Pressure

1 atmosphere $= 1.013 \times 10^5$ pascals
  $= 14.7$ lb in.$^{-2}$

1 pascal $\equiv 10$ dyn cm$^{-2} = 1.450 \times 10^{-4}$ lb in.$^{-2}$
  $= 7.501 \times 10^{-4}$ cm Hg

1 cm Hg $= 1.333 \times 10^4$ dyn cm^{-2}
  $= 1.316 \times 10^{-2}$ atmosphere $= 1.333 \times 10^3$
    pascals

1 in. $H_2O = 1.868$ mm Hg $= 249.1$ pascals

1 lb in.$^{-2} = 6.895 \times 10^3$ pascals
  $= 6.805 \times 10^{-2}$ atmosphere

1 lb ft$^{-2} = 47.88$ pascals

1 torr $\equiv 1$ mm Hg $= 133.3$ pascals

Viscosity and Flow Resistance

1 Pa s $\equiv 10$ poise

1 Pa s m$^{-3} \equiv 0.750 \times 10^{-8}$ torr s cm^{-3}

Energy

1 joule $\equiv 10^7$ ergs $= 0.2390$ calorie
  $= 0.7376$ ft lb

1 calorie $= 4.184$ joules

1 kcal $\equiv 10^3$ calories

1 joule $= 6.24 \times 10^{18}$ electron volts

1 electron volt $= 1.602 \times 10^{-19}$ joule

1 kWh $\equiv 3.6 \times 10^6$ joules

1 Btu $= 1.054 \times 10^3$ joules

1 ft lb $= 1.356$ joules

Mass Energy Conversion

1 amu $= 931 \times 10^6$ electron volts $\equiv 931$ MeV

Power

1 watt $\equiv 10^{-3}$ kilowatts $= 0.7376$ ft lb s^{-1}
  $= 1.341 \times 10^{-3}$ horsepower

1 horsepower $\equiv 550$ ft lb s$^{-1} = 7.457 \times 10^2$ watts

1 kilowatt $\equiv 10^3$ watts $= 1.341$ horsepower

Magnetic Field

1 gauss $= 10^{-4}$ tesla

LIFE
SCIENCE
PHYSICS

LIFE SCIENCE PHYSICS

JOSEPH W. KANE
MORTON M. STERNHEIM

Department of Physics and Astronomy
University of Massachusetts
Amherst, Massachusetts

JOHN WILEY & SONS,

New York   Santa Barbara   Chichester   Brisbane   Toronto

This book was printed by Murray Printing Co. and bound by Book Press.
It was set in 10 on 12 Times Roman by York Graphic Services, Inc.
The Designer was Raphael H. Hernandez.
Picture research was done by Kathy Bendo.

Library of Congress Cataloging in Publication Data:

Kane, Joseph W., 1938–
Life science physics.

Includes bibliographical references and index.
1. Physics.   I. Sternheim, Morton M., 1933–
joint author.   II. Title.
QC21.2.K33   530   78–503
ISBN 0-471-03137-2

Printed in the United States of America

10 9 8 7 6 5 4 3 2 1

To my mother.

JWK

To my wife, Helen, and my children, Laura, Amy, and Jeffrey.

MMS

PREFACE

Life Science Physics is intended for students with diverse backgrounds. It is based on our experience with a course taken by many students majoring in various life science areas, including zoology, premedical studies, botany, medical technology, forestry, and physical education, along with a scattering of students majoring in fields outside the life sciences. They usually were sophomores or juniors who had completed college level chemistry, mathematics, and biology courses, but some were freshmen, and a few were graduate students. Most had not studied physics in high school. Many of the students had been exposed to calculus, but few of them had confidence in their ability to use it seriously.

This book differs in several ways from conventional physics textbooks designed for science majors. First, the choice of basic physics topics to be included or emphasized has been determined by the interests and needs of life science majors. This has necessitated the inclusion of some topics no longer of great current interest to physicists, such as geometric optics, the mechanics of fluids, and acoustics. It has also resulted in minimizing historical material and areas of contemporary physics such as high energy physics and astrophysics, which have little direct impact on biology.

The second major difference is that many examples have been chosen from biological systems. This contrasts with the common use of examples with an engineering flavor. Thus, rigid body problems are frequently taken from biomechanics, resistance and capacitance examples from nerve conduction, heat transport and latent heat illustrations from the maintenance of body temperature, and so on. Instrumentation used in biomedical work is discussed whenever it illustrates physical principles.

The third and perhaps most significant novel feature of this book is the inclusion of entire sections and chapters devoted to the detailed application of physics to biological systems. These discussions motivate the student to learn physics by demonstrating its relevance to the life sciences, as well as measuring his or her understanding of the physical principles involved. They also show the overall unity of science by using concepts from physics, biology, and chemistry. However, no specific background in biology or chemistry is assumed.

Since we have included all the topics normally covered in physics courses for life science majors plus some unusual material, this book contains slightly more than can be covered in a standard two-semester or three-quarter course. There are 34 chapters that are divided into nine units. Four chapters cover material not usually found in any detail in introductory physics texts: Chapter Seventeen, The Mammalian Circulatory System; Chapter Twenty, Nerve Conduction; Chapter Thirty-two, Nuclear Magnetic Resonance; and Chapter Thirty-four, Ionizing Radiation. These chapters can be omitted or treated lightly without significant loss of continuity, as can some more traditional chapters such as Chapter Nine, Elastic Properties of Materials; Chapter Twenty-seven, Special Relativity; and Chapter Thirty-one, The Structure of Matter. In addition, most of the chapters end with one or more supplementary sections containing either biological applications or more traditional topics of secondary importance. This arrangement permits the instructor to select what will be included or emphasized and also helps the student to distinguish the basic principles of physics from relatively peripheral material.

Each chapter contains a generous set of exercises. Keyed to the sections in the book, these test the student's basic understanding of the material and aid in building confidence. Unkeyed problems are also included, with the few more difficult ones noted by asterisks. Exercises and problems are provided for the supplementary sections, so that they can be used as an integral part of the course.

Although we include a good deal of material not usually found in standard physics texts, we present the basic physics in a fairly conventional sequence and style. This is done because the standard arrangement seems to work best and because it permits the instructor to choose and organize the material to suit individual requirements and preferences. Although mechanics must be treated first and modern physics last, the material between (heat, fluids, electricity and magnetism, and wave motion) can be rearranged without causing major problems. This is true even though we stress the unity of physical concepts and point out their interconnections wherever possible.

Throughout this book, we use the minimum amount of formal mathematics consistent with an honest and precise treatment of the physics under discussion. This means that we use algebra rather freely and some geometry. Because some students are relatively weak in these skills when they begin studying physics, we proceed fairly slowly in the first few chapters and also provide a mathematical review in Appendix B. Differentiation is introduced in Chapter One, and it is used subsequently in definitions. The few derivations requiring calculus are placed at the very ends of chapters where they will not interrupt the text material. We avoid integration entirely, using instead the approach of guessing and then verifying by differentiation the solutions to problems such as the exponential decay law for a radionuclide. No examples, exercises, or problems require the use of calculus.

An important aspect of learning physics is developing certain patterns of thinking. Consequently, we stress the use of simple models for complex systems, of mathematical approximations, and of dimensional analysis. These are most evident in the biological applications but are present throughout the book. We also emphasize that physics is an experimental science, not an abstract intellectual discipline. Its implications relate continually to everyday experience as well as to work in other scientific fields.

Units are always a problem in an introductory text. The SI units are stressed, but British and cgs units are occasionally used to make contact with familiar quantities and to conform with the conventional usage in various applied areas.

This book could not have been written without the assistance of a very large number of people. Our greatest debt is to the many students who aided us in a variety of ways to learn what a life science physics course and textbook should be, and who suffered the inconvenience of using very preliminary versions of this material as a text. Norman C. Ford participated in the early stages of planning and writing this book but was forced to withdraw because of other commitments. A colleague, Stanley S. Hertzbach, taught from an early draft and offered useful advice. Helpful advice was also received from several reviewers, including Rubin Landau, Margaret McCarthy, Harvey Picker, Arnold Strassenberg, John Weir, Robert Williamson, and Steve Woods. Over a period of years, many graduate and undergraduate students offered valuable suggestions based on reading or teaching from parts of the manuscript or worked on the solutions to the exercises and problems. They include James Ledwell, David Long, Caroline Markey, Robert Meyer, Francesc Roig, Ernest Seglie, Thomas Slavkovsky, David Vetter, J. C. Wang and Jonathan Wainer.

The manuscript was typed by four capable and tolerant typists; the bulk of it was typed by Kathleen Ryan and the remainder by Lillian Camus, Linda Lisnerski, and Doris Maynard. Helen Sternheim helped with various aspects of manuscript preparation and provided advice and encouragement.

JOSEPH W. KANE
MORTON M. STERNHEIM

PROLOGUE: PHYSICS AND THE LIFE SCIENCE STUDENT

"Why should I study physics?" Sometimes asked with emotional overtones ranging from anguish to anger, this is one of the questions most frequently heard by physics teachers. It seems appropriate therefore to begin this book by attempting an answer.

One reason this question is asked so often is that many people who have not studied physics—and some who have—lack a clear notion of what physics is. Dictionaries are not much help. A typical short dictionary definition says that physics is the branch of science that deals with matter, energy, and their interactions. This is vague and general enough to include what is usually considered to be chemistry; in any case, it does not give any real feeling for what is involved. Longer dictionary entries usually expand the definition by noting that physics includes subfields such as mechanics, heat, electricity, and so forth. They give no clues as to why some subfields of science are included and others are not.

A better approach to defining physics is to ask what physicists are concerned about. Physicists attempt to understand the basic rules or *laws* that govern the operation of the natural world in which we live. Since their activities and interests evolve with time, the basic science called physics also changes with time. Many of the most active contemporary subfields of physics were un-dreamed of a generation or two ago. On the other hand, some parts of what are now considered to be chemistry or engineering were once considered to be physics. This is because physicists sometimes gradually abandon a field once the basic principles are known, leaving further developments and practical applications to others.

The fact that physics deals with the basic rules governing how the world works lets us see why people with varied interests may find the study of physics interesting and useful. For example, a historian who wants to understand the origins of our contemporary society will find significance in the story of the development of physics and its relationship to other human activities. Similarly, a philosopher concerned about concepts of space and time will profit greatly from understanding the revolutionary twentieth-century advances in physics. However, since we have written this book primarily for students majoring in the life sciences, we have not stressed the historical or philosophical aspects of physics. Rather, we have tried to make clear in every chapter the connection between physics and the life sciences.

Perhaps the most obvious impact of physics on biology and medicine is at the level of instrumentation. A knowledge of physics helps in the intelligent use of everything from light microscopes and centrifuges to electron microscopes and elaborate radiation detection systems used in nuclear medicine. Physics also enters biology in more fundamental ways. The physical laws governing the behavior of molecules, atoms, and atomic nuclei are the basis for all of chemistry and biochemistry. Physiology offers many examples of physical processes and principles: diffusion within cells, the regulation of the body temperature, the motion of fluids in the circulatory system, and electrical signals in nerve fibers are just a few. In comparative anatomy, the physics associated with an anatomical feature often helps to clarify the evolutionary process. Athletic activities ranging from running and jumping to karate can be studied and sometimes optimized with the aid of physical principles. In the course of developing and illustrating the basic principles of physics, we discuss all these life science applications and many others.

A few remarks about how one studies physics may be helpful. More than any other science, physics is a logical and deductive discipline. In any subfield of physics, there are just a few fundamental concepts or laws derived from experimental measurements. Once one has mastered these basic ideas, the applications are usually straightforward conceptually, even though the details may sometimes become complicated. Consequently, it is important to focus one's attention on the basic principles and to avoid memorizing a mass of facts and formulas.

Most of the basic laws of physics can be expressed rather concisely in the form of mathematical equations. This is a great convenience, since a tremendous amount of information is implicitly contained in a single equation. However, this also means that any serious attempt to learn or apply physics necessitates a willingness to use a certain amount of mathematics. High school algebra plus a bit of geometry is adequate for everything covered in this book, but a reasonable level of facility is required. A student who has become rusty at these mathematical skills may want to begin with the Mathematical Review in Appendix B. One post-high school mathematical technique, differentiation, is introduced in the first chapter. However, except in definitions, its use is restricted to a few derivations located in the Supplementary Topics. None of the exercises or problems require this mathematical tool.

In summary, we believe the student of the life sciences will benefit in two major ways from studying physics. The student will gain an understanding of the basic laws that govern everything in our world from the subatomic to the cosmic scale and will also learn much that will be important in his or her work in the life sciences. The study of physics as a basic science is not particularly easy, but we believe it is rewarding, particularly for students planning further training in related sciences. We hope that all who use this book will agree.

J. W. K.
M. M. S.

CONTENTS

Contents

Contents

Contents

UNIT ONE

THE GENERAL LAWS OF MOTION

In this unit and the next we discuss *mechanics:* the study of the motions of objects and of the forces that affect their motions. Its concepts and principles enter, directly and indirectly, into many areas of the physical and biological sciences. The laws of mechanics allow us to make predictions about such diverse phenomena as the motions of satellites, the movements of animals, and the strength and structure of both living and artificial systems. The laws of mechanics, applied to the motions of large numbers of atoms and molecules, provide an interpretation of the phenomena of heat and temperature. The properties of fluids—at rest and in motion—are understood in terms of the same laws; through them we can understand the flight of airplanes and animals or the flow of rivers and blood. Finally, with some twentieth-century modifications, the same mechanical laws play a central role in our present theories of atomic and nuclear phenomena.

The first two chapters of this unit are devoted to the concepts we require for a quantitative description of motion: *position, velocity,* and *acceleration.* In Chapter One we encounter these in the treatment of motion in a straight line; Chapter Two extends the same ideas to the description of motion in more than one dimension. The general laws of motion introduced by Newton, which relate the motions of objects to their causes, are discussed in Chapters Three to Five.

CHAPTER ONE

MOTION IN A STRAIGHT LINE

The most basic and obvious result of physical interactions is motion: a brick falls, an eardrum vibrates, a compass needle swings into line with a magnetic field, a meter needle moves on a scale, a radioactive nucleus emits a beta particle. Most of our understanding of nature is derived from our observations of motions and our efforts to relate them to their causes. Accordingly, we begin our study of physics by developing the ideas needed for a quantitative discussion of motion, starting in this chapter with the case of an object moving in a straight line.

Physics, like many other sciences, is largely based on quantitative measurements. These measurements must be correlated or interpreted in some way; often they are compared with theoretical predictions. To the extent that theory and experiment are in accord, we say that we have some understanding of the phenomena in question. A quantitative discussion of motion requires measurements of times and distances, so we must first consider the *standards, units,* and *errors* involved in physical measurements.

1.1 | MEASUREMENTS, STANDARDS, UNITS, AND ERRORS

Quantitative physical measurements must be expressed by numerical comparison to some agreed-upon set of standards. If you say a lecture lasted 53 minutes, you mean that it went on for the same length of time as it took for the wall clock to make some number of ticks. Here the quantity being measured has the *dimensions* of time, the *unit* for the measurement is the minute, and the clock is the *standard.* It is a *secondary standard,* since the minute is not defined by the properties of that one clock. All such measuring devices are calibrated directly or indirectly in terms of *primary standards* of length, time, and mass established by the international scientific community.

These primary standards are redefined from time to time as measurements become more precise. For example, the unit of length—the *meter*—was defined in 1889 as the length of a particular platinum-iridium bar kept under controlled conditions. This standard was discarded in 1960 because replication and preservation were inconvenient and subject to inaccuracies. The length standard is now based on the wavelength of the orange-red light emitted by atoms of krypton 86 in an electrical discharge tube. Standards have also been defined for the units of time and mass.

It is not accidental that standards have been set up for length, time, and mass. All physical quantities can be expressed in terms of some combination of these three fundamental dimensions, which we will denote at L, T, and M, respectively. For example, a velocity is a distance divided by a time, so its dimensions are L/T.

Systems of Units | In this text we emphasize a system of metric units that employs the meter, the *kilogram,* and the second as the units of length, mass,

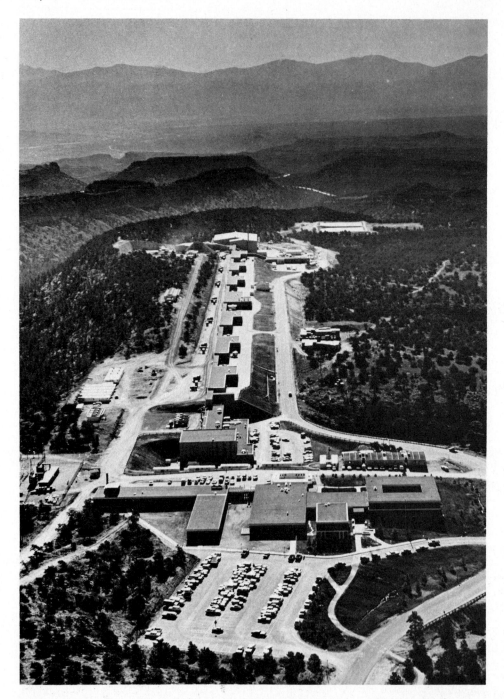

Figure 1.1.   The Clinton P. Anderson Los Alamos Meson Physics Facility, a half-mile long machine located high in the mountains of Northern New Mexico, accelerates large numbers of protons (hydrogen nuclei) to high velocities. When these protons strike a target they produce short-lived particles called mesons used both for basic physics research and for cancer therapy. Measurements made at this "meson factory" employ sophisticated electronic devices, but these are still calibrated indirectly in terms of the basic units of length, time, and mass. (Courtesy of Los Alamos Scientific Laboratory.)

TABLE 1.1

Representative lengths in meters

Atomic nucleus	10^{-15}
Sodium atom, diameter	10^{-11}
C—C bond	1.5×10^{-10}
DNA, diameter	2×10^{-9}
Microfilament, thickness	4×10^{-9}
Hemoglobin	7×10^{-9}
Cell membrane	10^{-8}
Small virus, diameter	2×10^{-8}
Small bacterium, diameter	2×10^{-7}
Visible light, wavelength	$4 - 7 \times 10^{-7}$
Mitochondria, diameter	$0.5 - 1.0 \times 10^{-6}$
Large bacterium, diameter	10^{-6}
Mammalian liver cell, diameter	2×10^{-5}
Sea urchin egg	7×10^{-5}
Giant amoeba, diameter	2×10^{-4}
Small crustacean	10^{-3}
Ostrich egg, diameter	4×10^{-2}
Mouse	10^{-1}
Human	$1 - 2 \times 10^{0}$
Blue whale	3×10^{1}
Brooklyn Bridge	10^{3}
Earth, diameter	1.3×10^{7}
Sun, diameter	1.2×10^{9}
Earth–sun, distance	1.3×10^{11}
Our galaxy, diameter	10^{22}
Distance to the farthest galaxies ever observed	10^{28}

and time, respectively. This system is known as the *Systeme Internationale (S.I.)* and is now the officially recognized set of scientific units. Unfortunately, history has bequeathed to us other systems of units that are still in use by scientists in many fields. One of these is the *c.g.s. system,* a metric system based on the centimeter, gram, and second; 1 centimeter is 0.01 meters, and 1 gram is 0.001 kilograms. Another system is the *British system,* familiar from everyday use in the English-speaking world, in which the unit of length is the foot; 1 foot equals 0.3048 meters. We will use c.g.s. and British units sparingly.

Representative lengths and times of various magnitudes are given in Tables 1.1 and 1.2, respectively. The standard prefixes used to define multiples of S.I. units are listed in Table 1.3. These prefixes make it possible, using powers of 10, to define units of a

convenient size for various purposes. (Powers of 10 are reviewed in Appendix B.1.) For example, the distance between two cities is usually measured in kilometers, where 1 kilometer is 10^3 meters. The dimensions of this book might be more conveniently expressed in centimeters than in meters.

Because many different kinds of units are in use, we frequently must convert quantities from one set of units to another. To see how this is done, let us suppose we need to convert 100 feet (100 ft) into the equivalent number of meters (m). From Appendix C,

$$1 \text{ ft} = 0.3048 \text{ m} \qquad (1.1)$$

Now we divide both sides by 1 ft, just as if the unit (feet) were an algebraic quantity:

$$\frac{1 \text{ ft}}{1 \text{ ft}} = \frac{0.3048 \text{ m}}{1 \text{ ft}}$$

The feet cancel on the left, so we have

$$1 = \frac{0.3048 \text{ m}}{1 \text{ ft}} = 0.3048 \text{ m ft}^{-1}$$

(*Note:* ft^{-1} is short for 1/ft.) Thus the quantity 0.3048 m ft^{-1} is just a way of writing the quantity 1. If

TABLE 1.2

Representative times in seconds

Nuclear events	$10^{-23} - 10^{-10}$
Atomic events: light absorption, electronic excitation	$10^{-15} - 10^{-9}$
Chemical events	$10^{-9} - 10^{-6}$
Chains of biochemical reactions	$10^{-8} - 10^{2}$
Quick contraction of striated muscle (wink)	10^{-1}
Fastest cell division	5×10^{2}
Typical bacterial generation time	3×10^{3}
Typical protozoan generation time	10^{5}
Small mammal generation time	4×10^{7}
Large mammal lifetime	$4 \times 10^{8} - 4 \times 10^{9}$
Lifetime of a lake	$10^{10} - 10^{12}$
Age of mammals	3×10^{15}
Age of vertebrates	10^{16}
Age of life	$>10^{17}$
Age of the earth	2×10^{17}

TABLE 1.3

Prefixes used to define multiples of S.I. units. These may be used with any of the basic S.I. units or of units derived from them

Fraction	Prefix	Symbol	Example
10^{-18}	atto	a	
10^{-15}	femto	f	
10^{-12}	pico	p	
10^{-9}	nano	n	1 nanosecond = 1 ns = 10^{-9} seconds
10^{-6}	micro	μ	
10^{-3}	milli	m	1 millimeter = 1 mm = 10^{-3} meters
10^{-2}	centi	c	1 centimeter = 1 cm = 10^{-2} meters
10^{-1}	deci	d	
10	deka	da	
10^{2}	hecto	h	
10^{3}	kilo	k	1 kilogram = 1 kg = 10^{3} grams
10^{6}	mega	M	
10^{9}	giga	G	
10^{12}	tera	T	

we multiply 100 ft by 1, nothing is changed, so we find

$$100 \text{ ft} = (100 \text{ ft})(1)$$
$$= (100 \text{ ft})(0.3048 \text{ m ft}^{-1})$$
$$= 30.48 \text{ m}$$

Notice that the units of feet and feet^{-1} cancel. This trick of multiplying by 1 eliminates any uncertainty as to whether we should multiply or divide by the conversion factor. For instance, we can divide 1 ft = 0.3048 m by 0.3048 m, giving

$$1 = \frac{1 \text{ ft}}{0.3048 \text{ m}} = 3.281 \text{ ft m}^{-1} \qquad (1.2)$$

Thus, 3.281 ft m^{-1} is another way of writing 1. However, if we multiply 100 ft by this factor, the units do not cancel properly.

Sometimes it is necessary to convert units for a quantity such as a velocity, which is measured in terms of two or more basic dimensions. We must then multiply by 1 for each of the conversions. This procedure is illustrated in the following examples.

Example 1.1

Convert a velocity of 60 mi h^{-1} (miles per hour) to feet per second (ft s^{-1}).

To carry out this conversion, we need a factor of 1 to convert hours to seconds and another to convert miles to feet. Since 1 h = 60 min = 3600 s, dividing by 3600 s gives

$$1 = \frac{1 \text{ h s}^{-1}}{3600}$$

Also, 1 mi = 5280 ft, so

$$1 = 5280 \text{ ft mi}^{-1}$$

Multiplying 60 mi h^{-1} by 1 twice gives

$$60 \text{ mi h}^{-1}(1)(1) = 60 \text{ mi h}^{-1}\left(\frac{1 \text{ h s}^{-1}}{3600}\right)(5280 \text{ ft mi}^{-1})$$
$$= 60\left(\frac{5280}{3600}\right) \text{ ft s}^{-1} = 88 \text{ ft s}^{-1}$$

Example 1.2.

A small swimming pool is 20 ft long, 10 ft wide, and 5 ft deep. Its volume is the product of these distances, or (20 ft)(10 ft)(5 ft) = 1000 ft^3. What is the volume in cubic meters (m^3)?

Here we must convert feet to meters three times, corresponding to changing the units for length, width, and depth. Using 1 ft = 0.3048 m or 1 = 0.3048 m ft^{-1},

$$1000 \text{ ft}^3(1)^3 = 1000 \text{ ft}^3(0.3048 \text{ m ft}^{-1})^3$$
$$= 1000(0.3048)^3 \text{ m}^3 = 28.3 \text{ m}^3$$

Types of Errors | Measurements and predictions are both subject to errors. Measurement errors are of two types, *random* and *systematic*. The meaning of these terms is best understood with the aid of an example: the time τ required for a weight on a string to swing back and forth once when released from a given point.

If someone uses a stopwatch to measure τ and repeats the experiment several times, each result will be slightly different from the others. Usually, most of the measurements will be close to the average of all the measurements. The variation in results about this average arises from the inability of the observer to start and stop the watch exactly the same way each time. The error introduced by this inability is *random* and can be reduced by taking the average of many measurements.

Even when many measurements are made, the average result for τ will be too small if the watch runs slow. This *systematic* error can be reduced by using a better watch or by comparing the watch used with a more accurate one and adjusting the results accordingly.

A systematic error may also result from the observer's reaction time. The observer may systematically start or stop the watch too early or too late. This error can be reduced by doing a more complex experiment. For example, the timing device may be started and stopped using a light beam and a photoelectric cell similar to those used in automatic door openers. Naturally, this apparatus will also have systematic and random errors, but they will be smaller than before.

Both systematic and random errors are present in all experiments. Reducing these errors generally requires increasingly elaborate apparatus and time-consuming procedures. High precision measurements and measurements of small effects require that great attention be given to identifying and reducing these errors.

Theoretical predictions usually have errors arising from various sources. A theoretical formula often contains measured quantities such as the mass of an electron or the speed of light, and there is some error associated with these measurements. For example, we will later obtain a formula for the period of a pendulum, the time τ in our discussion above. Theoretical predictions made with this formula depend on the accuracy of our present knowledge of the acceleration due to gravity. Also, like most theoretical formulas, this formula depends on the validity of several approximations. Among these are the absence of friction and air resistance and the assumption that the pendulum does not swing very far from its rest position.

In all careful scientific work the numerical accuracy must be stated with precision. However, it is customary in textbooks to avoid the difficulties of a complete error analysis in doing numerical examples by using the rules for *significant figures*. This means that in the statement "the length of a rod is 2.43 meters" the last digit (3) is somewhat uncertain; the exact length might turn out to be closer to 2.42 or 2.44 m. In the examples, exercises, and problems in this text, all numbers should be treated as known to three significant figures. For example, 2.5 and 3 should be interpreted in calculations as 2.50 and 3.00, respectively. Significant figures are reviewed in Appendix B.3.

1.2 | AVERAGE VELOCITY

The discussion of motion is based on measurements and calculations of *positions, velocities,* and *accelerations.* In this and the next few sections we use simple examples of moving cars to define and illustrate these concepts for motion along a straight line. We consider only *translational* motion: motion in which every part of an object moves in the same direction, and there is no rotation. Rotation will be discussed in a later chapter.

We begin by defining the average velocity. Suppose that a car is moving along a straight highway with marker posts every 100 m, and that it is observed to pass one of these posts every 5 s as in Fig. 1.2a. The *average velocity* of the car during a specific time interval is the distance traveled divided by the time elapsed:

$$\text{average velocity} = \frac{\text{distance traveled}}{\text{time elapsed}}$$

This definition is illustrated by the following example.

Example 1.3.
What is the average velocity of the car in Fig. 1.2a during the interval that the clock reading changes from 10 to 25 s?

From the figure, we see that the car travels $500 \text{ m} - 200 \text{ m} = 300 \text{ m}$ during the 15-s period, so

$$\text{average velocity} = \frac{300 \text{ m}}{15 \text{ s}} = 20 \text{ m s}^{-1}$$

Because the car in this example moves equal distances in equal times, the average velocity will be the same no matter what time interval is chosen. In this situation the motion is said to be *uniform,* and the driver will observe the speedometer reading to remain constant. Motion that is not uniform is said to be *accelerated.* In such cases the average velocity does depend on the time interval chosen. The speedometer reading of an accelerating car will be changing in time.

Often it is convenient to use graphs or algebraic formulas to describe the position and velocity of a moving object. For the car discussed above, we may designate the straight line along which it moves as an "x axis," and choose its position x to be zero at the

Figure 1.2. (a) The position of a car as observed at 5-second intervals. The speedometer reading is constant. (b) A graph of the position x versus time t.

post marked 0 meters. Also, we may use the symbol t to represent the observation times. If we arbitrarily take $t = 0$ when the first observation is made, then x is zero at $t = 0$. Also, x is 100 m when $t = 5$ s, x is 200 m when $t = 10$ s, and so on. Figure 1.2b shows these observations on a graph with time t along the horizontal coordinate and with the position x along the vertical coordinate. A straight line has been drawn connecting the points corresponding to the observations, since the car presumably moves steadily between observations.

The average velocity can now be defined more symbolically using this notation. Suppose that at some time which we will call t_1 the car is observed to be at position x_1 and that a later time t_2, it is located at x_2. The difference in positions

$$\Delta x = x_2 - x_1$$

is called the *displacement*. (The symbol Δ is the Greek letter "delta," and Δx is read "delta x." Δ usually represents a difference or change of the quantity that follows it.) The elapsed time between the observations is the difference

$$\Delta t = t_2 - t_1$$

In this notation, the average velocity $\bar{v}$ is the displacement divided by the elapsed time,

$$\bar{v} = \frac{\Delta x}{\Delta t} = \frac{x_2 - x_1}{t_2 - t_1} \qquad (1.3)$$

Note that this definition of the average velocity holds whether or not $\bar{v}$ is constant in time.

We will now repeat the preceding example to show how this notation is used.

Example 1.4.

   Using Eq. 1.3, again find the average velocity of the car in Fig. 1.2 during the period from $t = 10$ s to 25 s.
   Here t_1 is 10 s and t_2 is 25 s; from the graph, $x_1 = 200$ m and $x_2 = 500$ m. Thus

$$\bar{v} = \frac{x_2 - x_1}{t_2 - t_1} = \frac{500\text{ m} - 200\text{ m}}{25\text{ s} - 10\text{ s}} = \frac{300\text{ m}}{15\text{ s}} = 20\text{ m s}^{-1}$$

Describing the motion of an object in terms of its position along some coordinate axis (here, the highway with its labeled markers is our x axis) automatically distinguishes between positive and negative displacements. For example, suppose the car in Fig. 1.2a is moving downward in the picture instead of upward. In that case, x is decreasing with increasing t, and Eq. 1.3 yields a *negative* value of average velocity. All this means is that the object is moving in the opposite direction—from larger toward smaller values of x. The next example illustrates such a case.

Example 1.5.

   At $t_1 = 5$ s, a car is at $x_1 = 600$ m, and at $t_2 = 15$ s, it is at $x_2 = 500$ m. Find its average velocity.
   Using Eq. 1.3,

$$\bar{v} = \frac{x_2 - x_1}{t_2 - t_1} = \frac{500\text{ m} - 600\text{ m}}{15\text{ s} - 5\text{ s}}$$

$$= \frac{-100\text{ m}}{10\text{ s}} = -10\text{ m s}^{-1}$$

Note that $\bar{v}$ is negative, and the car is moving in the $-x$ direction even though the position x is positive.

When an object undergoes uniform motion, its x-t graph is a straight line, as in the example of Fig. 1.2. If the motion is accelerated, the graph is not a straight line, and the average velocity depends on the particular time interval chosen. For example, in Fig. 1.3, a car starting from rest travels a short distance in the first second and a longer distance in the next second as it speeds up. In this situation the average velocity will be less in the first second of the motion than in later seconds. This is illustrated by the following example.

Example 1.6.

   A car moves as shown in Fig. 1.3. Find its average velocity from $t = 0$ to $t = 1$ s and from $t = 1$ to $t = 2$ s.
   To calculate the average velocities, we need the positions at $t = 0$, 1, and 2 s. From Fig. 1.3, these are 0, 1, and 4 m, respectively. From $t = 0$ to 1 s, the average velocity is

$$\bar{v} = \frac{\Delta x}{\Delta t} = \frac{1\text{ m} - 0\text{ m}}{1\text{ s} - 0\text{ s}} = 1\text{ m s}^{-1}$$

From $t = 1$ to 2 s,

$$\bar{v} = \frac{\Delta x}{\Delta t} = \frac{4\text{ m} - 1\text{ m}}{2\text{ s} - 1\text{ s}} = 3\text{ m s}^{-1}$$

As anticipated, the average velocity is greater in the later time interval because the car is accelerating.

1.3 | INSTANTANEOUS VELOCITY

Most situations of interest involve accelerated rather than uniform motion. Since the average velocity depends on the time interval if the motion is accelerated, it is often more useful to characterize the motion by the *instantaneous velocity*—the velocity at a particular instant in time. For example, when we say that an accelerating car is moving at 10 m s^{-1}, we are referring to its instantaneous velocity at the present moment.

The instantaneous velocity is determined by computing the average velocity for an extremely small time interval. One procedure for doing this is illustrated by the following example.

Example 1.7.

   The motion of the car shown in Fig. 1.3 corresponds to the algebraic formula $x = bt^2$, where $b = 1$ m s^{-2}. Find its average velocity from 3 to 3.1 s, 3 to 3.01 s, and 3 to 3.001 s. (These time intervals are successively smaller, and the corresponding average velocities are increasingly good approximations for the instantaneous velocity at $t = 3$ s.)
   When $t = 3$ s, the position is $x = bt^2 = (1\text{ m s}^{-2})$ $\times$ $(3\text{ s})^2 = 9$ m; when $t = 3.1$ s, $x = (1\text{ m s}^{-2})(3.1\text{ s})^2 = 9.61$ m. Thus the average velocity from 3 to 3.1 s is

$$\bar{v} = \frac{\Delta x}{\Delta t} = \frac{9.61\text{ m} - 9\text{ m}}{3.1\text{ s} - 3\text{ s}} = 6.1\text{ m s}^{-1}$$

At $t = 3.01$ s, the position is $x = (1\text{ m s}^{-2})(3.01\text{ s})^2 = 9.0601$ m, so $\bar{v}$ from 3 to 3.01 s is

$$\bar{v} = \frac{9.0601\text{ m} - 9\text{ m}}{3.01\text{ s} - 3\text{ s}} = 6.01\text{ m s}^{-1}$$

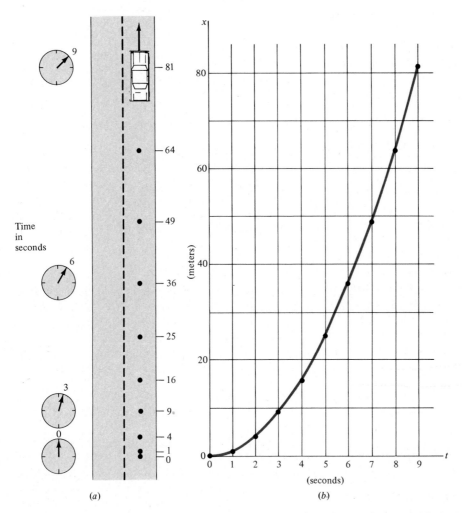

Figure 1.3.   (*a*) The positions of an accelerating car at 1-second intervals are represented by solid circles. (*b*) The position-time graph for the car.

Note that for the purposes of this example we have kept more than the usual number of significant figures. Finally, when $t = 3.001$ s, $x = (1$ m s$^{-2})(3.001$ s$)^2 = 9.006001$ m, so the average velocity from 3 to 3.001 s is

$$\bar{v} = \frac{9.006001 \text{ m} - 9 \text{ m}}{3.001 \text{ s} - 3 \text{ s}} = 6.001 \text{ m s}^{-1}$$

As the time interval Δt is made smaller and smaller, the average velocity $\bar{v}$ steadily approaches 6 m s^{-1}. Therefore, the instantaneous velocity v at $t = 3$ seconds is 6 m s^{-1}.

The above example shows us how to obtain the instantaneous velocity of an object. The average ve-

locity is computed for progressively shorter time intervals, and the value obtained for $\bar{v}$ when the interval becomes arbitrarily small is the instantaneous velocity v. Mathematically, the instantaneous velocity v is said to be the *limit* of the average velocity $\bar{v}$ as Δt approaches zero. The process of evaluating this limit is called *differentiation; v* is called the *derivative of x with respect to t* and is written as

$$v = \frac{dx}{dt} \tag{1.4}$$

The quantity dx/dt can be thought of as the ratio $\Delta x/\Delta t$ evaluated when both Δx and Δt are very small.

General expressions have been developed that often make it possible to find the formula for $v = dx/dt$ at *any* time if the algebraic relation between x and t is known. This eliminates the need to use the laborious process employed in the example to find the velocity at a *single* time. Several of these differentiation formulas are given in Appendix B.8.

In this textbook, we use differentiation in definitions of quantities such as the instantaneous velocity and in occasional derivations. These derivations are placed at the ends of some of the chapters and can be skimmed or omitted entirely without loss of continuity. Since none of the examples, exercises, or problems involve the use of differentiation, students who have not studied differentiation should encounter no special difficulties. For example, even though we make extensive use of the instantaneous velocity, we find ways to determine it without differentiating.

Like the average velocity, the instantaneous velocity can be either positive or negative; negative values correspond to motion toward decreasing values of x. The *speed* is the magnitude or size of the instantaneous velocity, so it is always positive or zero.

From now on, when we refer to the velocity or any other rate of change, we mean the instantaneous value unless the word "average" is explicitly used. Note that the distinction between the two kinds of velocities disappears for the special case of uniform motion.

Graphical Interpretation of Velocity | A
position-versus-time graph provides direct information about the velocity. As we have seen, a straight line graph corresponds to a constant velocity or uniform motion, while a curved graph corresponds to a changing velocity. Furthermore, the velocity is equal to a quantity called the *slope* of the x-t graph. In the straight line graph of Fig. 1.4, the slope of the line is

Figure 1.4. The x-t graph for an object undergoing uniform motion. The slope of the line is defined to be $\Delta x/\Delta t$, which is the velocity.

Figure 1.5. The slope of the x-t curve at P is equal to the slope $\Delta x/\Delta t$ of the straight line tangent to the curve at P. The velocity at P is equal to the slope of the curve at that point.

defined to be $\Delta x/\Delta t$, which is just the velocity. In the curved x-t graph of Fig. 1.5, the slope at point P is defined to be that of the straight line tangent to the curve at P. The velocity at P is equal to the slope, since the curve and the tangent line have the same rate of increase with time at that point in the motion.

1.4 | ACCELERATION

Like the position, the velocity can change with time. The rate at which the velocity changes is the *acceleration*. Again we can discuss either the average or instantaneous rate of change.

The *average acceleration* $\bar{a}$ from time t_1 to t_2, if the velocity changes by $\Delta v = v_2 - v_1$, is defined by

$$\bar{a} = \frac{\text{change in velocity}}{\text{time elapsed}} = \frac{\Delta v}{\Delta t} = \frac{v_2 - v_1}{t_2 - t_1} \quad (1.5)$$

If we measure velocity in meters per second and time in seconds, $\bar{a}$ will have the units of meters per second per second (usually abbreviated as m s^{-2} and read as "meters per second squared"). An average acceleration of 1 m s^{-2} corresponds to an average increase in the velocity of 1 m s^{-1} each second. We will illustrate this definition with an example.

Example 1.8.

A car accelerates from rest to 30 m s^{-1} in 10 s. What is its average acceleration?

From the definition,

$$\bar{a} = \frac{\Delta v}{\Delta t} = \frac{30 \text{ m s}^{-1} - 0}{10 \text{ s}} = 3 \text{ m s}^{-2}$$

This corresponds to an increase in the velocity of 3 m s^{-1} in each second of the 10-s second time interval.

The *instantaneous acceleration* a is defined as the average acceleration $\bar{a}$ for an extremely short time interval. Symbolically,

$$a = \frac{dv}{dt} \quad (1.6)$$

On a graph showing the velocity of an object versus time, a equals the slope of the v-t curve. This is illustrated by the following example.

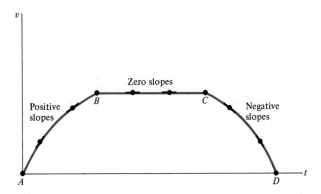

Figure 1.6. The velocity-versus-time graph for a car. The slope and acceleration are positive from A to B, zero from B to C, and negative from C to D.

Example 1.9.

Figure 1.6 shows the velocity-versus-time graph for a car. Describe its acceleration qualitatively.

From A to B, the velocity is increasing, the slope is positive, and the car is accelerating. However, the slope is becoming progressively smaller, so the acceleration is decreasing. From B to C, the velocity is constant and the slope and the acceleration are zero. From C to D, the velocity is decreasing, so the acceleration is negative. This *deceleration* increases in magnitude as the car slows.

1.5 | FINDING THE MOTION OF AN OBJECT

So far we have calculated velocities from position changes and accelerations from velocity changes. However, it is often true that it is the acceleration of an object that is either measured or predicted theoretically, and we wish to know the corresponding velocity and position changes. For example, when an animal jumps vertically from the earth's surface, it is subject to a constant acceleration due to gravity, which determines the motion of the animal while it is off the ground. In this section, we see how to find the subsequent motion of an object given its acceleration and its initial position and velocity.

Consider a car initially at rest at a traffic light, so $v = 0$ initially. The origin of the x axis can be chosen to be at the light, so initially $x = 0$ also. We set our clock at $t = 0$ as the light becomes green. If the car now accelerates with a constant or uniform acceleration a, we would like to find its velocity v and position x after a time t has elapsed.

Since the acceleration is constant, the average acceleration $\bar{a}$ equals the instantaneous acceleration a, and

$$a = \frac{\Delta v}{\Delta t}$$

or

$$\Delta v = a\,\Delta t$$

Because $v = 0$ when $t = 0$, we have $\Delta v = v$ and $\Delta t = t$, so

$$v = at \qquad (1.7)$$

This equation has an interesting and useful interpretation. If we look at the graph (Fig. 1.7a) of the acceleration from time 0 to time t, the area of the shaded rectangle is the product of its height a and width t, which is the velocity change at. *Thus the change in velocity equals the area under the* a-t *graph over the time interval chosen.* This result is quite general and is not restricted to constant acceleration situations.

To find the position x after a time t has elapsed, we use the definition of average velocity,

$$\bar{v} = \frac{\Delta x}{\Delta t}$$

or

$$\Delta x = \bar{v}\,\Delta t$$

Since x is zero at $t = 0$, we have $\Delta x = x$ and $\Delta t = t$, so $x = \bar{v}t$. In this uniform acceleration situation, we see from the v-t graph (Fig. 1.7b) that $\bar{v} = \frac{1}{2}v$, and $x = \bar{v}t$ becomes

$$x = \tfrac{1}{2}vt \qquad (1.8)$$

Since $v = at$, this can be rewritten as

$$x = \tfrac{1}{2}(at)t = \tfrac{1}{2}at^2 \qquad (1.9)$$

Again, these algebraic results have a direct graphical interpretation. The area of the shaded triangle under the v-t graph in Fig. 1.7b is $\frac{1}{2}$(base)(height) = $\frac{1}{2}vt$; according to Eq. 1.8, this is equal to x. *Thus the change in position is equal to the area under the* v-t *graph over the corresponding time interval.* This inter-

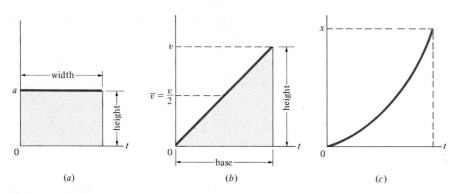

Figure 1.7. Motion with constant acceleration. (*a*) The acceleration *a* is constant. (*b*) The graph of $v = at$. The average velocity is $\bar{v} = v/2$. (*c*) The graph of $x = \frac{1}{2}at^2$.

pretation is also quite general and applies even if the acceleration is not constant.

Equation 1.7 for the velocity and Eq. 1.9 for the position completely describe the motion of an object starting from rest at time $t = 0$ at the origin with constant acceleration *a*. Equation 1.8 contains equivalent information and is useful in solving problems in which the final velocity but not the acceleration is stated. Another useful relation is obtained by eliminating the time. Rewriting Eq. 1.7 as $t = v/a$ and substituting this for t in Eq. 1.8 gives

$$x = \frac{1}{2}v\left(\frac{v}{a}\right) = \frac{v^2}{2a}$$

or

$$v^2 = 2ax \qquad (1.10)$$

These results for the constant acceleration problem are summarized in Table 1.4, along with the more complicated-looking equations obtained in the same way when the object begins to move from some arbitrary point x_0 with an initial velocity v_0. Note that in both cases we always set our clock so that the motion begins at time $t = 0$.

The use of the constant acceleration formulas is illustrated in the examples below and in the next section.

Example 1.10.

A car at rest at a traffic light accelerates at 2 m s^{-2} when the light turns green. After 4 s, what is its (a) velocity; (b) position?

(a) Let us choose our origin at the light, so $x_0 = 0$. Since we know the acceleration *a* and the time *t*, we can find the velocity using Eq. 1.7. Thus

$$v = at = (2 \text{ m s}^{-2})(4 \text{ s}) = 8 \text{ m s}^{-1}$$

The car has reached a velocity of 8 m s^{-1} after 4 s of its motion.

(b) Since we know the acceleration and the time, we can find the position x from Eq. 1.9,

$$x = \tfrac{1}{2}at^2 = \tfrac{1}{2}(2 \text{ m s}^{-2})(4 \text{ s})^2 = 16 \text{ m}$$

After 4 s the car has moved 16 m.

Note that we could also have found x from $x = \frac{1}{2}vt$ using our result for v obtained in part (a). Constant acceleration problems and physics problems in general can often be solved in two or more ways.

Example 1.11.

A car accelerates to 20 m s^{-1} with an acceleration of

TABLE 1.4

Motion with constant acceleration. The motion is assumed to start at time $t = 0$

Simpler Form ($v_0 = 0, x_0 = 0$)		General Form	
$x = \frac{1}{2}at^2$	(1.9)	$x = x_0 + v_0t + \frac{1}{2}at^2$	(1.11)
$v = at$	(1.7)	$v = v_0 + at$	(1.12)
$x = \frac{1}{2}vt$	(1.8)	$x = x_0 + \frac{1}{2}(v_0 + v)t$	(1.13)
$v^2 = 2ax$	(1.10)	$v^2 = v_0^2 + 2a(x - x_0)$	(1.14)

2 m s^{-2}. How far will it travel while it is accelerating if it is (a) initially at rest; (b) initially moving at 10 m s^{-1}?

(a) Here the initial velocity is zero and we know the final velocity $v = 20$ m s^{-1} and the acceleration, $a = 2$ m s^{-2}. We can take the origin to be the starting point, so $x_0 = 0$. Equation 1.10 contains known quantities plus the distance traveled, x, so solving for x,

$$x = \frac{v^2}{2a} = \frac{(20 \text{ m s}^{-1})^2}{2(2 \text{ m s}^{-2})} = 100 \text{ m}$$

We could also have solved this problem by using $v = at$ to find the time t and then using $x = \frac{1}{2}at^2$ to obtain x. (The reader should work out the problem in this way as an exercise.)

(b) Since the car is initially moving with velocity $v_0 = 10$ m s^{-1}, we must use the more general form of the constant acceleration equations to find the distance traveled. Solving Eq. 1.14 for x gives

$$x = x_0 + \frac{v^2 - v_0{}^2}{2a}$$
$$= 0 + \frac{(20 \text{ m s}^{-1})^2 - (10 \text{ m s}^{-1})^2}{2(2 \text{ m s}^{-2})} = 75 \text{ m}$$

The distance required to reach the desired velocity is shorter than in part (a) because the car is initially in motion.

Example 1.12

A car accelerates from rest with a constant acceleration of 2 m s^{-2} onto a highway where traffic is moving at a steady 24 m s^{-1}. (a) How long will it take for the car to reach a velocity of 24 m s^{-1}? (b) How far will it travel in that time? (c) The driver does not want the vehicle behind to come closer than 20 m nor force it to slow down. How large a break in traffic must he wait for?

(a) We set $x_0 = 0$ at the starting point. Then $v = at$ gives the time necessary for acceleration,

$$t = \frac{v}{a} = \frac{24 \text{ m s}^{-1}}{2 \text{ m s}^{-2}} = 12 \text{ s}$$

(b) Using Eq. 1.9, the distance traveled in 12 s is

$$x = \tfrac{1}{2}at^2 = \tfrac{1}{2}(2 \text{ m s}^{-2})(12 \text{ s})^2 = 144 \text{ m}$$

(c) Since the vehicle behind is moving at a constant velocity, its acceleration a is zero. We can use Eq. 1.11 with $v_0 = 24$ m s^{-1} and $t = 12$ s. Then we find that the oncoming vehicle travels a distance $x - x_0$ in 12 s, where

$$x - x_0 = v_0 t + \tfrac{1}{2}at^2$$
$$= (24 \text{ m s}^{-1})(12 \text{ s}) + 0 = 288 \text{ m}$$

Since the entering car travels 144 meters in this time, the oncoming vehicle gains $(288 - 144)$ m, or 144 m. If it is to come no closer than 20 m, then the break in traffic must be at least $(144 + 20)$ m, or 164 m.

The car in this example reaches 24 m s^{-1} or about 54 mi h^{-1} in 12 s, which is a fairly brisk acceleration. A less powerful car would take longer to reach this speed and would require a longer break in traffic.

The above examples illustrate the procedures for solving constant acceleration problems and physics problems in general. We identify which quantities are known and which are to be found and determine the equation or equations relating these quantities. If necessary, we then solve algebraically for one unknown in terms of known quantities. Numerical values are usually best substituted in the final step rather than at an earlier stage. This tends to minimize the arithmetic labor and also facilitates checking for errors.

A useful check on any problem is provided by the dimensions of the final result. If the unknown is a length, then the result should be in length units such as meters or feet; if not, an error has been made somewhere. Note, for example, that in parts (b) and (c) of Example 1.12 the time units cancel and the distances are in meters as required.

To summarize, we have found that given the initial position and velocity, we can find the velocity at a later time from the acceleration, and the position from the velocity. The equations listed in Table 1.4 can be used to find the motion of objects undergoing a constant acceleration. When the acceleration is not constant, the average acceleration may often be used in the constant acceleration equations to find an approximate description of the motion.

1.6 | THE ACCELERATION OF GRAVITY AND FALLING OBJECTS

Until now, our discussion of motion has been based on definitions and their consequences. Specifically, we have considered the mathematical relationships arising from the definitions of velocity and acceleration, but we have made no statements about how the natural world works. However, in order to discuss the motion of falling objects, we must use some information first obtained by Galileo from careful experimental observations.

We know from everyday experience that unsupported objects tend to fall toward the ground. The speed at impact often increases if the distance dropped increases. Thus, it is evident that falling objects undergo an acceleration, which we attribute to *gravity,* the gravitational attraction of the earth. Nevertheless, two essential aspects of this gravitational acceleration are not so readily observed.

Suppose gravity is the only factor affecting the motion of an object falling near the earth's surface, and air resistance is either absent or negligibly small. In this situation, it is found that:

1   *The gravitational acceleration is the same for all falling objects,* no matter what their size or composition.
2   *The gravitational acceleration is constant.* It does not change as the object falls.

Neither of these statements squares entirely with our everyday experience. Coins fall faster than bits of paper, which contradicts statement 1. Objects dropped from great heights reach a maximum or *terminal* velocity, which contradicts statement 2. However, both of these effects result from air resistance. A coin and scrap of paper fall together in a vacuum (Fig. 1.8), and an object falling at high altitude has a constant acceleration until entering the atmosphere.

The acceleration of gravity near the surface of the earth is denoted by g; it is approximately equal to

$$g = 9.8 \text{ m s}^{-2} = 32 \text{ ft s}^{-2} = 980 \text{ cm s}^{-2}$$

Small variations in g occur as a result of changes in latitude, elevation, and the density of local geological features.

In the examples below and in the problems and exercises at the end of the chapter we neglect such

(a)   (b)   (c)

Figure 1.8. (*a*) A coin and a piece of paper are released simultaneously from the top of a container. (*b*) If the container has air in it, the coin hits the bottom first. (*c*) If most of the air is pumped out of the container, both objects reach the bottom at the same time.

variations in g and assume that air resistance is unimportant. This allows us to use the constant acceleration equations to find the motion of falling objects.

Example 1.13

A ball is dropped from a window 84 m above the ground (Fig. 1.9). (a) When does the ball strike the ground? (b) What is the velocity of the ball when it strikes the ground?

In this type of problem we can choose the coordinate system so that the positive x direction is either up or down. The origin $x = 0$ can be at the window, ground level, or a mile underground; the choice is a matter of convenience. Let us choose the origin at the window level, so $x_0 = 0$, and positive values of x in the upward direction.

Since the ball is dropped with zero initial velocity, $v_0 = 0$. The gravitational acceleration is constant and in the $-x$ direction, so the equations of Table 1.4 can be applied using $a = -g = -9.8 \text{ m s}^{-2}$.

(a) The ball strikes the ground when $x = -84$ m. The time t at which this happens satisfies

$$x = \tfrac{1}{2}at^2$$

or

$$t^2 = \frac{2x}{a}$$

Thus $t = \sqrt{\dfrac{2x}{a}} = \sqrt{\dfrac{2(-84 \text{ m})}{-9.8 \text{ m s}^{-2}}} = 4.14 \text{ s}$

Notice that the positive root is used, since the ball hits the ground after it is released, not before.

(b) Using Eq. 1.7 with $t = 4.14$ s,

$$v = at = (-9.8 \text{ m s}^{-1})(4.14 \text{ s}) = -40.6 \text{ m s}^{-1}$$

The velocity is 40.6 m s^{-1} in the negative direction, or downward.

Example 1.14

Repeat the preceding example, choosing $x = 0$ at ground level and the $+x$ direction downward.

(a) With these choices, the initial height of the ball is $x_0 = -84$ m. The acceleration is $a = g = +9.8 \text{ m s}^{-2}$; it is positive because the positive direction is downward. The time when $x = 0$ and the ball strikes the ground can be found from Eq. 1.11. Using $v_0 = 0$, this equation becomes

$$x = x_0 + \tfrac{1}{2}at^2$$

Solving for t,

$$t = \sqrt{\frac{2(x - x_0)}{a}} = \sqrt{\frac{2(0 + 84 \text{ m})}{9.8 \text{ m s}^{-2}}} = 4.14 \text{ s}$$

15

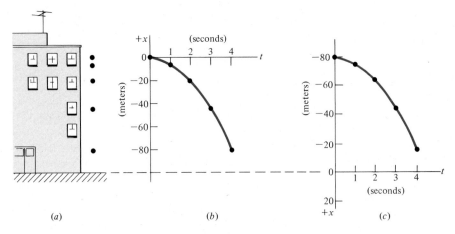

Figure 1.9. (*a*) A freely falling ball released with no initial velocity. Note that it falls through successively larger distances in each second. The corresponding *x-t* graphs use the notations of (*b*) Example 1.13 and (*c*) Example 1.14, respectively.

This agrees with our earlier result.

(b) When $t = 4.14\,\text{s}$,

$$v = at$$
$$= (9.8\,\text{m s}^{-2})(4.14\,\text{s}) = +40.6\,\text{m s}^{-1}$$

The velocity is $40.6\,\text{m s}^{-1}$ downward as before.

From this and the preceding example we see that the choices of the origin and of the positive *x* direction have no effect on our results except that we must correctly interpret our positions and signs.

When an object is thrown straight up, it experiences a constant downward gravitational acceleration. This is true at all times during the motion: while the object is going upward, at the instant it reaches its greatest height, and while it is coming down. However, the velocity is continually changing. Initially it is directed upward. As the object rises, the magnitude of the velocity uniformly decreases, and the *velocity is zero at the instant the object is at its highest point*. The object then begins to move downward, with the velocity steadily increasing in magnitude. This motion is illustrated by the *v-t* and *x-t* graphs in Fig. 1.10, where the $+x$ direction is taken to be upward. At *A* the object is moving upwards, at *B* it is instantaneously at rest at the highest point of its motion, and at *C* it is moving downward. Thus, *v* is initially positive, then zero for an instant, and afterwards negative. By contrast, the acceleration is equal to the slope of the *v-t* graph or to $-g$ at all times.

We can see from Fig. 1.10 that the motion is *sym-*

metric about its highest point (point *B*). The upward velocity 1 second before the peak and the downward velocity 1 second after the peak have the same magnitude. Consequently the time the object takes to traverse any given portion of the path will be the same going up and coming down.

These features of the motion are illustrated by the next example.

Example 1.15

A ball is thrown upward at $64\,\text{ft s}^{-1}$ from a window 192 ft above the ground. (a) How high does it go? (b) When does it reach its highest point? (c) When does it strike the ground?

Let us again choose the origin at the window and let $+x$ be upwards. Thus, $x_0 = 0$, $v_0 = 64\,\text{ft s}^{-1}$, and $a = -g = -32\,\text{ft s}^{-2}$.

(a) As the ball rises, its velocity decreases uniformly

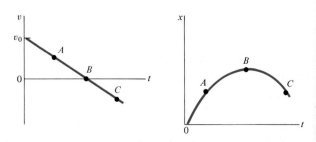

Figure 1.10. (*a*) *v-t* and (*b*) *x-t* graphs for an object thrown straight up from $x = 0$. The velocity is zero at the instant the object reaches its greatest height (point *B*), but the acceleration is always *-g*.

until $v = 0$ at the peak; we want to find the position x at which this occurs. From Eq. 1.14, using $v = 0$ and $x_0 = 0$,

$$x = \frac{v^2 - v_0{}^2}{2a} = \frac{0 - (64 \text{ ft s}^{-1})^2}{2(-32 \text{ ft s}^{-2})} = 64 \text{ ft}$$

Thus the ball reaches a maximum height 64 feet above the window, or $192 + 64 = 256$ ft above the ground.

(b) Using Eq. 1.12 and the fact that $v = 0$ at the peak height,

$$t = \frac{v - v_0}{a} = \frac{0 - 64 \text{ ft s}^{-1}}{-32 \text{ ft s}^{-2}} = 2 \text{ s}$$

The peak is reached in 2 s.

(c) At the ground $x = -192$ ft, so with Eq. 1.11, the time t when this occurs satisfies

$$x = v_0 t + \tfrac{1}{2} a t^2$$

$$-192 \text{ ft} = (64 \text{ ft s}^{-1})t + \tfrac{1}{2}(-32 \text{ ft s}^{-2})t^2$$

Dividing by 16 ft s^{-2} and rearranging terms, this becomes

$$t^2 - (4 \text{ s})t - 12 \text{ s}^2 = 0$$

This can be factored to

$$(t - 6 \text{ s})(t + 2 \text{ s}) = 0$$

This equation has two solutions, $t = 6$ s and $t = 2$ s. Since the ball cannot possibly hit the ground before it is thrown, $t = 6$ s is the correct result. (The extraneous root $t = -2$ s is not meaningless; it is the time at which the ball *would have had* to be thrown up *from the ground* in order to follow the same trajectory that it in fact did follow *after $t = 0$*.)

1.7 | MODELS IN PHYSICS

Real problems sometimes are so complex that an exact solution is either impossible or else requires very difficult measurements or calculations. A rough estimate of the correct solution can often be found with a *mathematical model* obtained by making simplifying assumptions and approximations.

We used a model in the preceding section when we considered falling objects and neglected air resistance. For a rock or a coin falling at low speeds, this is a good approximation. However, if the speed is high, air resistance is more important, and a model that neglects it can be misleading in its predictions. Also, as we saw, it would be very inaccurate to ignore air resistance for something as light as a scrap of paper.

Mathematical models can show what factors are important in a problem and provide a qualitative understanding that may not be obtainable from a more exact approach. Their predictions may indicate whether a more elaborate effort is worth attempting and may suggest lines of attack for such an effort. However, textbooks—including this one—and other sources are not always fully explicit in stating their assumptions and in distinguishing models from exact treatments. Thus, a moderate amount of skepticism is a healthy thing to acquire.

SUMMARY

The motion of an object along a straight line is described by its position, velocity, and acceleration. The velocity is the time rate of change of the position. The average velocity is the change in position or displacement divided by the elapsed time,

$$\bar{v} = \frac{\Delta x}{\Delta t}$$

The instantaneous velocity is the average velocity evaluated for an arbitrarily short time interval. Similarly, the average acceleration is the velocity change divided by the elapsed time,

$$\bar{a} = \frac{\Delta v}{\Delta t}$$

The instantaneous acceleration is the average acceleration evaluated for an arbitrarily short time interval.

Often the acceleration can be calculated theoretically or measured experimentally. If the initial position and velocity are known, their later values can then be found from the acceleration. In the special case where the acceleration is constant, we have found the equations of motion. These are useful for describing the motion of objects that experience the constant acceleration due to gravity.

(New York Public Library Picture Collection.)

GALILEO GALILEI
(1564–1642)

Galileo was born in Pisa. At 17 he began to study medicine, although he had also shown talent in music and art. His interests soon turned toward other branches of science, and he was appointed a professor of mathematics at the University of Pisa. There, between 1589 and 1592, he conducted his investigations of motion which are the basis for this chapter.

The Greek philosopher Aristotle (384–322 B.C.) had taught that heavy objects fall faster than light ones. Galileo performed a series of experiments on objects rolling down smooth inclines and concluded that, in the ideal frictionless case, all objects have the same acceleration. Furthermore, he showed that the distance varies as the time squared, which implies that the acceleration is constant. Galileo is considered to have shown the importance of experimentation in science.

In 1608, Galileo heard that two spectacle lenses used together would magnify a distant object, and he soon built a series of telescopes of increasing magnifying power. He discovered that our moon has mountain ranges, that Jupiter has moons, and that the sun has spots. But his observations also led to problems. Copernicus (1473–1543) had earlier doubted Aristotle's teaching that the earth is the center of the universe. He showed that the apparent motions of the sun, stars, and planets could be most simply explained by supposing that the earth itself is a planet that rotates daily on its axis and revolves annually about the sun. Galileo's observations supported this heretical view that the earth is not unique. Consequently, he ran into serious difficulties with the authorities.

Galileo's clash with Church dogma lasted for nearly two decades. Initially forbidden to discuss his ideas, he was later ordered to describe the Copernican ideas as only a theory. However, Galileo's analysis and presentation of the existing facts were so complete and convincing that, at age 70, he was tried for disregarding the earlier order. After the trial, he was kept under house arrest for the remaining 12 years of his life.

Checklist

Define or explain:

dimensions
standards
units
S.I.
c.g.s. system
British system
random errors
systematic errors
significant figures
translational motion
average velocity
uniform motion

accelerated motion
instantaneous velocity
slope
average acceleration
instantaneous
  acceleration
constant acceleration
  formulas
gravitational
  acceleration
terminal velocity
mathematical model

EXERCISES

The student should work out some of the exercises for each section to test his or her general understanding of the concepts before attempting the problems. Answers to most of the odd-numbered exercises and problems are given at the end of the book.

Section 1.1 | Measurements, Standards, Units, and Errors

1-1 An acre is 43,560 ft². How large is this in square meters (m²)?

1-2 Convert 40 mi h⁻¹ to meters per second (m s⁻¹).

1-3 A gallon is 231 cubic inches (in.³), and a liter is 1000 cm³. How many liters are there in a gallon?

1-4 A furlong is 220 yards, and a fortnight is 14 days. If a snail moves at 2 m h⁻¹, what is this in furlongs per fortnight?

1-5 A cell membrane is 70 angstrom (Å) units thick. If an angstrom unit is 10^{-10} m, what is the membrane thickness in inches?

1-6 If two quantities have different dimensions, can they be (a) multiplied; (b) added? Give examples to support your answers.

Section 1.2 | Average Velocity

1-7 A car travels 30 miles in 45 minutes on a straight highway. What is its average velocity in miles per hour (mi h⁻¹)?

1-8 A pilot wishes to fly 2000 km in 4 hours. What average velocity in meters per second is required to accomplish this?

1-9 Sketch the position-versus-time graph for a car that starts from rest and is driven 1 mile to a store. (Describe the motion in words.)

1-10 A car travels in a straight line at 40 mi h⁻¹ for 1 hour and at 60 mi h⁻¹ for 2 hours. (a) How far does it travel? (b) Find the average velocity.

1-11 A woman wants to drive 100 miles in 2 hours. If she averages 40 mi h⁻¹ for the first 1.5 hours, what average velocity must she maintain for the remaining time?

1-12 A man runs a 42-km marathon in 2.5 hours. Find the average velocity in (a) kilometers per hour (km h⁻¹); (b) miles per hour (mi h⁻¹); (c) feet per second (ft s⁻¹).

1-13 A sprinter runs the 100-m dash in 9.8 s. (a) What is the sprinter's average velocity? (b) Since the runner starts from rest, the velocity cannot be constant. Sketch the approximate position-versus-time graph for the runner. Explain the assumptions made.

1-14 Light travels at 3×10^8 m s⁻¹. A light-year is the distance light travels in 1 year, 365 days. Find the distance in kilometers to the nearest star, which is 4 light-years away from us.

1-15 A falling object moves so that its height x above the ground at time t is given by the equation $x = 200$ ft $- (16$ ft s⁻²$)t^2$, where x is in feet and t is in seconds. Find its average velocity from (a) $t = 0$ to $t = 2$ s; (b) $t = 2$ to $t = 4$ s.

1-16 A ball thrown straight upward has a height x above the ground, which is given by the equation $x = (19.6$ m s⁻¹$)t - (4.9$ m s⁻²$)t^2$ where x is in meters and t is in seconds. Find the average velocity from (a) $t = 0$ to $t = 2$ s; (b) $t = 2$ to 4 s.

1-17 From the position-versus-time graph of Fig. 1.11, find the average velocity from $t = 0$ s to (a) $t = 10$ s; (b) $t = 20$ s; (c) $t = 40$ s.

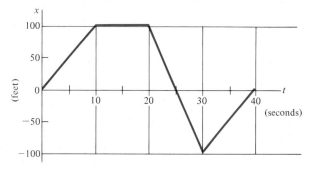

Figure 1.11. Exercises 1-17, 1-18, and 1-19.

Section 1.3 | Instantaneous Velocity

1-18   In Fig. 1.11, what is the instantaneous velocity at (a) $t = 5$ s;   (b) $t = 15$ s;   (c) $t = 25$ s; (d) $t = 35$ s?

1-19   Draw the instantaneous velocity-versus-time graph corresponding to Fig. 1.11.

1-20   Figure 1.12 shows the position of a pendulum versus time. In the interval $t = 0$ to τ, when is the velocity (a) zero; (b) positive; (c) negative?

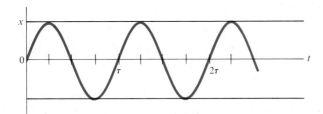

Figure 1.12.   The horizontal position of a pendulum versus time. Exercises 1-20 and, 1-21; Problem 1-47.

1-21   Figure 1.12 shows the position of a pendulum versus time. In the interval from $t = 0$ to τ, when does the velocity have its largest positive and negative values?

Section 1.4 | Acceleration

1-22   A car proceeds to pass another. Its speed increases from 50 to 100 km h^{-1} in 4 s. What is the average acceleration?

1-23   A car moves at a constant velocity of 50 m s^{-1} for 20 s. It then slows with a constant acceleration, coming to rest 10 s later. (a) Draw a velocity-time graph for the car. (b) Draw the acceleration-time graph for the car.

1-24   Draw the acceleration-versus-time graph corresponding to Fig. 1.13.

Figure 1.13.   Exercises 1-24 and 1-25.

1-25   When does the acceleration corresponding to Fig. 1.13 have (a) a maximum value; (b) a minimum value; (c) a value of zero?

Section 1.5 | Finding the Motion of an Object

1-26   A jet plane accelerates on a runway from rest at 4 m s^{-2}. After 5 s, find its (a) distance traveled; (b) velocity.

1-27   A racing car travels 90 m while it accelerates from rest to 30 m s^{-1}. What is its average acceleration?

1-28   A train traveling with a velocity of 30 m s^{-1} stops with a uniform acceleration in 50 s. (a) What is the acceleration of the train? (b) What is the distance traveled before coming to rest?

✓**1-29**   A car starts from rest at $t = 0$ and accelerates as shown in Fig. 1.14. Find its velocity at (a) $t = 10$ s, and (b) $t = 30$ s. (c) Draw the velocity-versus-time graph.

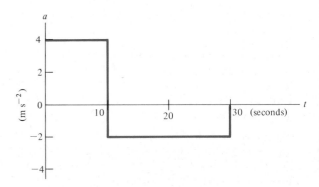

Figure 1.14.   Exercise 1-29.

1-30   Plot the position-versus-time graph corresponding to Fig. 1.15. (Assume the object starts from $x = 0$.)

1-31   Suppose a car can accelerate at 1 m s^{-2}. How large a break in traffic is needed to enter a highway where cars are moving at 20 m s^{-1} if the driver wants to avoid forcing the next car to slow down or to approach closer than 25 m?

1-32   A car moving at 15 m s^{-1} hits a stone wall. (a) A seat-belted passenger comes to rest in 1 m. What average acceleration does this person experience? (b) Another passenger without a seat belt strikes the windshield and comes to rest in 0.01 m. What average acceleration does this person experience?

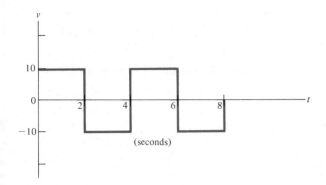

Figure 1.15. Exercise 1-30.

1-33 A dragster has a constant acceleration for 0.25 miles. Its speed at the end of this period is 245 mi h^{-1} = 359 ft s^{-1}. (a) What is the acceleration of the dragster? (b) How long does it require to cover the 0.25-mile distance?

Section 1.6 | The Acceleration of Gravity and Falling Objects

1-34 From what height must water fall to strike a turbine wheel with a vertical downward velocity of 30 m s^{-1}?

1-35 An antiaircraft shell is fired vertically upward with an initial velocity of 500 m s^{-1}. (a) Compute the maximum height of the shell. (b) How long does it take to reach this height? (c) When will the height be 1000 meters?

1-36 A model rocket is fired straight up with a constant acceleration of 50 m s^{-2} until the engine runs out of fuel after 4 s. Neglecting air resistance, find (a) the height of the rocket when the engine stops; (b) the maximum height reached; (c) the total time duration of the flight.

1-37 A rock falls from a cliff 60 m high. (a) Find the average velocity during the first 3 seconds of its fall. (b) At what instant of time is the instantaneous velocity equal to the average velocity of part (a)? (c) How long does it take for the rock to fall to the ground?

1-38 A boy standing beside a tall building throws a ball straight up with an initial velocity of 15 m s^{-1}. (a) How high will the ball rise? (b) How long will it take for the ball to reach its maximum height? (c) Another boy reaches out of a window 6 m above the initial position of the ball and at-

tempts to catch the ball. At what times will the ball pass him?

1-39 A rock dropped from the top of a tower strikes the ground in 4 s. (a) Find the velocity of the rock just before it strikes the ground. (b) Find the height of the tower.

1-40 A stone is thrown vertically downward from a bridge with an initial velocity of 10 m s^{-1}. It strikes the water in 3 s. (a) What is the velocity of the stone as it strikes the water? (b) What is the height of the bridge above the water?

1-41 A stone dropped from a bridge strikes the water in 5 s. (a) What is the velocity of the stone when it strikes the water? (b) What is the height of the bridge?

1-42 The hammer of a pile driver strikes the top of a pipe with a velocity of 7 m s^{-1}. From what height did the hammer fall?

1-43 A sandbag dropped from a balloon strikes the ground in 15 s. Which was the height of the balloon if it is initially (a) at rest in the air; (b) descending with a velocity of 20 m s^{-1}?

1-44 A box falls from an elevator that is ascending with a velocity of 2 m s^{-1}. It strikes the bottom of the elevator shaft in 3 s. (a) How long will it take the box to reach its maximum height? (b) How far from the bottom of the shaft was the box when it fell off the elevator? (c) What is the height of the elevator when the box is at its highest point?

PROBLEMS

The occasional problems preceded by an asterisk are the most difficult. Answers to most of the odd-numbered problems are given at the end of the book.

1-45 At the instant a bowler releases the ball, her hand is moving horizontally with a speed of 6 m s^{-1} relative to her body. If she is moving forward at a speed of 1 m s^{-1}, what is the velocity of the ball with respect to the floor?

1-46 When a bowling ball is released, a bowler's hand is moving relative to his forearm at 2.7 ft s^{-1}. The forearm is moving relative to the upper arm at 1.8 ft s^{-1}, and the upper arm is moving at 17.3 ft s^{-1} relative to the shoulder. (The velocities refer to the outermost ends of the respective parts of the body and are all horizontal.) If the bowler's shoulder is moving relative to the floor at 4.7 ft s^{-1}, how fast is the ball thrown?

1-47 The position-versus-time graph for a pendulum is shown in Fig. 1.12. Sketch the velocity-versus-time and acceleration-versus-time graphs for the motion.

1-48 A sled starting from rest slides down a hill with uniform acceleration. It travels 12 m in the first 4 s. When will the sled have a velocity of 4 m s^{-1}?

1-49 A dragster accelerates for one-quarter of a mile and then brakes to a stop in an additional one-half mile. (a) Sketch an approximate velocity-versus-time graph. (b) Sketch an approximate acceleration-versus-time graph.

1-50 Plane A, flying at 500 m s^{-1}, is 10,000 m directly behind plane B, moving in the same direction at 400 m s^{-1}. The pilot of plane A fires a missile which accelerates at 100 m s^{-2}. How long will it take for the missile to reach plane B? (Neglect the effects of the gravitational acceleration.)

1-51 A dog running at 10 m s^{-1} is 30 m behind a rabbit moving at 5 m s^{-1}. When will the dog catch up with the rabbit? (Both velocities remain constant.)

1-52 A rocket-powered experimental sled carrying a test pilot is brought to rest from 200 m s^{-1} over a distance d. If the pilot is not to be subjected to an acceleration greater than six times that of gravity, what is the minimum value of d?

1-53 The world record for the 100-yard dash is 9.0 s and for the 60-yard dash, it is 6.1 s. Assume a sprinter accelerates at a constant rate up to a maximum velocity that is maintained for the remainder of the race. (a) Find the acceleration. (b) What is the duration of the acceleration period? (c) What is the maximum velocity? (d) The record for the 220-yard dash is 19.5 s, while for the mile it is about 232 s. Are these times consistant with the assumptions made?

1-54 A sandbag is dropped from an ascending balloon that is 300 m above the ground and is ascending at 10 m s^{-1}. (a) What is the maximum height of the sandbag? (b) Find the position and velocity of the sandbag after 5 s. (c) How long does it take for the sandbag to reach the ground from the time it was dropped?

TABLE 1.5

Acceleration distances *d* and vertical heights *h* for several animals. All distances are in meters.

	Acceleration Distance (d)	Vertical Height (h)
Human	0.5	1.0
Kangaroo	1.0	2.7
Bushbaby	0.16	2.2
Frog	0.09	0.3
Locust	0.03	0.3
Flea	0.0008	0.1

SUPPLEMENTARY TOPICS

1.8 | VERTICAL JUMPING

We can use the equations for constant acceleration to analyze the relative jumping ability of various animals. Table 1.5 lists the heights recorded for some animals jumping vertically. Notice that the jumping height for the human is less than the record for the high jump, which is about 2 m. This is because a man 1.8 m tall (about 6 feet) is already in a position to clear a bar at about half his height by rotating his body to a horizontal position. The method of human jumping called the Western Roll (Fig. 1.16) is not used by other animals, so the heights listed in Table 1.5 are the appropriate ones to use for comparing jumping abilities.

Animals make standing jumps by folding their legs and extending them rapidly. Usually the acceleration distance d is somewhat shorter than the animal's legs. Once off the ground the animal experiences only the acceleration of gravity, so the constant acceleration formulas apply. We can also analyze the takeoff phase if we make the approximation that the acceleration a_t during takeoff is constant. This approximation is used in the following example.

Example 1.16.

Using the data in Table 1.5, find (a) the takeoff velocity v_t for a human and (b) the takeoff acceleration a_t.

(a) Let the coordinate x stand for the position of the person's midpoint. We will choose the $+x$ direction upward and take the origin $x = 0$ where the person is standing erect or just before leaving the ground, as in Fig. 1.17b. To find the takeoff velocity, we consider the airborne phase of the jump. During this phase, the

(a)

(b)

Figure 1.16. High jumpers using (*a*) the traditional Western Roll and (*b*) the newer Fosbury Flop techniques. In both cases the athlete keeps as much of his weight below the bar as possible. ((*a*) United Press International; (*b*) Wide World Photos.)

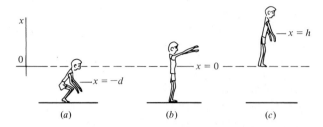

Figure 1.17. Positions during vertical jumping; (*a*) crouched, with $v = 0$; (*b*) fully extended at takeoff, with $v = v_t$; and (*c*) maximum height, with $v = 0$. The coordinate *x* indicates the height of the person's midpoint. The origin ($x = 0$) is at the height of the midpoint in the fully extended position shown in (*b*).

acceleration is $-g$ and the velocity changes from $v_0 = v_t$ to $v = 0$. The height changes by $(x - x_0) = h = 1$ m, so $v^2 = v_0^2 + 2a(x - x_0)$ becomes

$$0 = v_t^2 - 2gh$$

or

$$v_t^2 = 2gh$$

Thus

$$v_t = \sqrt{2gh} = \sqrt{(2)(9.8 \text{ m s}^{-2})(1 \text{ m})}$$
$$= 4.4 \text{ m s}^{-1}$$

(b) We will assume that during the takeoff phase the acceleration a_t is constant. The velocity increases from $v_0 = 0$ to $v = v_t$, and the height changes by $(x - x_0) = d = 0.5$ m. Thus, during the takeoff phase, $v^2 = v_0^2 + 2a(x - x_0)$ becomes

$$v_t^2 = 2a_t d$$

Comparing this with $v_t^2 = 2gh$, we find

$$a_t = \frac{h}{d} g$$

$$= \frac{1 \text{ m}}{0.5 \text{ m}}(9.8 \text{ m s}^{-2}) = 19.6 \text{ m s}^{-2}$$

The takeoff accelerations and times are vastly different for different animals. A human jumping with the same acceleration as a flea would reach a height of more than 50 m! Despite these great differences, we will see in Chapter Six that the total energy that a given amount of muscle can supply in a single jump is roughly the same in all animals.

EXERCISES ON SUPPLEMENTARY TOPICS

Section 1.8 | Vertical Jumping
1-55 A salmon jumps vertically out of the water at an initial velocity of 6 m s^{-1}. (a) How high will it jump? (b) How long will the salmon be out of the water?

1-56 How high will a woman jump if her takeoff velocity is the same as that of the flea?

1-57 From Table 1.5, compute the average takeoff acceleration and takeoff velocity for a locust, assuming the acceleration is constant. Compare these results with that for the human of Example 1.16.

1.58 As astronaut wearing a space suit can jump 0.5 m vertically at the surface of the earth. The gravitational acceleration on Mars is 0.4 times that on the earth. If his takeoff velocity is the same, how high can the astronaut jump on Mars?

1.59 If a human could achieve a takeoff acceleration equal to that of the flea, how high could she jump? (Assume the acceleration distance is still 0.5 meters.)

1-60. Compute the takeoff times for the human, bushbaby, and flea of Table 1.5.

Additional Reading

Alfred M. Bork and Arnold B. Arons, Resource Letter Col R-1 on Collateral Reading in Physics, *American Journal of Physics*, vol. 35, 1967, p. 1. A bibliography.

Marjorie Nicholson, Science and Literature, *American Journal of Physics*, vol. 33, 1965, p. 175. A bibliography.

Arnold B. Arons and Alfred M. Bork, *Science and Ideas*, Prentice-Hall, Inc., Englewood Cliffs, N.J., 1964. An anthology.

American Foundation for Continuing Education, *The Mystery of Matter*, Oxford University Press, New York, 1965. An anthology.

National Science Foundation, *Exploring the Universe*, McGraw-Hill Book Co., New York, 1963. An anthology.

W.F. Magie, *Source Book in Physics*, McGraw-Hill Book Co., New York, 1935. An anthology.

L.B. Macurdy, Standards of Mass, *Physics Today*, vol. 4, April 1951, p. 7.

D. Brouwer, The Accurate Measurement of Time, *Physics Today*, vol. 4, August 1951, p. 6.

Stillman Drake, *Discoveries and Opinions of Galileo*, Doubleday and Co., Garden City, New York, 1957.

Giorgio de Santillana, *The Crime of Galileo,* University of Chicago Press, Chicago, 1955.

R.B. Lindsay, Galileo Galilei, 1564–1642, and the Motion of Falling Bodies, *American Journal of Physics,* vol. 10, 1942, p. 285.

Sir James Gray, *How Animals Move,* Cambridge University Press, Cambridge, 1953. Pages 69–80 discuss jumping.

R. McNeill Alexander, *Animal Mechanics,* University of Washington Press, Seattle, 1968. Pages 28–33 discuss jumping.

Scientific American articles:

Herbert Butterfield, The Scientific Revolution, September 1960, p. 173.

P.A.M. Dirac, The Physicist's Picture of Nature, May 1963, p. 45.

Freeman J. Dyson, Mathematics in the Physical Sciences, September 1964, p. 128.

Allen V. Astin, Standards of Measurement, June 1968, p. 50.

Lord Ritchie-Calder, Conversion to the Metric System, July 1970, p. 17.

Barry N. Taylor, Donald N. Langenberg, and William H. Parker, The Fundamental Physical Constants, October 1970, p. 62.

J.E. Ravetz, The Origins of the Copernican Revolutions, October 1966, p. 88.

Stillman Drake, Galileo's Discovery of the Law of Free Fall, May 1973, p. 84.

Stillman Drake, The Role of Music in Galileo's Experiments, June 1975, p. 98.

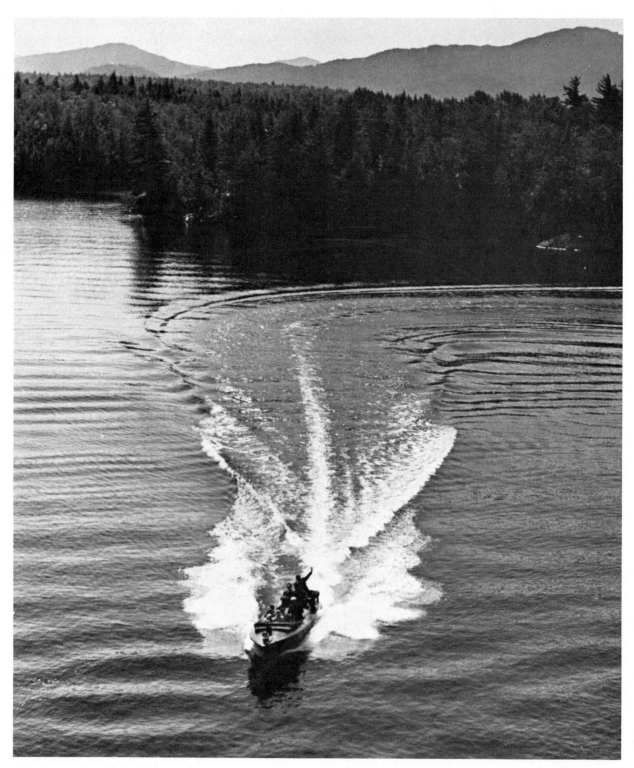

(Odie Monahan/FPG)

CHAPTER TWO

MOTION IN TWO DIMENSIONS

Although many of the principles of mechanics can be illustrated by objects moving in a straight line, their applications often involve more complex motions. Animals leaping forward, thrown or struck balls, and figure skaters all move in a vertical or horizontal plane. The definitions of position, velocity, and acceleration and their interrelations can be extended from the one-dimensional case to the more general situation if these quantities are represented by *vectors*. Vectors are mathematical objects having both a magnitude and a direction and are used to represent many other physical quantities in addition to the ones needed to describe motion. By contrast, some physical quantities, such as temperature and time, have no direction and are represented by ordinary numbers or *scalars*.

In the first section we introduce vectors and explain some of the rules for manipulating them. We then restate the definitions of the previous chapter in vector language and show that motion in a plane is equivalent to a pair of one-dimensional motions. This permits us to carry over the results of the preceding chapter and to discuss the motion of projectiles.

2.1 | AN INTRODUCTION TO VECTORS

In this section we introduce vectors and show how to add and subtract them. While vectors are written with an arrow over the symbol ($\vec{A}$), in print, they are denoted with boldface type, as in **A, s,** and **v.** The magnitude of the vector **A** is written as A or |**A**|. A vector is pictured in a diagram by an arrow with a length proportional to its magnitude and oriented to indicate its direction. For example, in Fig. 2.1, **A, B,** and **C** each have different directions, and **C** is longer or greater in magnitude than **A.** Also, |**A**| > |**B**|.

Addition of Vectors | The concept of vector addition is illustrated by an example involving two vectors representing displacements. Suppose someone walks a certain distance in one direction and then

(a)　　　　(b)　　　　(c)　　　　(d)

Figure 2.1. (*a*) The vectors **A** and **B** representing two displacements. (*b*) To add **B** to **A**, we place its tail at the head of **A**. (Moving a vector does not change it so long as the direction and magnitude are not altered.) (*c*) The sum **C** = **A** + **B** is a vector from the tail of the first vector, **A**, to the head of the second vector, **B**. **C** represents the net displacement. (*d*) The order in which the vectors are added does not matter. **A** + **B** = **B** + **A.**

Figure 2.2. (*a*) Three vectors, **A**, **B** and **C**. (*b*) and (*c*) Two possible orderings in computing the sum **D** of the three vectors.

turns and walks another distance in a second direction. The net change in position or displacement will depend on the magnitudes and directions of the two displacements. Let us denote the first displacement by **A** and the second by **B**. The net displacement **C** is the sum of **A** and **B**,

$$\mathbf{C} = \mathbf{A} + \mathbf{B}$$

The procedure for finding **C** is shown in Fig. 2.1. Note that the order of the vectors in the sum is irrelevant: **A** + **B** and **B** + **A** are the same thing. When there are three or more vectors to be summed, two vectors are added first, then the next, and so on. Again, the order of the vectors is irrelevant; for example, in Fig. 2.2, **A** + **B** + **C** = **B** + **A** + **C**.

The following example shows how the magnitude of **A** + **B** depends on the relative directions of **A** and **B**.

Example 2.1

A person walks 1 km due east. If the person then walks a second kilometer, what is the final distance from the starting point if the second kilometer is walked (a) due east; (b) due west; (c) due south?

We will call the first displacement **A** and the second **B**. Using the procedures outlined above, we construct the sum **C** = **A** + **B** for the three cases (Fig. 2.3).

(a) Since **A** and **B** are in the same direction, $C = A + B = 2A = 2$ km. **C** is directed due east.

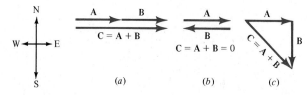

Figure 2.3. The sum of two displacement vectors that are equal in magnitude and (*a*) parallel; (*b*) opposite or antiparallel; and (*c*) perpendicular.

(b) Here, the vectors are opposite, so $C = A - B = 0$.

(c) From the Pythagorean theorem, $C^2 = A^2 + B^2 = 2A^2$, so

$$C = \sqrt{2}A = \sqrt{2}\ \text{km}$$

From the figure, **C** points toward the southeast.

Multiplication of a Vector By a Scalar

Multiplication of a vector by a scalar is defined so that the usual rules of algebra apply. If $2\mathbf{A} = \mathbf{A} + \mathbf{A}$ is to hold, then we must interpret $2\mathbf{A}$ as a vector in the same direction as **A** but with twice the length. For example, if **A** is a 2-mile displacement due north, than $2\mathbf{A}$ is a 4-mile displacement due north, and $5\mathbf{A}$ is a 10-mile displacement due north.

Vector Subtraction

Subtraction of vectors is also defined so that the usual algebraic rules apply. In order that $\mathbf{A} - \mathbf{A} = 0$ holds, $-\mathbf{A}$ must be interpreted as a vector equal in magnitude to A but opposite in direction. Thus if **A** is a 2-mile displacement due north, $-\mathbf{A}$ is a 2-mile displacement due south, and $-3\mathbf{A}$ is a 6-mile displacement due south. The vector difference $\mathbf{C} = \mathbf{B} - \mathbf{A} = \mathbf{B} + (-\mathbf{A})$ is evaluated by adding the vector $-\mathbf{A}$ to the vector **B** as in Fig. 2.4.

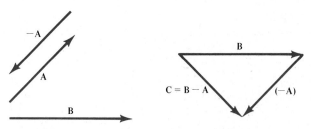

Figure 2.4. The vector $\mathbf{C} = \mathbf{B} - \mathbf{A}$ is found by adding $-\mathbf{A}$ to **B**.

Vector Components

A vector in a given plane is specified by two pieces of information, its magnitude and its direction. Equivalently, the vector can be specified by two other quantities, its *components* along a pair of perpendicular axes. These components are often useful in vector calculations.

The procedure for finding the components of a vector is shown in Fig. 2.5. The vector **A** is redrawn from the origin of the *x-y* axes, and the components of **A**, denoted by A_x and A_y, are constructed by

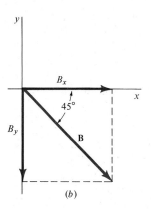

Figure 2.5. (*a*) A vector **A** and perpendicular *x* and *y* axes. (*b*) The components of **A** are A_x and A_y.

(*a*)

Figure 2.7. Example 2.2.

Figure 2.6. A right triangle.

drawing the dashed lines at right angles to the axes. Expressions relating the components to the magnitude of **A** and its angle θ with the *x* axis can be obtained using some properties of right triangles.

Figure 2.6 shows a right triangle with sides *a*, *b*, and *c*. These satisfy the Pythagorean relation,

$$a^2 + b^2 = c^2 \tag{2.1}$$

The sine, cosine, and tangent of the angle θ are defined by

$$\sin\theta = \frac{\text{opposite side}}{\text{hypotenuse}} = \frac{a}{c}$$

$$\cos\theta = \frac{\text{adjacent side}}{\text{hypotenuse}} = \frac{b}{c}$$

$$\tan\theta = \frac{\text{opposite side}}{\text{adjacent side}} = \frac{a}{b}$$

Thus, in Fig. 2.5*b*, the components A_x and A_y satisfy

$$A_x{}^2 + A_y{}^2 = A^2 \tag{2.2}$$

Also,

$$\cos\theta = \frac{A_x}{A}, \qquad \sin\theta = \frac{A_y}{A} \tag{2.3}$$

or

$$A_x = A\cos\theta, \qquad A_y = A\sin\theta \tag{2.4}$$

A_x is positive when it is directed in the $+x$ direction and negative when it is directed toward the $-x$ direction. Similarly A_y can be positive or negative, as we will see in the next example.

Example 2.2

Find the components of the vectors **A** and **B** in Fig. 2.7 if $A = 2$ and $B = 3$.

Using the trigonometric tables in Appendix A, we find that $\cos 30° = 0.866$ and $\sin 30° = 0.500$. Thus,

$$A_x = A\cos\theta = 2\cos 30° = 2(0.866) = 1.73$$
$$A_y = A\sin\theta = 2\sin 30° = 2(0.500) = 1.00$$

From Fig. 2.7*b*, B_x is positive and B_y is negative. With $\cos 45° = \sin 45° = 0.707$,

$$B_x = 3\cos 45° = 3(0.707) = 2.12$$
$$B_y = -3\sin 45° = -3(0.707) = -2.12$$

The sum of two or more vectors can be calculated conveniently in terms of components. To do this, we define the *unit vector* $\hat{\mathbf{x}}$ (read "*x*-hat"), a vector of length one in the $+x$ direction. Similarly, $\hat{\mathbf{y}}$ is a unit vector in the $+y$ direction. With these unit vectors, a vector **A** with components A_x and A_y can be written as

$$\mathbf{A} = A_x\hat{\mathbf{x}} + A_y\hat{\mathbf{y}}$$

If a vector **B** is also written as

$$\mathbf{B} = B_x\hat{\mathbf{x}} + B_y\hat{\mathbf{y}}$$

the sum $\mathbf{C} = \mathbf{A} + \mathbf{B}$ is

$$\mathbf{C} = (A_x + B_x)\hat{\mathbf{x}} + (A_y + B_y)\hat{\mathbf{y}} \tag{2.5}$$

*The components of **C** are the sums of the components of **A** and **B*** (Fig. 2.8). This is illustrated by the following example.

Example 2.3

In Fig. 2.9, $\mathbf{A} = 2\hat{\mathbf{x}} + \hat{\mathbf{y}}$ and $\mathbf{B} = 4\hat{\mathbf{x}} + 7\hat{\mathbf{y}}$. (a) Find the components of $\mathbf{C} = \mathbf{A} + \mathbf{B}$. (b) Find the magnitude

29

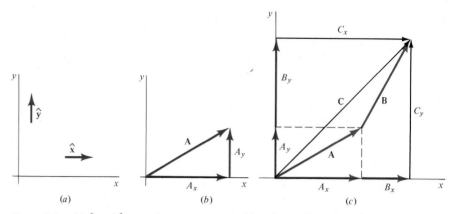

Figure 2.8. (*a*) $\widehat{\mathbf{x}}$ and $\widehat{\mathbf{y}}$ are unit vectors, vectors of length one directed along the coordinate axes. (*b*) A vector **A** can be constructed from its components A_x and A_y. (*c*) $\mathbf{C} = \mathbf{A} + \mathbf{B}$ in components is $C_x = A_x + B_x$, $C_y = A_y + B_y$.

of **C** and its angle θ with the x axis.

(a) Using Eq. 2.5,

$$\mathbf{C} = (2 + 4)\widehat{\mathbf{x}} + (1 + 7)\widehat{\mathbf{y}} = 6\widehat{\mathbf{x}} + 8\widehat{\mathbf{y}}$$

Thus $C_x = 6$, and $C_y = 8$.

(b) From the Pythagorean theorem,

$$C^2 = C_x{}^2 + C_y{}^2 = 6^2 + 8^2 = 100$$

so $C = 10$. From Fig. 2.9, we see that the angle θ satisfies

$$\tan \theta = \frac{C_y}{C_x} = \frac{8}{6} = 1.333$$

Using the table in Appendix A, we find $\theta = 53°$.

Figure 2.9.   Example 2.3.

2.2 | THE VELOCITY IN TWO DIMENSIONS

In two dimensions, position, velocity, and acceleration are represented by vectors. Their definitions are very similar to those for motion along a straight line,

and their x components are related to each other in the same way as are x, v, and a for straight line motion. Since this is also true for the y components, *a problem involving motion in a plane is effectively a pair of one-dimensional motion problems.*

Figure 2.10 shows an object moving in a plane labeled with x and y axes. The object could be a car, an animal, or a red blood cell, and it is represented symbolically by a point. If the displacement in a time interval Δt is denoted by the vector $\Delta \mathbf{s}$, then the *average velocity* of the object is

$$\overline{\mathbf{v}} = \frac{\Delta \mathbf{s}}{\Delta t} \tag{2.6}$$

We can write $\Delta \mathbf{s}$ in terms of its components in the x and y direction:

$$\Delta \mathbf{s} = \Delta x \widehat{\mathbf{x}} + \Delta y \widehat{\mathbf{y}} \tag{2.7}$$

Then the average velocity can also be written in terms of its components.

$$\overline{\mathbf{v}} = \frac{\Delta x}{\Delta t} \widehat{\mathbf{x}} + \frac{\Delta y}{\Delta t} \widehat{\mathbf{y}}$$

$$= \overline{v}_x \widehat{\mathbf{x}} + \overline{v}_y \widehat{\mathbf{y}} \tag{2.8}$$

Notice that

$$\overline{v}_x = \frac{\Delta x}{\Delta t}, \qquad \overline{v}_y = \frac{\Delta y}{\Delta t} \tag{2.9}$$

so each component of the average velocity looks like a one-dimensional average velocity.

Since the average velocity in a plane is equivalent to 2 one-dimensional average velocities, everything

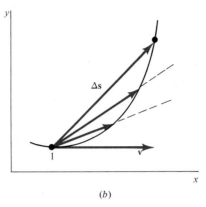

(a)   (b)

Figure 2.10. (a) An object moves along a path in a plane. At time t_1, it is at point 1, and at t_2 it is at point 2. The average velocity is parallel to $\Delta \mathbf{s}$. (b) As the time interval $t_2 - t_1$ becomes smaller, so does the displacement $\Delta \mathbf{s}$. The average velocity $\overline{\mathbf{v}} = \Delta \mathbf{s}/\Delta t$ approaches the instantaneous velocity $\mathbf{v}$ at t_1 which is tangent to the path at point 1.

stated in the preceding chapter about the average and instantaneous velocities holds here for each component. For example, the *instantaneous velocity* $\mathbf{v}$ is the average velocity for an extremely short time interval. In component form

$$\mathbf{v} = v_x \widehat{\mathbf{x}} + v_y \widehat{\mathbf{y}} \qquad (2.10)$$

Here v_x and v_y are the instantaneous rates of change of the x and y coordinates of the object. *At any instant,* $\mathbf{v}$ *is directed along the tangent to the path or trajectory of the object* (Fig. 2.10b). These ideas are illustrated by the following example.

Example 2.4

A car travels halfway around an oval racetrack at a constant speed of 30 m s^{-1} (Fig. 2.11). (a) What are its instantaneous velocities at points 1 and 2? (b) It takes 40 s to go from 1 to 2, and these points are 300 m apart. What is the average velocity of the car during this time interval?

(a) The instantaneous velocity is tangent to the path of the car, and its magnitude is equal to the speed. Thus, at point 1 the velocity is directed in the $+y$ direction and $\mathbf{v}_1 = 30$ m s$^{-1}\widehat{\mathbf{y}}$. Similarly, at point 2 the velocity is along the $-y$ direction, and $\mathbf{v}_2 = (30$ m s$^{-1})(-\widehat{\mathbf{y}}) = -30$ m s$^{-1}\widehat{\mathbf{y}}$.

(b) The average velocity is the displacement divided by the elapsed time. The displacement is entirely along the x direction, so $\Delta \mathbf{s} = (300$ m$)\widehat{\mathbf{x}}$. Since $\Delta t = 40$ s,

$$\Delta \overline{\mathbf{v}} = \frac{\Delta \mathbf{s}}{\Delta t} = \frac{300 \text{ m}}{40 \text{ s}}\widehat{\mathbf{x}} = 7.5 \text{ m s}^{-1}\widehat{\mathbf{x}}$$

The average velocity during this time interval is directed along the $+x$ axis. Its magnitude is less than the speed of 30 m s^{-1} because the car does not travel in a straight line.

2.3 | THE ACCELERATION IN TWO DIMENSIONS

The acceleration of an object moving in a plane is defined much as in the one-dimensional case. Suppose that in a time interval $\Delta t = t_2 - t_1$ the velocity changes by $\Delta \mathbf{v} = \mathbf{v}_2 - \mathbf{v}_1$. Then the *average acceleration* is

$$\overline{\mathbf{a}} = \frac{\Delta \mathbf{v}}{\Delta t} = \frac{\mathbf{v}_2 - \mathbf{v}_1}{t_2 - t_1} \qquad (2.11)$$

The *instantaneous acceleration* $\mathbf{a}$ is the average acceleration during an extremely short interval (Fig. 2.12). In components,

$$\mathbf{a} = a_x \widehat{\mathbf{x}} + a_y \widehat{\mathbf{y}} \qquad (2.12)$$

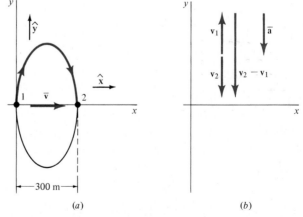

(a)   (b)

Figure 2.11. (a) A car goes halfway around an oval track. Its average velocity $\overline{\mathbf{v}}$ during this interval points in the $+x$ direction. (b) The average acceleration $\overline{\mathbf{a}}$ points in the $-y$ direction.

31

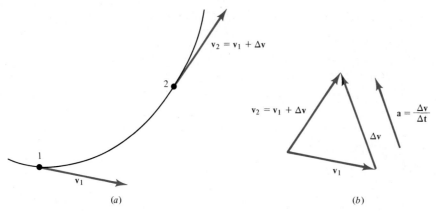

Figure 2.12. (a) The black curve represents the path of an object. The velocities at points 1 and 2 are vectors tangent to the trajectory at times t_1 and t_2. (b) The average acceleration $\bar{\mathbf{a}}$ from t_1 to t_2 is parallel to $\Delta\mathbf{v}$.

Here a_x and a_y are the rates of change of v_x and v_y.

The following example illustrates the calculation of the average acceleration.

Example 2.5

In the preceding example the velocity of the car changed from $\mathbf{v}_1 = 30 \text{ m s}^{-1}\hat{\mathbf{y}}$ to $\mathbf{v}_2 = -30 \text{ m s}^{-1}\hat{\mathbf{y}}$ in 40 s. What was the average acceleration of the car in that time interval?

The average acceleration is defined as the velocity change divided by the elapsed time:

$$\bar{\mathbf{a}} = \frac{\mathbf{v}_2 - \mathbf{v}_1}{\Delta t} = \frac{(-30 \text{ m s}^{-1}\hat{\mathbf{y}}) - (30 \text{ m s}^{-1}\hat{\mathbf{y}})}{40 \text{ s}}$$

$$= -1.5 \text{ m s}^{-2}\hat{\mathbf{y}}$$

Thus the average acceleration during the time the car goes from point 1 to point 2 is directed in the $-y$ direction, or downward in Fig. 2.11b.

This example illustrates an important point. A constant *velocity* means that the acceleration is zero, but a constant *speed* may or may not correspond to zero acceleration. An object moving in a curved path with constant speed has its velocity changing direction, so it is accelerating. We feel the effects of this acceleration when a car turns a corner quickly. *To reiterate, the acceleration is zero only when the speed and the direction of motion are both constant.*

2.4 | FINDING THE MOTION OF AN OBJECT

As we have stressed, a problem involving the motion of an object in a plane is equivalent to a pair of straight line motion problems. Consequently the x and y components of the position and velocity of an object can be calculated exactly as before if the acceleration and the initial position and velocity are known. For example, suppose an object is initially at rest at the origin and the acceleration is constant. In one dimension, the subsequent position and velocity in such a situation are given by $x = \frac{1}{2}at^2$ and $v = at$. Here, similar equations hold for the x and y motions separately:

$$x = \tfrac{1}{2}a_x t^2, \qquad y = \tfrac{1}{2}a_y t^2 \qquad (2.13)$$

and

$$v_x = a_x t, \qquad v_y = a_y t \qquad (2.14)$$

Similarly, if initially the object is at the position (x_0, y_0) and has velocity components (v_{0x}, v_{0y}), then we have the pairs of equations

$$x = x_0 + v_{0x}t + \tfrac{1}{2}a_x t^2,$$
$$y = y_0 + v_{0y}t + \tfrac{1}{2}a_y t^2 \qquad (2.15)$$

and

$$v_x = v_{0x} + a_x t, \qquad v_y = v_{0y} + a_y t \qquad (2.16)$$

Each of the other constant acceleration equations of Table 1.4 also leads to a pair of component equations for two-dimensional motion.

The use of these equations is illustrated in the next section.

2.5 | PROJECTILES

Kicked or thrown balls, jumping animals, and objects

Figure 2.13. Multiple exposure photographs show that a ball dropped from rest and one projected forward fall at the same rate. (Reproduced by permission of the publisher from PSSC Physics, fourth edition, 1976, by D. C. Heath and Company, Lexington, Massachusetts.)

dropped from windows are all examples of *projectiles*. *If air resistance is neglected, the motion of a projectile is influenced only by the constant gravitational acceleration.* Given the initial position and velocity, the subsequent position and velocity can be found using the constant acceleration equations.

To use these equations for the motion of a projectile, let us choose the *x* axis to be horizontal and the +*y* axis to be directed upward. Then $a_x = 0$ and $a_y = -g$, and Eqs. 2.15 and 2.16 become

$$x = x_0 + v_{0x}t, \qquad y = y_0 + v_{0y}t - \tfrac{1}{2}gt^2 \quad (2.17)$$

and

$$v_x = v_{0x}, \qquad v_y = v_{0y} - gt \quad (2.18)$$

The *x* and v_x equations show that the horizontal motion is uniform, and the *y* and v_y equations show the vertical motion is that of an object acting only under the influence of gravity.

The equations for the vertical motion can be used as in straight line motion to answer various questions. For example, the projectile strikes the ground when *y* equals the ground height, and the peak height occurs when $v_y = 0$. This approach can be applied to qual-

itative and quantitative projectile motion problems.

Two interesting examples of projectile motion are often shown as lecture demonstrations. In one of these, two steel balls are simultaneously released from a stand above the floor. Ball 1 is projected horizontally by a spring, while ball 2 is allowed to drop from rest. The problem is to predict which ball will hit the floor first.

Although this problem sounds hard initially, its solution is apparent once we realize that both balls have the same initial vertical velocity and position components and the same acceleration. Consequently, the equations for the vertical motion of the two balls are the same, and they must hit the floor simultaneously (Fig. 2.13). This illustrates clearly the idea that motion in the horizontal and vertical directions are independent.

The second demonstration is a bit more elaborate. A projectile is fired by a "cannon" pointed at a stuffed animal (Fig. 2.14). Just as the projectile leaves the cannon, the stuffed animal is released and falls. Somewhat surprisingly, the projectile strikes the animal in midair. The point is that if there were no gravitational acceleration, the projectile would travel in a straight line. However, since both the projectile and the stuffed animal experience the same gravitational acceleration, both fall at the same rate relative to where they would have been otherwise. Hence, if the cannon is aimed directly at the animal, the only effect of the gravitational acceleration is that they meet somewhat below where the animal was initially.

The following example illustrates the solution of quantitative projectile motion problems.

Figure 2.14. The projectile and stuffed animal fall together when simultaneously released. They collide at the point indicated with a cross.

Figure 2.15. A tennis ball is served horizontally.

Example 2.6

A tennis ball is served horizontally from 8 ft above the ground at 100 ft s^{-1} (Fig. 2.15). (a) The net is 40 ft away and 3 ft high. Will the ball clear the net? (b) Where will the ball land?

(a) To find the height of the ball at the net, we must first use the equations for the horizontal motion to find out when it will reach the net, where $x = 40$ ft. From this time, we can determine the height. Since $x_0 = 0$, Eq. 2.17 simplifies to $x = v_{0x}t$. Solving for t,

$$t = \frac{x}{v_{0x}} = \frac{40 \text{ ft}}{100 \text{ ft s}^{-1}} = 0.4 \text{ s}$$

Using $t = 0.4$ s, $y_0 = 8$ ft, and $v_{0y} = 0$, the vertical position is

$$y = y_0 + v_{0y}t - \tfrac{1}{2}gt^2$$
$$= 8 \text{ ft} - \tfrac{1}{2}(32 \text{ ft s}^{-2})(0.4 \text{ s})^2 = 5.44 \text{ ft}$$

Since the net is only 3 ft high, the ball easily clears it.

(b) The ball lands at the time t when $y = 0$. Once we have found this time, we can find the distance it has traveled from the horizontal motion. Substituting $y = 0$ and $v_{0y} = 0$ in $y = y_0 + v_{0y}t - \tfrac{1}{2}gt^2$, we find

$$0 = y_0 - \tfrac{1}{2}gt^2$$

Thus

$$t^2 = \frac{2y_0}{g} = \frac{2(8 \text{ ft})}{32 \text{ ft s}^{-2}} = \frac{1}{2}\text{s}^2$$
$$t = 0.707 \text{ s}$$

Hence the distance the ball travels before it lands is

$$x = v_{0x}t = (100 \text{ ft s}^{-1})(0.707 \text{ s}) = 70.7 \text{ ft}$$

Projectiles in Athletics | The preceding ex-
ample illustrates how the projectile motion formulas can be used to analyze a tennis serve. A serve should clear the net by a small but safe margin and land just inside the service court. If a beginner serves the ball rather slowly, the ball will not clear the net unless it is

served slightly above the horizontal direction. More advanced players can serve horizontally or slightly below the horizontal, since the ball moves faster and has a flatter trajectory. While a player can determine his or her own best serving angle by trial and error, the projectile motion formulas can be used to predict this angle given the initial speed. The advice given in textbooks on tennis is sometimes based on this kind of analysis. Many other athletic events involving projectiles that are thrown, kicked, or struck can be discussed using the projectile motion formulas.

SUMMARY

In two dimensions, position, velocity, and acceleration have both magnitude and direction, so they are represented by vectors. In diagrams, vectors are designated by arrows. To find the sum $\mathbf{C} = \mathbf{A} + \mathbf{B}$, we draw the tail of $\mathbf{B}$ from the head of $\mathbf{A}$; $\mathbf{C}$ is then a vector from the tail of $\mathbf{A}$ to the head of $\mathbf{B}$. Vector sums are often most readily evaluated by adding the components of the vectors along a conveniently selected pair of axes.

The velocity and acceleration in a plane are defined by restating the one-dimensional definitions in vector form: $\bar{\mathbf{v}} = \Delta\mathbf{s}/\Delta t$ and $\bar{\mathbf{a}} = \Delta\mathbf{v}/\Delta t$. Also, the x components of these quantities are related to each other exactly as in one dimension, as are the y components, so motion in a plane is effectively a pair of one-dimensional motions. Therefore, if we know the acceleration and the initial position and velocity, the subsequent motion can be found much as in one dimension.

Checklist

Define or explain:

vector	velocity in two
scalar	dimensions
addition of vectors	acceleration in two
subtraction of vectors	dimensions
vector components	projectile
unit vector	

EXERCISES

Section 2.1 | An Introduction to Vectors

2-1 Figure 2.16 shows a collection of vectors that can be combined in various ways. For example,

A + C = B. Find (a) **E + C;** (b) **A + F;**
(c) **A + D;** (d) **E + A;** (e) **E + 2A;** (f) **A − B;**
(g) **B − A;** (h) **C − A.**

Figure 2.16. Exercise 2-1.

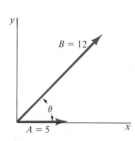

Figure 2.17. Exercises 2-2 and 2-4.

2-2   In Fig. 2.17, for what value of θ will **C = A + B** have (a) a minimum magnitude, and (b) a maximum magnitude? (c) Find **C** when $\theta = 90°$.

2-3   A vector has an x component of -10 and a y component of $+3$. (a) Draw a set of x-y axes and show the vector. (b) Calculate the magnitude and direction of the vector.

2-4   If $\theta = 72°$ in Fig. 2.17, find (a) the direction and magnitude of **C = A + B** by constructing a drawing using a ruler and a protractor; (b) the direction and magnitude of **C** using the component method.

2-5   For the vectors **A** and **B** in Fig. 2.18, find (a) **A + B;** (b) **B − A;** (c) **A − B.**

2-6   Using components for the vectors in Fig. 2.19, find the direction and magnitude of **E = A + B + C + D.**

2-7   Using components for the vectors in Fig. 2.19, find the direction and magnitude of **F = A − C + B − 2D.**

Figure 2.18. Exercise 2-5.

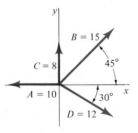

Figure 2.19. Exercises 2-6 and 2-7.

2-8   A woman walks 10 miles north, turns toward the northwest, and walks 5 miles further. What is her final position?

2-9   A ship sets out to sail 100 miles north but is blown by a severe storm to a point 200 miles east of its starting point. How far must it sail, and in what direction, to reach its intended destination?

2-10   One person walks northeast at 3 mi h^{-1}, and another heads south at 4 mi h^{-1}. How far apart are they after 2 hours?

2-11   For the vectors in Fig. 2.20, find the magnitude and direction of (a) **D = A + B + C;** (b) **E = A − B − C.**

Figure 2.20. Exercise 2-11.

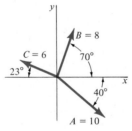

Figure 2.21. Exercise 2-12.

2-12   For the vectors in Fig. 2.21, find the magnitude and direction of (a) **D = A + B + C;** (b) **E = A − B − C.**

Section 2.2 | The Velocity in Two Dimensions

2-13   A car goes around a circular track 500 meters in diameter at a constant speed of 20 m s^{-1}. (a) How long does it take for the car to go halfway around the track? (b) What is its average velocity in that time interval?

2-14   A car goes around a circular track with a radius of 1000 meters at a constant speed of

10 m s^{-1}. (a) How long does it take the car to go once completely around the track? (b) What is the average velocity of the car over this time interval?

2-15   A ball is thrown at 30 m s^{-1} at an angle of 20° to the horizontal direction. Find the horizontal and vertical components of its initial velocity.

2-16   A plane flies for 3 hours and reaches a point 600 miles north and 800 miles east of its starting point. Find the direction and magnitude of its average velocity.

Section 2.3 | The Acceleration in Two Dimensions

2-17   A car initially traveling due north goes around a semicircle having a radius of 500 m at a constant speed of 20 m s^{-1}. (a) How long does this take? (b) What is the magnitude and direction of the average acceleration?

2-18   A rifle pointed at 30° to the horizontal fires a bullet at 250 m s^{-1}. If the bullet is accelerated uniformly in the barrel for 0.006 s, find (a) the magnitude of the acceleration; (b) its horizontal and vertical components.

2-19   A tennis ball is served by a player, bounces in the opponent's court, and is hit by the opponent toward the first player. Describe the direction and magnitude of the acceleration during each part of the motion.

Section 2.4 | Finding the Motion of an Object and Section 2.5 | Projectiles

2-20   A football kicked into the air from ground level hits the ground 30 m from where it started after 4 s. (a) Find its average velocity while in the air. (b) Find its average acceleration while in the air.

2-21   A ball is thrown horizontally at 20 m s^{-1} from a window 15 meters above the ground. (a) When will it hit the ground? (b) Where will it land?

2-22   A rifle pointed at 30° to the horizontal fires a bullet at 500 m s^{-1}. The rifle barrel is 0.7 m long. (a) Find the average acceleration in the rifle barrel. (b) Find the horizontal and vertical components of this acceleration.

2-23   A snowball is thrown from 2 m above the ground at a velocity of 10 m s^{-1} directed at 30° above the horizontal. (a) Find its horizontal and vertical position after 1 s. (b) Find its velocity components after 1 s.

2-24   (a) How long will the snowball of the preceding exercise be in the air? (b) Where will it land?

2-25   A baseball is hit at 40 m s^{-1} at angle of 30° to the horizontal. (a) How high will it go? (b) When will it reach that height? (c) What will be its horizontal distance from the batter at that time?

2-26   For the baseball in the preceding exercise, find (a) the distance it will travel; (b) the total time it will be in the air. (Neglect the fact that the ball is struck from slightly above ground.)

PROBLEMS

2-27   Suppose that in Fig. 2.14 the stuffed animal is initially 1 m above and 1.5 m to the right of the cannon that is pointed at the animal. The animal starts to fall as the cannonball is fired at 10 m s^{-1}. (a) What are the initial horizontal and vertical velocity components of the cannonball? (b) How long does it take for the horizontal coordinate of the ball to change by 1.5 m? (c) What are the vertical positions of the ball and the animal at that time?

2-28   One ball is thrown horizontally with a velocity v_0 from a height h, and another is thrown straight down with the same initial speed. (a) Which ball will land first? (b) Which ball will have the greater speed as it is about to land?

2-29   A quarterback throws a football so that it rises 4 ft in 20 yards and then begins to drop. What was the initial speed and direction of the football?

***2-30**   A man wishes to row across a river 0.5 miles wide. He points his boat toward the opposite shore and rows at 2 mi h^{-1} relative to the water. The current is 4 mi h^{-1}. (a) How long does it take him to cross the river? (b) Where does he land?

2-31   A tennis ball is served horizontally at a height of 8 ft, 40 ft from a net that is 3 ft high. (a) If it is to clear the net by at least 0.5 ft, what is its minimum initial velocity? (Neglect air resistance). (b) If it clears the net by 0.5 ft, where will it land?

2-32   Some books advise serving a tennis ball at an angle below the horizontal direction. To see if this is sound advice, suppose a ball is struck at an angle of 5° downward at a height of 8 ft with the relatively high speed of 100 ft s^{-1}. How high will it be when it reaches the net 40 ft away? (The net is 3 ft high. Neglect air resistance.)

SUPPLEMENTARY TOPICS

2.6 | PROJECTILES IN BIOMECHANICS

Many applications of projectile motion occur in athletics and in animal motion. Here we explore some further aspects of this subject.

In applications of projectile motion, it is convenient to have a formula for the horizontal distance traveled or *range, R*. To obtain this formula, consider a projectile launched from a flat surface (Fig. 2.22). The projectile lands at a time t when y returns to zero. The range can be found from the equation for x once this landing time is known.

With $y = y_0 = 0$ and $a_y = -g$, the equation $y = y_0 + v_{0y} - \frac{1}{2}gt^2$ becomes

$$0 = v_{0y}t - \frac{1}{2}gt^2$$

or

$$0 = (v_{0y} - \frac{1}{2}gt)t$$

The solutions of this equation are $t = 0$, which corresponds to the launch time, and

$$t = \frac{2v_{0y}}{g}$$

which is the time the projectile lands.

Substituting this time in $x = v_{0x}t$, the range is

$$R = \frac{2v_{0x}v_{0y}}{g}$$

If the initial velocity of the projectile is at a *launch angle* θ to the ground (Fig. 2.22), then $v_{0x} = v_0 \sin \theta$ and $v_{0y} = v_0 \cos \theta$. The range can then be expressed as

$$R = 2\frac{v_0^2}{g} \sin \theta \cos \theta \qquad (2.19)$$

Some interesting features of this result are illustrated by considering a kicked football. If it is kicked straight up, θ is 90°, cos 90° = 0, and the range is zero; the ball stays in the air for a while, but its horizontal range is zero. Similarly, if it is kicked straight ahead, θ is 0°, and sin 0° = 0. Again the range is zero, but now the ball immediately hits the ground.

Figure 2.22. A projectile with a velocity $\mathbf{v_0}$ launched at an angle θ to the horizontal has a range R.

Suppose now the football is kicked at 30°. The range is proportional to sin 30° cos 30° = (0.5) (0.866). If instead the ball is kicked at 60°, then the range is proportional to sin 60° cos 60° = (0.866) (0.5). The two products are equal, so the ranges are the same. However, the ball kicked at the steeper 60° launch angle will rise higher and stay in the air longer. Thus a punter can provide more time for his team to get downfield with the higher kick (Fig. 2.23). In general, projectiles launched from a flat surface at an angle θ and an angle 90° $- \theta$ at the same initial speed will have the same range, but the larger launch angles will result in higher paths with longer flight times.

A launch angle of 45° is a special case, since 90° $- \theta$ = 90° $-$ 45° = 45°. It turns out that this is the angle that gives the longest range for a given initial speed when the takeoff and landing heights are equal (Fig. 2.23). However, a shot put thrown from *above* ground level will have its longest range with a flatter trajectory or a *smaller* launch angle (Fig. 2.24). Conversely, an object thrown from *below* its landing point will have its greatest range for an angle *greater* than 45°.

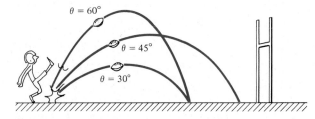

Figure 2.23. Footballs kicked with the same initial speed from ground level with launch angles of 30° and 60° have the same range. The maximum range occurs when the launch angle is 45°.

Horizontal Jumping | In Section 1.8, we saw that the constant acceleration formulas can be used to analyze vertical jumping by animals. Similarly, the

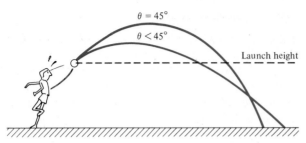

Figure 2.24. A shotput is thrown from above ground level. The trajectories for a launch angle of 45° and a smaller launch angle cross at a point below the launch height. The flatter trajectory has a longer range.

Figure 2.25. Frogs frequently jump with a launch angle of approximately 45°, the angle which produces the maximum range on flat ground.

projectile motion formulas can be used to discuss horizontal jumping, since they accurately describe the motion while the animal is in the air if air resistance is negligible.

Although a 45° launch angle produces the maximum range on flat ground for a given initial speed (Fig. 2.25), an animal may customarily jump at some other angle for reasons related to its needs or structure. For example, locusts often jump into the air and then start flying. In this case the range of the jump is clearly irrelevant, but the time duration may be significant. Whether or not they begin to fly, locusts usually jump at about 55°.

The calculation of the takeoff velocity is illustrated in the following example.

Example 2.7

What is the takeoff speed of a locust if its launch angle is 55° and its range is 0.8 m?

Since we know R and θ, we can find v_0 from

$$R = \frac{2v_0^2}{g} \sin \theta \cos \theta \qquad (2.19)$$

With $\sin 55° = 0.819$ and $\cos 55° = 0.574$,

$$v_0^2 = \frac{gR}{2 \sin \theta \cos \theta} = \frac{(9.8 \text{ m s}^{-2})(0.8 \text{ m})}{2(0.819)(0.574)}$$

$$= 8.3 \text{ m}^2 \text{ s}^{-2}$$

$$v_0 = 2.9 \text{ m s}^{-1}$$

EXERCISES ON SUPPLEMENTARY TOPICS
Section 2.6 | Projectiles in Biomechanics

2-33 A football is kicked at 20 m s^{-1} from ground level. Find its range if the launch angle is (a) 30°; (b) 60°; (c) 45°.

2-34 In the preceding example, how long will the football be in the air in each of the three cases?

2-35 An astronaut wearing his spacesuit can broad jump 2 m on the earth. How far can he jump on a planet where the gravitational acceleration is half that at the surface of the earth?

2-36 A girl wishes to throw a snowball at another child. If she can throw the snowball at 20 m s^{-1}, how far can she stand from the other child and still reach him?

2-37 A rifle is fired horizontally from the top of a tall mountain. Using a sketch, show the effect of the earth's curvature on the range of the bullet.

2-38 A kangaroo can jump 8 m. If it takes off at 45° to the horizontal, what is its takeoff speed?

2-39 A baseball thrown at 10° to the horizontal returns to its original height after 70 m. What was its original speed?

PROBLEMS ON SUPPLEMENTARY TOPICS

2-40 With the aid of a sketch, show that the maximum range for an object launched at a given speed from below its landing point occurs for a launch angle greater than 45°.

2-41 (a) Explain why the maximum range for a man doing a standing broad jump is not obtained for a takeoff angle of 45°. (b) Should the angle be less than or greater than 45°? Explain.

2-42 A frog can jump 0.9 m with a takeoff angle of 45°. (a) What initial velocity does this require? (b) With this same initial velocity directed vertically, how high could a frog jump? (c) The maximum height jumped by frogs is 0.3 m. What are some possible explanations for this difference?

2-43 A flea can jump 0.03 m. (a) If the takeoff angle is 70°, what is the initial velocity? (b) If the flea achieves this velocity in a takeoff distance of 8×10^{-4} m, what is its average acceleration during takeoff?

Additional Reading

Sir James Gray, *How Animals Move,* Cambridge University Press, Cambridge, 1953. Pages 69–80 discuss jumping.

R. McNeill Alexander, *Animal Mechanics,* University of Washington Press, Seattle, 1968. Pages 28–33 discuss jumping.

Scientific American articles:

Stillman Drake and James MacLachlan, Galileo's Discovery of the Parabolic Trajectory, March 1975, p. 102.

Cornelius T. Leondes, Inertial Navigation for Aircraft, March 1970, p. 80.

Graham Hoyle, The Leap of the Grasshopper, January 1958, p. 30.

Miriam Rothschild, et al., The Flying Leap of the Flea, November 1973, p. 92.

CHAPTER THREE

NEWTON'S LAWS OF MOTION

Having learned how to describe motion, we can now turn to the more fundamental question of what causes motion. An object is set into motion when it is pushed or pulled or subjected to a *force*. The discussion of forces and their effects is the central topic in mechanics.

Although there are many kinds of forces in nature, the effects of any force are described accurately by three general laws of motion first stated fully by Sir Isaac Newton (1643–1727). Guided by earlier astronomical observations and making several giant steps of intuition, Newton developed the laws of motion and also the expression for the gravitational attraction between two objects. He then showed that the orbital motions of the planets and the moon were in quantitative agreement with the predictions he made using these laws.

Newton's work represented a tremendous step forward in our understanding of the natural world and exerted a great influence on science and on the way people viewed science. For over two centuries, Newton's laws of motion served as the foundation of mechanics, with later workers finding full agreement between theory and experiment for a wide range of phenomena. Even though twentieth-century advances have shown that Newton's laws are inadequate at the atomic scale and at velocities comparable to the speed of light, 3×10^8 m s^{-1}, they provide an extremely accurate framework for discussing the motions of macroscopic objects at ordinary velocities. Thus, they are fully adequate for most applications in fields such as astronomy, biomechanics, geology, and engineering. The twentieth-century modifications of mechanics are discussed in later units.

3.1 | FORCE, WEIGHT, AND GRAVITATIONAL MASS

If we push or pull an object, we are exerting a force on it. Forces have both magnitudes and directions, so they are vector quantities. It is found that the *net* or *total* force on an object is the vector sum of all the forces acting on the object. For example, if two forces equal in magnitude but opposite in direction act on an object, there is no net force (Fig. 3.1a).

Figure 3.1. Forces have both magnitudes and directions and are vector quantities. (a) $\mathbf{F}_1$ and $\mathbf{F}_2$ are equal in magnitude but opposite, so their sum is zero. (b) Since $\mathbf{F}_1$ and $\mathbf{F}_2$ are equal, the net force is $\mathbf{F} = \mathbf{F}_1 + \mathbf{F}_2 = 2\mathbf{F}_1$.

Forces that are exerted only when two objects are in contact are referred to as *contact forces*. Examples are the force exerted by a compressed spring on an attached object, the upward force exerted by a table on a book resting upon it, and the force exerted on a bone by a contracting muscle. Other forces, including gravitational, magnetic, and electric forces, can be exerted between objects that are not in contact.

In order to make quantitative statements about forces, we must define a force unit. One way to do this is to use a spring to measure the gravitational force on an object adopted as a standard. If the spring is compressed, a pointer moves along a calibrated dial. When the standard is placed on a plate mounted on the spring, the pointer moves a certain distance. The standard object is then said to exert a force of one unit on the spring. The same reading will always be obtained if this procedure is performed at points on the earth's surface where the gravitational acceleration is the same. Other forces can then be measured by determining the compression of the spring once the dial has been calibrated using multiple copies of the standard object. In Section 3.5, we will see that the force unit can also be defined by measuring the acceleration of a standard object.

The unit of force is different in each of the three major systems of units. The S.I. unit is the *newton* (N), the c.g.s. unit is the *dyne* (dyn), and the British unit is the *pound* (lb). These units are related by

$$1\,N = 10^5\,dyn = 0.225\,lb$$

A particularly important force is the gravitational force on an object, which is referred to as its *weight* **w.** In the three force units, the weight of a fairly heavy man might be 1000 N or 225 lb or 10^8 dyn.

Closely related to the weight is the *gravitational mass m* of an object, which is defined as the weight divided by the acceleration of gravity:

$$m = \frac{w}{g} \tag{3.1}$$

The units of gravitational mass are the *kilogram* (kg), *gram* (g), and *slug* in the S.I., c.g.s., and British systems, respectively. For example, a man who weighs 1000 N on the earth has a gravitational mass of $w/g = 1000\,N/9.8\,m\,s^{-2} = 102\,kg$. Similarly, an object that weighs 64 lb has a gravitational mass of $64\,lb/32\,ft\,s^{-2} = 2$ slugs.

The mass, denoted dimensionally by M, completes the basic set of physical dimensions. All physical quantities can be written in terms of the dimensions of length, time, and mass. Table 3.1 shows the masses of some representative objects.

In this section we have introduced two major concepts, force and gravitational mass. A force is any push or pull on an object. The gravitational force on an object is referred to as its weight, and the gravitational mass of an object is its weight divided by the gravitational acceleration. We explore the relation-

ships among forces and masses in the following sections.

3.2 | NEWTON'S FIRST LAW

According to the Aristotelian view that dominated medieval ideas about motion, all objects move only if some force causes the motion to occur. Thus, a wagon detached from the horse that was pulling it would quickly come to rest because no force is pulling it along. The modern view is that the cart slows down and stops because of frictional forces acting upon it. This view is summarized in Newton's first law, which states that

> *Every object continues in a state of rest, or of uniform motion in a straight line, unless it is compelled to change that state by forces acting upon it.*

An equivalent statement of the first law is that if there is no force on an object, or if there is no net force when two or more forces act on the object, then

(*1*) *an object at rest remains at rest, and*
(*2*) *an object in motion continues to move with constant velocity.*

The first law holds true in the form stated only for measurements made by certain observers. A girl rid-

TABLE 3.1

Representative masses in kilograms

Electron	9×10^{-31}
Proton	2×10^{-27}
Oxygen atom	3×10^{-26}
Insulin molecule (a small protein)	10^{-23}
Penicillin molecule	10^{-18}
Giant amoeba	10^{-8}
Ant	10^{-5}
Hummingbird	10^{-2}
Dog	10^1
Human	10^2
Elephant	10^4
Blue whale	10^5
Oil tanker	10^8
Moon	7×10^{22}
Earth	6×10^{24}
Sun	2×10^{30}
Our galaxy	2×10^{41}

ing on a merry-go-round sees objects not subjected to any net force undergo rather complex motions, while a boy at rest on the ground sees them at rest or moving with constant velocity. Thus, Newton's first law as stated is true for the boy at rest but not for the girl. The point is that the girl is accelerating, since her velocity is changing, and Newton's first law as stated does not hold true for someone who is accelerating.

The first law leads us to define an *inertial* coordinate system or reference frame as one in which Newton's first law holds true as stated. Strictly speaking, the boy in the preceding example is not quite in an inertial frame, since he is standing on the earth, which is revolving daily on its axis, rotating annually about the sun, and moving with the solar system relative to the distant stars. Usually these effects can be ignored, and the earth can be treated as an inertial frame. However, the earth's daily revolution does affect the large-scale motions of the atmosphere and the oceans.

A coordinate system moving at a constant velocity relative to an inertial frame is itself an inertial frame. To see this, consider an observer standing on the ground and another in a car moving at a constant velocity. If both observers measure the velocity of a moving object, their measurements will differ by their *constant* relative velocity. Hence, they will agree on whether the motion of the object is accelerated, and Newton's first law will work equally well for both observers.

From this discussion, we see that cars or airplanes moving at constant velocities relative to the ground are inertial reference frames, while accelerating vehicles, merry-go-rounds, and playground swings are not. As a further illustration, consider the driver of an accelerating car. She feels the back of her seat exerting a force on her. Relative to the ground, she is accelerating, but relative to the car, she is not. The first law is consistent with the measurements of an observer at rest on the ground: the state of motion of the driver is being changed by forces acting on her. The first law does not work in the frame of the accelerating car, since there is a force on the driver, but she remains at rest.

3.3 | EQUILIBRIUM

When the state of motion of an object remains unchanged even though two or more forces are acting

upon it, the object is said to be in *equilibrium*. There are three types of equilibrium: *unstable, stable,* and *neutral.* These are best defined with the aid of an example.

Figure 3.2.   A ball at the crest of a hill is in unstable equilibrium. In the depression, it is in stable equilibrium. On the level plain, it is in neutral equilibrium.

Suppose a ball is placed exactly at the crest of a small hill (point *A* in Fig. 3.2). It may remain at rest there briefly, but if a slight breeze moves it a bit, it will rapidly accelerate down the hillside. This is unstable equilibrium: a small displacement leads to an unbalanced force that further increases the displacement from the equilibrium location. Conversely, in the depression at *B*, the ball will roll back if disturbed. This is stable equilibrium: a small displacement leads to an unbalanced force that restores the body to the equilibrium position. Finally, on the flat level area near *C*, there is still no unbalanced force, even if the ball is displaced slightly. This is neutral equilibrium.

From Newton's first law, the net force on an object must be zero in order that the translational motion of an object remain unchanged. This is one requirement for equilibrium. If we are dealing with an object that can rotate and not a point particle, then there is a second requirement that no twisting effects or *torques* act that would alter the rotational state of motion. This second requirement will be discussed in the next chapter.

To summarize, an object will be in translational equilibrium if the vector sum of the forces acting upon it is zero. The type of equilibrium is determined by noting how the forces change when the particle is disturbed slightly from its original state of rest or of uniform motion.

3.4 | NEWTON'S THIRD LAW

We turn now to Newton's third law, deferring the

Figure 3.3. The force on the person is equal in magnitude and opposite in direction to the force on (*a*) the wall; (*b*) the ground.

second law until the next section. The third law relates the forces that two objects exert on each other, and it is familiar in a general way from our everyday experiences. For example, suppose you are at rest in a swimming pool. If you push a wall with your legs, the wall exerts a force that propels you further into the pool. The *reaction* force the wall exerts on you is opposite in direction to the force you exert on the wall (Fig. 3.3). Similarly, in order for you to start walking forward, your feet must exert a backward force on the ground; the ground, in turn, pushes you forward.

Newton observed that whenever a person exerts a force on an object, the object exerts a force on the person that is *equal in magnitude and opposite in direction*. This relationship, which is referred to as the third law of motion, holds true whether or not the person or the object accelerates. The two forces acting between a person and an object, or between two objects, are called *action* and *reaction* forces. The general statement of the third law is

> *For every action, there is always an equal but opposite reaction.*

Stated more fully, if one object exerts a force **F** on a second, then the second object exerts an equal but opposite force −**F** on the first. For example, if you push a chair with a horizontal force of 10 N, the chair exerts a force on you of 10 N in the opposite direction. If the earth exerts a downward 1 lb gravitational force on a book, then the book exerts an upward 1 lb force on the earth.

It is important to realize that in each case the two forces—the action and reaction forces—act on different objects, so their effects do *not* cancel out. When you push the chair, it starts to move unless there is a sufficiently large frictional force exerted by the floor to prevent this. Or if a book is not subjected to any

force other than gravity, it will fall toward the earth. In other words, *only the forces acting on a particular object can affect its state of motion;* the forces exerted *by* an object will affect the motion of *other* objects.

There are many situations where forces are equal but opposite, but they are not action-reaction pairs in the sense of the third law. This often happens, as in the next example, where an object is in equilibrium and forces therefore cancel as a consequence of the *first law.*

Figure 3.4. (*a*) Since the woman is in equilibrium, the force **N** exerted by the floor on the woman is equal in magnitude to her weight and opposite in direction. (*b*) The reaction to the weight of the woman is a force −**w** on the earth; the reaction to the force **N** exerted by the floor is a force −**N** she exerts on the floor.

Example 3.1

A woman of weight **w** stands on a horizontal floor. Find the forces acting on her and identify the reactions to those forces (Fig. 3.4).

The woman remains at rest, so according to the first law, there must be no net force on her. Her weight **w** is directed downward, so there must be an equally large upward force exerted on her by the floor. This force is denoted by **N** in Fig. 3.4 since it is *normal* or perpendicular to the floor.

We see then that **N** = −**w** follows here from the first law. However, these equal and opposite forces are not an action-reaction pair. The reaction to the weight **w** is an upward gravitational force, −**w**, on the earth; the reaction to the normal force **N** is a downward force, −**N**, exerted by the woman on the floor (Fig. 3.4*b*).

Although it is customary to refer to action and reaction forces, the main point of the third law is that forces between two objects always occur in equal but opposite pairs. Which is the action or reaction is often a matter of choice, as in the case of gravitational forces.

3.5 | NEWTON'S SECOND LAW

When there is a net force acting on an object, the

object undergoes an acceleration in the same direction as the force. The acceleration and the force are also proportional in magnitude; if the force on a given object is doubled, so is the acceleration.

If two quantities are proportional, one is equal to some number or *proportionality constant* times the other. Thus, we can relate the force **F** and the acceleration **a** by *Newton's second law,*

$$\mathbf{F} = m\mathbf{a} \tag{3.2}$$

The proportionality constant m is called the *mass* of the object. (Strictly speaking, it is the *inertial mass,* but we will see in the next section that the inertial and gravitational masses of an object are equal.)

The mass of an object is a measure of the amount of matter it comprises, or in other words, of its inertia. The greater the mass of an object, the less effect a given force will have on its motion. Mass is related to but very different from weight. The weight of an object is the gravitational force on the object and is a vector quantity; mass is a scalar quantity.

Newton's second law $\mathbf{F} = m\mathbf{a}$ provides another way of defining the units of force in the various systems. A 1-N force acting on a 1-kg mass produces an acceleration of 1 m s^{-2}:

$$1 \text{ N} = (1 \text{ kg})(1 \text{ m s}^{-2}) = 1 \text{ kg m s}^{-2}$$

Similarly, in c.g.s. units, $1 \text{ dyn} = 1 \text{ g cm s}^{-2}$, and in British units $1 \text{ lb} = 1 \text{ slug ft s}^{-2}$. These definitions are equivalent to those given earlier.

To illustrate the use of these units and the second law, we may note that a 1000-kg car accelerating at 2 m s^{-2} must be subjected to a force $F = ma = (1000 \text{ kg})(2 \text{ m s}^{-2}) = 2000 \text{ N}$. Similarly, if a 10-lb force applied to a sled produces an acceleration of 4 ft s^{-2}, then its mass must be $m = F/a = (10 \text{ lb})/(4 \text{ ft s}^{-2}) = 2.5 \text{ slugs}$.

We consider many examples and applications of Newton's second law later in this and other chapters. However, we can see at this point that the second law explains why the equality of action and reaction forces is sometimes not immediately apparent. Since $\mathbf{F} = m\mathbf{a},$ the acceleration resulting from a given force varies inversely with the mass of an object. A person who steps off a cliff experiences an acceleration g resulting from the gravitational force exerted by the earth on the person. The earth experiences a force of equal magnitude, but its acceleration is very much

smaller because of its great mass. Similarly, when a small car and a large truck collide, the forces on each vehicle are equal in magnitude, but the acceleration of the car is much greater than that of the truck.

3.6 | THE SIGNIFICANCE OF NEWTON'S LAWS OF MOTION

Newton's three laws of motion are very fundamental statements about the physical world. All of mechanics may be viewed as Newton's laws applied in direct or indirect ways to a variety of forces and systems. Although we introduce many other quantities and concepts in our discussions of mechanics in the following chapters, none are so basic as Newton's laws of motion. Accordingly, before going on to discuss specific forces and examples of how one uses Newton's laws, we list these laws together here for emphasis and for convenient reference.

Newton's first law of motion: *When there is no net force on an object, (1) an object at rest remains at rest; and (2) an object in motion continues to move with a velocity that is constant in magnitude and direction.*

Newton's second law of motion: *The force **F** needed to produce an acceleration **a** is*

$$\mathbf{F} = m\mathbf{a} \tag{3.2}$$

where m *is the mass of the object.*

Newton's third law of motion: *For every action, there is an equal but opposite reaction. The action and reaction forces act on different objects.*

Note that in the form given, Newton's laws may be applied only with respect to inertial coordinate systems.

3.7 | GRAVITATIONAL FORCES

Newton's great triumph was his successful derivation of the observed laws of motion of the planets and the moon. The derivation was based on his three laws of motion and on a law he proposed for the gravitational force. This law is referred to as the *law of universal gravitation* and is regarded as a fundamental law of nature.

The law of universal gravitation states that all objects in the universe attract each other. For two spheres, or for two objects of any shape that are so small compared to their separation that they may be considered as point particles, the law has a simple form. If two spheres or particles have gravitational masses m and m', and their centers are separated by a distance r, the forces between the two spheres have a magnitude

$$F = \frac{Gmm'}{r^2}$$ (3.3)

G is called the *gravitational constant* and has a measured value of

$$G = 6.67 \times 10^{-11} \text{ N m}^2 \text{ kg}^{-2}$$

Figure 3.5.   Two spheres exert attractive forces on each other. **F** and **F'** are equal but opposite in accordance with Newton's third law of motion.

The gravitational forces are directed along the line connecting the centers of the two spheres (Fig. 3.5). The magnitude of the gravitational force varies as $1/r^2$, so Eq. 3.3 is referred to as an *inverse square law*.

As noted, Eq. 3.3 applies directly to spheres and to particles. For more complex objects, the forces between all the pairs of particles in the objects must be summed to find the net gravitational forces on the objects. However, these net forces are still equal but opposite as required by the third law of motion.

The gravitational force exerted by the earth on an object is relatively large because of the large mass of the earth. In contrast, the gravitational force between two objects of moderate mass is very small and hard to detect, as is shown quantitatively in the next example.

Example 3.2

The centers of two 10-kg spheres are separated by 0.1 m. (a) What is their gravitational attraction? (b) What is the ratio of this attraction to the weight of one of the spheres?

(a) Using Newton's law of gravitation, the forces between the spheres have a magnitude

$$F = G\frac{mm'}{r^2} = (6.67 \times 10^{-11} \text{ N m}^2 \text{ kg})\frac{(10 \text{ kg})(10 \text{ kg})}{(0.1 \text{ m})^2}$$

$$= 6.67 \times 10^{-7} \text{ N}$$

The forces are directed along the line connecting the centers of the spheres.

(b) The weight of one of the spheres is

$$w = mg = (10 \text{ kg})(9.8 \text{ m s}^{-2}) = 98 \text{ N}$$

Thus the ratio of the gravitational forces between the spheres to the weight of a sphere is

$$\frac{F}{w} = \frac{6.67 \times 10^{-7} \text{ N}}{98 \text{ N}} = 6.81 \times 10^{-8}$$

The small size of this ratio explains why we do not notice the gravitational attractions among objects of ordinary size.

Figure 3.6.   A person of mass m has a weight $w = GmM_E/R_E{}^2$ at the surface of the earth.

3.8 | WEIGHT

The weight of an object is the gravitational force it experiences. For an object near the surface of the earth, this force is mainly due to the earth's attraction.

Let us call the radius of the earth R_E and its mass M_E (Fig. 3.6). An object with a gravitational mass $\tilde{m}$ at the earth's surface is subjected to a gravitational force. From Eq. 3.3, this force is

$$F = G\frac{\tilde{m}M_E}{R_E{}^2}$$ (3.4)

The acceleration g resulting from this force can be found from Newton's second law, $\mathbf{F} = m\mathbf{a}$, which contains the inertial mass m:

$$g = \frac{F}{m} = \frac{1}{m}\left(G\frac{\tilde{m}M_E}{R_E{}^2}\right)$$

If we assume that the inertial and gravitational

45

masses are the same, then $\tilde{m}/m = 1$, and we find

$$g = G\,\frac{M_E}{R_E{}^2} \qquad (3.5)$$

This result says that the gravitational acceleration is the same for all objects. This is in agreement with experiments and justifies the assumption that the gravitational and inertial masses of an object are equal.

Note that the earth's radius R_E is 6400 km. Accordingly, the gravitational acceleration within a few meters or even a few kilometers of the surface will not differ appreciably from the value at the surface, 9.8 m s^{-2}.

Since the weight is directed toward the center of the earth, $\mathbf{g}$ is also in that direction. Thus

$$\mathbf{w} = m\mathbf{g} \qquad (3.6)$$

is the vector form of the relation between mass and weight given earlier.

As we have seen, mass and weight are related but quite different quantities. The mass of an object is an intrinsic property that is the same whether the object is in Chicago, on the moon, or in interstellar space. It is a measure of the amount of matter in the object and determines its inertia or its response to a force. The weight of an object varies from place to place and is the force resulting from gravity.

The mass of an object is determined *only* by the amount of matter present and is independent of its physical or chemical state. For example, if a cubic meter of oxygen gas at atmospheric pressure is cooled, it will liquefy and fill a volume of about 10^{-3} m^3. Nevertheless, it will still have the same number of molecules and the same mass. Similarly, when a volume of hydrogen gas and another volume of oxygen gas combine to form liquid water, the volume is reduced by a large factor, but again the mass is unchanged. The following example further illustrates the relationship between mass and weight.

Example 3.3

An astronaut weighs 180 lb on the earth. What is his weight on planet X, which has a radius $R_x = R_E/2$ and a mass $M_x = M_E/8$?

On the earth, his weight is

$$w_E = \frac{GmM_E}{R_E{}^2}$$

where m is his mass. On planet X, his mass is the same, but his weight is

$$w_x = \frac{GmM_x}{R_x{}^2} = \frac{Gm(M_E/8)}{(R_E/2)^2}$$
$$= \frac{4}{8}\,\frac{GmM_E}{R_E{}^2} = \frac{1}{2}\,w_E$$

Thus, his weight on planet X is $(\tfrac{1}{2})(180\text{ lb}) = 90\text{ lb}$.

3.9 | SOME EXAMPLES OF NEWTON'S LAWS

We now give several examples of how Newton's laws of motion are used. In each case, we employ a systematic procedure for relating the acceleration of an object or objects to the forces present.

1 For each object, we draw a careful sketch showing the forces acting *on* that object.
2 We then apply Newton's second law $\mathbf{F} = m\mathbf{a}$ to *each object separately*. If there are n forces $\mathbf{F}_1$, $\mathbf{F}_2, \ldots, \mathbf{F}_n$ acting on an object, the net force $\mathbf{F}$ is the sum of the forces, and we have

$$\mathbf{F} = \mathbf{F}_1 + \mathbf{F}_2 + \cdots + \mathbf{F}_n = m\mathbf{a}$$

In component form, this is

$$F_x = F_{1x} + F_{2x} + \cdots + F_{nx} = ma_x \quad (3.7)$$
$$F_y = F_{1y} + F_{2y} + \cdots + F_{ny} = ma_y \quad (3.8)$$

3 As in the earlier chapters, we usually do not substitute numerical values immediately. Instead we solve the equations for the unknown quantities symbolically and then substitute the numbers if they are given. This procedure facilitates checking for algebra and physics errors and often reduces the amount of arithmetic required.
4 We include the units of the numerical quantities and see that the final answer has the correct dimensions.

We use these steps in the examples below and in later chapters as well.

Example 3.4

An elevator has a mass of 1000 kg. (a) It accelerates upward at 3 m s^{-2}. What is the *tension* in the cable, that is, the force exerted by the cable on the elevator? (b) What is the tension if the acceleration is 3 m s^{-2} downward?

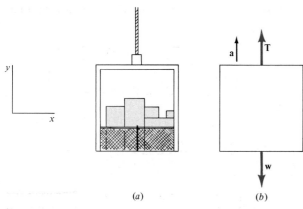

(a)   *(b)*

Figure 3.7. (*a*) The elevator and cable. (*b*) The force of the cable on the elevator, **T**, and the weight of the elevator **w**. When the elevator is accelerating upward, **T** is greater than **w**.

(a) The forces on the elevator are its weight **w** and the upward force **T** resulting from the cable (Fig. 3.7). Using Eq. 3.8 with two forces present,

$$F_{1y} + F_{2y} = ma_y$$

With $F_{1y} = T$ and $F_{2y} = -w = -mg$, this becomes

$$T - mg = ma_y$$

or

$$T = m(g + a_y) \qquad \text{(i)}$$

Then, with $a_y = 3$ m s^{-2} and $m = 1000$ kg,

$$T = (1000 \text{ kg})(9.8 \text{ m s}^{-2} + 3 \text{ m s}^{-2})$$
$$= 12{,}800 \text{ N}$$

Note that the tension T is greater than the weight mg. The cable must support the weight of the elevator and also provide the extra force needed for the acceleration.

(b) Equation i holds true no matter what the magnitude or sign of a_y. When **a** is directed downward, $a_y = -3$ m s^{-2}, and Eq. i gives

$$T = m(g + a_y)$$
$$= (1000 \text{ kg})(9.8 \text{ m s}^{-2} - 3 \text{ s}^{-2}) = 6800 \text{ N}$$

Note that T is now less than the weight mg, since the elevator is being allowed to accelerate downward.

Example 3.5

A child pulls a train of two cars with a horizontal force **F** of 10 N. Car 1 has a mass $m_1 = 3$ kg, and car 2 has a mass $m_2 = 1$ kg (Fig. 3.8). The mass of the string connecting the cars is small enough so it can be set equal to zero, and friction can be neglected. (a) Find the normal forces exerted on each car by the floor. (b) What is the tension in the string? (c) What is the acceleration of the train?

In this problem, there is a system of several objects. The three objects of interest are the two cars and the string connecting them. When Newton's second law **F** = m**a** is applied to each object, a set of equations results that must be solved simultaneously.

The fact that the string is massless ($m = 0$) implies that the net force **F** = m**a** on the string must be zero. Consequently, the forces exerted *by the two cars* on the ends of the string must cancel. The reactions to these forces that act *on the two cars* must also be equal but opposite. Because of this, we have labeled the string tension force by the same symbol **T** in the force diagrams for each of the cars.

(a) Since $a_y = 0$ for each car, $F_y = ma_y = 0$, so

$$N_2 - w_2 = 0$$

and

$$N_1 - w_1 = 0$$

Thus the normal forces exerted by the floor are

$$N_2 = w_2 = m_2 g = (1 \text{ kg})(9.8 \text{ m s}^{-2}) = 9.80 \text{ N}$$
$$N_1 = w_1 = m_1 g = (3 \text{ kg})(9.8 \text{ m s}^{-2}) = 29.4 \text{ N}$$

(b) For both cars, $a_x = a$, so for car 1, $F_x = m_1 a_x$ becomes

$$F - T = m_1 a \qquad \text{(i)}$$

Both T and a are unknowns, so we cannot find them from this one equation alone; as many equations as there are unknowns are needed to solve a problem. The second equation relating T and a is obtained by applying $F_x = m_2 a_x$ to car 2:

$$T = m_2 a \qquad \text{(ii)}$$

(a)   *(b)*   *(c)*

Figure 3.8. (*a*) The system in Example 3.5. (*b*) and (*c*) show the force diagrams of cars (2) and (1) respectively.

This gives $a = T/m_2$. Substituting this for a in Eq. i,

$$F - T = m_1 \left(\frac{T}{m_2} \right)$$

Solving for T,

$$T = \frac{F}{1 + \dfrac{m_1}{m_2}} = \frac{10 \text{ N}}{1 + \dfrac{3 \text{ kg}}{1 \text{ kg}}} = 2.5 \text{ N}$$

(c) Now we can find a from Eq. ii:

$$a = \frac{T}{m_2} = \frac{2.5 \text{ N}}{1 \text{ kg}} = 2.5 \text{ m s}^{-2}$$

Note that if we had considered the train as a single object of mass $m = m_1 + m_2 = (3 + 1)\text{kg} = 4 \text{ kg}$, we could immediately have found the acceleration from $a = F/m = 10 \text{ N}/4 \text{ kg} = 2.5 \text{ m s}^{-2}$. However, the normal and tension forces can only be found by considering the individual cars and not the system as a whole.

Example 3.6

A block of mass $m_1 = 20 \text{ kg}$ is free to move on a horizontal surface. A rope, which passes over a pulley, attaches it to a hanging block of mass $m_2 = 10 \text{ kg}$ (Fig. 3.9). Assuming for simplicity that the pulley and rope masses are negligible and that there is no friction, find (a) the forces on the blocks and (b) their acceleration. (c) If the system is initially at rest, how far has it moved after 2 s?

This problem is similar to the preceding problem of the two cars, although here one mass moves horizontally and one moves vertically. Again, the fact that the rope

and pulley are massless implies that the force exerted by the rope on m_1 is equal in magnitude to the force exerted by the rope on m_2. Accordingly we denote both forces by the same symbol **T**.

(a) We first apply $\mathbf{F} = m\mathbf{a}$ to the block on the surface. Since it has no vertical acceleration component, F_y must be zero. With $w_1 = m_1 g$,

$$N_1 - m_1 g = 0$$

Thus the normal force N_1 on block 1 due to the surface is

$$N_1 = m_1 g = (20 \text{ kg})(9.8 \text{ m s}^{-2}) = 19.6 \text{ N}$$

The system is accelerating with an unknown acceleration a. Block 1 has $a_x = a$, so $F_x = m_1 a_x$ becomes

$$T = m_1 a \tag{i}$$

Note that both T and a are unknown, so we cannot find either until we consider the motion of m_2.

We now apply Newton's second law to block 2. Since it is accelerating downward, $a_y = -a$, and $F_y = m_2 a_y$ becomes

$$T - w_2 = -m_2 a \tag{ii}$$

This equation also involves the two unknowns, T and a. Since the number of equations now equals the number of unknowns, we can solve for T and a. From Eq. i we have $a = T/m_1$. Substituting this in Eq. ii gives

$$T - w_2 = -m_2 \frac{T}{m_1}$$

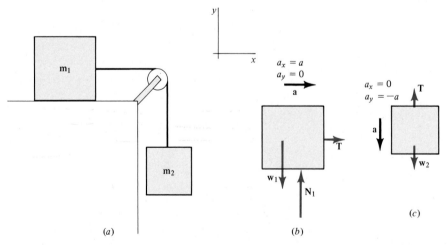

(a)   (b)   (c)

Figure 3.9. (b) and (c) are force diagrams for the two blocks when they are connected as shown in (a).

Solving for T,

$$T = \frac{w_2}{1 + \frac{m_2}{m_1}} = \frac{m_2 g}{1 + \frac{m_2}{m_1}} = \frac{(10 \text{ kg})(9.8 \text{ m s}^{-2})}{1 + \frac{10 \text{ kg}}{20 \text{ kg}}}$$

$$= 65.3 \text{ N}$$

(b) Using Eq. i, the acceleration is

$$a = \frac{T}{m_1} = \frac{65.3 \text{ N}}{20 \text{ kg}} = 3.27 \text{ m s}^{-2}$$

(c) Since the system is initially at rest and uniformly accelerated, the distance it moves in 2 s is

$$x = \tfrac{1}{2}at^2$$
$$= \tfrac{1}{2}(3.27 \text{ m s}^{-2})(2 \text{ s})^2 = 6.54 \text{ m}$$

Thus if we know the initial position and velocity, we can find the subsequent motion from the forces.

Example 3.7

A parachutist of weight **w** strikes the ground with her legs flexed and comes to rest with an upward acceleration of magnitude $3g$. Find the force exerted on her by the ground during landing (Fig. 3.10).

The forces on the woman are her weight **w** and a normal force **N** due to the ground. With $m = w/g$ and $a = 3g$, $F_y = ma_y$ becomes

$$N - w = ma = \left(\frac{w}{g}\right)(3g) = 3w$$

$$N = 4w$$

The force on her feet due to the ground is four times her weight. By contrast, if she is simply standing on the ground, the normal force equals her weight as seen in Example 3.1. Note that if she holds her legs more stiffly

during landing, she will come to rest with a greater acceleration in a shorter distance, and the forces on her feet will be greater.

3.10 | FRICTION

Friction is a force that always acts to resist the motion of one object sliding on another. Frictional forces are very important, since they make it possible for us to walk, use wheeled vehicles, and hold books. Microscopically, friction arises from many minute temporary bonds between the contact points of the two surfaces.

Frequently, we try to reduce the frictional forces that oppose some desired motion. This is often done with rollers or with wheels, since rolling peels the surfaces apart with a much smaller force than is required when they are torn apart by sliding.

Frictional forces in fluids are called *viscous forces*. They are often quite small compared to the friction between solid surfaces. Thus the use of lubricating liquids such as oil, which clings to the surfaces of metals, greatly reduces friction. Similarly, an air layer provides an almost frictionless support for a hovercraft vehicle or an air track demonstration apparatus.

When we walk or run, we are not conscious of any friction in our knees or other leg joints. These and many other mammalian joints are well lubricated by *synovial fluid*, which is squeezed through the cartilage lining the joints when they move (Fig. 3.11). This

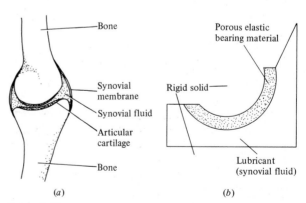

Figure 3.11. Human joints are lubricated by synovial fluid squeezed through the porous cartilage lining the joint. (*a*) A typical human joint. (*b*) A model approximately equivalent to the joint. (From Duncan Dowson, "Lubrication in Human Joints," Verna Wright, (ed), in *Lubrication and Wear in Joints,* Lippincott; Philadelphia, 1969.)

Figure 3.10. (*a*) A parachutist comes to rest with an acceleration of 3g. (*b*) The forces on her are a normal force **N** due to the ground and her weight **w**.

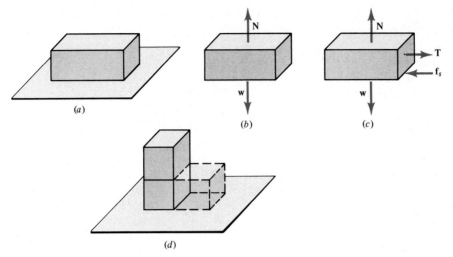

Figure 3.12. (*a*) A block at rest on a horizontal surface. (*b*) The force diagram. (*c*) The force diagram when a force **T** is applied. (*d*) The maximum static friction is independent of the contact area.

lubricant tends to be absorbed when the joint is stationary, increasing the friction and making it easier to maintain a fixed position. This is an excellent example of the clever biological engineering that nature has employed.

To make quantitative statements about friction, we consider a block at rest on a horizontal surface (Fig. 3.12*a,b*). Since the block is at rest, the first law requires that the net force on the block be zero. The vertical forces are the weight **w** and the normal force **N,** so we must have $N = w$. In the horizontal direction, there is no applied force and no motion, so the frictional force must also be zero, according to the first law.

Now suppose we apply a small horizontal force **T** to the right (Fig. 3.12*c*). If the block *remains at rest,* the friction force $\mathbf{f}_s$ can no longer be zero, since the first law requires that the net force be zero, or $f_s = T$. If **T** is gradually increased, $\mathbf{f}_s$ increases also. Eventually when **T** becomes large enough, the block begins to slide. *Thus there is a maximum possible static friction force $\mathbf{f}_s$(max).*

Experimentally, it is found that $\mathbf{f}_s$(max) has the following properties:

1 $\mathbf{f}_s$ (max) *is independent of the contact area.* For example, if we saw the block in half and stack one piece on the other (Fig. 3.12*d*), then $\mathbf{f}_s$(max) is unchanged.

2 *For a given pair of surfaces, f_s(max) is proportional to the normal force N.*

3 The number relating f_s(max) and N, called the *coefficient of static friction,* μ_s, is defined by

$$f_s(\text{max}) = \mu_s N \qquad (3.9)$$

μ_s depends on the nature of the two surfaces. Typically, for metals on metals, μ_s is between 0.5 and 1.0; for Teflon on metals, $\mu_s \approx 0.04$. Usually μ_s is less than 1 unless an adhesive is applied.

4 *The force necessary to keep an object sliding at constant velocity is smaller than that required to start it moving.* For example, it is easier to keep a heavy table or box moving than it is to get it started. Thus the *sliding or kinetic friction force f_k is less than f_s(max). It is independent of the contact area* and it satisfies

$$f_k = \mu_k N \qquad (3.10)$$

Here μ_k is the *coefficient of kinetic friction* and is determined by the nature of the two surfaces.

5 μ_k *is nearly independent of the velocity* and, since $f_k < f_s$(max),

$$\mu_k < \mu_s \qquad (3.11)$$

It is important to note that these properties of frictional forces are *not* fundamental statements about the physical world. Thus, they are quite different from Newton's laws of motion or the universal law of gravitation. Even though they give a good description of the frictional forces in common situations, they represent only an approximation to a complex problem. A more complete discussion would have to include the specific characteristics of the forces among the molecules making up the two surfaces. Friction is still an imperfectly understood phenomenon because of this complexity.

The following examples illustrates how friction is taken into account.

Example 3.8

A 50-newton block is on a flat, horizontal surface (Fig. 3.12). (a) If a horizontal force $T = 20$ N is applied and the block remains at rest, what is the frictional force? (b) The block starts to slide when T is increased to 40 N. What is μ_s? (c) The block continues to move at constant velocity if T is reduced to 32 N. What is μ_k?

(a) Since the block remains at rest when the force **T** is applied, the frictional force $\mathbf{f}_s$ must be equal but opposite to **T**. Consequently

$$f_s = T = 20 \text{ N}$$

(b) Since the block just begins to slide when the applied force is increased to 40 N, the maximum frictional force must be

$$f_s(\text{max}) = 40 \text{ N}$$

The vertical forces must add to zero, so the normal force N is equal to the weight w, or 50 N. Hence

$$\mu_s = \frac{f_s(\text{max})}{N} = \frac{40 \text{ N}}{50 \text{ N}} = 0.8$$

(c) Since the block moves with constant velocity when a 32-N force is applied, the net force must be zero. Consequently the frictional force f_k must equal the applied force, or

$$f_k = 32 \text{ N}.$$

Again the normal force must equal the weight, or 50 N, so

$$\mu_k = \frac{f_k}{N} = \frac{32 \text{ N}}{50 \text{ N}} = 0.64$$

As expected, the coefficient of kinetic friction μ_k is less than the coefficient of static friction μ_s.

Example 3.9

A block is at rest on an inclined plane (Fig. 3.13). The coefficient of static friction is μ_s. What is the maximum possible angle of inclination $\theta(\text{max})$ of the surface for which the block will remain at rest?

In this problem it is convenient to choose the coordinate axes as shown. By definition, the normal force **N** is perpendicular to the surface, or in the $-x$ direction. Similarly, the frictional force $\mathbf{f}_s$ is parallel to the surface or along the y direction. From Fig. 3.13c, the components of the weight **w** are

$$w_x = w \cos \theta, \qquad w_y = -w \sin \theta$$

When the block remains at rest, the first law requires that the x and y components of the forces cancel. Thus

$$f_s = w \sin \theta$$
$$N = w \cos \theta$$

Taking the ratio of these equations, the weight w cancels, leaving

$$\frac{f_s}{N} = \frac{\sin \theta}{\cos \theta} = \tan \theta$$

When the block is about to slip, $f_s = f_s(\text{max}) = \mu_s N$ and $\theta = \theta(\text{max})$. Thus we find

$$\mu_s = \tan \theta(\text{max})$$

An arrangement of this type provides a simple way to measure the coefficient of static friction. The angle is gradually increased until the block will not stay in place. For example, if $\theta(\text{max}) = 37°$, then $\mu_s = \tan 37° = 0.75$.

(a)

(b)

(c)

Figure 3.13. (a) A block on an inclined plane. (b) The forces on the block. (c) The components of the weight **w**.

(Culver Pictures)

SIR ISAAC NEWTON
(1642-1727)

Born in 1642, the year of Galileo's death, Newton made the crucial advances needed to complete our understanding of motion. He also made major contributions to optics and mathematics.

Newton was a thin frail baby and was raised by his grandmother after his widowed mother remarried when he was 2 years old. His difficult childhood may have contributed to his later psychotic tendencies. Throughout his brilliant career, he was extremely anxious when his work was published and irrationally violent whenever his ideas were challenged. He suffered at least two nervous breakdowns.

As an undergraduate at Cambridge (1661 to 1665), Newton soon mastered the literature of science and mathematics and began to enter unexplored regions. He formulated the binomial theorem and the basic concepts of calculus. During this period and the years immediately following, he also began to do research on optics and on planetary motion. He deduced that the force on a planet due to the sun must vary as $1/r^2$. Some 20 years later, he would extend this idea to the universal law of gravitation.

Although Newton's work was known only to a limited circle because of his reluctance to publish his research, he was appointed a professor at Cambridge in 1669. He developed the first reflecting telescope in order to circumvent the distortions inherent in lenses. When this telescope received an enthusiatic reception from the Royal Society of London, he was encouraged to present his other research in optics to the Society in 1672. Robert Hooke, the leading authority on optics, disagreed with some of Newton's ideas. This led to bitter disputes, with Newton finally withdrawing into isolation for some years.

Newton's greatest achievements were his advances in mechanics. Although many of his results were obtained quite early in his career, he did not present his theory of planetary motion until he was urged to do so in 1684 by Edmond Halley, an astronomer who had heard of Newton's work.

Newton's classic work, *Principia Mathematica,* appeared in 1687. Written in Latin, it contained the three laws of motion and the universal law of gravitation. This treatise constituted one of the foundations of modern science and made Newton internationally famous. It also effectively marked the end of Newton's active research, with his attention gradually turning to politics, theology, and scientific priority disputes.

Newton became master of the mint, a well-paying and normally undemanding job. However, he took the position seriously and was especially zealous in sending counterfeiters to the gallows. He also assumed the role of the leader of English science, becoming president of the Royal Society in 1703; in 1705, he became the first scientist to be knighted. Unfortunately, he repeatedly used his position to carry on bitter arguments with various scientists. The most prolonged of these was a 25-year battle with Leibniz (which ended with Newton's death in 1727) over credit for the development of the calculus. It is now agreed that Leibniz independently developed the calculus after Newton had, but before Newton published his results.

SUMMARY

Newton's three laws of motion allow us to predict the motion of an object from the forces acting upon it. The first law states that in an inertial coordinate frame, the object remains at rest or in motion with constant velocity unless a net force is applied. The second law states that the force **F** needed to produce an acceleration **a** satisfies $\mathbf{F} = m\mathbf{a}$, where m is the inertia or mass of the object. The third law states that if object A exerts a force on B, then B exerts an equal and opposite force on A.

All objects exert gravitational forces on each other. The forces between two spheres or point particles are proportional to the products of their masses and inversely proportional to the square of their separation. The mass of an object is the same everywhere in the universe; its weight $\mathbf{w} = m\mathbf{g}$ depends on the acceleration due to gravity at the object's location.

When the force applied to an object resting on a surface exceeds the maximum static friction force $\mu_s N$, it starts to move. The force of sliding or kinetic friction $\mu_k N$ is usually smaller than the maximum static friction force.

Checklist

Define or explain:

force

weight

gravitational mass

Newton's first law of motion

inertial coordinate system

equilibrium

Newton's third law of motion

Newton's second law of motion

inertial mass

law of universal gravitation

gravitational constant

coefficient of static friction

coefficient of kinetic friction

EXERCISES

Section 3.1 | Force, Weight, and Gravitational Mass

3-1   Find the direction and magnitude of the net force on the object in Fig. 3.14.

Figure 3.14.   Exercise 3-1.

3-2   Find the direction and magnitude of the net force on the object in Fig. 3.15.

Figure 3.15.   Exercise 3-2.   **Figure 3.16.**   Exercise 3-3.

3-3   Find the direction and magnitude of the net force on the object in Fig. 3.16.

3-4   A man weighs 160 lb. What is his mass in slugs?

3-5   A woman has a mass of 50 kg. What is her weight in newtons?

3-6   What is the weight of a 1 kilogram steak?

3-7   Find the weight of 500 grams of candy in (a) dynes, (b) newtons, and (c) pounds.

3-8   A candy bar weighs 1 ounce (1 oz) (16 oz = 1 lb). Find its mass in kilograms.

Section 3.2 | Newton's First Law and
Section 3.3 | Equilibrium

3-9   For which of the following observers does Newton's first law of motion hold true in the form stated in this chapter? (a) A person on a plane moving at constant speed in a constant direction. (b) A parachutist who has just stepped out of an airplane. (c) A parachutist who has reached terminal velocity and is falling at constant speed. (d) The pilot of a plane that is taking off from a runway. Explain your reasoning.

3-10   A person is in a car traveling around a circular track. Do the observations of this person agree with Newton's first law as stated in this chapter? Does your answer depend on whether the speed is constant? Explain.

3-11   A car is moving at a constant speed on a straight road, which is at an angle of 10° to the horizontal direction. (a) Is it in equilibrium? (b) What forces act on the car?

3-12   A 2000-kg airplane is in level flight at constant velocity. (a) What is the net force on the plane? (b) What is the upward lift force on the plane due to the air?

3-13   A round stone resting in the back of a pickup truck begins to roll toward the back as the truck begins moving. The driver is at rest with respect to the truck and concludes that a force is causing the stone to move toward the rear of the truck. (a) Is he correct? (b) What forces are acting on the stone? (c) Does the stone move forward or backward with respect to the ground?

Section 3.4 | Newton's Third Law

3-14   A boat in a flowing river is tied by a rope to a post on a dock. (a) Draw a diagram showing all of the forces acting on the boat. (b) Identify the horizontal forces acting on the boat and the associated reaction forces. (c) Identify the vertical forces acting on the boat and the associated reaction forces. (d) What is the net force on the boat?

3-15   A large airplane is pulled at constant velocity relative to the runway by a truck. The two are connected by an iron bar. (a) What are the forces on the airplane? (b) What are the forces on the truck? (c) What is the net force on the airplane? (d) What is the net force on the truck? (e) What is the net force on the iron bar? (f) Identify the action-reaction forces acting on the airplane, bar, and truck.

3-16   An airplane is flying horizontally with a constant velocity. The propellers are pushing backward on the air. (a) Is the airplane in equilibrium? (b) What forces are acting in the horizontal plane?

3-17   (a) A girl holds a ball motionless in her hand. Identify the forces acting on the ball and their reactions. (b) She throws the ball into the air. What are the forces on the ball while it is in the air? What are the reactions to these forces?

3-18   A car coasts to a stop on a flat, straight road. (a) What are the forces acting on the car? (b) What are the reactions to those forces?

Section 3.5 | Newton's Second Law

3-19   The takeoff acceleration of a woman jumping vertically is 20 m s^{-2}. Her mass is 50 kg. (a) How large a force does the ground exert on the woman? (b) What is the ratio of this force to her weight?

3-20   What acceleration is produced when a 100-N force is applied to a 10-kg rock?

3-21   What net force is needed to give a 1000-kg car an acceleration of 3 m s^{-2}?

3-22 A baseball of mass 0.15 kg is struck by a bat with a force of 5000 N. What is the acceleration of the ball?

Section 3.7 | Gravitational Forces

3-23 The moon is 3.9×10^5 km from the center of the earth. The mass of the moon is 7.3×10^{22} kg, and the mass of the earth is 6.0×10^{24} kg. How far from the earth's center are the gravitational forces on an object due to the earth and moon equal and opposite? (Assume the object is on the line connecting the earth and moon.)

3-24 The mass of the sun is 2.0×10^{30} kg, and the distance from the moon to the sun is 1.5×10^8 km. Using the data of the preceding example, find the ratio of the forces exerted by the earth and the sun on the moon.

Section 3.8 | Weight

3-25 The acceleration of gravity on the surface of Mars is 3.62 m s^{-2}. How much would a person who weighs 200 lb on earth weigh on Mars?

3-26 The mass of Mars is 6.42×10^{23} kg and the acceleration of gravity on its surface is 3.62 m s^{-2}. What is the radius of Mars?

3-27 The acceleration of gravity at the surface of a planet is half that on the surface of the earth. If the radius of the planet is half the radius of the earth, how is its mass related to the mass of the earth?

3-28 Planet Y has a radius one-third times that of the earth, and a mass $(1/3)^3 = 1/27$ times that of the earth. What is the weight of an astronaut with a mass of 70 kg?

3-29 If the earth is assumed to be a uniform sphere of radius 6.38×10^6 m, compute its mass from G and g.

Section 3.9 | Some Examples of Newton's Laws

3-30 An elevator of mass 900 kg accelerates upward at 3 m s^{-2}. What is the tension in the cable where it is attached to the elevator?

3-31 A horse can exert a horizontal force of 3.5×10^4 N on a rope that passes over a pulley to lift loads vertically. What is the acceleration of a load of weight (a) 3.5×10^4 N; (b) 3×10^4 N? (Neglect the masses of the rope and pulley.)

3-32 An elevator cable that is light in weight compared to the elevator car can support a weight of 2000 lb. If the elevator and occupants weigh 1750 lb, what is the maximum possible vertical acceleration of the elevator?

3-33 A 60-kilogram man hangs from a light cable suspended from a helicopter. Find the tension in the cable if the acceleration is (a) 5 m s^{-2} upward; (b) 5 m s^{-2} downward.

3-34 A human femur will fracture if the compressional force is 2×10^5 N. A person of mass 60 kg lands on one leg, so that there is a compressional force on the femur. (a) What acceleration will produce fracture? (b) How many times the acceleration of gravity is this?

3-35 A 55-kg woman wishes to slide down a stationary rope that will support a force of 400 N. What is the minimum acceleration of the woman if she safely slides down the rope?

3-36 An engine weighing 2×10^4 lb pulls a train weighing 10^5 lb on a level track with an acceleration of 1 ft s^{-2}. What would the acceleration be if the train weighed 6×10^4 lb?

3-37 In a collision, an automobile of mass 1000 kg stops with constant acceleration in 2 m from an initial speed of 20 m s^{-1}. (a) What is the acceleration of the car? (b) What is the force on the car during the collision?

Section 3.10 | Friction

3-38 A horse weighing 7500 N is able to exert a horizontal force of 6500 N on a load. What is the coefficient of static friction between the horse's feet and the ground? (Assume the force exerted by the horse is limited by its tendency to slip.)

3-39 A refrigerator weighs 1000 N. A horizontal force of 200 N is applied, but the refrigerator does not move. (a) What is the frictional force? (b) What can we conclude about the coefficient of static friction?

3-40 A box weighing 100 lb is at rest on a horizontal floor. The coefficient of static friction is 0.3. What is the minimum force needed to start it in motion?

3-41 A box weighing 100 lb is pushed on a horizontal floor. The coefficient of sliding friction is 0.2. What acceleration will result if a horizontal force of 40 lb is applied?

3-42 A sled weighing 1000 N is pulled along flat, snow-covered ground. The coefficient of static friction is 0.3, and the coefficient of sliding friction is 0.15. Find the force needed to (a) start the sled moving, and (b) keep it moving at constant velocity.

3-43 A stone boat of total weight 12,000 lb is used in a horse-pulling contest. The coefficient of static friction between the stone boat and the earth is 0.6, and the coefficient of sliding friction is 0.4. (a) What force must a pair of horses exert to start the stone boat moving? (b) What force must the horses exert to keep the stone boat moving at a constant velocity?

3-44 How can the adjustable inclined plane in Fig. 3.13 be used to measure the coefficient of kinetic friction for an object on the plane?

PROBLEMS

3-45 A 2000-lb automobile is traveling at 50 ft s^{-1} and skids to a stop with constant acceleration in 250 ft. What is the frictional force on the car?

3-46 A 0.5-kg ball is initially at rest. If a 10-N force is applied for 2 s, what is the final velocity of the ball?

3-47 A hockey player who weighs 160 lb comes to rest from 20 ft s^{-1} in 1 s. (a) What is his mass? (b) What is his average acceleration? (c) What force is required to provide this acceleration?

3-48 A runner's foot strikes the ground with a velocity of 10 m s^{-1} downward. If the effective mass of the foot and leg that is brought to rest is 9 kg, what is the force on the foot as it comes to rest with constant acceleration in (a) 0.03 m on a soft surface; (b) 0.005 m on a hard surface?

3-49 A woman of mass 55 kg steps off a table 1 m above the ground. (a) What is her velocity just before she strikes the ground? (b) If she lands on her feet with her body held rigid, she stops in 0.15 m. What average force is exerted upward on the woman during the impact? (c) If she flexes her legs and body during impact, she stops in 0.5 m. What average force does she now experience during impact?

3-50 A boy is fishing with a line that will sustain a maximum force of 40 N. If he hooks a 3-kg fish, which can exert a force of 60 N for several seconds,

what is the minimum acceleration with which the line must be played out during that time interval?

3-51 A subway train has three cars, each weighing 1.2×10^5 N. The frictional force on each car is 10^3 N, and the first car, acting as an engine, exerts a horizontal force of 4.8×10^4 N on the rails. (a) What is the acceleration of the train? (b) What is the tension in the coupling between the first and second cars? (c) What is the tension between the second and third cars?

3-52 If a 60-kg woman walks up a 45° hill at a constant velocity, what is the force she must exert parallel to the ground if (a) she walks straight up the hill; (b) she takes a zigzag path up the hill so the effective angle of ascent is 30°?

3-53 A 60-kg man pulls horizontally on a rope with a force of 600 N. The rope passes over a pulley and is attached to a post of mass 25 kg set in the ground. If the frictional force opposing the vertical motion of the post is 300 N, (a) what is the vertical acceleration of the post? (Assume the man moves as the post does.) (b) If the frictional force is the same, what is the maximum mass of a post that the man can pull from the ground?

3-54 The radius of the planet Venus is 6.1×10^3 km and that of the earth is 6.4×10^3 km. The mass of Venus is 82 percent of the earth's mass. What is the acceleration of gravity on the surface of Venus?

3-55 A wagon weighing 5×10^3 N is pulled along a muddy, horizontal road by a 500-kg horse. The coefficient of friction between the wagon wheels and the road is 0.2. (a) If the wagon is pulled at constant speed, what force must the horse exert on the ground to pull the wagon? (b) If the wagon is accelerated from rest to a velocity of 5 m s^{-1} in 5 s, what force must the horse exert on the ground?

3-56 A 60-kilogram man wishes to run on ice. The coefficient of static friction between his shoes and the ice is 0.1. What is his maximum possible acceleration?

3-57 A girl of mass 40 kg skis down a slope, which is at an angle of 37° with the horizontal. (Neglect air resistance.) If the coefficient of kinetic friction between her skis and the snow is 0.1, what is her acceleration?

3-58 In Fig. 3.17, the strings and pulleys are massless, and there is no friction. Find (a) the tensions in the strings, and (b) the acceleration of the system.

Figure 3.17. Problems 3-58 and 3-59.

3-59 Repeat the preceding problem if the coefficient of kinetic friction between the block on the surface and the surface is 0.1.

3-60 In Fig. 3.18 the string and pulley are massless, and there is no friction. Find (a) the tension in the string, and (b) the acceleration.

3-61 In the preceding problem, $m_1 = 2$ kg and $m_2 = 3$ kg. Find (a) the tension in the string, and (b) the acceleration. (c) If the system is released from rest, what are its velocity and position after 0.15 seconds?

* **3-62** Two people wish to push a food freezer weighing 400 lb up a ramp inclined at an angle of 37° with the horizontal. The coefficient of sliding friction between the freezer and ramp is 0.5.

(a) What minimum force must the people exert to slide the freezer up the ramp? (b) What acceleration will the freezer have if it is released and slides down the ramp? (c) If it slides 8 ft down the ramp and strikes a heavy object, coming to rest in 1 ft, what average force does it exert on the heavy object?

* **3-63** A man can exert a force of 700 N on a rope attached to a sled. The rope is at an angle of 30° with the horizontal. If the coefficient of kinetic friction between the sled and ground is 0.4, what is the maximum load on the sled that the man can pull at constant speed?

* **3-64** A box weighing 600 lb is at rest on a ramp at an angle of 37° with the horizontal. The coefficient of static friction between the box and ramp is 0.8. Find the minimum force required to move the box down the ramp if the force is applied (a) parallel to the ramp; (b) horizontally.

* **3-65** In a ski jump, the slope is initially at an angle of 45° to the horizontal. If the coefficient of kinetic friction between the skis and the snow is 0.1, find (a) a skier's acceleration, and (b) the velocity reached after 40 m on the ramp.

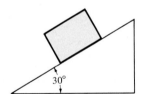

Figure 3.19. Problems 3-66 and 3-67.

3-66 If there is no friction, what is the acceleration of the block in Fig. 3.19?

* **3-67** If the coefficient of sliding friction is 0.2, what is the acceleration of the block in Fig. 3.19?

Additional Reading

E.N. daCosta Andrade, *Sir Isaac Newton: His Life and Work,* Science Study Series, Doubleday and Co., Garden City, New York, 1958.

Herbert Butterfield, *The Origins of Modern Science,* The Macmillan Co., New York, 1960. Chapters 8 and 10 discuss Newton's work.

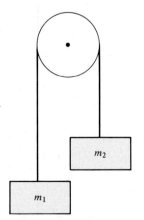

Figure 3.18. Problems 3-60 and 3-61.

J.W. Beams, Finding a Better Value for G, *Physics Today*, vol. 24, May 1971, p. 34.

R. McNeill Alexander, *Animal Mechanics*, University of Washington Press, Seattle, 1968. Pages 55–66 discuss friction in animals.

E.H. Freitag, The Friction of Solids, *Contemporary Physics*, vol. 2, 1961, p. 198.

Ernest Rabinowitz, Friction (Resource Letter F-1), *American Journal of Physics*, vol. 31, 1963, p. 897. A bibliography.

Scientific American articles:

I. Bernard Cohen, Newton, December 1955, p. 73.

D. Sciama, Inertia, February 1957, p. 99.

Carl Gans, How Snakes Move, June 1970, p. 82.

W.A. Heiskane, The Earth's Gravity, September 1955, p. 164.

George Gamow, Gravity, March 1961, p. 94.

R.H. Dicke, The Eötvös Experiment, December 1961, p. 84.

Clifford C. Will, Gravitational Theory, November 1974, p. 24.

F. Palmer, Friction, February 1956, p. 54.

Ernest Rabinowitz, Stick and Slip, May 1956, p. 109.

CHAPTER FOUR

STATICS

Statics is the study of the forces acting on an object that is in equilibrium and at rest. Even though no motion occurs, a number of interesting questions arise concerning these forces that can be answered with the aid of Newton's laws. For example, one can find the forces acting on various parts of engineering structures, such as bridges or buildings, or of biological structures, such as jaws, limbs, or backbones. Statics can be used to understand the force multiplication or *mechanical advantage* obtained with simple machines, such as the many levers found in the human body. It also addresses problems of balance and stability for objects as well as for animals. The range of questions answerable with statics makes it invaluable in fields as diverse as engineering, comparative anatomy, physical therapy, and orthodonture.

Before considering applications of statics, we must discuss the conditions for equilibrium of a *rigid body:* an object extended in space that does not change its shape or size, such as a baseball or a bone. The condition that the net force is zero is sufficient to ensure that a point particle at rest remains at rest; for a rigid body, this condition means that the body as a whole will not accelerate, or that it is in *translational equilibrium*. However, a rigid body will start to rotate if forces act so that there is a net turning effect, or *torque*. Thus the absence of a net torque is the second condition necessary for the equilibrium of a rigid body. Also, to consider balance and stability, we need to use the concept of the *center of gravity*. This is the

point at which the weight of a rigid body may be considered to be concentrated.

4.1 | TORQUES

Suppose an object is subjected to two equal but opposite forces. The net force is zero, so the object is in translational equilibrium. Nevertheless, it may not be in rotational equilibrium. As an example, consider a lunch counter stool with a freely rotating seat (Fig. 4.1). If a child applies equal but opposite forces $\mathbf{F}_1$ and $\mathbf{F}_2 = -\mathbf{F}_1$ to opposite sides of the seat, it will begin to rotate. Thus the seat does not remain at rest even though $\mathbf{F} = \mathbf{F}_1 + \mathbf{F}_2 = 0$, and there is no net force.

Clearly, in addition to $\mathbf{F} = 0$, we require another

Figure 4.1. Equal but opposite forces applied to opposite sides of the seat will cause it to start rotating. Hence, it is not in equilibrium, even though the net force is zero.

Figure 4.2.   The torque on the rigid body about point P has a magnitude $\Gamma = rF \sin \theta$.

equilibrium condition to exclude the possibility of rotational motion. The quantity that indicates the ability of a force to cause rotation is called the *torque*. *A rigid body is in rotational equilibrium when there is no net torque acting on it.* The torque Γ depends on the force **F**, the distance **r** from a point on the axis of rotation to where the force acts, and the angle θ between **r** and **F** (Fig. 4.2). The torque has a magnitude

$$\Gamma = rF \sin \theta \qquad (4.1)$$

This expression for the torque can be illustrated in various ways. For example, suppose we need to un-

screw a large nut that is rusted into place (Fig. 4.3). To maximize the torque, we use the longest wrench available and exert as large a force as possible. When $\theta = 90°$, $\sin \theta = 1$, which is the largest value of $\sin \theta$. Consequently, we should pull at right angles to the wrench. Note that when θ is $0°$ or $180°$, $\sin \theta$ is zero, and there is no torque.

Another example is provided by the screen door in Fig. 4.4. A spring opposes the rotation of the door and permits it to open through an angle ϕ which increases with the torque. We observe that the torque is greatest when the force is applied as far as possible from the hinge and at right angles to the door.

The definition $\Gamma = rF \sin \theta$ allows one to compute the magnitude of a torque. It also shows that the dimensions of a torque are force times length, so that the S.I. torque unit is a newton meter (N m). However, the definition is incomplete, since rotations have a direction associated with them. When an object such as a wrench or door is constrained to rotate

Figure 4.3.   The torque has a magnitude $\Gamma = rF \sin \theta$. Accordingly it is a maximum when **r** and **F** are at right angles as in (*a*). The torque is zero when **r** and **F** are parallel ($\theta = 0°$) as in (*b*) or opposite ($\theta = 180°$) as in (*c*).

Figure 4.4.   A top view of a screen door hinged at P. A spring resists opening of the door, so that the deflection angle ϕ increases with the applied torque. When a force **F** is applied opposite or parallel to **r**, no torque results. The greatest torque occurs when **F** is perpendicular to **r** and **r** is as large as possible.

BAH, trying to compute trajectory of spendings

Figure 4.5. $C = A \times B$ has a magnitude $AB \sin \theta$. Its direction is given by the right hand rule.

about a given axis, we must say whether a torque will tend to produce clockwise or counterclockwise rotations. Some objects, such as balls and gymnasts, are free to rotate about any axis. In such cases, *the torque is regarded as a vector directed along the axis about which it tends to cause rotation.*

To write the definition of torque in vector notation, we define the *vector product* or *cross product* of two vectors. The vector product of **A** and **B** is a vector **C**, which is perpendicular to both **A** and **B** and denoted by

$$C = A \times B \qquad (4.2)$$

The magnitude of **C** is (see Fig. 4.5)

$$C = AB \sin \theta \qquad (4.3)$$

where θ is the angle between **A** and **B**.

Since **C** is perpendicular to both **A** and **B**, it points along the direction perpendicular to the plane defined by **A** and **B**. You can determine whether **C** points toward or away from this plane with the aid of the *right-hand rule* as follows:

1 Point the fingers of your right hand along the first vector (**A** in Fig. 4.5).
2 Orient your arm so that you can bend your hand at the wrist and rotate your palm forward, through an angle of less than 180°, until your fingers point toward **B**.

3 Your thumb now points in the same direction as $C = A \times B$.

From the definition of the cross product, it follows that $A \times B$ and $B \times A$ are in opposite directions, or $A \times B = -B \times A$. Thus the order of the factors in a cross product is significant; this product does not obey the *commutative principle.* This contrasts with ordinary algebra, where $x \cdot y = y \cdot x$ holds, and with vector addition, where $A + B = B + A$. These operations are commutative.

In terms of this notation the torque Γ can be written as

$$\Gamma = r \times F \qquad (4.4)$$

We note that the magnitude of Γ is $\Gamma = rF \sin \theta$ as before. The direction of Γ is given by the right-hand rule and indicates the axis about which rotation will tend to occur.

To illustrate the right-hand rule, suppose **r** is in the $+x$ direction and **F** is in the $+y$ direction (Fig. 4.6*a*). Using the right-hand rule, we point the fingers of our right hand in the $+x$ direction. When our palm faces the $+y$ direction, and our thumb is out of the page, we can rotate our fingers 90° toward the $+y$ direction. Thus $\Gamma = r \times F$ *is out* of the page. Γ is shown in the figure as a circled dot representing the tip of an arrow.

Suppose instead that **r** is in the $-x$ direction and **F** is in the $+y$ direction (Fig. 4.6*b*). Pointing our fingers toward $-x$ with the palm toward $+y$, we can rotate 90° from **r** to **F**. In this case our thumbs and $\Gamma = r \times F$ are into the page. Here Γ is shown as a circled cross symbolizing the tail of an arrow.

The results of this discussion can be expressed in a slightly different form when the plane of the rotation is known (Fig. 4.7). Torques that tend to produce a *counterclockwise rotation* of the object are directed outward from the page and are conventionally taken

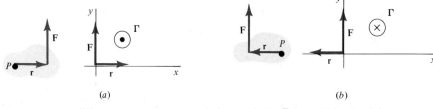

(a)   *(b)*

Figure 4.6. (*a*) When torques are computed about point P, $\Gamma = r \times F$ is out of the page. It is symbolized by a dot representing the tip of an arrow. (*b*) Γ is into the page and symbolized by a cross representing the tail of an arrow.

Figure 4.7. (*a*) Counterclockwise torques are taken to be positive. (*b*) Clockwise torques are taken to be negative.

Figure 4.9. The torque due to a couple is the same about every point.

to be *positive*. Similarly, torques causing *clockwise rotations* are directed into the page and are conventionally taken to be *negative*.

The magnitude of the torque, $\Gamma = rF \sin \theta$, can be rewritten as $\Gamma = r_\perp F$ where the subscript $\perp$ means perpendicular (Fig. 4.8). That is, the magnitude of $\boldsymbol{\Gamma}$ is the product of F and the perpendicular distance from P to the *line of action* of $\mathbf{F}$. The perpendicular distance $r_\perp$ is called the *lever arm*. The torque can also be written as $\Gamma = rF_\perp$, the distance r times the component of $\mathbf{F}$ perpendicular to $\mathbf{r}$.

Couples | A pair of equal but opposite forces with

different lines of action is called a *couple*. The pair of forces applied to the lunch counter seat in Fig. 4.1 is an example of a couple. Couples do not exert a net force on an object even though they do exert a net torque. They have the interesting property that the net torque is independent of the choice of the point P from which distances are measured. This is seen explicitly in the following example.

Example 4.1

Two equal and opposite forces act on an object with different lines of action (Fig. 4.9). Find the net torque on the object resulting from these forces.

If we compute the torque about the point P in the figure, the torque resulting from the force at x_2 is $\Gamma_2 = -x_2 F$. (The minus sign indicates a clockwise torque.) The torque resulting from the force at x_1 is $\Gamma_1 = x_1 F$. The net torque is

$$\begin{aligned} \Gamma = \Gamma_1 + \Gamma_2 &= x_1 F - x_2 F \\ &= (x_1 - x_2)F = -lF \end{aligned}$$

The minus sign means that the net torque tends to cause clockwise rotation. Note that only the distance l between the lines of action of the forces appears in the result, so that the torque is independent of the location of the point P.

In the next section we see how torques are used to determine whether a system is in rotational equilibrium.

4.2 | EQUILIBRIUM OF RIGID BODIES

From the discussion of the preceding section, we see that there are two conditions for the equilibrium of a rigid body:

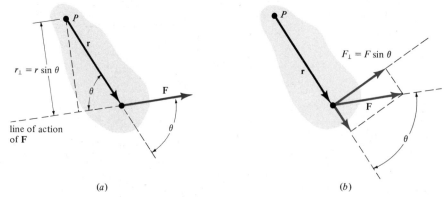

Figure 4.8. The magnitude of the torque about P is $\Gamma = rF \sin \theta$. This can be rewritten as (*a*) $\Gamma = r_\perp F$, where $r_\perp = r \sin \theta$ is the lever arm. (*b*) Also, $\Gamma = rF_\perp$, where $F_\perp = F \sin \theta$.

(a)

(b)

(c)

Figure 4.10. (a), (b) When two children balance on a see-saw, the heavier child must be closer to the pivot. (c) Another equilibrium configuration.

1 *The net force on the object must be zero,*

$$\mathbf{F} = 0 \qquad (4.5)$$

2 *The net torque on the object computed about any convenient point must be zero,*

$$\mathbf{\Gamma} = 0 \qquad (4.6)$$

A familiar application of these conditions is provided by two children balancing on a seesaw (Fig. 4.10). If their weights are unequal, they soon find that the heavier child must be closer to the pivot. For example, if one child is twice as heavy as the other, he or she must sit half as far from the pivot. Let us see how this comes about from the equilibrium conditions.

Example 4.2

Two children of weights w_1 and w_2 are balanced on a board pivoted about its center (Fig. 4.10). (a) What is the ratio of their distances x_2/x_1 from the pivot? (b) If $w_1 = 200$ N, $w_2 = 400$ N, and $x_1 = 1$ m, what is x_2? (For simplicity, we assume the board to be weightless; this will not affect the result.)

(a) According to the first equilibrium condition, the force $\mathbf{N}$ exerted by the support must balance their weights so that the net force is zero:

$$N - w_1 - w_2 = 0, \qquad N = w_1 + w_2$$

This gives us no information about their positions. However, we have not yet used the condition that the net torque is zero. We will compute torques about the pivot point, P. Then the lever arm is zero for $\mathbf{N}$, since its line of action passes through the pivot, and it has no torque. The torques resulting from the weights are $\Gamma_1 = x_1 w_1$ and $\Gamma_2 = -x_2 w_2$. Thus, $\mathbf{\Gamma} = \Gamma_1 + \Gamma_2 = 0$ requires

$$x_1 w_1 - x_2 w_2 = 0$$

or

$$\frac{x_2}{x_1} = \frac{w_1}{w_2}$$

This is the condition relating the positions x_1 and x_2 when the seesaw is balanced. Either position may be chosen arbitrarily; the other is then determined by this condition.

(b) If $w_1 = 200$ N, $w_2 = 400$ N, and $x_1 = 1$ m, then we must have

$$x_2 = x_1 \frac{w_1}{w_2} = (1 \text{ m}) \frac{(200 \text{ N})}{(400 \text{ N})} = 0.5 \text{ m}$$

This is consistent with our initial statement that if one child is twice as heavy as the other, he or she must sit half as far from the pivot.

In the example above we chose to compute the torques about the pivot point. However, the equilibrium condition says that the torques calculated about *any* point must add up to zero. In the next example we will solve the same problem, computing torques about a different point. We will find that the result is the same.

Example 4.3

Again find x_2/x_1 for the seesaw of the preceding example, calculating torques about the point P_1, where the child of weight w_1 is seated.

To proceed, we redraw the force diagram of Fig. 4.10 as in Fig. 4.11. Computing torques about P_1, $\mathbf{N}$ and $\mathbf{w}_2$ produce torques $x_1 N$ and $-(x_1 + x_2)w_2$, respectively; $\mathbf{w}_1$

Figure 4.11. Calculation of the torques' about P_1.

produces no torque, since its lever arm is zero. In equilibrium the sum of these torques must be zero so

$$-(x_1 + x_2)w_2 + x_1N = 0 \qquad \text{(i)}$$

But the forces must also add up to zero, so

$$N - w_1 - w_2 = 0$$

or

$$N = w_1 + w_2$$

Substituting for N in Eq. i, we find

$$-(x_1 + x_2)w_2 + x_1(w_1 + w_2) = 0$$

Cancelling some terms, we obtain our earlier result

$$x_2w_2 = x_1w_1 \quad \text{or} \quad \frac{x_2}{x_1} = \frac{w_1}{w_2}$$

This example illustrates that we can often simplify our calculations with a clever choice of the point about which we compute the torques. For example, using the pivot eliminated the unknown normal force **N** from the torque equation. The same final results are always obtained no matter what point is used to calculate torques.

We now use the equilibrium conditions to find the forces on a human forearm.

Example 4.4

A model for the forearm in the position shown in Fig. 4.12 is a pivoted bar supported by a cable. The weight **w** of the forearm is 12 N and can be treated as concentrated at the point shown. Find the tension **T** exerted by the biceps muscle and the force **E** exerted by the elbow joint.

The tension **T** and the weight **w** have no horizontal components. Since the net horizontal force must be zero, the force **E** exerted by the joint cannot have a horizontal component. We assume tentatively that **E** is directed downward; a negative result will indicate it points the opposite way.

Applying the condition $\mathbf{F} = 0$,

$$T - E - w = 0$$

This contains both unknowns, T and E. Calculating torques about the pivot, **E** produces no torque. **w** produces a torque $-(0.15 \text{ m})w$, and **T** produces a torque $(0.05 \text{ m})T$. Thus, $\boldsymbol{\Gamma} = 0$ becomes

$$-0.15w + 0.05T = 0$$

or

$$T = 3w = 3(12 \text{ N}) = 36 \text{ N}$$

The first equation then gives

$$E = T - w = 36 \text{ N} - 12 \text{ N} = 24 \text{ N}$$

E is positive, so it is directed downward as initially assumed.

Note that both the tension **T** exerted by the muscle and the force **E** exerted by the elbow are considerably larger than the weight they are supporting. This is because, relative to the joint, the lever arm for the weight is larger than that for the muscle. When a weight is held in the hand, it is even further from the joint, and the additional forces the muscle and joint must provide are correspondingly larger.

4.3 | THE CENTER OF GRAVITY

The torque about any point produced by the weight of an object is equal to that due to a concentrated object of the same weight placed at a point called the *center of gravity* (C.G.) (Fig. 4.13). This fact simplifies the mechanics of both static and moving objects and was implicitly assumed in Example 4.4 when we treated the weight of the forearm as concentrated at a point. The centers of gravity of uniformly dense symmetric objects are at their geometric centers (Fig. 4.14). For less symmetric objects the C.G. can be calculated mathematically or located experimentally.

Figure 4.12. (*a*) The forearm is supported by the biceps muscle and pivoted at the elbow. (*b*) The forearm can be considered as a pivoted bar supported by a cable. The pivot represents the elbow joint, and the cable represents the biceps. (Adapted from Williams and Lissner.)

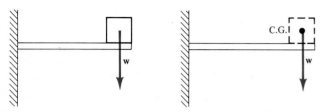

Figure 4.13. The torque produced by the weight of a rigid body is equal to that due to a concentrated object of the same weight placed at the center of gravity.

Figure 4.14. The center of gravity of a uniformly dense symmetric object is at its geometric center. Note that the C.G. of the doughnut is at the center of the hole and hence not in the object itself.

A suspended object always hangs so that its center of gravity is directly below the point of suspension, since in this position the torque resulting from the weight about that point is zero (Fig. 4.15). This observation provides a way of locating the C.G. experimentally. If an object is suspended from a point P_1, it comes to rest with the C.G. on the vertical line through P_1. Suspending the object from a second point P_2, the C.G. lies on the vertical line through P_2. The C.G. is then at the intersection of the two lines (Fig. 4.16).

We now show how to find the center of gravity mathematically, starting with the simplest possible system: two point weights on a weightless rod (Fig. 4.17). We take the C.G. to be a distance x_1 from w_1 and x_2 from w_2. Then the net torque about the C.G. due to the two weights must be equal but opposite. This is similar to the seesaw of Example 4.2, and we can use the balance condition obtained there, $x_2/x_1 = w_1/w_2$. For example, if $w_1 = w_2$, then $x_1 = x_2$; the C.G. is at the midpoint, as is always true for a symmetric object. If $w_2 = 2w_1$, then $x_2 = x_1/2$; the C.G. is closer to the heavier weight, as we might have expected.

Another way of finding the C.G. of two weights leads to a formula that can immediately be adapted

to any number of weights. Figure 4.18 shows the same two weights on a weightless rod located on an x axis. The C.G. is at an unknown point X. From the definition, a weight $w = w_1 + w_2$ concentrated at X will produce a torque equal to the sum of the torques due to w_1 and w_2. The individual torques about the origin are $\Gamma_1 = -x_1 w_1$ and $\Gamma_2 = -x_2 w_2$. Thus the total torque is

$$\Gamma = \Gamma_1 + \Gamma_2 = -x_1 w_1 - x_2 w_2$$

A single weight w at X gives a torque $\Gamma = -Xw$. Equating the two expressions for Γ, we find the center of gravity is at

$$X = \frac{x_1 w_1 + x_2 w_2}{w}.$$

Suppose, for example, that $w_1 = w_2$. Then $w = w_1 + w_2 = 2w_1$, and

$$X = \frac{x_1 w_1 + x_2 w_1}{2w_1} = \frac{x_1 + x_2}{2}$$

Figure 4.15. An object will hang so that its center of gravity is below the point of suspension, P.

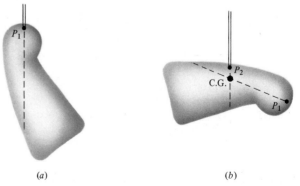

(a)   (b)

Figure 4.16. (a) The center of gravity lies on the vertical line through P_1. (b) The C.G. also lies on the vertical line through P_2, so it is at the intersection of the two lines.

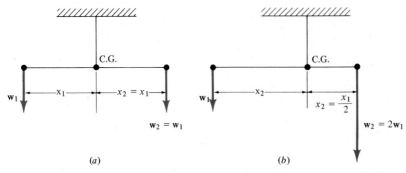

Figure 4.17.   The center of gravity of two point weights. (*a*) Equal weights. (*b*) Unequal weights.

Figure 4.18.   The center of gravity of two point weights on a weightless rod is at X.

Thus the center of gravity X is midway between the two weights, as was found before.

If there are more than two weights, the center of gravity is found in the same way. The result is

$$X = \frac{x_1 w_1 + x_2 w_2 + x_3 w_3 + \cdots}{w} \qquad (4.7)$$

where

$$w = w_1 + w_2 + w_3 \cdots \qquad (4.8)$$

The following example illustrates how this result is used when there are three weights.

Example 4.5

A weightless plank 4 m long has one concrete block at the left end, another at the center, and two blocks at the right end (Fig. 4.19). Where is the center of gravity? Let us call the weight of a block w_0; the numerical value is not needed since it will cancel out. We can choose our origin arbitrarily, so we will take it at the left end. Then the total weight is $w = w_1 + w_2 + w_3 = 4w_0$, and Eq. 4.7 becomes

$$X = \frac{x_1 w_1 + x_2 w_2 + x_3 w_3}{w}$$

$$= \frac{0 + (2\text{ m})w_0 + (4\text{ m})(2w_0)}{4w_0} = 2.5\text{ m}$$

Thus the C.G. is between the center of the plank and the heavier end.

The equation for the center of gravity, Eq. 4.7, contains weights in both the numerator and denominator. If we substitute $w = mg$ for each weight, the factors of g cancel out. Then, X involves the masses instead of the weights, and it is called the *center of mass* (C.M.). There is no difference between the C.G. and C.M. as long as **g** has the same direction and magnitude for each weight.

Although we have only discussed weights placed at several points on a straight line, the procedure for finding the center of gravity for more complicated arrangements is basically the same. If weights are at several points in a plane, then the C.G. is at a point (X, Y) in the plane. Equation 4.7 is then used to find X, and an analogous equation involving the y coordinates of the weights is used to find Y.

4.4 | STABILITY AND BALANCE

The number and position of an animal's legs have

Figure 4.19.   The center of gravity is between the center of the plank and the heavier end.

Figure 4.20. The beam is (a) in equilibrium; (b) not in equilibrium. The base area defined by the supports is shown in color.

apparently been partially determined by its requirements for stability and balance. The basic idea is illustrated by the beam in Fig. 4.20. If its center of gravity is between the two supports, the torques about the C.G. due to N_1 and N_2 are opposite and cancel, so the beam is in equilibrium. However, when the center of gravity is to the left of the supports, the torques about the C.G. due to N_1 and N_2 are both positive. Since the net torque is not zero, the beam topples over. Thus, an object is balanced only when its *center of gravity lies above the base area* defined by its supports.

An animal standing on four legs is analogous to a table. A table placed on a surface that is gradually tilted will topple over when the center of gravity is no longer over the portion of the surface defined by the tips of the four legs (Fig. 4.21). The shorter the legs for a given tabletop, the larger the angle θ at which this occurs and the greater its stability; a low table is

more stable than a high one. Similarly, the centers of gravity of automobiles, ships, and even vases must be kept low for good stability to be achieved. Thus, we can see that rats and squirrels, which have relatively short legs, are adapted well for living on steep slopes or on tree branches. The long-legged horse and antelope are proportioned for efficient running on nearly flat terrain.

If a quadruped lifts one leg, it will remain balanced if its center of gravity is over the triangular base area formed by its remaining three legs. By moving its legs in the correct sequence, it also can walk slowly while always keeping three feet on the ground and the center of mass over the triangle they define (Fig. 4.22). This sequence is right front, left rear, left front, right rear, and it is used by all four-legged animals and by human infants. Of course, humans, birds, and some animals can balance on one or two feet, but their large feet make it relatively easy to do this.

Figure 4.21. A four-legged table will tip when a vertical line from the center of mass passes just outside the base area formed by the legs. Since $\theta_a < \theta_b$, a long-legged table (a) is less stable than a short-legged one (b).

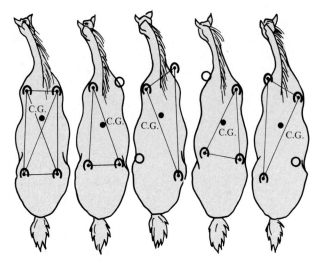

Figure 4.22.   The diagram of a quadruped walking as seen from above. The open circle represents the foot that is off the ground. Notice that the center of gravity is always within the triangle formed by the three feet on the ground.

When a quadruped runs rapidly, it can happen that only one or two legs are on the ground at one time. The tendency to tip diagonally forward or to roll sideways is quickly countered when the other feet are set down. Thus, brief periods of instability are required for fast motion of quadrupeds and bipeds (Fig. 4.23).

Three feet on the ground is the minimum for stability for animals with small feet, so insects that have six legs can move three at a time and still be stable at all times. Since they have very small masses, even a slight gust of wind would tip them over if they had periods of instability. The need to be stable in a moderate wind also explains why their legs do not point nearly vertically as in mammals, but are instead splayed outward.

Figure 4.23.   A sprinter, at the start of and even during a race, has his center of mass well forward of his feet as shown in this drawing. This means that he is in a very unstable position. He maintains his balance by bringing his legs forward just in time to keep from falling. This extreme position aids the athlete in exerting a larger force on the ground to increase his acceleration.

Figure 4.24.   (*a*), (*b*), and (*c*) show the relative positions of the applied force $\mathbf{F}_a$, the load force $\mathbf{F}_L$, and the fulcrum for class I, II, and III levers, respectively. The load force $\mathbf{F}_L$ shown here is equal and opposite to the force produced by the lever on the load, $-\mathbf{F}_L$.

The fact that animals usually have the minimum number of legs consistent with stability is apparently related to considerations of strength and weight. A single leg weighing as much as two thinner legs is better able to withstand torques tending to bend the leg. Accordingly, the portion of the body weight composing the legs is minimized by keeping the number of legs as small as possible.

4.5 | LEVERS; MECHANICAL ADVANTAGE

Levers, pulley systems, and screw jacks are all examples of *machines*. In each case, a force $\mathbf{F}_a$ is applied and a load force $\mathbf{F}_L$ is balanced. The *mechanical advantage* (M.A.) of the machine is defined as the ratio of the magnitudes of these forces,

$$\text{mechanical advantage} = \text{M.A.} = \frac{F_L}{F_a}$$

Levers | A lever in its simplest form is a rigid bar used with a fulcrum (Fig. 4.24). Three classes of levers are defined according to the relative positions of $\mathbf{F}_L$, $\mathbf{F}_a$, and the fulcrum. The following example illustrates the effects of a lever.

Example 4.6
Suppose the load $\mathbf{F}_L$ on a class I lever (Fig. 4.24*a*) has a magnitude of 2000 N. A person exerts a force $F_a = 500$ N to balance the load. (a) What is the ratio of the distances x_a and x_L? (b) What is the mechanical advantage of this lever?

(a) To find x_a/x_L, we compute the torques about the fulcrum. The torque due to $\mathbf{F}_a$ is $\Gamma_a = -x_a F_a$, and the torque due to $\mathbf{F}_L$ is $\Gamma_L = x_L F_L$. For balance, these must sum to zero, so

$$x_L F_L - x_a F_a = 0$$

and

$$\frac{x_a}{x_L} = \frac{F_L}{F_a} = \frac{2000 \text{ N}}{500 \text{ N}} = 4$$

(b) The mechanical advantage of the lever used in this way is then

$$\text{M.A.} = \frac{F_L}{F_a} = \frac{x_a}{x_L} = 4$$

For all classes of levers the mechanical advantage can be expressed as a ratio of distances from the fulcrum, as in the preceding example. If the forces are at right angles to the lever, in equilibrium the ratio of the magnitudes of the load force to the applied force is

$$\frac{F_L}{F_a} = \frac{x_a}{x_L}$$

Hence, for all classes of levers,

$$\text{M.A.} = \frac{x_a}{x_L} \qquad \text{(forces} \perp \text{lever)} \qquad (4.10)$$

With the forces at right angles to the lever the M.A. of class III levers is always less than 1, and the M.A. of class II levers is always greater than 1. Class I levers can have an M.A. larger or smaller than 1.

For all levers the M.A. given in Eq. 4.10 is an ideal value. Real machines always have frictional forces present that reduce the actual mechanical advantage below the ideal value.

4.6 | MUSCLES

Many examples of levers are found in the bodies of animals. Muscles provide the forces for using these levers.

A muscle is composed of thousands of long, thin fibers. When a muscle is stimulated by an electrical pulse from the nervous system, it contracts briefly or *twitches*, thereby exerting a force. A series of pulses sent to a muscle causes a series of twitches in the fibers. The twitches are close together in time but occur at different times in different parts of the muscle, so the apparent result is a smooth contraction of the muscle. If the frequency of the twitches is increased, the tension in the muscle increases up to a state of maximum tension. Further nerve impulse rate increases cause no further increase in tension.

The maximum tension of a muscle is proportional to its cross-sectional area at the widest point. This maximum tension also depends on the length of the muscle, which can be varied. The greatest tension can be achieved when the muscle is only slightly elongated relative to its resting or undisturbed state and is about 30 to 40 newtons per square centimeter of cross-sectional area. The maximum tension possible drops rapidly if the muscle is substantially elongated or shortened. To see an example of this, bend your wrist forward as far as you can and try to make a tight fist. Most people either fail to close their fingers completely or else do so with a rather weak grip.

4.7 | LEVERS IN THE BODY

The evolutionary development of the limbs and other skeletal structures in animals has been strongly affected by the animals' needs. We recall that when the forces are perpendicular to a lever, its mechanical advantage is $F_L/F_a = x_a/x_L$. Thus, short limbs with small values of x_L will have relatively large mechanical advantages and be able to exert large forces (Fig. 4.25). However, the distance that the end of a limb moves is proportional to its length x_L, so rapid motion requires a long limb. Consequently, a compromise must be made between strength and speed of movement. For example, the foreleg of a fast-moving horse has a mechanical advantage of 0.08. The armadillo, which is a burrowing animal, has a foreleg with an M.A. of about 0.25. Hence, it cannot move as quickly, but it has the strength needed for burrowing.

Figure 4.25.   A limb being flexed may be represented by a pivoted bar supported by a cable. The cable represents a muscle and the pivot represents a joint. The M.A. of the limb is x_a/x_L, while the distance d that the end of the limb moves is proportional to x_L.

The Spinal Column | The human spinal column is made up of 24 *vertebrae* separated by fluid-filled *disks*. When a person bends, the spine is effectively a lever with a very small M.A. Hence, bending

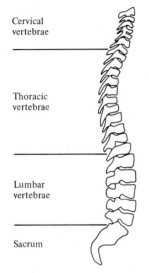

Figure 4.26.   The anatomy of the spinal column.

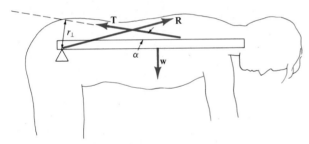

Figure 4.27.   Force diagram for the spine of a person bending over with the back horizontal.

over to pick up even a light object produces a very large force on the *lumbrosacral* disk, which separates the last vertebra from the *sacrum*, the bone supporting the spine (Fig. 4.26). If weakened, this disk can rupture or be deformed, causing pressure on nearby nerves and severe pain.

To understand why this force is so large, we can use a model that treats the spine as a pivoted rod. The pivot corresponds to the sacrum and exerts a force **R** (Fig. 4.27). The various muscles of the back are equivalent to a single muscle producing a force **T** as shown. When the back is horizontal, the angle α is 12°. **w** is the weight of the torso, head, and arms, about 65 percent of the total body weight.

Notice that because α is small, the line of action of **T** passes close to the pivot, so its lever arm $r_\perp$ is small. However, the weight **w** acts at right angles to the

spine, and its lever arm is much longer. Hence for their torques to balance, the muscle force **T** must be much greater than the weight. Because **T** is large, its horizontal component is also large. In equilibrium, the force **R** due to the sacrum must have an equal but opposite horizontal component, so this force due to the sacrum is also much larger than the weight.

If this calculation is carried out in detail, the numbers obtained are impressively large. For a 170-lb man, **T** and **R** are each close to 500 lb! If the man is also lifting a 40-lb child, so that there is an extra 40-lb weight at the right end of the bar in Fig. 4.27, **T** and **R** are each about 750 lb! Such forces in the muscles and on the disk are potentially quite hazardous.

Since even bending over without lifting a load puts great stress on the spine, it should be avoided. If, instead, one flexes the knees but keeps the back vertical, then the centers of gravity of all the weights lie almost directly above the sacrum. Consequently their torques about the sacrum are small, and the muscles need not exert any appreciable force. The force on the disk is then approximately the total weight supported. For the 170-lb man, this weight is about 110 lb for the body alone and 150 lb with a 40-lb load. This is a far safer way to lift even a light object (Fig. 4.28).

SUMMARY

The quantity that indicates the ability of a force to cause rotation is the torque Γ. If a force **F** acts at a distance **r** from a point on the axis of rotation, the torque about that point has a magnitude $rF \sin \theta$, where θ is the angle between **r** and **F**. The torque is a vector quantity and is directed perpendicular to the plane of **r** and **F** or along the rotational axis. In the

Figure 4.28.   (*a*) Incorrect and (*b*) correct ways to lift a weight.

vector or cross product notation, $\boldsymbol{\Gamma} = \mathbf{r} \times \mathbf{F}$, the direction of $\boldsymbol{\Gamma}$ is then given by a right-hand rule. Alternatively, if the motion is in a specific plane, counterclockwise torques are conventionally taken as positive and clockwise torques as negative.

For a rigid body to be in equilibrium, two vector equations must be satisfied:

$$\mathbf{F} = 0$$
$$\boldsymbol{\Gamma} = 0$$

The first of these is the condition that the net force is zero and ensures that there is no change in the translational motion. The second is the condition that the net torque about some convenient point is zero; it ensures that the rotational motion remains the same.

For objects at rest or in motion, the entire weight and mass of the object can be considered to be located at a point called the center of gravity or center of mass. This fact facilitates the understanding of stability and balance.

Checklist

Define or explain:

torque
lever arm
vector product or cross
  product
couple

equilibrium conditions
center of gravity
center of mass
machines
mechanical advantage

EXERCISES

Section 4.1 | Torques

4-1   Find the magnitude and sign of the torque due to each of the weights in Fig. 4.29 relative to point P.

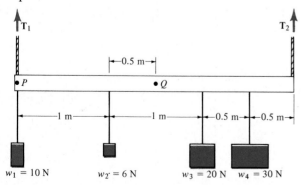

$w_1 = 10\ N$ $w_2 = 6\ N$ $w_3 = 20\ N$ $w_4 = 30\ N$

Figure 4.29.   Exercises 4-1, 4-2, and 4-7.

4-2   Find the magnitude and sign of the torque due to each of the weights in Fig. 4.29 relative to point Q.

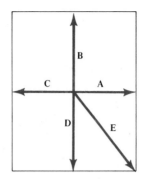

Figure 4.30.   The vectors are drawn from the center of a rectangle. Exercise 4-3.

4-3   In Fig. 4.30, consider the vector products $\mathbf{A} \times \mathbf{A}$, $\mathbf{A} \times \mathbf{B}$, $\mathbf{A} \times \mathbf{C}$, $\mathbf{A} \times \mathbf{D}$, and $\mathbf{A} \times \mathbf{E}$. (a) Which of these vector products are zero? (b) Which are directed into the page? (c) Which are directed out of the page? (d) Which are equal to each other in magnitude and direction?

Figure 4.31.   Exercises 4-4 and 4-6.

4-4   In Fig. 4.31, find the torques due to $\mathbf{F}_1$ and $\mathbf{F}_2$ relative to point P.

4-5   A cyclist applies a downward force $\mathbf{F}$ of magnitude 100 N to the pedal of her bicycle (Fig. 4.32). (a) Find the magnitude and direction of the torques in each position shown. (b) In which position is the torque a maximum?

Section 4.2 | Equilibrium of Rigid Bodies

4-6   The bar in Fig. 4.31 is pivoted at point P. Will it tend to start rotating if it is initially at rest? Explain, and indicate in which direction it would rotate if your answer is yes.

Figure 4.32.   Exercise 4-5.

4-7   A weightless bar supported by two vertical ropes has four weights hung from it (Fig. 4.29). Find the tensions T_1 and T_2 in the ropes.

4-8   Two children balance on a weightless seesaw. One weighs 160 N and is seated 1.5 m from the fulcrum. The second is seated 2 m on the other side of the fulcrum. What is the weight of the second child?

Figure 4.33.   Exercise 4-9.

4-9   Find the forces F_1 and F_2 on the tooth in Fig. 4.33. (In orthodonture, forces applied to the teeth lead to forces on the supporting bones. Gradually the bone tissue breaks down and permits the tooth to rotate or translate. New bone tissue grows in the space left behind. The forces must be small enough to avoid damaging the root of the tooth.)

4-10   Figure 4.34 shows the forearm considered in Example 4.4 when the person is holding a 12-N weight ($\mathbf{w}_1$) in the hand ($\mathbf{w}$ is the weight of the forearm). (a) Find the force $\mathbf{T}$ exerted by the biceps muscle and the force $\mathbf{E}$ exerted by the elbow joint.

(b) In Example 4.4, with $\mathbf{w}_1 = 0$, we found $T = 36$ N and $E = 24$ N. Why are these forces more than twice as large here?

Figure 4.34.   Exercise 4-10, Problem 4-24.

4-11   Children with weights w_1 and w_2 balance on a seesaw. The weight w of the seesaw can be considered to act at its center of gravity, which is directly over the pivot. In terms of w, w_1, and w_2, (a) find the force exerted by the pivot, and (b) find the ratio x_2/x_1 of the children's distances from the pivot.

Section 4.3 | The Center of Gravity

4-12   Three weights are positioned on a weightless rod, as shown in Fig. 4.35. Where is their center of gravity?

Figure 4.35.   Exercise 4-12.

4-13 Two weights are hung from the ends of a horizontal meter stick. If the weight at $x = 0$ is 10 N and the center of gravity is at $x = 0.8$ m, what is the weight at $x = 1$ m? (Neglect the weight of the stick.)

4-14 A woman's forearm has a mass of 1.1 kg and her upper arm has a mass of 1.3 kg. When her arm is held straight out, the C.G. of the forearm is 0.3 m from the shoulder joint, and the C.G. of the upper arm is 0.07 m from the shoulder joint. What is the position of the center of gravity of the entire arm with respect to the shoulder joint?

4-15 An 80-kilogram hiker carries a 20-kg pack. The center of gravity of the hiker is 1.1 m above the ground when he is not wearing the pack. The C.G. of the pack is 1.3 m from the ground when it is worn. How far above the ground is the C.G. of the hiker and the pack?

Section 4.4 | Stability and Balance

4-16 At what angle θ will the table in Fig. 4.36 tip over?

Figure 4.36. Exercise 4-16.   **Figure 4.37.** Exercise 4-18.

4-17 Ships returning to their home ports without a cargo are sometimes loaded with rocks in their holds or water in their tanks. Why is this done?

4-18 An amusing toy consists of a figure holding a curved pole with weights at either end (Fig. 4.37). It is stable in the position shown and will not fall when pushed gently. Explain why. (*Hint:* Where is the center of gravity?)

Section 4.5 | Levers; Mechanical Advantage

4-19 A man places a 2-m long bar under a boulder weighing 4500 N. He uses a fulcrum 0.2 m from the point where the bar touches the rock (Fig. 4.38). What force **F** must he exert to lift the rock?

Figure 4.38. Exercise 4-19.

Figure 4.39. Exercise 4-20.

4-20 An oar is held 0.4 m from the oarlock (Fig. 4.39). If it contacts the water at an average of 1.4 m from the oarlock, what is its mechanical advantage?

4-21 Figure 4.40 shows a pair of tweezers. What is its mechanical advantage?

4-22 Figure 4.41 shows a pair of pliers. (a) What is its mechanical advantage? (b) If a force $F = 10$ lb is applied, what force is exerted on the object?

4-23 Give examples of class I levers with M.A.s equal to 1, less than 1, and greater than 1.

Section 4.7 | Levers in the Body

4-24 Figure 4.34 shows the forearm represented as a pivoted bar. T is the force exerted by the

Figure 4.40. Exercise 4-21.   **Figure 4.41.** Exercise 4-22.

biceps muscle. (a) What class of lever does this represent? (b) What is the mechanical advantage of the forearm for supporting its own weight, w? (c) What is its M.A. for supporting a load w_1 held in the hand? (d) If the muscle contracts 1 centimeter, how far will the load in the hand move?

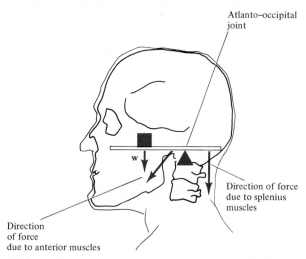

Figure 4.42.   Muscles moving and supporting the head. Exercise 4-25.

4-25   The head pivots about the atlanto-occipital joint (Fig. 4.42). The splenius muscles attached behind the joint support the head. (a) What class of lever does this represent? (b) The anterior muscles produce forward motions of the head. What class of levers does their action represent? (c) Which muscles have the larger mechanical advantage? Speculate on the reasons for this.

PROBLEMS

4-26   Find the tension in the ropes in Fig. 4.43.
4-27   In Fig. 4.44, an object is supported by a hinged, weightless rod and a cable. Find the ten-

Figure 4.43.   Problem 4-26.   **Figure 4.44.**   Problem 4-27.

sion in the cable and the force exerted by the hinge.
4-28   In Fig. 4.45, an object is supported by a hinged, weightless rod and cable as shown. If $w = 1000$ N, find (a) the tension in the cable, and (b) the force on the rod at the hinge.

Figure 4.45.   Problems 4-28 and 4-29.

4-29   In Fig. 4.45, the hinged rod and cable are weightless. The cable will break when the tension exceeds 2000 N. What is the maximum weight w that can be supported?
*4-30   The tension T at each end of the chain in Fig. 4.46 is 20 N. What is the weight of the chain?

Figure 4.46.   Problem 4-30.

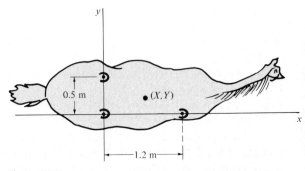

Figure 4.47.   A horse viewed from above standing on three legs. Problem 4-31.

4-31 A horse stands with its left forefoot off the ground (Fig. 4.47). The left rear and right front legs each support 300 lb of its weight, which is 1000 lb. (a) What force is exerted by the right hind leg? (b) Find the position (X, Y) of the center of gravity.

4-32 What is the position of the center of mass of the three masses of Fig. 4.48?

Deltoid muscle

Humerus

T

(a)

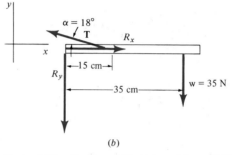

$\alpha = 18°$

T

R_x

x

R_y

←15 cm→

←35 cm→

$w = 35$ N

(b)

Figure 4.51. Problem 4-36. (Adapted from Williams and Lissner.)

Figure 4.48. Problem 4-32.

Figure 4.49. Problem 4-33.

1 m

M

C.G. (man alone)

1.5 m

1.1 m

0.3 m

* **4-33** The man in Fig. 4.49 has a mass of 100 kg. His arms are held straight out to the side, and in one hand he holds a mass M. (a) Find the horizontal and vertical positions of the C.G. of the man plus the mass M. (Choose the origin to be midway between his feet). (b) What is the maximum mass the man can hold without falling over?

* **4-34** A four-legged table has a top of mass 20 kg. The legs are positioned at the corners of the top and have a mass of 2 kg each. The dimensions of the table are shown in Fig. 4.50. At what angle θ will the table tip over?

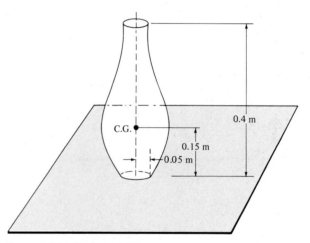

C.G.

0.4 m

0.15 m

0.05 m

Figure 4.52. Problem 4-37.

0.8 m

1 m

θ

Figure 4.50. Problems 4-34 and 4-35.

4-35 The table in Fig. 4.50 has massless legs. At what angle θ will it tip over?

* **4-36** The deltoid muscle raises the upper arm to a horizontal position (Fig. 4.51). (a) Find the tension T exerted by the muscle and the components R_x and R_y of the force exerted by the shoulder joint. (b) What is the mechanical advantage of the muscle for lifting the arm?

* **4-37** A vase 0.4 m tall has its center of gravity 0.15 m from the bottom, which is a circle of radius 0.05 m (Fig. 4.52). How far can the top of the vase be pushed to the side without toppling it?

4-38 Show that the M.A. of a class III lever is always less than 1, assuming the forces are perpendicular to the lever.

4-39 Show that the M.A. of a class II lever is always greater than 1, assuming the forces are perpendicular to the lever.

Figure 4.53.   Problem 4-40.

***4-40**   In Fig. 4.53, the weight of the upper body is $w = 110$ lb. Find the force T exerted by the spinal muscles and the components R_x and R_y of the force R exerted by the pivot (sacrum) if the weight held w_1 is (a) zero; (b) 40 lb.

SUPPLEMENTARY TOPICS

4.8 | THE JAWS OF ANIMALS

Mechanics enables us to understand why many anatomical structures have evolved to their present state, since the mechanical functions of bones, muscles, and joints largely determine their sizes and shapes. This is illustrated nicely by the development of the lower jaws of mammals.

It is often advantageous for an animal to be able to bite very hard. The biting force depends on the magnitude, direction, and point of application of the forces exerted by the muscles closing the jaw. This leads to certain optimal shapes and sizes of jaws. In addition, the bones of the jaw joint connecting the upper and lower jaws must be strong enough to prevent fractures and dislocations. From fossil records, we know that mammals evolved from mammal-like reptiles. As this occurred, the muscles attached to the lower jaw progressively *increased* in size, while the bones forming the jaw joint steadily *decreased* in size. This apparent paradox can be explained in terms of the changes in the direction and point of application of the muscular forces.

Figure 4.54 shows the basic differences between the lower jaws of a primitive reptile and a typical present-day mammal. The first is a simple bar, with the muscles pulling upward at a point close to the joint. The mammalian jaw has a large bump or projection called the *coronoid process*. Attached to this is the *temporalis* muscle, which pulls *backward* as well as upward (force **T** in Fig. 4.54). The *masseter* and *pterygoideus* muscles pull *forward* as well as upward (force **M**).

A primitive reptile biting with an upward force $-\mathbf{B}$ on food between its back teeth experiences an equal but opposite reaction force **B** downward on the jaw. Since the muscular force **M** is applied close to the joint, static equilibrium can be achieved only if a large downward force **R** is exerted by the joint.

Calculating torques about the point 0, the net torque is zero if

$$x_B B - x_R R = 0 \qquad \text{or} \qquad R = \frac{x_B}{x_R} B$$

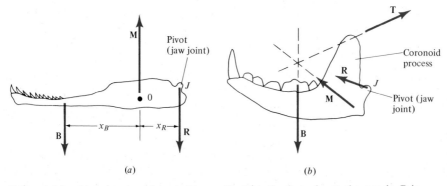

Figure 4.54.   (*a*) Lower jaw of a primitive reptile. **M** is the force due to the muscle. **B** is the reaction force from the object being bitten, and **R** is the force due to the jaw joint at *J*. (*b*) A mammalian jaw. Muscle forces are shown as **T** and **M**. As explained in the text, the force **R** due to the jaw joint can be zero if the lines of action of the three forces **T**, **B**, and **M** intersect as shown here.

Since the net force on the jaw must be zero, $M - B - R = 0$, and the required muscle force is

$$M = B + R = B \left(1 + \frac{x_B}{x_R}\right)$$

For example, if $x_B = 2x_R$ and $B = 1\,\text{N}$, then $R = 2\,\text{N}$, and $M = 3\,\text{N}$. Thus the force B on the food is smaller than the forces M and R exerted by the muscle and the joint respectively. Clearly the strength of the joint is a limiting factor in how hard the reptile can bite and in how large a muscle it can safely employ.

In the mammalian jaw, the force $\mathbf{M}$ is applied further from the joint and another muscular force, $\mathbf{T}$, is also present (Fig. 4.55). If the lines of action of $\mathbf{T}$, $\mathbf{M}$, and $\mathbf{B}$ all cross at a point, their torques about this point are zero. Then the second equilibrium condition, $\mathbf{\Gamma} = 0$, requires that the line of action of $\mathbf{R}$ pass through this point as well. Furthermore, when the forces also satisfy $\mathbf{T} + \mathbf{M} + \mathbf{B} = 0$, *no force $\mathbf{R}$ need be supplied by the joint to satisfy the requirement* $\mathbf{F} = 0$. If $\mathbf{T} + \mathbf{M} + \mathbf{B}$ is not zero, or if their lines of action do not exactly meet at one point, a force, $\mathbf{R}$, will have to be supplied by the joint, but $\mathbf{R}$ will still be much smaller than in the reptile. Thus, a smaller joint structure is adequate, and the strength of the joint does not limit the size of the muscle the animal can have.

Example 4.7

To illustrate the superiority of the mammalian jaw, suppose that the muscle forces $\mathbf{T}$ and $\mathbf{M}$ of Fig. 4.55 are both at $\theta = 45°$ to the horizontal. If there is to be no force $\mathbf{R}$ supplied by the joint, how is $\mathbf{M}$ related to $\mathbf{T}$, and how large a force, $\mathbf{B}$, is exerted on the food? (Assume the

lines of action of $\mathbf{B}$, $\mathbf{T}$, and $\mathbf{M}$ all cross at a point, so the second equilibrium condition $\mathbf{\Gamma} = 0$ is satisfied.)

Since the forces have both x and y components, $\mathbf{F} = 0$ can be used in component form as $F_x = 0$ and $F_y = 0$. With $F_x = 0$, we obtain

$$T \cos \theta - M \cos \theta = 0$$

so $M = T$. With $F_y = 0$,

$$T \sin \theta + M \sin \theta - B = 0$$

Using $M = T$ and $\sin \theta = \sin 45° = \sqrt{2}/2$,

$$B = (T + M) \sin \theta = 2T \sin \theta = \sqrt{2}T$$

Thus the force B exerted by the jaw on the food is greater than either of the muscular forces T and M, and the force due to the joint is zero. (This conclusion is also evident from the lengths of the vectors in Fig. 4.55b). By contrast, we found that in the reptile, B is smaller than the muscular force or the joint force.

Just as we have compared mammalian and more primitive jaws, one can also compare different mammals. Carnivores use their powerful front teeth to seize or tear their prey, while herbivores grind their food sideways between their molars. The weight of a carnivore's temporalis muscle is one-half to two-thirds of the total weight of the jaw-closing muscles. However, in herbivores, this muscle weighs only about one-tenth of the total. It is left as an exercise to show why this is the proper adaptation for the needs of these two types of animals.

4.9 | CENTER OF GRAVITY OF HUMANS

Information about the center of gravity of humans is useful in many applications. The center of gravity of a freely falling object follows the same trajectory as a simple particle, even though the object may be rotating or changing its shape. This simplifies the analysis of jumping, gymnastics, and other athletic activities. In physical therapy, an amputee fitted with an artificial limb that is lighter than the natural limb has his or her center of gravity shifted. This must be taken into account in planning the person's rehabilitation.

In Section 4.3, we described how the center of gravity can be located by suspending an object from two different points. Another technique is better suited for live humans and animals (Fig. 4.56). A board of length l is supported at its ends by knife

(a)   (b)

Figure 4.55.   Forces on the mammalian jaw when no force is supplied by the jaw joint.

Figure 4.56. A method for finding the center of gravity of a human.

edges resting on scales adjusted to read zero with the board alone. When the person stands on the board, the scales read w_1 and w_2, respectively.

The condition $\mathbf{\Gamma} = 0$ for the net torque on the board can be used to find X. Calculating torques about point P,

$$-Xw_1 + (l - X)w_2 = 0$$

or

$$X = \frac{lw_2}{w_1 + w_2} \qquad (4.11)$$

The measurement is repeated twice more with the subject first turning 90° and then lying down. In this way all three coordinates of the center of gravity are determined.

The detailed measurement of the masses, sizes, and centers of gravity of body segments is difficult, and the results vary from individual to individual. Data for a typical man are given in Figs. 4.57 and 4.58 and Table 4.1 and are used in some of the problems.

4.10 | PULLEY SYSTEMS

Pulleys, like levers, are simple machines that are used in many situations. A single pulley is used to change the direction of a force, while combinations of pulleys can be used to reduce the force needed to lift a heavy load.

If friction in the bearings is negligible, the equilibrium tension in the rope or cable is the same on either side of the pulley. This property is used in discussing some typical pulley arrangements in the following examples. We assume in these examples that the friction is negligible and that the pulleys and ropes are massless.

Example 4.8

What applied force $\mathbf{F}_a$ is necessary to lift the weight $\mathbf{w}$ in Fig. 4.59?

The forces on pulley 1 are shown in Fig. 4.59b. The rope is continuous, and the tension on both sides of the pulleys is the same. If the weight is being raised at constant speed, the system is in equilibrium. Hence, $2F_a - w = 0$, and $F_a = w/2$. The force required is only half the weight, and the mechanical advantage is

$$\text{M.A.} = \frac{w}{F_a} = 2$$

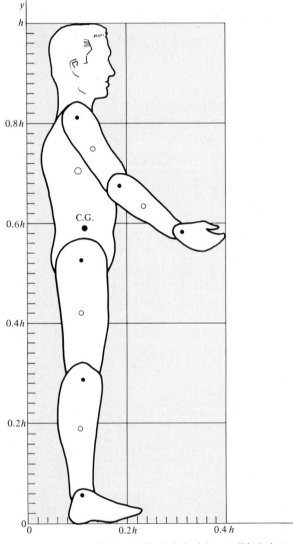

Figure 4.57. The limbs, position of the joints (solid circles), and the position of the centers of gravity (open circles) of several body segments for a typical man. (Adapted from Williams and Lissner.)

Figure 4.58.   The man shown in Fig. 4.57 is now bending over so that his back is nearly horizontal. Note that his center of gravity is still over his feet. (Adapted from Williams and Lissner.)

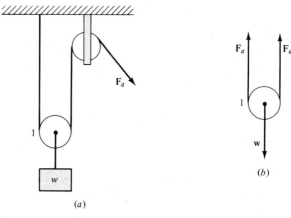

(a)

(b)

Figure 4.59.   The tension in the rope is the same throughout the system, so the two forces on either side of pulley 1 are equal.

From the preceding two examples, we can infer a rule for the mechanical advantage of pulley systems used to lift weights. *The mechanical advantage of the system is equal to the number of parallel ropes supporting the pulley to which the load is attached:* 2 in Fig. 4.59 and 4 in Fig. 4.60. Note that this rule does *not* apply when, as in the next example, the forces applied to the load are not all parallel.

Example 4.9

What applied force $\mathbf{F}_a$ is necessary to raise the weight $\mathbf{w}$ with the pulley system of Fig. 4.60? What is the mechanical advantage of this system?

Again the tension in each vertical segment of rope is the same, so $4F_a - w = 0$, and $F_a = w/4$. The mechanical advantage for this system is

$$\text{M.A.} = \frac{w}{F_a} = 4$$

Example 4.10

Leg traction is applied to a patient's leg as shown in Fig. 4.61. What horizontal force is exerted on the leg?

The sum of the forces on each pulley is zero, since the pulleys are at rest. From Fig. 4.61b, the horizontal forces that act on the pulley attached to the foot satisfy

$$2w \cos \theta - F_L = 0$$

or

$$F_L = 2w \cos \theta$$

TABLE 4.1

Masses and centers of gravity of body segments of the man in Figs. 4.57 and 4.58. His total mass is *m* and his height is *h*. For example, if his mass is 70 kg, then the mass of his trunk and head is 0.593*m* = 0.593(70 kg) = 41.5 kg

| Segment | Mass | Center of Gravity Position For Segment | | | |
| | | Fig. 4.57 | | Fig. 4.58 | |
		x	y	x	y
Trunk and head	0.593m	0.10 h	0.70 h	0.26 h	0.52 h
Upper arms	0.053m	0.14 h	0.75 h	0.35 h	0.45 h
Forearms and hands	0.043m	0.24 h	0.64 h	0.34 h	0.29 h
Upper legs	0.193m	0.12 h	0.42 h	0.11 h	0.40 h
Lower legs and feet	0.118m	0.10 h	0.19 h	0.17 h	0.18 h

This force can be changed by altering either w or θ. Since $\cos\theta$ varies from 1 to 0 as θ goes from 0° to 90°, any force from zero to $2w$ can be obtained by choosing the correct angle. When θ is large, $\cos\theta$ is small. Then the weight w and the tension in the rope are much larger than the force F_L exerted on the foot.

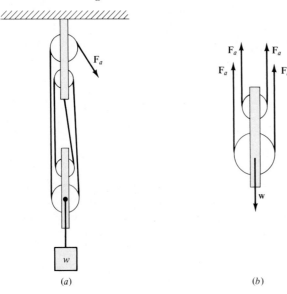

(a)   (b)

Figure 4.60.   Example 4.9.

EXERCISES ON SUPPLEMENTARY TOPICS
Section 4.8 | The Jaws of Animals
4-41   A snake exerts a muscle force $M = 5$ N (see Fig. 4.54a). M acts at a distance of 0.03 m from the joint, and the resulting bite force is 2 N. Find (a) the distance from the joint to the line of action of the bite force, and (b) the force exerted by the jaw joint.

4-42   (a) In a typical herbivore, the maximum magnitude of the force T is one-tenth the maximum magnitude of the force M in Fig. 4.62. Assuming that there is no force at the joint, would you expect the animal to exert the largest biting force near the front or back of the jaw? (b) In a carnivore the maximum value of T is about twice that of M. Would you expect the maximum biting force to be exerted further from or closer to the jaw joint than in the herbivore? Explain.

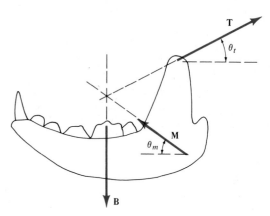

Figure 4.62.   Exercise 4-42; Problems 4-49 and 4-50.

Section 4.9 | Center of Gravity of Humans
4-43   A board 4 m long is used to find the C.G. as in Fig. 4.56. When a person is on the board, the scale readings are $w_1 = 200$ N and $w_2 = 600$ N. What is the position of the C.G. of the person?

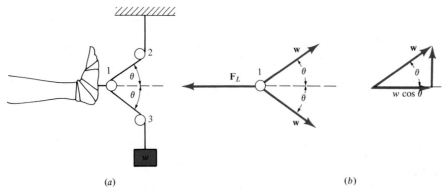

(a)   (b)

Figure 4.61.   (a) A system of pulleys used to apply a force in leg traction. The magnitude of this force can be adjusted by changing the angle θ. Pulley 1 is attached to the foot, and pulleys 2 and 3 are mounted on a rigid frame that is not shown. (b) The forces on pulley 1.

4-44 Using the data of Table 4.1, find the C.G. of the man in Fig. 4.57.

4-45 Using the data of Table 4.1, find the C.G. of the man in Fig. 4.58.

Section 4.10 | Pulley Systems

4-46 What force F must be applied in Fig. 4.63 to lift the load?

4-47 In Fig. 4-61, a 50-N force is to be applied to the leg. If a 10-kg mass is hung from the cable, what angle θ should be used?

4-48 Suppose the rope in Fig. 4.60 is pulled at 0.25 m s^{-1}. How fast will the load rise?

PROBLEMS ON SUPPLEMENTARY TOPICS

4-49 A mammal bites so that the muscle force **M** of Fig. 4.62 has a magnitude of 30 N. What is the force **B** of the bite? (Assume $\theta_t = \theta_m = 45°$.)

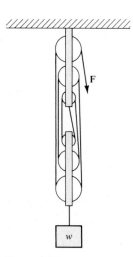

Figure 4.63. Exercise 4-46.

*__4-50__ In a particular carnivore the magnitude of the force T is 1.3 times the magnitude of M (Fig.

4.62). There is no force at the jaw joint. If $\theta_m = 60°$, find (a) θ_t, and (b) the ratio B/M.

Additional Reading

John M. Cooper and Ruth B. Glassow, *Kinesiology,* 4th ed., C.V. Mosby Co., St. Louis, 1976. Balance and body levers in humans.

Signi Brunnstrom, *Clinical Kinesiology,* 3rd ed., F.A. Davis Co., Philadelphia, 1972. Balance and body levers in humans.

Katherine F. Wells and Janet Wessel, *Kinesiology,* 5th ed., W.B. Saunders Co., Philadelphia, 1971. Balance, body levers in humans.

Marian Williams and Herbert R. Lissner, *Biomechanics of Human Motion,* W.B. Saunders Co., Philadelphia, 1962. Physical therapy applications.

John J. Hay, *The Biomechanics of Sports Techniques,* Prentice-Hall, Englewood Cliffs, N.J., 1973.

Sir James Gray, *How Animals Move,* Cambridge University Press, Cambridge, 1953. Chapter 3 discusses balance and stability.

R.A.R. Tricker and B.J.K. Tricker, *The Science of Movement,* Mills and Boon Ltd., London, 1967. Chapters 3, 4, 15, and 16 discuss balance and stability.

R.C. Thurow, *Edgewise Orthodontics,* 3rd ed., The C.V. Mosby Co., St. Louis, 1972. Applications of statics.

R. McNeill Alexander, *Animal Mechanics,* University of Washington Press, Seattle, 1968. Pages 5–13 discuss the jaws of animals.

A.W. Crompton, On the Lower Jaw of Diarthrognathus and the Origin of the Mammalian Lower Jaw, *Proceedings of the Zoological Society of London,* vol. 140, 1963, p. 697–753.

Scientific American article:

Milton Hildebrand, How Animals Run, May 1960, p. 148.

CHAPTER FIVE

CIRCULAR MOTION

Newton's laws of motion permit us to find the motion of any object if we know the forces acting upon the object and its initial position and velocity. In the preceding chapter, we considered objects that are in equilibrium and remain at rest. Earlier, we discussed the uniformly accelerated motion that occurs when the net force is constant in magnitude and direction. In this chapter, we consider another kind of motion frequently encountered, motion in a circular path.

We will see that when an object moves in a circular path at a constant speed, its acceleration is directed toward the center of the circle. The force required to produce this acceleration can be provided in various ways. Friction provides the required force for a car on a flat, circular track; gravity for an artificial satellite orbiting the earth; and electrical forces for an electron orbiting an atomic nucleus. In this chapter, we discuss these and other examples of circular motion.

5.1 | CENTRIPETAL ACCELERATION

Consider a car going around a circular track at a constant speed (Fig. 5.1a). The car is said to be undergoing *uniform circular motion,* and it has no acceleration component along the direction of motion. However, the velocity vector **v** is continually changing direction. Figure 5.1b shows that the velocity change Δ**v** that occurs in a short time Δt points toward the center of the circle. The acceleration **a** is Δ**v**/Δt evaluated for a very small Δt, so **a** also points toward the center of the circle or is directed *radially inward.* We will show in Section 5.7 that for motion in a circle of radius r with speed v, this acceleration toward the center or *centripetal acceleration* has a magnitude of

$$a_r = \frac{v^2}{r} \tag{5.1}$$

The subscript r reminds us the acceleration is radial. The following example illustrates the concept of radial acceleration.

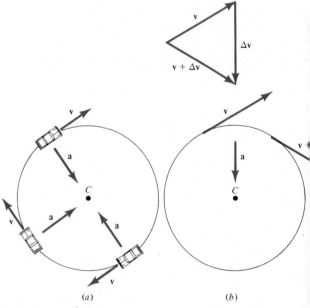

Figure 5.1. A car going around a circular track at a constant speed has an acceleration directed toward the center C of the circle. Its tangential acceleration component is zero since the speed is constant.

Figure 5.2.   Forces on a car on a flat circular track. Not shown are the forces due to the air and the road along the direction of motion.

Example 5.1

A car goes around a flat, circular track of radius 200 m at a constant speed of 30 m s^{-1}. What is its acceleration?

Since the speed is constant, there is no tangential acceleration. However, because the velocity vector is changing direction, there is an acceleration toward the center of the circle. This acceleration has a magnitude of

$$a_r = \frac{v^2}{r} = \frac{(30 \text{ m s}^{-1})^2}{200 \text{ m s}^{-2}} = 4.5 \text{ m s}^{-2}$$

In the absence of any net force, an object moves in a straight line at a constant speed in accordance with Newton's first law. If the object moves in a circular path, it has a radial acceleration $a_r = v^2/r$, so some force must be producing this acceleration. From the second law, the force must be equal to the mass times the acceleration. *Thus the force F needed to produce an acceleration a_r is*

$$F = ma_r = \frac{mv^2}{r} \qquad (5.2)$$

In the case of a car on a flat track, the centripetal acceleration results from a frictional force exerted on the tires by the road (Fig. 5.2). Other forces also act on the car: its weight, directed downwards; an equal but opposite upward normal force due to the road; a backward force due to air resistance; and a forward force exerted by the road on the tires. However, the only force that can produce a radial acceleration is the frictional force acting perpendicular to the motion. Accordingly, the maximum frictional force possible sets a limit to the centripetal acceleration of the car. This limit is determined by the coefficient of static friction, not the coefficient of kinetic friction, since the tire is momentarily at rest relative to the

road at the point of contact. If the driver tries to negotiate a curve too rapidly, the maximum frictional force will be exceeded and the car will skid. Once a skid begins, the car is sliding rather than rolling, so the frictional force is then determined by the coefficient of kinetic friction and is reduced in magnitude. This reduction in the frictional force makes it difficult to maintain control of the vehicle and can lead to an accident. The force needed to produce a centripetal acceleration is found in the next example.

Example 5.2

The car in the preceding example travels on a flat, circular track of radius 200 m at 30 m s^{-1} and has a centripetal acceleration $a_r = 4.5$ m s^{-2}. (a) If the mass of the car is 1000 kg, what frictional force is required to provide the acceleration? (b) If the coefficient of static friction μ_s is 0.8, what is the maximum speed at which the car can circle the track?

(a) Since the mass is 1000 kg and the acceleration is 4.5 m s^{-2}, Newton's second law gives

$$F = ma_r = (1000 \text{ kg})(4.5 \text{ m s}^{-2}) = 4500 \text{ N}$$

This is the frictional force required to circle the track at 30 m s^{-1}.

(b) From Fig. 5.2, the normal force **N** is equal in magnitude to the weight $\mathbf{w} = m\mathbf{g}$. Thus the maximum frictional force possible is $\mu_s N = \mu_s mg$, and the maximum velocity satisfies

$$\frac{mv^2}{r} = \mu_s mg$$

or

$$v = \sqrt{\mu_s rg}$$

Note that the mass has cancelled, so the maximum velocity on this track is the same for any car as long as the coefficient of static friction is the same. Substituting the numerical values,

$$v = [(0.8)(200 \text{ m})(9.8 \text{ m s}^{-2})]^{1/2}$$
$$= 39.6 \text{ m s}^{-1}$$

If the driver attempts to exceed 39.6 m s^{-1}, the car will not be able to continue on the circular course, and it will skid as in Fig. 5.3.

Vector Notation | It is useful to write the expression for the centripetal acceleration in vector form. In Fig. 5.3, **r** is a vector directed from the center of the circle, and $\hat{\mathbf{r}} = \mathbf{r}/r$ is a unit vector in that

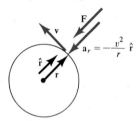

Figure 5.3. (*a*) A car on a flat circular track of radius *r* requires an acceleration $a_r = v^2/r$. (*b*) If the road cannot supply a frictional force mv^2/r, then the car will tend to move more nearly in a straight line and will skid.

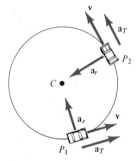

Figure 5.4. The vector $\hat{\mathbf{r}} = \mathbf{r}/r$ is parallel to $\mathbf{r}$ and has unit magnitude. $\mathbf{a}_r$ and $\mathbf{F} = m\mathbf{a}_r$ are directed opposite to $\mathbf{r}$ and to $\hat{\mathbf{r}}$.

direction. The acceleration $\mathbf{a}_r$ is directed opposite to $\mathbf{r}$, so

$$\mathbf{a}_r = -\frac{v^2}{r}\hat{\mathbf{r}} \tag{5.3}$$

Similarly the force $\mathbf{F} = m\mathbf{a}_r$ producing the acceleration can be written in vector form as

$$\mathbf{F} = -\frac{mv^2}{r}\hat{\mathbf{r}} \tag{5.4}$$

The equation $\mathbf{a}_r = -(v^2/r)\hat{\mathbf{r}}$ is also applicable in some cases other than uniform circular motion. When a car moves on a circular track with a variable speed, it has a tangential acceleration component $\mathbf{a}_T$ along the direction of motion that is equal in magnitude to the rate of change of the speed (Fig. 5.5). However, its acceleration component perpendicular to the motion is still v^2/r. Also, if the car travels on a road that is not circular, any small segment of the road can be considered as part of a circle. The radius of that circle is called the *radius of curvature* at the point *P* (Fig. 5.6). Again v^2/r is the acceleration component perpendicular to the motion.

5.2 | EXAMPLES OF CIRCULAR MOTION

In the preceding section, we considered in some detail

the example of a car on a flat, circular track. We will now discuss some other examples of circular motion.

Banked Turns | Good highways usually have banked or slanted curves, so that the normal force exerted by the road on the car has a horizontal component. This horizontal component can provide part of or all the force needed to produce the centripetal acceleration, reducing the role of the frictional force. The road is then much safer, especially under slippery conditions.

To examine this idea in detail, we consider a car on a banked curve. We suppose that the driver wants to drive at the correct speed v, so that no frictional force is required to provide the centripetal acceleration. The problem, then, is to find what that speed should be.

Figure 5.7 shows the relevant forces acting on the car: its weight $\mathbf{w} = m\mathbf{g}$ and the normal force $\mathbf{N}$. (The forces along the direction of motion play no role

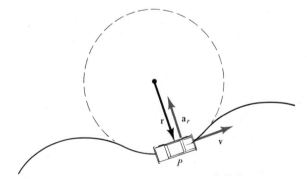

Figure 5.5. A car going around a circular track with an increasing speed. It has a centripetal acceleration $a_r = v^2/r$ and a tangential acceleration a_T equal to the rate of change of the speed. (Note that both $\mathbf{v}$ and $\mathbf{a}_r$ are larger at P_2 than at P_1.)

Figure 5.6. At *P* the road has a radius of curvature *r*. The centripetal acceleration at *P* is $a_r = v^2/r$.

Figure 5.7. A car on a banked curve is moving at the correct speed v so that the horizontal component N_x of the normal force provides the centripetal acceleration and no frictional force is required.

here.) Since the horizontal component of **N** must provide all the centripetal acceleration,

$$N \sin \theta = \frac{mv^2}{r}$$

There is no vertical acceleration component, so the net vertical force is zero, and

$$N \cos \theta = mg$$

Dividing the first equation by the second, the masses cancel out, leaving

$$\frac{\sin \theta}{\cos \theta} = \frac{v^2}{rg}$$

Since $\sin \theta / \cos \theta = \tan \theta$, this gives

$$v^2 = rg \tan \theta \qquad (5.5)$$

This equation gives the speed v at which the car can negotiate a curve banked at an angle θ without any assistance from frictional forces. At any other speed, a frictional force will be needed to supplement the horizontal component of the normal force. On a very slippery day, a car going faster will tend to slide out of the curve and a slow car will tend to slide toward the inside of the curve. The use of this result is illustrated by the following example.

Example 5.3
A curve of radius 900 m is banked, so no friction is required at a speed of 30 m s^{-1}. What is the banking angle θ?

Solving Eq. 5.5 for $\tan \theta$,

$$\tan \theta = \frac{v^2}{rg} = \frac{(30 \text{ m s}^{-1})^2}{(900 \text{ m})(9.8 \text{ m s}^{-2})} = 0.012$$

$$\theta = 6°.$$

As we have seen, on a banked road a component of

the normal forces provides part of or all the force accelerating a car toward the center of the curve. Banking also supplies the entire force needed for a bird or an airplane to turn. A bird is held aloft by aerodynamic lift forces perpendicular to the surface of its wings. When one wing or the tip of one wing is rotated about its long axis, the lift forces become unbalanced, and the bird tilts or banks. In airplanes, banking is accomplished with ailerons, which are movable surfaces on the trailing edges of the wings. Banking causes the aerodynamic lift force to have a horizontal component and to make the bird or airplane turn (Fig. 5.8).

This is exactly the same situation as in the banked curve, and Eq. 5.5 can therefore be used to relate the banking angle, the speed, and the radius of the turn. For example, birds such as swallows, which can maneuver very rapidly, experience accelerations of a few g's.

Motion in a Vertical Circle | Swinging a water-filled bucket in a vertical circle provides a simple but dramatic example of circular motion. We will see why spilling can be avoided if the pail is swung fast enough.

Figure 5.9 shows a bucket of total mass m swinging in a circle of radius R. At the bottom of the circle (point A), the tension **T** in the rope is opposite to the weight $\mathbf{w} = m\mathbf{g}$, so $\mathbf{F} = m\mathbf{a}_r$ gives

$$T - mg = \frac{mv^2}{R}, \qquad T = mg + \frac{mv^2}{R}$$

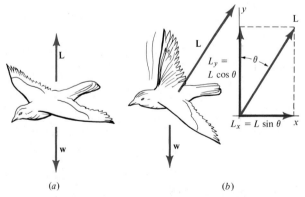

Figure 5.8. (a) Forces on a bird include its weight **w** and a lift force **L**. (The forward thrust and backward drag forces are not shown.) (b) When the bird banks, the lift force **L** has a horizontal component.

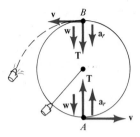

Figure 5.9. Forces acting on a pail swung in a vertical circle. If the rope breaks at the top of the circle, the pail follows the dotted path.

Thus the tension exceeds mv^2/R by the weight mg.

At the top of the circle (point B), $\mathbf{T}$, $\mathbf{w}$, and $\mathbf{a}_r$ are all downward, so

$$T + mg = \frac{mv^2}{R}, \qquad T = -mg + \frac{mv^2}{R}$$

Thus the tension here is *less* than mv^2/R by an amount equal to the weight. If v^2/R is less than g, then T is negative, corresponding to a force opposite to the assumed direction. However, a rope can only pull, not push, so this cannot happen. Consequently, if the pail is swung too slowly, the pail and its contents will drop below the circular path; to avoid wet clothing, we must have $T \geq 0$ at the top of the circle, or

$$\frac{v^2}{R} \geq g$$

For example, if $R = 1$ m, v must satisfy

$$v \geq \sqrt{gR} = \sqrt{(9.8 \text{ m s}^{-2})(1.0 \text{ m})} = 3.13 \text{ m s}^{-1}$$

It is natural to ask why the pail and the water do not fall when they reach the top of the circle. The answer is *that they do fall, but not fast enough*. One way to see this is to consider what would happen if the rope broke as the pail reached the top of the circle. It would then follow the dashed curve in Fig. 5.9 and fall less rapidly than when the rope is intact. Thus the rope forces the pail and its contents to fall even faster than they do under the influence of gravity alone.

We might note that if the pail is swung fast enough to break the rope, it will most likely happen at the bottom of the circular path rather than at the top, because the tension is greater than mv^2/r at the bottom and less than mv^2/r at the top. Also, the gravita-

tional force speeds up the pail as it falls, so v is greater at the bottom than at the top. This further increases the tension at the bottom.

5.3 | EFFECTIVE WEIGHT

When we stand in an elevator that is accelerating upward, we feel heavier than usual. Similarly, when the acceleration is downward, we have the feeling that our weight is reduced. Our perception of our weight is determined by the forces exerted on our bodies. These forces can be measured by a spring scale. We define the *effective weight* $\mathbf{w}^e$ of an animal or an object as the total force that an object exerts on a spring scale. While such a measurement may not always be feasible, this definition will be useful in understanding and calculating the effective weight of an accelerating object. The following example illustrates how the effective weight is found.

Example 5.4

A woman of mass m stands on a spring scale in an elevator. Find the effective weight of the woman if the elevator is accelerating upward at $0.2g$.

The forces acting on the woman are her weight $\mathbf{w} = m\mathbf{g}$ and the force $\mathbf{S}$ due to the spring scale (Fig. 5.10).

Using Newton's second law, $\mathbf{F} = m\mathbf{a}$,

$$S - mg = ma, \qquad S = mg + ma$$

Since by definition the effective weight is equal in magnitude to the spring scale force S,

$$w^e = m(g + a) = mg(1 + 0.2) = 1.2mg$$

The effective weight of the woman is then 1.2 times her normal weight, mg.

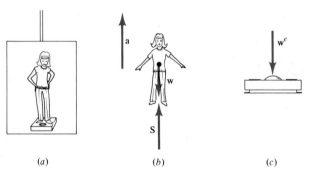

Figure 5.10. (*a*) A woman accelerating upward in an elevator. (*b*) The forces on the woman. (*c*) The woman exerts a force $\mathbf{w}^e$ on the spring scale. By Newton's third law, $\mathbf{S} = -\mathbf{w}^e$.

If the acceleration were downward, a similar calculation would show that the woman's effective weight is less than mg.

From Fig. 5.10, it is clear how to find the effective weight $\mathbf{w}^e$ in general. By definition, $\mathbf{w}^e = -\mathbf{S}$, where $\mathbf{S}$ is the force exerted by the scale. From Newton's second law, $\mathbf{S} + m\mathbf{g} = m\mathbf{a}$, or $-\mathbf{S} = m\mathbf{g} - m\mathbf{a}$. Thus,

$$\mathbf{w}^e = m\mathbf{g} - m\mathbf{a} \qquad (5.6)$$

An object in free fall has an acceleration $\mathbf{a}$ equal to $\mathbf{g}$, so its effective weight is zero. Since an artificial satellite in orbit around the earth is in free fall, an astronaut in such a satellite is weightless. The following example illustrates the calculation of $\mathbf{w}^e$ when the vectors are not all parallel.

Example 5.5

On a carnival ride, a girl of mass m stands in a cylindrical cage of radius R. The cage is spun about its cylindrical axis so that her speed is v (Fig. 5.11). What is her effective weight?

From Eq. 5.6, $\mathbf{w}^e = m\mathbf{g} - m\mathbf{a}$. Taking horizontal and vertical components,

$$w^e_x = \frac{mv^2}{R}, \qquad w^e_y = -mg$$

Using the Pythagorean relation (Fig. 5.11c),

$$w^e = m\sqrt{g^2 + \left(\frac{v^2}{R}\right)^2}$$

Thus the girl's effective weight is greater than mg.

The sedimentation of material in a liquid occurs because the weight of the material exceeds that of an equal volume of liquid. The sedimentation rate is proportional to the acceleration of gravity g, which can in effect be increased by spinning the material rapidly in a centrifuge. Ultracentrifuges used in research laboratories achieve accelerations of up to 500,000 g.

5.4 | ELECTRIC CHARGES AND FORCES

We have seen that gravitational forces are responsible for keeping the planets in their nearly circular orbits around the sun. *Electric* forces, which are in some ways very similar to gravitational forces and in other ways quite different, are responsible for keeping electrons in their orbits in atoms. Although we will defer a full discussion of electric forces until Chapter Eighteen, we introduce them here and discuss a simple model of an atom.

The gravitational force is a fundamental force of nature. Related to this force is the concept of mass, which plays a basic role in the description of the physical world. An equally basic concept is that of *charge*. Just as two masses produce forces on each other, so also do two charges produce forces on one another. However, an important difference is that there are two kinds of charges, described in mathematical terms as positive and negative. Although two masses always attract each other, like charges (both positive or both negative) repel each other, while unlike charges attract each other. A further complication is that charges may produce two types of forces. Charges at rest exert forces on each other

Figure 5.11.   Forces on a girl in a spinning cage.

called *electric forces,* but charges in motion exert additional forces on each other called *magnetic forces.*

Charge and the Structure of Matter |

While it requires sophisticated mathematics to fully describe atoms and molecules, a simple atomic model (Fig. 5.12) gives some insight into charges, their properties, and the role they play in nature. There are slightly more than 100 naturally occurring or artificially produced atomic species or *elements.* The typical atomic radius is 10^{-10} m. The model pictures atoms as rather like a solar system, with negatively charged *electrons* orbiting a heavy, positively charged *nucleus.* The attractive electric force between the negative electrons and the positive nucleus holds the atom together, just as the gravitational attraction holds the solar system together. When an atom has its full complement of electrons, it is electrically neutral, that is, its total charge is zero. If one or more electrons are added or taken away the atom is said to be *ionized.*

The constituents of the massive nucleus were not completely identified until the 1930s. These constituents, collectively called *nucleons,* are *neutrons* that are uncharged and *protons* that have a positive charge. The proton charge is denoted by e and is exactly equal in magnitude to the electron charge, $-e$. A neutron or proton is approximately 1800 times as massive as an electron.

A typical nuclear radius is 10^{-14} m, some 10,000 times smaller than that of the atom. One may well ask how a nucleus can have many positively charged protons so close together if they repel each other. The explanation depends on the existence of still another fundamental force of nature that does not depend upon charge. If two protons approach each other, they electrically repel each other more and more strongly until they are close enough for the very strong *nuclear force* to take effect. This force overwhelms the electric repulsion and holds the protons together.

Atoms combine in various ways, all of which depend on the electric forces produced by charges, to form molecules and macroscopic objects that, in their normal state, are electrically neutral. Thus, common objects are not ordinarily charged but their existence and solidity depend on charges being present in their microscopic structure.

Figure 5.12. A simple model of an atom. Negatively charged electrons orbit a small, massive nucleus under the influence of electrical forces. The nucleus, which is not shown to scale, contains two kinds of nucleons: positively charged protons and neutral neutrons.

Two remarkable facts concerning charges have emerged from physics research up to the present. Using particle accelerators, complex nuclear reactions have been produced in which particles are created or destroyed. In this way, many short-lived particles have been discovered that are not ordinarily present in matter. Every one of these particles has an electric charge that is an exact integer multiple of the electronic charge: $0, \pm e, \pm 2e, \ldots$. There has been much speculation that particles called *quarks* having a fraction of the electronic charge may exist, but so far they have not been observed. Thus, it presently seems that charges cannot exist in units other than that found on the electron. The second observation is that when particles are created or destroyed, the net charge remains constant. This is referred to as *conservation of charge.* For example, if a positive charge is destroyed, a negative charge is also destroyed, leaving the net charge unchanged. The reasons for these characteristics of charge are not known.

Coulomb's Law |

The first detailed studies of the forces between charges were made in 1784 by Charles Augustin de Coulomb (1736–1806). He found that the electric force, like the gravitational force, is inversely proportional to the distance squared. However, because there are two kinds of charge, the force can be either attractive or repulsive.

Two point charges, q_1 and q_2, exert equal and opposite electric forces on each other. If the vector distance to q_1 from q_2 is $\mathbf{r}$, then the *force experienced by q_1 due to q_2* is given by Coulomb's law,

$$\mathbf{F}_{12} = \frac{kq_1q_2}{r^2}\,\widehat{\mathbf{r}} \qquad (5.7)$$

Here $\widehat{\mathbf{r}}$ is a unit vector pointing from q_2 to q_1, and k is an experimentally determined constant.

The S.I. unit of charge is the *coulomb* (C). In S.I. units, it is found that

$$k = 9.0 \times 10^9 \text{ N m}^2 \text{ C}^{-2} \qquad (5.8)$$

and that the magnitude of the charge on a proton or electron is

$$e = 1.60 \times 10^{-19} \text{ C}$$

The coulomb is a large unit of charge. For example, according to Coulomb's law, the force between a pair of 1 C charges 1 m apart is

$$F = (9 \times 10^9 \text{ N m}^2 \text{ C}^{-2})\frac{(1 \text{ C})(1 \text{ C})}{(1 \text{ m})^2} = 9 \times 10^9 \text{ N}$$

This is about 1 million tons! Consequently isolated charges as large as a coulomb are rarely encountered.

The direction of the electric force depends on the relative signs of the charges q_1 and q_2 (Fig. 5.13). If the two charges have the same sign (both positive or both negative), then q_1q_2 is positive, so that $\mathbf{F}_{12}$ is directed along $\widehat{\mathbf{r}}$, and the charges repel each other. If the charges have opposite signs, with one positive and the other negative, then q_1q_2 is negative; $\mathbf{F}_{12}$ is then directed along $-\widehat{\mathbf{r}}$, and the charges attract. In each case, Newton's third law holds true: the force $\mathbf{F}_{21}$ on q_2 due to q_1 equals $-\mathbf{F}_{12}$.

The gravitational force law (see Chapter Three) can be written in a vector form similar to Eq. 5.7 for the electric force. The gravitational force on a particle of mass m_1 due to a particle of mass m_2 is

$$\mathbf{F}_{12} = -\frac{Gm_1m_2}{r^2}\,\widehat{\mathbf{r}} \qquad (5.9)$$

Figure 5.13. (*a*) Like charges, both positive or both negative, repel each other. (*b*) Unlike charges, one positive and one negative, attract each other.

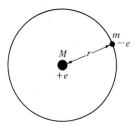

Figure 5.14.   A simple model of the hydrogen atom. A negative electron moves in a circular orbit around a massive positive proton.

Again $\widehat{\mathbf{r}}$ is a unit vector pointing from particle 2 toward particle 1; the minus sign indicates the force is toward particle 2, or attractive. Since both the gravitational and electric forces vary as $1/r^2$, their ratio for a pair of particles of given mass and charge is independent of the separation.

The next two examples illustrate some features of the simplest atom, hydrogen, which has a single electron orbiting a proton.

Example 5.6

In a simple model of the normal undisturbed hydrogen atom, an electron moves around a proton in a circular orbit of radius 5.29×10^{-11} m. The proton mass is $M = 1.67 \times 10^{-27}$ kg and the electron mass is $m = 9.11 \times 10^{-31}$ kg (Fig. 5.14). What are the electric and gravitational forces exerted on the electron by the proton?

Since the proton and electron have opposite charges, $+e$ and $-e$, the electric force is attractive, and its magnitude is

$$F = \frac{ke^2}{r^2}$$

$$= (9 \times 10^9 \text{ N m}^2 \text{ C}^{-2})\frac{(1.6 \times 10^{-19} \text{ C})^2}{(5.29 \times 10^{-11} \text{ m})^2}$$

$$= 8.23 \times 10^{-8} \text{ N}$$

The gravitational force is also attractive; from Chapter Three, its magnitude is

$$F_G = \frac{GmM}{r^2}$$

$$= (6.67 \times 10^{-11} \text{ N m}^2 \text{ kg}^{-2})$$

$$\times \frac{(1.67 \times 10^{-27} \text{ kg})(9.11 \times 10^{-31} \text{ kg})}{(5.29 \times 10^{-11} \text{ m})^2}$$

$$= 3.63 \times 10^{-47} \text{ N}$$

Comparison of these two forces shows that in this situation the electric force is about 10^{39} times stronger than

Chapter 5   Circular Motion

the gravitational force. Thus, in atomic physics the gravitational force can be completely ignored.

Example 5.7

Using the numbers given in the preceding example, find the speed of the electron in the hydrogen atom.

The centripetal acceleration of the electron results from the electric force, so the speed v satisfies

$$\frac{mv^2}{r} = \frac{ke^2}{r^2}$$

or

$$v = \sqrt{\frac{ke^2}{mr}}$$

Substituting the numerical values from the preceding example,

$$v = \left[\frac{(9 \times 10^9 \text{ N m}^2 \text{ C}^{-2})(1.6 \times 10^{-19} \text{ C})^2}{(9.11 \times 10^{-31} \text{ kg})(5.29 \times 10^{-11} \text{ m})}\right]^{1/2}$$
$$= 2.19 \times 10^6 \text{ m s}^{-1}$$

This is a large velocity; it is about 1 percent of the speed of light, 3×10^8 m s^{-1}.

We will see later that this simple model of the hydrogen atom and its generalization for atoms with many electrons correctly predicts some, but not all, of the observed properties of atoms. The failure of Newtonian mechanics to account for a variety of atomic phenomena led to revolutionary changes early in this century in our view of the world at atomic and substomic levels.

SUMMARY

An object undergoing uniform circular motion has a centripetal acceleration $a_r = v^2/r$ directed toward the center of the circle. This centripetal acceleration is produced by a net force acting on the object, which is equal to $m\mathbf{a}_r$.

The effective weight of an object with an acceleration $\mathbf{a}$ is $m(\mathbf{g} - \mathbf{a})$. A freely falling object and its contents are weightless.

The electric force between two charges is proportional to the product of the charges and inversely proportional to their separation squared:

$$F_{12} = \frac{kq_1q_2}{r^2}$$

The force is attractive for unlike charges and repulsive for like charges. Electric forces keep electrons in atomic orbits.

Checklist

Define or explain:

uniform circular motion
centripetal acceleration
banked curve
effective weight
electric charge
electric force

nucleus
nucleons, neutrons,
  protons
conservation of charge
Coulomb's law

EXERCISES

Section 5.1 | Centripetal Acceleration

5-1   A woman runs on a circular track of radius 100 m at a speed of 8 m s^{-1}. What is her acceleration?

5-2   A boy rides a bicycle at 10 m s^{-1} on a flat curve of radius 200 m. (a) What is his acceleration? (b) If the boy and the bike have a total mass of 70 kg, how large a force is needed to provide this acceleration?

5-3   A racing car rounds a turn at 60 m s^{-1}. If the force needed to provide the centripetal acceleration is equal to the weight of the car, what is the radius of the turn?

5-4   A man sits without a seat belt in a car. He tends to slide to the left as the car makes a right turn. Is there a force pushing the man to the left? Explain.

5-5   The speeds of centrifuges are limited in part by the strength of the materials used in their construction. A centrifuge spins a 10 g = 10^{-2} kg sample at a radius of 0.05 m at 60,000 revolutions per minute. (a) What force must the centrifuge exert on the sample? (b) What is the mass of a sample at rest with a weight equal to this force?

5-6   A child sits 4 m from the center of a merry-go-round, which turns completely each 10 s. What is the child's acceleration?

5-7   A jet fighter plane flying at 500 m s^{-1} pulls out of a dive on a circular path. What is the radius of the path if the pilot is subjected to an upward acceleration of $5g$?

5-8   The radius of the earth's orbit about the sun is 1.5×10^8 km, and its period is 365 days. What is the centripetal acceleration of the earth?

5-9   A trained pilot can pull out of a dive on a circular path with an upward acceleration of $5.5g$ at the bottom of the path. An untrained pilot can perform the same maneuver at the same speed but only with an acceleration of $3g$. What is the ratio of the minimum radii of the paths in which the two pilots can fly?

5-10 A centrifuge used for testing human tolerance to acceleration has a gondola at a distance of 16 m from the vertical axis of rotation. What speed is required to produce a horizontal acceleration of $11g$? $(16 m) \frac{11(9.8 ms^{-2})}{16 m} = v^2$

5-11 A car is traveling around a flat curve of radius 0.25 km. The coefficient of static friction between the tires and the road is 0.4. At what speed will the car begin to skid?

5-12 A woman of mass 60 kg runs around a flat, circular track of radius 200 m at 6 m s^{-1}. (a) What is her acceleration? (b) What force produces this acceleration? (c) How large is this force?

Section 5.2 | Examples of Circular Motion

5-13 At what angle should the track of the preceding exercise be banked so that there is no frictional force required?

5-14 A racing car travels around a curve of radius 4000 ft. If the frictional force is zero and the speed is 200 ft s^{-1}, at what angle is the curve banked?

5-15 A curved track has a radius of 336 m and is banked at 35°. At what speed is the frictional force zero?

5-16 Why is it not safe to drive on steeply banked curves at low speeds under very slippery conditions?

5-17 A swallow flies in a horizontal arc of radius 15 m at 18 m s^{-1}. (a) What is its acceleration? (b) What is the banking angle?

5-18 A curve is banked so that no friction is required at 60 mi h^{-1}. If a car is moving at 40 mi h^{-1} on the curve, in which direction is a frictional force exerted by the road?

Section 5.3 | Effective Weight

5-19 An airplace climbs and then turns downward in a circular arc of radius R. If its speed is 400 m s^{-1}, at what radius will the pilot experience weightlessness at the top of the arc?

5-20 An ultracentrifuge spins samples so that their effective weight is 10^5 times their normal weight. If the sample is at a radius of 0.05 m, how many revolutions per minute does the machine make?

5-21 A bird of weight w flies at 15 m s^{-1} in a horizontal circle of radius 15 m. What is the bird's effective weight?

5-22 A ferris wheel of radius 16 m rotates uniformly once every 20 s. (a) What is the centripetal acceleration? (b) What is the effective weight of a 45-kg rider at the highest point of the ride? (c) What is the rider's effective weight at the lowest point of the ride?

Section 5.4 | Electric Charges and Forces

5-23 Objects can be given a net charge by rubbing, as is observed by people walking on woolen rugs in dry weather. How many electrons must be transferred to give an object a net charge of $+10^{-6}$ C? Must electrons be added or taken away?

5-24 A gram of hydrogen contains about 6×10^{23} atomic electrons. What fraction of the electrons must be removed to give the sample a net charge of 10^{-3} C?

5-25 Two identical charges 0.1 m apart exert electrical forces of 10 N on each other. (a) What is the magnitude of one of the charges? (b) Find the ratio of this charge to that on an electron.

5-26 Cosmic ray protons strike the upper atmosphere at an average rate of 1500 protons per square meter each second. How much charge is received by the earth in a 24-hour day? (The radius of the earth is 6.4×10^6 m.)

5-27 A kilogram of molecular hydrogen contains 3.01×10^{26} molecules, each consisting of two hydrogen atoms. (a) What is the total charge on the electrons in 1 kg of hydrogen? (b) What is the total charge on the protons? (c) If all the electrons were removed and placed 1 m away from the protons, what would be the electrical force between them?

5-28 In a nucleus the minimum separation between two protons is about 10^{-15} m. (a) Find the electrical force between two protons at this distance. (b) Find the ratio of this force to the force between a proton and an electron separated by 10^{-10} m.

5-29 A salt crystal is made up of Na$^+$ ions, which are lacking one electron, and Cl$^-$ ions, which have one extra electron. What is the force between a Na$^+$ ion and a Cl$^-$ ion separated by 5×10^{-10} m?

5-30   In a simple model of the normal hydrogen atom, the radius of the circular electronic orbit is 5.29×10^{-11} m and the speed of the electron is 2.19×10^{6} m s^{-1}. Find (a) the acceleration of the electron, and (b) the number of orbits completed per second.

PROBLEMS

5-31   A car travels on a curve of radius 100 m banked at 20° at a speed such that no friction is needed. (a) What is the speed of the car? (b) Find the ratio of the normal force to the weight.

***5-32**   A car of weight w travels on a curve of radius 200 m banked at 10°. (a) At what speed is no friction required? (b) What frictional force is required if the car travels 5 m s^{-1} faster than this speed?

***5-33**   A curve of radius 300 m is banked at an angle of 10°. (a) At what speed is no friction required? (b) If the coefficient of friction is 0.8, what are the maximum and minimum speeds at which the curve can be traveled?

5-34   A bird of mass 0.3 kg flies in a horizontal curve of radius 20 meters at 15 m s^{-1}. (a) What is the banking angle? (b) What is the lift force exerted by the air on the bird?

5-35   The earth's radius is 6400 km and it rotates on its axis once every 24 hours. (a) What is the centripetal acceleration at the equator? (b) If a man weighs 700 N at the North Pole, what is his effective weight at the equator? (c) The earth is actually not quite spherical; it is flattened slightly at the poles and broadened at the equator. What qualitative effect does this have on your answer to part (b)?

5-36   A 2-kg rock is tied to a light rope 1 meter long and is swung in a horizontal circle. The rope is at an angle of 30° to the horizontal. (a) What is the tension in the rope? (b) What is the speed of the rock?

5-37   An airplane moving at 400 m s^{-1} can safely be subjected to a 8g acceleration when banking. How long will it take for the plane to turn 180°?

SUPPLEMENTARY TOPICS

5.5 | SATELLITES

Like a water pail swing in a vertical circle, an artificial satellite in a circular earth orbit has an acceleration due to gravity toward the earth. It is falling just fast enough to stay in that orbit (Fig. 5.15). We can find a formula relating the orbital radius r to the *period T*, the time needed for one full orbit.

If the satellite mass is m and the earth mass is M_E, the gravitational force on the satellite is, from Eq. 5.9,

$$\mathbf{F} = -\frac{GmM_E}{r^2}\hat{\mathbf{r}}$$

Thus, $\mathbf{F} = m\mathbf{a}_r$ becomes

$$-\frac{GmM_E}{r^2}\hat{\mathbf{r}} = -\frac{mv^2}{r}\hat{\mathbf{r}}$$

In the period T the planet travels a distance $2\pi r$, so its

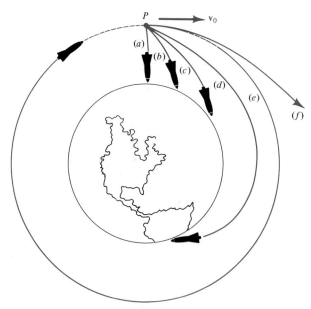

Figure 5.15.   Rockets placed at P with their engines off experience a gravitational acceleration toward the earth. Trajectory (a) corresponds to a rocket with no initial velocity; it falls straight down. As the rocket is given in increasingly large initial velocity $\mathbf{v}_o$ in the direction shown, the trajectory changes as indicated by paths (b), (c), and so on. Trajectory (e) is a closed circular orbit, corresponding to an artificial earth satellite. If the initial velocity is still larger, the rocket will escape from the earth (f). Note that the rocket is always falling, just as a thrown rock falls. In a circular orbit, it moves parallel to the surface just fast enough so that its fall carries it around the earth.

speed is $v = 2\pi r/T$, and the equation above can be rewritten as

$$\frac{GmM_E}{r^2} = \frac{m}{r}\left(\frac{2\pi r}{T}\right)^2$$

Solving for T^2, we find the relation between T and r,

$$T^2 = Cr^3 \qquad (5.10)$$

The constant C is

$$C = \frac{4\pi^2}{M_E G} \qquad (5.11)$$

Notice that C is independent of the satellite mass m. Hence the motion of the moon and of all artificial earth satellites will satisfy $T^2 = Cr^3$ with the *same* value of C.

The relation $T^2 = Cr^3$ also holds true for the planets in their nearly circular orbits about the sun. (In this case M_E in the constant C must be replaced by the solar mass, and r must be replaced by the average orbital radius.) This was one of the three laws of planetary motion discovered by Kepler in the early part of the seventeenth century from the careful analysis of observations made by earlier workers. Newton showed that all three laws could be derived using the gravitational force law and the equations of motion.

In the next example, we calculate the orbital radius of a communications satellite that has a period of exactly 1 day and consequently is always above the same point on the earth.

Example 5.8

The moon has a period $T_m = 28$ days and an orbital radius $r_m = 3.8 \times 10^5$ km. What is the orbital radius r_s of a satellite that has a period $T_s = 1$ day?

We can apply $T^2 = Cr^3$ to both the artificial satellite and the moon:

$$T_s^2 = Cr_s^3$$
$$T_m^2 = Cr_m^3$$

If we take the ratio of these equations, C cancels, leaving

$$\frac{T_s^2}{T_m^2} = \frac{r_s^3}{r_m^3}$$

or

$$r_s = r_m\left[\frac{T_s}{T_m}\right]^{2/3} = (3.8 \times 10^5 \text{ km})\left[\frac{1 \text{ day}}{28 \text{ day}}\right]^{2/3}$$
$$= 4.2 \times 10^4 \text{ km}$$

5.6 | PHYSIOLOGICAL EFFECTS OF ACCELERATION

The advent of high-speed aviation and, more recently, of space travel has spurred considerable research into the physiological reactions of humans to acceleration. Some representative accelerations and their time durations are given in Table 5.1.

Most of us have experienced moderate vertical accelerations in high-speed elevators. The effects we feel are related to the fact that most, but not all, of the body is fairly rigid. The blood flows in distensible vessels, so when the body is accelerated upward, the blood accumulates in the lower body. When the acceleration is downwards, the blood volume increases in the upper body. Also, the internal organs of the body are not rigidly held in place, and there may be unpleasant feelings produced by the displacement of these organs during acceleration.

The ability of a person to withstand an acceleration depends on both the magnitude and the duration of the acceleration. Because of the inertia of the blood and distensible organs, the effects on them of moderate accelerations (a few g's) are unimportant if the acceleration lasts for only a small fraction of a second. The limit of tolerance is then some tens of g's and is set by the structural strength of the vertebrae. As the time duration increases, so also do the dangers.

Extensive studies have been made of the circulatory disturbances of pilots subjected to accelerations lasting a few seconds or longer. Pilots in an airplane pulling out of a dive for several seconds may experience two successive types of *blackouts*. A visual blackout occurs first at about $3g$ as a result of reduced blood pressure in the retina, which is very sensitive to oxygen deprivation. The blood pressure drops because the heart has difficulties in pumping the blood with its increased effective weight. By modifying the pilot's position, training him to tense his abdominal muscles, and equipping him with a flight suit that reduces blood pooling in the lower body, the visual blackout threshold can be increased to about $5g$. In addition, the reduced supply of blood to the brain leads to a complete blackout or unconsciousness at about $6g$. Since many high performance planes can sustain about $9g$ when pulling out of a dive, the limits of human tolerance may easily be exceeded.

When a plane climbs and then circles downward,

TABLE 5.1

The approximate duration and magnitude of some brief accelerations in multiples of the gravitational acceleration, $g = 9.8$ m s^{-2}

Type of Acceleration	Acceleration in Multiples of g	Duration (seconds)
Elevators		
Fast service	0.1–0.2	1–5
Comfort limit	0.3	
Emergency stop	2.5	
Automobiles		
Comfortable stop	0.25	5–8
Very unpleasant	0.45	3–5
Maximum possible	0.70	3
Crash (possibly survivable)	20–100	0.1
Aircraft		
Normal takeoff	0.5	10–20
Catapult takeoff	2.5–6	1.5
Crash landing (possibly survivable)	20–100	
Seat ejection	10–15	0.25
Humans		
Parachute opening	8–33	0.2–0.5
Parachute landing	3–4	0.1–0.2
Fall into firemen's net	20	0.1

Adapted from D.E. Goldman and H.E. von Gierke in Harris & Crede, (eds), *Shock and Vibration Handbook*, McGraw-Hill, New York, 1961, Chapter 44.

the plane and pilot are both more susceptible than they are at the bottom of a dive. An engorgement of the vessels in the head causes a reduction of heart activity and hence a reduction of the oxygen supplied to the retina and brain. Planes are usually not designed to withstand as large a stress in this maneuver as in a dive, so they also will be structurally sound only under lower g stresses. An illustration of the upward effective weight attained during some maneuvers is provided by the next example.

Example 5.9

An airplane of mass m is flown at a speed of 300 m s^{-1} in a vertical circle (Fig. 5.16). What is the minimum radius r of the circle so that the upward effective weight does not exceed $3mg$?

Since the effective weight $\mathbf{w}^e$ is now upward and equal to $3mg$ in magnitude, $\mathbf{w}^e = -3m\mathbf{g}$. According to Eq. 5.6, $\mathbf{w}^e = m\mathbf{g} - m\mathbf{a}$, so

$$-3m\mathbf{g} = m\mathbf{g} - m\mathbf{a}$$

or $\mathbf{a} = 4\mathbf{g}$. The acceleration here is the centripetal acceleration, so we have

$$\frac{v^2}{r} = 4g$$

Thus the radius that will lead to an upward effective weight of $3mg$ is

$$r = \frac{v^2}{4g} = \frac{(300 \text{ m s}^{-1})^2}{4(9.8 \text{ m s}^{-2})} = 2300 \text{ m}$$

A smaller radius will lead to a larger effective weight.

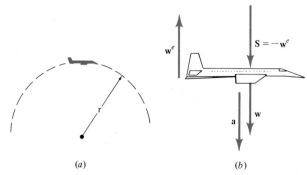

Figure 5.16. S is the downward force exerted by the air on the airplane.

CHARLES AUGUSTIN de COULOMB (1736–1806)

HENRY CAVENDISH (1731–1810)

(The Bettmann Archive.)

(Radio Times Hulton Picture Library.)

Although we know it today as Coulomb's law, the expression for the electrostatic force between two charges was first discovered by Henry Cavendish. This fact is not surprising when one learns about Cavendish's bizarre personality.

Charles Augustin de Coulomb was a French nobleman who began a career as a military engineer and gradually became interested in scientific research. At the beginning of the French Revolution, he wisely retired to the safety of a provincial town to concentrate on his experiments. In 1777, he competed for a prize offered by the French Academy of Sciences for the improvement of magnetic compasses. He found that if he suspended a compass needle from a fine hair or thread, the torque exerted on the needle was proportional to the angle through which it had rotated. This *torsion balance* principle made it possible for him to measure electrostatic forces accurately and to obtain the force law. His results were published in the normal fashion, and he received the credit for this important discovery.

Unknown to Coulomb, his English contemporary, Cavendish, had already done the same kinds of electrostatic experiments with a torsion balance. Cavendish was an extraordinarily eccentric person: shy, absentminded, and a recluse to the point of insisting on dying alone. He never completed his studies at Cambridge because he could not bear to face his professors at the required examinations. He avoided all people, especially women, as much as possible.

Cavendish came from a wealthy family and never cared for nor needed to worry about money. Devoting himself to scientific research for his entire life, he made many major discoveries. However, he published very few of them and had no concern about receiving credit for his advances. In particular, his electrical researches anticipated many discoveries of the following decades but remained unknown until James Clark Maxwell examined Cavendish's notes decades after his death. Cavendish did publish some of his early work on the properties of hydrogen gas, but his experiments in which he discovered the inert gas now called argon were ignored until they were repeated a century later.

The most important experiment performed by Cavendish was, in effect, a measurement of the mass of the earth. From Newton's law of universal gravitation, it follows that the gravitational acceleration at the surface of the earth is $g = GM_E/R_E^2$, where G is the gravitational constant and M_E and R_E are the earth's mass and radius respectively. Since g and R_E are easily measured, a determination of either G or M_E determines the other quantity. Cavendish used a torsion balance to measure the small gravitational force exerted on two small spheres at the ends of a pivoted rod by two other large spheres. This gave him a value for G and therefore for the mass of the earth.

5.7 | DERIVATION OF THE RADIAL ACCELERATION FORMULA

In Fig. 5.17, an object moves with a constant speed v in a circular path of radius r. The position vector **r** is the vector from the center of the circle to the object at time t; **v** is the velocity vector at this time. Both vectors are constant in magnitude but not in direction. As the object goes once around the path, the position vector rotates through a full circle, as does the velocity vector. *Hence the position vector* **r** *and the velocity vector* **v** *rotate at the same rate, and their fractional changes in any time must be the same.* Suppose, for example that in a short time, Δt, the velocity changes by $\Delta \mathbf{v}$ and the position by $\Delta \mathbf{r}$ (Fig. 5.17). Then we must have

$$\frac{\Delta v}{v} = \frac{\Delta r}{r}$$

(This equation can also be obtained using similar triangles as shown in Fig. 5.17). Dividing by Δt,

$$\frac{1}{v}\frac{\Delta v}{\Delta t} = \frac{1}{r}\frac{\Delta r}{\Delta t}$$

This states that the rate of change of **v** is to v as the rate of change of **r** is to r. If Δt is very small, then $\Delta v/\Delta t$ becomes the acceleration a_r, and $\Delta r/\Delta t$ becomes the velocity v. Thus, we obtain

$$\frac{a_r}{v} = \frac{v}{r}$$

Multiplying by v yields the desired result, $a_r = v^2/r$.

EXERCISES ON SUPPLEMENTARY TOPICS

Section 5.5 | Satellites

5-38 An artificial earth satellite moves in a circular orbit with a radius one-fourth that of the moon. What is its period?

5-39 An artificial satellite is to be placed in an orbit around the sun so that its period is 8 earth years. By definition, the radius of the earth's orbit is 1 astronomical unit (A.U.). What is the radius of the satellite's orbit in A.U.?

5-40 The distance from the sun to the earth is 1 astronomical unit. What is the length of a "year" on a planet 9 A.U. from the sun?

PROBLEMS ON SUPPLEMENTARY TOPICS

5-41 The orbital radius of the moon is 3.8×10^5 km and its period is 28 days. (a) Find its acceleration, a_m. (b) The gravitational acceleration at the earth's surface, which is 6400 km from its center, is $g = 9.8$ m s^{-2}. Using the $1/r^2$ dependence of the gravitational force law, what gravitational acceleration g' would one expect at the radius of the moon's orbit? (c) Compare g' and a_m. (This was one of Newton's original tests of the universal law of gravitation.)

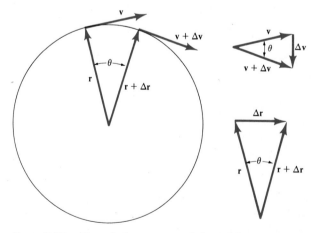

Figure 5.17. The velocity vector v and the position vector r rotate at the same rate, so the angle θ is the same in both triangles. Thus the triangles are similar, and $\triangle v/v = \triangle r/r$.

5-42   Suppose the gravitational force were proportional to $1/r^3$ instead of $1/r^2$. What would be the relation between the period of a planet and its orbital radius? Compare this result with the observed relation, $T^2 = Cr^3$.

Additional Reading

D.E. Goldman and H.E. von Gierke, in Cyril M. Harris and Charles E. Crede (eds.), *Shock and Vibration handbook,* McGraw-Hill Book Co., New York, 1961, Chapter 44, Effects of Shock and Vibration on Humans.

Otto Glasser (ed.), *Medical Physics,* vol. I, The Year Book Publishers, Inc. Chicago, 1944, p. 22. Effects of acceleration.

J.L.E. Dreyer, *Tycho Brahe: A Picture of Scientific Life and Work in the Sixteenth Century,* Dover Publications Inc., New York, 1963. Brahe and his observations of planetary motions.

Scientific American articles:

J.W. Beams, Ultrahigh-Speed Rotation, April 1961, p. 134.

Terence A. Rogers, The Physiological Effects of Acceleration, February 1962, p. 60.

Gerald Feinberg, Ordinary Matter, May 1967, p. 126.

Thomas C. Van Vlandern, Is Gravity Getting Weaker? February 1976, p. 44.

UNIT TWO

ADDITIONAL TOPICS IN MECHANICS

The principles described in Unit One form the basis for analyzing all mechanics problems. In particular, Newton's laws of motion tell us how to use the forces on objects to predict their motion. There are, however, large classes of problems for which it is either difficult or inconvenient to use Newton's laws directly. For example, it is difficult to determine all the forces when a skier descends an uneven slope or when two objects collide. In the rotational motion of solid objects, it is not the forces as much as the torques that are important. Nevertheless, there are convenient ways to handle many problems of these types.

In the first two chapters of this unit, we discuss the concepts of work, energy, and momentum. In these discussions, we carefully separate the forces present into several categories. This separation can often be accomplished even though we may not have detailed knowledge of the forces. Under certain conditions, we then obtain the very convenient result that the energy or momentum or both are constant and hence said to be conserved.

In Chapter Eight, we discuss the rotation of solid objects. Many of the results of this chapter are analogous to those developed for translational motion. For example, a quantity called the angular momentum is sometimes constant or conserved. Under such circumstances the understanding of many rotational phenomena is greatly simplified.

In Chapter Nine, we see how the strength of an object depends on its shape and composition. Construction materials, bones, and trees are obvious cases where such questions are important.

The final chapter of this unit describes mechanical vibrations and oscillations. We can predict when such motion may occur and identify some of its important features, such as the frequency.

This unit concludes our formal development of mechanics, but it is by no means the end of its usefulness to us. Throughout this text, we make explicit or implicit use of the principles of mechanics. Since these first two units are based almost entirely on Newton's laws, the general range of application of mechanics is as broad as the various forces that appear in nature. Even in our discussion of atoms and molecules, where Newton's laws break down, many of the principles we have discussed will survive.

CHAPTER SIX

WORK, ENERGY, AND POWER

Work, energy, and power are words with a variety of meanings in everyday usage. However, to the scientist these terms have very specific definitions. In this chapter, we consider these definitions and the relationships among work and various kinds of energy for mechanical systems. Although these relationships are obtained from Newton's laws, they can often be readily used when the forces are unknown or when the systems are so complex as to make the direct use of Newton's laws impracticably difficult.

This chapter also provides us with our first look at a *conservation law*. We find that under certain conditions the mechanical energy of a system is *constant* and is then said to be *conserved*. This fact provides a very powerful tool for understanding and solving certain mechanical problems. However, it is now known that a much broader energy conservation law is valid in nature. If we calculate or measure the total energy—mechanical, electrical, thermal, and so on—then this total energy is constant, even though any one type need not be conserved. What is observed in nature is an interchange of energy from one form to another, with the *sum* remaining constant. This total energy conservation law became fully understood when Einstein showed that matter and energy are two forms of the same quantity. He saw that not only may energy be converted from one form to another but it may also be converted to matter and vice versa.

The idea of energy conservation will be used throughout this textbook. Because of its general nature, it is a useful concept in an enormous range of applications. In Chapters Seven and Eight, we find examples of two other mechanical quantities that obey conservation laws, the momentum and the angular momentum. Like the energy, these provide the means for understanding many otherwise complex systems.

6.1 | WORK

In this section, we define the work done on an object by a force and show how it is calculated. It will become clear in the following sections that the concept of work plays a fundamental role in the analysis of many mechanical problems.

Figure 6.1 shows a constant force **F** acting on an object as it moves a distance **s**. The *work done by the force* is defined as its component F_s along the displacement times the magnitude of the displacement,

$$W = F_s s \qquad (6.1)$$

If **F** is at an angle, θ, to **s**, as in Fig. 6.2, then

$$F_s = F \cos \theta \qquad (6.2)$$

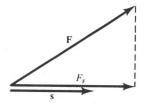

Figure 6.1. The work done by the force **F** during a displacement **s** is $W = F_s s$.

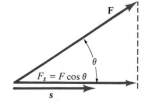

Figure 6.2. The component of **F** along **s** is $F_s = F \cos \theta$. The work is then $W = Fs \cos \theta$.

and the work can be written as

$$W = Fs \cos \theta \qquad (6.3)$$

The S.I. unit of work is the *joule* (J). Since work has dimensions of force times distance, a joule is a new-ton-meter. The c.g.s. unit is the dyne-centimeter or *erg*; the British unit is the foot-pound.

Notice that our definition of work differs in some ways from the nontechnical meaning of the word. According to Eq. 6.3, we do twice as much work in pushing an object along the floor if we double its weight or the distance it moves. This is consistent with the everyday notion of work. However, this is not so if we stand in one spot supporting a heavy weight. We would consider that we are working rather hard, but since there is no displacement, we conclude that no work is done on the weight.

However, work is being done in the body as nerve impulses repeatedly trigger muscle fiber contractions. Unlike a bone or a steel post, a muscle fiber cannot sustain a static load. Instead, it must repeatedly relax and contract, doing work in each contraction. We are unaware of this process because of the large number of fibers and the rapidity of the contractions.

In the following examples we show how the work done by a force acting on an object is computed.

Example 6.1

A 600-N force is applied by a man to a dresser that moves 2 m. Find the work done if the force and displacement are (a) parallel; (b) at right angles; (c) oppositely directed (Fig. 6.3).

(a) When **F** and **s** are parallel, $\cos \theta = \cos 0° = 1$, and

$$W = Fs \cos \theta = (600 \text{ N})(2 \text{ m})(1) = 1200 \text{ J}$$

The man does 1200 J of work on the dresser.

(b) When **F** is perpendicular to **s**, $\cos \theta = \cos 90° = 0$, and $W = 0$. No work is done when the force is at right angles to the displacement.

(c) When **F** and **s** are opposite, $\cos \theta = \cos 180° = -1$, and

$$W = Fs \cos \theta = (600 \text{ N})(2 \text{ m})(-1) = -1200 \text{ J}$$

In this case the work done by the force is negative, so *the object is doing work on the man.*

Example 6.2

A horse pulls a barge along a canal with a rope in which the tension is 1000 lb (Fig. 6.4). The rope is at an angle of 10° with the towpath and the direction of the

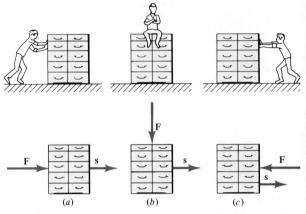

Figure 6.3. A force is applied (*a*) parallel to the motion, (*b*) perpendicular to the motion, and (*c*) opposite to the motion. In each case the work done by the force **F** is different. In part (*c*) we may imagine that the dresser is being slowed and brought to rest. In (*a*) the man does work on the dresser; in (*b*) he does no work, and in (*c*) the dresser does work on the man.

barge. (a) How much work is done by the horse in pulling the barge 100 ft upstream at a constant velocity? (b) What is the net force on the barge?

(a) The work done by the constant force **T** in moving the barge a distance **s** is given by

$$W = Ts \cos \theta$$

where θ is the angle between **T** and **s**. Using $\cos 10° = 0.985$,

$$W = (1000 \text{ lb})(100 \text{ ft})(0.985) = 9.85 \times 10^4 \text{ ft lb}$$

(b) Since the barge moves at a constant velocity, the sum of all the forces on it must be zero. There must be another force acting that is not shown in Fig. 6.4, a force exerted on the barge by the water that is equal and opposite to **T**.

In this example the net work done by all the forces acting on the barge is zero because the net force is zero. The work done by the force due to the water is -9.85×10^4 ft lb. In other words, the barge does 9.85×10^4 ft lb of work on the water.

Figure 6.4. (*a*) A horse pulling a barge at constant velocity. (*b*) The tension in the rope is **T**.

Figure 6.5. The area under the curve is the work done by the constant force **F** acting through a displacement **s**.

In our defining equation, $W = F_s s = Fs \cos \theta$, for work, we assumed that the force **F** was constant. In many situations, this is, at best, only approximately true. If the force varies in magnitude or direction with respect to the displacement, the correct procedure is to consider the work done in a series of many small displacements. In *each* piece, we compute $\Delta W = \bar{F}_s s$ where $\bar{F}_s$ is the average force in this part of the motion. The sum of all these small terms then gives the net work done.

We do not need to compute such sums in this textbook. However, we will make use of a graphical method for computing the work. We note that if we graph F_s for a constant force versus s for a given motion, then the rectangular area under the curve is the product $F_s s$ (Fig. 6.5). This equals the work done by the force. This result is true in general: *the area under the graph of F_s versus s for any motion is the work done.*

6.2 | KINETIC ENERGY

The *kinetic energy* of an object is a measure of the work an object can do by virtue of its motion. As we show below the translational kinetic energy of an object of mass m and velocity **v** is $\frac{1}{2}mv^2$.

Objects can also have kinetic energy associated with rotational motions or the motions of their constituent molecules. These kinds of kinetic energy are dealt with in later chapters.

The work done on an object and its kinetic energy obey the following fundamental principle:

The total work done on an object by all the forces acting upon it, including frictional and gravitational forces, is equal to the change in the kinetic energy of the object.

This work-energy principle can be derived quite generally from Newton's laws of motion. However, it is more instructive for our purposes to see how it comes about in a simple situation.

Figure 6.6. A force **F** does work on an object as it moves a distance **s**. The velocity changes from v_0 to **v**.

Consider an object of mass m subjected to a constant force **F** (Fig. 6.6). The object moves a distance **s** parallel to **F**. Since its acceleration $\mathbf{a} = \mathbf{F}/m$ is constant, the initial velocity v_0 and final velocity **v** satisfy $v^2 = v_0^2 + 2as$ or

$$as = \tfrac{1}{2}v^2 - \tfrac{1}{2}v_0^2$$

From Newton's second law, $\mathbf{F} = m\mathbf{a}$, the work done by the force **F** is $W = Fs = mas$. With our expression for as, this becomes

$$W = \tfrac{1}{2}mv^2 - \tfrac{1}{2}mv_0^2 \qquad (6.4)$$

The final kinetic energy K and the initial kinetic energy K_0 of the object are defined by

$$K = \tfrac{1}{2}mv^2 \qquad \text{and} \qquad K_0 = \tfrac{1}{2}mv_0^2 \qquad (6.5)$$

With Eq. 6.4,

$$W = K - K_0 \qquad (6.6)$$

*Therefore the work done by **F** is equal to the change in the kinetic energy of the object.* Note that work and kinetic energy have the same dimensions and units.

One consequence of Eq. 6.6 is that if work is done on the object, its kinetic energy increases. Conversely, if the object does work on an external agency, its kinetic energy decreases. This is illustrated by the work that is done on a person who slows down or stops a moving object. The following examples should serve to clarify these ideas.

Example 6.3

A woman pushes a toy car, initially at rest, toward a child by exerting a constant horizontal force **F** of magnitude 5 N through a distance of 1 m. (a) How much work is done on the car? (b) What is its final kinetic energy? (c) If the car has a mass of 0.1 kg, what is its final speed? (Assume no work is done by frictional forces.)

(a) When the woman pushes the car, she exerts a force on it to the right (Fig. 6.7a). The displacement

Figure 6.7. (a) A woman pushes a toy car to the right. She exerts a horizontal force **F** on the car, which is parallel to its displacement **s**. (b) A child stopping the car exerts a force **F′** on the car. **F′** is opposite to the displacement **s′** of the car as it is brought to rest.

$s = 1$ m is also to the right, so the work done on the car is

$$W = Fs = (5 \text{ N})(1 \text{ m}) = 5 \text{ J}$$

(b) The initial kinetic energy K_0 is zero, so the final kinetic energy is

$$K = W = 5 \text{ J}$$

(c) The final kinetic energy is $K = mv^2/2$, so with $m = 0.1$ kg, the speed is

$$v = \sqrt{\frac{2K}{m}} = \sqrt{\frac{(2)(5 \text{ J})}{0.1 \text{ kg}}} = 10 \text{ m s}^{-1}$$

Example 6.4

In the preceding example the woman releases the toy car with a kinetic energy of 5 J. It moves across the floor and reaches the child who stops the car by exerting a constant force F' opposite to its motion. The car stops in 0.25 m. Find F' (Fig. 6.7b) if no work is done on the car by frictional forces.

While the car is moving toward the child, no work is done on the car, and its kinetic energy remains 5 J until it reaches the child. The initial kinetic energy K_0 is 5 J, and the final kinetic energy K is zero, since the car comes to rest, so

$$W = K - K_0 = 0 - 5 \text{ J} = -5 \text{ J}$$

The work done is also given by the product of F', s', and the cosine of the angle between them, which is 180° here. Thus, we can also write

$$W = F's' \cos 180° = -F's'$$

so

$$F' = -\frac{W}{s'} = -\frac{(-5 \text{ J})}{(0.25 \text{ m})} = 20 \text{ N}$$

The negative sign for the work done, $W = -5$ J, indicates that the car does work on the child. These two examples show that positive work done on an object gives it kinetic energy, which is then available to do work, in this case on the child.

6.3 | POTENTIAL ENERGY AND CONSERVATIVE FORCES

The work-energy relation of the preceding section includes the work done by *all* the forces acting on the object. However, it is useful in most cases to separate the work done into three distinct categories: the work done by gravitational forces, by frictional forces, and by the other forces acting on the object, which we will call *applied forces*. In this section, we examine the work done on an object by gravitational forces. We will see that the work done by gravity on the object can be automatically taken into account by introducing another form of energy called *potential energy*. Forces that can be taken into account in this way are said to be *conservative forces*.

The idea of potential energy can be introduced by again using the example of a toy car. Suppose a child pushes a car up an inclined board and that the work done by friction is negligible (Fig. 6.8). The forces on the car are its weight **w**, the force **F** applied by the child parallel to the board, and a normal force **N**. The coordinate x measures the distance of the car from the bottom of the board. At the bottom, the velocity is v_0 and the kinetic energy is $K_0 = \frac{1}{2}mv_0^2$. At x, the velocity is v and the kinetic energy is $K = \frac{1}{2}mv^2$. The work done by the applied force **F** is the product of F and the displacement, $W_a = Fx$. The component of the gravitational force acting along the plane is $mg \sin \theta$, and it is opposite to the motion up the plane. Thus the work done by the gravitational force is $W_g = -mgx \sin \theta$. The normal force does no work, since **N** is perpendicular to the motion. From the work-energy principle, the total work done, $W_a + W_g$, equals the change in kinetic energy, so

$$\boxed{Fx - mgx \sin \theta = K - K_0} \qquad (6.7)$$

This result can be rewritten if we note in Fig. 6.8b that $x \sin \theta = h - h_0$, the height change of the car. Then $mgx \sin \theta$ becomes $mg(h - h_0)$, which depends only on the *difference of the heights* of the two points and not on the angle θ. Because of this, it is useful to

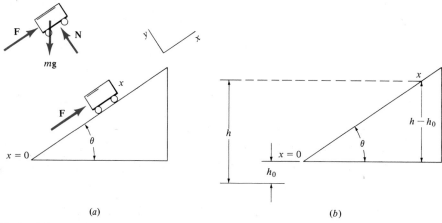

Figure 6.8. (*a*) The toy car is subject to its weight, a normal force, and an applied force **F**. (*b*) By definition, $\sin \theta = (h - h_0)/x$, so $x \sin \theta = (h - h_0)$.

define the *potential energy* of the car at the bottom of the board and at x by

$$\mathcal{U}_0 = mgh_0 \quad \text{and} \quad \mathcal{U} = mgh \qquad (6.8)$$

$W_a = Fx$ is the work done by the *applied force*, so Eq. 6.7 becomes

$$W_a = (K - K_0) + (\mathcal{U} - \mathcal{U}_0) \qquad (6.9)$$

When no work is done by frictional forces, the work done by the applied force equals the change in the kinetic energy plus the change in the potential energy. Notice that it is only the difference in the potential energies that is important. This means that the reference level used for measuring heights can be chosen for convenience.

The value of the potential energy concept is most clearly seen when the applied force is zero. Then Eq. 6.9 becomes

$$K + \mathcal{U} = K_0 + \mathcal{U}_0 \qquad (6.10)$$

The sum $K + \mathcal{U}$ is the same at the bottom of the board and at x. Since x is arbitrary, the sum must be constant everywhere on the plane. In other words, *in the absence of any forces that do work other than the gravitational force, the total mechanical energy* E defined by

$$E = K + \mathcal{U} \qquad (6.11)$$

is constant or *conserved*. As the car moves up the frictionless plane, it gradually slows, losing kinetic energy but gaining potential energy. The potential energy is converted back into kinetic energy once the

car comes to rest and begins to slide downward. Thus *potential energy is energy associated with the position or configuration of a mechanical system which can, at least in principle, be converted into kinetic energy or be used to do work.*

The following example shows how energy conservation can be used to solve an otherwise complex problem.

Example 6.5

A woman skis from rest down a hill 20 m high (Fig. 6.9). If friction is negligible, what is her speed at the bottom of the slope?

The forces acting on the skier are her weight and a normal force due to the ground. The effects of the weight (the gravitational force) are contained in the potential energy, and the normal force does no work because it is perpendicular to the displacement. Thus, no work is done by applied forces, and the total energy $E = K + \mathcal{U}$ is constant.

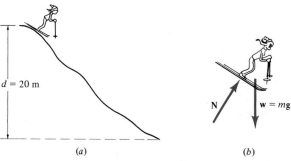

Figure 6.9. The forces on the skier are a normal force **N** and the weight $\mathbf{w} = m\mathbf{g}$.

Since we can choose the reference level for measuring potential energy as we wish, we choose the bottom of the slope as the level at which $\mathcal{U} = 0$. The kinetic energy at the top is $K_0 = 0$, since she starts from rest; her potential energy there is $\mathcal{U}_0 = mgd$. Her final kinetic energy at the bottom of the hill is $K = \frac{1}{2}mv^2$, and her final potential energy is $\mathcal{U} = 0$. Thus $K + \mathcal{U} = K_0 + \mathcal{U}_0$ becomes

$$\tfrac{1}{2}mv^2 + 0 = 0 + mgd$$

Her speed v at the bottom of the slope is then found from

$$v = \sqrt{2gd} = \sqrt{2(9.8 \text{ m s}^{-2})(20 \text{ m})}$$
$$= 19.8 \text{ m s}^{-1}$$

This example shows the advantage of the energy conservation approach to solving mechanics problems. We cannot use $\mathbf{F} = m\mathbf{a}$ directly unless we know the exact shape of the slope so that we can compute the force. Even with this information the calculation would be difficult. Energy conservation immediately tells us the velocity at any height.

Conservative Forces

The gravitational force has the interesting property that when an object moves from one point to another, the work done by the force does not depend on the choice of path. For example, in Fig. 6.10 the work done by gravity when an object moves from B to C is $-mg(h - h_0)$. No work is done by gravity when the object moves horizontally from A to B, so the total work done by gravity along the path ABC is $-mg(h - h_0)$. When the block is moved vertically from A to C, the work done by gravity again is $-mg(h - h_0)$. Therefore the work is the same for both paths.

Any force that has the property that the work it does is the same for all paths between any two given points is said to be a *conservative* force. Gravitational, electrical, and spring forces are examples of conservative forces; friction and many other forces are not conservative. The effects of *any* conservative force can always be taken into account by introducing a suitable potential energy term.

6.4 | DISSIPATIVE FORCES

We have seen that the work done by conservative forces can be handled conveniently by the introduction of the potential energy concept. We now deal with the second special category of forces, the frictional forces, which can dissipate mechanical energy.

Figure 6.10. The work done by gravity is the same for the paths *ABC* and *AC*. Whenever the work done by a force is the same for all paths, the force is said to be conservative, and its effects can be included in the potential energy.

Since the work done by a frictional force depends on the path, friction is not conservative. Also, friction always opposes the motion of an object, so it always does negative work. The energy expended by an object against the frictional forces is usually converted into heat energy and hence is lost as mechanical energy.

We can now restate the work-energy relation in a form that reflects the division of the forces into conservative, dissipative, and applied forces. If the work done by the applied forces is W_a, and the energy dissipated by friction is denoted by Q,

$$\boxed{W_a = (K - K_0) + (\mathcal{U} - \mathcal{U}_0) + Q} \quad (6.12)$$

This is the central result of this chapter:

The work done by the applied forces acting on the system is equal to its change in kinetic and potential energies plus the energy lost or dissipated.

It should be emphasized that the content of Eq. 6.12 is exactly that of the equation $W = K - K_0$, which we discussed earlier in this chapter. The difference is that we have divided the work W done by *all* the forces acting on a system into the categories of work done by conservative, dissipative, and applied forces.

The frictional force on a moving object was written earlier as the coefficient of kinetic friction times the normal force. (Since work is done only on moving objects, it is the coefficient of kinetic friction that must always be used. When an object rolls, the point of contact between the object and the supporting surface is instantaneously at rest. In this special case the frictional force does no work.) The frictional force is always directed opposite to the motion, so if an object moves a distance s against a frictional force $\mu_K N$, the energy lost is

$$Q = \mu_K N s \quad (6.13)$$

We illustrate this in the following example.

Example 6.6

The skier in Example 6.5 reaches the flat ground at the bottom of the slope with a speed of 19.8 m s^{-1} and then, by turning her skis sideways, quickly comes to a stop. If the coefficient of kinetic friction is 2.5, how far will she skid before coming to a halt?

Since the ground is level, there is no potential energy change and all her kinetic energy, $\frac{1}{2}mv^2$, must be dissipated. The normal force is equal and opposite to her weight, so the work done by the frictional force over a distance x is $Q = \mu_K mgx$. Thus

$$\frac{1}{2}mv^2 = \mu_K mgx$$

and

$$x = \frac{v^2}{2g\mu_K} = \frac{(19.8 \text{ m s}^{-1})^2}{2(9.8 \text{ m s}^{-2})(2.5)} = 8 \text{ m}$$

The coefficient of friction in this example is large because the skier is actually deforming and moving snow as she stops. Her energy is then dissipated partially as heat and partially as energy necessary to disturb the snow.

6.5 | OBSERVATIONS ON WORK AND ENERGY

In this section we discuss the general result,

$$W_a = K - K_0 + \mathcal{U} - \mathcal{U}_0 + Q \quad (6.12)$$

The total mechanical energy of an object was defined as $E = K + \mathcal{U}$, where the kinetic energy is $K = \frac{1}{2}mv^2$, and the potential energy $\mathcal{U}$ may be characterized as the energy by virtue of position. Equation 6.12 contains the change of mechanical energy from one position to another, so it can be rewritten as

$$W_a = \Delta E + Q \quad (6.14)$$

where

$$\Delta E = E - E_0 = (K + \mathcal{U}) - (K_0 + \mathcal{U}_0) \quad (6.15)$$

As we saw, *if no work is done by applied forces and no energy is lost via dissipative forces, $W_a = 0$ and $Q = 0$. Then $\Delta E = 0$ and*

$$K + \mathcal{U} = K_0 + \mathcal{U}_0 \quad (6.16)$$

This result is referred to as *mechanical energy conservation*. Under these circumstances, the total mechanical energy or the sum of potential and kinetic energies remains constant, although either may change at the expense of the other.

In many situations, dissipative forces are present which convert mechanical energy into heat, sound, or other forms of energy. The heat and noise generated by a saw or drill are examples of this. Heat represents energy transferred to the random motion of the molecules of a substance, increasing their average energy. As we will see later in Chapter Eleven, increasing the average molecular energy is equivalent to raising the temperature.

Q is positive when mechanical energy is dissipated and negative when energy is added from some external source in such a way that the mechanical energy is increased. This happens, for example, in a steam engine, where heated steam expands and does work. Similarly, sound waves cause portions of the inner ear to vibrate. Here, sound energy is transformed into mechanical energy of the structure of the ear.

Forms of Energy | Energy exists in many forms in addition to mechanical and heat energy. A hot object transmits energy to its surroundings not only by direct contact but also by the emission of electromagnetic waves that travel with the speed of light. The chemical bonds in molecules arise from electric forces. These chemical bonds can be broken or altered, causing the chemical energy to be released. For example, when a fossil fuel and oxygen are brought together at an elevated temperature, chemical changes occur that release energy. Similarly, the body uses food to synthesize molecules that later break down to provide energy as needed.

The energy associated with the forces that bind atomic nuclei together is very large. Some of this is released in a nuclear weapon or power reactor when uranium nuclei *fission* into smaller nuclei. Energy is also released when hydrogen nuclei combine or *fuse* to form a larger nucleus. Fusion supplies the energy in stars and in hydrogen bombs. In a more controlled form, fusion may supply much of our future energy if present research programs are successful. Table 6.1 lists the energies associated with various phenomena.

Conservation of Total Energy | We see that energy occurs in many forms. It has been found experimentally that energy can be changed into different forms, but it is never created or destroyed. *This is the principle of conservation of total energy.* Historically, failures of energy conservation have led to the identification of new forms of energy. Since energy

TABLE 6.1

Approximate energy associated with various events and phenomena

Description	Energy (J)
Radio energy emitted by galaxy during its lifetime	10^{55}
Rotational energy of the Milky Way	10^{52}
Energy release in a supernova explosion	10^{44}
Rotational energy of the sun	10^{36}
Rotational energy of the earth	10^{29}
Annual solar energy incident on the earth	10^{24}
Annual food requirements of the human race	10^{19}
Annual U.S. fuel energy usage	
Energy release during Krakatoan eruptions of 1883	10^{18}
Energy release of 15-megaton fusion bomb	10^{17}
Annual electrical energy output of large generating plant	10^{16}
Thermal energy in 1 ton of coal	10^{10}
Kinetic energy of large jet aircraft	10^{9}
Average daily food energy intake of a human adult	10^{7}
Kinetic energy of a home run in baseball	10^{3}
Work done by human heart per beat	0.5
Discharge of a single neuron	10^{-10}
Typical energy of a proton in a nucleus	10^{-13}
Typical energy of an electron in an atom	10^{-18}
Energy to break one bond in DNA	10^{-20}

occurs in so many forms, energy conservation is an underlying concept that unifies all of science. Although mechanical energy conservation occurs only under special circumstances, the total energy is always conserved. In this chapter, our discussion is mainly limited to mechanical energy, but the concepts developed here will appear often in the remainder of the book.

6.6 | SOLVING PROBLEMS USING WORK AND ENERGY

We have used simple examples to infer a general result

$$W_a = \Delta E + Q$$

The following examples illustrate the procedure for using this result. In each case we follow a series of steps to solve a problem:

1 We draw a force diagram showing all the forces acting on the object of interest.
2 We identify the applied forces that do work, the forces that dissipate energy, and the forces, such as the gravitational force, that may be included in the potential energy.
3 We calculate the work done by applied forces W_a, the energy dissipated Q, and the change in energy ΔE between two points in the motion. These terms are then related by $W_a = \Delta E + Q$.

Example 6.7

Two identical balls of mass m are thrown from a window a height h above the ground. The initial speed of each ball is v_0 but they are thrown in different directions (Fig. 6.11.). What is the speed of each ball as it strikes the ground? (Neglect air resistance.)

Each ball is subjected only to the gravitational force $m\mathbf{g}$ after it is released. The gravitational force is conservative and is included in the potential energy. No applied forces act, so $W_a = 0$, and in the absence of frictional forces, $Q = 0$ also. Thus, for each ball, $\Delta E = 0$. We choose the ground as the reference level for computing the potential energy so initially $\mathfrak{U}_0 = mgh$ for each. When each strikes the ground $\mathfrak{U} = 0$ and the initial kinetic energy of each ball is $K_0 = \frac{1}{2}mv_0^2$, so $\Delta E = 0$ becomes

$$0 = K - K_0 + \mathfrak{U} - \mathfrak{U}_0$$

or

$$0 = K - \tfrac{1}{2}mv_0^2 + 0 - mgh$$

Thus $K = \frac{1}{2}mv^2$ is also the same for each ball,

$$\tfrac{1}{2}mv^2 = \tfrac{1}{2}mv_0^2 + mgh$$

or

$$v = \sqrt{v_0^2 + 2gh}$$

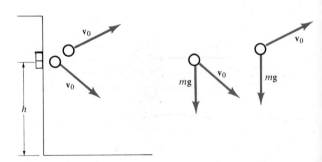

Figure 6.11. Two identical balls thrown in different directions with the same initial speed. The force on each is $m\mathbf{g}$ downward.

Thus, while the two balls do not strike the ground at the same time, they do have the same speed when they hit.

F

mg

N

Figure 6.12. The forces acting on the skier in Example 6.8.

Example 6.8

A 40-kg girl is pulled to the top of a ski slope by a rope tow. The top of the tow is higher than the bottom by 100 m. She is initially at rest, and she gets off the tow moving at 1 m s^{-1}. If the friction is negligible, how much work does the tow do on the girl? (See Fig. 6.12.)

Taking the bottom of the tow as our reference point for the potential energy, her initial energy is $E_0 = K_0 + \mathcal{U}_0 = 0$. When she has risen a height h and has a velocity v, her energy is

$$E = K + \mathcal{U} = \tfrac{1}{2}mv^2 + mgh$$

With $Q = 0$, the work W_a done by the rope tow is

$$
\begin{aligned}
W_a &= E - E_0 = \tfrac{1}{2}mv^2 + mgh \\
&= \tfrac{1}{2}(40\,\text{kg})(1\,\text{m s}^{-1})^2 + (40\,\text{kg})(9.8\,\text{m s}^{-2})(100\,\text{m}) \\
&= 20\,\text{J} + 39200\,\text{J} = 39220\,\text{J}
\end{aligned}
$$

Note that most of the work of the tow goes into raising the skier. Very little is needed to increase her kinetic energy.

Example 6.9

In the pole vault, an athlete uses a pole to convert the kinetic energy of running into potential energy when the pole is vertical (Fig. 6.13). A good sprinter runs at a speed of 10 m s^{-1}. Disregarding the additional height the athlete gains by using his arms to raise his center of mass well above the position of his hands on the pole, how high can the athlete raise his center of gravity?

Initially, his C.G. is about 1 m above the ground. Just before the athlete begins to use the pole, $\mathcal{U}_0 = 0$ and $K_0 = \tfrac{1}{2}mv^2$, where $v = 10$ m s^{-1}. At the top of an ideal jump, $v = 0$, so $K = 0$ and $\mathcal{U} = mgh$, where h is the height of his center of gravity above its initial position. Since only the force of gravity acts on the airborne jumper, $W_a = Q = 0$ and

$$mgh = \tfrac{1}{2}mv^2$$

Then

$$h = \frac{v^2}{2g} = \frac{(10\,\text{m s}^{-1})^2}{2(9.8\,\text{m s}^{-2})} = 5.1\,\text{m}$$

His center of gravity is then about 6.1 m above the ground. The present world pole vault record is near 5.5 m.

6.7 | GRAVITATIONAL POTENTIAL ENERGY

We showed in Section 6.3 that the constant gravitational force on an object near the surface of the earth is conservative. This means that the work it does is not dependent on the path between two points and that the effects of the force can be included in the potential energy. It can also be shown that the more general gravitational force between two spheres with masses m and M, respectively, is conservative. This force (see Chapter Three) is given by

$$\mathbf{F} = -\frac{GmM}{r^2}\hat{\mathbf{r}} \qquad (6.17)$$

Figure 6.13. Photo of a pole vaulter at equally spaced time intervals. Note that the vaulter is nearly at rest at the top of the jump where all his kinetic energy has been converted to potential energy. (Dr. Harold Edgerton, MIT, Cambridge, Ma.)

where r is the distance between their centers. We also show in Section 6.12 that if the potential energy is chosen to be zero when the masses are infinitely far apart, then the gravitational potential energy of the two masses at a separation r is,

$$\mathfrak{U} = -\frac{GmM}{r} \qquad (6.18)$$

$\mathfrak{U}$ is negative, since work must be done to increase the separation of the masses to infinity where $\mathfrak{U} = 0$. This result must be used whenever the distance of an object above the earth's surface is a significant fraction of the earth's radius. Its use is illustrated in the following example.

Example 6.10

Find the kinetic and total mechanical energies of an earth satellite of mass m in a circular orbit of radius r.

The kinetic energy is $K = \frac{1}{2}mv^2$, and the potential energy is $\mathfrak{U} = -GM_E m/r$. To find K in terms of G, M_E, m, and r, we use Newton's second law $\mathbf{F} = m\mathbf{a}$ for a circular orbit,

$$\frac{GM_E m}{r^2} = \frac{mv^2}{r}$$

or

$$mv^2 = \frac{GM_E m}{r}$$

Thus the kinetic energy can be written as

$$K = \frac{1}{2}mv^2 = \frac{GM_E m}{2r} = -\frac{1}{2}\mathfrak{U}$$

Since $\mathfrak{U}$ is negative, K is positive and half as large as $\mathfrak{U}$ in magnitude. The total energy is

$$E = K + \mathfrak{U} = -\tfrac{1}{2}\mathfrak{U} + \mathfrak{U} = \tfrac{1}{2}\mathfrak{U}$$

$$= -\frac{GM_E m}{2r}.$$

Example 6.11

A projectile is fired vertically from the surface of the earth. What is the minimum velocity required for the projectile to escape the earth's gravitation force? (Neglect the gravitational effects of the sun and the other planets.)

At the surface of the earth the velocity is v_0, so

$$E_0 = K_0 + \mathfrak{U}_0 = \frac{1}{2}mv_0{}^2 - \frac{GM_E m}{R_E}.$$

If the projectile is to escape permanently from the earth, it will eventually reach a very large value of r, so that

$\mathfrak{U} = 0$. If it has the minimum energy needed to do this, its velocity and kinetic energy will also be zero at this distance. Thus the minimum total energy needed to escape the earth is $E = K + \mathfrak{U} = 0$. Since the mechanical energy is conserved

$$\frac{1}{2}mv_0{}^2 - \frac{GM_E m}{R_E} = 0$$

Then

$$v_0 = \sqrt{\frac{2GM_E}{R_E}}$$

This is the minimum velocity necessary at the earth's surface to escape from the earth.

We can simplify the result somewhat by noticing that the weight of an object of mass m at the surface of the earth is $mg = GM_E m/R_E{}^2$. Using this, we have

$$v_0 = \sqrt{\frac{2GM_E}{R_E}} = \sqrt{2gR_E}$$

The escape velocity for any planet can be found from its gravitational acceleration and radius. For the earth, $g = 9.8 \text{ m s}^{-2}$ and $R_E = 6.4 \times 10^6 \text{ m}$, so the escape velocity is

$$v_0 = \sqrt{2(9.8 \text{ m s}^{-2})(6.4 \times 10^6 \text{ m})}$$
$$= 1.12 \times 10^4 \text{ m s}^{-1}$$

Earlier in this chapter, we used $\mathfrak{U} = mgh$ for the gravitational potential energy, while we have used $\mathfrak{U} = -GM_E m/r$ here. It is left as a problem to show that if an object is at a small height h above the earth's surface, the potential energy is

$$\mathfrak{U} = -\frac{GM_E m}{(R_E + h)} \simeq -\frac{GM_E m}{R_E}\left(1 - \frac{h}{R_E}\right) (h \ll R_E)$$

(The symbol $\simeq$ means approximately equal.)

Using $g = GM_E/R_E{}^2$, this becomes

$$\mathfrak{U} = -mgR_E + mgh$$

Since we always discuss potential energy differences, the constant term $-mgR_E$ drops out of all calculations. Thus, as long as we are near the earth's surface, we can use $\mathfrak{U} = mgh$.

6.8 | THE ELECTRICAL POTENTIAL ENERGY

The gravitational force is not the only conservative force that we encounter. The electric force is also conservative. In Chapter Five, we saw that the force on a charge q due to a second charge Q a distance r away is

$$\mathbf{F} = k\frac{qQ}{r^2}\hat{\mathbf{r}} \qquad (6.19)$$

Here $k = 9 \times 10^9$ N m^2 C^{-2}. Except for a minus sign and the fact that qQ can be positive or negative, this is identical in mathematical form to the gravitational force law. Then we can immediately say that *the electrical force is conservative*. If we replace $-GmM$ by kqQ in the gravitational potential energy formula, we obtain an expression for the potential energy of a charge q in the presence of a second charge Q,

$$\mathfrak{U} = \frac{kqQ}{r} \qquad (6.20)$$

Because the electric force is conservative the change in potential energy if the charge q is moved from r_1 to r_2 is the same *no matter what path is taken*.

When discussing electrical energies, it is convenient to define the *electric potential V* as the potential energy per unit charge,

$$V = \frac{\mathfrak{U}}{q} \qquad (6.21)$$

This is a general definition. When the potential energy results from a single point charge Q, the electric potential is

$$V = k\frac{Q}{r} \qquad (6.22)$$

The S.I. unit of electric potential is called the *volt* (V); $1 \text{ V} = 1 \text{ J C}^{-1}$.

The contribution of the electrical energy to the total energy is illustrated in the next example.

Example 6.12

In a hydrogen atom the electron orbits about a proton under the influence of the electric force. If the orbit is circular with radius $a_0 = 5.3 \times 10^{-11}$ m, what is the energy of the electron? (The electron and proton charges are $-e = -1.60 \times 10^{-19}$ C and $+e = +1.60 \times 10^{-19}$ C respectively.)

In Example 6.10 the energy of an orbiting satellite due to gravitational forces was calculated to be $E = -GM_E m/2r$. The result holds true for this problem also if we replace $GM_E m$ by $(-kqQ)$. Then $E = +kqQ/2r$. With $q = -e$, $Q = e$, and $r = a_0$, we have the energy of the electron,

$$E = -\frac{1}{2}\frac{ke^2}{a_0}$$

Using $k = 9 \times 10^9$ N m^2 C^{-2}, this is equal to

$$E = -\frac{1}{2}\frac{(9 \times 10^9 \text{ N m}^2 \text{ C}^{-2})(1.6 \times 10^{-19} \text{ C})^2}{5.3 \times 10^{-11} \text{ m}}$$
$$= -2.17 \times 10^{-18} \text{ J}$$

Often, energies of electrons or other atomic particles are most conveniently expressed in units of *electron volts* (eV). An electron volt is the kinetic energy acquired when a charge, e, is accelerated by a potential difference of one volt,

$$1 \text{ eV} = (1.60 \times 10^{-19} \text{ C})(1 \text{ V})$$
$$= 1.60 \times 10^{-19} \text{ J}$$

In these units, the energy of the electron in the hydrogen atom is -13.6 eV.

Example 6.13

In the cathode tube of an oscilloscope or a television picture tube, electrons are accelerated from rest through a potential difference of 20,000 V. What is their velocity? (The electron mass is 9.11×10^{-31} kg.)

From the definition of the electric potential, the *change* in potential energy is $\Delta\mathfrak{U} = e\,\Delta V$ where $\Delta V = 20,000$ V. Then from energy conservation,

$$\tfrac{1}{2}mv^2 = e\,\Delta V$$

Hence,

$$v = \sqrt{\frac{2e\,\Delta V}{m}}$$
$$= \sqrt{\frac{2(1.60 \times 10^{-19} \text{ C})(20,000 \text{ V})}{9.11 \times 10^{-31} \text{ kg}}}$$
$$= 8.38 \times 10^7 \text{ m s}^{-1}$$

Note that, as in gravitational problems, we can find the final velocity from the potential energy difference alone and need not know anything about the path taken in the tube.

6.9 | POWER

We are often less concerned with the net amount of work done than with the rate at which it is done. For example, we may remove snow from a driveway either with a hand shovel or with a snowplow. The same work is done, but the plow does it much faster. We say that the plow is more powerful. In this section, we describe the relationship between work and power.

111

When an amount of work ΔW is done in a time Δt, the *average power* is defined as the average rate of doing work,

$$\bar{\mathcal{P}} = \frac{\Delta W}{\Delta t} \qquad (6.23)$$

The instantaneous power $\mathcal{P}$ is found by considering smaller and smaller time intervals, so

$$\mathcal{P} = \frac{dW}{dt} \qquad (6.24)$$

From this definition, we see that the S.I. unit of power is a joule per second, which is called a *watt* (W). In the British system the unit of power is the foot-pound per second; an alternative unit is the *horsepower* (hp), where

$$1 \text{ hp} = 550 \text{ ft lb s}^{-1} = 746 \text{ W}$$

The work done by a force $\mathbf{F}$ acting through a small displacement $\Delta \mathbf{s}$ in a short time Δt is $\Delta W = F_s \Delta s$. Dividing by Δt gives the power,

$$\mathcal{P} = F_s \frac{\Delta s}{\Delta t}$$

Since $\Delta s / \Delta t$ is the velocity, the power is also given by,

$$\mathcal{P} = F_s v \qquad (6.25)$$

Thus the power is the force times the velocity. The relation between work and power is seen in the next example.

Example 6.14

A 70-kg man runs up a flight of stairs 3 m high in 2 s. (a) How much work does he do? (b) What is his average power output?

(a) The work done, ΔW, is equal to his change in potential energy, mgh. Thus

$$\Delta W = mgh = (70 \text{ kg})(9.8 \text{ m s}^{-2})(3 \text{ m})$$
$$= 2060 \text{ J}$$

(b) His average power is the work done divided by the time,

$$\bar{\mathcal{P}} = \frac{\Delta W}{\Delta t} = \frac{2060 \text{ J}}{2 \text{ s}} = 1030 \text{ W}$$

Example 6.15

A 500-lb piano is raised by a hoist at a constant velocity of 2 ft s^{-1}. What is the power expended by the hoist in horsepower?

Since the force is parallel to the velocity,

$$\mathcal{P} = Fv = (500 \text{ lb})(2 \text{ ft s}^{-1}) = (1000 \text{ ft lb s}^{-1})$$
$$= (1000 \text{ ft lb s}^{-1}) \left(\frac{1 \text{ hp}}{550 \text{ ft lb s}^{-1}} \right)$$
$$= 1.82 \text{ hp}$$

6.10 | JUMPING; SCALING LAWS IN PHYSIOLOGY

In Chapter 1, we related the heights reached by jumping animals to the velocities and accelerations involved. Here we describe a procedure called *scaling*, which permits one to draw some general conclusions about jumping. We begin by reanalyzing the vertical jump of a human in terms of work and energy.

Figure 6.14 shows a man jumping. Initially he crouches, lowering his center of gravity a distance d called the acceleration distance. We shall measure potential energy from this reference level. As he accelerates and straightens, he does work to raise his potential and kinetic energies. At the instant of takeoff, his potential energy is $\mathcal{U}_0 = mgd$. If his upward velocity at this instant is v_0, his kinetic energy is $\frac{1}{2}mv_0^2$. Thus, in reaching the erect takeoff position, he has done work W where

$$W = mgd + \tfrac{1}{2}mv_0^2$$

From takeoff to the top of the jump, the only force acting on the man is his weight. Thus, during the airborne part of the jump, his mechanical energy is constant. At the top of the jump, $K = 0$ and $\mathcal{U} = mg(h + d)$. This must equal the energy at takeoff so

$$W = mgd + \tfrac{1}{2}mv_0^2 = mg(h + d). \qquad (6.26)$$

Thus the total energy that the human must supply to make the jump is $W = mg(h + d)$. In our subsequent discussion, we neglect d compared to h and use

$$W \simeq mgh$$

This is only a fair assumption for humans, since $d/h \simeq \frac{1}{2}$, but it is a good approximation for smaller animals. Notice that Eq. 6.26 also gives us the takeoff velocity

$$v_0 = \sqrt{2gh}$$

Having found the total energy in terms of the height reached in the jump, we might hope to determine the energy expended by various muscles and

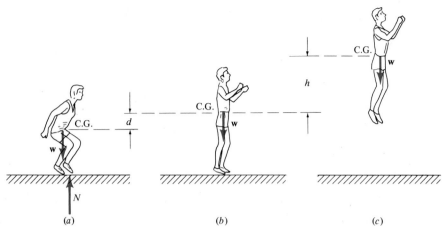

Figure 6.14. (*a*) A man preparing to jump. The forces on the man are his weight **w** down and the normal force **N** of the ground acting upward. (*b*) The man at takeoff. (*c*) During the airborne phase the only force on the man is **w.** The total change in the position of his center of gravity during the jump is $d + h$.

compare these numbers for various animals. Unfortunately, the large number of bones and muscles involved makes such an analysis very complicated, a difficulty characteristic of many biological problems. However, we can make some progress in comparing the jumping abilities of different animals with the aid of a procedure called *scaling*. Scaling assumes that the basic characteristics of a biological system change or *scale* with its overall size in a simple but plausible way. This procedure, which can be used in many physiological problems, leads to predicted *scaling laws* for the jumping ability versus the size of an animal, which can be compared to the data.

The simplest scaling model assumes that the mass m of an animal is proportional to its volume and that this volume is proportional to the cube of some characteristic length l. Thus the scaling assumption is $m = cl^3$ when c is a constant. For example, the ratio of the characteristic lengths of a 0.02-kg mouse and a 700-kg cow would be

$$\frac{l_{\text{cow}}}{l_{\text{mouse}}} = \left(\frac{m_{\text{cow}}}{m_{\text{mouse}}}\right)^{1/3} = \left(\frac{700 \text{ kg}}{0.02 \text{ kg}}\right)^{1/3} = 32.7$$

This may, for example, represent the relative lengths of the legs.

Using this length, we note that the volume of an animal or any of its organs is proportional to l^3, the surface area of the body and the cross-sectional areas of the muscles are proportional to l^2, and the length of the limbs is proportional to l.

Comparing the jumping ability of animals we notice the interesting fact that the heights of the jumps are not so very different for animals that are extremely different in size. Thus the rat kangaroo, about the size of a rabbit, jumps nearly as high as a large kangaroo. Locusts and fleas jump to roughly the same height. We may ask, what characteristics do they share that would explain this result?

Two possibilities come immediately to mind.

1   The energy supplied per unit mass of muscle is the same for all animals.
2   The power supplied per unit mass of muscle is the same for all animals.

We examine the first assumption here, leaving the second as a problem (Problem 6–63). The first assumption says that an animal of mass m should be able to do an amount of work W proportional to its mass, $W \propto l^3$. But we saw that the work done during a jump of height h is $W \simeq mgh$, so $mgh \propto l^3$. Since $m \propto l^3$, we also have $l^3gh \propto l^3$ or $h = $ constant. Thus h does not depend on l.

This prediction, that the height reached is independent of the size of an animal, is roughly correct experimentally. By contrast, the predictions obtained with the second hypothesis disagree with the data. Thus, we may tentatively conclude that the *energy expended per unit mass is about the same for all animals of the same general type.* Another conclusion follows by noting that the takeoff velocity $v_0 = \sqrt{2gh}$

113

is roughly independent of size, the takeoff time $t = d/\bar{v} = d/(\frac{1}{2}v_0)$ scales with l. Hence the average power expended per unit mass, which is the work done per unit mass, $W/m \simeq gh$, divided by the time, must vary as $1/l$. This predicts that larger animals expend energy at a slower rate.

Comparisons between mammals and insects are complicated by the different ways they use their leg muscles. Mammals use muscle contractions directly, but insects employ a catapult arrangement. For example, a flea has an elastic material called *resilin* in its knee joint. The flea gradually bends its leg, stretching the resilin, and locks the knee in position. In a jump, the knee is unlocked, and the resilin contracts rapidly, straightening the leg. Thus, insects employ stored *elastic* potential energy and use their muscles somewhat indirectly.

The discussion of scaling laws as applied to jumping shows how interesting qualitative insights can sometimes be obtained in complex biological systems. The model we have used assumed that all body dimensions scale in exactly the same way with the mass of the animal. However, it is possible to make other scaling assumptions. Another approach to scaling is described in Chapter Nine.

SUMMARY

Many problems can be most easily solved using the concepts of work and energy. The relationship between these quantities is contained in the two most important statements of this chapter.

1 The total work W done by *all* forces acting on an object equals the change in kinetic energy of the object, $K - K_0$:

$$W = K - K_0$$

In this equation, $W = F_s s$, where s is the displacement of the object, and F_s is the component of the net force along the direction of the displacement. $K_0 = \frac{1}{2}mv_0^2$ and $K = \frac{1}{2}mv^2$ are the kinetic energies of the object before and after the displacement, respectively.

2 If the total work W done is categorized according to the type of force doing the work, the general relationship has the more useful form

$$W_a = (K - K_0) + (\mathfrak{U} - \mathfrak{U}_0) + Q$$

W_a is the work done by the applied forces. Q is the energy lost as a result of dissipative or frictional forces. $\mathfrak{U}_0$ and $\mathfrak{U}$ are the initial and final potential energies, and K_0 and K are the corresponding kinetic energies.

The gravitational and electric forces are conservative forces; the work done by these forces does not depend on the path. Their effects can always be included in the potential energy.

Under the special but not uncommon condition that there are no applied forces and no dissipative forces, both $W_a = 0$ and $Q = 0$, and we say that the mechanical energy is conserved:

$$K + \mathfrak{U} = K_0 + \mathfrak{U}_0$$

When mechanical energy is conserved, $E = K + \mathfrak{U}$ is constant. K and $\mathfrak{U}$ may each change, but the sum must remain constant.

The power associated with any force is the rate at which work is done by that force.

Checklist

Define or explain:

work
joule
kinetic energy
potential energy
conservative force
total mechanical energy
dissipative force
applied force

mechanical energy
   conservation
gravitational potential
   energy
electric potential
volt
electron volt
power
watt

EXERCISES

Section 6.1 | Work

6-1   A child pulls a toy car with a 10-N force at a 20° angle to the horizontal (Fig. 6.15). If the car moves a distance of 6 m, how much work does the child do?

Figure 6.15. Exercise 6.1.

6-2   A woman pushes horizontally on a chair with a force of 300 N. Compute the work done on the chair if (a) it moves 2 m parallel to the force; (b) it moves 1 m opposite to the force; (c) the chair does not move.

6-3   A woman exerts a constant horizontal force of 200 N on a child and a tricycle. The tricycle moves 2 m, and the work done by the woman is 100 J. What is the angle between the force and the displacement of the tricycle?

6-4   A motorcycle comes to a skidding stop in 5 m. During stopping, the force on the cycle due to the road is 200 N and is directly opposed to the motion. (a) How much work does the road do on the cycle? (b) How much work does the cycle do on the road?

6-5   A girl pulls a box weighing 40 N a distance of 10 m across the floor at constant speed. How much work does she do if the coefficient of sliding friction is 0.2?

Section 6.2 | Kinetic Energy

6-6   A 1000-kg automobile has a velocity of 40 km h^{-1}. What is its kinetic energy?

6-7   What is the kinetic energy of a 0.25-kg stone moving at 10 m s^{-1}?

6-8   A baseball of mass 0.15 kg is thrown at 30 m s^{-1}. (a) What is its kinetic energy? (b) If it is thrown by a man who exerts a constant force over a distance of 1.5 m, what force does he exert?

6-9   A 200-lb man is in a car traveling at 60 mi h^{-1} = 88 ft s^{-1}. (a) Find his kinetic energy. (b) The car strikes a concrete wall and comes to rest after the front of the car has collapsed 3 ft. The man is wearing a seat belt and harness. What is the average force exerted by the belt and harness during the crash?

6-10   Show that kinetic energy has dimensions of force times distance.

6-11   A 200-kg swordfish swimming at 5 m s^{-1} rams a wooden yacht which is at rest. Its sword penetrates the yacht, and the fish is stopped in 1 m. (a) What is the initial kinetic energy of the fish? (b) How much work is done by the fish?

6-12   Fishing line is usually sold as being capable of withstanding a certain force. What strength line is necessary to handle a 10-lb salmon swimming at 10 ft s^{-1} if it is stopped in 0.5 ft?

6-13   A baseball is thrown from center field to second base, and its velocity diminishes from 20 m s^{-1} to 15 m s^{-1}. If its mass is 0.15 kg, how much energy was lost due to air resistance? (Assume the initial and final heights are the same.)

Section 6.3 | Potential Energy and Conservative Forces

6-14   A swimmer dives off a board into a pool, swims to the edge, and climbs back onto the board. Identify and discuss the kinds of forces present and the work they do.

6-15   Mountain roads usually wind back and forth up the mountain side rather than going straight up. Why?

6-16   A child on a swing reaches a maximum height of 6 ft above its lowest position. What is the speed of the swing at the lowest point? (Neglect frictional forces.)

6-17   A horizontal force **F** applied to a block of mass m pushes it at constant speed a distance l along a plane at an angle θ to the horizontal (Fig. 6.16). If the plane is frictionless, how much work is done by the force **F**?

Figure 6.16.   Exercise 6-17.   **Figure 6.17.**   Exercise 6-21.

6-18   A beer can is dropped from a window 100 ft above the ground. How fast will it be moving when it lands? (Neglect air resistance.)

6-19   A car moving at 82 mi h^{-1} = 120 ft s^{-1} hits a concrete wall. From what height would a car have to be dropped onto concrete to achieve the same result?

6-20   A boy sits on a ferris wheel. How much work is done on him by gravitational forces when the wheel makes one complete turn?

6-21   A 4.5-N salmon swims a distance of 5 m up a fish pass at constant velocity. It experiences a 1.3-N dissipative force from the water. The fish rises a distance of 0.5 m while swimming up the pass. (Fig. 6.17). (a) How much work must the fish do to overcome the dissipative force? (b) What is the change in the fish's potential energy? (c) How much total work does the fish do in swimming up the pass?

6-22   A baseball thrown straight up reaches a height of 50 m. What was its initial velocity? (Neglect air resistance.)

Figure 6.18.   Water going over a dam. It has zero velocity at the top and drops a distance h. Exercise 6-23.

6-23   Water flows over a dam and falls a distance h in a smooth stream, as shown in Fig. 6.18. Assume that the water has zero velocity at the top of the falls. What is the velocity at the base of the falls?

Section 6.4 | Dissipative Forces; Energy Loss

6-24   A hockey puck with an initial velocity of 4 m s^{-1} slides on ice. The coefficient of kinetic friction is 0.1. How far will the puck slide before stopping?

6-25   A box with an initial speed of 3 m s^{-1} slides to rest on a horizontal floor in 0.5 m. What is the coefficient of kinetic friction?

6-26   At some amusement parks, one can slide down a ramp, as shown in Fig. 6.19. (a) What is the velocity at the bottom? (Assume the ramp is frictionless.) (b) What distance l is necessary to stop if the coefficient of friction at the bottom is 0.5?

6-27   A sled slides 100 m down a hill that slopes at an angle of 30° with the horizontal direction. The sled attains a final velocity of 20 m s^{-1} at the base of the slope. What fraction of the energy was lost due to friction?

Figure 6.19.   Profile of the amusement park slide. l is the stopping distance. Exercise 6-26.

Section 6.5 | Observations on Work and Energy and Section 6.6 | Solving Problems Using Work and Energy

6-28   A 50-kg woman climbs a mountain 3000 m high. (a) How much work does she do? (b) A kilogram of fat supplies about 3.8×10^7 J of energy. If she converts fat into mechanical energy with a 20 percent efficiency rate, how much fat will she consume in the climb?

6-29   A dieter lifts a 10-kg mass a distance of 0.5 m 1000 times. (a) How much work does he do? (b) Fat supplies 3.8×10^7 J of energy per kilogram, which is converted to mechanical energy with a 20 percent efficiency rate. How much fat will the dieter use up?

6-30   A man lifts a 20-kg mass 0.5 m vertically. (a) How much work does he do? (b) When he lowers the mass to its original height, how much work is done on the man? (c) What happens to this energy?

Section 6.7 | Gravitational Potential Energy

6-31   How much energy is required to put a spaceship of mass m in orbit around the earth at a distance from the surface equal to the earth's radius?

6-32   The average center-to-center distance from the earth to the moon is 3.84×10^5 km. What is the smallest distance from the center of the earth at which the gravitational potential energies of the earth and moon are equal? The mass of the moon is approximately $1.2 \times 10^{-2} M_E$.

6-33   What is the escape velocity from the moon? (Use the data of Appendix D.2).

Section 6.8 | The Electrical Potential Energy

6-34   A proton is fired directly toward a second proton, which is held stationary. The velocity of the incident proton is 2×10^6 m s^{-1}. How close will the protons be when they both are at rest?

6-35   A charge q is placed between two charged plates that have a 10 V potential difference as shown in Fig. 6.20. (a) Which way will a positive charge accelerate? Explain. (b) Which way will a negative charge accelerate?

6-36   A beam of electrons with velocity v is injected into a region between two plates (Fig. 6.21). If the electrons reach the second plate with zero velocity, what is the potential difference between the two plates? Which is at the higher potential?

6-37 According to Bohr's theory of the hydrogen atom, the only possible radii for the electron are a_0, $2a_0$, $3a_0$, ... where $a_0 = 5.3 \times 10^{-11}$ m. How much energy in electron volts must be supplied to make the transition from the orbit of radius a_0 to the orbit with radius na_0, $n = 2, 3, \ldots$? (This energy can be supplied by electromagnetic radiation.)

Section 6.9 | Power

6-38 Two teams of students pull a rope in a tug of war. Team A is gaining over B, since the rope is moving in its direction at a constant rate of 0.01 m s^{-1}. The tension in the rope is 4000 N. How much power is expended by team A?

6-39 If a man can do mechanical work at a rate of 8 W kg^{-1} of body mass during sustained activity, how fast can he run up stairs?

6-40 A girl of mass 40 kg climbs a rope 8 m long at constant speed in 15 s. What power does she expend during the climb?

6-41 An elevator motor produces 2000 W of power. How fast can it lift a 1000-kg load?

6-42 Cruising speeds for fish 0.3 m long are about 0.35 m s^{-1}. The average power expended is about 4.5 W kg^{-1} of body mass. Assume the fish has a mass of 0.4 kg. (a) What is the average power expended by the cruising fish? (b) What is the average force exerted by the fish on the water? (c) How much work does the fish do in 10 minutes?

6-43 A motor operating a water pump can produce 1000 W of power. If the kinetic energy change is negligible, how many kilograms of water per second can it pump from a well 20 m deep?

6-44 Water is pumped from a well 200 m below the surface. (a) How much work must the motor do

to raise 1 kg of water? (Assume the kinetic energy change is negligible.) (b) If the pump delivers 10 kg min^{-1}, what is its power output?

6-45 How much does it cost to keep a 100-W bulb running per 24-hour day if electrical energy costs 5 cents per kilowatt-hour (kWh)? (1 kWh = 1000 W $\times$ 1 hour.)

Section 6.10 | Jumping; Scaling Laws in Physiology

6-46 Using the data in Table 1.5, calculate the work and power per unit mass in jumping for (a) a kangaroo, and (b) a locust.

6-47 Which of the animals listed in Table 1.5 does the most work per unit mass in jumping vertically? Which does the least?

6-48 Show that if the rate at which oxygen is absorbed by an animal and supplied to its tissues varies as the surface area of the arteries, then its power output per unit mass should vary as l^{-1}. What does this suggest about the pulse rate?

6-49 How does the rate of heat loss per unit body mass in a cold environment vary with the characteristic length l of an animal? (The rate of heat loss is proportional to the surface area.)

PROBLEMS

*6-50 If a salmon encounters a falls when swimming upstream, it will attempt to ascend the falls in one of two ways. If it can swim fast enough, it will swim up the falls. If it cannot, it will jump from the base of the falls into the falls at a height where the water is moving slowly enough so that it can swim up to the top.

Assume that the salmon has a maximum swimming speed of 5 m s^{-1} in still water and that the water at the top of the falls and at the base of the falls is at rest. (a) What is the maximum height of a falls that the salmon can swim up without jumping? (b) If the falls is 1 m high, what is the velocity of the fish with respect to the ground when it begins to swim upward at the bottom of the falling stream? (c) If the falls is 2 m high, what is the minimum height in the stream to which the fish must jump in order to swim up the rest of the way? (d) In order to jump the distance necessary in part (c), what must the fish's initial velocity be when it leaves the water?

Figure 6.20. Exercise 6-35.   **Figure 6.21.** Exercise 6-36.

*6-51   On a certain river, a falls 6 m high is encountered by the salmon of Problem 6-50. To facilitate the progress of the salmon upstream, a fish pass is constructed. It is a series of sloping flow passages with connecting resting pools where the water velocity is virtually zero. If we refer to each pool with its flow passage as a step, what is the minimum number of steps necessary to allow the salmon to ascend the falls without jumping?

*6-52   The potential energy of an object of mass m at a height h above the earth's surface is $\mathcal{U} = -GM_Em/(R_E + h)$. R_E is the earth's radius and M_E its mass. Using Eq. B.15 in Appendix B, show for $h \ll R_E$ that

$$\mathcal{U} \simeq -\frac{GM_Em}{R_E}\left(1 - \frac{h}{R_E}\right).$$

*6-53   Show that for a rocket fired in the vicinity of the earth, the minimum velocity to escape the solar system is $v_s = \sqrt{2}C/T$, where C is the circumference of the earth's orbit and T is 1 year, the period of the earth's rotation about the sun. (Neglect the energy needed to escape the earth.)

6-54   Using the formula in the preceding problem, compare the escape velocities needed to escape the earth and to escape the solar system.

6-55   Assuming all planets have the same density or mass to volume ratio, show how the escape velocity varies with the radius of the planet.

6-56   Water is sometimes pumped up to a reservoir at times of low demand for electric power and used to drive turbines during peak demand periods. Assuming 80 percent efficiency, how much water is needed to produce 100 megawatts (MW) of power for 2 hours if the water falls 40 m? (1 MW = 10^6 W.)

6-57   The engine delivers 200 hp to the propeller of a single-engine plane cruising at a constant 170 mi h^{-1} in level flight. What is the drag force due to the air?

6-58   The Horseshoe Falls on the Niagara River in Canada is about 50 m high and 800 m wide. The water is moving at 10 m s^{-1} and has a depth of 1 m as it goes over the falls. (a) What volume of water goes over the falls every second? (b) What is the change in potential energy of this volume of water? (One cubic meter of water has a mass of 10^3 kg.) (c) If this potential energy could be converted di-

rectly into electrical energy, how much electrical power would be produced? (d) The total electrical power-generating capacity in the United States is about 5×10^{11} W. What percentage of this power could be produced by harnessing the Horseshoe falls with 80 percent efficiency?

6-59   An automobile engine delivers a maximum usable power of 110 hp. Neglecting dissipative forces, if the car weighs 3000 lb, how fast can it travel uphill at constant speed if the road is at an angle of 6° with the horizontal?

6-60   The muscle force is proportional to the cross-sectional area of the muscle. Show that if the speed of contraction is constant, the power per kilogram is proportional to l^{-1}.

*6-61   Show that if the power output of an animal of length l varies as l^2, the speed with which it can run uphill varies as l^{-1}.

6-62   Show that if the consumption of oxygen for a mammal of characteristic length l varies as l^2, the time it can remain under water without breathing varies as l.

6-63   Show that if the power expended per unit mass of muscle were the same for all animals of the same type, the height jumped would vary as $l^{2/3}$, where l is the characteristic size of the animal.

SUPPLEMENTARY TOPICS

6.11 | RUNNING

In our discussion of jumping, we assumed that all the work done by the muscles is converted into mechanical energy and that no energy is dissipated. This cannot be true for an animal running at constant speed on flat ground. In this case the mechanical energy remains constant. However, in each stride, energy is expended by the muscles in accelerating the legs and in lifting the center of gravity of the body. This energy is dissipated as the legs slow down and the center of gravity falls. A variety of dissipative forces are involved, making any model that relies on a direct analysis based on Newton's laws very difficult to use for running.

Another kind of model proposed recently by a mathematician, J.B. Keller, sidesteps the details of the forces. The runner is assumed to exert a net force f which varies with the time but never exceeds a maximum force, F_{max}. The power expended by the runner is the force times the velocity, fv. A dissipative force, D, opposes the motion and is assumed to be proportional to the velocity, so $D = cv$. The runner is able to draw on an initial reserve of stored energy E_0, but once that is exhausted the speed is limited by the rate σ (assumed constant) at which additional energy is supplied by metabolic processes.

Given these assumptions, the mathematical problem is to find the force f that the runner must exert to run a given race as fast as possible. The four physiological parameters of this model—the quantities F_{max}, c, E_0, and σ—are chosen by comparing the calculated times with the men's world records for several distances. If the model then predicts many other records correctly, it can be considered successful, and its physiological parameters may be of use in studying other activities.

Keller's solution to this problem states that for races shorter than 291 m, a runner should exert the maximum force F_{max} throughout the race. In longer events, he should accelerate as fast as possible for a second or two and then run at constant speed until he slows slightly in the last few steps. These results are consistent with the usual advice to maintain a steady pace. However, runners often finish long races with a burst of speed rather than by slowing down. The competitive aspects of racing may be the reason for this.

Table 6.2 gives the four physiological parameters found by comparison with the men's world records. Figure 6.22 shows that the predicted average speed-versus-distance curve agrees very well with the data at all distances up to about 10,000 m. The striking initial rise and later decrease in this speed as the distance increases is correctly predicted.

TABLE 6.2

Physiological parameters in S.I. units for the Keller model of running for a man of mass 80 kg. These constants are all proportional to the body mass. For example, for a runner of mass 100 kg, the parameters should be multiplied by 100/80 = 1.25

Maximum force	$F_{max} = 976$ N
Coefficient of the dissipative force	$c = 89.7$ N s m^{-1}
Stored metabolic energy	$E_0 = 193,000$ J
Rate of metabolic energy conversion	$\sigma = 3330$ W

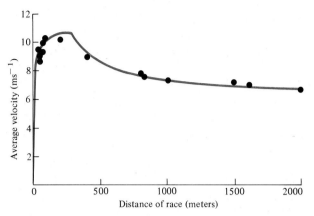

Figure 6.22. The average velocity for running races of various lengths. The dots represent average speeds for men's world records. The curve is the theoretical prediction using the values of the four parameters in Keller's model which give the best overall fit to the world records. (From Joseph B. Keller, *Physics Today.* © American Institute of Physics.)

The use of the physiological parameters is illustrated by the following example.

Example 6.16

The world record for the 5000-m run is 796.6 s, corresponding to an average velocity of 6.28 m s^{-1}. Find the power dissipated at this speed by a runner with a mass of 80 kg, and compare this with the rate at which additional energy is supplied by metabolic processes.

The power dissipated is $\mathcal{P}_d = Dv = (cv)v$. From Table 6.2, $c = 89.7$ N m^{-1}, and

$$\mathcal{P}_d = cv^2 = (89.7 \text{ N s m}^{-1})(6.28 \text{ m s}^{-1})^2$$
$$= 3538 \text{ W}$$

Energy is supplied at a rate of $\sigma = 3330$ W, so the runner is expending stored energy. The difference between the power dissipated and the power supplied is 208 W. In 797 s, this corresponds to an energy expenditure of $E = (208 \text{ W})(797 \text{ s}) = 166,000$ J. This is close to the total stored energy $E_0 = 193,000$ J, indicating as expected that the runner has used nearly all his stored energy.

It remains to be seen whether this particular model correctly describes the most important features of running. In any event, it shows how a simple model incorporating energy considerations can illuminate a very complex problem.

6.12 | DERIVATION OF THE GRAVITATIONAL POTENTIAL ENERGY FORMULA

When the separation between m and M is changed by

119

Δ**r**, the work ΔW_g done by the gravitational force is minus the potential energy change $\Delta \mathcal{U}$. The minus sign corresponds to the fact that when an object near the earth rises, its potential energy increases; the gravitational force does negative work. Thus, if $\bar{F}_r$ is the average gravitational force component in the **r** direction (Fig. 6.23),

$$\Delta W_g = \bar{F}_r \, \Delta r = -\Delta \mathcal{U}$$

Dividing by Δr and letting Δr become very small,

$$F_r = -\frac{d\mathcal{U}}{dr}$$

We can check that this relationship is satisfied by Eq. 6.18 by using Eq. B.23 in Appendix B,

$$\frac{d}{dr}\left(\frac{1}{r}\right) = -\frac{1}{r^2}$$

Therefore the choice $\mathcal{U} = -GmM/r$ gives

$$F_r = GmM \frac{d}{dr}\left(\frac{1}{r}\right) = -\frac{GmM}{r^2}$$

The minus sign indicates that **F** is opposite to **r,** so this is equivalent to Eq. 6.17.

EXERCISES ON SUPPLEMENTARY TOPICS

Section 6.11 | Running

6-64   (a) According to Keller's model, how large is the dissipative force on an 80-kg man running at 8 m s^{-1}? (b) What is his power expenditure?

6-65   Metabolizing a gram of fat yields about 8000 J of mechanical energy. If all his energy is supplied from metabolizing fat, how much does an 80-kg runner consume at 7 m s^{-1} according to Keller's model?

PROBLEMS ON SUPPLEMENTARY TOPICS

6-66   In Keller's model, the power expenditure of a runner equals the power input when $v^2 = \sigma/c$. Calculate the velocity at which this occurs, and compare it to the world record average velocity for a 5000-m race, which is 6.28 m s^{-1}.

6-67   (a) Calculate the maximum force a 60-kg runner can exert. (b) In a 2000-m race, the world record average speed is 6.64 m s^{-1}. What is the dissipative force? (c) Explain the difference.

Figure 6.23.   Since the gravitational force **F** is opposite to **r,** $F_r = -\mathbf{F}$. When m moves away from M, the gravitational force does negative work, and the potential energy increases.

6-68   (a) Suppose a runner exerted the maximum possible force and did not have any additional energy supply. How far could he run using his stored energy? (b) What period of time is needed for the metabolic processes to supply an amount of energy equal to the stored reserves?

Additional Reading

Elmer L. Offenbacher, Physics and the Vertical Jump, *American Journal of Physics,* vol. 38, 1970, p. 829.

Sir James Gray, *How Animals Move,* Cambridge University Press, Cambridge, 1953. Pages 69–80 discuss jumping.

R. McNeill Alexander, *Animal Mechanics, University of Washington Press, Seattle,* 1968. Pages 28–33 discuss jumping.

E.F. Adolph, Quantitative Relations in the Physiological Constitutions of Mammals, *Science,* vol. 109, 1949, p. 579. Scaling in biology.

Walter R. Stahl, Similarity and Dimensional Methods in Biology, *Science,* vol. 137, 1962, p. 205.

A. Gold, Energy Expenditure in Animal Locomotion, *Science,* vol. 181, 1973, p. 275.

John Maynard Smith, *Mathematical Ideas in Biology,* Cambridge University Press, Cambridge, 1968. Scaling, running.

Joseph B. Keller, A Theory of Competitive Running, *Physics Today,* vol. 26, September 1973, p. 42.

Knut Schmidt-Nielson, Locomotion: Energy Cost of Swimming, Flying, and Running, *Science,* vol. 177, 1972, p. 222.

Knut Schmidt-Nielsen, *How Animals Work,* Cambridge University Press, London, 1972. Applications of physical principles to physiology; scaling.

Galileo Galilei, *Dialogues Concerning Two New Sciences,* translated by Henry Crew and Alfonso de Salvio, The Macmillan Co., New York, 1914; Dover Publications, New York 1954. The first discussion of scaling in biology.

David Pilbeam and Stephen Jay Gould, Size and Scaling in Human Evolution, *Science,* vol. 186, 1974, p. 892.

V.A. Tucker, The Energetic Cost of Moving About, *American Scientist,* vol. 63, 1975, p. 413. Advantages of flying, swimming, bicycling compared to walking and running.

Scientific American articles:

Milton Hildebrand, How Animals Run, May 1960, p. 149.

Henry W. Ryder, Harry J. Carr, and Paul Herget, Future Performances in Footracing, June 1976, p. 109.

CHAPTER SEVEN

IMPULSE AND MOMENTUM

Many common events can be regarded as collisions even though we may not think of them as such. A baseball player hitting or catching a ball, a blow struck in a boxing match, and the action of seat belts on passengers during an automobile accident are all examples of collisions. Although collisions are not unusual in our experience, they are difficult to analyze using the methods we have described so far in this textbook. In particular, since the forces acting during a collision are often difficult to determine, Newton's second law is hard to use directly. Furthermore, in many collisions, mechanical energy is not conserved, and the methods of energy conservation described in the previous chapter are also frustrated.

However, as we see in this chapter, such problems can be understood and described in terms of a quantity called the *momentum*. The momentum of an object is the product of its mass and velocity. The momentum has a useful and surprising characteristic: when two objects collide, the momentum of each may change, but the *total* momentum of the system of two objects may remain constant or be conserved. In many collisions the total momentum of the system is either exactly conserved or, as in automobile collisions, may be regarded as conserved to a very good approximation.

The change in the momentum of an object is very closely related to the forces on the object. Because of this, we can sometimes find the average forces acting during a very complex event by measuring the momentum change.

In this chapter, we describe and analyze a variety of events where momentum is conserved or, if it is not, where the forces may be found when the momentum changes. The forces acting when momentum changes occur are contained in a quantity called the *impulse.*

7.1 | IMPULSE AND MOMENTUM

With the aid of an example, we now define *impulse* and *momentum* and find the relationship between them. Figure 7.1 shows a man pushing a dresser with casters. We will suppose that the frictional forces on the dresser are negligible. With Newton's laws, we can find a relation among the forces acting on the dresser, its velocity change, and the time the force acts.

The force $\mathbf{F}$ acts on the dresser for a time t. During this period the velocity of the dresser changes from $\mathbf{v}$ to $\mathbf{v}'$. By definition, the average acceleration $\bar{\mathbf{a}}$ is the velocity change divided by the time t, $\bar{\mathbf{a}} = (\mathbf{v}' - \mathbf{v})/t$. However, from Newton's second law, the average acceleration and force are related by $\bar{\mathbf{F}} = m\bar{\mathbf{a}}$, so

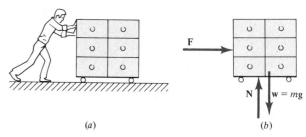

(a)   (b)

Figure 7.1. (a) A man pushes a dresser, which moves frictionlessly on casters. (b) The forces on the dresser are $\mathbf{F}$, the weight $\mathbf{w} = m\mathbf{g}$, and the normal force $\mathbf{N}$.

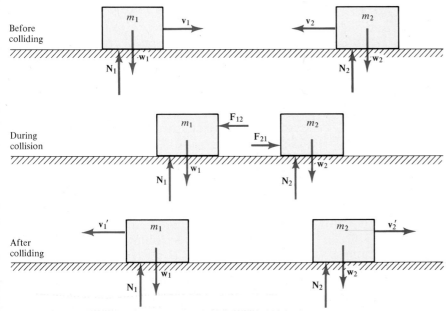

Figure 7.2. Two objects slide frictionlessly towards each other. Their initial momenta are $\mathbf{p}_1 = m_1\mathbf{v}_1$ and $\mathbf{p}_2 = m\mathbf{v}_2$. During the collision the objects exert equal and opposite forces on each other. After the collision the momenta of the two objects are $\mathbf{p}'_1 = m_1\mathbf{v}'_1$ and $\mathbf{p}'_2 = m_2\mathbf{v}'_2$.

$$\bar{\mathbf{F}} = m\left(\frac{\mathbf{v}' - \mathbf{v}}{t}\right)$$

or

$$\bar{\mathbf{F}}t = m\mathbf{v}' - m\mathbf{v} \qquad (7.1)$$

The product of the average force and the time $\bar{\mathbf{F}}t$ is called the *impulse*. The product $m\mathbf{v}$, denoted by $\mathbf{p}$, is the *momentum*. Equation 7.1 means that *the impulse equals the change in momentum*,

$$\bar{\mathbf{F}}t = \mathbf{p}' - \mathbf{p} \qquad (7.2)$$

We see that the momentum of an object has the dimensions of force times time or mass times velocity. The S.I. momentum unit is a kilogram-meter per second (kg m s^{-1}). The following example illustrates how Eq. 7.2 may be used.

Example 7.1

A baseball, initially at rest, is struck with a bat. The velocity of the 0.15-kg ball just after it is hit is 40 m s^{-1}. If the impact time is 10^{-3} s, what is the average force on the ball?

The initial momentum $\mathbf{p}$ of the ball is zero, since it starts from rest; the final momentum is $\mathbf{p}' = m\mathbf{v}'$. Thus, from $\bar{\mathbf{F}}t = \mathbf{p}' - \mathbf{p}$, the average force $\bar{F}$ on the ball is

$$\bar{F} = \frac{mv'}{t} = \frac{(0.15 \text{ kg})(40 \text{ m s}^{-1})}{(10^{-3} \text{ s})} = 6000 \text{ N}$$

This is more than 1300 lb.

7.2 | MOMENTUM CONSERVATION

The concept of momentum is most useful when two or more objects act on one another. For example, in Fig. 7.2 we show two objects that collide. Friction is negligible, and each object experiences a gravitational force and a normal force. During the collision, they also exert forces on each other. These forces, $\mathbf{F}_{12}$ and $\mathbf{F}_{21}$, are an action-reaction pair, so they are equal and opposite and $\mathbf{F}_{12} + \mathbf{F}_{21} = 0$.

If the collision time is Δt, the change in momentum of each object can be computed from the average forces $\bar{\mathbf{F}}_{12}$ and $\bar{\mathbf{F}}_{21}$. For m_1, $\bar{\mathbf{F}}_{12}\,\Delta t = \mathbf{p}'_1 - \mathbf{p}_1$, and for m_2, $\bar{\mathbf{F}}_{21}\,\Delta t = \mathbf{p}'_2 - \mathbf{p}_2$. But $\bar{\mathbf{F}}_{12} + \bar{\mathbf{F}}_{21} = 0$, so if we add these two equations, we find

$$\mathbf{p}'_1 - \mathbf{p}_1 + \mathbf{p}'_2 - \mathbf{p}_2 = 0$$

or

$$\mathbf{p}_1 + \mathbf{p}_2 = \mathbf{p}'_1 + \mathbf{p}'_2 \qquad (7.3)$$

This equation means that the net momentum of the two objects before and after the collision is the same, or the total momentum of the system is conserved.

This is important because it means that we can

Figure 7.3. (a) A neutron collides with a stationary nitrogen atom. (b) The neutron is absorbed and a single object of mass $m + M$ is formed.

relate the velocities of the objects before and after the collision *without knowing anything about the forces between them during the collision.* The fact that these forces are always equal and opposite is sufficient to obtain the result. The following example illustrates this idea.

Example 7.2

A neutron moving at 2700 m s^{-1} collides head-on with a nitrogen atom at rest and is absorbed. The neutron and nitrogen masses are $m = 1.67 \times 10^{-27}$ kg and $M = 23.0 \times 10^{-27}$ kg, respectively. What is the final velocity of the combined object? (Fig. 7.3.)

The only forces affecting the motion during the collision are those between the neutron and atom. Thus the net momentum is constant. Before the collision, $p = mv$ and afterwards $p' = (m + M)v'$, where v' is the final velocity. Then $mv = (m + M)v'$, so

$$v' = \frac{mv}{(m + M)} = \frac{(1.67 \times 10^{-27} \text{ kg})(2700 \text{ m s}^{-1})}{(1.67 + 23.0) \times 10^{-27} \text{ kg}}$$

$$= 183 \text{ m s}^{-1}$$

In order to know when momentum is conserved, we must understand the role of *internal* and *external* forces. In Fig. 7.2b the normal and gravitational forces are external forces. Since there is no vertical motion, $\mathbf{N}_1 + \mathbf{w}_1 = 0$ and $\mathbf{N}_2 + \mathbf{w}_2 = 0$, and there is no net impulse as a result of these forces. The force $\mathbf{F}_{12}$ is an external force on m_1, and $\mathbf{F}_{21}$ is an external force on m_2. However, if m_1 and m_2 are considered as *one system*, we can say that $\mathbf{F}_{12}$ and $\mathbf{F}_{21}$ are internal forces of the system. Thus, in Fig. 7.2, there is *no net external force acting on the system.* The momentum of m_1 alone is not conserved nor is that of m_2. However, the net momentum of the system is conserved.

Whenever there is no net external force acting on a system, its momentum is conserved. Momentum is therefore always conserved for an isolated system, one subjected only to internal forces.

In many situations of interest to us, the momentum is only approximately conserved because there are net external forces acting. However, the impulses resulting from these external forces are often small enough to be neglected. For example, when two cars collide, they exert huge forces on each other for a very short time, and the impulse on each is substantial. An external force such as friction may also be present, but its magnitude is not large enough to contribute any appreciable impulse in the brief time of the collision. Hence, only the internal forces between the cars are really important, and momentum conservation can be used to analyze the collision.

Even though momentum is always conserved when a collision occurs within an isolated system, mechanical energy may or may not be conserved. A collision is said to be *elastic* if the mechanical energy is conserved and *inelastic* if it is not. The neutron-nitrogen atom collision of Example 7.2 illustrates inelastic collisions.

Example 7.3

We found in example 7.2 that when a neutron of mass 1.67×10^{-27} kg, moving at 2700 m s^{-1}, is absorbed by a stationary nitrogen nucleus of mass 23.0×10^{-27} kg, the resulting object moves at 183 m s^{-1}. How much kinetic energy is lost in the collision?

Before the collision the kinetic energy is that of the neutron alone,

$$K_0 = \tfrac{1}{2}mv^2 = \tfrac{1}{2}(1.67 \times 10^{-27} \text{ kg})(2700 \text{ m s}^{-1})^2$$
$$= 6.1 \times 10^{-21} \text{ J}$$

After the collision,

$$K = \tfrac{1}{2}(m + M)v'^2$$
$$= \tfrac{1}{2}(1.67 + 23.0) \times 10^{-27} \text{ kg } (183 \text{ m s}^{-1})^2$$
$$= 0.41 \times 10^{-21} \text{ J}$$

Thus, 5.7×10^{-21} J, or about 93 percent of the initial kinetic energy is lost.

The following two examples further illustrate the use of momentum conservation.

Example 7.4

A cannon is mounted inside a railroad car, which is initially at rest but can move frictionlessly (Fig. 7.4). It fires a cannonball of mass $m = 5$ kg with a horizontal velocity $v = 15$ m s^{-1} at an end wall. The total mass of the cannon and railroad car is $M = 15{,}000$ kg. (a) What is the velocity **V** of the car while the cannonball is in flight? (b) If the cannonball becomes embedded in the wall, what is the velocity of the car and ball after impact?

(a) When fired, the cannon exerts a force to the right on the ball. The ball exerts an equal and opposite force on the cannon, so the car and cannon recoil to the left. The net momentum is conserved because the external frictional force is negligible in this example. The momentum before firing is zero, so after firing the momentum of the ball to the right must be equal in magnitude to that of the car and cannon to the left. Thus $mv = MV$ and

$$V = \frac{mv}{M} = \frac{(5 \text{ kg})(15 \text{ m s}^{-1})}{(15{,}000 \text{ kg})}$$
$$= 5 \times 10^{-3} \text{ m s}^{-1}$$

The recoil speed of the car and cannon is very small because of their large mass.

(b) As the ball becomes embedded in the wall, it exerts a force on the wall to the right in Fig. 7.4. The wall, in turn, exerts a force to the left on the ball. The ball and car both stop moving when this happens, since the net momentum is still zero. Meanwhile the car will have rolled to the left as the ball traveled to the right.

In Problem 7-20, it is found that the ball in the preceding example moves further to the right than does the car to the left. Because of this, the position of the center of gravity of the entire system remains unchanged.

> *The motion of the center of gravity of a system is unaffected by the internal forces and is determined entirely by the external forces acting on the system.*

In the example, the initial velocity of the center of gravity was zero, and the net external force due to gravity and the rails was zero. Therefore the center of gravity remained at rest. When a net external force

Figure 7.4. The cannon fires a ball toward the right. The cannon is mounted firmly to the floor of the railroad car.

(a)   (b)

Figure 7.5. (a) The trajectory of an artillery shell. (b) The motion of the center of gravity of a system is determined solely by the external forces. Therefore the center of gravity of the shell continues to follow the original trajectory after the shell explodes. None of the fragments need follow this path.

$\mathbf{F}_{ext}$ does act on a system, the center of gravity has an acceleration equal to $\mathbf{F}_{ext}/M$, where M is the total mass of the system. Thus the center of gravity moves under the influence of $\mathbf{F}_{ext}$ in exactly the same fashion as a single object of mass M.

Example 7.5

An artillery shell explodes in the air. Neglecting air resistance, what can be said about the subsequent motion of its fragments?

Before the explosion, the shell moves in the trajectory described in our discussion of projectiles in Chapter Two. After the explosion the fragments fly off in many directions, and we cannot determine anything about their individual paths without more information. However, their center of gravity is unaffected by the internal forces of the explosion, so it continues to follow the original trajectory (Fig. 7.5). Similarly, divers or gymnasts may execute complicated maneuvers, but their centers of gravity follow simple trajectories (Fig. 7.6).

7.3 | HIGHWAY SAFETY

Automobile accidents are complex events involving many variables. However, many questions about the safety of passengers involved in such accidents can be explored using the concepts of impulse and momentum.

To visualize what happens in an accident, suppose that a car hits a stone wall or a tree head-on at an appreciable speed. The front end of the car will collapse, and the passenger compartment will come to rest in a meter or so. The time required for this to occur is typically a few tenths of a second. A passenger who is wearing a seat belt or shoulder harness will therefore come to rest in a few tenths of a second. However, an unrestrained passenger will slide forward

Figure 7.6. The diver's center of gravity moves in a projectile trajectory after he leaves the board.

(a)	*(b)*	*(c)*

Figure 7.7. A passenger in a vehicle during a collision, *(a)* without seat belts, *(b)* with a lap belt only, *(c)* with a lap belt and shoulder harness.

at roughly the original speed of the car, striking the windshield or dashboard (Fig. 7.7). Such a passenger will come to rest in a very short time and experience much larger forces during the impact. The force may also be applied over a very small area, increasing the severity of the injuries. The following example illustrates the difference between the two situations.

Example 7.6

A car traveling at 10 m s^{-1} (about 22 mi h^{-1}) collides with a tree. (a) An unrestrained passenger strikes the windshield head first and comes to rest in 0.02 s. The contact area between the head and windshield is 6×10^{-4} m^2 (roughly 1 in^2), and the mass of the head is

5 kg. Find the average force and the force per unit area exerted on the head. (b) A passenger of mass 70 kg wearing a shoulder harness comes to rest in 0.5 s. The area of the harness in contact with the passenger is 0.1 m^2 (about 1 ft^2). Find the average force and the average force per unit area.

(a) Again we use $\bar{\mathbf{F}}t = \mathbf{p}' - \mathbf{p}$. The final momentum is zero, since the windshield is stationary, and the initial momentum is the head mass times the velocity. Thus the magnitude of the average force is

$$\bar{F} = \frac{p}{t} = \frac{(5 \text{ kg})(10 \text{ m s}^{-1})}{0.02 \text{ s}} = 2500 \text{ N}$$

The average force per unit area is

$$\frac{\bar{F}}{A} = \frac{2500 \text{ N}}{6 \times 10^{-4} \text{ m}} = 4.16 \times 10^7 \text{ N m}^{-2}$$

This is a very large force per unit area and is certain to cause serious injury.

(b) The average force is found from the change in momentum of the entire body as the speed of the car changes from 10 m s^{-1} to zero. Thus the magnitude of $\bar{F}$ is

$$\bar{F} = \frac{p}{t} = \frac{(70 \text{ kg})(10 \text{ m s}^{-1})}{0.5 \text{ s}} = 1400 \text{ N}$$

This is much smaller than the force exerted on the head of the unrestrained passenger above. The average force per unit area is

$$\frac{\bar{F}}{A} = \frac{1400 \text{ N}}{0.1 \text{ m}^2} = 1.4 \times 10^4 \text{ N m}^{-2}$$

Since this is smaller than the force per unit area on the unrestrained passenger by a factor of 3000, the chance of serious injury is much smaller for this passenger.

Improving the safety of passengers in auto collisions involves increasing the time over which the momentum changes or increasing the area over which the decelerating forces are applied. In the first category are improved passenger restraints and front-end designs that permit more gradual slowing of the passenger compartment. The removal of sharp objects and the use of inflatable air bags are examples of attempts to increase the contact area.

SUMMARY

The central point of this chapter is Eq. 7.2 and the statements that follow from it,

$$\bar{\mathbf{F}}t = \mathbf{p}' - \mathbf{p} \tag{7.2}$$

The product of the average force and the time during which it acts is called the impulse. The momentum **p** of an object with mass m and velocity **v** is $\mathbf{p} = m\mathbf{v}$. The impulse equals the change in the momentum.

When two objects collide, each suffers a change in momentum. However, the total momentum is constant or conserved if no net external force acts on the two objects. Consequently, the total momentum of an isolated system is always conserved. Momentum conservation is particularly useful in collision problems, even though energy may or may not be conserved. If the energy is also conserved, the collision is said to be elastic; if not, the collision is called inelastic.

Checklist
Define or explain:

impulse external forces
momentum elastic collision
momentum conservation inelastic collision
internal forces

EXERCISES

Section 7.1 | Impulse and Momentum

7-1 A golf ball initially at rest is struck with an average force of 2600 N for 1.25×10^{-3} s. What is the final velocity of the 0.047-kg ball?

7-2 If the momentum of a car increases by 9×10^4 kg m s^{-1} in 12 s, what is the average force accelerating the car?

7-3 (a) What is the change in momentum of a car that experiences a force of 6000 N for 6 s? (b) If the mass of the car is 1000 kg, what is its change in velocity?

7-4 A pilot ejected from a plane suffers an average acceleration of 12 g's for a period of 0.25 s. (a) If the mass of the pilot is 70 kg, what is the average force he experiences? (b) Find the magnitude of the change in his velocity. Neglect air resistance.

Section 7.2 | Momentum Conservation

7-5 A propeller-driven airplane pushes air toward the rear of the plane as it moves forward. Describe the force on the plane in terms of the change in momentum of the air.

7-6 A man sits on a chair fixed to a cart which is initially at rest on frictionless rails. (a) If he throws a sandbag off the side of the cart, will the cart move? Explain. (b) If he throws a sandbag off the back of the cart, will the cart move? Explain.

7-7 A man seated at the stern of a sailboat attempts to move forward by blowing on the sails. Explain what will happen.

7-8 A rocket of mass M explodes into two pieces just as it reaches its maximum height. One piece of mass m_1 *stops and then falls* vertically to the earth. What is the velocity of the other piece of mass m_2 just after the rocket separates? (The speed of the rocket at its maximum height is v_0.)

7-9 A cart of mass m on a frictionless air track has an initial velocity $\mathbf{v}_0$ toward a cart of mass M which is at rest. What are the final velocities of the two carts after they collide if the carts lock together?

7-10 A 1-g $= 10^{-3}$-kg bullet has a horizontal velocity of 200 m s^{-1}. It strikes and becomes embedded in a 1-kg block of wood which rests on a smooth table top. What is the velocity of the block and bullet after impact? (Neglect friction.)

Section 7.3 | Highway Safety

7-11 A vehicle with velocity v_0 collides with (a) a stationary vehicle of the same mass, (b) a solid wall, and (c) a vehicle of the same mass moving in the opposite direction with speed v_0. Which is(are) the most severe as far as the effect on passengers? Explain.

7-12 A truck of mass 4500 kg moving at 10 m s^{-1} strikes the back of a car that is at rest. The car and occupants have a mass of 950 kg. (a) What is the speed of the car just after impact if the car and truck lock together? (b) If the impact lasts 0.3 s, what is the average force on a 60-kg passenger?

7-13 In a head-on collision with a stationary obstacle, the stopping distance of a vehicle of mass 1500 kg is 0.3 m. If the vehicle was traveling at a speed of 13 m s^{-1} just before impact, what is the change in momentum?

7-14 A 2000-lb Volkswagon and a 4000-lb Ford are each moving at 40 mi h^{-1} when they collide head-on. Find their final velocities just after the collision, (a) if the collision is elastic; (b) if they remain locked together.

PROBLEMS

7-15 A man is sitting on a sled on an ice

pond so that the frictional forces are zero. He has a machine gun that fires bullets of mass 1.3×10^{-2} kg with a muzzle velocity of 800 m s^{-1}. (a) What is the momentum of each bullet? (b) What is the average force on the man, per bullet, if that force is experienced for 0.2 s? (c) What is the speed of the man on the sled after he fires 100 bullets? Assume that the mass of the man, sled, and gun is 90 kg, and neglect the loss of mass because of the release of 100 bullets.

7-16   A neutron of mass m and velocity v collides elastically with an atomic nucleus of mass M which is initially at rest. (a) If the collision is head-on, how much kinetic energy does the neutron lose? (b) Show that the maximum kinetic energy loss by the neutron occurs when $m = M$. (This is why hydrogen and other light elements are used to slow down neutrons in nuclear reactors.)

7-17   An empty railroad gondola car rolls initially at 5 m s^{-1} and then collects 1000 kg of water in a storm in which the rain falls vertically. If there is no friction and the empty car has a mass of 12,000 kg, what is its final speed?

7-18   A freight car with a mass of 30,000 kg rolls at 5 m s^{-1} under a grain elevator where 10,000 kg of grain is dropped vertically into the car. (a) Assuming frictional forces between the car and rails are negligible, what is the final velocity of the car? (b) How much kinetic energy is lost by the car, and how is this energy transformed?

***7-19**   A billiard ball strikes an identical billiard ball and is deflected 45° from its original direction. Show that if the collision is elastic, the other ball must move at 90° to the first and with the same speed. (*Hint:* Momentum is a vector quantity. Momentum is conserved separately in directions parallel and perpendicular to the initial motion.)

7-20   (a) If the distance d in Fig. 7.4 is 6 m, how far does the cannonball of Example 7.4 travel before striking the wall? (b) How far do the car and cannon move? (c) Show that the center of gravity of the entire system remains stationary when the ball is fired.

***7-21**   A bullet of mass 10^{-2} kg and initial horizontal velocity of 250 m s^{-1} strikes and is embedded in a 1-kg wooden block. The wooden block hangs on the end of a long string. (a) What is the velocity of the block and bullet after impact? (b) How high will the block and bullet swing upward?

***7-22**   Explain how the toy pictured in Fig. 7.8 operates. The toy is composed of steel balls hung on strings. When one ball is initially raised and released, the last ball at the other end swings up. When two balls are initially displaced, two on the other end swing upwards.

Figure 7.8.   When the raised ball is released it swings downward and collides elastically with the second ball. After a sequence of collisions, the right-hand ball swings upward while the others remain at rest. Problem 7-22.

7-23   A person lies quietly on a horizontal table that is supported by air jets. Thus the person and table are essentially isolated. During a heart contraction, blood is accelerated from rest toward the head. Shortly after leaving the heart through the aorta, the upward flowing blood goes around a curve and flows toward the feet. Since reaction forces occur when the heart ejects blood and when the blood changes direction and momentum, sensitive measuring devices can measure the resultant acceleration and displacement of the body. Given the description of blood flow described above, what sequence of body displacements would one expect to observe? (The apparatus described is called a ballistocardiograph.)

SUPPLEMENTARY TOPICS

7.4 | MOMENTUM AND THE USE OF THE BODY

In many athletic activities, one tries to maximize the momentum transfer. For example, in boxing, punches thrown by a simple extension of the arm are not nearly as effective in imparting momentum to an opponent as those thrown in conjunction with body movement. However, in karate, large momentum transfers are often attained by high-speed motion of the limbs instead of by entire body movements.

Contact sports are not the only area where momentum transfer is important. For example, the pri-

Figure 7.9. A shot-putter in action. Ideally, the momentum of the body parts is very small just after release. The momentum of the body is transferred to the ball.

mary aim of a shot-putter is to transform the low-velocity motion of the mass of the entire body into high-velocity motion of the smaller ball (Fig. 7.9). Momentum also plays an important role in activities where a ball is struck.

Table 7.1 lists the typical velocities and times measured for good athletes in various sports using balls. The striker is the instrument used to hit the ball, such as a baseball bat, tennis racket, or foot. To obtain useful information from these data using impulse and momentum, we must recognize that the mass of the striker, and hence its momentum, is somewhat ambiguous. Using tennis as an example,

the striker is a racket of mass 0.4 kg. However, in using the racket, the arm and part of the body can be considered as part of the striker; the racket acts as an extension of the body. The effective striker mass depends on what part of the body is used and on how it is used. When a person swings a tennis racket primarily through wrist action, the effective striker mass is small, and the swing cannot be carried through firmly.

It must also be remembered that the person involved in striking the ball will usually be in contact with the ground. This means that the ball and striker cannot truly be regarded as parts of an isolated system with no external forces. When a tennis racket hits a ball, by Newton's third law the ball exerts a force on the racket, the racket exerts a force on the body, and the body exerts a force on the ground. Therefore, if a racket imparts momentum to a ball, momentum is also transferred to the earth.

To avoid the difficult details of this situation, it is convenient to define the *effective mass* of the striker. That is, we pretend the striker and ball are two parts of a system on which no external forces act, so the total momentum of the striker and ball can be considered constant. This effective mass is not just a trick to allow us to compute a meaningless number. A detailed study of the motions of an athlete can aid in understanding how to increase the effective mass of the striker and thereby achieve higher ball velocities.

We denote the mass of the ball by m and its final velocity by $\mathbf{v}'$; its initial velocity is zero. The effective mass of the striker is denoted by M and its initial and

TABLE 7.1

Masses of balls, velocities of strikers before and after impact, and the impact times during which the balls and strikers are in contact.

Ball	Ball Mass (kg)	Ball Velocity (m s^{-1}) Before	After	Striker Velocity (m s^{-1}) Before	After	Impact Time (s)
Baseball (hit from rest)	0.15	0	39	31	27	1.35×10^{-3}
Football (punt)	0.42	0	28	18	12	8×10^{-3}
Golf ball (driver)	0.047	0	69	51	35	1.25×10^{-3}
Handball (serve)	0.061	0	23	19	14	1.35×10^{-2}
Soccer ball (kick)	0.43	0	26	18	13	8×10^{-3}
Squash ball (serve)	0.032	0	49	44	34	3×10^{-3}
Softball (hit from rest)	0.17	0	35	32	22	3×10^{-3}
Tennis ball (serve)	0.058	0	51	38	33	4×10^{-3}

Figure 7.10. A batter has his feet firmly set and his body is moving toward the ball on impact. This increases the effective striking mass of the bat.

final velocities by **V** and **V′**. If the motion is in a straight line and momentum is conserved,

$$MV = mv' + MV' \qquad (7.5)$$

Assuming m, v', V, and V' are known, we can solve this equation for the effective mass of the striker,

$$M = \frac{mv'}{V - V'} \qquad (7.6)$$

Notice that this effective mass can be increased by performing the motion in such a way as to minimize the change in velocity of the striker during the impact (Fig. 7.10).

Example 7.7

(a) Using the data of Table 7.1 for a tennis serve, calculate the effective mass of the striker. (b) What is the average force on the tennis ball during impact?

(a) With the data of Table 7.1, the effective mass of the striker is

$$M = \frac{mv'}{V - V'} = \frac{(0.058 \text{ kg})(51 \text{ m s}^{-1})}{(38 - 33) \text{ m s}^{-1}} = 0.59 \text{ kg}$$

This is greater than the mass of the racket alone, which is 0.4 kg.

(b) During impact the average force on the ball $\bar{F}$ is found from $\bar{F}t = m(v' - v)$. With $v = 0$ and the impact time of 4×10^{-3} s,

$$\bar{F} = \frac{m(v' - v)}{t} = \frac{(0.058 \text{ kg})(51 \text{ m s}^{-1})}{4 \times 10^{-3} \text{ s}} = 740 \text{ N}$$

7.5 | MOMENTUM IN KARATE

In karate, the high momentum of a strike is achieved through rapid motion of the limbs. In the forward punch, the fist moves in an approximate straight line,

and impact is supposed to occur when the speed of the fist is at a maximum (Fig. 7.11). This occurs when the arm is about 70 percent extended (Fig. 7.12). Because of the way the punch is thrown, the striking mass is less than or at most equal to the mass of the arm. In the following example, we estimate the force and energy transferred in the forward punch.

Example 7.8

The forward punch shown in Fig. 7.11 is delivered to an opponent's head. Impact occurs when the arm extension is 70 percent; the fist and head move together after impact. (a) Using the data in Fig. 7.12, what is the average force of the blow? (b) How much kinetic energy is lost? Assume the arm mass is $m_a = 7$ kg, the head mass is $m_h = 6$ kg, and the impact time is 10^{-3} s.

(a) From Fig. 7.12, the fist speed at 70 percent arm extension is $v = 5.5$ m s^{-1}. The initial momentum is then $m_a \mathbf{v}$. After impact the two masses move together at a speed v'. Since momentum is conserved, $m_a v = (m_a + m_h)v'$ and

$$v' = \frac{m_a v}{m_a + m_h} = \frac{(7 \text{ kg})(5.5 \text{ m s}^{-1})}{(7 \text{ kg} + 6 \text{ kg})} = 2.96 \text{ m s}^{-1}$$

Now that we have the initial and final velocities of the arm, we can use $\bar{F}t = \mathbf{p}' - \mathbf{p}$ to find the average force on the fist,

$$\bar{F} = \frac{m_a v' - m_a v}{t}$$

$$= \frac{(7 \text{ kg})(2.96 \text{ m s}^{-1}) - (7 \text{ kg})(5.5 \text{ m s}^{-1})}{(10^{-3} \text{ s})}$$

$$= -1780 \text{ N}$$

Figure 7.11. A forward punch in karate. The dashed line shows the path of the fist.

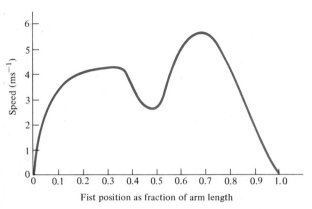

Figure 7.12. The speed of the fist versus the fractional arm extension in a forward karate punch. The data are taken from the study of high speed motion pictures. (From J. D. Walker, *American Journal of Physics*, vol. 43, October 1975, p. 845.)

The force on the opponent is equal in magnitude but opposite to the force on the fist.

(b) The initial kinetic energy is that of the arm,

$$K_0 = \tfrac{1}{2}m_a v^2 = \tfrac{1}{2}(7 \text{ kg})(5.5 \text{ m s}^{-1})^2 = 106 \text{ J}$$

After impact the kinetic energy is that of the arm plus the opponent's head,

$$K = \tfrac{1}{2}(m_a + m_h)v'^2$$
$$= \tfrac{1}{2}(7 \text{ kg} + 6 \text{ kg})(2.96 \text{ m s}^{-1})^2 = 57 \text{ J}$$

The difference, $K - K_0 = 49$ J, is energy dissipated in doing damage to the opponent.

In karate the objective is to disable an opponent, possibly by fracturing a bone. This requires rapid impact and large energy transfers. In boxing the intent is to gradually disable an opponent with a succession of blows. The impact time is longer, especially when gloves are worn. The energy transfer per blow is then reduced proportionately.

EXERCISES ON SUPPLEMENTARY TOPICS

Section 7.4 | Momentum and the Use of the Body

7-24 What is the average force on the foot in a football punt?

7-25 Using the data of Table 7.1, find the average force on the hand of a handball player when serving.

7-26 Discuss, in terms of impulse and momentum, the difference between boxing with bare knuckles and with gloves.

7-27 What is the effective mass of the striker in a soccer kick?

7-28 Use the data in Table 7.1 for a baseball to find (a) the kinetic energy transferred to the baseball, and (b) the mechanical energy lost.

7-29 A fielder catches a baseball of mass 0.15 kg moving at 40 m s⁻¹. If the fielder's glove moves backward 0.1 m during the catch, what is the average force on the fielder's glove?

7-30 (a) In a football punt, what is the effective mass of the striker? (b) What percentage of the body weight of an 80-kg person is this?

7-31 Discuss how the concept of effective striker mass applies to boxing techniques.

7-32 (a) From the data of Table 7.1, compute the effective striker mass for a baseball hit from rest. (b) Compare the force on the baseball with that of the karate blow in Example 7.8 and that on the head in Example 7.6.

PROBLEMS ON SUPPLEMENTARY TOPICS

7-33 An object of mass m and velocity v_0 collides with a second object of mass M that is originally at rest. If the objects move together after the collision and momentum is conserved, show that (a) the final velocity of m and M is

$$\frac{1}{1 + \dfrac{M}{m}} \, v_0$$

(b) Show that the ratio of the energy lost in the collision ΔK to the original kinetic energy $K_0 = \tfrac{1}{2}mv_0^2$ is

$$\frac{\Delta K}{K_0} = \frac{1}{1 + \dfrac{m}{M}}$$

(c) Draw sketches of the graphs of the final velocity versus m/M and the ratio $\Delta K/K_0$ versus m/M.

7-34 Using the results of Problem 7-33, compare boxing and karate. In boxing the bodies are fully involved, so $m \simeq M$. In karate the striking mass is usually that of only one limb, but the object struck is much more massive, so $m \ll M$.

7-35 A world record shot-putter can launch a 7.27-kg ball with a speed of 15 m s⁻¹. If the mass of the athlete is 115 kg, what is the speed of his center of gravity just before releasing the ball? (Assume that his momentum is zero after release.)

Additional Reading

James G. Hay, *The Biomechanics of Sports Techniques,* Prentice-Hall, Englewood Cliffs, N.J., 1973. Momentum conservation in athletics.

Stanley Plagenhoef, *Patterns of Human Motion: A Cinematographic Analysis,* Prentice-Hall, Englewood Cliffs, N.J., 1971. Momentum conservation in athletics.

Jearl D. Walker, Karate Strikes, *American Journal of Physics,* vol. 43, October 1975, p. 845.

F.I. Ordway, Principles of Rocket Engines, *Sky and Telescope,* vol. 14, 1954, p. 48.

Scientific American article:

H.W. Lewis, Ballistocardiography, February 1958, p. 89. Momentum conservation and heart action.

CHAPTER EIGHT

THE ROTATION OF RIGID BODIES

Until now, we have been primarily concerned with translational motion. However, situations in which rotational motion occurs are just as familiar. A combination of the two kinds of motion can also occur, as illustrated by the wheel of a car. Its geometric center is in translational motion at the velocity of the vehicle. However, the point on the edge of the wheel that is in contact with the ground is momentarily at rest, while the point directly opposite at the top of the wheel is moving with twice the velocity of the car. Each part of the wheel has a different velocity at any given time.

Rotational motion is most readily described in terms of *angular* position, velocity, and acceleration. The equations describing rotational motion in terms of these quantities are similar in form to those for translation, and many of our results for translation have rotational analogues. However, despite the mathematical similarities between translation and rotation, we will encounter some rather startling kinds of rotational motion.

8.1 | ANGULAR VARIABLES

In this section, we introduce the *angular variables* that most conveniently represent rotational motion. To define angular variables, we examine a wheel spinning about its axis (Fig. 8.1). At the time shown, *all points* along the colored spoke have moved through the same angle θ from the vertical direction. However, points near the rim have traveled further than points near the axis. Because of this, it is more con-

Figure 8.1. A wheel of radius R rotates about its axle. The spoke shown in color was initially vertical and the wheel has rotated through an angle θ. The point at a distance r from the center has moved a distance $s = r\theta$.

venient to describe rotational motion in terms of angles rather than distances.

Angles can be measured either in degrees or in *radians* (rad). Referring to Fig. 8.1, we define the *angular position θ in radians by*

$$\theta = \frac{s}{r} \tag{8.1}$$

This result for θ is the same for any point on the colored spoke of the wheel. The value of s for a point a distance r from the rotation axis is $r\theta$. For a complete rotation of the wheel, s is equal to the circumference $2\pi r$, and $\theta = 2\pi$ rad. Since a full circle is 360°, 2π rad = 360°, and 1 rad = 360°/2π = 57.3° (Fig. 8.2). Note that $\theta = s/r$ implies that θ is dimensionless; it is the ratio of two lengths. Thus the radian is not a unit in quite the same sense as the meter or foot. We use radians as a reminder of how the angle is defined. We mention this point again below.

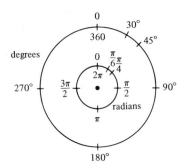

Figure 8.2. Typical angles in degrees and radians. An object that rotates through one and one-half revolutions is said to have rotated through $2\pi + \pi = 3\pi$ radians. 1 radian = 57.3°.

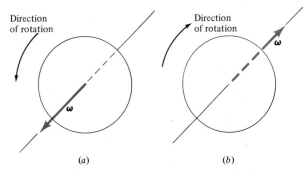

Figure 8.3. (a) For counterclockwise rotations ω is directed out of the page. (b) ω is directed into the page for clockwise rotations.

The angular velocity ω (omega) has a magnitude equal to the rate of change of the angle. The average angular speed is defined by $\bar{\omega} = \Delta\theta/\Delta t$, where $\Delta\theta$ is the change in angle in the time Δt. As Δt becomes arbitrarily small, the angular speed at a given instant in radians per second is

$$\omega = \frac{d\theta}{dt} \tag{8.2}$$

The vector ω is conventionally taken to be directed along the axis of rotation. It is directed out of the page, parallel to the rotation axis, if the rotation is counterclockwise (Fig. 8.3a). If the rotation is clockwise, as in Fig. 8.3b, ω is directed into the page. ω *is always directed along the rotation axis.*

A convenient method of identifying the direction of ω is to curl the fingers of the right hand around the rotation axis in the direction of rotation. The right thumb then points in the direction of ω (Fig. 8.4). The following example illustrates these ideas.

Figure 8.4. Curling the fingers of the right hand in the direction of rotation, the thumb points perpendicular to the disk in the direction of ω.

Example 8.1

A disk rotates clockwise with a constant angular velocity at two revolutions per second (Fig. 8.3b). (a) What is the direction and magnitude of the angular velocity? (b) What is the rotation angle after 4 s?

(a) The disk rotates twice or through 4π rad each second, so its angular velocity is

$$\omega = \frac{\Delta\theta}{\Delta t} = \frac{4\pi \text{ rad}}{1 \text{ s}} = 4\pi \text{ rad s}^{-1}$$

The direction of ω is into the page since the motion is clockwise.

(b) In 4 s, the disk rotates through an angle

$$\Delta\theta = \omega \, \Delta t = (4\pi \text{ rad s}^{-1})(4 \text{ s}) = 16\pi \text{ rad}$$

The speed v of points on a rotating object can be related to ω. From the definition, the displacement of a point a distance r from the rotation axis is $s = r\theta$, so $ds/dt = r \, d\theta/dt$. $v = ds/dt$ is the speed of the point and $\omega = d\theta/dt$ is the angular speed, so

$$v = r\omega$$

r has units of meters and ω is in radians per second (rad s^{-1}), but the product $r\omega$ has units of rad m s^{-1} = m s^{-1}; the radians are dropped in the final result.

The speed of points on a rotating object is proportional to the distance r from the rotation axis. The high speeds sometimes encountered when objects rotate are illustrated in the following example.

Example 8.2

The maximum speed of the blades on rotary lawn mowers is limited to reduce the hazard from flying stones and other debris. A presently available model has a rotation rate of 3700 revolutions per minute and a blade 0.25 m in radius. What is the speed at the tip of the blade?

To use $v = r\omega$, we need to find ω. Converting 3700 rev min⁻¹ to rad s⁻¹,

$$\omega = 3700 \, \frac{\text{rev}}{\text{min}} \left(\frac{2\pi \, \text{rad}}{\text{rev}} \right) \left(\frac{1 \, \text{min}}{60 \, \text{s}} \right) = 387 \, \text{rad s}^{-1}$$

Then the velocity at the blade tip is $v = r\omega = (0.25 \, \text{m})(387 \, \text{rad s}^{-1}) = 97 \, \text{m s}^{-1}$, which is nearly 200 mi h⁻¹.

The *angular acceleration* $\boldsymbol{\alpha}$ (alpha) is defined as the rate of change of the angular velocity,

$$\boldsymbol{\alpha} = \frac{d\boldsymbol{\omega}}{dt} \tag{8.4}$$

The units of $\boldsymbol{\alpha}$ are radians per second per second. If the orientation of the rotation axis does not change, $\boldsymbol{\alpha}$ points along the axis and is either parallel or opposite to $\boldsymbol{\omega}$. For example, if the disk of Fig. 8.5 is increasing its rotation rate, $\boldsymbol{\alpha}$ and $\boldsymbol{\omega}$ are parallel. If the disk is slowing down, $\boldsymbol{\alpha}$ is opposite to $\boldsymbol{\omega}$.

When an object rotates, points on the object experience a radial or centripetal acceleration (Chapter Four). If the object has an angular acceleration, points on the object have both a tangential and a radial acceleration (Fig. 8.5). The magnitude of the tangential acceleration of a point is $a_T = dv/dt$, so from $v = r\omega$, $dv/dt = r \, d\omega/dt$ and

$$a_T = r\alpha \tag{8.5}$$

In Chapter Four, we found that the magnitude of the radial acceleration is $a_r = v^2/r$. Using $v = r\omega$, the vector $\mathbf{a}_r$ can be written as

$$\mathbf{a}_r = -\frac{v^2}{r}\hat{\mathbf{r}} = -\omega^2 r \hat{\mathbf{r}} \tag{8.6}$$

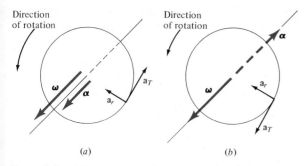

Direction of rotation

Direction of rotation

(a) *(b)*

Figure 8.5. In *(a)* the angular velocity of the disk is increasing, so α and ω are parallel. In *(b)* the angular velocity is decreasing so α and ω are antiparallel. The tangential and radial accelerations $\mathbf{a}_T$ and $\mathbf{a}_r$ of a point on the disk are also shown.

The minus sign indicates that $\mathbf{a}_r$ is directed opposite to $\hat{\mathbf{r}}$ or radially inward.

Relatively large accelerations often accompany rotational motion. For example, suppose a woman with an arm of length 0.6 m swings her arm rapidly in a vertical circle at, say, 1.5 revolutions per second = 3π rad s⁻¹ (Fig. 8.6). The radial acceleration of a point on her hand will be $a_r = \omega^2 r = (3\pi \, \text{rad s}^{-1})^2(0.6 \, \text{m}) = 53 \, \text{m s}^{-2}$. This is an inward acceleration of more than 5 g.

Figure 8.6. A woman swinging her arm in a vertical circle. Points on her arm and hand will experience an inward radial acceleration $\mathbf{a}_r$.

8.2 | NEWTON'S LAWS FOR ROTATIONAL MOTION

In earlier chapters, we described how, by using Newton's laws, we can find the forces and the acceleration for objects in translational motion. Once the acceleration is known, the subsequent motion can be predicted. Similarly for rotational motion, we would like to be able to find the angular acceleration $\boldsymbol{\alpha}$ from which we can predict the subsequent motion.

For translational motion, we found the acceleration from the forces using $\mathbf{F} = m\mathbf{a}$. For rotational motion, we show that it is not the forces but the torques exerted by these forces that determine the angular acceleration. We use a simple example to show how this occurs.

Figure 8.7 shows a mass m at the end of a string swinging on a frictionless plane in a horizontal circle. The mass is subjected to two forces, the tension $\mathbf{T}$ in the string directed toward the center C, and the force $\mathbf{F}_a$ applied at right angles to the string. These two forces produce radial and tangential accelerations; $\mathbf{T} = m\mathbf{a}_r$ and $\mathbf{F}_a = m\mathbf{a}_T$.

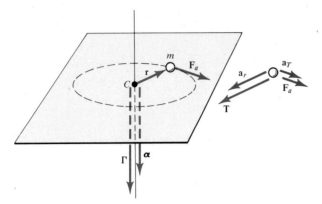

Figure 8.7. The force $\mathbf{F}_a$ exerts a torque on m about the point C. It also produces tangental, $\mathbf{a}_T$, and centripetal, $\mathbf{a}_r$, accelerations.

The line of action of $\mathbf{T}$ passes through point C, so it produces no torque about that point and has no effect on the rotation. On the other hand, $\mathbf{F}_a$ does produce a torque Γ about point C, $\Gamma = F_a r$. From Newton's second law, $F_a = ma_T$, and $a_T = r\alpha$. Thus the torque due to F_a can be written as $\Gamma = F_a r = (mr\alpha)r = mr^2\alpha$. The quantity mr^2 is the *moment of inertia, I,* of the point mass. The vector form of the torque equation is then

$$\Gamma = I\alpha \qquad (8.7)$$

This is the rotational version of Newton's second law. In this form, Eq. 8.7 holds true for any rigid body and states that:

For any object rotating about a fixed axis, the net torque on the object is equal to the moment of inertia of the object times the angular acceleration.

Moment of Inertia
Despite the fact that $\Gamma = I\alpha$ is similar in form to $\mathbf{F} = m\mathbf{a}$, it is important to realize that both the torque Γ and the moment of inertia I depend on the position of the axis of rotation. We will also find that I depends on the shape and mass of the rotating object.

To calculate the moment of inertia of a complex object, we must mentally separate the object into N small pieces of mass $m_1, m_2, \ldots, m_N$. Then each piece is a distance $r_1, r_2, \ldots, r_N$ from the axis of rotation. The moment of inertia of the first piece is $m_1 r_1^2$, that of the second is $m_2 r_2^2$, and so on. The net moment of inertia is the sum of all such terms

$$I = m_1 r_1^2 + m_2 r_2^2 + \cdots + m_N r_N^2 \qquad (8.8)$$

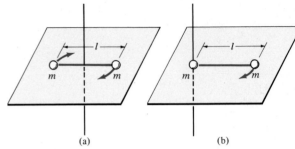

(a)   (b)

Figure 8.8. Two point masses on a massless bar with axis of rotation (a) through the center; (b) through an end.

From this we see that the moment of inertia is large when the pieces are far from the rotation axis. The calculation of the moment of inertia is illustrated in the following example.

Example 8.3
Two equal point masses m are at the ends of a massless thin bar of length l (Fig. 8.8). Find the moment of inertia for an axis perpendicular to the bar through (a) the center, and (b) an end.

(a) For an axis through the center, each mass is a distance $l/2$ from the axis. Summing the mr^2 terms for each of the two masses,

$$I = m\left(\frac{l}{2}\right)^2 + m\left(\frac{l}{2}\right)^2 = \frac{ml^2}{2}$$

(b) For an axis through an end, the mass at that end has $r = 0$, while the other mass is at a distance l, so

$$I = 0 + ml^2 = ml^2$$

Thus we see that the moment of inertia depends on the position of the rotation axis.

Example 8.4
Find the moment of inertia of a bicycle wheel of radius R if the mass m is concentrated in the rim and tire. (The spokes are considered massless.)

If we break up the tire into small pieces, each one is the same distance R from the axle. Thus the moment of inertia is just the total mass m times R^2

$$I = mR^2$$

This arrangement yields the maximum moment of inertia for a wheel of a given mass and radius. It is this large moment of inertia that gives a bicycle some of its stability. Any torques caused, for example, by a rough road will only produce small angular accelerations, since $\alpha = \Gamma/I$ and I is large.

The moment of inertia is often required for objects such as rods or cylinders whose mass is distributed in

TABLE 8.1

Moments of inertia. The mass of each object is taken to be *m*. The dotted line is
the rotation axis.

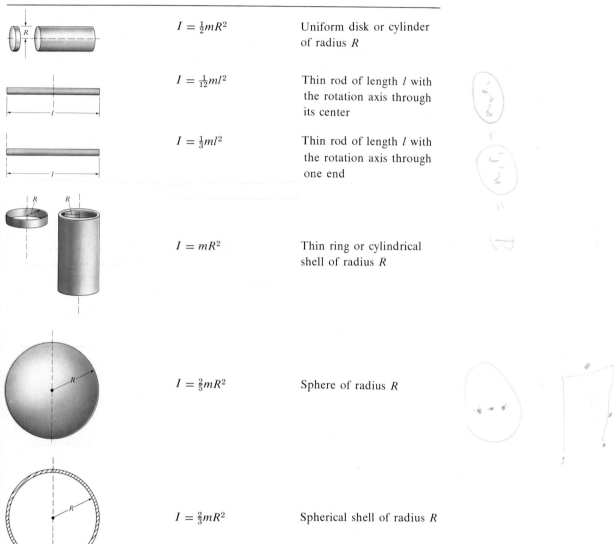

	$I = \frac{1}{2}mR^2$	Uniform disk or cylinder of radius R
	$I = \frac{1}{12}ml^2$	Thin rod of length l with the rotation axis through its center
	$I = \frac{1}{3}ml^2$	Thin rod of length l with the rotation axis through one end
	$I = mR^2$	Thin ring or cylindrical shell of radius R
	$I = \frac{2}{5}mR^2$	Sphere of radius R
	$I = \frac{2}{3}mR^2$	Spherical shell of radius R

a continuous fashion. If this distribution is known in detail, then *I* can be calculated mathematically. Except for a few special cases such as the examples above, this requires mathematics beyond the scope of this textbook. Some of the results obtained are listed in Table 8.1.

For irregular objects such as a bone or forearm, it is usually necessary to determine the amount of inertia experimentally. Such experimental results are often expressed by giving the mass *m* of the object and *radius of gyration k* defined by

$$I = mk^2 \qquad \text{or} \qquad k = \sqrt{\frac{I}{m}} \qquad (8.9)$$

A point mass *m* at a distance *k* from the rotation axis would have the same moment of inertia as the actual object. For example, the radius of gyration of the sphere in Table 8.1 is found from

$$k = \sqrt{\frac{I}{m}} = \sqrt{\frac{\frac{2}{5}mR^2}{m}} = R\sqrt{\frac{2}{5}} = 0.63R$$

The following example shows how $\Gamma = I\alpha$ can be used in a problem involving both rotation and translation.

Example 8.5

A wheel is on a horizontal axle of radius $r = 0.01$ m supported by frictionless bearings (Fig. 8.9). A 5-kg block attached to a string is wrapped around the axle and released from rest. It falls a distance of 1 m in 10 s. (a) What is the acceleration of the block? (b) What is the tension in the string? (c) What is the moment of inertia of the wheel and axle?

(a) The forces on the falling mass are constant, so its acceleration a is constant. Since the block falls from rest, we can use $x = \frac{1}{2}at^2$. With $x = 1$ m and $t = 10$ s,

$$a = \frac{2x}{t^2} = \frac{2(1\text{ m})}{(10\text{ s})^2} = 0.02\text{ m s}^{-2}$$

(b) Using Newton's second law for the falling mass, $mg - T = ma$. Solving for T,

$$T = m(g - a) = (5\text{ kg})(9.8 - 0.02)\text{ m s}^{-2} = 48.9\text{ N}$$

(c) The line of action of the bearing force $\mathbf{N}$ passes through the rotation axis, so $\mathbf{N}$ gives zero torque; the magnitude of the torque due to $\mathbf{T}$ is rT. The acceleration a of the block is also the tangential acceleration of a point on the edge of the axle. Thus $a = r\alpha$ and $\Gamma = I\alpha$ becomes

$$rT = I\frac{a}{r}$$

Solving for I,

$$I = \frac{r^2 T}{a} = \frac{(0.01\text{ m})^2(48.9\text{ N})}{0.02\text{ m s}^{-2}} = 0.245\text{ kg m}^2$$

This example illustrates one method of determining I experimentally. By measuring the time for the mass to fall we can find I. This procedure is often used in the introductory physics laboratory.

8.3 | ANGULAR MOTION AND ANGULAR MOMENTUM

In the preceding sections, we have shown that the angular position, velocity, and acceleration are useful in describing the rotation of rigid bodies. Further-

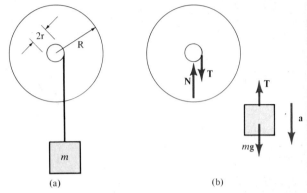

Figure 8.9. (a) A wheel and axle mounted on frictionless bearings have a block of mass m hanging from a string wrapped around the axle. (b) The forces on the axle and on the falling mass. $\mathbf{T}$ is the tension on the string, and $\mathbf{N}$ is the force exerted by the bearings.

more, we found that the analog of Newton's second law for rotation is $\Gamma = I\alpha$. From these results, it follows quite directly that all the principles that we have discussed for translation have rotational analogs. These analogs are summarized in Tables 8.2 and 8.3.

In Table 8.2, we list the equations describing the motion of an object with a constant angular acceleration α along the axis of rotation. The other rotational relations and their translational counterparts are given in Table 8.3.

For example, the kinetic energy is $\frac{1}{2}mv^2$ for translation and $\frac{1}{2}I\omega^2$ for rotation; m is replaced by I, and $\mathbf{v}$ by ω. The angular momentum at any instant (the analog of $\mathbf{p} = m\mathbf{v}$ for translation) is the product of the momentum of inertia and the angular velocity,

$$L = I\omega \qquad (8.10)$$

The relationship between the angular impulse $\Gamma \Delta t$ and the change in angular momentum is especially important.

$$\Gamma \Delta t = \mathbf{L}_f - \mathbf{L}_i \qquad (8.11)$$

A torque Γ acting during a time Δt changes the angular momentum from an initial value of $\mathbf{L}_i$ to a final value $\mathbf{L}_f$.

Just as the conservation of momentum plays an important role in the analysis of many phenomena, the conservation of angular momentum is also an important principle. (In this text, momentum always refers to the quantity $\mathbf{p} = m\mathbf{v}$. If we are referring to $L = I\omega$ we always use the words *angular momen-*

TABLE 8.2

Equations for constant angular acceleration α along the axis of rotation and their translational motion analogs. In using these equations, one direction along the rotation axis is defined as positive and the other as negative; θ, ω, and α can each be positive or negative. The motion is assumed to begin at $t = 0$

Constant Linear Acceleration a	Constant Angular Acceleration α
$x = x_0 + v_0 t + \frac{1}{2}at^2$	$\theta = \theta_0 + \omega_0 t + \frac{1}{2}\alpha t^2$
$v = v_0 + at$	$\omega = \omega_0 + \alpha t$
$x = x_0 + \frac{1}{2}(v + v_0)t$	$\theta = \theta_0 + \frac{1}{2}(\omega + \omega_0)t$
$v^2 = v_0{}^2 + 2a(x - x_0)$	$\omega^2 = \omega_0{}^2 + 2\alpha(\theta - \theta_0)$

TABLE 8.3

Translational and rotational motion comparisons

Quantity	Translation	Rotation	Relationship
Position, displacement	x, s	θ	$x = r\theta$
Velocity	$\mathbf{v}$	ω	$v = r\omega$
Acceleration	$\mathbf{a} = \mathbf{a}_r + \mathbf{a}_T$	α	$a_T = r\alpha$
			$a_r = \omega^2 r$
Net force, torque	$\mathbf{F}$	$\mathbf{\Gamma}$	$\mathbf{\Gamma} = \mathbf{r} \times \mathbf{F}$
Mass, moment of inertia	m	I	$I = m_1 r_1{}^2 + m_2 r_2{}^2 + \cdots$
Momentum, angular momentum	$\mathbf{p} = m\mathbf{v}$	$\mathbf{L} = I\omega$	
Newton's second law	$\mathbf{F} = m\mathbf{a}$	$\mathbf{\Gamma} = I\alpha$	
	$= \dfrac{\Delta \mathbf{p}}{\Delta t}$	$= \dfrac{\Delta \mathbf{L}}{\Delta t}$	
Equilibrium	$\mathbf{F} = 0$	$\mathbf{\Gamma} = 0$	
Kinetic energy	$K = \frac{1}{2}mv^2$	$K = \frac{1}{2}I\omega^2$	
Impulse	$\mathbf{F}\,\Delta t = \mathbf{p}_f - \mathbf{p}_i$	$\mathbf{\Gamma}\,\Delta t = \mathbf{L}_f - \mathbf{L}_i$	
Work	$W = F_s s$	$W = \Gamma_\theta \theta$	
Power	$\mathcal{P} = F_s v$	$\mathcal{P} = \Gamma_\theta \omega$	
Uniform acceleration	See Table 8.2		

tum.) As we see from $\mathbf{\Gamma}\,\Delta t = \mathbf{L}_f - \mathbf{L}_i$, the *angular momentum is constant or conserved if the external torque acting on the system is zero:*

$$\mathbf{L}_f = \mathbf{L}_i; \qquad I_f \omega_f = I_i \omega_i \qquad (\mathbf{\Gamma} = 0) \quad (8.12)$$

Many of us have had the experience of spinning on a stool. When the friction is small, the torque tending to change the angular momentum and angular velocity is also small. Angular momentum is approximately conserved. If one starts spinning with their legs folded and then extends them, the moment of inertia increases and the angular velocity decreases (Fig. 8.10). This follows from $I_f \omega_f = I_i \omega_i$. If I_f is greater than I_i, ω_f is less than ω_i. Conversely, if spinning is initiated with the legs extended, folding the legs will cause one to spin faster.

Figure skaters also spin very rapidly. They start a spin with their limbs extended and increase their angular velocity by pulling their limbs in closer to their bodies (Fig. 8.11).

In both of these examples the angular momentum is conserved, but there is a change in the rotational

Figure 8.10. Changing the moment of inertia results in a change in angular velocity if the angular momentum is conserved.

 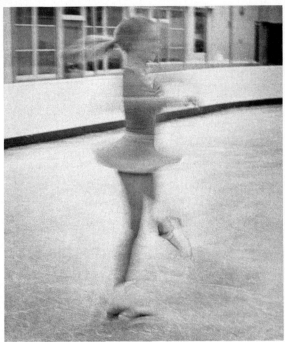

Figure 8.11. A figure skater can increase her angular speed by folding her arms. (Photos by David Leonardi.)

kinetic energy. When the limbs are extended or retracted, there are equal and opposite forces acting on the body and limbs. These two forces produce no net torque, but they each do work. Thus, when we retract our limbs to spin faster, we do work, increasing our kinetic energy.

Example 8.6

A figure skater begins to spin at 3π rad s^{-1} with her arms extended. (a) If her moment of inertia with arms folded is 60 percent of that with arms extended, what is her angular velocity when she folds her arms? (b) What is the fractional change in her kinetic energy?

(a) If we assume that the ice is nearly frictionless, angular momentum is conserved: $I_f\omega_f = I_i\omega_i$. With $I_f = 0.6I_i$ and $\omega_i = 3\pi$ rad s^{-1}, we have

$$\omega_f = \frac{I_i}{I_f}\omega_i = \frac{3\pi \text{ rad s}^{-1}}{0.6} = 5\pi \text{ rad s}^{-1}$$

(b) Her initial and final kinetic energies are $K_i = \frac{1}{2}I_i\omega_i^2$ and $K_f = \frac{1}{2}I_f\omega_f^2$. To find the fractional change, we compute

$$
\begin{aligned}
\frac{\Delta K}{K_i} &= \frac{K_f - K_i}{K_i} \\
&= \frac{\frac{1}{2}(0.6I_i)(5\pi \text{ rad s}^{-1})^2 - \frac{1}{2}I_i(3\pi \text{ rad s}^{-1})^2}{\frac{1}{2}I_i(3\pi \text{ rad s}^{-1})^2} \\
&= \frac{2}{3}
\end{aligned}
$$

The skater increases her kinetic energy by 67 percent when she folds her arms. This is equal to the work she has to do to bring her arms in.

Conservation of angular momentum does *not* mean that the angular position of an object must remain the same in the absence of external torques. One example of this is given in Fig. 8.12.

A more interesting example of this principle is provided by a falling cat, which can always land on its feet if the fall is long enough. Looking at Fig. 8.13, in (b) the cat prepares to turn about a horizontal axis by drawing the front paws close to this axis, diminishing the moment of inertia of the fore part. The hind legs are left extended, so the rear part has a larger moment of inertia. In (c) the cat rotates the fore part in one direction and the rear part in the opposite direction as required by angular momentum conservation. In (d) the fore legs are extended and the hind legs drawn in, so the rear portion has the smaller moment of inertia. Thus in (e) the cat rotates

Figure 8.13. The sequence of motions performed by a cat falling from rest when initially upside down. The sketches are from photos at 1/20 second intervals. The motions are described in the text. (Adapted from Tricker and Tricker.)

the hind portion with relatively little counter-rotation of the fore part.

The tail of the cat also plays a role in this motion. The tail is suddenly coiled and then unwound or straightened and rotated so that its angular momentum is in one direction, and the body then acquires an angular momentum in the opposite direction. However, if a cat has no tail, its ability to land upright is only slightly impaired.

8.4 | SENSORY PERCEPTION OF ANGULAR MOTION

If a person is blindfolded and put on a frictionless chair that can rotate, the following results are observed.

1   If the chair is turned 90° clockwise and stopped, the person can identify the direction of rotation and the fact that it has stopped. Reversal of the process yields the same results.

Figure 8.12. Top view of a person sitting in a swivel chair that turns easily. If an arm is moved repeatedly along the dotted path, the swivel chair gradually rotates clockwise. The chair is in motion only when the hand is moving along the circular parts of the path, since the hand's angular momentum is zero when moving along a line through the axis of rotation. Holding a heavy weight in the hand makes the chair rotate faster.

Rotation of the chair

141

2 If the chair is rapidly accelerated and the rotation is continued, the person can correctly identify the direction of rotation for about 20 s. After 20 s the answers are uncertain.

3 If after about 30 s the chair is slowed to a lower angular velocity, the person will generally indicate that motion has stopped.

4 If the chair is then stopped, the person usually perceives rotation in the opposite direction.

To interpret these results we must examine the structure of the inner ear. The inner ear is composed of two portions (Fig. 8.14). One portion, the *cochlea*, contains the auditory response elements. The other portion is comprised of three *semicircular canals*, which primarily detect motion of the head and have little auditory function.

The structure of one of the semicircular canals is shown in Fig. 8.15. The canal contains a watery fluid called the *endolymph* and has a projection somewhat like a swinging door, called the *cupola*, which senses relative motion of the fluid. To understand the operation of the canal, consider what happens to a bucket of water suspended from a rope when it is set into rotation (Fig. 8.16). The water is initially at rest and has zero angular momentum. It remains at rest when the bucket starts spinning, since the frictional force between the wall of the bucket and the water is small. However, the torque resulting from this force gradually increases the angular momentum of the water, and after some seconds the water is rotating with the same angular velocity as the bucket. If the

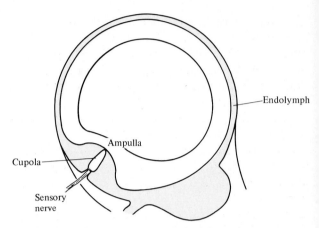

Figure 8.15. A semicircular canal of a human. The endolymph is a fluid that can flow around the canal. The ampulla is a swelling into which the cupola projects. The cupola can block the entire canal, but it is elastic and will bend when the fluid moves. When the cupola is pushed aside by the flowing fluid, the sensory nerve detects this motion and the information is transmitted to the brain.

bucket is now stopped, the water continues to rotate for some time. The *relative* motion of the water and bucket is now opposite to that when the bucket was first set into motion.

In the inner ear, it takes a second for the endolymph to start moving at about the speed of the canal, so initially the cupola is deflected by its motion relative to the fluid. This deflection is sensed by the nerve, and information is transmitted to the brain. Once the canal, fluid, and cupola are all rotating at the same

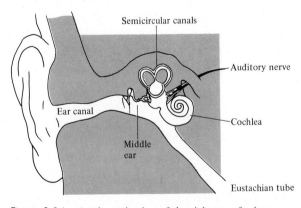

Figure 8.14. A schematic view of the right ear of a human. The three mutually perpendicular semicircular canals function as indicators of rotations about three perpendicular axes in space.

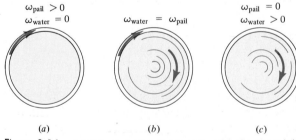

Figure 8.16. (*a*) A bucket of water is spun, with the water initially at rest. (Small scraps of paper on the surface make its motion more readily visible.) The water remains nearly at rest for a few seconds, and the pail is moving *clockwise* relative to the water. (*b*) After a while the frictional torques have increased the angular velocity of the water and it has the same angular velocity as the bucket. (*c*) When the bucket is stopped the water continues to move for some time until slowed by the frictional torques. Relative to the water, the pail is now moving *counterclockwise*.

rate, the cupola starts to return to its normal position. However, the elastic or spring force on the cupola is so weak that it takes about 20 s for this return to be completed. The perception of rotation persists during this 20-s period, but afterward the ear can no longer detect rotation.

If the rotation rate is now reduced, the canal slows while the fluid continues to move briefly at the higher speed. Thus the cupola is deflected in the opposite direction, and the subject perceives a change in the direction of rotation. If this slower rotation continues for about 20 s, the cupola once more returns to its normal position, and the subject again perceives no rotation. Finally, if the rotation is stopped, the cupola is again deflected, and the subject perceives a rotation opposite to the original direction.

When a subject is put through the same sequence but is not blindfolded, the eyes do not normally stare directly outward at passing objects. Instead, the eyes focus briefly on a given object, then fix on another object, and so on. This is called the *nystagmus reflex*. The flicking of the eyes will continue as long as the sensory nerve detects a bending of the cupola. When the cupola resumes its normal position, the reflex action of the eyes stops and the subject, if untrained, sees objects streaking by. Then balance is lost.

When the rotation is stopped, the cupola is again deflected and the nystagmus reflex acts, this time in the opposite direction. Objects at rest appear to be moving steadily and then moving back again. The postural muscles are then trying to respond to nonexistent motion, resulting in staggering.

SUMMARY

The equations describing rotational motion are very similar in appearance to those for translational motion if a proper choice of angular variables is made. Measuring angles in radians leads to convenient definitions of the angular position, velocity, and acceleration. Newton's second law for rotational motion has the form $\Gamma = I\alpha$. Γ is the torque acting, and I is the moment of inertia. The moment of inertia of an object is proportional to its mass, but it also depends on how the mass is distributed with respect to the rotation axis.

The angular momentum of an object with a moment of inertia I and angular velocity ω is $\mathbf{L} = I\omega$. A torque will produce a change in the angular momentum according to $\Gamma \Delta t = \mathbf{L}_f - \mathbf{L}_i$. If the net torque is zero, the angular momentum is constant or conserved. This explains a number of situations where changes in the moment of inertia result in changes in angular velocity. If $\mathbf{L} = I\omega$ is constant, an increase in I results in a decrease in ω and vice versa.

Checklist

Define or explain:

angular position	moment of inertia
radian	Newton's second law for
angular velocity	rotation
angular acceleration	radius of gyration
tangental acceleration	angular momentum
centripetal acceleration	angular momentum
torque	conservation

EXERCISES

Section 8.1 | Angular Variables

8-1 A point on a bicycle wheel moves a distance $s = 1$ m. If this point is 0.4 m from the wheel axis, through what angle has the wheel rotated in (a) radians, and (b) degrees?

8-2 A wheel rotates through two complete revolutions. Through what angle has the wheel rotated in (a) degrees, and (b) radians?

8-3 The following angles are given in radians. Find the corresponding angles in degrees, and sketch these angular coordinates on circles as in Fig. 8.2. (a) $\theta = \pi/3$ rad, (b) $\theta = 3\pi/4$ rad, (c) $\theta = 9\pi/4$ rad.

8-4 Figure 8.17 shows a right triangle with two of the angles given. (a) Find these two angles in radians. (b) The sum of the internal angles of any triangle is 180°. Find the third angle in Fig. 8.17 in radians.

Figure 8.17. Exercise 8-4.

8-5 Figure 8.18 shows a sphere rotating about a vertical axis. What is the direction of the angular velocity in (*a*) and in (*b*)?

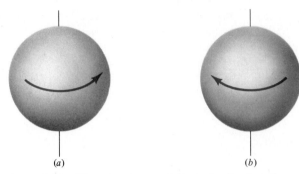

Figure 8.18. The colored arrows indicate the direction of rotation of the sphere about a vertical axis. Exercises 8-5 and 8-6.

8-6   If the sphere in Fig. 8.18*a* rotates at a constant rate, what is the direction and magnitude of ω if the angular displacement is (a) π rad in 0.4 s? (b) 270° in 0.6 s?

8-7   A bicycle rider moves past us from left to right. If her speed is 5 m s^{-1}, (a) what is the direction and magnitude of the angular velocity of one of the wheels, which is 0.4 m in radius? (b) The rider is increasing her speed as she passes, and her acceleration is 1 m s^{-2}. What is the angular acceleration of a wheel?

8-8   (a) Find the radial acceleration at the edge of a 12-in. phonograph record rotating at 78 rev min^{-1}. (b) The record comes to a stop with a uniform angular acceleration in 2 s. Find the tangential acceleration at the edge and the angular acceleration.

8-9   An ultracentrifuge produces a radial acceleration of 300,000 times the acceleration of gravity at a distance of 0.05 m from the rotation axis. What is the angular velocity in radians per second and revolutions per minute?

8-10   A ferris wheel at an amusement park rotates once every 20 s. Its radius is 10 m. (a) What is its angular velocity in radians per second? (b) What is the radial acceleration of a passenger?

8-11   (a) What is the angular velocity in radians per second of the earth's rotation about its own axis? (b) What is the radial acceleration of a person at the equator? (The mean radius of the earth is 6.4×10^6 m.)

Section 8.2 | Newton's Laws for Rotational Motion

8-12   A grinding wheel, a disk of uniform thickness, has a radius of 0.08 m and a mass of 2 kg.

(a) What is its moment of inertia? (b) How large a torque is needed to accelerate it from rest to 120 rad s^{-1} in 8 s?

8-13   Assuming that the earth is a uniform sphere, find its moment of inertia about an axis through its center. (The mean radius of the earth is 6.4×10^6 m. Its total mass is 6×10^{24} kg.)

8-14   Two masses m_1 and m_2 hang on a pulley of mass M (Fig. 8.19). The pulley is a solid cylinder of radius R and rotates without friction. What is the tangential acceleration of the wheel if $M = m_2$ and $m_1 = \frac{1}{2}m_2$?

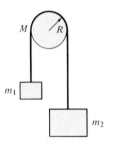

Figure 8.19. Exercises 8-14 and 8-15.

8-15   Two masses m_1 and m_2 hang on a frictionless pulley of mass M (Fig. 8.19). If all the mass of the pulley can be considered to be at its rim, and $M = 2m_2 = 3m_1$, find the acceleration of the masses m_1 and m_2.

8-16   Assume the disk of Fig. 8.5*b* has an initial angular velocity of $\omega_0 = 10$ rad s^{-1} directed as shown. The angular acceleration is 2.5 rad s^{-2} directed opposite to ω_0. (a) How long does it take for the disk to stop turning? (b) Through what angle has the disk turned in this time?

8-17   A car accelerates uniformly from rest to 20 m s^{-1} in 15 s. The wheels have a radius of 0.3 m. (a) What is the final angular velocity of the wheels? (b) What is the angular acceleration of the wheels? (c) What is their angular displacement during the 15-s interval?

8-18   The flywheel of an automobile engine is turning at 700 revolutions per minute. The accelerator is depressed, and in 6 s the speed is 3500 rev min^{-1}. (a) Find the initial and final angular velocities in rad s^{-1}. (b) Find the average angular acceleration. (c) Assuming constant angular acceleration, find the angular displacement during the 6 second acceleration period. (d) Find the tangential acceleration of a point on the flywheel of the engine that is 0.2 m from the axis.

Section 8.3 | Angular Motion and Angular Momentum

8-19   A bicycle wheel has a radius of 0.36 m, and the bicycle is moving at 6 m s⁻¹. The mass of the wheel is 2 kg. (a) What is the angular velocity of the wheel? (b) Assuming the mass of the wheel is entirely at its edge, find its kinetic energy. (c) What is the angular momentum of the wheel?

8-20   An acrobat holds a long pole and walks on a tightrope. (a) Using angular momentum arguments, explain which way he should tip the pole if he starts tipping toward the right side. (b) What effect will weights at the ends of the pole have?

8-21   A tricycle wheel of mass 2 kg has a moment of inertia of 0.04 kg m². (a) What is its radius of gyration? (b) If the wheel rotates once per second, what is the angular momentum of the wheel? (c) What is the kinetic energy of the wheel?

8-22   A 3-kg cylinder of radius 0.2 m is rotating about its axis at 40 rad s⁻¹. (a) Find the kinetic energy and angular momentum if the cylinder is solid. (b) Find the kinetic energy and angular momentum if the cylinder is a thin shell.

8-23   Why does a girl walking on the top of a fence keep her arms outstretched?

8-24   A car is at rest with the engine idling. If the accelerator is suddenly depressed, the left side of the car drops slightly and the right side rises. Which way is the engine turning?

8-25   Refinements in materials technology have opened the possibility of using the energy stored in flywheels to operate vehicles. (a) If a flywheel is a solid cylinder of mass 1000 kg and radius 1 m, what is its kinetic energy when it rotates at 1000 rad s⁻¹? (b) How long can it supply energy to a vehicle at an average power of 20 kW?

8-26   Why must a single engine plane be trimmed for normal horizontal flight by raising one wing flap or aileron and lowering the opposite one? Is this necessary in planes with two engines?

PROBLEMS

8-27   On a moving bicycle (Fig. 8.20), the angular momentum of the spinning wheels is directed to the rider's left. If the bicycle is ridden "no-handed,"

what is the effect of the rider leaning to her left? Explain.

Figure 8.20.   Problem 8-27.

8-28   A girl of mass 50 kg standing at the center of a turntable is spun at 1.5 revolutions per second. In each hand she holds a 6-kg mass, and initially these are close to her body. Estimate her rotational velocity once she has extended her arms outward. (Neglect the moment of inertia of the turntable.)

8-29   An insect of mass 10^{-3} kg walks around the edge of a phonograph turntable of mass 0.5 kg and radius 0.15 m. If the insect comes back to the spot on the record it started from, how far has the record turned? (Neglect friction in the bearings.)

8-30   If the earth's polar ice caps melt, what will happen to the length of the day?

8-31   Show that the angular momentum of a particle of mass m and velocity v, moving in a circle of radius r, has a magnitude $L = mvr$.

***8-32**   In the Bohr model of the hydrogen atom, the only possible electronic orbits are those for which the angular momentum is $nh/2\pi$, where n is an integer and h is Planck's constant. (Use $L = mvr$ from Problem 8-31.) (a) What relation must the radius and linear momentum satisfy in this model? (b) We saw in Chapter Six that the kinetic energy of an electron in an orbit of radius r is $\frac{1}{2}mv^2 = ke^2/2r$. Show that the possible radii of the orbits are $n^2h^2/(4\pi^2kme^2)$. (c) What are the kinetic and total energies for these radii?

***8-33**   For some purposes, a diatomic molecule can be thought of as a dumbbell, with two masses at the ends of a rigid rod and a moment of inertia I. According to Bohr's model for such a molecule, such a dumbbell can only have an angular momentum which is an integer multiple of $h/2\pi$, where h is a number called Planck's constant; the angular momentum is said to be *quantized*. What are the possible kinetic energies of such a dumbbell?

8-34   A single engine airplane is viewed from the front. The propeller begins to rotate counterclockwise. (a) Are the forces due to the ground on the two wheels, one mounted under each wing, the same? If not, which is larger, that on the left wheel or the right? (b) Once the propeller is turning with a constant angular velocity, which wheel experiences the largest force due to the earth?

8-35   A man seated on a stool that rotates freely holds a 5-kg mass in each hand. With his hands in his lap, the moment of inertia about a vertical axis is 20 kg m². With his arms outstretched the total moment of inertia is 35 kg m². Initially the man has an angular velocity of 3 rad s⁻¹ with his arms in his lap. (a) What is the angular velocity of the man with his arms extended? (b) If the man drops the weights while his arms are extended, what is his angular velocity?

8-36   A high diver of height h wishes to do a forward double flip dive in a tuck position, with her arms around her folded legs. She leaves the board in an erect position with an angular velocity of 2.0 rad s⁻¹. Her radius of gyration is $0.25h$ when she is erect and is $0.1h$ in the the tuck position. (a) What is her angular velocity in the tuck position? (b) What is the minimum time necessary for two complete turns? (c) If the diver initially has no vertical translational velocity, what is the minimum height of the diving board above the water if the dive is to be successful? (d) Does a tall person have any advantage in performing this dive?

8-37   Figure 8.21 shows a helicopter in flight. The large propeller which provides the lift force rotates clockwise when viewed from above. Explain, using angular momentum and torque, the function of the small propeller on the tail of the aircraft. This small propeller spins about a horizontal axis.

Figure 8.21.   Problem 8–37. (Courtesy of Sikorsky Aircraft, Division of United Technologies.)

Figure 8.22. A toy gyroscope. Its three pairs of pivots (1, 2, 3) isolate it from any external torques. (Courtesy of Sperry Division, Sperry Rand Corporation.)

SUPPLEMENTARY TOPICS

8.5 | GYROSCOPIC MOTION

The relationship between the torque and angular momentum leads to some startling motions. The behavior of a gyroscope is an example of this (Fig. 8.22). A gyroscope is constructed so that a spinning wheel is supported beneath its center of gravity and is isolated from any external torques. Thus the angular momentum of the spinning wheel continues to point in one direction even if the frame is tilted or rotated.

Gyroscopes of very sophisticated design have virtually no friction and are used in inertial guidance systems for planes, rockets, and ships. Three such gyroscopes can be used to indicate three fixed directions in space. From measurements of a vehicle's acceleration relative to these axes, the velocity and position changes of the vehicle can be computed electronically.

Figure 8.23 shows a toy top on a pedestal. The wheel spins on an axle, which is supported at one end. If the top is placed as shown and released, it will fall if the wheel is not spinning. However, if the wheel is spinning rapidly the top does not fall. Instead it *precesses:* the axis slowly turns in the horizontal plane. To analyze this motion, we must see how the torque affects the angular momentum of the wheel.

The gravitational force **w** exerts a torque on the top about the end resting on the pedestal. If the wheel is not spinning, the initial angular momentum L_i is zero. The torque produces an angular momentum L_f in a time Δt; $\mathbf{\Gamma} \Delta t = \mathbf{L}_f$. Since the torque $\mathbf{\Gamma} = \mathbf{r} \times \mathbf{w}$ and the resulting angular momentum are directed perpendicular to **r** and **w,** the top rotates clockwise and falls.

When the wheel is spinning, the situation changes. Now there is an initial angular momentum $\mathbf{L}_i$ along the top axis. $\mathbf{\Gamma}$ is perpendicular to $\mathbf{L}_i$ and the final angular momentum after a short time is $\mathbf{L}_f = \mathbf{L}_i + \mathbf{\Gamma} \Delta t$. Since $\mathbf{\Gamma} \Delta t$ is perpendicular to $\mathbf{L}_i$, $\mathbf{\Gamma}$ acts to change the direction but not the magnitude of the angular momentum. Thus $\mathbf{L}_f$ and $\mathbf{L}_i$ have the same magnitudes but different directions (Fig. 18.23c). The torque causes the angular momentum and the top to precess in a circle.

Tops are fascinating devices to experiment with, and many different kinds of motion can be produced. When the end of the axis away from the pivot is released with exactly the right initial horizontal velocity, the precessional motion of the axis described above occurs. If, instead, the end of the axis is simply dropped, the top initially begins to fall, and it then gradually acquires the sideways precessional velocity. The result is a vertical bobbing motion called *nutation,* which occurs along with the horizontal precession. If the top is spinning fast enough, the magnitude of the vertical or nutational displacement is small initially and rapidly diminishes as a result of frictional effects. Only the precessional motion is observed after a brief interval has elapsed.

Precession of the Equinoxes | The earth rotates about its axis with a period of one day. The axis of rotation precesses very slowly resulting in what is called the precession of the equinoxes.

When the earth was formed, somewhat more mass accumulated along its equator than at the poles be-

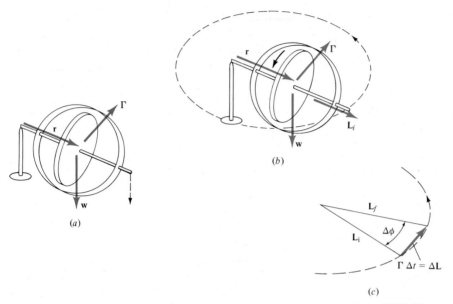

Figure 8.23. (a) The torque produced by **w** causes the top to rotate clockwise and fall. The wheel is not spinning. (b) and (c) When the wheel is spinning the torque causes the angular momentum to precess in a horizontal plane. The end of the top moves along the dashed circle shown.

cause of the centripetal acceleration. The moon, and to some extent the sun, exert unequal forces on the opposite sides of the earth (Fig. 8.24). The net gravitational torque on a spherical planet is zero, so the net torque on the earth results entirely from the unequal forces on the *equatorial bulge* and is quite small.

Because the torque is small, the precession is very slow, with one complete cycle lasting 25,800 years. One effect of this precession is to cause a slow change in the time of year when a star is seen in a given position. Over the last 2000 years this time has shifted about 1 month. A second result is the gradual shifting of the seasons. After 12,900 years winter will occur in that position of the earth's orbit where summer occurs now.

PROBLEMS ON SUPPLEMENTARY TOPICS

8-38 A boy holds a bicycle wheel that is spinning rapidly. The wheel is in the horizontal plane. Viewed from above, the rotation is counterclockwise. (a) Which way does the angular momentum point? (b) The boy tries to turn the plane of rotation of wheel by pushing to his right on the upper end of the axle and to his left on the lower end. What is the direction of the torque? (c) What happens to the wheel?

8-39 (a) If a torque applied to the earth reduced the length of the day by 1 hour, how large an angular impulse would be required? (b) If this torque arose from a couple applied at opposite sides of the equator acting for 1 hour, how large a force would be required? (Use the data of Appendix D.2.)

*__8-40__ A weight is hung from the free end of the axis of a top similar to the top in Fig. 8.23. What

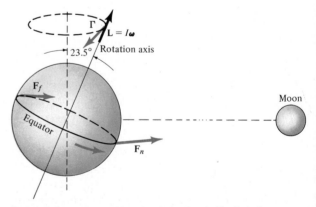

Figure 8.24. The earth's axis of rotation is tilted as shown. The gravitational forces exerted on the equatorial bulge are F_n and F_f. F_n is larger than F_f because the near side of the earth is closer to the moon. The precession is along the dotted curve.

Figure 8.25. A flywheel mounted on a horizontal axle viewed from the rear of the ship. The rotation of the wheel is such that its angular momentum is directed as shown. Problem 8-42.

qualitative effect will this have upon its rate of precession?

8-41 "A top will fall to the floor if released when it is not spinning but will precess if it is spinning." Since this statement does not state any minimum rotational velocity, it suggests that an arbitrarily small rotational speed is sufficient to prevent the top from falling. Explain how nutation resolves this apparent paradox.

8-42 In large ships flywheels are often used to reduce the side to side rolling caused by waves. If a large wave approaches the ship from the left in Fig. 8.25, how does the flywheel effect the ship's subsequent motion?

Additional Reading

R.A.R. Tricker and B.J.K. Tricker, *The Science of Movement,* Mills and Boon Ltd, London, 1966. Applications of angular momentum to balance and motion.

R.L. Page, *The Mechanics of Swimming and Diving,* The Physics Teacher, Vol. 14, p. 72, 1976.

Scientific American articles:

Alfred Gessow, The Changing Helicopter, April 1967, p. 38.

James E. McDonald, The Coriolis Effect, May 1952, p. 72.

Richard P. Post and Stephen F. Post, Flywheels, December 1973, p. 17.

Cornelius T. Leondes, Inertial Navigation for Aircraft, March 1970, p. 80.

CHAPTER NINE

ELASTIC PROPERTIES OF MATERIALS

We have discussed the motion of objects using the implicit assumption that these objects never changed in size or shape. However, an object made from any real material will always be deformed at least slightly and may even break when forces or torques are applied. For example, a steel or wooden beam will bend when a weight is hung from it, and a bone will twist and perhaps fracture when subjected to a torque.

Although materials are held together by complicated forces among the molecules, the effects of these forces can be categorized quite adequately using a few measured quantities. With these quantities one can determine the size and shape of a steel beam needed to safely support a given load or the torque a particular bone can withstand without breaking.

The first portion of this chapter is devoted to the description of the *stresses* which produce deformations or *strains* in a material. We find that the strain depends on the way in which the stress is applied. Using experimentally determined parameters, we can then discuss the strength of materials and the optimal design of objects. We also apply our results to find the relation between the lengths and radii of columns. This is used to discuss tree sizes and to develop a very interesting scaling hypothesis relating the structure and function of animals.

9.1 | GENERAL ASPECTS OF STRESS AND STRAIN

If a certain force stretches a rubber band a given distance, then a force twice as great is needed to produce the same elongation in two such bands, or equivalently, in a *single* band of twice the cross-sectional area. Thus the deformations of materials are determined by the *force per unit area, and not by the total force*. Because of this it is useful to define the *stress* σ in a bar of cross sectional area A (Fig. 9.1a) subjected to a force **F** as the ratio of the force to the area,

$$\sigma = \frac{F}{A} \tag{9.1}$$

The stress is opposed by the intermolecular forces

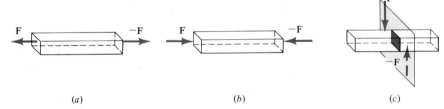

Figure 9.1. A bar subjected to (*a*) tension; (*b*) compression; (*c*) shear forces.

within the material. For example, if the leg of a table is supporting a 100-N weight, then the intermolecular forces must exert an upward 100-N force on the top layer of molecules in the leg.

Three kinds of stress are commonly defined (Fig. 9.1). *Tension stress* is the force per unit area producing elongation of an object. *Compression stress* acts to compress an object. *Shear stress* corresponds to the application of scissorslike forces.

The change in the length of the bar under tension or compression stress is proportional to its length. For example, if a bar of length l subjected to a tension force **F** stretches a distance Δl, then each half of the bar stretches $\frac{1}{2}\Delta l$. The *strain ε* is the *fractional change in length* (Fig. 9.2),

$$\varepsilon = \frac{\Delta l}{l} \qquad (9.2)$$

From this definition, we see that ε is dimensionless and does not depend on the length of the bar. There are three kinds of strains: tension, compression, and shear. Any deformation of an object can be considered as a combination of these three strains. Equation

Figure 9.2. A bar of length l and area A subjected to a tension stress has its length increased by Δl. The ratio $\Delta l/l$ is the strain, ε.

Figure 9.3. An experiment to measure the relationship between stress and strain.

9.2 defines both tension and compression strains. We describe shear strains later in the chapter.

The relation between the stress and the strain for a material under tension can be found experimentally. A bar clamped tightly at each end is gradually stretched, and the applied force F needed to do this is recorded at intervals (Fig. 9.3). The fractional change in length is then the strain, and the force per unit area is the stress. Typical results obtained in this way are shown in Fig. 9.4. Analogous graphs for compression and shear stresses can also be obtained.

For small values of the strain, the stress-strain graph (Fig. 9.4) is a straight line; the stress σ is linearly proportional to the strain ε. This is called the *linear region* for a material. Beyond the *linear limit A*, the stress is no longer linearly proportional to the

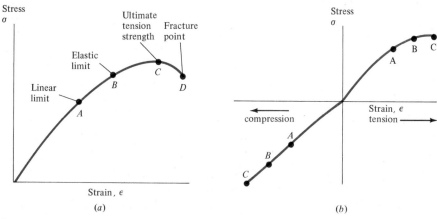

Figure 9.4. (*a*) The stress associated with a given tension strain for a ductile metal. (*b*) Stress-strain graph for a brittle material such as bone. The fracture point D (not shown) is now very close to ultimate strength point C. Note that the slopes in the linear portion of the curve are not the same for tension and compression.

151

strain. However, from *A* to the *elastic limit* or *yield point B,* the object still returns to its original dimensions when the applied force *F* is removed. The deformation up to *B* is said to be elastic. If the applied force is further increased, the strain increases rapidly. In this region, if the applied force is removed, the object does not return completely to its original dimensions; it retains a permanent deformation. The highest point *C* on the stress-strain graph is the *ultimate tension strength* σ_t of the material or its maximum stress. Beyond this point, additional strain is produced even by a reduced applied force, and *fracture* occurs at point *D*. From *B* to *D* the material is said to undergo *plastic deformation*. If the ultimate tension strength and fracture points *C* and *D* are close together, as in Fig. 9.4b, the material is *brittle;* if they are far apart, as in Fig. 9.4a, the material is said to be *ductile*.

When discussing the properties of materials, it must be remembered that all materials exhibit the phenomenon of *fatigue*. After many cycles of applying and removing a load, their ultimate strength gradually diminishes, and the material finally fails, even under weak stresses. For example, a paper clip bent back and forth several times eventually breaks easily. The effects of fatigue must be considered in situations as diverse as designing a bridge and developing pins to insert into bone fractures.

The reasons for material fatigue are not completely understood, but it is believed that after repeated deformations the internal molecular structure of the material is changed. These changes result in decreased intermolecular forces and hence in a reduction in strength of the material.

9.2 | YOUNG'S MODULUS

The elastic deformations of a solid are related to the associated stresses by quantities called *elastic moduli*. In the linear region of the stress-strain graph for tension or compression, the slope equals the stress-to-strain ratio and is called the *Young's modulus E* of the material:

$$E = \frac{\sigma}{\varepsilon} \qquad (9.3)$$

For homogeneous materials such as steel, the Young's moduli for compression and tension are usu-

ally equal. For inhomogeneous materials such as concrete or bone, the moduli for compression and tension are different. Table 9.1 lists representative Young's moduli, ultimate tension strengths σ_t and ultimate compression strengths σ_c for various materials.

The following example illustrates the relationships we have described.

Example 9.1

(a) If the minimum cross-sectional area of the femur of a human adult is 6×10^{-4} m², what is the compressional load at which fracture occurs? (The femur is the main bone in the upper leg.) (b) Assuming the stress-strain relationship is linear until fracture, find the strain at which the fracture occurs.

(a) From Table 9.1, the ultimate compression strength σ_c for bone is 17×10^7 N m⁻². This is the force per unit area that will lead to fracture, and the total force is found by multiplying by the cross-sectional area of the bone. Thus

$$F = \sigma_c A = (17 \times 10^7 \text{ N m}^{-2})(6 \times 10^{-4} \text{ m}^2)$$
$$= 1.02 \times 10^5 \text{ N}$$

This force is large; it is about 15 times the weight of a 70-kg person. However, it is readily exceeded if one falls several meters and lands in a rigid position.

(b) Using the definition of Young's modulus, $E = \sigma/\varepsilon$, with $E = 0.9 \times 10^{10}$ N m⁻² from Table 9.1,

$$\varepsilon = \frac{\sigma}{E} = \frac{17 \times 10^7 \text{ N m}^{-2}}{0.9 \times 10^{10} \text{ N m}^{-2}} = 0.019$$

Thus the bone is reduced in length by 1.9 percent at the load that will cause fracture. The experimental value of the strain at fracture is slightly larger, since the stress-strain curve is not linear (Fig. 9.4b). It flattens out as shown in Fig. 9.4b as the ultimate strength is approached.

The linear stress-strain region of Fig. 9.4 is also called the *Hooke's law* region. In this region, since the stress is linearly related to the strain, the force is linearly related to the elongation. This can be seen using the definition of Young's modulus rewritten as $\sigma = E\varepsilon$. With the definitions of the stress $\sigma = F/A$ and the strain $\varepsilon = \Delta l/l$, this becomes

$$\frac{F}{A} = E \frac{\Delta l}{l}$$

Thus, in tension or compression the force on an object is proportional to its elongation,

TABLE 9.1

Young's moduli and ultimate strengths for representative materials. All quantities have units of N m^{-2}

Material	Young's Modulus, E	Ultimate Tension Strength, σ_t	Ultimate Compression Strength, σ_c
Aluminum	7×10^{10}	2×10^{8}	
Steel	20×10^{10}	5×10^{8}	
Brick	2×10^{10}	4×10^{7}	
Glass	7×10^{10}	5×10^{7}	11×10^{8}
Bone (along axis)			
Tension	1.6×10^{10}	12×10^{7}	
Compression	0.9×10^{10}		17×10^{7}
Hardwood	10^{10}		10^{8}
Tendon	2×10^{7}		
Rubber	10^{6}		
Blood vessels	2×10^{5}		

$$F = k\,\Delta l \qquad (9.4)$$

k is called the *spring constant,* and

$$k = \frac{EA}{l} \qquad (9.5)$$

As long as an object under stress is in the linear region, the object acts like a spring; k is large for strong springs. From the definition of k, we see that increasing the cross-sectional area and decreasing the length both serve to strengthen the spring properties of the object. Spring forces will be discussed further in Chapter 10.

9.3 | BENDING STRENGTH

Almost all mechanical structures ranging from beams to tree trunks and human limbs are subjected to various kinds of stresses. When the stress is a simple compression or tension stress, the shape of the object is unimportant, since the deformation depends only on the cross-sectional area. However, the ability of an object to resist bending or to bend without breaking depends not only on the composition but also on the shape of the object. For example, a hollow tube made from a given amount of material is stronger than a solid rod of the same length constructed from an equal amount of material. Similarly, there is a definite relationship between the lengths and radii of tree trunks and of animal limbs imposed by their shape and composition. In this section, we see how humans and nature design structures for both strength and lightness.

Figure 9.5 shows a bar of length l and rectangular cross section with sides a and b. Placed on two supports, it bends somewhat under its own weight. When

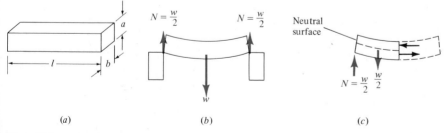

(a)   (b)   (c)

Figure 9.5. (a) and (b) A rectangular bar will bend under its own weight when supported at two points. (c) The left half of the bar experiences forces from one support and its weight. It also experiences forces from the other half of the bar (shown in black).

153

we look at the left half of the bar (Fig. 9.5c), we notice that the vertical force from the left support and the weight of this half are equal and opposite. However, these forces have different lines of action, so they form a couple that rotates the half-bar clockwise. Since the bar is in equilibrium, the right half must exert forces that produce an equal but opposite torque.

From Fig. 9.5c we note that the upper part of the bar is being compressed and the lower part is under tension. The *neutral surface* suffers no change in length. This means that the strength of the bar depends on the elastic properties of the bar.

The upper and lower surfaces of the bar are distorted most, so the largest internal forces will appear at these surfaces (Fig. 9.6b). These forces produce a torque opposing that of the weight and support. The further from the neutral surface they act, the greater will be their contribution to the torque. Thus, with thick bars we can obtain large torques with relatively small internal forces, making it possible to support large loads.

This idea can be put into a more symbolic form. A somewhat complex mathematical analysis shows that when the bar of Fig. 9.5 bends with a radius of curvature R, the *internal torque* Γ in the bar is given by (Fig. 9.7)

$$\Gamma = E \frac{I_A}{R} \tag{9.6}$$

E is Young's modulus for the material, and I_A is called the *area moment of inertia*. For the rectangular bar the area moment of inertia is

$$I_A = \frac{a^3 b}{12} \tag{9.7}$$

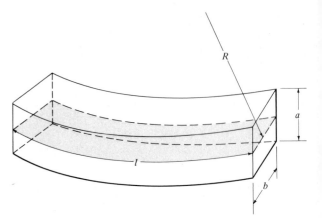

Figure 9.7. A beam of length l bends with a radius of curvature R.

Notice that I_A increases rapidly as a is increased, since a enters to the third power.

Table 9.2 lists the area moments of inertia for several common structures. We use these results in the following example to show that thick boards resist bending better than thin ones.

Example 9.2

Two identical wooden 2 × 6 boards are supported at each end (Fig. 9.8). Each supports its own weight, but one is resting with its wide side down and the other with its narrow side down. Which board bends most and what is the ratio of the radii of curvature for the two boards?

Since each board supports its own weight, the internal torque of each must be the same. Thus, from Eq. 9.6, $I_{A1}/R_1 = I_{A2}/R_2$, where R_1 and R_2 are the radii of curvature of the two boards. Using the definition of the area moment of inertia for each of the two rectangular boards, we see that $I_{A1} = (2 \text{ in.})^3(6 \text{ in.})/12 = 4 \text{ in.}^4$ and $I_{A2} = (6 \text{ in.})^3(2 \text{ in.})/12 = 36 \text{ in.}^4$. Then

$$\frac{4 \text{ in.}^4}{R_1} = \frac{36 \text{ in.}^4}{R_2}$$

or

$$\frac{R_2}{R_1} = 9$$

The radius of curvature of the board with its narrow side down is nine times that of the other. Since a large radius of curvature implies little bending, board 2 does not bend as much as board 1. Consequently, board 2 is less likely to break when a heavy weight is placed upon it.

The results above suggest that to construct strong, light structural members, most of the material should be located as far as practical from the neutral surface.

Figure 9.6. (a) The shaded plane, called the neutral surface, suffers no change in dimension as the bar bends. (b) Detailed internal forces acting on the left half of the bar. The pairs of forces at the upper and lower edges produce the largest torques because they are large and far apart.

Neutral surface

(a)   (b)

TABLE 9.2

Area moments of inertia for vertical loading

Cross section		I_A
Rectangle:	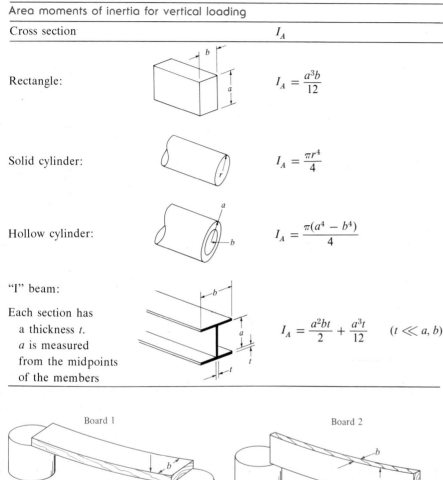	$I_A = \dfrac{a^3 b}{12}$
Solid cylinder:		$I_A = \dfrac{\pi r^4}{4}$
Hollow cylinder:		$I_A = \dfrac{\pi(a^4 - b^4)}{4}$

"I" beam:
 	
Each section has
  a thickness t.
  a is measured
  from the midpoints
  of the members
		$I_A = \dfrac{a^2 b t}{2} + \dfrac{a^3 t}{12}$ $(t \ll a, b)$

Board 1

$b = 6$ in.
$a = 2$ in.

Board 2

$b = 2$ in.
$a = 6$ in.

Figure 9.8. Identical boards resting flat and on edge.

An I beam (Fig. 9.9) is better able to withstand bending torques due to forces in any direction perpendicular to its length than a beam with a square cross section made from the same amount of material. By contrast the two beams can support the same applied *compressional* force, since their cross-sectional areas are the same. Similarly, a hollow tube has more bending strength than a solid rod of the same length and weight. This is why metal chair and table legs are usually hollow (Fig. 9.10).

From this discussion, it would seem advantageous to make structural members with very large diameters and very thin walls. However, a limit to how far one may go in this direction is imposed by the tendency of thin-walled structures to *buckle*. Figure 9.11 shows an experiment one can readily perform to illustrate this point. One sheet of $8\frac{1}{2} \times 11$ paper is rolled into a cylinder a single layer thick and fastened with tape. A second sheet is rolled into a cylinder an inch or so in diameter and taped. When the cylinders

Figure 9.9. (*a*) An I beam is constructed so that most of the material is near the outer surfaces during bending in any direction. (*b*) Brackets with an *L*-shaped cross section are designed for the same purpose. This bracket could be used as a wall mounted support for a shelf.

Figure 9.10. The legs of both chairs have the same length and are constructed from the same amount of material. Those of (*a*) are solid cylinders, while those of (*b*) are hollow cylinders.

Figure 9.11. (*a*) A large radius, thin walled paper tube buckles easily. (*b*) A narrower tube made from an identical sheet of paper with thicker walls will not buckle under the same load.

are stood on end and this textbook is placed on top, the larger cylinder immediately buckles and collapses, while the smaller one is able to support the load. Thus the thinner walls of the larger cylinder are unable to withstand a force applied approximately

along its axis. We will consider buckling in more detail in the next section.

Nature has made extensive use of the principle that hollow structures are stiffer than solid ones of the same cross-sectional area. Bones are generally hollow. For example, in the human femur, the ratio of the inner and outer radii is about 0.5, and the cross-sectional area is only 78 percent of that of a solid bone with the same bending strength. Smaller mammals and birds have bones with relatively thinner walls. For example, the ratio of the inner to outer radii for the humerus of a swan is 0.9, and the cross-sectional area is 38 percent of that of a solid bone with the same strength. The danger of collapse by buckling in this thin-walled bone is reduced by thin, reinforcing struts of bone placed across the interior of the humerus.

9.4 | BUCKLING STRENGTH AND STRUCTURAL DESIGN IN NATURE

In nature the failure of structural members usually results from large torques of various types rather than from simple compression or tension stresses. For example, except when someone falls from a great height, fractures of limb bones are usually the result of bending or twisting. We noted in the preceding section that a thin tube will readily buckle if a force is applied along its axis. More generally, any beam or column may buckle under such a force. Here we discuss the buckling strength of a cylindrical column and illustrate our results by seeing how nature seems to have used buckling strength as a criterion in designing the approximately cylindrical trunks of trees.

To understand how buckling occurs, consider the long, thin, cylindrical column in Fig. 9.12. It is held almost but not exactly vertically, so that its center of gravity is not quite over the center of the base, point *P*. The weight, therefore, exerts a torque about *P* that causes the column to bend. If the material is strong enough, the bending stops when the torques due to the internal forces in the material become large enough to balance the torque due to the weight. However, if the column is very tall and thin, as it bends the torques due to the weight will grow faster than will the torques due to the internal forces. The column will then buckle and collapse.

Figure 9.12. A cylindrical column tilted so that its weight is not directly over the center of its base, point P.

More generally, any vertical column supporting a load or even just its own weight will eventually buckle if its radius is held fixed and its height is increased. This *critical height* is determined by the Young's modulus of the material. This is because Young's modulus determines the internal forces for a given deformation.

For a solid cylinder of radius r supporting only its own weight, the critical height will be shown in Section 9.7 to be

$$l_{cr} = cr^{2/3} \qquad (9.8)$$

Here c is a constant that depends on the weight per unit volume and the Young's modulus of the material composing the column. This result implies, for example, that if a column is just barely stable against buckling, doubling its radius does not permit doubling its height. This is illustrated in the following example.

Example 9.3

Two columns are made of the same material. One has a radius r_1 and the other has a radius of $2r_1$. If both columns can just support their own weight without buckling, what is the ratio of their lengths?

The length of the column of radius r_1 is $l_1 = cr_1^{2/3}$. The other column has a length $l_2 = c(2r_1)^{2/3}$. The ratio l_2/l_1 is

$$\frac{l_2}{l_1} = \frac{(2r_1)^{2/3}}{r_1^{2/3}} = (2)^{2/3} = 1.59$$

Thus the column with twice the radius can only be about 1.6 times as long.

The Height of Trees | The result $l_{cr} = cr^{2/3}$ is
quite general. For example, with suitable values of the constant c, it holds true for tapered columns, hollow columns, and columns supporting loads. It

must also hold true for trees; that is, the maximum height a tree can have and be stable against buckling must vary as $r^{2/3}$. Whether buckling is, in fact, the limiting factor in determining the height of trees can be investigated by comparing their measured heights and radii (Fig. 9.13). While there is considerable scatter in the data, the results support the idea that buckling strength is the key factor in determining the proportions of trees. We will see in Section 9.6 that there is evidence that buckling strength also determines the body proportions of animals.

Figure 9.13. Data for North American trees. The dashed line is that of $l = cr^{2/3}$, with $c = 34.9$ chosen to fit the data. The solid line is the theoretical result for a tapered column that is just on the verge of buckling. Presumably no data points appear above this line since such trees would buckle under their own weight. (From T. McMahon, *Science,* vol. 179, pp. 1201–1204, March 23, 1973. Copyright 1973 by the American Association for the Advancement of Science.)

9.5 | SHEARING AND TWISTING TORQUES

Thus far, we have only considered compression and tension stress. However, it is also common to have forces acting on an object that cause shearing or twisting. In this section, we describe shear forces qualitatively and twisting torques more analytically.

A situation involving shearing forces is shown in Fig. 9.14. A short, fat bar is clamped rigidly at one end, and a force is applied parallel to the face of the bar at its free end. The shearing forces cause a deformation of the bar without bending.

In Fig. 9.14 the original height of the bar is h and

Figure 9.14. A short clamped bar subjected to shearing forces deforms without bending.

the bar is deformed through an angle α. The upper surface of the bar moves a distance δ with respect to the base. The shear stress on the upper face of the bar is $\sigma_s = F/A$ and the shear strain is defined as

$$\varepsilon_s = \frac{\delta}{h} = \tan\alpha$$

The relation between σ_s and ε_s defines the *shear modulus G*:

$$\sigma_s = G\varepsilon_s \qquad (9.9)$$

TABLE 9.3

Shear moduli for some materials in N m^{-2}

Material	Shear Modulus, G
Aluminum	2.4×10^{10}
Bones (long)	10^{10}
Copper	4.2×10^{10}
Glass	2.3×10^{10}
Hardwood	10^{10}
Steel	8.4×10^{10}
Tungsten	11.4×10^{10}

Table 9.3 gives the shear moduli for some materials. Usually the shear modulus is between one-third and one-half of Young's modulus for the material.

Even though an object is subjected only to shearing forces, compression or tension stresses may result on various planes in the object. For example, the cube in Fig. 9.15 is in translational and rotational equilibrium under the action of four shearing forces of equal magnitude acting along the faces. The cube is deformed slightly by these forces as shown. If we examine an imaginary cut along a diagonal plane as in Fig. 9.15b, then the shaded part of the cube must experience a compressive force **F** in the direction shown, since this part is in equilibrium. Similarly, for the plane in Fig. 9.15c, the shaded part experiences a tension force. The net result is that the square shown on the side of the cube in Fig. 9.15d becomes a rectangle when the cube is deformed by shearing forces.

From this discussion it is clear that stress and strain are complex quantities except in rather special cases. As we saw, externally applied shear forces result in both compressive and tension forces inside the material. Full specification of the stress must include the three components of the force on each of three perpendicular planes at each point in the object. Such complete analyses of stress-strain relations are beyond the scope of this textbook.

We now briefly discuss the effect of a torque directed along the axis of a cylinder arising from twisting forces. Such torques occur when a skier's leg is twisted in a fall or when power is transmitted by a rotating shaft. Figure 9.16 shows a cylinder fixed at one end. A couple is applied at the free end, so that there is a torque directed along the axis. If the result-

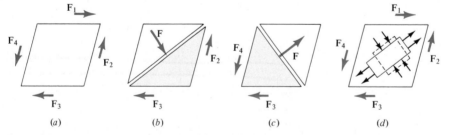

Figure 9.15. (*a*) A cube subjected to four shearing forces of equal magnitude. The cube deforms slightly but it is in static equilibrium. (*b*) The parts formed by an imaginary cut along a diagonal plane are subjected to compressive stresses. (*c*) The other diagonal plane forms parts that experience tension stresses. (*d*) The dotted lines show a square on the side of the cube before the shearing forces are applied. A square drawn with this orientation is deformed into a rectangle by the shearing forces, even though the cube face itself is deformed from a square into an equilateral parallelogram or rhombus.

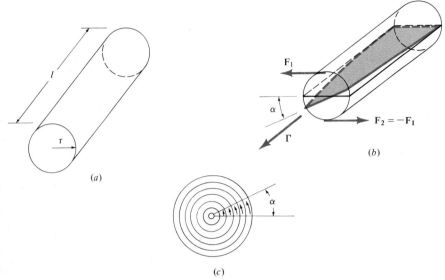

Figure 9.16. (*a*) A cylinder of length *l* and radius *r*. (*b*) The far end of the cylinder is held fixed by a constraint (not shown), and the near end is subjected to forces producing a torque along the axis. The resulting twist of the planes in the cylinder increases linearly with the distance from the fixed end if the applied torque is not too large. (*c*) The near end. Adjacent cylindrical layers deform more as one goes further from the cylinder axis. This is a shearing deformation.

ing deformation is not too large, it is found that a plane drawn along the axis of the cylinder becomes twisted, as in Fig. 9.16*b*. The angle of twist increases linearly with the distance from the fixed end, so that the radial lines remain straight, as do lines originally drawn along the outside of the cylinder parallel to the axis.

To find the relation between the torque $\mathbf{\Gamma}$ and the deformation α, it is necessary to compute the stress σ_s at various distances from the cylinder axis. Because adjacent cylindrical layers are subjected to shear forces (Fig. 9.16*c*), the stress σ_s and strain ε_s are related by $\sigma_s = G\varepsilon_s$. When the torques on each layer are computed and summed, it is found that

$$\Gamma = GI_p \frac{\alpha}{l} \qquad (9.10)$$

This is similar in form to the result for bending torques, but I_p is the *polar moment of inertia*. For a cylinder of radius *r*,

$$I_p = \frac{\pi r^4}{2} \qquad \text{(solid cylinder)} \qquad (9.11)$$

Since the polar moment of inertia increases with the fourth power of the radius, doubling the radius of a cylinder increases its resistance to twisting by a factor of $2^4 = 16$.

If an object is subjected to an increasingly large twisting torque, it will eventually fracture. The torques and corresponding angles at which this happens for several human limb bones are listed in Table 9.4. Under the right conditions, a relatively small force can lead to a fracture. For example, a ski tip is about a meter from the heel of a skier's boot, so a 100-N force at the tip produces a torque of 100 N m. According to the table, this is sufficient to break the

TABLE 9.4

The breaking torque and breaking angle for twisted bones in humans

Bone	Breaking Torque (N m)	Breaking Angle of Twist
Leg		
Femur	140	1.5°
Tibia	100	3.4°
Fibula	12	35.7°
Arm		
Humerus	60	5.9°
Radius	20	15.4°
Ulna	20	15.2°

tibia, the larger of the two bones in the lower leg.

The toe of the boot is about 0.3 m from the heel. If the foot and ski do not rotate, an applied torque of 100 N m requires that the skier exert a force of 100 N m/0.3 m = 330 N at the toe (Fig. 9.17). If the toe release is set to open at a lower force level, the risk of injury is reduced.

Twisting fractures in bones or metal cylinders normally are not breaks at right angles to the axis but instead are spiral fractures (Fig. 9.18). The explanation for this phenomenon is suggested by our discussion of the stresses and strains in a cube subjected

Figure 9.17. An overhead view of a skier's boot and ski. If the ski and foot do not rotate, the net torque about the heel must be zero. Thus a 100-N force applied to the ski results in a 330-N force exerted on the ski by the toe.

Figure 9.18. A spiral fracture of a tibia. (Courtesy of Dr. Robert C. Runyon.)

only to shearing forces. We saw that for some surfaces in the cube, there was pure tension or compression stress, and a square drawn appropriately on a face of the cube was deformed into a rectangle. Similarly, when a rod is twisted, planes drawn at 45° angles to the axis experience pure tension or compression. Fracture occurs when the smaller of the ultimate strengths for tension and compression is exceeded on one of these planes (Fig. 9.19).

SUMMARY

Objects subjected to forces or torques change their shape and may break. The fractional change in size or shape is the strain ε, and the force per unit area producing the deformation is the stress σ. For small applied forces or torques, the stress and strain in a material are usually linearly related. The proportionality constant relating stress and strain in the linear region is Young's modulus for compression or tension,

$$\sigma = E\varepsilon$$

In the presence of a shear stress, the relationship is

$$\sigma_s = G\varepsilon_s$$

where G is the shear modulus.

When a beam bends with a radius of curvature R due to a torque Γ, they are related by

$$\Gamma = E\frac{I_A}{R}$$

I_A is the area moment of inertia. If the area moment of inertia of the beam is large, the radius of curvature for a given torque is also large, and the beam does not bend much.

A column of radius r supporting its own weight has a maximum or critical length

$$l_{cr} = cr^{2/3}$$

c depends on Young's modulus for the material and on the weight per unit volume. If the length exceeds l_{cr}, the column will buckle. This criterion appears to be a limiting factor in the growth of trees.

A twisting torque Γ applied to a uniform cylinder of length l is related to the angle α through which the cylinder twists by

$$\Gamma = GI_p\frac{\alpha}{l}$$

(a) (b) (c)

Figure 9.19. (*a*) When a cylinder is twisted, lines parallel to the axis are twisted, but lines in planes perpendicular to the axis are unaffected. (*b*) When the cylinder is twisted, one set of parallel lines drawn at 45° to the axis is pushed together and the other set is stretched out. (*c*) A single square from part (*b*) becomes deformed into a rectangle. The stress is pure tension along the long sides and pure compression along the short sides.

I_p is the polar moment of inertia and depends on the radius of the cylinder.

Checklist

Define or explain:

tension stress and strain	Young's modulus
compression stress and strain	spring constant
	neutral surface
linear limit	internal torque
elastic limit	area moment of inertia
ultimate tension strength	critical height
	shear stress and strain
brittle	shear modulus
ductile	polar moment of inertia

EXERCISES

Section 9.1 | General Aspects of Stress and Strain

9-1 A tension force of 100 N is applied to the 0.1-m² face of a bar. What is the stress in the bar?

9-2 A tension stress of 2×10^6 N m^{-2} is applied to a bar 0.05 m² in cross section. What is the applied force?

9-3 A 0.4-m pipe under compressional stress changes length by 0.005 m. What is the strain in the pipe?

9-4 The largest tension strain that can occur before fracture in aluminum is 0.003. What is the maximum change in length of a 1-m aluminum pipe?

Section 9.2 | Young's Modulus

9-5 (a) 100-kg mass is suspended from the end of a vertical, 2-m long steel bar with a cross-sectional area of 0.1 m². Find the stress and strain in the bar.

(b) How much does the bar stretch? (c) What is the maximum mass that can be suspended from this bar?

9-6 A hardwood post with dimensions 10 cm by 15 cm by 3 m supports a load of 1000 N along its length. (a) Find the stress and strain in the post. (b) What is its change in length?

9-7 If the minimum cross-sectional area of a human femur is 6.45×10^{-4} m², what is the tension load at which fracture occurs?

9-8 A sheet of glass has an area of 0.5 m² and is 0.005 m thick. (a) If it is placed horizontally, what is the uniformly distributed load at which fracture occurs? (b) What is the change in thickness if half that load is applied?

9-9 A vertical steel post is 3 m long and has a radius of 0.1 m. It is supporting a load of 10^5 N. (a) Find the stress and strain in the post. (b) Find the change in length.

9-10 The average cross-sectional area of a woman's femur is 10^{-3} m², and it is 0.4 m long. The woman weighs 750 N. (a) What is the length change of this bone when it supports half of the weight of the woman? (b) Assuming the stress-strain relationship is linear until fracture, what is the change in length just prior to fracture? (c) Is the answer to part (b) an overestimate or underestimate?

9-11 What is the spring constant of a human femur under compression of average cross-sectional area 10^{-3} m² and length 0.4 m?

Section 9.3 | Bending Strength

9-12 A cylindrical rubber rod is 0.5 m long and has a radius of 0.005 m. (a) What is its area moment of inertia? (b) What torque is exerted by the internal elastic forces on the ends of a rod when it is bent into a circle?

161

9-13   A cylindrical steel rod 2 m long has a radius of 0.01 m. If it is loaded so that it bends elastically with a 20-m radius of curvature, what is the torque due to this load?

9-14   A board is 1 in. by 6 in. in cross section. (a) Compute the area moments of inertia for loads parallel to the longer dimension and for loads parallel to the shorter dimension. (b) What is the ratio of the radii of curvature for the two deflections if equal loads are applied to the board in both orientations?

Section 9.4 | Buckling Strength and Structural Design in Nature

9-15   Using $l = cr^{2/3}$ with the experimental value $c = 34.9$ m$^{1/3}$, find the height of a tree of trunk radius $\frac{1}{8}$ m. Comment on whether your answer seems reasonable.

9-16   Give an argument as to why the data of Fig. 9.13 do not fall on a single straight line.

9-17   A uniform column will buckle under its own weight when

$$l = \left(\frac{2E}{w_0}\right)^{1/3} r^{2/3}$$

For hardwood $E = 10^{10}$ N m^{-2} and $w_0 = 5900$ N m^{-3}. Compare the length of a uniform column of wood of radius $\frac{1}{8}$ m with that of an actual tree of the same trunk radius for which $l = cr^{2/3}$ and $c = 34.9$ m$^{1/3}$.

Section 9.5 | Shearing and Twisting Torques

9-18   Two bones of equal radius are subjected to equal twisting torques. If one is longer than the other, which will fracture first?

9-19   A steel bar is clamped in a vise, so that a cube with sides 0.01 m protrudes above the jaws of the vise as in Fig. 9.14. (a) If a force of 100 N is applied along the top face of the cube, what are the stress and strain? (b) What is the horizontal displacement of the top face?

PROBLEMS

9-20   Estimate the twisting torques on the femur of a football player when he makes a sharp pivot on one foot. Does your result have any bearing on the types of cleats and playing surfaces that should be used?

Figure 9.20.   Problem 9-21.

9-21   The I beam in Fig. 9.20 is constructed from plates 0.003 m thick and is 0.3 by 0.4 m. (a) Calculate its area moment of inertia for loads applied vertically. (b) Calculate the area moment of inertia for a beam with square cross section having the same weight per unit length.

9-22   A 100-N bar is 5 m long. What is the magnitude of the torque on one half due to forces from the other half when it is supported horizontally? (*Hint:* See Fig. 9.5c and assume that the bar bends very little.)

*9-23   Two cylinders are constructed, one solid with radius r and one hollow with radii $a = 2r$ and $b = 3r/2$. If both cylinders are subjected to the same load, perpendicular to their long axes, what is the ratio of their radii of curvature?

9-24   A rod of radius a is replaced by a hollow tube of the same length with inner radius a. (a) If the tube is to have the same area moment of inertia as the rod, what must its outer radius be? (b) What is the ratio of the weights of the tube and rod?

9-25   The inner radius of a bone is half the outer radius. What would be the outer radius of the bone if, for a given torque, it were to twist through the same angle as a steel rod of the same length with a 1-cm radius? (The polar moment of inertia of a hollow cylinder is $\pi(a^4 - b^4)/2$, where a and b are the outer and inner radii, respectively.)

SUPPLEMENTARY TOPICS
9.6 | STRUCTURE AND FUNCTION

The buckling strength criterion was employed in Section 9.4 to relate structure and size in trees. It can also be used to relate structure and physiological function in animals with a different form of the scaling concept introduced in Chapter Six. We will see that this different scaling approach leads to much better agreement with observations.

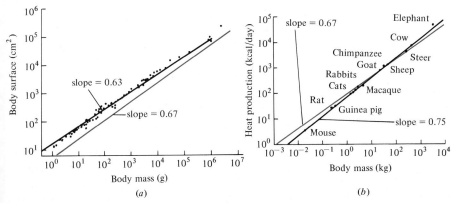

Figure 9.21. (a) The body surface area of mammals of various size. (b) The metabolic rates of various mammals versus mass. The black lines are the predictions of the buckling strength model; the colored lines are those of the simple scaling model. (From T. McMahon, Science, vol. 179, pp. 1201–1204, March 23, 1973. Copyright 1973 by the American Association for the Advancement of Science.)

The simple version of scaling adopted earlier assumes that all of an animal's body dimensions scale with a single characteristic length l. Thus the body volume, and hence the mass m, varies as l^3; the length varies as $l \propto m^{1/3}$. Any area A then varies as l^2 or as $m^{2/3} = m^{0.67}$, so it is predicted that the body surface area will vary as $m^{0.67}$. It is also predicted that the *metabolic rate*—the rate at which food energy is used in the body—will vary in the same way, since the rate of oxygen absorption must vary as the surface area of the lungs, and the rate of heat loss must vary as the body surface area.

The actual scaling laws found by experiment are not always in exact agreement with the predictions of this simple model (Fig. 9.21). For example, one failure of the simple scaling procedure was observed by Kleiber in 1932. He found that the rate of heat production in mammals ranging in size from mice to elephants is not proportional to $m^{0.67}$ but to $m^{0.75}$. This difference is small, but it is sufficient to raise doubts about the validity of the model.

Recently McMahon suggested that one should take account of the fact that most body segments are cylindrical and are perhaps built to withstand buckling. If this is true, the length l and radius r of each body segment are related as we found in Section 9.4, $l \propto r^{2/3}$. Because the volume of a cylindrical body segment is $\pi r^2 l$, the mass m should be proportional to $r^2 l$. Using $r \propto l^{3/2}$ we find $m \propto r^2 l = (l^{3/2})^2 l = l^4$. This is significantly different from the other scaling assumption, $m \propto l^3$.

In this model, we see that lengths should scale as $l \propto m^{1/4}$. Also because $l \propto r^{2/3}$, $r^{2/3} \propto m^{1/4}$ or $r \propto m^{3/8}$. Thus, our scaling assumption is that lengths and radii of body segments scale with mass as

$$l \propto m^{1/4}, \qquad r \propto m^{3/8} \qquad (9.12)$$

The first result follows from $m \propto l^4$. The second comes from $m \propto l^4 \propto (r^{2/3})^4 \propto r^{8/3}$. We can use these results to find how the body surface area of mammals with cylindrical body segments scales with their mass.

To find the surface area, we note that most cylindrical body segments are connected to other segments at either one or both ends. The surface area of one segment will then be proportional to the area of the sides of the cylinder. Thus, $A_{\text{surf}} = 2\pi r l$. Using the relations of Eq. 9.12, $A_{\text{surf}} \propto rl$, so

$$A_{\text{surf}} \propto m^{3/8} m^{1/4} = m^{5/8}$$

$\frac{5}{8} = 0.625$ is very close to 0.63 and fits the surface area data in Fig. 9.21a quite well.

To find the metabolic rate, we calculate not the surface area but the power used to flex a muscle. This power $\mathcal{P}$ is the force exerted, F, times the velocity of muscle contraction, v. Since all mammalian muscles are found to exert the same force per unit area $\sigma = F/A$, we write the force as $F = \sigma A$. A is the cross-sectional area of the muscle. Thus,

$$\mathcal{P} = Fv = \sigma Av$$

For voluntary muscle fiber, it has been found experimentally that the velocity of muscle contraction is also the same for all mammals. Thus σ and v do not depend on the mass and $\mathcal{P} \propto A$. The cross-sectional area A depends on the square of the radius of the muscle, $A \propto r^2$, so

$$\mathcal{P} \propto A \propto r^2 \propto m^{0.75}$$

This is just the result shown in Fig. 9.21b, if we assume that the power expended and the heat production scale in the same way. The rate of work done by the heart muscle should also vary as $m^{0.75}$ and, because metabolic processes use oxygen absorbed through the lung wall, the lung area should depend on $m^{0.75}$. Both of these results have been established experimentally.

From these results, we can also find how pulse rates scale. The metabolic rate, and hence the oxygen demand of the body, is proportional to $m^{0.75}$. The volume of blood pumped per heartbeat is proportional to the heart volume or to m. The blood pumped per second will vary as mf, where f is the heart rate. Then $mf \propto m^{0.75}$ or $f \propto m^{-0.25}$. This means that larger animals should have lower pulse rates. Again, this has been observed experimentally.

To summarize, the assumption that the cylindrical segments of animals have their shape determined by the buckling criterion gives rise to scaling laws that are in good agreement with a variety of experiments. Even if further study should show that this model is not adequate to explain other kinds of data, it does illustrate clearly how physical principles can influence the relationship between size and function in biological systems.

9.7 | DERIVATION OF $l_{cr} = cr^{2/3}$

We consider the particular example of a uniform column of radius r and length l which bends under its own weight with a radius of curvature R (Fig. 9.22). The torque due to the weight w is $\Gamma = wd$ and must be counteracted by a torque acting at the base of the column equal to $\Gamma = EI_A/R$. For a cylinder, $I_A = \pi r^4/4$. Thus, when the column is just at the buckling point,

$$\frac{E\pi r^4}{4R} = wd \qquad (9.13)$$

We now need to find expressions for w and d.

If the weight per unit volume is w_0, the total weight is w_0 times the volume or $w = w_0(\pi r^2 l)$. The distance d can be found from the colored triangle in Fig. 9.22.

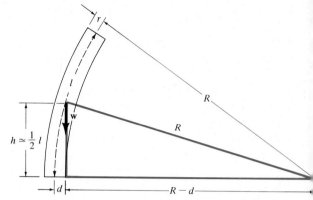

Figure 9.22. A column of length l and radius r, supported at its base, bends with a radius of curvature R.

The sides of the triangle have lengths R, $R - d$, and h. If the radius of curvature is large compared to l, then $h \simeq \tfrac{1}{2}l$. Using the Pythagorean theorem

$$(R - d)^2 + \left(\frac{l}{2}\right)^2 = R^2$$

Squaring and neglecting the term d^2, we find $d = l^2/8R$.

Using our results for w and d in Eq. 9.13,

$$\frac{E\pi r^4}{4R} = w_0(\pi r^2 l)\frac{l^2}{8R}$$

or

$$l = \left(\frac{2E}{w_0}\right)^{1/3} r^{2/3}$$

This is the critical length l_{cr} of Eq. 9.8 for a uniform column. Here, $c = (2E/w_0)^{1/3}$, where E is Young's modulus and w_0 is the weight per unit volume of the column.

EXERCISES ON SUPPLEMENTARY TOPICS

Section 9.6 | Structure and Function

9-26 If the properties of animals did not depend on buckling strength but on compressional strength, how would the cross-sectional area of the legs of animals depend on body weight?

9-27 How does the time an animal can stay under water scale with body mass? Use the scaling hypothesis of Section 9.6. (Assume the oxygen used varies with volume.)

9-28 How does the heat loss to the environment scale with body mass? (Use the scaling hypothesis of Section 9.6.) Are your results consistent with the fact that small animals are not generally found in arctic areas?

9-29 If all mammals lived the same number of heartbeats, which animals would live longest?

PROBLEMS ON SUPPLEMENTARY TOPICS

9-30 Find how the speed with which an animal can run uphill varies with its mass, assuming (a) the simplest scaling law; (b) the buckling strength criterion.

9-31 Assume that as a human grows from an infant to an adult, the bones grow according to the relationship $l \propto r^{2/3}$. (a) If the breaking angle during twisting is always the same, are adults or infants more susceptible to fractures due to twisting forces? (b) How does the twisting torque scale with body mass?

9-32 If the proportions of animal body segments are determined by the structural buckling strength criterion, how does the height jumped vary with the mass? Compare your result with that of Chapter 6.

9-33 The column of Fig. 9.22 will fracture when the stress at the outer (and inner) edges reaches the maximum, σ_t. Show that this occurs when $\sigma_t = Er/R$.

9-34 Using the results of Problem 9-33, find the maximum radius of curvature for a hardwood tree with $r = 0.25$ m. Use $E = 10^{10}$ N m^{-2} and $\sigma_t = 10^8$ N m^{-2}.

9-35 What is the radius of a cylindrical steel rod that fractures when the radius of curvature is 4 m? (Use the results of Problem 9-33).

Additional Reading

Harold M. Frost, *An Introduction to Biomechanics,* Charles C Thomas, Springfield, Ill., 1967. Brief introduction to the properties of biological materials.

Hiroshi Yamada, in F. Gaynor Evans (ed.), *Strength of Biological Materials,* Williams and Wilkins Co., Baltimore, 1970. Results of extensive research.

T. McMahon, Size and Shape in Biology, *Science,* vol. 179, 1973, p. 1201. Buckling strength and scaling.

R. McNeill Alexander, *Animal Mechanics,* University of Washington Press, Seattle, 1968. Chapters 3 and 4 discuss elastic properties, strength.

Francis Gaynor Evans, *Stress and Strain in Bones,* Charles C Thomas, Springfield, Ill., 1957.

Haywood Blum, Physics and the Art of Kicking and Punching, *American Journal of Physics,* vol. 45, 1977, p. 61. Karate and the strength of materials.

Scientific American articles:

John J. Gilman, Fracture in Solids, February 1960, p. 94.

Carl W. Condit, The Wind Bracing of Buildings, February 1974, p. 93.

Francis P. Bundy, Superhard Materials, August 1974, p. 62.

Thomas A. McMahon, The Mechanical Design of Trees, July 1975, p. 92.

CHAPTER TEN

VIBRATIONAL MOTION

When an object moves back and forth repeatedly over the same path, it is said to be oscillating or vibrating. Some familiar examples are a child on a swing, a clock pendulum, and a violin string. Oscillations also play an important role in many physical phenomena outside the field of mechanics. For example, the molecules of solids vibrate about their equilibrium positions and the currents in electrical circuits can reverse directions or oscillate. There are also many biological examples of oscillations, such as the production of sound by the human vocal cords and the motions of insect wings.

Although the physical nature of vibrating systems can vary greatly, the equations describing small oscillations of an object about an equilibrium position often relate its acceleration, velocity, and displacement in exactly the same special way. A vibration of this kind is called *simple harmonic motion,* and its mathematical description is always the same except for differences in the symbols used.

Simple harmonic motion is characterized by several quantities. The *amplitude* is the maximum displacement of the oscillating object from equilibrium. A complete oscillation back and forth, which returns the system to its original state, is called a *cycle*. The *period* τ is the time needed for one full cycle. The *frequency f* of the oscillation is the number of cycles in a unit time. The period and frequency are related by $f\tau = 1$.

In this chapter, we first find the general relations describing simple harmonic motion and discuss several examples of this motion. We then discuss the

effects of a frictional force that dissipates energy and of external forces that vary in time and provide energy. We find that there is a special frequency of the external force for which the oscillations have the greatest amplitude. Examples of this *resonance* phenomenon occur in molecular, mechanical, biological, and other systems.

10.1 | SIMPLE HARMONIC MOTION; AN EXPERIMENT

In Chapter Nine, we found that certain objects, when stretched or compressed, exert a force opposing this action that is directly proportional to the length of stretch or compression. Simple coil springs have this property. When a weight placed on such a spring is moved from equilibrium and released, the resulting motion is referred to as *simple harmonic motion*. This means that the position, velocity, and acceleration change in time in a specific way.

Figure 10.1. An experimental arrangement in which the shadows of a hanging weight on a spring and of a peg mounted on a wheel coincide. Both shadows undergo simple harmonic motion.

The way these quantities change in time can be determined with the aid of an experiment involving a weight hung on a spring and a vertical rotating wheel with a peg mounted at its edge. Both the hanging weight and the rotating peg are illuminated by a spotlight, and their shadows fall on a screen (Fig. 10.1). Once the weight is set in motion, the speed of the wheel is adjusted until the two shadows move up and down the same number of times per second. When this is done, it is seen that the motion of the two shadows is identical, so that both shadows must have the same equations relating their position, velocity, and acceleration. Thus, we can use our knowledge of the circular motion of the peg on the wheel to obtain the mathematical description of the simple harmonic motion of the oscillating weight.

The peg is at a radius R and is rotating at an angular speed ω measured in radians per second. Starting the clock when the peg is at its highest point, at time t the peg has rotated through an angle $\theta = \omega t$. Since the wheel turns through 2π radians in one period τ, we can also write $2\pi = \omega\tau$. The frequency f, the number of cycles per second, is defined by $f = 1/\tau$, so

$$\omega = \frac{2\pi}{\tau} = 2\pi f \qquad (10.1)$$

Thus, we can also write the angle as $\theta = 2\pi ft$. If time is measured in seconds, then f is measured in cycles per second or *hertz* (Hz).

From Fig. 10.2, the position of the peg's shadow is $x = R\cos\theta$ or

$$x = R\cos(2\pi ft) \qquad (10.2)$$

We have chosen the origin of the x axis so that $x = 0$ when the shadow is at the midpoint of the motion. Since $\cos(2\pi ft)$ is always between $+1$ and -1, x varies back and forth between the values R and $-R$; R is the *amplitude* of the oscillation.

The velocity $\mathbf{v}_t$ of the rotating peg is always tangential to its motion and has a magnitude $v_t = \omega R = 2\pi fR$. Figure 10.3 shows $\mathbf{v}_t$ for the same position of the object as in Fig. 10.2. From the figure, the *velocity v of the shadow* is

$$v = -v_t\sin\theta = -2\pi fR\sin(2\pi ft) \qquad (10.3)$$

The minus sign means that v is directed downward when the object is in the position shown.

Figure 10.2. The position of a peg mounted on a wheel turning with constant angular speed ω. The object is a distance R from the axis of rotation. The position x of the shadow is negative when the shadow is below the horizontal dashed line.

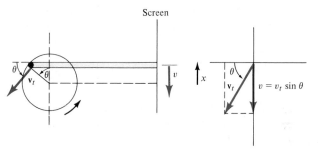

Figure 10.3. The instantaneous velocity $\mathbf{v}_t$ of the rotating object has a magnitude $v_t = \omega R$. The velocity of the shadow is v. v is negative when it is in the $-x$ direction.

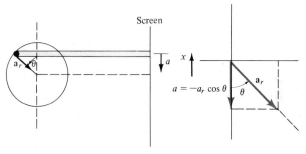

Figure 10.4. The acceleration $\mathbf{a}_r$ of the rotating object has a magnitude $a_r = (2\pi f)^2 R$ and the shadow has an acceleration a.

To find the acceleration of the shadow, we recall that the acceleration of an object undergoing uniform circular motion is directed radially toward the center of the circle and has a magnitude

$$a_r = \omega^2 R = (2\pi f)^2 R \qquad (10.4)$$

From Fig. 10.4, the acceleration of the shadow is

$$a = -a_r\cos\theta = -(2\pi f)^2 R\cos(2\pi ft) \qquad (10.5)$$

Again the minus sign means that when the shadow is in the position shown, the acceleration a is in the negative x direction.

In summary, the *shadow of an object* rotating with

constant angular speed at a distance R from the center of rotation moves according to the equations

$$\left.\begin{array}{l} x = R\cos(2\pi ft) \\ v = -(2\pi f)R\sin(2\pi ft) \\ a = -(2\pi f)^2 R\cos(2\pi ft) \end{array}\right\} \qquad (10.6)$$

From these equations, we obtain the important result

$$a = -(2\pi f)^2 x \qquad (10.7)$$

This means that the acceleration is opposite to the direction of the displacement and proportional to its magnitude. We will see shortly that this relationship occurs also for a weight on a spring and a swinging pendulum. *Equation 10.7 can be considered as a definition of simple harmonic motion;* whenever the motion of an object satisfies this equation, it is executing simple harmonic motion with a *characteristic frequency f.* An alternative derivation of these results is given in Section 10.9.

10.2 | THE WEIGHT ON A SPRING

In the experiment just described, the shadows of the rotating peg and the hanging weight on a spring move identically. Thus the experiment shows that the weight executes simple harmonic motion. This can also be seen mathematically if Newton's second law of motion is applied to the weight.

A spring is designed so that the applied force and the change in length are proportional to each other.

Thus a force $F = kx$ is needed to stretch a spring by an amount x; k is called the spring constant, and x is the change from the equilibrium length of the spring with no force applied.

In Fig. 10.5, a weight $w = mg$ is hung from a spring with spring constant k. In equilibrium, the force exerted by the spring on the weight is $F_s = kd$, where d is the amount by which the spring is stretched from the equilibrium length it has when no weight is attached. When the weight remains at rest, $kd = mg$. Now if the weight is displaced upward by a distance x from this equilibrium position, the spring force is $F_s = k(d - x)$. Using $F = ma$, we have $F_s - w = ma$, or

$$k(d - x) - mg = ma$$

Using $kd = mg$, this becomes

$$a = -\frac{k}{m}x \qquad (10.8)$$

This is identical to the equation defining simple harmonic motion, $a = -(2\pi f)^2 x$, provided that $(2\pi f)^2 = k/m$. We conclude that the weight on the spring executes simple harmonic motion with a frequency

$$f = \frac{1}{\tau} = \frac{1}{2\pi}\sqrt{\frac{k}{m}} \qquad (10.9)$$

This shows that if the mass is increased, the spring oscillates less rapidly; if the spring is made stiffer so k

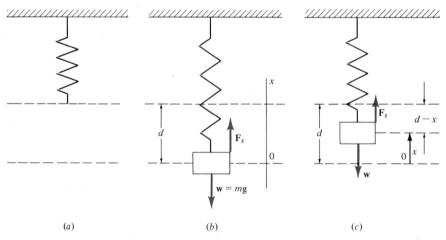

(a)   (b)   (c)

Figure 10.5. (*a*) A spring with constant k. (*b*) A weight w is hung from the spring. It is in equilibrium and remains at rest when $kd = mg$. (*c*) The weight is displaced a distance x from the equilibrium position. The force F_s due to the spring is $k(d - x)$.

is increased, the spring oscillates more rapidly. The following example shows how the spring constant and frequency of a spring and mass may be found.

Example 10.1

When a steel ball of mass 0.1 kg hangs at rest from a spring, the spring stretches 0.2 m. (a) What is the spring constant? (b) Find the characteristic frequency and period of oscillation when the ball is set in motion.

(a) Since the ball is in equilibrium, $kd = mg$, and the spring constant is

$$k = \frac{mg}{d} = \frac{(0.1 \text{ kg})(9.8 \text{ m s}^{-2})}{(0.2 \text{ m})} = 4.9 \text{ N m}^{-1}$$

(b) The characteristic frequency is

$$f = \frac{1}{2\pi}\sqrt{\frac{k}{m}} = \frac{1}{2\pi}\sqrt{\frac{4.9 \text{ N m}^{-1}}{0.1 \text{ kg}}} = 1.1 \text{ Hz}$$

The period is $\tau = 1/f = 0.91$ s.

10.3 | THE PHYSICAL PENDULUM

An object swinging back and forth also undergoes simple harmonic motion. Such an object is called a *physical pendulum*. We will find that the physical pendulum serves as a useful model for the analysis of certain kinds of body motions. A *simple pendulum* is an idealized example of a physical pendulum, consisting of a massless rod with a point mass at its end. Results for the simple pendulum can be obtained directly from those for the physical pendulum.

We consider an object of arbitrary shape that can swing frictionlessly about an axis A (Fig. 10.6). The

object has mass m and moment of inertia I about A. The distance from A to the center of mass C.G. is d. The object is initially displaced and released. Figure 10.6 shows the object when the displacement is θ, which is taken to be positive when the C.G. is to the right of the vertical line through A. Since the motion is not linear but angular, we must compute torques and angles rather than forces and distances.

The torque is $\Gamma = rF\sin\theta$, where r is the length of the line from A to C.G., and $F\sin\theta$ is the component of the force perpendicular to this line. From the figure we see that $r = d$ and $F = mg$. The torque is then $\Gamma = -mgd\sin\theta$, where the minus sign indicates that the torque is clockwise. Using the angular form of Newton's second law, $\mathbf{\Gamma} = I\boldsymbol{\alpha}$, we have $-mgd\sin\theta = I\alpha$ or

$$\alpha = -\frac{mgd}{I}\sin\theta$$

This is not the equation for harmonic motion even though α is the angular acceleration associated with the displacement θ. However, at small angles $\sin\theta$ and θ are almost equal. For example, if θ is measured in radians, $\sin\theta$ and θ differ by only about 1 percent at 15° and by about 5 percent at 30°. Thus for *small angles* we have

$$\alpha = -\frac{mgd}{I}\theta$$

This has the correct form for harmonic angular motion, $\alpha = -(2\pi f)^2\theta$, if we identify the characteristic frequency as

$$f = \frac{1}{\tau} = \frac{1}{2\pi}\sqrt{\frac{mgd}{I}} \qquad (10.10)$$

A meter stick swinging about one end is an example of a physical pendulum.

Example 10.2

A meter stick is suspended from one end and released. Its initial angle with the vertical direction is $\pi/24 \text{ rad} = 7.5°$ (Fig. 10.7). When does the meter stick first become vertical?

According to Table 8.1, the moment of inertia I for a rod of mass m and length l rotating about an axis through its end is $\frac{1}{3}ml^2$; the distance d from the C.G. to the axis of rotation is $\frac{1}{2}l$. With Eq. 10.10,

$$f = \frac{1}{2\pi}\sqrt{\frac{mgd}{I}} = \frac{1}{2\pi}\sqrt{\frac{mgl/2}{ml^2/3}} = \frac{1}{2\pi}\sqrt{\frac{3g}{2l}}$$

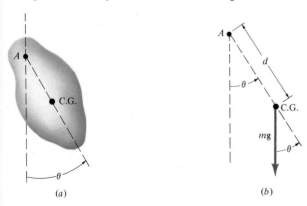

(a)   *(b)*

Figure 10.6. *(a)* An object of mass m and moment of inertia I with respect to the axis at A. *(b)* A schematic diagram showing the force $\mathbf{w} = m\mathbf{g}$. The distance from A to the center of gravity (C.G.) is d. At equilibrium the C.G. is on the vertical dotted line beneath A.

$$= \frac{1}{2\pi} \sqrt{\frac{3(9.8 \text{ m s}^{-2})}{2(1 \text{ m})}} = 0.61 \text{ Hz}$$

Notice that the frequency does not depend on the mass. The meter stick is vertical after one-quarter of a period or $t = \tau/4 = 1/4f$. Thus

$$t = \frac{1}{4f} = \frac{1}{4(0.61 \text{ s}^{-1})} = 0.41 \text{ s}$$

Figure 10.7.   A meter stick pivoted at one end.

Simple Pendulum | The simple pendulum is

an idealized system that is also sometimes a good approximation to a real system. It consists of a point mass m at one end of a massless rod of length l pivoted about the other end. Its moment of inertia is ml^2 and $d = l$ (Fig. 10.8).

Using the results for the physical pendulum, the frequency of the simple pendulum is

$$f = \frac{1}{\tau} = \frac{1}{2\pi} \sqrt{\frac{mgl}{ml^2}} = \frac{1}{2\pi} \sqrt{\frac{g}{l}} \quad (10.11)$$

The following example describes a simple pendulum composed of a rock on a string.

Example 10.3

A small rock swings at the end of a string 1 m long. What is its characteristic frequency?

Using the preceding results,

$$f = \frac{1}{2\pi} \sqrt{\frac{g}{l}} = \frac{1}{2\pi} \sqrt{\frac{9.8 \text{ m s}^{-2}}{1 \text{ m}}} = 0.5 \text{ Hz}$$

Note that this frequency is less than that of the meter stick in Example 10.2. This is because most of the meter stick mass is closer to the axis of rotation.

The result, $f = (1/2\pi) \sqrt{g/l}$ for a simple pendulum, is particularly simple and interesting. Since g is nearly constant over the earth's surface, f depends only on the length of the simple pendulum. The mass is not important except that it must be much larger than

Figure 10.8.   (*a*) A simple pendulum is a point mass at the end of a massless rod. (*b*) The period of a simple pendulum is independent of its mass.

that of the rod or string, so that we may assume $I = ml^2$. One way of observing this is to notice that the period of oscillation of different people on identical swings is nearly independent of their weight. The dependence of the period on length may also be verified roughly by observing the way the period of a weight tied to a string varies as the length of the string is changed.

The implications of the variation of period with length are often important in our daily activities. Normal walking speeds are very nearly those determined by considering the legs as swinging rigid rods. In running, the leg is bent during its forward motion. This reduces its length and hence its period. The bent leg moves forward more quickly than a rigid leg, reducing the effort required. Furthermore, the arms are moved in opposition to the legs to provide balance but are kept bent to reduce their period and the effort required to move them back and forth.

10.4 | ENERGY IN SIMPLE HARMONIC MOTION

In the preceding section, we discussed the pendulum that executes simple harmonic motion. It is convenient to define the potential energy to be zero when the pendulum is at its lowest point, so its energy is entirely kinetic at this position. When the pendulum is at its highest point the velocity is zero, and the energy is entirely potential. Since no forces other than gravity do work, the total mechanical energy is conserved, and the maximum potential energy is equal to the maximum kinetic energy. As the pendulum swings back and forth, there is a continual conversion of kinetic to potential energy and then back to kinetic energy (Fig. 10.9).

Similarly, when a mass oscillates on a spring, the total energy is constant, and there is a continual interchange of potential and kinetic energy. Figure 10.10 shows a mass on a spring sliding on a frictionless table; the origin is chosen at the equilibrium point where the spring force is zero. Again, it is convenient to define the potential energy to be zero at the equilibrium point. As the mass passes through $x = 0$, its energy is entirely kinetic. At the points of maximum displacement, the mass is momentarily at rest, and the kinetic energy is zero. All the energy is then potential energy stored in the spring.

To find the potential energy at position x, we note that the spring force there is $F = kx$ and $F = 0$ at $x = 0$. The average spring force as the mass moves from x to 0 is then $\overline{F} = \frac{1}{2}kx$, and the work done by the spring is $W = \overline{F}x = \frac{1}{2}kx^2$. This must be equal to the potential energy at x, so

$$\mathcal{U} = \tfrac{1}{2}kx^2 \qquad (10.12)$$

The following example illustrates this result.

Example 10.4

A mass of 2 kg on a spring is extended 0.3 m from the equilibrium position and released from rest (Fig. 10.10). The spring constant $k = 65$ N m^{-1}. (a) What is the initial potential energy of the spring? (b) What is the maximum speed of the mass after it is released?

(a) Initially the displacement is 0.3 m, so

$$\mathcal{U}_0 = \tfrac{1}{2}kx^2 = \tfrac{1}{2}(65 \text{ N m}^{-1})(0.3 \text{ m})^2 = 2.9 \text{ J}$$

(b) The energy is totally kinetic when the spring and mass pass through the unstretched position, $x = 0$. The kinetic energy there is $\frac{1}{2}mv^2$ and is equal to the initial potential energy. Thus

$$\tfrac{1}{2}mv^2 = \mathcal{U}_0$$

or

$$v = \sqrt{\frac{2\mathcal{U}_0}{m}} = \sqrt{\frac{2(2.9 \text{ J})}{(2 \text{ kg})}} = 1.7 \text{ m s}^{-1}$$

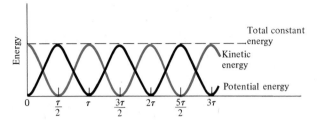

Figure 10.9. The total mechanical energy of a pendulum is constant, although the kinetic and potential energies vary with time. The same is true for a mass on a spring.

Figure 10.10. The magnitude of the force on the mass when the spring is stretched a distance x is $F = kx$.

10.5 | DAMPED OSCILLATIONS

Most real situations involving vibrational motion cannot be described precisely by the equations of simple harmonic motion because of the presence of dissipative forces such as friction or air resistance. For example, a child on a swing, a clock pendulum, and a violin string all gradually come to rest unless energy is supplied to replace the losses. In this section, we briefly describe the reduction in amplitude or *damping* caused by dissipative forces. In the next section, we discuss what happens when energy is fed into a system by an external source.

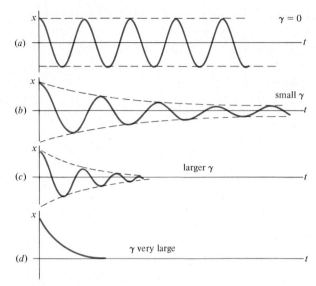

Figure 10.11. The displacement of a weight on a spring versus time. (*a*) No dissipative forces. The displacement is alternately positive and negative with constant amplitude. (*b*) With a dissipative or damping force, the amplitude gradually diminishes. (*c*) With more damping, the amplitude diminishes more rapidly. (*d*) If the damping is very large, no oscillations occur.

Dissipative forces typically depend on the velocity, but the exact dependence varies and can be quite complicated for some systems. Often it is assumed that the dissipative force F_d is linearly proportional to v, or $F_d = -\gamma v$. (The qualitative behavior of an oscillator does not greatly depend on the exact form of the force law.) Here γ is the *damping constant,* and the minus sign indicates that the damping force opposes the motion.

The effect of including the damping force in the equation of motion for a weight on a spring is illustrated in Fig. 10.11. If γ is zero, oscillations continue with the same amplitude indefinitely. When a small amount of damping is present, the oscillations steadily decrease in amplitude until they are negligibly small. If γ is larger, then the oscillations diminish faster. Oscillations cannot occur at all when γ is very large; if displaced, the weight returns to its equilibrium position without oscillating.

The theory of damped oscillations has some interesting biological applications. For example, it provides a way of measuring the friction present in the joints of mammalian limbs. As we noted in Chapter Three, this friction is quite small because of the lubricating effects of synovial fluid. If you sit so that your lower leg can swing freely from the knee and set it in motion, it will gradually come to rest if no muscular forces are exerted. Measurement of the rate at which the amplitude diminishes leads to information about the frictional forces. In a normal knee, there is very little friction, and the oscillations damp out slowly, much as in Fig. 10.11*b*.

An example of strongly damped motion is provided by the cupola in the inner ear of vertebrates. We learned in Chapter Eight that this is a swinging door type of structure that is easily displaced from its equilibrium position by the relative motion of the endolymph fluid and the semicircular canal. Here the spring constant k is very small, and the damping due to the fluid is comparatively large. As a result, the cupola returns very slowly over a period of about 20 s to its equilibrium position, and it does not oscillate at all. This is similar to the situation shown in Fig. 10.11*d*.

Another structure in the ear of vertebrates, the *otolith,* provides an example of an intermediate amount of damping. All vertebrates have two or three otoliths, which provide information about acceleration and tilting of the animal. The otolith is made of calcium carbonate and is about three times as dense as water. It is connected by springlike tissue to a cavity filled with watery fluid (Fig. 10.12*a*). When the head tilts or accelerates, the otolith moves relative to the fluid, and this motion is detected by sensory nerves. Although the otolith does go past the new equilibrium position when the head tilts or accelerates, it oscillates only a few times before coming to rest. This is the kind of situation shown in Fig. 10.11*c*. Thus, evolution has arranged that, in a very short time, information about the amount of tilt or acceleration will be available to the brain without significant ambiguity due to continued oscillations.

10.6 | FORCED OSCILLATIONS AND RESONANCE

We saw in the preceding section that unless energy is supplied, the amplitude of an oscillator usually decreases in time because of frictional effects. To overcome such losses, clock pendulums are driven by coiled springs, and children on swings pump their feet. When energy is fed into a vibrating system, it is said to be undergoing *forced oscillations.*

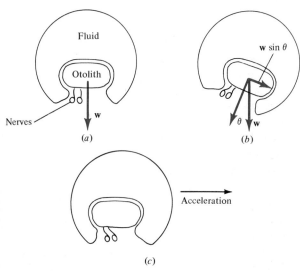

Figure 10.12. (*a*) The otolith is denser than the surrounding fluid. Tilting (*b*) or accelerating (*c*) causes the otolith to move. Nerve endings adjacent to the otolith sense its motion. Note that in (*b*) and (*c*) the side portions of the fluid filled chamber change size. (Adapted from Alexander.)

A singer holding a note of a certain frequency can set up vibrations in a glass. If the singer persists, the energy absorbed by the glass may cause large enough vibrations so that the glass will shatter (Fig. 10.13). This is only true of well-made glassware. In poorer quality glasses, which are less homogeneous in composition, various portions of the glass have different characteristic frequencies, and no single frequency will cause shattering.

Soldiers crossing bridges break step because the regular march step may be at just the right frequency to set the bridge vibrating and perhaps cause its ultimate destruction. A spectacular example of a bridge being set into motion and collapsing is that of the Tacoma Narrows Bridge in Washington. The wind set the bridge oscillating with a steadily increasing amplitude (Fig. 10.14).

Insect wings may vibrate at up to 120 times per second, although only about three nerve impulses per second trigger the wing muscles. The nerve impulses arrive at just the right frequency to maintain the natural vibrational motion of the wing.

In all these examples, there are both dissipative forces that reduce the vibrations and external forces that supply energy. Depending on the physical circumstances, there may be a balance of these two energies, so that the amplitude of the motion is constant, as with the clock or the insect wing. Sometimes energy enters the system faster than it is dissipated, and disaster follows, as with the glass and the Tacoma Narrows Bridge. Finally, if the energy does not enter the system at very nearly the right frequency, little or no vibration occurs, since the energy supplied is immediately dissipated.

As we have seen, energy is most effectively supplied to an oscillator when the external force acts at the correct frequency, which is usually close to the frequency of the oscillator with no external force. This phenomenon is called *resonance,* and the optimum frequency is referred to as the *resonant* frequency. A child pumping a swing or a parent pushing from behind learns to apply forces at just the right interval to attain the maximum amplitude. Similarly, people trying to push a car stuck in snow or mud are most successful when they allow the car to rock back and forth and time their pushes appropriately.

Figure 10.13. A trained singer can break a glass by sustaining the right note. (Ella Fitzgerald in an ad for Memorex tape. Courtesy Memorex Corporation.)

Figure 10.14. Winds caused the Tacoma Narrows Bridge to collapse after several hours of increasing vibrations. After this disaster, which occurred in 1940, some similar bridges were substantially modified. (Wide World Photos.)

SUMMARY

Any motion in which the displacement x and acceleration a are related by

$$a = -(2\pi f)^2 x$$

is simple harmonic motion. f is the natural or characteristic frequency of the motion.

A common example of simple harmonic motion is a physical pendulum that swings back and forth at relatively small angles. If the pendulum has a mass m and moment of inertia I, and d is the distance from the rotation axis to the center of gravity, the characteristic frequency of the motion is

$$f = \frac{1}{2\pi} \sqrt{\frac{mgd}{I}}$$

For a simple pendulum with all the mass concentrated at a distance l from the pivot,

$$f = \frac{1}{2\pi} \sqrt{\frac{g}{l}}$$

A mass attached to a spring undergoes simple harmonic motion. If the mass is m and the spring constant is k, the frequency of the motion is

$$f = \frac{1}{2\pi} \sqrt{\frac{k}{m}}$$

The total mechanical energy of an object undergoing simple harmonic motion is constant. There is a repeated interchange between the kinetic and potential energy portions. If dissipative forces are present, the energy is not constant, and the motion is said to be damped. The rate at which the motion dies out is proportional to the magnitude of the dissipative force.

If an external driving force is also present, the motion is again quite similar. However, the amplitude of the motion depends on the frequency of the external force. When the frequency of the external force is equal to the natural frequency of the oscillator, the amplitude is largest, and the system is at resonance.

Checklist
Define or explain:

harmonic motion	spring constant
amplitude	physical pendulum
cycle	simple pendulum
frequency	damping
period	forced oscillations
spring motion	resonance

EXERCISES

Section 10.1 | Simple Harmonic Motion; An Experiment

10-1 In the experiment comparing the motion of the shadows of the peg and the weight on a spring, the angular frequency of the wheel is $\omega = 0.5$ rad s^{-1}. Find (a) the frequency f, and (b) the period of the motion.

10-2 In an experiment comparing the shadows of the peg and weight, $f = 1.5$ Hz. (a) Find ω. (b) If the amplitude is 0.2 m, what is the position at $t = \frac{1}{3}$ s?

10-3 In an experiment comparing the shadows

of a rotating peg and a weight, $f = 0.5$ Hz, and the amplitude is 0.1 m. Find the positions and velocities of the shadows at (a) $t = 0$ s, (b) $t = 0.5$ s, and (c) $t = 1$ s.

Section 10.2 | The Weight on a Spring

10-4 Masses m and M are suspended from two identical springs of spring constant k. When set in motion, the characteristic frequency of M is three times that of m. What is the ratio M/m?

10-5 The characteristic frequency of a mass on a spring is 5 Hz. What is the acceleration of the mass when the displacement is 0.15 m?

10-6 A spring stretches 0.05 m when a 0.3-kg mass is suspended from it. (a) What is the spring constant? (b) What is the characteristic frequency of this mass on the spring?

10-7 The period of a 0.75-kg mass on a spring is 1.5 s. What is the spring constant?

Section 10.3 | The Physical Pendulum

10-8 A small weight swings at the end of a string. If the period is 1 s, how long is the string?

10-9 A simple pendulum has a 1.5-s period on Earth. When it is set swinging at the surface of another planet, the period is found to be 0.75 s. What is the acceleration of gravity on this planet?

10-10 A steel ball on the end of a cable is used in demolition work. The period of swing is found to be 7 s. What is the length of the cable? (Neglect the mass of the cable.)

10-11 (a) Estimate the moment of inertia of your lower leg and foot when pivoted at the knee. (b) What is your estimate of the characteristic frequency of the lower leg and foot when pivoted at the knee? (c) How does your estimate in part (b) compare to your observations?

10-12 A uniform rod is suspended at one end. The characteristic period of its swing is 2 s. What is the length of the rod?

Section 10.4 | Energy in Simple Harmonic Motion

10-13 A 0.5-kg mass on a spring has a period of 0.3 s. The amplitude of the motion is 0.1 m. (a) What is the spring constant? (b) What is the potential energy stored in the spring at maximum displacement? (c) What is the maximum speed of the mass?

10-14 A 0.05-kg mass is hung from a massless rubber band and stretches it 0.1 m. (a) What is the spring constant of the rubber band? (b) What is the characteristic frequency of oscillation of the system? (c) What is the period of the oscillation? (d) If the mass is pulled 0.05 m below the equilibrium position and released, what will be the energy associated with the oscillations?

Section 10.5 | Damped Oscillations

10-15 The otolith in a fish has a mass of 0.022 g $= 2.2 \times 10^{-5}$ kg, and the effective spring constant is 3 N m^{-1}. (a) What is the characteristic frequency of the otolith? (b) Is the frequency consistent with the idea that the otolith should respond rapidly to accelerations or changes in orientation?

10-16 The characteristic frequency of a man's lower leg and foot, when swung at the knee, is 1.3 Hz. The motion is damped out after six swings. (a) What is the period of the motion? (b) How long does the leg swing?

10-17 The otolith of a fish has a mass of 0.1 g $= 10^{-4}$ kg and a spring constant at 3 N m^{-1}. (a) Find the characteristic frequency. (b) Assuming the damped otolith comes to rest in a time equal to twice its natural period, how long is the otolith in motion after the fish's head is tilted?

PROBLEMS

10-18 Using sense organs in their legs, spiders can detect vibrations in their webs when their prey is captured. When trapped in one web, a 1-g $= 10^{-3}$-kg insect causes the web to vibrate at 15 Hz. (a) What is the spring constant of the web? (b) What would be the frequency of a 4-g insect caught in the web?

10-19 During the motion of a weight on a spring, the position, velocity and acceleration are given by

$$x = x_0 \cos(2\pi ft)$$
$$v = -2\pi f x_0 \sin(2\pi ft)$$
$$a = -(2\pi f)^2 x_0 \cos(2\pi ft)$$

Evaluate x, v, and a for (a) $t = 0$, and (b) $t = 1/(2f)$. (c) In words, describe how x, v, and a change with time between $t = 0$ and $t = 1/(2f)$.

10-20 Young's modulus for bone is $E = 1.6 \times 10^{10}$ N m^{-2}. The tibia is 0.2 m long and has an average cross-sectional area of 0.02 m². (a) What is the spring constant of the bone? (b) A man weighs 750 N. How much is the bone compressed if it supports half his weight? (c) If the bone oscillates lengthwise with half the body weight on it, what is the characteristic frequency of oscillation?

10-21   An instructor wishes to prepare a class demonstration of simple harmonic motion. She has only a 2-kg mass available but has a selection of springs. (a) What spring constant should she choose to have a period of 2 s? (b) She selects the spring by measuring the extension of the available springs with the 2-kg mass. What extension is she seeking?

10-22   The characteristic frequency of the simple pendulum of a clock is 0.7 Hz. The pendulum is 0.5 m long. How should the length be changed to change the frequency to 0.8 Hz?

10-23   The human leg can be approximated by a cylinder. (a) Estimate the characteristic frequency of your legs when pivoted from the hip with the knee locked. (b) If normal walking were performed with the legs swinging at their natural frequency, how far could you walk in 1 hour?

10-24   (a) If one assumes that the arm is a uniform rod, estimate the characteristic frequency of the extended arm. (b) Estimate the frequency if the forearm is at a right angle with the upper arm as in running.

10-25   A ball on a 30-m cable is used for demolition work. If the maximum angular displacement is 20°, what is the velocity of the ball at its lowest point? (Neglect the mass of the cable.)

SUPPLEMENTARY TOPICS

10.7 | THE EFFECTS OF VIBRATION ON HUMANS

People have always experienced vibrational motion in activities such as walking or running. However, they are now routinely subjected to vibrational motion when they travel in cars or airplanes, operate tractors, use power tools, or work near machinery. The effects of vibration may range from mild annoyances to injury or death, depending on the amplitude, frequency, and duration of the vibration. Considerable research has been directed at measuring these

effects, so that safer and more comfortable vehicles and machines can be designed.

Discussions of the effects of vibration normally use the peak acceleration a_{max} rather than the amplitude as a variable. According to Eq. 10.7, the acceleration is related to the displacement by $a = -(2\pi f)^2 x$. When x is equal to the amplitude A or to $-A$, the magnitude of the acceleration is

$$a_{max} = (2\pi f)^2 A \qquad (10.13)$$

Laboratory experiments are usually done with platforms or seats that vibrate at a single frequency, although practical vibrational motion situations often involve a combination of oscillations with many different frequencies. Figure 10.15 shows the reactions to vertical oscillations for subjects sitting on a hard seat. The greatest sensitivity occurs at 6 to 7 hertz; larger accelerations can be tolerated at higher or lower frequencies.

These ideas are illustrated in the following example.

Example 10.5

What peak acceleration is perceived as alarming at 6 Hz? What is the corresponding vibrational amplitude?

From Fig. 10.15, the peak acceleration at 6 Hz which is perceived as alarming is about 0.65 g, where g is the acceleration due to gravity. With Eq. 10.13, the amplitude is

$$A = \frac{a_{max}}{(2\pi f)^2} = \frac{(0.65)(9.8 \text{ m s}^{-2})}{[2\pi(6 \text{ s}^{-1})]^2}$$
$$= 4.5 \times 10^{-3} \text{ m}$$

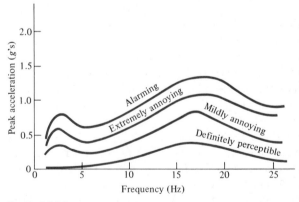

Figure 10.15.   The average reactions of men to simple harmonic motion. The subjects are on hard seats that vibrate vertically. (Adapted from S. Lippert, *Vibration Research*.)

Figure 10.16. (*a*) A model that treats the body as a single mass describes the motion adequately for frequencies up to 2 Hz. The device resembling a piston in a fluid symbolizes the damping. (*b*) A more elaborate model that takes into account the relative motion of the body parts. (Adapted from R. R. Coermann et. al., *Aerospace Med.,* vol. 31, p. 443, 1960.)

Mathematical models can correlate the observations of the response of the body to vibrations (Fig. 10.16). At frequencies up to 2 Hz, the body reacts as though it were a single mass attached to a spring with some damping present. Above 2 Hz, there is relative motion of the body parts, and a much more complex model is needed. The body as a whole has a resonance at about 6 Hz, but the abdominal mass resonates at 3 Hz, the pelvis at 5 and 9 Hz, the head relative to the shoulders at 20 Hz, and the eyeballs in their sockets at 35 and 75 Hz (Fig. 10.17). Vibration well below the "alarming" level causes physiological changes in the circulatory and nervous systems and impairs coordination, vision, and speech. High levels of vibration can cause serious or even fatal lung, heart, intestinal, and brain damage.

Above 20 Hz it is relatively easy to protect people

from excessive vibration using cushioned seats and various simple suspension arrangements. However, tractors and trucks have most of their vibration at 1 to 7 Hz, and peak accelerations up to 1*g* sometimes occur. Springs and hydraulic shock absorbers can be used to construct seats that reduce these vibrations to acceptable levels.

It is useful to know that when a person's legs are slightly bent, considerably less vibration is transmitted to the upper body than when the person stands or sits erect. Operators of farm machinery sometimes stand in this way to reduce the vibration to a level lower than that they experience when sitting. Similarly, if one rides a bicycle on a rough road, the discomfort can be reduced by supporting part or all of the weight on the pedals rather than on the seat.

10.8 | QUANTITATIVE DESCRIPTION OF RESONANCE

In this section, we sketch the quantitative description of forced oscillations and resonance. We use a specific model system in our discussion.

Suppose that a motor with suitable linkages exerts a force on a mass hanging from a spring and that this external force acts alternatively up and down at a frequency f_0 according to the expression $F_0 \cos(2\pi f_0 t)$.

Figure 10.17. The amplitude of the motion of various body parts can exceed that of the original vibration. This often happens when the frequency is close to the resonant frequency of a particular part. Here a subject is seated on a test platform built to resemble an automobile. The figure shows the maximum displacements of the car and the subject. Note that the subject's head moves a distance *h* which is larger than the distance *t* moved by the test car.

If the mass is m, the spring constant k, and the damping force $-\gamma v$, then from Newton's second law of motion we find

$$F_0 \cos(2\pi f_0 t) - kx - \gamma v = ma$$

If we wait long enough so that the amplitude reaches a constant value, the motion of the spring looks like simple harmonic motion. However, the frequency of oscillation is that of the external force and not the characteristic oscillator frequency f. The amplitude of the motion in this case depends on the frequency. One finds for the amplitude A,

$$A = \frac{F_0}{(2\pi)^2 m Z} \qquad (10.14)$$

where

$$Z = \sqrt{(f_0{}^2 - f^2)^2 + \left(\frac{f}{2\pi m}\right)^2 \gamma^2},$$
$$f = \frac{1}{2\pi}\sqrt{\frac{k}{m}} \qquad (10.15)$$

We can interpret these results as follows. If the damping constant γ is small, then as f_0 approaches f, A becomes very large (Fig. 10.18). As γ increases, the A graph becomes broader and less sharply peaked. Thus the greater the damping, the less pronounced is the resonance effect.

At zero frequency $(f_0 = 0)$ the amplitude is

$A = F_0/m(2\pi f)^2$. The weight will tend to vibrate at its normal frequency f, but the equilibrium position is slightly shifted by the now constant force F_0. At large frequencies, $A \simeq 0$. The weight cannot respond to the rapid changes in the external force. No energy is absorbed and the oscillator remains at rest.

To summarize, when both an external force and a dissipative force act on an oscillating system, the system oscillates at the frequency of the external force with an amplitude that depends on the natural frequency f and the external force frequency f_0. When the external force is applied at or near the characteristic frequency of the oscillator, the amplitude becomes large if the damping constant is small.

10.9 | DERIVATION OF THE EQUATIONS FOR HARMONIC MOTION

The derivation given in Section 10.1 of the expressions for the position, velocity, and acceleration of an oscillator utilized our knowledge of circular motion. Another derivation can be given more concisely using differentiation. The basic problem is to find an equation for the position such that the corresponding acceleration is related to it by Eq. 10.7. Let us try a solution for x of the form $x = R \cos \omega t$, where R and ω are constants. The velocity is dx/dt, so differentiating x (see Appendix B, Eq. B.27)

$$v = -\omega R \sin \omega t$$

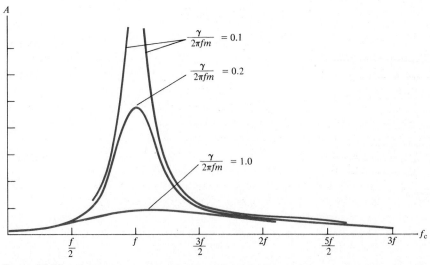

Figure 10.18. Graphs of A versus the external force frequency f_0 for several values of γ. The resonance peak at f becomes broader and less pronounced as γ is increased.

Since the acceleration is dv/dt, differentiating v gives

$$a = -\omega^2 R \cos \omega t$$

The expressions for x and a then satisfy

$$a = -\omega^2 x.$$

With $\omega = 2\pi f$, this is just Eq. 10.7.

EXERCISES ON SUPPLEMENTARY TOPICS
Section 10.7 | The Effects of Vibrations on Humans

10-26   When a man is placed on a shaking platform, his head is observed to oscillate relative to the shoulders with an amplitude that is greatest at 20 Hz. Estimate the spring constant for this motion.

10-27   A tractor seat is mounted on springs. When a 70-kg adult sits on the seat, the characteristic frequency is 7 Hz. What is the characteristic frequency when a child of mass 25 kg sits on the seat? (Neglect the mass of the seat itself.)

10-28   A 50-kilogram woman sits on a spring-mounted seat. The spring is compressed 5×10^{-3} m. (Neglect the mass of the seat itself.) (a) What is the spring constant? (b) What is the characteristic frequency of the motion? (c) Would this be an appropriate size spring for use in a vehicle?

10-29   What is the spring constant of a tractor seat if the resonant frequency is 2 Hz with a 70-kg person seated on it? (Neglect the mass of the seat.)

PROBLEMS ON SUPPLEMENTARY TOPICS

10-30   A human can endure vibration at 4 Hz with a maximum acceleration of $4g$ for a brief period. What is the maximum displacement of the body under these conditions?

10-31   Since various parts of the bodies of animals have characteristic frequencies of vibration, we can think of these parts as being connected by springs (Fig. 10.16b). The spring is formed by the flexible connections of these parts of the body. In Chapter Nine, we learned that the spring constant is pro-portional to the cross-sectional area divided by the length of the spring material. (a) Using the scaling hypothesis of Section 9.6, $l \propto r^{2/3}$, show that the spring constant $k \propto m^{1/2}$, where m is the body mass. (b) Show that the characteristic frequencies should scale as $f \propto m^{-1/4}$.

10-32   The abdomen and thorax of a 60-kg human has a resonance at about 3 Hz. (a) Using the results of Problem 10-31, what would you expect the corresponding characteristic frequency to be in a 20-g $= 2 \times 10^{-2}$-kg mouse? (b) Experimentally the frequency in mice is between 18 and 25 Hz. How does this compare to the result in part (a)?

10-33   When a person stands erect on a vibration table, there is a resonant vibration of the entire body, including the head, at about 2 Hz. The effect on the upper body is only slightly reduced by standing with the legs slightly bent. When the frequency is increased, the legs become much better absorbers. The amplitude at the head is only about 30 percent of that of the table at 5 Hz. Explain this in terms of the high-frequency portion of Fig. 10.18.

10-34   The otolith of a fish has a mass of 0.3 g $= 3 \times 10^{-4}$ kg and an effective spring constant $k = 3$ N m^{-1}. The damping constant $\gamma = 1.5 \times 10^{-2}$ N m^{-1} s. (a) What is the characteristic frequency of the otolith? (b) What is the ratio $\gamma/(2\pi fm)$? (c) Does the otolith have a fairly sharp resonance?

Additional Reading:

C.M. Harris and C.E. Crede (eds.), *Shock and Vibration Handbook,* vol. 3, McGraw-Hill Book Co., New York, 1961, D.E. Goldman and H.E. von Gierke, Effects of Shock and Vibration on Man, Chapter 44.

S. Lippert (ed.), *Human Vibration Research,* The Macmillan Co., New York, 1963.

C.H. Bachman, Some Observations on the Process of Walking, *The Physics Teacher,* September 1976, p. 360. Pendulum motion and walking.

R. McNeill Alexander, *Animal Mechanics,* University of Washington Press, Seattle, 1968. Chapter 7.

UNIT THREE

HEAT

The study of heat and the thermal properties of matter is actually a study of energy and energy transfer. Heat phenomena can be interpreted on a molecular basis; for example, a warm substance has a greater degree of molecular motion than a cold one. The temperature can then be thought of as a measure of the kinetic energy of this motion.

In the first chapter of this unit (Chapter Eleven), we develop the molecular basis for thermal phenomena and temperature. This is accomplished by considering a simplified model of real gases, the *ideal gas*. This model is a good approximation to a real gas at low pressures and moderate temperatures. Surprisingly enough, the ideal gas model also gives a very good description of the phenomena of diffusion and osmosis.

Historically, thermal phenomena were first described on the basis of observations made on macroscopic quantities of material. After developing the molecular basis for heat and temperature, we shift our attention to the behavior and description of macroscopic systems. Thus, in Chapter Twelve, we describe the ability of substances to absorb heat; the *heat capacity;* the subject of phase changes, such as that from a liquid to a solid; and how heat is transferred from one object to another. All these topics are extremely important in scientific work and in our everyday lives.

Long before the relationship between molecular energies and thermal phenomena was understood or even recognized, a great deal of effort went into the study of what is now called *thermodynamics.* Recognizing that the thermal properties of materials could be characterized by such general quantities as the pressure, volume, and temperature, it was found that very general and amazingly important predictions could be made about how systems exchange thermal energy and do work. At the basis of these studies are the first and second laws of thermodynamics, which are covered in Chapter Thirteen. As an example of the generality of thermodynamics, we show how Carnot, in 1824, predicted the maximum possible efficiency for a heat engine. Such engines are used in automobiles and in the production of electricity in power plants. Carnot's result shows that the ~~optimal~~ maximum operation of such engines depends only on the operating temperatures. Thermodynamics has proven to be a powerful and useful approach to many problems. This is particularly evident to students of chemistry, since thermodynamics is frequently used to study chemical reactions.

CHAPTER ELEVEN

TEMPERATURE AND THE BEHAVIOR OF GASES

The concept of temperature plays an important role in the physical and biological sciences. As we learn in this chapter, this is because the temperature of an object is directly related to the average kinetic energy of the atoms and molecules composing the object. Since natural processes often involve energy changes, the temperature plays the role of a label for these changes.

In our everyday experience, this same idea holds true. Our perception of hot and cold is actually a measure of how rapidly energy exchange occurs between objects. Touching something hot results in a rapid and sometimes damaging transfer of energy into our bodies.

In this chapter, we first discuss several temperature scales. The measurement of temperature is based on the variation of physical properties of materials with temperature. *Dilute gases*—those with average intermolecular separations that are very large compared to the molecular dimensions, so that molecular forces are unimportant—are sometimes used to measure temperature. This is because their pressure and temperature are related accurately by a simple expression called the *ideal gas law*. The ideal gas law can be derived by applying Newton's laws of motion to a model that depicts gas molecules as noninteracting spheres. This analysis also makes clear the relationship between temperature and energy and provides a framework for discussing the processes of diffusion and osmosis.

11. 1 | TEMPERATURE SCALES

Many physical quantities always have the same value at a given temperature. For example, the length of a rod varies with the temperature, but it has the same value every time it is placed in a container of ice and water. Because of the reproducibility of experiments of this nature, these properties may be used to define a temperature scale.

One common thermometer uses the volume of a fixed mass of mercury to indicate the temperature. A fine glass tube is attached to a larger bulb. The bulb and part of the tube are filled with mercury, and the rest of the tube is evacuated and sealed. As the temperature increases, the volume of mercury increases faster than that of the bulb, so the mercury rises in the tube.

To calibrate the thermometer, one usually chooses two reference temperatures and divides the interval between them into some number of equal steps. Thus we might take the freezing and boiling points of water at normal atmospheric pressure as our reference temperatures and divide the interval between them into 100 equal steps. We would then have the *Celsius* (centigrade) temperature scale if we set the freezing temperature equal to 0°C and the boiling temperature equal to 100°C. This scale is in everyday use in most of the world and is widely used in scientific work. The *Fahrenheit* (°F) scale used in the United States was originally defined by setting the lowest temperature obtained with a prescribed ice-water–salt mixture equal to $0°F$ and the temperature of the human body to $96°F$. Because of the variability of the body temperature, this scale was later redefined, so that water freezes at $32°F$ and boils at $212°F$. The relationship between the Celsius temperature T_C and the Fahrenheit temperature T_F is given exactly by the equation

$$T_C = \tfrac{5}{9}(T_F - 32°F) \tag{11.1}$$

For example, normal body temperature is $98.6°F$. On the Celsius scale, this is

$$\begin{aligned} T_C &= \tfrac{5}{9}(T_F - 32°F) \\ &= \tfrac{5}{9}(98.6°F - 32°F) \\ &= 37.0°C. \end{aligned}$$

The procedure given above for defining a temperature scale does not clarify the meaning of temperature and depends on the choice of materials. For example, suppose the length of a steel rod were chosen to measure the temperature, with the same calibration procedure used again. Although the steel and mercury thermometers would necessarily agree at the two reference points, they might differ slightly at intermediate temperatures. In the following sections, we will see how these difficulties can be avoided by using the properties of gases to define a temperature scale.

11.2 | THE IDEAL GAS LAW

The properties of gases play an important role in our understanding of temperature and thermal phenomena for two reasons. First, it is relatively easy to study a model of a gas in which the molecules do not interact with one another. This *ideal gas model* can be used to find the relationship between the pressure, temperature, and volume of the gas called the *equation of state*. Second, it is a fortuitous fact that all real gases behave exactly like the ideal gases if they are sufficiently dilute.

In this section, we describe a dilute gas in terms of its pressure, volume, temperature, and the number of moles of gas present. We see that all dilute gases obey a very simple equation of state called the *ideal gas law*. In the next section, we describe the molecular basis for this law.

Number of Moles | Even a relatively small sample of a pure gas contains a large number of molecules. Some gases such as helium (He) and argon (A) are monatomic; that is, their molecules are single atoms. Molecules of polyatomic gases such as oxygen (O_2), nitrogen (N_2), and ammonia (NH_3) contain two or more atoms.

The masses of atoms and molecules are tabulated using a scale defined so that the mass of a ^{12}C atom is exactly 12 atomic mass units (amu). It has been determined that

$$1 \text{ amu} = 1.660 \times 10^{-27} \text{ kg} \tag{11.2}$$

Atomic masses for the elements are tabulated in Appendix D.3. The *molecular mass* (often imprecisely termed the molecular weight) of a molecule is the sum of the masses of its constituent atoms in atomic mass units. Molecular masses are calculated in the following example.

Example 11.1

Find the molecular masses of carbon dioxide (CO_2) and molecular hydrogen (H_2). The atomic masses of H, O, and C are 1.008, 15.999, and exactly 12, respectively.

$$\begin{aligned} M(CO_2) &= M(C) + 2\,M(O) \\ &= 12.00 \text{ amu} + 2(15.999 \text{ amu}) \\ &= 43.998 \text{ amu} \end{aligned}$$

$$M(H_2) = 2\,M(H) = 2(1.008 \text{ amu}) = 2.016 \text{ amu}$$

A *gram-mole* or more simply, a mole, of a substance is an amount whose mass in grams is *numerically equal* to the molecular mass in atomic mass units. Thus, 1 mole of CO_2 has a mass of 43.998 g, and 20.16 g of H_2 is 10 moles. Because of its definition, 1 mole of CO_2 contains exactly the same number of molecules as does 1 mole of H_2 or 1 mole of any other substance. The number of molecules in 1 mole is called *Avogadro's number* N_A, where

$$N_A = 6.02 \times 10^{23} \text{ molecules mole}^{-1} \tag{11.3}$$

Pressure | Figure 11.1 shows a gas in a cylinder fitted with a piston of area A. The magnitude of the force F the gas exerts on the piston divided by A is the *pressure P:*

$$P = \frac{F}{A} \tag{11.4}$$

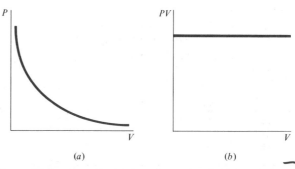

Figure 11.1. The pressure on the piston is equal to the force F exerted on it by the gas divided by the area A of the piston.

The gas exerts a force per unit area equal to P in magnitude and normal to the walls of its container at every point. Methods for measuring pressures will be discussed in Chapter Fourteen.

The S.I. unit of pressure is the pascal; $1 \text{ Pa} = 1 \text{ N m}^{-2}$. Normal atmospheric pressure is

$$1 \text{ atmosphere} = 1 \text{ atm} = 1.013 \times 10^5 \text{ Pa}$$

The relationship between force and pressure is used in the following example.

Example 11.2

A gas at a pressure of 10 atm is in a cubical container of side 0.1 m. If the pressure outside is atmospheric pressure, what is the net force on one wall of the container?

The force due to the gas inside is

$$F_i = P_i A = (10 \text{ atm})(1.013 \times 10^5 \text{ Pa atm}^{-1})(0.1 \text{ m})^2$$
$$= 1.013 \times 10^4 \text{ N}$$

The force on the outside due to the atmosphere is

$$F_a = P_a A = (1.013 \times 10^5 \text{ Pa})(0.1 \text{ m})^2$$
$$= 0.1013 \times 10^4 \text{ N}$$

The net outward force is the difference

$$F_i - F_a = 1.013 \times 10^4 \text{ N} - 0.1013 \times 10^4 \text{ N}$$
$$= 0.912 \times 10^4 \text{ N}$$

In this example the force on the wall is proportional to the difference between the internal pressure P_i and the atmospheric pressure P_a. The difference $P_i - P_a$ is called the *gauge pressure* of the gas. For example, the gauge on a tire pump registers the gauge pressure. In this textbook, the term "pressure" without any modifier always means the absolute pressure, not the gauge pressure.

Equation of State |
Having defined the pressure and the mole, we can now discuss the equation of state for a dilute gas, which is called the ideal gas law. The ideal gas law relates the pressure, volume, number of moles, and temperature of a dilute gas.

The full ideal gas law was deduced from several different observations. Robert Boyle (1627–1691), a contemporary of Isaac Newton, found that if the temperature of a dilute gas is held fixed, the product of the pressure P and volume V is a constant:

$$PV = \text{constant} \qquad \text{(temperature constant)} \qquad (11.5)$$

This is called *Boyle's law*. It states that if the volume is decreased, the pressure increases, a familiar fact to anyone who has used a hand air pump (Fig. 11.2).

A second set of experiments was performed early in the nineteenth century in France. Jacques Charles (1746–1823) and Joseph Gay-Lussac (1778–1850) found that if the temperature of a dilute gas is changed, but the pressure is held fixed, the volume of the gas changes in a simple way (Fig. 11.3). This result could be summarized algebraically as *Charles' law*,

$$\frac{V}{T_C + 273.15} = \text{constant} \qquad (P \text{ constant}) \qquad (11.6)$$

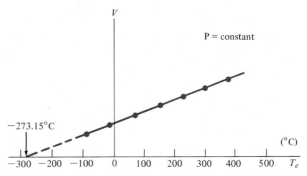

(a)   (b)

Figure 11.2. From Boyle's law, for a gas at fixed temperature, (a) P is inversely proportional to V, or (b) PV is constant.

Figure 11.3. The volume of a gas decreases uniformly as the temperature decreases. The solid line is drawn through actual data points and the dashed line is the extension down to zero volume. This dashed line intersects the temperature axis at $-273.15°C$.

Here T_C is the Celsius temperature.

Because of this universality of behavior of all dilute gases, it was suggested that such gases be used to define a temperature scale. This scale was defined such that the temperature T is,

$$T = T_c + 273.15 \qquad (11.7)$$

T is called the *Kelvin* or *absolute temperature* (see Table 11.1). The increments on this scale are the same as those of the Celsius scale, but $T = 0$ K (zero degrees Kelvin) represents a minimum temperature. Below $T = 0$ K, the volume of an ideal gas would be negative and hence physically impossible. Extensive study has confirmed that no process can be used to carry the temperature below this *absolute zero* of temperature.

With this temperature scale, we can conveniently combine Boyle's law and Charles' law by noting that PV/T is always constant. For constant T this is Boyle's law, and for constant P it is Charles' law. The constant can be determined from experiment. It is found that for n moles of gas,

$$\boxed{PV = nRT} \qquad (11.8)$$

This result is the *ideal gas law*. R is the *universal gas constant*,

$$R = 8.314 \text{ J mole}^{-1} \text{ K}^{-1}$$
$$= 0.08207 \text{ liter atm mole}^{-1} \text{ K}^{-1}$$
$$(1 \text{ liter} = 10^{-3} \text{ m}^3 = 10^3 \text{ cm}^3)$$

Defining a temperature scale using the ideal gas law has the advantage of not depending on the properties of any one material such as mercury or steel. Gas thermometers are maintained at standards laboratories and are used to calibrate more convenient thermometers, such as mercury thermometers. Another important advantage of defining temperature in this way is that it enables us to relate temperature to molecular quantities, as shown in the next section.

Standard conditions for a gas are defined to be $P = 1$ atm and $T_C = 0°C$. The following example illustrates an important property of ideal gases at standard conditions.

Example 11.3

What is the volume of one mole of an ideal gas at standard conditions?

From Eq. 11.7, $T = T_C + 273.15° = 273.15$ K. The ideal gas law then gives

TABLE 11.1

Kelvin temperatures at which representative physical and biological phenomena occur

Description	T
Absolute zero	0
Melting point of nitrogen	67
Gasoline freezes	123
Dry ice (CO_2 freezes)	195
Water freezes	273.15
Hybernating squirrel	275
Body temperature in man	310
Body temperature in birds	315
Water boils	373.15
Fireplace fire	1,100
Gold melts	1,336
Gas flame (stove)	1,900
Surface of sun	6,000
Center of earth	16,000
Center of sun	10^7

$$V = \frac{nRT}{P}$$
$$= \frac{(1 \text{ mole})}{1 \text{ atm}} (0.082 \text{ liter atm mole}^{-1} \text{ K}^{-1})(273.15 \text{ K})$$
$$= 22.4 \text{ liters}$$

Thus under standard conditions, a mole of an ideal gas occupies 22.4 liters.

In summary, all real gases satisfy the ideal gas law to a high degree of accuracy if they are sufficiently dilute. The ideal gas law also provides a good approximation to the behavior of real gases at moderate pressures and temperatures.

11.3 | TEMPERATURE AND MOLECULAR ENERGIES

The ideal gas law, $PV = nRT$, was originally obtained from several types of experiments. However, it is also possible to construct a theoretical model of a gas that yields the ideal gas law. A by-product of this model is the direct identification of the average kinetic energy of the gas molecules with the Kelvin temperature.

In the ideal gas model, we regard the molecules as spheres that never collide with each other but that do collide with the container walls. These collisions are assumed to be elastic, so the molecules lose no energy, but they do change direction. The change in direction involves a change in momentum of the molecules, and this means that there is a reaction force on the container walls. The average force per unit area exerted by the molecules on the walls is the pressure of the gas.

In Section 11.6, we use this model to derive the expression for the pressure on the walls due to the colliding molecules. The result of that calculation is that the product of the pressure and volume, PV, is related to the average kinetic energy of the molecules, $(K)_{ave}$, by

$$PV = \tfrac{2}{3}nN_A(K)_{ave} \qquad (11.9)$$

Here, n is the number of moles of gas present, N_A is Avogadro's number, m is the mass of a molecule, and

$$(K)_{ave} = \left(\frac{mv^2}{2}\right)_{ave} \qquad (11.10)$$

If we compare these results of the model with the ideal gas law, $PV = nRT$, we have

$$nRT = \tfrac{2}{3}nN_A(K)_{ave}$$

Thus

$$(K)_{ave} = \frac{3}{2}\left(\frac{R}{N_A}\right)T = \frac{3}{2}k_BT \qquad (11.11)$$

This is an extremely important result. It gives us a direct interpretation of the absolute temperature in terms of the average kinetic energy of a molecule of a gas. $k_B = (R/N_A)$ is called *Boltzmann's constant*

$$k_B = 1.38 \times 10^{-23}\ \text{J K}^{-1}$$

The thermal energy k_BT is a ubiquitous factor in the natural sciences. By knowing the temperature we have a direct measure of the energy available for initiating chemical, physical, and biological processes. We can also use Eq. 11.11 to identify what is called the root mean square (rms) speed for the molecules at a given temperature. With the rms speed defined by $(K)_{ave} = mv^2_{rms}/2$,

$$v_{rms} = \sqrt{\frac{2(K)_{ave}}{m}} = \sqrt{\frac{3k_BT}{m}} \qquad (11.12)$$

Individual molecules of a gas can have almost any speed. However, as shown in Fig. 11.4, there is a speed at which the largest number will be found for a given temperature. The peaks in the curves occur at

speeds that are slightly below v_{rms}. Thus, in an approximate sense we can think of v_{rms} as a typical speed of a molecule in the gas. These relationships allow us to compute the average molecular energy and the rms speed of molecules of any ideal gas at a given temperature.

Example 11.4

(a) What is the average kinetic energy of a hydrogen molecule at $27°C = 300$ K in joules and in electron volts? (b) What is the rms speed?

(a) From $(K)_{ave} = \tfrac{3}{2}k_BT$,

$$\begin{aligned}(K)_{ave} &= \tfrac{3}{2}k_BT \\ &= \tfrac{3}{2}(1.38 \times 10^{-23}\ \text{JK}^{-1})(300\ \text{K}) \\ &= 6.21 \times 10^{-21}\ \text{J}\end{aligned}$$

Since 1 electron volt $= 1.6 \times 10^{-19}$ J, the energy is also

$$(K)_{ave} = 0.039\ \text{eV}$$

(b) In Ex. 11.1, we found that the molecular mass of H_2 is

$$\begin{aligned}2.016\ \text{amu} &= (2.016\ \text{amu})(1.66 \times 10^{-27}\ \text{kg amu}^{-1}) \\ &= 3.35 \times 10^{-27}\ \text{kg.}\end{aligned}$$

Thus, using $v_{rms} = \sqrt{2(K)_{ave}/m}$,

$$\begin{aligned}v_{rms} &= \sqrt{\frac{2(6.21 \times 10^{-21}\ \text{J})}{(3.35 \times 10^{-27}\ \text{kg})}} \\ &= 1930\ \text{m s}^{-1}\end{aligned}$$

We have calculated $(K)_{ave}$ and v_{rms} for the molecule H_2 in Example 11.4. Our calculations are correct for the translational motion, but such a molecule can also rotate and vibrate, and each of these motions has kinetic energy. In the equations we have developed

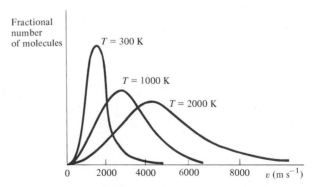

Figure 11.4. The distribution of molecular speeds of diatomic hydrogen/gas H_2. At higher temperatures the maximum speed is greater and there is a greater spread.

$(K)_{ave}$ should be interpreted as the translational kinetic energy only.

Two processes that depend on the thermal energy and are very important in biological systems are *osmosis* and *diffusion*. Both may be understood from the kinetic theory of an ideal gas and are described in the next two sections.

11.4 | DIFFUSION

When perfume is sprayed into still air, the scent eventually spreads to all parts of a room. Similarly, a drop of dye placed in a solvent will gradually spread throughout the container, even though we are careful not to disturb or stir the liquid. The rather slow process by which the molecules spread out evenly is called *diffusion*.

We may visualize the diffusion process for a typical gas by referring to Fig. 11.5. A small amount of helium gas is released at point A in an air-filled container. At a given moment, we draw an imaginary hemispherical surface so that most of the helium atoms are inside the surface, but a few are outside. These helium atoms are in constant random motion, bouncing off air molecules, other helium atoms, and the walls. A certain number of helium atoms will cross the surface from the inside outward and some from the outside inward. Because most of the helium is inside, more helium atoms will pass outward than inward, increasing the number outside. We see that there is a net drift of helium atoms away from A into the remainder of the container. *This flow from higher to lower concentrations is diffusion.*

The average distance the helium atoms diffuse increases with the time, but in a somewhat surprising fashion. Consider the path of a typical atom as illustrated in Fig. 11.6. The atom makes many collisions and changes direction repeatedly. Often the effect of one straight line section or *step* is nearly cancelled by a later step. On the average, the distance l from the initial point 0 increases, but much more slowly than the number of steps, N_s. When statistical methods are applied to such a path, it is found that l increases as the square root of the number of steps, or $l^2 \propto N_s$. Since the number of steps is proportional to the time, l^2 is proportional to t. This result for the case of *random motion* is quite different from the relationship $l \propto t$ which holds true for straight line motion.

Figure 11.5. A container of air has some helium atoms released into it at A. The helium atoms spread out, and at some instant, we draw an imaginary hemispherical surface approximately separating the regions of high and low concentrations of helium.

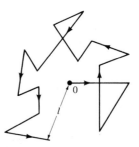

Figure 11.6. The random motion of an atom. On the average the distance l is proportional to square root of the number of straight line steps.

Conventionally, instead of $l^2 \propto t$, one writes an equation for the mean squared displacement x^2_{rms} in one direction,

$$x^2_{rms} = 2Dt \qquad (11.13)$$

where D is called the *diffusion constant*. Similar equations hold for y^2_{rms} an z^2_{rms}. The value of D depends on the nature of the diffusing atom or molecule and the choice of the solvent or medium (Table 11.2). The following numerical example illustrates how slow the diffusion process is in a typical situation.

Example 11.5

How long will it take a hemoglobin molecule to diffuse 1 cm $= 10^{-2}$ m in water?

Using Eq. 11.13 with $D = 6.9 \times 10^{-11}$ m^2 s^{-1} and $x = 10^{-2}$ m, we have

$$t = \frac{x^2_{rms}}{2D} = \frac{(10^{-2} \text{ m})^2}{2(6.9 \times 10^{-11} \text{ m}^2 \text{ s}^{-1})} = 7.24 \times 10^5 \text{ s}$$

This is 201 hours or about 8.4 days!

11.5 | DILUTE SOLUTIONS; OSMOTIC PRESSURE

The behavior of real gases can often be closely approximated by the ideal gas law, which is derived with a model that depicts the gas molecules as small, non-interacting particles. Somewhat surprisingly, the *osmotic pressure of dilute solutions* can also be accurately predicted using a form of the ideal gas law.

Figure 11.7 illustrates the meaning of osmotic pressure. The outer vessel contains water, and the inner one is filled initially to the same height with a solution of water and sugar. Water molecules can pass freely through the membrane separating the two vessels, but the larger sugar molecules cannot. Because the membrane is <u>permeable</u> to water molecules and *impermeable* to sugar molecules, it is called *semipermeable*.

Since the concentration of water molecules is greater in the outer vessel, there is a net diffusion of water molecules into the inner vessel. <u>One may think of this flow of water molecules as resulting from a pressure difference between the pure water and the solution. This pressure difference is called the *osmotic pressure,* denoted by</u> π.

Gradually the water molecules move into the inner vessel, raising the level of the solution. This flow will stop when the upward osmotic pressure at the membrane is equal to the downward pressure of the raised solution (Fig. 11.7b).

The osmotic pressure can be predicted on the basis of a surprisingly simple model. We associate a pressure $P_w^{\,o}$ with the pure water in the outer vessel and a pressure $P_w^{\,i}$ with the water inside. The sugar inside has a pressure P_s, so the total pressure inside is $P_w^{\,i} + P_s$. Since the net force on the membrane is zero, the inner and outer pressures must be equal, $P_w^{\,o} = P_w^{\,i} + P_s$. The pressure difference driving the water molecules, the osmotic pressure π, is then

$$\pi = P_w^{\,o} - P_w^{\,i} = P_s \qquad (11.14)$$

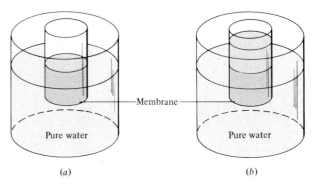

Figure 11.7. (*a*) Initially the levels of the water outside the container and the sugar solution inside are the same. (*b*) The membrane is not permeable to sugar molecules. Water molecules enter the container and in equilibrium the level of the solution inside is higher than that of the pure water.

Experimentally, for dilute solutions this pressure is found to obey the ideal gas law,

$$\pi = P_s = \frac{nRT}{V} \qquad (11.15)$$

where n is the number of moles of sugar, and V is the initial volume of the fluid in the inner vessel. This surprising result can be interpreted as meaning that, in the dilute solution, the sugar molecules behave like an ideal gas. It appears that because the sugar molecules seldom encounter one another, although they are almost constantly in collision with water molecules, they behave like noninteracting particles. This is just the basic assumption of the ideal gas model.

Sometimes it is convenient to rewrite the equation for the osmotic pressure in terms of the *concentration* of the solute. If n moles of solute are in a volume V, the concentration is,

$$c = \frac{n}{V} \qquad (11.16)$$

The S.I. units of c are moles per cubic meter. The osmotic pressure is then given by

$$\pi = cRT \qquad (11.17)$$

Here c is the solute concentration: the sugar in Fig. 11.7. The following example illustrates the use of this result and shows the role that osmosis plays in maple trees.

Example 11.6

Sap rises in the early spring in maple trees as a result of the osmotic pressure difference between the sugar

TABLE 11.2

Representative values of the diffusion constant D at $20°C = 293$ K

Molecule	Solvent	$D(m^2\ s^{-1})$
Hydrogen (H_2)	Air	6.4×10^{-5}
Oxygen (O_2)	Air	1.8×10^{-5}
Oxygen (O_2)	Water	1.0×10^{-9}
Glucose ($C_6H_{12}O_6$)	Water	6.7×10^{-10}
Hemoglobin	Water	6.9×10^{-11}
DNA	Water	1.3×10^{-12}

solution (sap) within the tree and the water in the ground around the roots. The sap contains 1 percent by weight of sucrose ($C_{12}H_{22}O_{11}$) in water. Assuming the temperature is 27°C find (a) the concentration in moles m^{-3}; (b) the osmotic pressure; and (c) the height the sap rises.

(a) The molecular mass of sucrose is $12(12) + 22(1) + 11(16) = 342$ amu, so a mole has a mass of 342 g. One cubic meter of solution has a mass nearly equal to that of 1 m^3 of water, or 10^3 kilograms. The mass of the sucrose is 1 percent of that, or 10 kg = 10^4 g. Thus 1 m^3 of solution contains 10^4 g/(342 gram-mole^{-1}) = 29.2 moles of sucrose. Its concentration is 29.2 moles m^{-3}.

(b) The osmotic pressure is

$$\pi = cRT$$
$$= (29.2 \text{ moles m}^{-3})(8.314 \text{ J mole}^{-1} \text{ K}^{-1})(300 \text{ K})$$
$$= 7.28 \times 10^4 \text{ N m}^{-2}$$

(c) A column of water of height h and cross-sectional area A has a volume hA. A cubic meter of water has a mass of 1000 kg and a weight of 9800 N. The weight of the column is then $W = 9800hA$. The pressure under such a column is the weight per unit area, or 9800h. This column is supported by the osmotic pressure so, omitting the units,

$$\pi = 9800h$$

Then

$$h = \frac{\pi}{9800} = \frac{7.28 \times 10^4}{9800} = 7.43 \text{ m}$$

The concentration of sugar in maple trees can be higher than 1 percent, and it may be that osmosis is largely responsible for the spring flow of sap in maples. In most other trees the concentration of solutes that will not pass through the root wall is much smaller. In these cases, osmotic pressure is insufficient to account for the movement of fluids in trees. We discuss another mechanism for this fluid transport in Chapter Sixteen.

Osmosis is extremely important in understanding a wide variety of biological processes. Both plant and animal tissues are composed of cells containing complex solutions. Solutes that can pass through the cell membrane are present along with impermeable solutes. The fluids surrounding the cells are also complex solutions but with different compositions. In equilibrium the total osmotic pressures due to imper-meable molecules and ions must be the same inside and outside the cell or the difference in osmotic pres-

sures will cause water to enter or leave the cell, along with dissolved permeable materials.

To illustrate this, suppose a person drinks a large quantity of water. Water enters the blood, thereby reducing the concentration of solutes relative to the body tissue. Consequently the body tissues take on water. The flow of water into the kidneys is also increased because of the increased osmotic pressure difference. The kidneys then excrete more dilute urine until the blood concentration returns to equilibrium values. By contrast, a person with a high fever may lose a great deal of water from the tissues and hence from the blood until it is impossible for the kidneys to absorb water and dissolved permeable solutes.

Fluids given intravenously to a patient are usually adjusted so their concentrations of impermeable solutes and osmotic pressures balance those of the tissues. Such solutions are called *isotonic*, If a cell is placed in a solution with a lower concentration of impermeable solutes, there will be a tendency for water to enter the cell. A red blood cell has relatively rigid walls. When it is placed in pure water, an influx of water molecules occurs, which raises the internal pressure, since the cell cannot expand appreciably. Equilibrium would occur at about 8 atm, but the cells will usually rupture before this pressure is reached. Cells that can change volume readily will expand or contract when placed in solutions that are not isotonic.

SUMMARY

Temperature is a measure of the average translational mechanical energy of the molecules in a substance. Several temperature scales are in use, but the one that most directly characterizes the energy is the Kelvin scale. Because it is directly related to molecular energies, the Kelvin scale starts at absolute zero.

Dilute real gases behave as ideal gases. The pressure, volume, number of moles, and temperature of an ideal gas are related by

$$PV = nRT$$

The model that describes an ideal gas as a collection of noninteracting molecules leads to the identification of the Kelvin temperature with the average kinetic energy per molecule,

$$(K)_{ave} = \tfrac{3}{2}k_B T$$

Molecules diffuse slowly from regions of high to low concentrations. The root mean squared diffusion distance in one direction, x_{rms}, is related to the time by

$$x_{rms}^2 = 2Dt$$

Diffusion also plays a role in the osmotic flow of fluids through semipermeable membranes. The permeable fluid will flow between regions of different concentration as if there were a pressure difference across the membrane equal to the osmotic pressure,

$$\pi = cRT$$

where c is the concentration of the impermeable solute.

Checklist

Define or explain:

Celsius (centigrade)	Boltzmann's constant
Fahrenheit	root mean square speed
molecular mass	diffusion
mole	diffusion constant
Avogadro's number	random motion
pressure	semipermeable
dilute gas	membrane
ideal gas	concentration
Kelvin temperature	isotonic

EXERCISES

Section 11.1 | Temperature Scales

11-1 What temperature on the Celsius scale corresponds to 100° on the Fahrenheit scale?

11-2 What temperature on the Fahrenheit scale corresponds to 50° on the Celsius scale?

11-3 At what temperature are the readings on a Fahrenheit and a Celsius thermometer the same?

11-4 What temperature on the Celsius scale corresponds to 105°F, a dangerously high temperature in humans?

Section 11.2 | The Ideal Gas Law

Refer to Appendix D.3 for atomic masses where needed.

11-5 Find the molecular mass of HCl.

11-6 In a crude experiment the molecular mass of NH_3 is found to be 17.5 amu. If the accepted value is used for the hydrogen mass, what is the atomic mass of nitrogen from this experiment?

11-7 What is the mass of two moles of H_2?

11-8 What is the mass of 0.3 moles of ammonia, NH_3? (The molecular mass of NH_3 is 17.03 amu.)

11-9 How many molecules are in 3 moles of sucrose?

11-10 How many molecules are in 0.7 moles of mercury?

11-11 How many moles of carbon tetrachloride (CCl_4) will contain 9.5×10^{23} molecules?

11-12 What is the mass of 6.02×10^{23} atoms of magnesium?

11-13 What is the mass of 17.4×10^{23} molecules of carbon dioxide? (The molecular mass of CO_2 is 44.0 amu.)

11-14 How many molecules are there in 18 g of hydrogen chloride, HCl? (The molecular mass of HCl is 34.46 amu.)

11-15 What is the force exerted by the atmosphere on a field 50 m by 100 m?

11-16 What is the force due to the atmosphere on one side of a door of area 2 m²?

11-17 If the difference in pressure on the two sides of a closed door of area 2 m² is 0.01 atmosphere, what is the net force on the door? Do you think you could open it by hand?

11-18 What pressure on the 12-m² base of an automobile of mass 1300 kg would be necessary to support it?

11-19 Estimate the pressure you exert on the ground (a) when standing erect on both feet; (b) when lying down.

11-20 A gas occupies 5 m³ at a pressure of 1 atm. What is the pressure when the volume is 1.5 m³ if the temperature is held constant?

11-21 The pressure of a gas changes from 1.5 atm to 0.3 atm at constant temperature. What is the ratio of the final and initial volumes?

11-22 The volume of a gas doubles at constant pressure. What is the final temperature if the gas was initially at 30°C?

11-23 If the temperature of a gas is raised from 0°C to 100°C at constant pressure, by what factor is its volume changed?

11-24 What is the volume of 3 moles of an ideal gas at $P = 2$ atm and $T = 300$ K?

11-25 What is the temperature of 1 mole of an ideal gas with a pressure of 0.3 atm in a volume of 3 m³?

11-26 One mole of an ideal gas occupies 2.24×10^{-2} m³ at standard conditions. What will the pressure be if the volume is increased to 1 m³ at constant temperature.

Section 11.3 | Temperature and Molecular Energies

11-27 If the average kinetic energy of a gas is doubled and the volume remains constant, what is the change in pressure?

11-28 What is the ratio of the rms speeds of hydrogen gas (H_2) and oxygen gas (O_2) if both gases are at the same temperature?

11-29 Natural uranium is composed of 99.3 percent ^{238}U, with a mass of 238 amu, and 0.7 percent ^{235}U, with a mass of 235 amu. It is the ^{235}U that is mainly used in reactors and weapons. The two isotopes are separated using diffusion processes. This method utilizes the fact that the rms speed of the gas UF_6 is different for the two isotopes. What is the ratio of their speeds at 37°C?

11-30 Suppose that all the translational molecular kinetic energy of 1 mole of an ideal gas at a temperature of 300 K could be used to raise a 1-kg mass. How high would the block be raised?

11-31 The rms speed of the molecules in an ideal gas with molecular mass 32.0 amu is 400 m s⁻¹. (a) What is the average translational kinetic energy? (b) What is the temperature of the gas?

Section 11.4 | Diffusion

11-32 What average distance will O_2 molecules diffuse in air in 1 hour? Assume $T_C = 20°C$.

11-33 How long will it take glucose molecules to diffuse an average distance of 1 mm = 10^{-3} m in water at 20°C?

11-34 If a solute diffuses through water an average distance of 10^{-2} m in 6 hours, what is its diffusion constant?

PROBLEMS

11-35 How many molecules are in 500 g of sucrose, $C_{12}H_{22}O_{11}$?

11-36 What is the molecular mass of tributyrin (fat), $C_3H_5O_3(OC_4H_7)_3$?

11-37 How many molecules are in 100 g of ethyl alcohol, C_2H_5OH?

11-38 A cylinder contains 0.02 m³ of oxygen at a temperature of 25°C and a pressure of 15 atm. (a) What volume does this gas occupy at 25°C and a pressure of 1 atm? (b) A man is breathing pure oxygen through a face mask at a rate of 0.008 m³ min⁻¹ at atmospheric pressure. How long will the cylinder of gas last?

11-39 The rms speed for an ideal monatomic gas at 300 K is 299 m s⁻¹. What is the atomic mass of the atoms? Identify the gas from the periodic table in Appendix D.3.

11-40 In example 11.4, the rms speed of hydrogen molecules (H_2) at 27°C (300 K) was calculated to be 1930 m s⁻¹. At what temperature would hydrogen molecules have an rms speed sufficient to escape from the earth (1.1×10^4 m s⁻¹)?

*11-41 The walls of blood capillaries are impermeable to proteins. The two major protein groups in blood plasma are

Protein Group	Concentration	Ave. Molecular Mass
Albumin	0.045 g m⁻³	69,000 amu
Globulin	0.025 g m⁻³	140,000 amu

(a) Calculate the concentration of each protein group in moles m⁻³. (b) Calculate the osmotic pressure of blood plasma at a temperature of 310 K due to the proteins dissolved in it. (c) How tall a column of water can this pressure support?

11-42 The alveoli of the lungs are small air sacs about 10^{-4} m in radius. The membrane walls of the sacs that separate the air space from the blood capillaries are about 0.25×10^{-4} m thick. The capillaries themselves have a radius of about 5×10^{-6} m. (a) Assuming that O_2 diffuses through the wall and blood as it does through water, what average time is required for O_2 to diffuse from the center of an alveolus to the center of a capillary? (b) How does this compare with the time, 0.1 s, that the blood is in transit around a alveolus?

11-43 The osmotic pressure of ocean water is 22 atm at 300 K. What is the salt concentration? (*Hint:* NaCl dissociates in water into Na⁺ and Cl⁻ ions.)

*11-44 A solution of sodium chloride (NaCl) in water of 160 moles m^{-3} concentration is isotonic with blood cells. What is the osmotic pressure in atmospheres in the cells if the temperature is 300 K? (*Hint:* NaCl dissociates in water into Na$^+$ and Cl$^-$ ions.)

*11-45 The osmotic pressure of a red blood cell is 8 atms. The cell is placed in a water solution containing 100 moles m^{-3} of a solute to which the cell membrane is impermeable. Will the cell tend to expand, contract, or remain the same size? Explain. (Assume $T = 300$ K.)

Figure 11.8. A single molecule moving in a container of area A and length l. The molecule collides elastically with the walls and with the piston. The piston must be held in place with a force **F** to offset the force due to the molecule.

SUPPLEMENTARY TOPICS

11.6 | MODEL DERIVATION OF THE IDEAL GAS LAW

Spherical molecules that collide only with the walls of a container are shown in Fig. 11.8. The volume of the container is $V = Al$, and there are nN_A molecules inside.

Each time a molecule hits the piston, its x component of velocity reverses itself (Fig. 11.9). Since the y component of the velocity is unchanged, the momentum change in one collision is $2mv_x$.

We can find the force on the piston due to one molecule from Newton's second law in the form $F\Delta t = \Delta p$, where Δp is the momentum change and Δt the time interval. The time interval Δt between impacts with the piston is the time for the molecule to travel the length of the container and back, so $2l = v_x \Delta t$. Thus the force is

$$F_1 = \frac{\Delta p}{\Delta t} = \frac{2mv_x}{\left(\frac{2l}{v_x}\right)} = \frac{mv_x^2}{l}$$

For the nN_A molecules actually present, the total force on the piston is nN_A times the average value of mv_x^2/l,

$$F = \frac{nN_A}{l}(mv_x^2)_{\text{ave}}$$

The pressure is the force per unit area. Using $V = Al$,

$$P = \frac{F}{A} = \frac{nN_A}{V}(mv_x^2)_{\text{ave}}$$

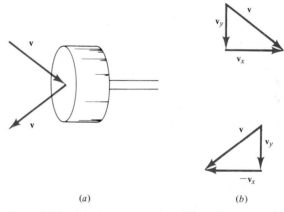

(*a*)   (*b*)

Figure 11.9. (*a*) when a molecule collides with a piston, its speed does not change, but the molecule does change directions. (*b*) $\mathbf{v}_y$ is the same before and after the collision, but $\mathbf{v}_x$ is reversed.

Since m is the same for each molecule, we need to compute $(v_x^2)_{\text{ave}}$. This average can be more conveniently written by noticing that $(v_x^2)_{\text{ave}}$ should be just the same as $(v_y^2)_{\text{ave}}$ and $(v_z^2)_{\text{ave}}$. Since v^2, the total velocity squared, is the sum of the squares of its components, we can use $(v_x^2)_{\text{ave}} = \frac{1}{3}(v^2)_{\text{ave}}$ and then write

$$PV = \frac{2nN_A}{3}\left(\frac{mv^2}{2}\right)_{\text{ave}}$$
$$= \tfrac{2}{3}nN_A(\text{K})_{\text{ave}} \qquad (11.9)$$

This is the result we quoted earlier.

Additional Reading

R.E. Wilson, Standards of Temperature, *Physics Today,* vol. 6, January 1953, p. 10.

D.K.C. MacDonald, *Near Zero: The Physics of Low Temperatures,* Science Study Series, Doubleday and Co., Garden City, N.Y., 1961. Paperback.

Mark W. Zemansky, *Temperatures Very High and Very Low,* Momentum Series, D. Van Nostrand and Co., Princeton, N.J., 1964. Paperback.

Scientific American articles:

Marie Boas Hall, Robert Boyle, August 1967, p. 84.

Arthur K. Solomon, The State of Water in Red Cells, February 1971, p. 88.

Reuben Hersh and Richard J. Griego, Brownian Motion and Potential Theory, March 1969, p. 66.

CHAPTER TWELVE

THERMAL PROPERTIES OF MATTER

Most scientific measurements depend on temperature either because the magnitude of the quantity measured is temperature-dependent or because the measuring apparatus itself has characteristics that depend on the temperature. We mentioned examples of the first effect in the last chapter: the pressure in a gas, the length of a rod, and the volume of a quantity of mercury all change as the temperature changes. An example of the dependence of instruments on temperature is the fact that most clocks run at different rates at different temperatures. Accurate time measurements therefore require knowledge of the thermal properties of matter.

Because the properties of matter depend on temperature, the exchange of thermal energy is extremely important. Since biological processes can only function properly over a small temperature range, both man and nature have devised ways of either limiting or improving the means by which this energy is transferred. For example, we insulate our homes, and our bodies respond to high temperatures by perspiring.

In this chapter, we investigate how the properties of materials depend on temperature. We begin with the thermal expansion of solids and liquids and then discuss how the temperature of an object changes as thermal energy or *heat* is added or withdrawn. This leads us to a discussion of the transfer of thermal energy, and we use our ideas to study temperature control in warm-blooded animals.

The term "heat flow" should properly be used only to describe a transfer of energy that occurs by virtue of a temperature difference; the quantity of energy transferred is the heat. However, during the historical development of the subject it was very common to refer to a "quantity of heat." This informal usage has persisted in much of the literature and is occasionally found in this textbook also.

12.1 | THERMAL EXPANSION

When a substance is heated, its volume usually increases, and each dimension increases correspondingly. This increase in size can be understood in terms of the increased kinetic energy of the atoms or molecules. The additional kinetic energy results in each molecule colliding more forcefully with its neighbors. The molecules effectively push each other further apart and the material expands.

At the macroscopic level, we can find a convenient relation between the change in length of an object and the temperature change. Suppose the original length of the object in Fig. 12.1 is l, and a small increase in length Δl occurs when the temperature increases by a small amount ΔT. If we divide the object into two equal parts, each part will have length $l/2$ and will expand by $\Delta l/2$. Consequently, the change in length Δl is directly proportional to the length l. In addition, we find from experiment that if we double the temperature change by raising the temperature by $2\,\Delta T$, the expansion also doubles. A single equation expressing both proportionalities is

$$\Delta l = \alpha l \, \Delta T \qquad (12.1)$$

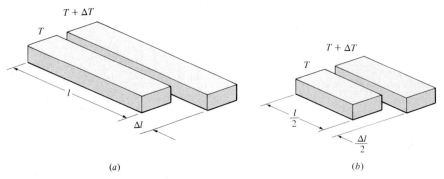

$T + \Delta T$

T

l

Δl

(a)

$T + \Delta T$

T

$\dfrac{l}{2}$

$\dfrac{\Delta l}{2}$

(b)

Figure 12.1. The expansion of a bar is proportional to its length. When the bar is halved in length, the expansion is also halved.

The constant α is the *coefficient of linear expansion*. It is a property of a given material and depends somewhat on the temperature. Typical values of α are given in Table 12.1. We see that α has the dimensions of inverse temperature, so its units are K^{-1}. *Since only the change in temperature is important, we can measure ΔT in Kelvin or Celsius degrees*. We use this fact throughout this chapter. Since α depends somewhat on temperature, Eq. 12.1 is exact only for very small temperature changes. However, it is often quite accurate if an average value of α is used even if ΔT is 100°C or more.

The following example illustrates the importance of thermal expansion.

Example 12.1

The roadbed of the Golden Gate Bridge is 1280 m long. During a certain year the temperature varies from −12°C to 38°C. What is the difference in the lengths at those temperatures if the road is supported by steel girders? For steel, $\alpha = 1.27 \times 10^{-5}$ K^{-1}.

With $\Delta T = 38°C - (-12°C) = 50°C = 50$ K,

$$\Delta l = \alpha l \, \Delta T$$
$$= (1.27 \times 10^{-5} \text{ K}^{-1})(1280 \text{ m})(50 \text{ K})$$
$$= 0.81 \text{ m}$$

This substantial change in the roadbed length must be allowed for in the design of the bridge. If the structure could not alter its length with changes in temperature, huge forces would be developed and severe damage would result.

Two interesting applications of thermal expansion are the *thermostat* and the *ultramicrotome*. A thermostat has two metal strips with different coefficients of linear expansion attached to each other. When they

TABLE 12.1

Coefficient of linear expansion for various materials

Material	Temperature (°C)	α (K^{-1})
Aluminum	−23	2.21×10^{-5}
	20	2.30×10^{-5}
	77	2.41×10^{-5}
	527	3.35×10^{-5}
Diamond	20	1.00×10^{-6}
Celluloid	50	1.09×10^{-4}
Glass (most types)	50	$8.3 \ \times 10^{-6}$
Glass (Pyrex)	50	$3.2 \ \times 10^{-6}$
Ice	−5	5.07×10^{-5}
Steel	20	1.27×10^{-5}
Platinum	20	$8.9 \ \times 10^{-6}$

are heated the unequal expansion causes the strips to bend; if they bend far enough, they open or close a switch that may, for example, control a heating system or air conditioning.

The *ultramicrotome* is an apparatus designed to make very thin tissue slices for use in microscopes. A

Brass

Steel

Figure 12.2. Brass and steel strips attached to each other expand at different rates when heated. ($\alpha_{\text{brass}} > \alpha_{\text{steel}}$).

Figure 12.3. A schematic view of an ultramicrotome. The mechanism for turning the holder is not shown.

sample, mounted on a rotating metal arm, passes a knife edge (Fig. 12.3). If the metal arm is heated at a constant rate, it will expand uniformly, and a thin specimen slice will be cut on each turn. The metal arm can be expended as slowly as 10^4 Å (10^{-6} m) per minute.

Volume Expansion

When an object is heated, all of its dimensions increase. Thus, while each length changes according to $\Delta l = \alpha l \Delta T$, the change in volume ΔV accompanying a temperature change ΔT is

$$\Delta V = \beta V \Delta T \tag{12.2}$$

Here β is the *coefficient of volume expansion*.

The coefficient β can be related to α for materials that expand at the same rate in all directions. Consider a cube of volume $V = l^3$ that is warmed from T to $T + \Delta T$ (Fig. 12.4). If the sides change from l to $l + \Delta l$ the *change in volume* is $\Delta V = (l + \Delta l)^3 - l^3$. In all cases of interest to us, the change in length Δl is much less than the original length. If $(l + \Delta l)^3$ is multiplied out, it is a very good approximation to keep only the first two terms, which are $l^3 + 3l^2 \Delta l$. Then *the change in volume* is just $\Delta V = 3l^2 \Delta l = 3V \Delta l/l$. Using $\Delta l = \alpha l \Delta T$, we have

$$\Delta V = 3\alpha V \Delta T$$

Comparing this with $\Delta V = \beta V \Delta T$, we see that

$$\beta = 3\alpha \tag{12.3}$$

WATER

In discussing thermal expansion, special mention must be made of liquid water because it is one of the very few substances with a negative coef-

Figure 12.4. When the temperature is increased each dimension of a cube changes from l to $l + \Delta l$.

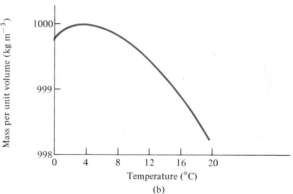

Figure 12.5. (*a*) Coefficient of volume expansion for water. (*b*) The mass per unit volume of water.

ficient of volume expansion at some temperatures. Figure 12.5 shows both β and the mass per unit volume of water plotted against the temperature. β varies as the temperature changes and even reverses sign at 3.98°C. Thus, as T rises from 0°C, water contracts up to 3.98°C and then expands as the temperature increases further; water has its greatest mass per unit volume at 3.98°C.

This characteristic of water is extremely important for aquatic life. As the air temperature decreases in early winter, the surface water of lakes cools. When this surface water reaches 3.98°C, it sinks to the bottom; the warmer, less dense water from beneath floats to the surface. The cool descending water carries oxygen with it. Once the entire lake has experienced this mixing and has reached 3.98°C, further cooling occurs at the surface and ice forms. The lower density ice floats, so that lakes freeze from the surface downward. Aquatic life survives the winter in the freshly oxygenated water beneath the ice.

12.2 | HEAT CAPACITY

When an object at one temperature is placed near or in contact with another object at a higher temperature, energy is transferred to the cooler object. This object then experiences a rise in temperature. The ratio of the amount of energy transferred to the temperature change is called the *heat capacity*.

When energy transfer occurs because of a temperature difference, we say that thermal energy or *heat* is being transferred. Alternatively, energy may be transferred to a substance by doing work on the substance as, for example, by stirring a liquid or compressing a gas.

When work is done on or heat is added to a substance, there is an increase in the *internal energy U*, which is the total energy of all the molecules. For monatomic gases such as He or Ne in which the molecules consist of a single atom, the internal energy is just the sum of the translational kinetic energies. A polyatomic gas molecule may also rotate; it will then have additional kinetic energy. Such a molecule may also have energy associated with vibrations of the constituent atoms about their equilibrium positions. In solids and liquids, interactions among the molecules likewise contribute to the internal energy.

Suppose the internal energy of n moles of a substance changes by a small amount ΔU and the temperature changes by ΔT when heat is added or when work is done on the substance. The *molar heat capacity* is then defined as

$$C = \frac{1}{n}\frac{\Delta U}{\Delta T}$$ (12.4)

The units of C are joules per mole per degree Kelvin (J mole^{-1} K^{-1}).

The heat capacity is an important characteristic of

materials. Perhaps the best example of this is water, which has a relatively high heat capacity. This is one reason that water is used to cool engines and to store heat in solar energy heating systems. The heat capacity is also important because it gives us a convenient opportunity to compare theory with experiment for many materials. For example, we can calculate heat capacities from the ideal gas model and compare them with measured values for real gases.

The heat capacity of a monatomic ideal gas can be found from what we have learned. We found in Chapter Eleven that the average kinetic energy per molecule in such a gas is

$$(K)_{ave} = \frac{3}{2}\frac{R}{N_A}T$$

R is the gas constant, and N_A is Avogadro's number. For n moles or nN_A molecules, the internal energy U is the sum of the kinetic energies of the molecules $U = nN_A(K)_{ave} = 3nRT/2$. If the gas does not change its volume, the heat added ΔQ is equal to the internal energy change, $\Delta Q = \Delta U = 3nR\,\Delta T/2$. With the definition of the molar heat capacity, we find

$$C = \frac{1}{n}\frac{\Delta U}{\Delta T} = \frac{3R}{2} \quad \text{(ideal monatomic gas)} \quad (12.5)$$

The molar heat capacity of a monatomic gas is $C = \frac{3}{2}R = 12.47 \text{ J K}^{-1}$. This prediction agrees very well with measurements on real monatomic gases.

A closely related quantity is the *specific heat capacity c*. This is the heat required for a unit temperature change in a unit mass of a substance. It is related to C by

$$c = \frac{C}{M}$$ (12.6)

Here M is the mass of 1 mole. For example, helium gas has a molecular mass of 4 amu; 1 mole is 4 g = 4×10^{-3} kg, and

$$c = \frac{12.47 \text{ J mole}^{-1} \text{ K}^{-1}}{4 \times 10^{-3} \text{ kg mole}^{-1}}$$
$$= 3.12 \times 10^3 \text{ J kg}^{-1} \text{ K}^{-1}$$

In terms of c, the heat required for a temperature change ΔT in a mass m is

$$\Delta Q = mc\,\Delta T \quad \text{(constant volume)} \quad (12.7)$$

Since c depends on the temperature, this equation

is exact only for very small ΔT. However, like the thermal expansion formula, it is usually adequate for large values of ΔT if the average specific heat capacity is used. Equation 12.7 holds true only when the volume is held fixed. More heat is required if the gas is allowed to expand as it is heated, because the gas does some work in expanding. In practice, heat capacities of solids and liquids are often measured at constant pressure, and c is then found indirectly.

While we have been able to calculate the heat capacity of a monatomic gas from first principles, the problem of calculating the energy required to raise the temperature of a polyatomic gas, a liquid, or a solid is more difficult. For example, we noted that for polyatomic molecules the rotational and vibrational energies must be included in the internal energies. Modern theory predicts that at low temperatures the molar heat capacity of any gas results entirely from the translational motion, and C is $\frac{3}{2}R$. At intermediate temperatures rotations also contribute and C is $\frac{5}{2}R$ for a diatomic gas. At high temperatures, vibrational, rotational, and translational motions all contribute to the specific heat, and C is $\frac{7}{2}R$ for a diatomic gas. The molar heat capacity of hydrogen gas H_2 versus temperature is shown in Fig. 12.6, where the three contributions are clearly seen.

As a gas is cooled and becomes a liquid, further difficulties in making accurate calculations are encountered. For example, the heat capacity of steam is only half as large as that of liquid water. The difference arises from a change in the effects of the forces among molecules. In the gas, the molecules are far apart, and the forces are negligible. In a liquid, the

TABLE 12.2

Specific heat capacities at constant pressure at 25°C, except where noted, of various substances in kcal kg⁻¹ K⁻¹ and in J kg⁻¹ K⁻¹. (In this textbook we approximate the specific heat capacity of liquid water as 1.0 kcal kg⁻¹ K⁻¹)

| | Specific Heat Capacity, c | |
Substance	kcal kg⁻¹ K⁻¹	J kg⁻¹ K⁻¹
Aluminum	0.215	898
Steel	0.107	447
Diamond	0.124	518
Lead	0.031	130
Copper	0.092	385
Helium (gas)	1.24	5,180
Hydrogen (H_2) (gas)	3.41	14,250
Iron	0.106	443
Nitrogen (N_2) (gas)	0.249	1,040
Oxygen (O_2) (gas)	0.219	915
Water (liquid)	0.998	4,169
Ice ($-10°$ to $0°$C)	0.500	2,089
Steam ($100°$ to $200°$C)	0.470	1,963

molecules are closer to each other, so their mutual attraction becomes significant, adding to the internal energy and making specific heat calculations very difficult. Oddly enough, in the solid form, specific heats are again relatively easy to calculate. Specific heats of representative substances are given in Table 12.2.

Heat capacities are measured using *calorimeters* (Fig. 12.7). Heat ΔQ is supplied by an electrical heater to a well-insulated calorimeter that holds the sample, and a thermometer measures the temperature rise ΔT. A sample with mass m and specific heat

Figure 12.6. Molar heat capacity of hydrogen gas H_2. At low temperatures the heat capacity is due to the translational kinetic energy. At higher temperatures the rotational kinetic energy must be included, and the vibrational kinetic energy is important at still higher temperatures.

Figure 12.7. A calorimeter.

capacity c absorbs an amount of heat equal to $mc\,\Delta T$. The container also absorbs some heat; if its mass is m_c and its specific heat capacity is c_c, this heat equals $m_c c_c\,\Delta T$. Adding the heat absorbed by the sample and the container, we have

$$\Delta Q = mc\,\Delta T + m_c c_c\,\Delta T$$

so

$$c = \frac{\Delta Q - m_c c_c\,\Delta T}{m\,\Delta T}. \qquad (12.8)$$

The use of the calorimeter is illustrated in the following example.

Example 12.2

There are 0.1 kg of carbon in a calorimeter at 15°C. The container has a mass of 0.02 kg and is made of aluminum. The addition of 892 J of heat energy brings the temperature to 28°C. What is the specific heat capacity of carbon? Assume the specific heat capacity of aluminum in this temperature range is 900 J kg^{-1} K^{-1}.

Substituting in Eq. 12.8, with $\Delta T = 28°C - 15°C = 13$ K.

$$C = \frac{\Delta Q - m_c c_c\,\Delta T}{m\,\Delta T}$$

$$= \frac{892\text{ J} - (0.02\text{ kg})(900\text{ J kg}^{-1}\text{ K}^{-1})(13\text{ K})}{(0.10\text{ kg})(13\text{ K})}$$

$$= 506\text{ J kg}^{-1}\text{ K}^{-1}.$$

In our discussion of heat capacities based on the ideal gas model, the relationship between heat and energy is quite clear. Historically, scientists studied heat long before any molecular theory was developed. Since it was not realized that heat is just another form of energy, a separate set of units was developed to measure heat. The gram-calorie (cal) was defined as the heat required to heat 1 g of water from 14.5°C to 15.5°C. Later, it was discovered that the energy represented by 1 calorie is equivalent to 4.18 J. The first suggestion of this equivalence was made in 1842 by Julius Robert Mayer (1814–1878), a German physician, on the basis of physiological observations. Mayer also suggested that energy can be neither created nor destroyed. His work was largely ignored, and it was not until the work of James Prescott Joule (1818–1889) several years later that the equivalence of heat and energy was accepted. The principle of the conservation of energy was also independently proposed by Hermann von Helmholtz (1821–1894).

Although 125 years have elapsed in which scientists could have adopted a common energy unit for both heat and work, present custom is still to measure mechanical and electrical energies in joules and heat energies in calories. To make matters worse, American engineers frequently use the British thermal unit (BTU), defined as the heat required to heat 1 lb of water from 63°F to 64°F, and the kilowatt-hour (kWh), which is equal to 3.6×10^6 J. Further, nutritionists measure the energy supplied by food in kilocalories (kcal), which are 10^3 times as large as the gram-calorie. Conversion factors for these energy units may be found in Appendix C.

12.3 | PHASE CHANGES

Most substances can exist in solid, liquid, or gas *phases*. For example, water may be ice, liquid, or steam. A transition from one of these phases to another is called a *phase change*. Many other kinds of phase changes occur in nature. A solid may change from one crystalline structure to another; at low temperatures, a material may become magnetic or lose its electrical resistance. All these changes take place very abruptly at a sharply defined temperature.

The temperature at which a phase change occurs usually depends on additional variables such as pressure. This is illustrated by the *phase diagram* for water (Fig. 12.8). At the temperature and pressure corresponding to point A, water can exist only as ice. If we keep the pressure fixed and add heat, the temperature

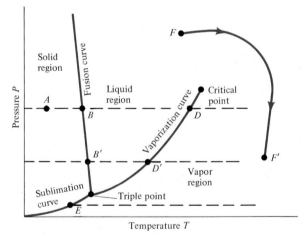

Figure 12.8. Phase diagram for water. The lines indicate the temperature and pressures at which phase changes occur.

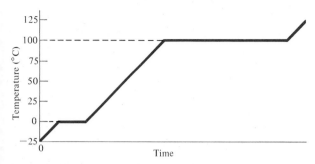

Figure 12.9. The temperature of a sample of water versus time. Heat is added at a uniform rate and the pressure is held constant at one atmosphere.

rises until point B is reached. Now as more heat is added the temperature does not rise. Instead, the ice gradually melts into water, and the *temperature remains constant until all the ice is melted* (Fig. 12.9). As more heat is added, the temperature of the liquid again steadily increases until point D is reached. *Here again the temperature remains constant* until all the liquid has been converted into water vapor. Additional heat will now increase the temperature of the gas.

If we repeat this experiment at a lower pressure, the phase changes occur at B' and D', which correspond to different temperatures than do B and D. If we repeat the experiment once more at an even lower pressure, we find that ice *sublimes* at E directly into vapor without passing through the liquid phase.

Sublimation has become an important part of preserving food. Food is frozen and placed in a low-pressure enclosure. As heat is added the ice sublimates and the vapor is drawn off. This freeze-drying process does not damage the food and preserves its shape and taste. The food can later be reconstituted by adding water.

Two points with special significance are noted on the diagram. At the *triple point,* liquid, solid, and vapor may all exist together. At the *critical point,* a kilogram of liquid and a kilogram of gas have the same volume, and the distinction between the two phases vanishes. Then if the pressure and temperature are adjusted so that a sample passes from point F to F' along the path shown in Fig. 12.8, the change of phase from liquid to vapor is never observed.

The energy absorbed or liberated in a phase change is called the *latent heat.* At atmospheric pressure, the *latent heat of fusion L_f* needed to melt ice is 79.7 kcal kg^{-1}. The *latent heat of vaporization L_v* needed to boil water at atmospheric pressure is 539 kcal kg^{-1}. Some other latent heats are given in Table 12.3. The heat ΔQ needed to change the phase of a mass m is

$$\Delta Q = Lm \qquad (12.9)$$

where L is the appropriate latent heat. The role of the latent heat of fusion is illustrated in the following example.

Example 12.3
How much heat is required to melt 5 kg of ice at 0°C?
Since the latent heat of fusion L_f is 79.7 kcal kg^{-1}, $Q = Lm$ gives

$$Q = L_f m = (79.7 \text{ kcal kg}^{-1})(5 \text{ kg}) = 398 \text{ kcal}$$

The latent heat and specific heat capacity are both important in determining the equilibrium state of a system. This is shown in the following example.

TABLE 12.3

Latent heats at atmospheric pressure

Substance	Melting Point (°C)	Latent Heat of Fusion (kcal kg^{-1})	Boiling Point (°C)	Latent Heat of Vaporization (kcal kg^{-1})
Helium			−268.9	5
Nitrogen	−209.9	6.09	−195.8	48
Ethyl alcohol	−114	24.9	78	204
Mercury	−39	2.82	357	65
Water	0	79.7	100	539
Silver	96	21.1	2193	558
Lead	327	5.86	1620	218
Gold	1063	15.4	2660	377

Example 12.4

Ice cubes at $-10°C$ with a total mass of 0.045 kg are placed in a 0.3-kg cup of tea at 30°C. What is the final equilibrium temperature T? (The heat capacity of tea is the same as water.)

Let us suppose provisionally that all the ice melts. If this is correct, the final temperature we calculate must not be below 0°C. The heat needed to bring the ice to 0°C is

$$Q_1 = m_{ice}c_{ice}\,\Delta T$$
$$= (0.045\ \text{kg})(0.5\ \text{kcal kg}^{-1}\ \text{K}^{-1})(10°C)$$
$$= 0.225\ \text{kcal}$$

The heat needed to melt the ice is

$$Q_2 = L_f m_{ice}$$
$$= (79.7\ \text{kcal kg}^{-1})(0.045\ \text{kg})$$
$$= 3.59\ \text{kcal}$$

The heat needed to warm the melted ice from 0°C to the final temperature T_f is

$$Q_3 = m_{ice}c\,\Delta T$$
$$= (0.045\ \text{kg})(1\ \text{kcal kg}^{-1}\ \text{K}^{-1})T_f$$
$$= 0.045\ \text{kcal K}^{-1}\ T_f$$

Now the heat lost by the tea originally at 30°C is (using $c_{tea} = 1\ \text{kcal kg}^{-1}\ \text{K}^{-1}$).

$$Q_4 = m_{tea}c_{tea}\,\Delta T$$
$$= (0.3\ \text{kg})(1\ \text{kcal kg}^{-1}\ \text{K}^{-1})(30°C - T_f)$$
$$= 9\ \text{kcal} - 0.30\ \text{K}^{-1}\ T_f$$

Since the heat lost must equal the heat gained, we have $Q_1 + Q_2 + Q_3 = Q_4$ or, suppressing the units,

$$0.225 + 3.59 + 0.45\ T_f = 9 - 0.30\ T_f.$$

Solving for T_f, we find $T_f = 14.2°C$. The final temperature is greater than zero, so our original assumption that all the ice melts is correct.

The following example illustrates how to compute the mass of ice melted when the final temperature is 0°C.

Example 12.5

Replace the tea in the preceding example by 0.045 kg of water at 30°C. What is the equilibrium condition?

If we assume again that the ice melts completely, we find that T_f is negative, violating our hypothesis. Melting 1 kilogram of ice liberates 79.7 kcal, enough to cool $79.7\ \text{kg}/30 = 2.66\ \text{kg}$ of water from 30° to 0°C. Since there are equal amounts of ice and water, it is clear without further calculation that not all the ice melts and the equilibrium temperature is 0°C. We must therefore solve for the mass m_1 of ice that is melted.

As in the preceding example, heating the ice to 0°C requires $Q_1 = 0.225$ kcal. Melting a mass m_1 of ice requires

$$Q_2 = L_f m_1 = (79.7\ \text{kcal kg}^{-1})m_1$$

The heat lost by the water originally at 30°C is

$$Q_3 = m_{water}c_{water}\,\Delta T$$
$$= (0.045\ \text{kg})(1\ \text{kcal kg}^{-1}\ \text{K}^{-1})(30°C)$$
$$= 1.35\ \text{kcal}$$

Equating the heat gained to the heat lost, we have

$$0.225 + 79.7\ m_1 = 1.35$$

so $m_1 = 0.0141$ kg.

12.4 | HEAT CONDUCTION

Heat transfer always occurs from regions of higher to regions of lower temperature, so that two objects isolated from their surroundings gradually approach a common temperature. In this section, we discuss heat *conduction* between objects in contact. Heat transfer by *convection* and *radiation* is considered in the following two sections.

When two objects at temperatures T_1 and T_2 are connected by a rod, their temperature difference $\Delta T = T_2 - T_1$ will steadily diminish (Fig. 12.10). If we cut the rod lengthwise so that each section has a cross section of $A/2$, then half the heat flow will be conducted in each section. Thus the rate at which heat flows from the hotter to the colder object must be proportional to the cross-sectional area A.

The rate of heat flow also depends on ΔT and the length l. If we double ΔT and double the length at the same time, the heat flow is unchanged. If the temperature difference is doubled and the length is un-

Figure 12.10. A rod of cross-sectional area A and length Δl conducts heat from the higher to the lower temperature object.

TABLE 12.4

Thermal conductivities in kcal s^{-1} m^{-1} K^{-1}

Substance	Thermal Conductivity, κ
Silver	0.10
Copper	0.095
Aluminum	0.057
Steel	0.019
Ice	4×10^{-4}
Glass, concrete	2×10^{-4}
Water	1.4×10^{-4}
Animal muscle, fat	5×10^{-5}
Wood, asbestos	2×10^{-5}
Felt, rock wool	1×10^{-5}
Air	5.7×10^{-6}
Down	4.6×10^{-6}

changed, the heat flow doubles. The same thing happens if ΔT is unchanged and the length is halved. Thus the heat flow must depend on the ratio $\Delta T/l$, which is called the *temperature gradient*. We can summarize these observations algebraically in an equation for the *heat flow* $H = \Delta Q/\Delta t$,

$$H = \kappa A \frac{\Delta T}{l} \tag{12.10}$$

where κ is a proportionality constant called the *thermal conductivity*. Equation 12.10 is exact when ΔT is very small. However, the equation for heat flow becomes more complicated if κ varies with temperature or when the geometry is not so simple. In this chapter, we assume that κ is constant in the temperature range of interest.

Representative values of κ are given in Table 12.4. Typically, metals are good thermal conductors. Their conductivities are greater by factors of 10^3 or 10^4 than those of thermal insulators such as asbestos or rock wool.

One of the most important insulators is air. Insulation in homes and in the material for warm clothing utilize this fact. The fibers of the material trap air in the material, and this air acts as an insulator. Storm or double-pane windows use air trapped between two glass panes to reduce conductive heat losses.

Body tissue is also a good insulator. When the environment is warm, the interior body temperature is quite uniform (Fig. 12.11a). Because body tissues are poor conductors, the inner core of the body can be kept warm in a cold environment (Fig. 12.11b).

Example 12.6

A person walking at a modest speed generates heat at a rate of 0.07 kcal s^{-1}. If the surface area of the body is 1.5 m^2 and if the heat is assumed to be generated 0.03 m below the skin, what temperature difference between the skin and interior of the body would exist if the heat were conducted to the surface? Assume the thermal conductivity is the same as that for animal muscle, 5×10^{-5} kcal s^{-1} K^{-1}.

Despite the dissimilarity between humans and the rod in Fig. 12.10, we may still apply Eq. 12.10 to a small section of tissue. Summing over the sections is approximately equivalent to using the total body surface area for the area A.

Solving $H = \kappa A \, \Delta T/l$ for ΔT,

$$\Delta T = \frac{lH}{\kappa A}$$

$$= \frac{(0.03 \text{ m})(0.07 \text{ kcal s}^{-1})}{(5 \times 10^{-5} \text{ kcal s}^{-1} \text{ m}^{-1} \text{ K}^{-1})(1.5 \text{ m}^2)}$$

$$= 28°C$$

Since the actual temperature difference is only a few degrees (Fig. 12.11), we can conclude that heat is not removed from the body by conduction through tissue from the interior to the exterior of the body. In fact, the flow of warm blood from the interior of the body to the cooler exterior is the major factor in body heat transport.

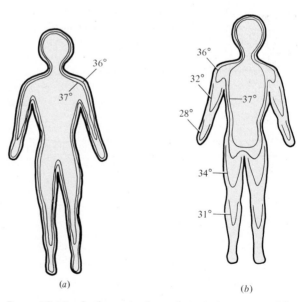

Figure 12.11. Isotherms (surfaces of constant temperature) in the body for (a) a warm environment and (b) a cold environment. (Adapted from Aschoff and Wever, Naturwissenschafter, vol. 45, p. 477, 1958.)

12.5 | HEAT TRANSFER BY CONVECTION

Although some heat is transported by conduction in liquids and gases, a much larger quantity may be carried by the motion of the fluid itself. This is called *convection*. In Fig. 12.12*a*, the liquid near the heat source is heated and expands slightly, becoming lighter than the overlying cooler fluid. It then rises and is replaced by cooler, heavier fluid. When the warmer fluid arrives at the cooler region of the container, it cools, contracts, and begins to sink again. Had the container been heated from the top, convection would not have occurred, and the bulk of the fluid would have been heated by the much slower conduction process.

A hot water or steam radiator provides another illustration of convection (Fig. 12.12*b*). Air near the radiator is heated and rises, while air near the outside walls and the windows is cooled and sinks. This establishes the flow pattern shown.

There are many difficulties in developing a quantitative theory of convection. For example, a given surface loses heat faster when it is vertical than when it is horizontal. Despite such difficulties, we can make some progress using an approximate formula. In still air the rate of convective heat transfer for a surface area A is given approximately by the empirical formula,

$$H = qA\,\Delta T \qquad (12.11)$$

Here ΔT is the temperature difference between the surface and the air distant from the surface. The *convective heat transfer constant q* depends on the shape and orientation of the surface and to some extent on ΔT. For a naked human, we use the average value $q = 1.7 \times 10^{-3}\,\text{kcal s}^{-1}\,\text{m}^{-2}\,\text{K}^{-1}$. Heat loss by convection is important for humans, as shown in the following example.

Example 12.7

In a warm room, a naked resting person has a skin temperature of 33°C. If the room temperature is 29°C and the body surface area is 1.5 m², what is the rate of heat loss due to convection?

Using $q = 1.7 \times 10^{-3}\,\text{kcal s}^{-1}\,\text{m}^{-2}\,\text{K}^{-1}$, we have

$$
\begin{aligned}
H &= qA\,\Delta T \\
&= (1.7 \times 10^{-3}\,\text{kcal s}^{-1}\,\text{m}^{-2}\,\text{K}^{-1})(1.5\,\text{m}^2)(33°\text{C}-29°\text{C}) \\
&= 0.010\,\text{kcal s}^{-1}
\end{aligned}
$$

Figure 12.12. Paths of convective flow in (*a*) a container of liquid; (*b*) in a room heated by a radiator.

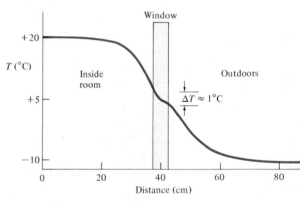

Figure 12.13. Air and glass temperatures near a window on a cold windless day. For clarity, the window thickness has been exaggerated.

A resting person in this situation will generate heat at about twice this rate. Thus, under these moderate conditions, convection provides the mechanism for about 50 percent of the body's heat loss. If there is a breeze or if the room temperature is lower, convective heat losses will increase correspondingly.

Convection plays a significant role in many everyday experiences. In a warm room the air a centimeter or so away from a cold windowpane feels rather cold. The air in contact with the glass has the same temperature as the pane; the air temperature rises perceptibly for some distance moving away from the window (Fig. 12.13). A similar gradual decrease in air temperature occurs outside if there is no wind. However, because the conduction of heat across the glass is very efficient, only a small temperature difference exists across the window.

If there is a wind, the nearly stagnant layer of

TABLE 12.5
Effect of moving air on temperature

Wind Speed (mi h⁻¹)	Actual Temperature, °F												
Calm	15	10	5	0	−5	−10	−15	−20	−25	−30	−40	−45	−55
	Effective Temperature, °F												
5	10	5	0	−5	−10	−15	−20	−25	−30	−35	−45	−50	−60
10	0	−10	−15	−20	−25	−35	−40	−45	−50	−60	−65	−70	−75
15	−10	−20	−25	−30	−40	−45	−50	−60	−65	−70	−80	−85	−90
20	−15	−25	−30	−35	−45	−50	−60	−65	−75	−80	−85	−95	−100
25	−20	−30	−35	−45	−50	−60	−65	−75	−80	−90	−95	−105	−110
30	−25	−35	−40	−50	−55	−65	−70	−80	−85	−95	−100	−110	−115

☐ —*Moderate danger*   ▦ —*Great danger*

warm air just outside the window is removed more rapidly than when convective forces alone are acting. This results in a lower temperature at the outside window surface and hence increases the rate of heat loss from the window. This is equivalent to having a lower outside temperature and is referred to in weather reports as the *wind chill factor*. The *effective temperature* decreases rapidly as the wind velocity increases, as shown in Table 12.5. Skiers and snow-mobiliers who create their own wind must be conscious of the potential hazards. Exposed flesh can freeze in about a minute at an effective temperature of −25°F, which can, for example, be achieved by moving at 20 mi h⁻¹ in air at 10°F. Effective temperatures below −75°F are extremely dangerous, since freezing can occur in seconds.

12.6 | RADIATION

Conduction and convection require the presence of some material, be it solid, liquid or gas. However, we know that heat can also be transmitted through a vacuum, since the sun's energy traverses millions of kilometers of space before reaching the earth. The process by which this occurs is called *radiation*. Radiant heat transfer also occurs in transparent media.

The term "radiation," as used in this chapter, is another word for electromagnetic waves. These are waves that are of electric and magnetic origin and can carry energy. In a hot object the atomic charges oscil-

late rapidly, sending out energy as electromagnetic waves, somewhat like ripples on a pond. These waves travel with the speed of light, $c = 3 \times 10^8 \text{ m s}^{-1}$. Visible light, radio waves, and x-rays are all examples of electromagnetic waves (Fig. 12.14). The energy carried by the waves depends on the motion of the charges and hence on the temperature. These electromagnetic waves are discussed further in later chapters.

A wave is characterized by its wavelength λ and frequency *f*. The wavelength is the distance between successive wave crests; the frequency is the number of crests passing a given point each second and is equal to the frequency of vibration of the charge producing

Figure 12.14. Frequencies and wavelengths of various kinds of electromagnetic waves. The scales are logarithmic.

205

the electromagnetic wave. The distance between crests, λ, times f, the number of crests per second passing a point, must equal the velocity of the wave, so

$$f\lambda = c \qquad (12.12)$$

For example, red light has a wavelength of about 7×10^{-7} m, which corresponds to a frequency

$$f = \frac{c}{\lambda} = \frac{3 \times 10^8 \text{ m s}^{-1}}{7 \times 10^{-7} \text{ m}} = 4.2 \times 10^{14} \text{ Hz}$$

When a surface is exposed to radiation, it may absorb part or all of the incident energy. The fraction of the radiation absorbed is the *emissivity e* of the surface. A perfect absorber is called a *blackbody* and has $e = 1$; for a perfect reflector, $e = 0$. All objects have emissivities between 0 and 1. The emissivity usually varies with the wavelength. Sunlight is most intense in the visible light region, which corresponds approximately to wavelengths from 4×10^{-7} m to 7×10^{-7} m. The darkest human skin has an emissivity of 0.82 for visible light and the lightest skin has $e = 0.65$. The radiation from objects at or near typical room or body temperatures is predominantly in the longer infrared wavelengths. All human skin has an emissivity of nearly 1 at these wavelengths.

In principle, every object at a nonzero temperature emits some radiation at all wavelengths. However, the amount of energy radiated at each wavelength depends on the temperature (Fig. 12.15). An object at 800°C looks red since it emits some radiation in the longest wavelength or red portion of the visible spectrum but very little in the blue portion. An object heated to 3000°C looks white because it is emitting substantial amounts of radiation throughout the visible light range. Similarly, very hot stars appear relatively blue, while cooler ones are somewhat red.

Two formulas express algebraically some of the information in Fig. 12.15. First, the wavelength λ_{max} at which the radiation is most intense depends on the Kelvin temperature T according to the *Wien displacement law,*

$$\lambda_{max} = \frac{2.898 \times 10^{-3}}{T} \text{ m K} \qquad (12.13)$$

The temperature of the sun determines the wavelength at which most of its radiation reaches us. This is shown in the following example.

Figure 12.15. The heat energy radiated per second by 1 m^2 of surface. As the temperature increases, λ_{max} shifts to shorter wavelengths (vertical dashed lines).

Example 12.8

The sun has a surface temperature of 6000 K. What is the wavelength of maximum radiation?

From the Wien displacement law,

$$\lambda_{max} = \frac{2.898 \times 10^{-3}}{6000 \text{ K}} \text{ m K} = 4.83 \times 10^{-7} \text{ m}$$

Thus the maximum solar radiation is in the visible portion of the spectrum as noted earlier. It is interesting that the eyes of animals are most sensitive very near the wavelength at which there is the most radiation available for seeing.

The second characteristic of radiation illustrated in Fig. 12.15 is the fact that as the temperature increases, the total energy radiated from a surface increases rapidly. The rate at which energy radiates from a surface of area A at temperature T was given by Josef Stefan (1853–1893) in 1879 and is called *Stefan's law,*

$$H = e\sigma A T^4 \qquad (12.14)$$

Here

$$\sigma = 5.67 \times 10^{-8} \text{ W m}^{-2} \text{ K}^{-4}$$
$$= 1.36 \times 10^{-11} \text{ kcal s}^{-1} \text{ m}^{-2} \text{ K}^{-4}$$

is Stefan's constant. Note that the emissivity, which was introduced in the discussion of absorption, also plays a role in emission. Here, e is the emissivity of the surface averaged over the wavelengths in which the radiation is emitted. The maximum energy is emitted by a blackbody, for which e is 1.

Radiation losses are important in animals. This is illustrated in the following example.

Example 12.9

The person in Example 12.7 had a skin temperature of $33°C = 306$ K and was in a room where the walls were at $29°C = 302$ K. If the emissivity is 1 and the body surface area is 1.5 m², what is the rate of heat loss due to radiation?

We must consider two competing processes. First, the person at a temperature $T_2 = 33°C = 306$ K will radiate heat at a rate

$$H_{out} = e\sigma A T_2{}^4$$
$$= (1)(1.36 \times 10^{-11} \text{ kcal s}^{-1} \text{ m}^{-2} \text{ K}^{-4})$$
$$\times (1.5 \text{ m}^2)(306 \text{ K})^4$$
$$= 0.179 \text{ kcal s}^{-1}$$

This is over 15,000 kcal per day, or roughly six times the typical daily heat output of a human; a person losing this much heat would quickly freeze. However, heat is also received from the surroundings. If these are at temperature $T_1 = 29°C = 302$ K, heat will be received at the rate

$$H_{in} = e\sigma A T_1{}^4$$
$$= (1)(1.36 \times 10^{-11} \text{ kcal s}^{-1} \text{ m}^{-2} \text{ K}^{-4})$$
$$\times (1.5 \text{ m}^2)(302 \text{ K})^4$$
$$= 0.170 \text{ kcal s}^{-1}$$

Thus the net rate of heat loss is $(0.179 - 0.170)$kcal s^{-1} = 0.009 kcal s^{-1}. This is roughly equal to the convective loss rate.

In almost all radiation problems, one must compute the difference between the rates at which energy is emitted at two temperatures. Thus if one temperature is $T + \Delta T$ and the other is T, the difference is $\Delta H = e\sigma A[(T + \Delta T)^4 - T^4]$. In many such problems ΔT is very much less than T, and it is a good approximation to write $(T + \Delta T)^4 - T^4 \simeq 4T^3 \Delta T$. Then

$$\Delta H = 4e\sigma A T^3 \Delta T \qquad (\Delta T \ll T) \quad (12.15)$$

The fractional error introduced by the approximation leading to this result is of the order $\Delta T/T$. In the preceding example, $\Delta T/T = 0.013$, so the error would be about 1.3 percent if the approximate formula were used. The following example illustrates a situation where this approximate result can be used accurately.

Example 12.10

Comparing two patches of skin of area A on a person's chest, the radiation rate is found to differ by 1 percent. What is the difference in skin temperature?

Suppose the lower temperature patch is at $T = 37°C = 310$ K, and the temperature of the other is $T + \Delta T$. Then the fractional difference divided by the heat loss at 310 K is

$$\frac{\Delta H}{H} \simeq \frac{4e\sigma A T^3 \Delta T}{e\sigma A T^4} = \frac{4\Delta T}{T} = 0.01$$

Solving for ΔT,

$$\Delta T = 0.01 \frac{T}{4} = (0.01)\frac{(310 \text{ K})}{4} = 0.775 \text{ K}$$

Thus, a temperature change of less than 1 K alters the radiation rate by 1 percent.

The preceding example illustrates the physical principle underlying *thermography*, a technique with uses in many fields. In medical applications, special infrared detectors are used to produce a photographic record of the infrared radiation emitted by a patient (Fig. 12.16). In photos of this type, areas with higher temperatures appear darker; differences as small as $0.1°C$ are detectable. The photographs show, for example, how circulation is reduced by smoking. Thermography is often used in preliminary screenings to identify patients who may have breast cancers, thyroid tumors, or other diseases.

It is also possible to measure accurately the weak radiation emitted by the body at microwave wavelengths, which are longer than those of infrared radiation. Recent experiments show that such measurements may help doctors to detect tumors up to 10 cm below the surface. By contrast, infrared radiation is absorbed more readily in body tissues, so thermography is sensitive only to tumors closer to the skin. Both techniques are noninvasive and intrinsically safe, since the patient is not subjected to any external source of radiation or other probe.

SUMMARY

Many properties of materials depend on temperature. Since temperature is indicative of the internal or thermal energy of the material, this dependence on temperature is usually a reflection of the different

207

Figure 12.16. Thermograms showing warmer areas as darker. (*a*) Joint inflammation accompanying rheumatoid arthritis. (*b*) Increased peripheral circulation is shown in a foot after treatment (right). (Courtesy of AGA Medical Division.)

states of motion of the molecules. For example, most objects will experience a fractional change in length $\Delta l/l$ proportional to the temperature change ΔT,

$$\frac{\Delta l}{l} = \alpha \, \Delta T$$

Similarly, fractional changes in volume of uniform materials are given by

$$\frac{\Delta V}{V} = \beta \, \Delta T = 3\alpha \, \Delta T$$

where α is the linear expansion coefficient, and β is the volume expansion coefficient.

When heat is added to an object, its temperature will change or it will undergo a phase change. If n moles of a substance experience an internal energy change ΔU and a temperature change ΔT, the molar heat capacity is given by

$$C = \frac{1}{n} \frac{\Delta U}{\Delta T}$$

If heat ΔQ is added to a substance but no temperature change occurs, then a mass m of the substance will experience a phase change where

$$\Delta Q = Lm$$

Here L is the latent heat for the phase change.

Heat energy is exchanged between objects at different temperatures by one or more of three processes: conduction, convection, and radiation. If a sample of material of cross-sectional area A has a temperature difference ΔT along its length, the rate of heat transfer from the high to low temperature end by conduction is

$$H = \kappa A \frac{\Delta T}{l}$$

where κ is the thermal conductivity of the material.

An object may transfer heat by convection. If the object has a surface area A and is at a temperature ΔT above that of a fluid that can circulate, the heat loss rate is given approximately by

$$H = qA\,\Delta T$$

Convective heat transfer increases dramatically if the fluid is forced to move. For example, convective losses are much greater on windy days than on calm days.

Heat transfer by radiation occurs through the emission and absorption of electromagnetic radiation. An object at a temperature T will emit most of its radiation at wavelengths near

$$\lambda = \frac{2.898 \times 10^{-3}}{T}\,\text{m}$$

The total rate at which energy is emitted depends on the fourth power of the temperature,

$$H = e\sigma AT^4$$

Here A is the surface area, e the emissivity of the surface, and $\sigma = 5.67 \times 10^{-8}$ W m^{-2} K^{-4} is Stefan's constant. The heat lost by radiation is the difference between that emitted by the object and that absorbed from the object's surroundings.

Checklist

Define or explain

linear expansion	triple point
thermostat	critical point
volume expansion	latent heat
internal energy	thermal conductivity
molar heat capacity	convection
specific heat capacity	emissivity
calorimeter	blackbody
phase change	Wien displacement law
sublimation	Stefan's law

EXERCISES

Section 12.1 | Thermal Expansion

12-1 A steel railroad track is 20 m long at 20°C. How much longer is it at 40°C?

12-2 A steel railroad track is 30 m long at 0°C. How much shorter is it at −20°C?

12-3 An aluminum meter stick is exactly 1 m long at 20°C. How much shorter is it at 0°C? Use $\alpha = 2.30 \times 10^{-5}$ K^{-1}.

12-4 For Pyrex glass, β is about one-third that of ordinary glass. What does this imply about thermal stresses?

12-5 Why does heating a jar lid make it easier to open?

12-6 The surface area of a solid increases with temperature according to the formula $\Delta A = 2\alpha A\,\Delta T$. By how much does the area of a rectangular steel plate 0.5 m by 2.5 m change when it is heated from 0°C to 40°C?

Section 12.2 | Heat Capacity

12-7 How many calories are needed to heat 0.15 kg of helium gas from 20°C to 80°C at constant volume?

12-8 A 10-kg mass is dropped to the ground from a window 50 m high. Assuming that all the kinetic energy is converted to heat, how many kilocalories of heat are produced?

12-9 A snowball is dropped from a rooftop 20 m above the ground. If its temperature is −10°C, and if we suppose that none of the heat generated is lost to the surroundings, what is its final temperature?

12-10 A container with a mass of 0.6 kg is at a temperature of 20°C. When 2.5 kg of boiling water is poured into it, the final temperature is 90°C. What is the specific heat capacity of the container?

12-11 A calorimeter of mass 0.4 kg and a specific heat capacity of 0.15 kcal kg^{-1} K^{-1} contains a sample with a mass of 0.55 kg. If 2450 J of energy is supplied electrically and the temperature rises 4°C, what is the specific heat capacity of the sample?

12-12 A 5-kg meteorite hits the ground at 2000 m s^{-1}. How many calories of heat are liberated if all its kinetic energy is converted to heat?

Section 12.3 | Phase Changes

12-13 How much heat is needed to melt a 10-kg block of ice that is initially at −10°C?

12-14 How much heat is required to heat 1 kg of water from 20°C at atmospheric pressure to boiling and to convert it entirely into steam?

12-15 If 0.15 kg of ice at 0°C is added to 0.25 kg of water at 20°C, (a) does all the ice melt? (b) What is the final temperature?

Section 12.4 | Heat Conduction

12-16   A picnic jug contains 1.3 kg of water and 0.6 kg of ice. If 8.5×10^{-3} kcal of heat enters through the insulation each second, how long does it take for all the ice to melt?

12-17   A cabin wall is made of wood 0.05 m thick and has an area of 12 m². If the outside surface of the wall is at 0°C and the inside surface is at 20°C, how much heat per second is lost through the wall?

12-18   How much heat per second will be conducted by a copper rod 4 m long with a cross-sectional area of 0.015 m² if one end is at 250°C and the other is at 40°C?

12-19   How much heat per second is conducted across a wooden wall of area 25 m² and thickness 0.1 m if the temperature inside is 20°C and the temperature outside is −10°C?

Section 12.5 | Heat Transfer by Convection

12-20   How much heat will a naked person with surface area 1.4 m² lose by convection in air at 0°C? Assume the average q is 1.7×10^{-3} kcal s⁻¹ m⁻² K⁻¹ and that the skin temperature is 30°C.

12-21   A naked person with a surface area of 1.8 m², a skin temperature of 31°C, and an average q of 1.7×10^{-3} kcal s⁻¹ m⁻² K⁻¹ loses 0.03 kcal s⁻¹ by convection. What is the air temperature?

12-22   A windowpane is at 10°C and has an area of 1.2 m². If the outside air temperature is 0°C, how much heat lost by convection per second? The q for the window is 10^{-3} kcal s⁻¹ m⁻² K⁻¹.

Section 12.6 | Radiation

12-23   What is the wavelength of maximum radiation intensity for a surface at 37°C?

12-24   What is the wavelength of maximum radiation intensity for a surface at 1000°C?

12-25   What is the wavelength of maximum radiation intensity for a surface at 2000°C?

12-26   Calculate the error made if Eq. 12.15 rather than Eq. 12.14 were used if $T + \Delta T = 319$ K and $T = 270$ K.

12-27   Estimate the error made if Eq. 12.15 rather than Eq. 12.14 were used if $T + \Delta T = 297$ K and $T = 293$ K.

12-28   A naked person with surface area 1.8 m² and skin temperature 33°C is in a room at 10°C.

(Assume $e = 1$.) (a) How much heat does the person radiate per second? (b) What is the person's net loss of heat due to radiation?

12-29   The temperature usually drops faster on clear nights than on cloudy nights. Why?

12-30   An object is at 300 K, and an identical object is at 900 K. What is the ratio of the energies radiated? (Assume the emissivity is the same for both objects?)

PROBLEMS

*12-31   The pendulum of a large clock consists of a thin steel rod with a heavy weight at the end. At 20°C the rod is 1.22 m long and the clock keeps accurate time. (a) By how much does the length change if the temperature rises to 40°C? (b) Does the clock run fast or slow? (c) What is the fractional change in the period of the pendulum? (*Hint:* See Chapter Ten.) (d) What is the error of the clock in seconds per day?

12-32   (a) A car with a full 20-gallon steel gasoline tank is parked in the sun. If its temperature rises 30°C, how much gasoline spills out? (For gasoline, $\beta = 9.50 \times 10^{-5}$ K⁻¹ and for steel, $\beta = 3.81 \times 10^{-5}$ K⁻¹). (b) At what time of day is gasoline cheapest?

12-33   (a) Show that the coefficient of volume expansion β for an ideal gas is $1/T$, where T is the Kelvin temperature. (Assume that the pressure remains constant.) (b) What is β at 20°C?

12-34   Assuming that no heat is transferred to or from the surroundings, how much does the temperature of a river rise in going over a 100-ft falls? (Assume the velocity of the water is the same above and below the falls.)

12-35   A 75-kg man utilizes energy at the rate of 2400 kcal per day. Suppose that 10 percent of this energy is used for work and 90 percent is waste heat. If his body had no way of releasing that heat, by how much would his temperature rise per hour on the average? (The specific heat of animal tissue is approximately equal to that of water.)

12-36   How fast must a lead bullet travel so that it melts when fired into a thick slab of wood, assuming that none of the heat is lost to the wood? (Assume initially that $T_C = 20$°C.)

12-37   Using the data in Fig. 12.6, find the number of joules of energy required to heat 1 kg of hydrogen gas from (a) 30 to 40 K, and from (b) 260 to 270 K.

12-38   A steel wool pad is rubbed on a steel frying pan with a force component of 10 N along the direction of motion. Each stroke is 0.1 m long, and there are 0.8 strokes per second. (a) How much heat is generated per second? (b) If none of the heat is lost to the surroundings and the combined mass of the pan plus pad is 1.25 kg, by how much will their temperature rise after 1 minute?

12-39   How much heat is needed to melt a mole of solid N_2 initially at its melting point?

*12-40   The prevailing winds in the western United States are from the west. In many areas the region east of mountain ranges is hot and dry. Suggest a reason for this. (*Hint:* Typically, there is relatively great precipitation over the mountains. What effect does this have on the temperature of the air.)

*12-41   A pot has a copper bottom layer 0.005 m thick and an inner steel layer 0.002 m thick. The inside of the pot is at 100°C and the outside of the bottom is at 103°C. (a) What is the temperature of the copper-steel junction? (b) How much heat is conducted per second if the area of the bottom is 0.04 m²?

*12-42   (a) How much heat per second is conducted across a wooden wall of area 20 m² and thickness 0.003 m if the temperature difference is 40°C? (b) How much heat is conducted per second if there is a layer of rock wool 0.004 m thick on one side of the wall?

12-43   A pot with a steel bottom 0.01 m thick rests on a hot stove. The area of the bottom is 0.1 m². The water inside the pot is at 100°C, and 0.05 kg are evaporated every 3 minutes. Find the temperature of the lower surface of the boiler, which is in contact with the stove. (Assume that no heat is lost to the room.)

12-44   A picnic chest has dimensions 0.5 × 0.3 × 0.35 m. It is insulated with a 0.02-m thick layer of a material with conductivity of 10^{-5} kcal s^{-1} m^{-1} K^{-1}. (a) If the temperature difference across the layer is 35°C, how much heat per second will be conducted? (b) How many kilograms of ice will melt each hour in the chest?

12-45   A homeowner turns down the thermostat from 23°C to 18°C. If the average outside temper-

ature is 0°C, estimate the percentage reduction in fuel consumption.

12-46   The average outside temperature is 0°C; inside a house the temperature is 23°C. To reduce fuel bills by 10 percent, what thermostat setting should be chosen?

12-47   A surface at 20°C faces a cloudless night-time sky. (a) What is the rate of radiation per square meter of the surface? (Assume $e = 1$.) (b) The amount of radiation from the sky is very small by comparison. Why?

12-48   The surface of the sun is at about 6000 K. (a) How much power does it radiate per square meter of surface area? (b) How much power reaches the upper atmosphere of the earth per square meter? (The diameter of the sun is 1.39×10^6 km, and the mean radius of the earth's orbit is 1.49×10^8 km.)

12-49   A man having surface area 1.8 m² wears a garment 0.01 m thick with a thermal conductivity of 10^{-5} kcal s^{-1} m^{-1} K^{-1}. (a) If his skin temperature is 34°C and the outside of his garment is at −10°C, what is his rate of heat loss? (b) Is he adequately dressed? Explain.

12-50   A girl having a surface area of 1.2 m² wears a down jacket and pants 0.03 m thick. If her skin temperature is at 34°C and she can safely lose 72 kcal h^{-1} by conduction, what is the lowest temperature for which the clothing is adequate? (Assume that the outside of the clothing is at the same temperature as the air.)

SUPPLEMENTARY TOPICS

12.7 | TEMPERATURE REGULATION IN WARM-BLOODED ANIMALS

Most biological processes are temperature-dependent, so the body temperature of an animal must be kept within a narrow range. Warm-blooded animals such as birds and mammals control their own temperatures by regulating the heat loss from their bodies. By contrast, cold-blooded animals are dependent on their environment for the maintenance of their body temperature. For example, snakes are often found sunning themselves on sun-warmed rocks. Many insects must beat their wings before takeoff to raise the temperature of their flight muscles. In this section, we are concerned only with the ways in which warm-blooded animals regulate their temperatures on a short-term basis.

In addition to the mechanisms used daily, biological adaptations to changing climatic conditions also occur on seasonal and evolutionary time scales. The heavy coats of animals, bird migration, and hibernation are all seasonal adaptations. On a longer time scale, nature has favored the development of larger animals in the coldest climates. Such animals have large ratios of volume to surface area, and the corresponding heat production to heat loss ratios are high.

In warm-blooded animals, the main object of temperature regulation is to keep vital organs and muscles at a nearly ideal temperature. Since heat loss occurs either at the body surface or through water evaporation from the lungs, the object is to adjust the flow of heat from the organs and muscles to the lung and body surfaces. The blood plays an important role in this process, since it carries heat from the internal to the external portions of the body. Also, the body's thermostat, the *hypothalamus,* which is located in the brain, uses the blood temperature to monitor the system. The hypothalamus acts to keep itself at a nearly constant temperature. In doing so, the temperature of other internal organs may vary much more widely. This is analogous to a house with a single thermostat in one room. The temperature in that room will be nearly uniform, while the temperature in other rooms may fluctuate considerably.

The source of body heat is the chemical metabolism of food. A resting 70-kg human generates about 70 kcal hr^{-1}; during heavy exercise the rate may be up to 20 times greater. Depending on the air temperature and the clothing worn, the heat generated may be needed to overcome convective and radiative losses, or it may be a waste product to be disposed of by the body.

A warm-blooded animal has a number of mechanisms to use in controlling its temperature. To raise its temperature, the body reduces the blood flow through the capillaries nearest the skin surface. Flesh is a poor conductor of heat, so this is effective in reducing heat losses. Also, body hair can be fluffed up to increase insulation. (Even humans have a vestige of this manifested as goose bumps; the body is attempting to fluff almost nonexistent hair). Finally, heat production may be increased by shivering.

The body is cooled by convection and radiation from the skin and by the evaporation of sweat from the skin and of water from the lungs. If the interior temperature begins to increase, the body first increases blood flow near the skin surface to increase convective and radiative losses and then, if necessary, uses the evaporative loss mechanism. Humans and horses, among other animals, sweat from glands located over much of the body and thus benefit from evaporation over a large surface area. As much as 1.5 kg of sweat per hour may be evaporated from the human body. The latent heat of vaporization for sweat at 37°C is about equal to that of water at the same temperature, 580 kcal kg^{-1}. Note that this is larger than the 539 kcal kg^{-1} necessary to vaporize water at 100°C. Under many conditions the evaporation of sweat is the primary cooling mechanism used by the body.

Furry animals such as dogs do not sweat but take major advantage of evaporative cooling in the lungs. They exhale large volumes of air by panting. Panting itself generates heat but fortunately less than is lost by evaporation.

The various contributions to the rate of heat loss and production for a typical human adult are approximately as follows. If the area A is given in square meters, the skin temperature T_s and air temperature T_a are given in degrees Celsius, and the rate r of sweating is in units of kilograms per hour, then

H_m = heat generated by metabolism: 70 to 1400 kcal h^{-1}

H_c = heat lost by convection (still air): 6.1 $A(T_s - T_a)$ kcal h^{-1}

H_r = heat lost by radiation: 5.6 $A(T_s - T_a)$ kcal h^{-1}

H_s = heat lost by evaporation of sweat: 580r kcal h^{-1}

H_l = heat lost by evaporation from the lungs: 9 kcal h^{-1}

The value of H_l is given for normal breathing rates; it increases in proportion to the rate of breathing. Typically, H_l is a small fraction of the heat loss, and we neglect its variation.

If the body temperature is held constant, then the heat losses are related by

$$H_m = H_c + H_r + H_s + H_l \qquad (12.16)$$

The following example illustrates the importance of sweating during moderate levels of activity.

Example 12.11

A person generates heat at a rate of 200 kcal h^{-1}. How much sweat is produced per hour if the surface area is 1 m^2? Assume the body temperature is 37°C and that of the air is 28°C.

Using the results given above,

$$H_s = H_m - H_c - H_r - H_l$$
$$= 200 \text{ kcal h}^{-1} - 6.1 \,(1.0)(9°\text{C})\text{kcal h}^{-1}$$
$$-5.6(1.0)(9°\text{C})\text{kcal h}^{-1} - 9 \text{ kcal h}^{-1}$$
$$= 85.7 \text{ kcal h}^{-1}$$

Since the heat removed by sweating is 580r kcal h^{-1}, the number of kilograms of sweat required per hour is

$$r = \frac{H_s}{580 \text{ kcal h}^{-1}} = \frac{85.7 \text{ kcal h}^{-1}}{580 \text{ kcal h}^{-1}} = 0.15$$

PROBLEMS ON SUPPLEMENTARY TOPICS

12-51 A person produces 150 kcal h^{-1} of waste heat. If all this heat is dissipated by evaporation of sweat, how much sweat per hour is required?

12-52 A naked person with a surface area of 1.5 m^2 and a skin temperature of 40°C is in a sauna at 85°C. (a) How much heat does the person absorb by radiation from the walls, assuming emissivity 1? (b) How much heat does the person radiate to the surroundings? (c) How much sweat must be evaporated per hour, assuming that no heat is transferred by convection? (Neglect the metabolic heat production.)

Additional Reading

T.C. Ruch and H.D. Patton (ed.), *Physiology and Biophysics,* vol. 3, W.B. Saunders, Philadelphia, 1973. Chapter 5 discusses temperature regulation in humans.

Scientific American articles:

Robert L. Sproull, The Conduction of Heat in Solids, December 1962, p. 92.

David M. Gates, Heat Transfer in Plants, December 1965, p. 76.

G. Yale Eastman, The Heat Pipe, May 1968, p. 38.

G. Neugebauer and Eric E. Becklin, The Brightest Infrared Sources, April 1973, p. 28.

Jacob Gershon-Cohen, Medical Thermography, February 1967, p. 94.

David Turnbull, The Undercooling of Liquids, January 1965, p. 38.

Laurence Irving, Adaptations to Cold, January 1966, p. 94.

T.H. Benzinger, The Human Thermostat, January 1961, p. 134.

Bernd Heinrich and George A. Bartholomew, Temperature Control in Flying Moths, June 1972, p. 70.

Francis G. Carey, Fishes with Warm Bodies, February 1973, p. 36.

CHAPTER THIRTEEN

THERMODYNAMICS

Thermodynamics is the study of the transformation of energy from one form to another and from one system to another. It is particularly useful when the microscopic properties of a system are complex and one wishes to predict and understand the results of experiments involving only a few macroscopic parameters.

In this chapter, we discuss the first and second laws of thermodynamics. The first law is a generalization of the fundamental result of Chapter Six, that the work done *by* a system equals its change in mechanical energy plus the energy that *leaves* the system.

The second law of thermodynamics can be phrased in several ways. In one form it is a statement that a quantity called the *entropy* tends to increase in all real processes. The entropy change of a system is related to the heat entering or leaving a system divided by the absolute temperature of the system. Microscopic theories of matter have shown that the entropy of a system is also closely related to the randomness or disorder of its constituents. We use the second law to obtain the ultimate limits of efficiency of thermal energy conversion processes. Such processes are used in both fossil and nuclear electrical power-generating plants.

13.1 | MECHANICAL WORK

Work can be done on or by a system in many ways. A gas may be compressed or allowed to expand against a piston. A liquid may be stirred, and a solid may be pounded with a hammer. Electric charges may be brought near a material, so that the electric forces alter the arrangements of charges inside the material. Thus the kinds of work that can be done are as varied as the forces that can be exerted on a system. In this section, we obtain an expression for the work done by a substance or system when its volume changes.

When a substance or system expands or contracts, the work W done by the system can be related to the volume change of the material. It is conventional in thermodynamics to take W as positive when work is done *by* the system. Note by contrast that in Chapter Six, we took W to be positive when work was done *on* the system.

It is most convenient to develop our ideas using the example of a gas. Figure 13.1 shows a gas at a pressure P in a closed cylinder. A movable piston of cross-sectional area A forms one end of the enclosure. The gas exerts a force $F = PA$ on the piston. When the piston moves a small distance Δx parallel to the force, the work done by the gas is $\Delta W = F \Delta x = PA \Delta x$.

Since $\Delta V = A \Delta x$ is the change in volume of the gas,

$$\Delta W = P \Delta V \tag{13.1}$$

Figure 13.1. The work done by a gas or a piston during a small displacement Δx is $\Delta W = F \Delta x = P \Delta V$.

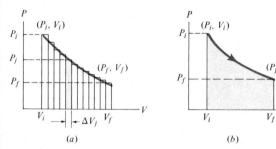

Figure 13.2. (*a*) The work done by the system is the sum of the areas of the small segments. In each segment, P_j is assumed constant, and for the colored segment shown $\Delta W_j = P_j \Delta V_j$. (*b*) The total work is exactly equal to the total colored area under the *P* versus *V* curve. *W* is positive in this example. (*c*) A process in which work is done on the system, so *W* is negative.

A large displacement can be regarded as composed of a series of very small displacements Δx_j, so that during each displacement the force $F_j = P_j A$ can be considered constant. Then the total work *W* done by the system is the sum of all the terms $P_j \Delta V_j$ (Fig. 13.2),

$$W = P_1 \Delta V_1 + P_2 \Delta V_2 + \cdots \qquad (13.2)$$

The work done by the system is equal to the area under the *PV* curve. This result is valid for a gas in a container of any shape and also for volume changes in liquids and solids. Equation 13.2 has a simple form if the work is done in an *isobaric process, that is, at constant pressure*. If the initial and final volumes of the system are denoted by V_i and V_f, then the work done is

$$W = P(V_f - V_i) \qquad (13.3)$$

For an isobaric process, the work done by the system is positive if $(V_f - V_i)$ is positive; work is done on the system if $(V_f - V_i)$ is negative.

The following example describes an isobaric process.

Example 13.1

A gas at a pressure of 2 atm $= 2.02 \times 10^5$ Pa is heated and is allowed to expand against a frictionless piston at constant pressure. If the volume change is 0.5 m³, how much work is done by the gas?

Using Eq. 13.3,

$$W = P(V_f - V_i) = (2.02 \times 10^5 \text{ Pa})(0.5 \text{ m}^3)$$
$$= 1.01 \times 10^5 \text{ J}$$

13.2 | THE FIRST LAW OF THERMODYNAMICS

The first law of thermodynamics relates the change in the internal energy of a system, the heat transferred to the system, and the work done by the system. We can understand its content by again using a gas as an example.

We consider a container of gas fitted with a movable piston (Fig. 13.3). If we add heat *Q* to the system but do not allow the piston to move, the temperature and hence the internal energy *U* of the gas will increase. We can also change the internal energy by doing work on the gas. Thus, if we insulate the container walls and push the piston in, we compress the gas. The work done on the system equals the change in internal energy (Fig. 13.3*b*), since no heat enters or leaves the gas.

More generally we may add heat *Q* to the gas, have the gas do work *W*, and the difference is the change in internal energy of the gas, $\Delta U = U_f - U_i$. U_f and U_i are the final and initial energies of the gas. This is the *first law of thermodynamics*

$$U_f - U_i = Q - W \qquad (13.4)$$

Q is positive if heat is *added* to the system, and *W* is positive if work is *done by* the system. Although we have used a gas in this example, this result is generally true for all systems and does not depend on the presence of a gas.

From our presentation of the concepts of heat and

215

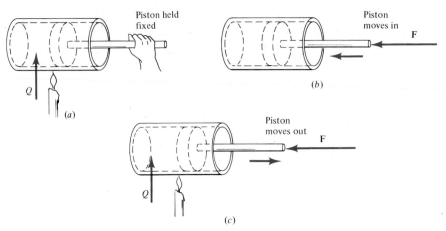

Figure 13.3. (a) With the piston held fixed, the heat added, Q, equals the internal energy increase ΔU. (b) The walls are insulated. The force **F** does work on the piston and hence on the gas. This work is equal to the internal energy increase. (c) Heat Q is added to the gas, and it expands against the external force **F**. The gas does work W. The difference between Q and W is the change in internal energy of the gas ΔU.

temperature, the first law seems an almost obvious statement of the conservation of energy. Nevertheless, the first law forms one of the most fundamental cornerstones of thermodynamics. Historically, the first law was by no means obvious. Until the work of Mayer, Joule, Helmholtz, and Rumford in the 1840s, heat was regarded as a material substance inside an object. This substance, called caloric, could flow from one object to another. Many observations were satisfactorily explained with this idea. Mayer was the first to suggest that heat and internal energy were closely related. Joule then showed that one could produce as much heat as one wished by doing work. The idea of a fixed amount of caloric being contained in a substance died with these developments. Heat, work, and internal energy were then identified as different manifestations of the same quantity, energy.

Two statements about the physical world are contained in the first law. First, heat and work are to be treated on an equal basis. Second, because the same change in internal energy can be obtained by adding heat or having work done on the system or by a combination of these processes, *the internal energy change is independent of how the change was achieved*. The difference between the initial and final internal energies of the system must depend *only* on its initial and final *states*, that is, on the change in quantities such as the temperature, pressure, and volume. This idea is illustrated in Fig. 13.4. Suppose a system undergoes two processes, represented by curves (1) and (2), in which the initial and final pressures and volumes are (P_i, V_i) and (P_f, V_f). Along path (1) the area beneath the curve is larger than that beneath curve (2). Hence, more work is done by the

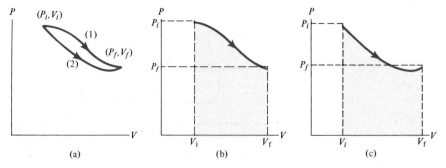

Figure 13.4. (a) A system can change from a state (P_i, V_i) to state (P_f, V_f) by many paths, two of which are shown. (b) The work done by the system is the area beneath the PV curve and is greater for process (1) than for process (2) shown in (c).

system in process (1) than in (2). From the first law, since $U_f - U_i$ is the same for both processes, more heat must be added to the system in process (1) than in (2) to achieve the same final state.

Two types of processes will be of particular importance in our later discussion of heat cycles. One is an *isothermal* or constant temperature process. In practice, such processes are difficult to achieve, but they may be visualized for an ideal gas. The internal energy of an ideal gas depends only on the temperature. We can picture adding heat very slowly to such a gas and allowing it to expand, doing work. If the process is slow, the temperature, and hence the internal energy, will remain constant. Thus, for an isothermal process in an ideal gas, $Q = W$.

A second, more easily achieved process is an *adiabatic* process. This is a process in which no heat enters or leaves the system. Thus, $U_f - U_i = -W$ for an adiabatic process. This condition is common because we can insulate systems to minimize heat transfer. Alternatively, processes may occur so rapidly that no heat flow occurs. If we pinch the hose on a bicycle pump, pushing the plunger once results in a sudden increase in the internal energy of the air. It will take a few seconds before heat flow from the inside to outside of the pump will be noticed. This flow results from the fact that the air inside is at a higher temperature.

13.3 | THE SECOND LAW OF THERMODYNAMICS

The first law of thermodynamics is useful in understanding the flow of energy during a given process. However, it yields no information about which energy-conserving processes are possible nor does it allow us to predict what state a system will be in under a given set of conditions. The second law can be used to answer these questions.

For example, suppose fuel is burned and the heat produced is supplied to a steam engine. The first law requires that the work done by the engine plus the heat rejected by it to the surroundings must equal the heat added, since the internal energy of the engine does not change. However, the first law gives no clue as to the ratio of the work done to the heat supplied, that is, the *efficiency* of the engine. The second law makes it possible to calculate the efficiency for an idealized engine and to set limits on the efficiencies of real engines.

A second example of its use is in chemical reactions. When a reaction occurs the first law allows one to predict how much energy is absorbed or liberated. However, the second law allows one to predict, for given conditions of temperature and pressure, what the equilibrium state of this system will be.

In this section, we discuss the microscopic and macroscopic forms of the second law. The efficiency of heat engines is discussed in the next section. Applications to chemical systems are treated in detail in chemistry and biochemistry texts.

Microscopic Form of the Second Law |

The microscopic form of the second law is a statement about the probable behavior of a large number of molecules or other particles. It states that systems tend to evolve from highly ordered, relatively improbable configurations to more disordered, more statistically probable configurations. Equivalently, systems tend toward states of maximum molecular disorder or chaos. For example, Fig. 13.5 shows two ways in which the molecules of a gas may move. Both pictures show a gas with the same internal energy, but Fig. 13.5a represents a highly ordered situation unlike Fig. 13.5b. The second law says that a situation *similar* to the disordered state is *more probable* than one similar to the ordered state. Thus, it is possible to imagine a great many pictures similar to 13.5b, but only a few like Fig. 13.5a. Poker players will recognize an equivalent idea: there are only a few hands with a royal flush, but many hands with no value at all, so the odds strongly favor hands of the latter type. Similarly, from this point of view, parking in a tight space at a curb is more difficult than moving onto the street, because there are many more moving configurations than parked ones.

(a) (b)

Figure 13.5 Two patterns of molecular motion in a gas. (a) Highly ordered. (b) Much less ordered or more chaotic.

JULIUS ROBERT MAYER
(1814 – 1878)

JAMES PRESCOTT JOULE
(1818 – 1889)

HERMAN VON HELMHOLTZ
(1821 – 1894)

(Brown Brothers)

(Radio Times Hulton Picture Library)

(Culver Pictures)

Until nearly the middle of the nineteenth century, heat was generally regarded as a fluid called *caloric*. This weightless fluid, which could flow from one object to another but could neither be created nor destroyed, provided a rather complete explanation of many experiments of this period. The three men pictured above are now linked historically with the demise of the caloric theory and with extending the conservation of energy principle to include thermal phenomena.

Mayer was the first to suggest that the energies associated with gravity, electricity, heat, and motion are intimately connected; when one kind of energy disappears, an equivalent amount of other types appears. However, Mayer and his ideas both suffered from neglect. Educated in Germany as a physician, Mayer became interested in physics and published his observations in 1842. His article was written in a metaphysical style and was not viewed as very convincing. Part of the resistance to Mayer's ideas arose because it was not even clear that he understood Newton's laws, much less the concepts involved in a theory of energy conservation.

While the caloric theory was dying even before Mayer's work, the real acceptance of heat as another form of energy came as a result of the independent work of Joule and Helmholtz in 1847. Joule was an English brewery owner and amateur scientist who had built a reputation as a careful and ingenious experimenter. By 1847, Joule had very accurately shown that mechanical work, in this case the work required to turn a paddle wheel immersed in water, is equivalent to heat, since the temperature of the water increased.

While Joule provided the most convincing experimental evidence for the equivalence of mechanical energy and heat, Helmholtz systematically developed the concept of the conservation of energy in one of the most important scientific papers of the nineteenth century.

Helmholtz was also a physician, serving as a surgeon in the Prussian army. In 1849, he became a professor of physiology in Konigsberg. His work on energy conservation, stimulated by his observations on muscle motion, was only one example of his remarkable ability to understand the physics behind biological systems. Helmholtz invented the ophthalmoscope for peering into the eye and also developed the ophthalmometer for measuring the curvature of the eye. In addition, he revived a theory of color vision credited to Young and expanded on it. Helmholtz's study of the ear included the role played by the bones of the middle ear and the cochlea and went on into the difficult but important subject of the quality of sounds. His book, *Sensations of Tone,* is a cornerstone of physiological acoustics. He was the first to measure the speed of a nerve impulse. In his later life, his interests and research led directly to the experimental discovery of electromagnetic waves by one of his students.

While Joule and Helmholtz pursued successful scientific careers, Mayer did not fare so well. Mayer was deeply affected by his lack of recognition and attempted suicide in 1849. After a period of mental illness from which he never fully recovered, he lived out his life in obscurity.

Macroscopic Form of the Second Law

The second law was first formulated as a statement about large or macroscopic systems. This form is easier to use for many applications, although its physical interpretation is perhaps more subtle than that of the molecular chaos discussed above. The two forms have been shown to be equivalent using microscopic statistical theories.

The macroscopic form of the second law is a statement that a quantity called the *entropy* tends to assume a maximum value. Like the internal energy, the entropy of a system depends only on its state and not on how that state is achieved.

The definition of entropy involves the concept of *reversible* and *irreversible* processes. A reversible process is one in which the system can be returned to its original state with no net change in either the system or its surroundings. For example, if no friction, turbulence, or other dissipative effect is present, the adiabatic expansion of a gas is reversible (Fig. 13.6). This is because an adiabatic compression can return the system to its original state. The work done on the gas during the compression is equal to that done by the gas during the expansion; the net work done by the gas and its surroundings is zero.

Most natural processes are irreversible. When heat is transferred between objects at different temperatures, the heat can be returned to the higher temper-

ature object, but to do so requires that the surroundings do work, as in a refrigerator. Thus the surroundings must be modified to return the system to its original state. Reversible processes, like frictionless systems in mechanics, are an idealization only approximately realized in real systems.

We can now define the entropy of a system. If a small quantity of heat ΔQ is added to a system at a Kelvin temperature T during a reversible process, the entropy change of this system is defined by

$$\Delta S = \frac{\Delta Q}{T} \qquad \text{(reversible process)} \qquad (13.5)$$

If ΔQ is large, it can be divided into many small amounts ΔQ_i such that the temperature T_i is nearly

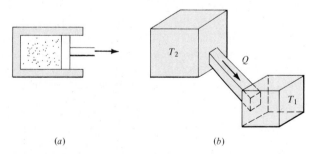

(a)   *(b)*

Figure 13.6. In a reversible process the system and surroundings can both be restored to their original states. (*a*) A frictionless adiabatic expansion is reversible. (*b*) Heat transfer between objects at different temperatures is irreversible.

constant during the transfer of ΔQ_i. Then, in a reversible process, the total change in entropy is found by summing the small entropy changes $\Delta Q_i/T_i$. Note that when heat *leaves* a system, ΔQ is *negative* and so is the associated entropy change of the system. For an irreversible process, the entropy change of an isolated system can be evaluated by considering reversible processes that would bring the system to the same final state.

We can now give the macroscopic form of the second law. For any process, *the total entropy of a system plus its surroundings may never decrease:*

$$\Delta S(\text{total}) \geq 0 \qquad (13.6)$$

The total entropy change is zero for a reversible process and positive for an irreversible process. This is called the *second law of thermodynamics.* Microscopically this is equivalent to saying that the molecular disorder of a system and its surroundings remains constant if the process is reversible and increases if it is not.

It is possible to derive Eq. 13.6 starting from either of two experimental observations. One is that heat never flows spontaneously from a colder to a hotter object. This is called the Clausius form of the second law. The second observation is that it is impossible to extract heat from an object and convert it *entirely* into work. This is the Kelvin form of the second law. The path from either verbal form of the second law to the entropy or molecular chaos forms is complex and beyond the scope of this textbook. We can, however, see that Eq. 13.6 is indeed satisfied in some simple processes. The following examples also show how one computes entropy changes in reversible and irreversible processes.

Example 13.2

Find the entropy changes of the system and of its surroundings for a reversible adiabatic process.

In an adiabatic process the heat entering a system is zero. Since the process is reversible, by Eq. 13.5 the entropy change of the system is zero. Similarly, since the surroundings transfer no heat, their entropy change is also zero. Thus $\Delta S(\text{total}) = 0$ as required for reversible processes by the second law.

Example 13.3

Heat is reversibly withdrawn from 10^{-2} kg of liquid water at 0°C until it is entirely converted into an ice cube at the same temperature. (a) What is the entropy

change of the water? (b) What is the net entropy change of the water and surroundings?

(a) When a mass m of water is converted to ice, heat $\Delta Q = -L_f m$ must be removed, where L_f is the latent heat of fusion. The temperature is constant at $T = 0°C = 273$ K so the entropy change is

$$\Delta S = \frac{-L_f m}{T} = -\frac{(80 \text{ kcal kg}^{-1})(10^{-2} \text{ kg})}{273 \text{ K}}$$

$$= -0.0029 \text{ kcal K}^{-1}$$

(b) The process occurs reversibly, so the net entropy change of the water plus surroundings is zero. Since the entropy of the water decreases, the entropy of the surroundings must *increase* by the same amount, $\Delta S = 0.0029 \text{ kcal K}^{-1}$.

In this example we can see the relationship between entropy and order. When the water solidifies to an ordered solid phase, the entropy decreases. Since the net change in entropy is zero, the entropy of the surroundings increases. Conversely, when the ice melts, its entropy increases. If the melting occurs reversibly, the entropy of the surroundings decreases.

Example 13.4

Two large objects are isolated from their surroundings. They are at temperatures T_1 and T_2 with $T_2 > T_1$ and are placed in thermal contact. A small quantity of heat Q is transferred by conduction, leaving their temperatures nearly unchanged. Find the entropy changes.

The entropy change of the surroundings is zero, since the system is isolated. This is an irreversible process, so we calculate the entropy changes for reversible paths leading to the same final states. We may, for example, consider the case where heat Q is reversibly removed from the object at T_2 by placing a cylinder of gas in contact with it and allowing the gas to expand isothermally. This gives an entropy change $\Delta S_2 = -Q/T_2$. Similarly, a second cylinder can be placed in contact with the cooler object and heat transferred reversibly. The entropy change of the cooler object is then $\Delta S_1 = Q/T_1$. Thus the total entropy change of the two objects is

$$\Delta S(\text{total}) = -\frac{Q}{T_2} + \frac{Q}{T_1} = Q\left(\frac{1}{T_1} - \frac{1}{T_2}\right)$$

$$= Q\frac{(T_2 - T_1)}{T_1 T_2}$$

Since T_2 is greater than T_1, this is positive as predicted for an irreversible process.

Although the second law of thermodynamics in the form of Eq. 13.6 states that no *net* decrease in entropy can occur, it is not true that the entropy of a *system* cannot be reduced. When water is frozen, the entropy of the water is lowered. However, if careful consideration is taken of the surroundings, it is found that the *total entropy of the system plus its surroundings remains constant or increases.*

13.4 | THE CARNOT THEOREM AND CONVERSION OF ENERGY

If one applies the second law of thermodynamics to a reversible heat engine, one finds that the efficiency of the engine in converting heat into mechanical work is always less than 100 percent. Real heat engines, such as automobile engines and electric power-generating plants, always have some friction or turbulence and are therefore irreversible. Their efficiencies are necessarily still lower. This remarkable property of heat engines was discovered by Sadi Carnot (1796–1832) in 1824 and is called *Carnot's theorem.*

Carnot's theorem shows that the conversion of heat energy into other forms of energy is qualitatively different from other kinds of energy conversion. For example, a swinging pendulum can completely convert mechanical potential energy into mechanical kinetic energy and back again. The kinetic energy of moving water in a turbine can be transformed into electrical energy by a generator; the efficiency is limited only by friction and turbulence and can be quite high in practice. Electrical energy can also be converted into mechanical energy by a motor; again there is no theoretical limit on the efficiency, and the actual efficiencies can be 90 percent or better. Also, chemical energy can be converted into electrical energy with very high efficiency, as in a fuel cell where hydrogen and oxygen combine and produce an elec-

trical current. Electrical energy can also be transformed into chemical energy without any fundamental limitation. In the human body the conversion of chemical energy from food into mechanical work is usually done at an efficiency of up to about 30 percent. This is not because of any second law limitation. It occurs because the body loses energy in converting foods into compounds that the body can utilize effectively at the cellular level.

Any heat engine can be thought of as acting in a *cycle* in which a substance at a high temperature is allowed to do work, so that its internal energy and temperature decrease. The cycle is completed by adding heat and returning the substance to its original state.

Carnot's theorem is proved by considering a particular reversible heat engine, called a *Carnot engine* or a *Carnot cycle.* This cycle consists of four reversible processes, which are illustrated in Fig. 13.7 for the special case of an ideal gas. However, the *calculated efficiency is independent of the material* used in the engine; we only use the ideal gas since it is easy to discuss.

The path from *a* to *b* is an isothermal expansion in which heat Q_2 is absorbed from a reservoir at T_2. The path *bc* is an adiabatic expansion; no heat is transferred. The path *cd* is an isothermal compression in which heat Q_1 leaves the system at a lower temperature T_1. Finally the material is returned to its original state by an adiabatic compression *da*; again, no heat is transferred.

In expanding, the gas does work equal to the area under the curve *abc* (Fig. 13.7*b*). When it is compressed, work must be done on the gas equal to the smaller area under the curve *cda* (Fig. 13.7*c*). The net

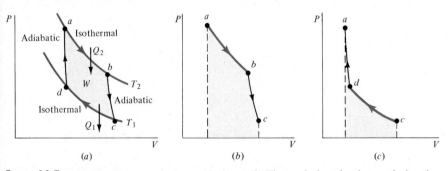

Figure 13.7.   (*a*) The Carnot cycle for an ideal gas. (*b*) The work done by the gas during the expansion equals the colored area. (*c*) The work done on the gas during the compression equals the colored area.

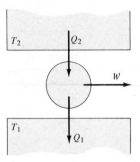

Figure 13.8. The work done in a complete cycle equals the heat absorbed at T_2 less the heat rejected at T_1.

work W done by the gas over one full cycle therefore equals the shaded area enclosed by the path (Fig. 13.7a). Since the gas at the end of the cycle is in its original state, the net change in internal energy is zero. The first law therefore requires that the net work done by the gas must equal the heat absorbed minus the heat rejected, or

$$W = Q_2 - Q_1 \qquad (13.7)$$

This is shown symbolically in Fig. 13.8.

Since the Carnot cycle is reversible, the total entropy change of the system is zero. For the reservoir at T_2 the entropy change is $\Delta S_2 = -Q_2/T_2$. The entropy change of the lower temperature reservoir is $\Delta S_1 = Q_1/T_1$. The gas returns to its original state, so its entropy change is zero. Thus

$$\Delta S(\text{total}) = \Delta S_1 + \Delta S_2 = -\frac{Q_2}{T_2} + \frac{Q_1}{T_1} = 0$$

or

$$\frac{Q_1}{Q_2} = \frac{T_1}{T_2}$$

The efficiency is defined as the work done divided by the heat absorbed, $e = W/Q_2$. Using $W = Q_2 - Q_1$, we find

$$e = \frac{W}{Q_2} = 1 - \frac{Q_1}{Q_2} = 1 - \frac{T_1}{T_2} \qquad (13.8)$$

This ratio of the work done to the heat supplied is the *efficiency* of the Carnot engine. According to Eq. 13.8, the efficiency is always less than 1 unless the colder reservoir is at absolute zero, which is, in fact, unattainable even in principle.

The following example shows the efficiency of a Carnot engine operating between two easily achieved temperatures.

Example 13.5

A Carnot engine operates between 100°C and 0°C. What is its efficiency?

Converting the temperatures to the Kelvin scale, $T_1 = 273$ K and $T_2 = 373$ K. Then

$$e = 1 - \frac{T_1}{T_2} = 1 - \frac{273}{373} = 0.268$$

Only 26.8 percent of the heat supplied is converted into mechanical work; the remainder is rejected into the colder reservoir.

As we noted earlier, the efficiency of a Carnot engine is independent of the substance used. This is clear from the derivation, because the calculation of the efficiency used only the fact that the cycle consisted of two isothermal and two adiabatic processes and did not rely on any properties of the material. This means that one cannot hope to improve the efficiency of an engine by an ingenious choice of materials if that engine is already fairly close to the Carnot efficiency for the temperatures of its cycle. Carnot also showed that *no heat engine operated cyclically between two temperature reservoirs is more efficient than that described by the Carnot cycle.*

Real engines always have some losses resulting from friction and turbulence and must be less efficient than Carnot engines operating between the same temperatures. To maximize the efficiency, T_2/T_1 is made as large as practicable. High performance auto engines have large ratios of maximum to minimum cylinder volumes or compression ratios in order to achieve correspondingly large gas temperature ratios in the cylinders.

The temperatures and pressures of the steam boilers in electric power-generating plants are also very high. Modern fossil fuel electric-generating plants are about 40 percent efficient; the excess heat lost when the steam condenses is given off into a lake or river or transferred to the atmosphere with cooling towers. Present commercial nuclear power plants are at best 34 percent efficient, so the thermal pollution problems they pose are somewhat larger. These efficiencies should be compared with the ideal Carnot efficiencies of 52 percent and 44 percent, respectively, for the two types of plants.

The lower efficiency of the nuclear plants is related

Figure 13.9. The major components of a steam electric power generating plant. The heat source may be coal or oil or a uranium-fueled reactor.

to limitations on the high temperatures allowable for the uranium in the reactor. The uranium oxide in the center of the fuel rods must be kept well below melting temperature. This limits the temperature at the outside of the fuel rods. In all power plants the maximum temperature is limited by design and materials problems, and the low temperature exhaust temperature is determined by the available cooling water in lakes and rivers. The components of a major steam generating plant are shown in Fig. 13.9.

Example 13.6

A nuclear power plant generates 500 MW = 5×10^8 W at 34 percent efficiency. The waste heat goes into a river, such as the Connecticut River, with an average flow of 3×10^4 kg s^{-1}. How much does the water temperature rise?

If the plant operates at 34 percent efficiency, it must have a total heat input of 1.47×10^9 W. Of this, 66 percent or 9.7×10^8 W must go into the river. The specific heat of water is $c = 1$ kcal kg^{-1} K^{-1} = 4.18×10^3 J kg^{-1} K^{-1}. In 1 second the heat $\Delta Q = 9.7 \times 10^8$ J will raise the water temperature by ΔT, where $\Delta Q = mc\,\Delta T$. With $m = 3 \times 10^4$ kg,

$$\Delta T = \frac{\Delta Q}{mc}$$

$$= \frac{9.7 \times 10^8 \text{ J}}{(3 \times 10^4 \text{ kg})(4.18 \times 10^3 \text{ J kg}^{-1}\text{ K}^{-1})}$$

$$= 7.7°C$$

This temperature rise could cause serious damage to the organisms in the river. In a case such as this, cooling towers are used to dissipate heat to the atmosphere rather than the river. A 40 percent efficient fossil fuel plant of the same output power would raise the river temperature by almost 6°C.

The expression $e = 1 - T_1/T_2$ for the efficiency of a Carnot engine provides a useful way of interpreting the second law. If heat is transferred from a hotter object to a colder one, reducing their temperature difference, then the efficiency of a Carnot engine operating between these two heat reservoirs decreases. This means that less work can be done with a given heat input and that the energy stored in these objects is therefore less available. For example, with tanks of water at 0°C and 100°C, we can operate an engine; this cannot be done if we mix these tanks together so that they are at a common temperature. The statement that systems tend toward states of maximum entropy therefore implies that *energy tends to become less and less available for doing mechanical work.*

SUMMARY

Thermodynamics is the study of the transformation of energy from one form to another. The first law of thermodynamics states that the internal energy change of a system is equal to the heat added to the system minus the work done by the system. In this chapter we have only described systems that do mechanical work.

The second law of thermodynamics is expressed in terms of the entropy. In its microscopic form, it is a statement that systems tend to evolve toward increasing disorder. The macroscopic statement of the second law, while it is somewhat subtle in its content, can be used to make very general statements about processes and their efficiencies. In particular, the maximum efficiency of a heat engine operating with high and low temperatures of T_2 and T_1, respectively, is

$$e = 1 - \frac{T_1}{T_2}$$

Checklist

Define or explain:

first law of	reversible and
thermodynamics	irreversible processes
adiabatic process	second law of
isothermal process	thermodynamics
disordered state	Carnot's theorem
entropy	Carnot efficiency

Figure 13.10. Exercises 13-1 and 13-3.

EXERCISES

Section 13.1 | Mechanical Work

13-1 In Fig. 13.10, how much work is done by the system in the process (a) along path (2) from A to B? (b) If the system is returned from B to A along the same path, how much work is done by the system?

13-2 A gas does work in an isobaric process at $P = 10^5$ Pa. How much work is done by the gas if (a) $V_i = 10^{-2}$ m³, and $V_f = 2.24 \times 10^{-2}$ m³; (b) $V_i = 2 \times 10^{-2}$ m³, and $V_f = 0.5 \times 10^{-2}$ m³?

Section 13.2 | The First Law of Thermodynamics

13-3 In Fig. 13.10, a system undergoes a process from point A to point B. How much work is done by the system on (a) path (1), and (b) path (2)? (c) If process (1) is carried out adiabatically, what is the change in the internal energy?

Figure 13.11. $P_1 = 1$ atm, $P_2 = 3$ atm, $V_1 = 0.02$ m³, and $V_2 = 0.10$ m³. Exercise 13-4.

13-4 In undergoing the process from A to B in Fig. 13.11, the increase in internal energy of a substance is 3×10^5 joules. How much heat is absorbed by the system?

13-5 Find the change in internal energy of the system when (a) a system absorbs 2000 J of heat and produces 500 J of work; (b) a system absorbs 1100 J of heat and 400 J of work is done on it.

Section 13.3 | The Second Law of Thermodynamics

13-6 A new deck of playing cards has cards of each suit together and ordered numerically. The deck is then shuffled. Has its entropy changed? If so, why?

13-7 The second law of thermodynamics states that the disorder of the universe is always either constant or increasing. (a) How is this compatible with the fact that plants and animals develop into highly ordered systems? (b) How can the second law of thermodynamics be compatible with the theory of evolution, which states that more highly ordered complex living things evolve from simpler species?

13-8 A system at a constant temperature of 300 K absorbs 10^4 J of heat, and no work is done. (a) What is the change in entropy of the system? (b) What is the change in its internal energy?

13-9 A helium-filled balloon is punctured, and the gas spreads uniformly throughout a room. (a) Is the process reversible? Explain. (b) Has the entropy of the helium and the air in the room increased or decreased? Explain.

Section 13.4 | The Carnot Theorem and Conversion of Energy

13-10 What is the maximum efficiency of a heat engine working between 100°C and 400°C?

13-11 An internal combustion engine using natural gas and air as a working substance has a temperature in the spark ignited firing chamber of 2150 K and an exhaust temperature of 900 K. The difference between the heat supplied and the work done by the engine in each second is 4.6×10^6 J. (a) What is the ideal Carnot efficiency of this engine? (b) How much work is actually done per second if the heat input is 7.9×10^6 W? (c) What is the actual efficiency of this engine?

13-12 If a heat engine operates at 40 percent efficiency and it absorbs 10^4 W from the high temperature reservoir, at what rate does it do mechanical work?

13-13 If fossil fuel and nuclear generating plants operate with efficiencies of 40 per-cent and 30 percent efficiency, respectively, and the low temperature reservoir is at 300 K for both, what is the minimum temperature of the steam produced by the fuel in each case?

13-14 If the power plant of example 13.6 were a fossil plant operating at 40 percent efficiency, what would the temperature rise of the water be as a result of the waste heat?

PROBLEMS

13-15 A system goes from state *a* to *c* via path *abc* in Fig. 13.12. 10^5 J of heat flow into the system, and 4×10^4 J of work are done. (a) How much heat flows into the system along path *adc* if 10^4 J of work are done? (b) If the system returns to state *a* from state *c* via the zigzag path, the work is 2×10^4 J. How much heat enters or leaves the system? (c) If $U_a = 10^4$ J and $U_d = 5 \times 10^4$ J, what is the heat absorbed along the paths *ad* and *dc*?

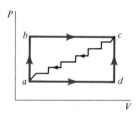

Figure 13.12. Problem 13-15.

13-16 Eight liters of air at room temperature and atmospheric pressure are compressed isothermally to a volume of 3 liters. The air then expands adiabatically to a volume of 8 liters. Show the process on a *P-V* diagram.

13-17 An engine operates on the cycle shown in Fig. 13.13, a temperature-entropy diagram. What is the efficiency of the engine?

*13-18 Draw a schematic *T-S* diagram for a Carnot engine operating between temperatures T_2 and T_1, where $T_2 > T_1$.

*13-19 Six poker chips are each painted white on one side and yellow on the other. (a) In how many possible ways can the chips be displayed in each combination from all white, one white and five yellow, . . . , to all yellow? (b) Which combination is most probable? (c) What is the implication of part (b) in terms of entropy and disorder?

*13-20 If a Carnot cycle is reversed, it represents an ideal cooling system. An amount of heat Q_1 is then taken in from a reservoir at a temperature T_1,

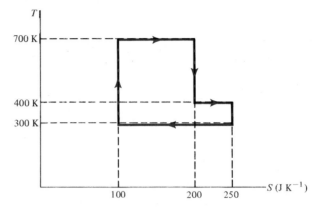

Figure 13.13. Problem 13-17.

and Q_2 is released into a higher temperature reservoir at T_2. Show that the work W necessary to operate the system is given by

$$W = Q_2(1 - T_1/T_2)$$

13-21 About 2×10^5 J of heat must be removed from a tray of water to produce ice cubes. (a) How long will it take to remove this much heat in a 200-W input Carnot refrigerator operating between inner and outer temperatures of 270 K and 310 K, respectively? (Use the result of the preceding problem.) (b) Is this time consistent with your estimate of how long it takes to freeze an ice tray? Explain the difference, if any.

13-22 The motor in an ideal (Carnot) refrigerator delivers 200 W of useful power. The freezing compartment of the refrigerator is at $T_1 = 270$ K, and the room air is at $T_2 = 300$ K. Using the formula in Problem 13-20, find the maximum amount of heat that can be removed from the refrigerator in 1 minute.

13-23 A nuclear boiling water reactor heats steam to 285°C, and the cooling water is at 40°C. The actual operating efficiency of the power plant is 34 percent. (a) What is the ideal efficiency of the plant? (b) If the plant produces 500×10^6 W of electric power, what is the power lost as compared to ideal operation?

SUPPLEMENTARY TOPICS

13.5 | HUMAN METABOLISM

All living things require energy to sustain the life process. Green plants obtain their energy directly from the sun though the process of photosynthesis. Plants such as mushrooms, which do not utilize photosynthesis, as well as animals require food capable of providing chemical energy. In all cases, living plants and animals operate within the constraints described by thermodynamics.

The first law of thermodynamics provides a convenient scheme for cataloging the factors that enter the complex subject of human metabolism. Suppose that in a time Δt, a person does an amount of mechanical work ΔW. This can be directly measured in activities such as bicycling, shoveling snow, or pushing a cart. Heat will usually leave the body, so ΔQ will be negative. It can be measured by finding the amount of heat that must be removed to keep the temperature constant in a room in which the person is working. From the first law, the internal energy change ΔU will be given by $\Delta U = \Delta Q - \Delta W$. Dividing by Δt, this becomes a relation among the rates of change,

$$\frac{\Delta U}{\Delta t} = \frac{\Delta Q}{\Delta t} - \frac{\Delta W}{\Delta t} \qquad (13.9)$$

The rate of change of internal energy can be measured accurately by observing the rate at which a person uses oxygen in converting food into energy and waste materials. For example, a mole (180 g) of glucose, which is a typical carbohydrate, combines with 134.4 liters of oxygen gas in a series of steps to form carbon dioxide and water. In the process, 686 kcal of energy are released. The *calorific equivalent of oxygen* is defined as the ratio of the energy released to the oxygen consumed. For glucose, this ratio is 686 kcal/134.4 liters = 5.10 kcal liter^{-1}. The *energy content per unit mass* is defined as the energy released divided by the mass. For glucose, this ratio is 686 kcal/180 g = 3.81 kcal g^{-1}.

Table 13.1 lists the average energy content per unit mass and calorific equivalent of oxygen for the carbohydrates, proteins, and fats usually consumed. The calorific equivalent of oxygen of all of these is the same to within a few percent. Thus, an average value of 4.83 kcal liter^{-1} is used in converting measured oxygen consumption rates to rates of internal energy change. For example, if a person consumes oxygen at the high rate of 100 liters h^{-1}, the rate of internal energy change is (100 liters h^{-1})(4.83 kcal liters^{-1}) = 483 kcal h^{-1}.

Basal Metabolic Rate | All animals, including humans, use internal energy even while sleeping. The rate of energy consumption while resting but awake is called the basal metabolic rate. It is about 1.2 W kg^{-1} for an average 20-year-old man and 1.1 W kg^{-1} for a woman of the same age. This corresponds to about 1700 kcal per day and 1400 kcal per day for a 70-kg man and a 60-kg woman, respectively. Much of the energy consumed by a resting person is converted directly into heat. The remainder is used to do work in the body and then converted into heat.

Food materials are not used directly in the body. Instead they are converted into materials such as ATP (adenosine triphosphate), which can be used in the tissues. About 55 percent of the internal energy is lost as heat in this conversion. The remaining 45 percent is available to do internal work in the body organs or to enable the skeletal muscles to contract and do work on an external object.

When a person is performing an activity such as running up stairs or doing housework, the metabolic rate increases (Table 13.2). Part of the increased conversion of internal energy is needed to provide the mechanical work done by the person. The remainder results from the increased internal demands of the body. For example, in shoveling, the metabolic rate is about eight times the basic metabolic rate, but little mechanical work is actually done. The metabolic energy is used primarily by the skeletal muscles in changing and maintaining the body position.

The following example illustrates some of the ideas we have developed to this point.

TABLE 13.1

The average energy content per unit mass of food and the calorific equivalent of oxygen for a typical diet

Food	Energy Content per Unit Mass (kcal g^{-1})	Calorific Equivalent of Oxygen (kcal liter^{-1})
Carbohydrate	4.1	5.05
Protein	4.2	4.46
Fat	9.3	4.74
Ethanol	7.1	4.86
Standard average		4.83

TABLE 13.2

Approximate metabolic rates per unit mass of a
20-year-old man during various activities

Activity	$-\dfrac{1}{m}\dfrac{\Delta U}{\Delta t}$ (W kg^{-1})
Sleeping	1.1
Lying awake	1.2
Sitting upright	1.5
Standing	2.6
Walking	4.3
Shivering	Up to 7.6
Bicycling	7.6
Shoveling	9.2
Swimming	11.0
Lumbering	11.0
Skiing	15.0
Running	18.0

Example 13.7

(a) How much internal energy is used by a 65-kg man when bicycling for 4 hours? (b) If this energy is obtained by the metabolism of body fat, how much fat is used in this period?

(a) From Table 13.2, the metabolic rate while cycling is 7.6 W kg^{-1}. A 65-kg man then uses energy at a rate of (7.6 W kg^{-1})(65 kg) = 494 W. Four hours is 1.44×10^4 s, so the net energy usage is

$$-\Delta U = (494 \text{ W})(1.44 \times 10^4 \text{ s})$$

$$= 7.1 \times 10^6 \text{ J}$$

$$= (7.1 \times 10^6 \text{ J})\frac{(1 \text{ kcal})}{(4.18 \times 10^3 \text{ J})}$$

$$= 1700 \text{ kcal}$$

(b) The energy equivalent of fat is 9.3 kcal g^{-1} so the mass of fat needed to supply the necessary energy is

$$\text{Mass of fat} = \frac{(1700 \text{ kcal})}{(9.3 \text{ kcal g}^{-1})}$$

$$= 180 \text{ g} = 0.18 \text{ kg}$$

To appreciate this result, it is useful to compare it with the energy equivalent of the food required by a fairly sedentary man during a 24-hour day, which is 2500 kcal. Hence the exercise of bicycling for 4 hours will use up about two-thirds of the daily energy required by a sedentary man. This indicates that limiting food intake is a more practical way to decrease body weight than is exercise for most people.

The Efficiency of Food Utilization | The effi-

ciency of a human using the chemical energy in food to do useful work can be defined in several ways. The usual convention is to compare the measured rate at which mechanical work is done with the actual metabolic rate during the activity minus the basal metabolic rate. The efficiency e in percent is then

$$e = \frac{100 \dfrac{\Delta W}{\Delta t}}{\left| \dfrac{\Delta U}{\Delta t} - \dfrac{\Delta U}{\Delta t}_{\text{basal}} \right|}\% \qquad (13.10)$$

The denominator is the magnitude of the difference in actual and basal metabolic rates. The efficiency would be 100 percent if all the additional energy were converted to mechanical work. Table 13.3 lists some measured efficiencies.

The following example illustrates the calculation of efficiency for hiking up a mountain.

Example 13.8

A 20-year-old woman of mass 50 kg climbs a mountain 1000 m high in 4 hours. Her metabolic rate per unit mass during this activity is 7 W kg^{-1}. (a) What is the difference between her actual and basal metabolic rate? (b) How much work is done in the climb? (c) What is her efficiency?

(a) Since the woman's basal rate is 1.1 W kg^{-1}, the difference per unit mass is $(7 - 1.1)$W kg^{-1} = 5.9 W kg^{-1}. The total difference in rates is this times her mass or

$$\left| \frac{\Delta U}{\Delta t} - \frac{\Delta U}{\Delta t}_{\text{basal}} \right| = (50 \text{ kg})(5.9 \text{ W kg}^{-1})$$

$$= 295 \text{ W}$$

TABLE 13.3

Maximum efficiencies for physical work

Activity	Efficiency in Percent
Shoveling in stooped posture	3
Weight lifting	9
Turning a heavy wheel	13
Climbing ladders	19
Climbing stairs	23
Cycling	25
Climbing hills with a 5° slope	30

Adapted from E. Grandjean, *Fitting the Task to the Man; An Ergonomic Approach*, Taylor and Francis, London, 1969.

(b) The work done during the climb is equal to the change in the woman's potential energy so

$$\Delta W = mgh = (50 \text{ kg})(9.8 \text{ m s}^{-2})(1000 \text{ m})$$
$$= 4.9 \times 10^5 \text{ J}$$

The rate at which she does work in 4 hours = 1.44×10^4 s is

$$\frac{\Delta W}{\Delta t} = \frac{4.9 \times 10^5 \text{ J}}{(1.44 \times 10^4 \text{ s})} = 34 \text{ W}$$

(c) Her efficiency is found from the definition

$$e = \frac{100 \dfrac{\Delta W}{\Delta t}}{\left| \dfrac{\Delta U}{\Delta t} - \dfrac{\Delta U}{\Delta t}_{\text{basal}} \right|} = \frac{100(34 \text{ W})}{295 \text{ W}} = 11.5\%$$

Thus the woman is utilizing food energy for mechanical work at 11.5 percent efficiency. In general the efficiencies in human activities are below 30 percent.

It is important to notice that the rate at which work can be done depends on how long the activity is continued. A person in good condition can produce a power of nearly 21 W kg^{-1} during bicycle racing, but only for about 5 or 6 seconds. When doing work for a period of as long as 5 hours, the maximum metabolic rate is 6 or 7 W kg^{-1}. For a person doing physical labor the annual average metabolic rate should be 4 W kg^{-1} or less to be safe.

EXERCISES ON SUPPLEMENTARY TOPICS
Section 13.5 | Human Metabolism
13-24 A woman of mass 55 kg produces heat at a rate of 1.1 W kg^{-1} while lying at rest on a hot day. If her body temperature is constant, (a) what is the rate at which her internal energy changes? (b) How much internal energy will she use up in 8 hours? (c) If all of this energy is produced by metabolizing carbohydrates, what mass of carbohydrate is consumed?

13-25 A woman with a normal diet uses internal energy at a rate of 3 W kg^{-1} and has a mass of 50 kg. (a) What is her rate of oxygen consumption? (b) How much oxygen does she use in 8 hours?

13-26 A 60-kg man shovels dirt with an efficiency of 3 percent, and his metabolic rate is 8 W kg^{-1}. (a) What is his power output? (b) How much work does he do in 1 hour? (c) How much waste heat does his body produce in 1 hour?

13-27 A resting 45-kg woman has a normal basal

metabolic rate. (a) What volume of oxygen does she use in 1 hour? (b) If she walks for 1 hour, how much oxygen does she consume?

13-28 If a 70-kg, 20-year old man uses 1 liter of oxygen per minute, (a) what is his metabolic rate? (b) If he could do work at 100 percent efficiency, what would his mechanical power output be?

13-29 Experimentally it was found by J.P. Joule in 1846 that in 24 hours a horse was capable of doing work equivalent to raising a weight of 10^8 N to a height of 0.3 m. The hay and corn consumed during this period were equivalent to a reserve of internal energy of 1.2×10^8 J. (a) What fraction of this internal energy was used by the horse to do mechanical work? (b) If the horse's actual efficiency was 30 percent, what is the basal metabolic rate of the horse?

PROBLEMS ON SUPPLEMENTARY TOPICS
13-30 The basal metabolic rate for most humans steadily decreases after age 20, decreasing by about 20 percent by age 70. (a) Will an older person become cold on a winter day more quickly than a younger person? Why? (b) If a 70-year-old man does work at the same rate as a younger man and both work with the same efficiency, which man will have the higher metabolic rate? Explain.

13-31 If a 70-kg person's diet yields a food energy equivalent of 1.25×10^7 J, how much work can that person do at 15 percent efficiency with a metabolic rate of 250 W before all of the food energy is used up?

13-32 If a 90-kg man exercises with a metabolic rate of 7.5 W kg^{-1}, how long must he continue to use up 1 kg of fat?

13-33 A 70-kg sprinter does work at a rate of 820 W during a bicycle dash lasting 11 s. If the efficiency is 20 percent and only carbohydrates are consumed, what mass of carbohydrate is used?

13-34 A 45-kg woman runs up a flight of stairs 5 m high in 3 s. (a) What is her mechanical power output? (b) If the woman's basal metabolic rate is 1 W kg^{-1} and she works with 10 percent efficiency, what is her metabolic rate when ascending the stairs? (c) What is the total amount of oxygen consumed by the woman during the climb?

13-35 If the woman of Problem 13-34 descends the stairs, her change in potential energy is negative. Is any metabolic energy required in this process? Explain.

13-36 In Chapter Nine the buckling strength scaling model led to the conclusion that the metabolic rate of an animal of mass m should vary as $m^{0.75}$. This also means that the metabolic rate per unit mass varies as $m^{0.75}/m = m^{-0.25}$. The basal metabolic rate of a 60-kg man is 1.2 W kg^{-1}. What is the basal metabolic rate per unit mass of a 960-kg horse?

13-37 Using the scaling law described in Problem 13-36, what is the expected basal metabolic rate of a 6400-kg elephant if that of a 0.04-kg mouse is 0.3 W?

13-38 A hummingbird requires an expenditure of 0.06 W to hover. The measured oxygen consumption rates for a hummingbird at rest and hovering are 5×10^{-6} liters s^{-1} and 35×10^{-6} liters s^{-1}, respectively. What is the efficiency of the hummingbird while hovering?

13-39 A dieter consumes 2500 kcal per day and uses 3000 per day. If the deficit is made up by the use of stored body fat, in how many days will the dieter lose 1 kg?

Additional Reading:

W.F. Magie, *A Source Book in Physics,* McGraw-Hill Book Co., New York, 1935, pp. 196–211. Excerpts from papers by Mayer and Joule.

V.V. Raman, Where Credit is Due—The Energy Conservation Principle, *Physics Teacher,* vol. 13, 1975, p. 80.

T.C. Ruch and H.D. Patton (eds.), *Physiology and Biophysics,* vol. 3, W.B. Saunders, Philadelphia, 1973. Chapter 5 by Arthur C. Brown discusses human metabolism.

George B. Benedek and Felix M.H. Villars, *Physics: With Illustrations from Medicine and Biology,* vol. I, Addison-Wesley Publishing Co., Reading, Mass., 1973, pp. 5–115. Human metabolism.

Knut Schmidt-Nielsen, *Animal Physiology,* 3rd ed., Prentice-Hall, Inc., Englewood Cliffs, N.J., 1970. Human and animal metabolism.

I. Prigogine, G. Nicolis, and A. Babloyantz, Thermodynamics of Evolution, *Physics Today,* November 1972, p. 23; December 1972, p. 38. Difficult but interesting articles.

Scientific American articles:

Mitchell Wilson, Count Rumford, October 1960, p. 158.

J.W.L. Köhler, The Stirling Refrigeration Cycle, April 1965, p. 119.

Graham Walker, The Stirling Engine, August 1973, p. 80.

C.M. Summers, The Conversion of Energy, September 1971, p. 148.

Lynwood Bryant, Rudolf Diesel and His Rational Engine, August 1969, p. 108.

Freeman Dyson, Energy in the Universe, September 1971, p. 50.

W. Ehrenberg, Maxwell's Demon, November 1967, p. 103.

Stanley W. Angrist, Perpetual Motion Machines, January 1968, p. 114.

O.V. Lounasmaa, New Methods for Approaching Absolute Zero, December 1969, p. 26.

John R. Clark, Thermal Pollution and Aquatic Life, March 1969, p. 18.

James B. Kelley, Heat, Cold and Clothing, February 1956, p. 107.

R.E. Newell, The Circulation of the Upper Atmosphere, March 1964, p. 62.

C.B. Chapman and J.H. Mitchell, The Physiology of Exercise, May 1965, p. 88.

J.R. Brett, The Swimming Energetics of Salmon, August 1965, p. 80.

Vance A. Tucker, The Energetics of Bird Flight, May 1969, p. 70.

David M. Gates, The Flow of Energy in the Biosphere, September 1971, p. 88.

Bernd Heinrich, The Energetics of the Bumblebee, April 1973, p. 97.

Eugene S. Ferguson, The Measurement of the Man-Day, October 1971, p. 96.

Rudolfo Margaria, The Sources of Muscular Energy, March 1972, p. 84.

Salvador E. Luria, Colicins and the Energetics of Cell Membranes, December 1975, p. 30.

Jules Janick, Carl H. Noller and Charles L. Rhykerd, The Cycles of Plant and Animal Nutrition, September 1976, p. 74.

The September 1970 issue of *Scientific American* deals with *The Biosphere,* and has several articles relating to energy cycles and life.

UNIT FOUR

FLUIDS

Fluids play a unique role in our lives and in our study of science, mainly because of their characteristic of flowing and the fact that they conform to the shape of their container. In this context, we can regard both liquids and gases as fluids.

Animals transport nutrients and remove wastes via the fluids of their circulatory systems. Similarly, the transport of materials in plants occurs in fluids. The flight of birds and planes involves fluid motion, as do the weather, waves, and ocean currents.

All these phenomena can be described by applying the principles of mechanics to fluids. However, because fluids do not remain in a fixed shape and because they may be compressed, a complete analysis can become very complicated. In order to simplify matters, in this unit we assume that fluids are *incompressible;* that is, they remain constant in density. For most liquids, this is a good first approximation. For gases, we must realize that our methods will only apply to those situations where the pressure and temperature variations are small.

In the first chapter of this unit (Chapter Fourteen), we also assume that there are no frictional forces between segments of a fluid moving with respect to one another. This is fine for fluids at rest and for some applications when fluids move. However, we will see in Chapter Fifteen that these frictional or *viscous* effects are often important. In Chapter Sixteen, we discuss some important properties arising from the intermolecular forces in fluids, and finally in Chapter Seventeen, we apply our knowledge to some aspects of the human circulatory system.

CHAPTER FOURTEEN

THE MECHANICS OF NONVISCOUS FLUIDS

In this chapter, we begin to apply the laws of mechanics to fluids. In mechanics, we found that if we could identify the forces acting on an object or objects, we could then predict the subsequent motion or describe the state of equilibrium. For our discussion of fluids, we use the same philosophy, but we need to take account of the fact that a given mass of fluid does not have a fixed shape. This complication is handled by using the *density* and *pressure* instead of the mass and force concepts we used previously.

Once we have defined the density and pressure, we proceed to describe fluids at rest and nonviscous (frictionless) fluids in motion. For fluids at rest, we first develop an understanding of why an object may either sink or float in a fluid. For moving fluids, we develop *Bernoulli's equation,* which puts work and energy concepts into a form suitable for fluids. This is the central point of this chapter and can be used, for example, to understand why fluids in connected containers tend to have the same surface levels and how fluids flow from one place to another.

An important condition on this discussion is the assumption that the fluid is incompressible: a given mass of fluid always occupies the same volume though its shape may change. This condition is described mathematically by the *continuity equation,* which simply states that the amount of fluid entering a tube must equal the amount leaving; this equation plays an important part in our discussions.

14.1 | DENSITY AND PRESSURE

The direct use of the concepts of mass and force is inconvenient when we apply mechanics to fluids. A particular mass of fluid may be deformed as it moves, so keeping track of it would present overwhelming geometrical problems. This difficulty is avoided nicely by describing fluids in terms of the *density* ρ and *pressure P,* neither of which is a vector nor dependent on geometry.

Density | A given mass of an incompressible fluid will occupy a fixed volume, regardless of its shape. Hence, we define the *density* ρ of a fluid as its mass to volume ratio,

$$\rho = \frac{\text{mass}}{\text{volume}} = \frac{m}{V} \qquad (14.1)$$

In S. I. units, densities are measured in kilograms per cubic meter (kg m^{-3}). Table 14.1 lists the densities of some common fluids.

Relative Density | The *relative density* or *specific gravity* is the ratio of the density of a substance to that of water at 0°C. The relative density can be conveniently found from Table 14.1, since the density of water at 0°C is 1000 kg m^{-3}. For example, the relative density of mercury is the dimensionless number $13,600/1000 = 13.6$. From the conversion factors given in the table title, we also see, for example, that a cubic foot of mercury weighs $(13.6)(62.5 \text{ lb}) = 850 \text{ lb}$.

Pressure | In Chapter Eleven the pressure in a gas was defined as the force per unit area exerted on the walls of the vessel containing the gas. Here we give an

TABLE 14.1

Densities of some fluids at atmospheric pressure. Conversion factors are $1 \text{ kg m}^{-3} = 10^{-3} \text{ g cm}^{-3} = 1.94 \times 10^{-3} \text{ slug ft}^{-3}$. (At 0°C, 1 ft³ of water weighs 62.5 lbs; 1 cm³ of water has a mass of 1 g)

Fluid	Density (kg m⁻³)	Temperature (°C)
Hydrogen (H₂)	0.0899	0
Helium (He)	0.178	0
Nitrogen (N₂)	1.25	0
Carbon dioxide (CO₂)	1.98	0
Oxygen (O₂)	1.43	0
Air	1.29	0
	1.20	20
	0.95	100
Water, pure	1000.	0
	958.	100
Sea water	1025.	15
Alcohol, ethyl	791.	20
Chloroform	1490.	20
Ether	736.	0
Linseed oil	930.	0
Glycerin	1260.	0
Mercury	13600.	0
Whole blood	1059.5	25
Blood plasma	1026.9	25

equivalent definition that is also applicable to points within a fluid.

Figure 14.1a shows a fluid at rest *in equilibrium.* Suppose we wished to remove a spherical portion of the fluid in such a way as to leave the remaining fluid undisturbed. Then we would have to apply, in some manner, forces such as those shown in Fig. 14.1b. *These forces would replace the actual effect of the removed fluid* and would be perpendicular or normal to the spherical surface at each point.

    (a)   (b)

Figure 14.1. (a) A cross section of a segment of fluid in equilibrium with a spherical portion. (b) Some of the forces necessary to maintain equilibrium when the portion is removed.

The sum of the *magnitudes* of the normal forces divided by the surface area is the *average pressure* $\bar{P}$ on the surface of the sphere:

$$\bar{P} = \frac{\text{magnitude of normal forces on surface}}{\text{surface area}}$$

$$= \frac{F_N}{A} \tag{14.2}$$

To obtain the *pressure P at a point,* we shrink the imaginary sphere until its radius and area are arbitrarily small.

Pressure has units of force per unit area. As mentioned previously, the S.I. unit of pressure is the pascal; $1 \text{ Pa} = 1 \text{ N m}^{-2}$. At sea level the normal atmospheric pressure in various units is

$$\begin{aligned} 1 \text{ atm} &= 1.013 \times 10^5 \text{ Pa} \\ &= 14.7 \text{ lb in}^{-2} \\ &= 1.01 \text{ bars} \\ &= 760 \text{ torr} \\ &= 760 \text{ mm Hg} \end{aligned}$$

The *bar* and *millibar* are used extensively in meteorology. The *torr* or the *millimeter of mercury* (mm Hg) is used in medicine and physiology. Normal atmospheric pressure will support a column of mercury of height 760 mm = 0.76 m.

14.2 | BULK MODULUS

In our discussion of fluids, we assume that they are incompressible. In this section, we discuss this point and see how accurate the assumption is.

We learned in Chapter Nine that when a solid bar is compressed by exerting forces on its ends, its fractional length change is proportional to the force exerted per unit area. Similarly, if a material is subjected to uniform inward forces on all sides, its volume decreases. The resistance of a material to such a deformation is characterized by its *bulk modulus, K.* We will see that the volume or density changes that occur in liquids are usually very small and may be neglected in most situations.

Figure 14.2 shows a solid object of volume V and density ρ. It is immersed in a liquid confined to a cylinder that is fitted with a piston. If a force is applied to the piston, the pressure on the solid changes

233

Figure 14.2. A method of applying inward forces on all sides of a solid object.

by ΔP. This is related to the volume change ΔV or the density change $\Delta \rho$ by

$$\Delta P = -K \frac{\Delta V}{V} = K \frac{\Delta \rho}{\rho} \qquad (14.3)$$

The signs reflect the fact that an increase in pressure produces an increase in density but a decrease in volume. The bulk modulus K is positive. Typical values of K for solids and liquids are given in Table 14.2. The small size of the volume and density changes is illustrated by considering the following extreme example.

Example 14.1

The pressure 2 km below the surface of the ocean is 200 times atmospheric pressure. At this depth, what is the change in volume of a mass of water that occupies 1 m^3 at the surface?

According to Table 14.2, the bulk modulus of water is 0.22×10^{10} N m^{-2}. The pressure change ΔP is $(200 - 1) = 199$ atm, where 1 atm is 1.01×10^5 Pa. Solving $\Delta P = -K \Delta V/V$ for ΔV, we find

$$\Delta V = -\frac{V \Delta P}{K}$$

$$= -\frac{(1 \text{ m}^3)(199)(1.01 \times 10^5 \text{ Pa})}{0.22 \times 10^{10} \text{ N m}^{-2}}$$

$$= -9.13 \times 10^{-3} \text{ m}^3$$

In this extreme example, the volume of water is reduced by less than 1 percent, and the density is increased by a corresponding amount. This shows that volume changes associated with moderate pressure changes in liquids are rather small. The same is

TABLE 14.2

Bulk moduli for some solids and liquids (Bulk moduli for gases are given in Table 24.1)

Materials	Bulk Modulus, K (N m^{-2})
Liquids	
Ethyl alcohol	0.09×10^{10}
Mercury	2.5×10^{10}
Water	0.22×10^{10}
Solids	
Aluminum	7.0×10^{10}
Copper	$14. \times 10^{10}$
Glass	3.7×10^{10}
Steel	$16. \times 10^{10}$

true for solids. For gases the bulk moduli are much smaller, and the volume changes may be substantial. Thus when we apply our results for incompressible fluids to gases, we must be careful to discuss only those situations where appreciable density and volume changes do not occur.

14.3 | ARCHIMEDES' PRINCIPLE

An object floating or submerged in a fluid experiences an upward or *buoyant* force due to the fluid. To understand this buoyant force B, consider a segment of fluid of volume V, density ρ_0, and weight $w_0 = \rho_0 g V$ (Fig. 14.3a). The segment is in equilibrium, so from Newton's first law, B must be equal but opposite to its weight. Thus

$$B = \rho_0 g V$$

The buoyant force is merely the force exerted by the remainder of the fluid to keep the segment at rest.

Now suppose that the imaginary segment is re-

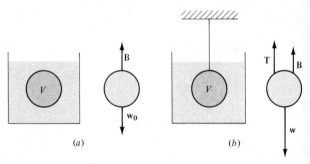

Figure 14.3. (a) An imaginary segment of fluid with the forces acting on it. (b) An object suspended by a string in a fluid and the forces acting on it.

placed by an object of volume V suspended on a string. The density ρ of the object is greater than that of the fluid. The forces on the object are its weight $w = \rho g V$, the tension in the string T, and the buoyant force B. The fluid does not distinguish between the object and the fluid it replaces, so again $B = \rho_0 g V$. Since the object is in equilibrium, $T = w - B$, or

$$T = (\rho - \rho_0)gV \qquad (14.4)$$

The tension in the string is reduced by the weight of the displaced fluid. The principle that *the buoyant force on the object is equal to the weight of the displaced fluid* was deduced by Archimedes (287–212 B.C.) and is called *Archimedes' principle.* Archimedes' principle provides a convenient way to determine densities. This is illustrated in the next example.

Example 14.2

A piece of metal of unknown volume V is suspended from a string. Before submersion the tension in the string is 10 N. When the metal is submerged in water, the tension is 8 N. What is the density ρ of the metal?

Before submersion the tension is $T_i = \rho g V$. After submersion the tension is $T_f = (\rho - \rho_0)gV$, where the density of water is $\rho_0 = 10^3$ kg m^{-3}. Dividing the second equation by the first to eliminate V gives

$$\frac{T_f}{T_i} = \frac{\rho - \rho_0}{\rho}$$

Solving for ρ,

$$\begin{aligned}
\rho &= \frac{\rho_0 T_i}{T_i - T_f} \\
&= \frac{(1000 \text{ kg m}^{-3})(10 \text{ N})}{(10 \text{ N} - 8 \text{ N})} \\
&= 5000 \text{ kg m}^{-3}
\end{aligned}$$

An object less dense than a fluid will float partially submerged. If a portion V_S of its volume V is submerged, then the buoyant force is $\rho_0 g V_S$. This must equal the weight $\rho g V$ of the object, so $\rho_0 g V_S = \rho g V$, or

$$\frac{\rho}{\rho_0} = \frac{V_S}{V} \qquad (14.5)$$

Thus the ratio of the densities is equal to the fraction of the volume submerged. This is illustrated in the following example of an iceberg.

Example 14.3

The density of ice is 920 kg m^{-3} while that of sea water is 1025 kg m^{-3}. What fraction of an iceberg is submerged?

The fraction is

$$\frac{V_S}{V} = \frac{\rho}{\rho_0} = \frac{920 \text{ kg m}^{-3}}{1025 \text{ kg m}^{-3}} = 0.898$$

Almost 90 percent of the iceberg is submerged.

14.4 | THE EQUATION OF CONTINUITY; STREAMLINE FLOW

Suppose an incompressible fluid completely fills a channel such as a pipe or an artery. Then if more fluid enters one end of the channel, an equal amount must leave the other end. This principle, which can be put into various mathematical forms, is called the *equation of continuity.* It will be quite useful in many of our discussions.

If the fluid enters one end at a rate Q_1, it must leave the other end at a rate Q_2, which is the same rate. Thus the equation of continuity can be written as

$$Q_1 = Q_2 \qquad (14.6)$$

For example, if 1 m^3 s^{-1} enters, then a 1 m^3 s^{-1} must leave.

This can be put into a more useful form if all the fluid in the channel is moving with a uniform velocity v. Consider a section of a tube with a constant cross-sectional area A (Fig. 14.4). In a time Δt, the fluid moves a distance $\Delta x = v \, \Delta t$, and the volume of fluid leaving the tube is $\Delta V = A \, \Delta x = Av \, \Delta t$. Alternatively, ΔV equals the flow rate Q times the time interval Δt, or $\Delta V = Q \, \Delta t$. Comparing these expressions for ΔV, we see that

$$Q = Av \qquad (14.7)$$

The flow rate equals the cross-sectional area of the channel times the velocity of the fluid.

For a channel whose cross section changes from A_1 to A_2, this result together with $Q_1 = Q_2$ gives another form of the equation of continuity,

$$A_1 v_1 = A_2 v_2 \qquad (14.8)$$

Figure 14.4. The flow rate in a tube is $Q = Av$.

The product of the cross-sectional area and the velocity of the fluid is constant. If at some point, A decreases, v must increase. For example, if the area is halved, then the velocity doubles.

Usually the flow velocity is not uniform in a channel. For example, in the next chapter we will encounter situations where the fluid near the walls of the channel is moving at a lower speed than the fluid near its center. The equation of continuity still holds for such cases if it is written in terms of the average flow velocity $\bar{v}$. The flow rate is $Q = A\bar{v}$, and at two points in a channel, $A_1\bar{v}_1 = A_2\bar{v}_2$.

Streamline Flow
A type of fluid flow, called *streamline flow,* is most readily discussed quantitatively and is important in many applications. It is best defined by imagining a simple experiment. Suppose we use a very fine eyedropper to inject some ink into a fluid, without interrupting the flow (Fig. 14.5*a*). If the ink line formed in this way does not disperse or mix but remains narrow and well defined, the flow is called *streamline.* If several eyedroppers injected ink side by side, the ink pattern might look like Fig. 14.5*b*. Finally, we can imagine a *tube* of streamlines as in Fig. 14.5*c*. The fluid in this tube is undergoing streamline flow. If instead, the ink lines swirl and mix, the flow is said to be *turbulent* (Fig. 14.5*d*).

The tube is a useful concept because, by definition, the streamlines do not cross one another; no fluid flows into or out of the sides of the tube. Accordingly the equation of continuity can be applied to this flow tube, and *the product Av is the same at all points in the*

flow tube. This property of streamline flow is used in the next section.

14.5 | BERNOULLI'S EQUATION

We now consider Bernoulli's equation. First derived by Daniel Bernoulli (1700–1782) in 1738, it states the consequences of the principle that the work done on a fluid as it flows from one place to another is equal to the change in its mechanical energy. Bernoulli's equation can be used under the following conditions:

1. The fluid is *incompressible;* its density remains constant.
2. The fluid does not have appreciable frictional effects; it is *nonviscous.* Consequently, no mechanical energy is lost due to friction.
3. The flow is *streamline, not turbulent.*
4. The velocity of the fluid at any point does not change during the period of observation. (This is called the *steady-state* assumption.)

In this section, we see how Bernoulli's equation is obtained by using the relation between work and mechanical energy. Its applications are discussed in the following sections.

Consider the fluid in a section of a flow tube with a constant cross section A (Fig. 14.6*a*). According to the equation of continuity, the product Av remains constant. Thus the velocity v does not change as the fluid moves through the tube, and its kinetic energy re-

Figure 14.5. (*a*) A single streamline. (*b*) A group of adjacent streamlines. (*c*) A flow tube. The walls of the tube are composed of streamlines. (*d*) Turbulent flow.

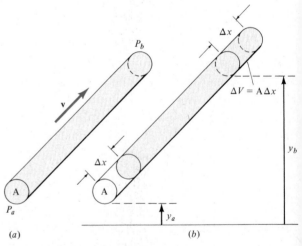

Figure 14.6. (*a*) The fluid in a section of flow tube with a constant cross section has the same velocity everywhere. (*b*) The fluid has moved a short distance Δx.

mains the same. However, the potential energy changes as the fluid rises.

The net force on the fluid in the tube due to the surrounding fluid is the cross-sectional area A times the difference in pressures on the ends, or $(P_a - P_b)A$. If the fluid in the section moves a short distance Δx, then the work done on it is the product of the force and the displacement $(P_a - P_b)A\,\Delta x$. Since $A\,\Delta x$ is the volume ΔV of the fluid leaving the section (Fig. 14.6b), the work done on the fluid is

$$W = (P_a - P_b)\,\Delta V$$

This work done on the fluid must equal the increase $\Delta\mathcal{U}$ in its potential energy. $\Delta\mathcal{U}$ can be calculated if we note that the fluid leaving the section has a mass $\rho\,\Delta V$ and a potential energy $(\rho\,\Delta V)gy_b$, while the fluid entering at the bottom of the section has a potential energy $(\rho\,\Delta V)gy_a$. Thus $\Delta\mathcal{U} = \rho g\,\Delta V(y_b - y_a)$. Equating this to W, we have

$$P_a - P_b = \rho g(y_b - y_a)$$

or

$$P_a + \rho gy_a = P_b + \rho gy_b \quad (v = \text{constant}) \quad (14.9)$$

Thus the pressure P plus the *potential energy per unit volume* ρgy of the fluid is the same everywhere in the flow tube if the velocity remains constant.

More generally, if the cross-sectional area of the flow tube changes, the fluid velocity v and *kinetic energy per unit volume* $\frac{1}{2}\rho v^2$ will also change. The work done on the fluid must then be set equal to the change in the potential plus kinetic energy of the fluid. The result is *Bernoulli's equation*,

$$P_a + \rho gy_a + \tfrac{1}{2}\rho v_a{}^2 = P_b + \rho gy_b + \tfrac{1}{2}\rho v_b{}^2 \quad (14.10)$$

The pressure plus the total mechanical energy per unit volume, $P + \rho gy + \frac{1}{2}\rho v^2$, is the same everywhere in a flow tube.

Bernoulli's equation is the main result of this chapter. As is clear from its derivation, it is a restatement of the relationship between work and energy in a form suitable for use in the physics of fluids. Many examples of its uses are seen in this chapter and in those following.

14.6 | STATIC CONSEQUENCES OF BERNOULLI'S EQUATION

We first examine the implications of Bernoulli's

equation when the fluid is at rest, so $v = 0$ and $P + \rho gy$ is constant.

Fluid at Rest in a Container |
Suppose fluid is at rest in a container shaped like the one in Fig. 14.7. We can find the pressure at a point B in terms of the pressure at the surface and the depth. To do this, we calculate $P + \rho gy$ at points A and B, choosing the y axis so that $y = 0$ at the bottom of the container. At A the pressure is atmospheric pressure, P_{atm}; at B the pressure is P_B. Thus, $P_{\text{atm}} + \rho gh = P_B + \rho gy_B$ or, with $h - y_B = d$,

$$P_B = P_{\text{atm}} + \rho gd \quad (14.11)$$

This result shows that the pressure at a depth d in a fluid at rest is equal to the surface pressure plus the potential energy density change ρgd corresponding to this depth. Equation 14.11 can also be interpreted as a statement of the force on a unit area a distance d below the surface of the liquid. The force per unit area P_B is the sum of two terms: P_{atm}, the pressure due to the atmosphere, and ρgd, the pressure due to the weight of the liquid above the point B.

Calculating $P + \rho gy$ at points B and D gives $P_B + \rho gy_B = P_D + \rho gy_D$. Or, since $y_B = y_D$,

$$P_B = P_D$$

Thus the pressure at the same depth at two places in a fluid at rest is the same. In particular, since points A and E are both at atmospheric pressure, the surface of the liquid is at the same height at both points. Thus the surfaces of liquids at rest in connected containers of any shape must be at the same height if they are open to the atmosphere.

These ideas are illustrated in the next two examples.

Figure 14.7. Fluid at rest in a container. As is explained in the text the level of the surface of the fluid in each of the segments is the same.

(The Bettmann Archive.)

ARCHIMEDES
(287–212 B.C.)

Archimedes was a mathematical and scientific genius whose equal did not appear until Newton, 2000 years later. Born the son of an astronomer, his relationship with King Hieron II of Syracuse afforded him both independent means and the attention of a man of power.

Archimedes is best remembered for his ingenious machines, which captured the popular imagination, played a role in the wartime defense of Syracuse from the Romans, and in the end resulted in larger-than-life stories about him. For example, the seige of Syracuse by the Romans during the Carthaginian war lasted 3 years, largely because of the defenses developed by Archimedes. It is said that he developed giant reflecting mirrors that focused sunlight on the Roman ships, setting them on fire. Giant cranes were also used to lift and upset ships. The war ended badly for Archimedes, as he is said to have been stabbed to death by one of the conquering Roman soldiers. At the time, he was doing geometry in the sand and refused to be disturbed.

Many of the stories about Archimedes' inventions are distorted because he himself considered them beneath the dignity of the pure scientist and never left any written record of them. His written testaments include treatises on geometry, where he approached the foundations of calculus, equilibrium and the center of gravity, and hydrostatics.

The principle of hydrostatics that carries his name grew from a problem concerning the gold content of a crown made for Hieron. Archimedes was asked to determine whether the crown was pure gold or was adulterated with silver. After much thought, he realized while at the public baths that he could measure the volume of the crown very accurately by placing it in water and measuring the amount displaced. He then compared the weight of the crown with the weight of an equal volume of gold. This discovery apparently had at least two immediate results. The idea so excited him that Archimedes dashed home naked shouting "Eureka, eureka," and the goldsmith was executed because the crown was not pure.

Archimedes fully understood and demonstrated the importance of the lever. He is said to have arranged a lever and pulley system with which the king himself pulled a large ship onto the shore. His work on the lever and on the importance of the center of gravity laid much of the groundwork for modern mechanics.

238

Example 14.4

What is the pressure on a swimmer 5 m below the surface of a lake?

Using $d = 5$ m and $\rho = 1000$ kg m^{-3}, we find

$$P_B = P_{atm} + \rho g d$$
$$= 1.01 \times 10^5 \text{ Pa} + (1000 \text{ kg m}^{-3})(9.8 \text{ m s}^{-2})(5 \text{ m})$$
$$= 1.50 \times 10^5 \text{ Pa}$$

Example 14.5

The pressure 1 m above a floor is measured to be normal atmospheric pressure, 1.01×10^5 Pa. How much greater is the pressure at the floor if the temperature is $0°C$?

Here, $d = 1$ m. From Table 14.1 the density of air at atmospheric pressure and $0°C$ is 1.29 kg m^{-3}. Thus

$$P_B = P_{atm} + \rho g d$$
$$= 1.01 \times 10^5 \text{ Pa} + (1.29 \text{ kg m}^{-3})(9.8 \text{ m s}^{-2})(1 \text{ m})$$
$$= (1.01 \times 10^5 + 12.6) \text{ Pa}$$

Thus the pressure at the floor is greater by 12.6 Pa or by about 1 part in 10^4, which is negligible for most purposes. The pressure change is small, unlike the preceding example, because the density of the air is very small compared to that of a typical liquid. The properties of gases were discussed in Chapter Eleven, with the assumption that the pressure in a gas-filled container is the same everywhere. This is usually a good approximation. However, if the pressure of a gas is measured at two very different heights, a considerable difference will be observed. For example, the atmospheric pressure at Aspen, Colorado, which is 2500 m above sea level, is about 80 percent of the sea level pressure.

The Manometer | The open-tube manometer

is a U-shaped tube used for measuring gas pressures. It contains a liquid that may be mercury or, for measurements of low pressures, water or oil. One end of the tube is open to the atmosphere, and the other end is in contact with the gas in which the pressure is to be measured (Fig. 14.8). The manometer can also

be used to measure pressures in a liquid provided that the liquid does not mix with the manometer fluid.

Measuring heights from the base of the U tube, $P + \rho g y$ is $P + \rho g y_1$ at the left surface of the column and $P_{atm} + \rho g y_2$ at the right surface. Equating these,

$$P + \rho g y_1 = P_{atm} + \rho g y_2$$

or

$$P = P_{atm} + \rho g(y_2 - y_1)$$
$$= P_{atm} + \rho g h \tag{14.12}$$

Thus, a measurement of the height difference h of the two columns determines the gas pressure P. In a blood pressure gauge, the sphygmomanometer, the pressure measured is that in an air sac wrapped around the upper arm. We return to a discussion of such measurements in Chapter Seventeen.

The pressure P in Eq. 14.12 is the *absolute pressure*. The difference between this and atmospheric pressure, $P - P_{atm}$, is the *gauge pressure*. The gauge pressure is then exactly equal to $\rho g h$.

Blood Pressure Measurements by Cannulation | In many experiments with anes-

thetized animals, the blood pressure in an artery or vein is measured by the direct insertion into the vessel of a *cannula*, which is a small glass or plastic tube containing saline solution plus an anticlotting agent. The saline solution, in turn, is in contact with the fluid in a manometer. It is necessary to have the surface of contact between the saline solution and the manometer fluid either at the same level as the insertion point of the cannula or to correct for the height difference (Fig. 14.9). Calculating $P + \rho g y$ at suitable points

Figure 14.8. The open tube manometer.

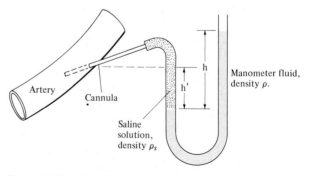

Figure 14.9. Measurement of blood pressure by cannulation.

Figure 14.10. (a) Venturi tube. (b) Enlarged view of the region where column 1 connects to the flow tube.

(Problem 14-42), it follows that the blood pressure P_B is given by

$$P_B = P_{atm} + \rho gh - \rho_s gh' \qquad (14.13)$$

Mercury is commonly used as the manometer fluid in measurements of arterial pressures. However, pressures in the veins are relatively low, and the use of mercury as a manometer fluid would give poor accuracy because h would be very small. Consequently the saline solution is used as the manometer fluid.

Physiologists often use electromanometers in which the manometer fluid pushes against a membrane instead of rising up a tube. The flexing of the membrane is proportional to the pressure, and the extent that the membrane is flexed is translated into an electrical signal. This signal drives a recorder that automatically gives a continuous record of the blood pressure.

14.7 | FLOW METERS

In this section we apply the complete Bernoulli equation to two kinds of flow meters. With minor modifications, they can be used to measure the flow in blood vessels, the airspeed of planes, and many other flow rates.

The Venturi Tube | In a *venturi tube,* fluid flows through different cross-sectional areas in different portions of the tube. When the tube narrows, the fluid velocity increases. Then, according to Bernoulli's equation, the pressure drops. Measurement of this pressure change determines the fluid velocity. The pressure measurement can be made with narrow vertical columns inserted into the main tube (Fig. 14.10a) or with electrical sensors.

Before using Bernoulli's equation, we must clear up a difficulty that arises because the liquid in the columns is at rest while the liquid in the tube is moving. Bernoulli's equation cannot be applied directly to relate the pressures at points C and D in Fig. 14.10b because the fluid at the two points is not in the same streamline. However, if the pressures were unequal, fluid would flow from one point to the other. Since this does not occur, $P_C = P_D$. Thus the pressure in the columns is the same as the pressure in the streamline.

Bernoulli's equation requires that $P + \rho gy + \frac{1}{2}\rho v^2$ is the same everywhere in a flow tube. Applying Bernoulli's equation to points at the same height in the flow stream just below the columns,

$$P_1 + \tfrac{1}{2}\rho v_1{}^2 = P_2 + \tfrac{1}{2}\rho v_2{}^2$$

From the continuity equation, $A_1 v_1 = A_2 v_2$, or

$$v_2 = \frac{A_1}{A_2} v_1 \qquad (14.14)$$

Using this expression for v_2, the preceding equation can be written as

$$P_1 - P_2 = \frac{1}{2}\rho v_1{}^2 \left[\frac{A_1{}^2}{A_2{}^2} - 1 \right] \qquad (14.15)$$

Thus, a measurement of $P_1 - P_2$ and knowledge of the areas determines v_1; v_2 can also be found using Eq. 14.14.

The following example shows how this flow meter can be used to measure the velocity of the blood in an artery.

Example 14.6

The flow of blood through a large artery in a dog is diverted through a venturi flow meter. The wider part of the flow meter has an area $A_1 = 0.08$ cm^2, which equals the cross-sectional area of the artery. The narrower part of the flow meter has an area $A_2 = 0.04$ cm^2. The pressure drop in the flow meter is 25 Pa. What is the velocity v_1 of the blood in the artery?

The ratio of the areas A_1/A_2 is dimensionless and has the value $0.08/0.04 = 2$. From Table 14.1, the density of whole blood is 1059.5 kg m^{-3}. Dropping the units, Eq. 14.15 becomes

$$25 = \tfrac{1}{2}(1059.5)v_1{}^2(2^2 - 1)$$

Solving for v_1,

$$v_1 = \sqrt{\frac{(2)(25)}{(1059.5)(2^2 - 1)}} = 0.125 \text{ m s}^{-1}$$

Figure 14.11. A Prandtl tube in a constant velocity flow stream. The right arm of the U-tube connects to the chamber opening at B. The left connects to the opening at A, where the fluid has zero velocity.

The Prandtl Tube | Figure 14.11 shows a

Prandtl tube inserted in a flow stream. It interrupts the flow pattern very little except at point A, where the fluid has zero velocity. At point B the velocity is assumed to be the streamline flow velocity v. From Bernoulli's equation, neglecting the small difference in heights at A and B,

$$P_A - P_B = \frac{1}{2}\rho v^2$$

where ρ is the fluid density. If the manometer fluid has density ρ_m, then $P_C = P_D$ gives

$$P_A + \rho g(h + x) = P_B + \rho g x + \rho_m g h$$

or

$$P_A - P_B = (\rho_m - \rho)gh$$

Comparing the two expressions for $P_A - P_B$,

$$\tfrac{1}{2}\rho v^2 = (\rho_m - \rho)gh \qquad (14.16)$$

Thus, a manometer reading gives a direct measurement of the flow velocity. As in the venturi tube, electromanometers may be employed instead of the open-tube manometer.

SUMMARY

The fundamental variables used in describing fluids are the density and pressure. The density is the mass per unit volume. The pressure is the force per unit area that a segment of fluid exerts on adjacent segments. If the pressure is increased, the density decreases; these changes are related by the bulk modulus.

The definition of the pressure leads directly to Archimedes' principle: the buoyant force on an object in a fluid is equal to the weight of the fluid displaced.

If a fluid may be treated as incompressible, the flow satisfies the continuity equation. This equation states that when fluid flows in a tube, the rate of flow must be constant even if the tube dimensions change, so the product Av is constant.

With the additional assumption that frictional or viscous forces are unimportant, we can equate the work done on a fluid to its change in mechanical energy. The result is Bernoulli's equation, which states that $P + \rho g y + \frac{1}{2}\rho v^2$ is constant during steady-state flow everywhere in a streamline flow tube. This equation can be applied to fluids in motion or at rest.

Checklist

Define or explain:

density	turbulent flow
relative density	flow tube
pressure	Bernoulli's equation
pascal	manometer
bulk modulus	absolute pressure
buoyant force	gauge pressure
Archimedes' principle	venturi tube
equation of continuity	Prandtl tube
streamline flow	

EXERCISES

Section 14.1 | Density and Pressure

14-1   Find the density of gasoline if 5 kg has a volume of 7.35×10^{-3} m^3.

14-2   Battery acid has a density of 1290 kg m^{-3} and contains 35 percent sulphuric acid by weight.

How many kilograms of sulphuric acid are in 10^{-3} m^3 of battery acid?

14-3   Find the volume of 1 kg of air at 0°C and atmospheric pressure.

14-4   What is the mass of 1 liter $= 10^{-3}$ m^3 of whole blood?

14-5   Find the weight in pounds of 9.13 ft^3 of chloroform.

14-6   A force of 25 N is exerted on the piston of a syringe. The piston has an area of 10^{-4} m^2. If no fluid escapes, what is the increase in pressure in the fluid of the syringe?

14-7   The pressure at the bottom of a cylindrical water tank is 3 atm. (a) Find the pressure in pascals. (b) If the tank has a radius of 10 m, what force does the water exert on the bottom of the tank?

Section 14.2 | Bulk Modulus

14-8   One cubic meter of water at the surface of the ocean has a mass of 1025 kg. What is the mass of 1 m^3 of water at a depth where the pressure is 300 atm? (Assume that the temperature is constant.)

14-9   A steel rod of length 1 m and radius 0.01 m is dropped into the ocean. Find the change in volume of the rod when it has reached the ocean floor if the pressure there is 250 atm. (Assume that the temperature is constant.)

Section 14.3 | Archimedes' Principle

14-10   A 160-lb man just floats in fresh water with virtually all of his body below the surface. What is his volume?

14-11   An object weighs 100 N in air and 75 N in water. What is the relative density of the object?

14-12   A tank of water on a scale weighs 50 lb. A 3-lb salmon is placed in the tank and swims around. What is the reading on the scale?

14-13   A balloon has a capacity of 0.1 m^3. What weight can it lift when it is filled with helium? Use the densities of helium and air at 0°C of Table 14.1.

14-14   A log weighing 40 lb is dropped into a river at 0°C. If the relative density of the log is 0.8, what will be the volume of the log above the surface?

Section 14.4 | The Equation of Continuity; Streamline Flow

14-15   The radius of a water pipe decreases from 0.2 to 0.1 m. If the average velocity in the wider portion is 3 m s^{-1}, find the average velocity in the narrower region.

14-16   A garden hose with a cross-sectional area of 2 cm^2 has a flow of 200 cm^3 s^{-1}. What is the average velocity of the water?

14-17   A blood vessel of radius r splits into four vessels, each with radius $r/3$. If the average velocity in the larger vessel is v, find the average velocity in each of the smaller vessels.

Section 14.5 | Bernoulli's Equation

14-18   Can Bernoulli's equation be used to describe the flow of water through a rapids in a stream? Explain.

14-19   A baseball is thrown by a pitcher and curves as it approaches the batter. Can Bernoulli's equation be applied to this problem using a reference frame (a) moving with the ball; (b) fixed with respect to the ground? Explain.

14-20   If water moves through pipes of constant cross section from the basement to the first floor, will the pressure stay the same? Explain.

Section 14.6 | Static Consequences of Bernoulli's Equation

14-21   Deep-sea photographs have been made at depths of 8000 m. (a) What is the pressure at this depth? (b) What is the force on the camera window if it measures 0.1 by 0.15 m?

14-22   What is the pressure difference between the heart and brain of a giraffe if the brain is 2 m above the heart? (Assume that the velocity of the blood is the same in both locations.)

14-23   Estimate the drop in the pressure of the atmosphere as one goes from sea level to the top of a hill 500 m high if the temperature is 0°C.

14-24   How high can water rise in the pipes of a building if the gauge pressure at the ground floor is 2×10^5 Pa?

14-25   A submarine dives to a depth of 100 m in sea water. What gauge pressure is necessary to expel water from its ballast tanks?

14-26   A certain pressure can support a column of pure water 0.7 m high. The same pressure will support a column of saline solution 0.6 m high. What is the density of the saline solution?

14-27   A large artery is cannulated, and a saline

solution of density 1300 kg m^{-3} is used as the manometer fluid. What is the blood pressure (gauge pressure) if the height difference in the manometer tubes is 0.67 m?

Section 14.7 | Flow meters

14-28 A venturi tube has a radius of 1 cm in its narrower portion and 2 cm in its wider sections. The velocity of water in the wider sections is 0.1 m s^{-1}. Find (a) the pressure drop, and (b) the velocity in the narrower portion.

14-29 A Prandtl tube flow meter is used to measure the aortic blood velocity of a dog. If water is the manometer fluid, what is the height difference in the manometer tube when the blood velocity is 0.1 m s^{-1}?

14-30 Suppose a venturi tube is equipped with vertical columns as in Fig. 14.10. Show that

$$g(y_1 - y_2) = \frac{1}{2}v_1^2\left(\frac{A_1^2}{A_2^2} - 1\right)$$

PROBLEMS

14-31 A block of oak weighs 90 N in air. A lead weight weighs 130 N when immersed in water. When attached together, they weigh 100 N in water. What is the density of the wood?

14-32 What is the initial acceleration of an iron ball of specific gravity 8 (a) when placed in water, and (b) when placed in mercury? (c) What is the direction of the acceleration in each case?

14-33 A *hydraulic jack* is constructed as shown in Fig. 14.12. The cross-sectional areas of the pistons are A_1 and A_2. Show that in equilibrium the forces F_1 and F_2 are related by

$$\frac{F_1}{A_1} = \frac{F_2}{A_2}$$

Piston area, A_2

Piston area, A_1

Figure 14.12. A hydraulic jack. Problems 14-33 and 14-34.

14-34 A hydraulic lift similar to that in Fig. 14.12 with pistons of cross-sectional areas 1500 cm^2 and 75 cm^2 is used to lift a dentist's chair of weight 1500 N. (a) What force is necessary on the small piston to raise the chair? (b) What distance must the small piston be moved to raise the chair 0.1 m?

14-35 A tank contains oxygen gas at 0°C. The pressure at the bottom of the tank is 100 atm. If the tank is 1 m tall, what is the pressure at the top? (Assume the average density of the oxygen is 143 kg m^{-3}.)

14-36 A *barometer* is constructed by filling a long tube with mercury and then inverting it, open side down, in a container of mercury exposed to the atmosphere (Fig. 14.13). (a) Show that atmospheric pressure P_{atm} is equal to $\rho g h$, where ρ is the density of mercury. (b) What is the height h when $P_{atm} = 1.01 \times 10^5$ Pa?

Figure 14.13. A mercury barometer. Problem 14-36.

14-37 During a whole blood transfusion, the needle is inserted in a vein where the pressure is 15 torr. At what height must the blood container be placed, relative to the vein, so that the blood just enters the vein? (1 torr = 133.3 Pa).

* **14-38** A *siphon tube* of cross-sectional area 3×10^{-4} m^2 is used to drain a tank of water. The tube is initially filled with water, and with the ends closed, one end is placed in the tank 0.25 m below the water surface. The other end is outside the tank and at a distance of 0.5 m below the immersed end. (a) What is the initial velocity of the water flowing out of the tube when the ends are opened? (b) Is the flow continuous? (c) What is the velocity of the water when the water surface in the tank is 0.1 m above the immersed end?

* **14-39** A siphon tube is filled with gasoline and closed at each end. One end is inserted in a gasoline tank 0.3 m below the surface of the gasoline.

The other end is placed 0.2 m below the end in the tank and both ends of the tube are opened. The tube has an inner cross-sectional area of 4×10^{-4} m². The density of gasoline is 680 kg m^{-3}. (a) What is the initial velocity of the gasoline in the tube? (b) What is the initial rate of flow?

14-40   An artery or vein may become partially blocked or *occluded,* when some material reduces the vessel radius over a small portion of its length. (a) Does the velocity in the occluded region change? (b) Does the pressure in the occluded region change? (c) Describe the possible sequence of events in the flexible vessel if the pressure inside the vessel becomes very small compared to the pressure outside.

14-41   A sealed bottle has water in it, and two tubes enter the bottle through the seal as shown in Fig. 14.14. Tube *B* is attached to a pump at one end, and the other end is above the water surface. If tube *A* is immersed 0.15 m below the water surface and the other end is open to the air, what is the minimum pressure obtained inside the bottle? What will occur when this pressure is reached? (Bottles of this type are used for draining unwanted fluids from the body, such as those that collect in and around the lungs in some diseases.)

Figure 14.14.   Problem 14-41.

14-42   The blood pressure in an artery is measured by cannulation (Fig. 14.9). The cannula contains saline solution of density ρ_s, and the manometer fluid has density ρ. Show that the blood pressure is given by Eq. 14.13.

SUPPLEMENTARY TOPICS

14.8 | THE FLIGHT OF ANIMALS AND AIRPLANES

A full discussion of flight involves a combination of elaborate mathematical theories and practical experimentation because the flow patterns about wings are very complex. However, with the aid of Bernoulli's equation, we can derive some qualitative results applicable to airplanes and to birds, insects, and other flying animals.

If we watch a plane from the ground, the air at any point is disturbed briefly as the plane passes and then returns to its original state. Since the fluid motion is not constant in time, we may not apply Bernoulli's equation to our observations. However, to an observer on the plane, the fluid flow is steady in time. *Thus, Bernoulli's equation holds true provided we apply it in a coordinate frame at rest relative to the plane.*

Fig. 14.15*a* shows the flow of air of initial velocity v and density ρ about a stationary airplane wing. The streamlines above the wing are closer together than those below the wing. According to the continuity equation, this indicates that the velocity v_a of the air above the wing is greater than that below, v_b (Fig. 14.16). Since Bernoulli's equation says that $P + \rho g y + \frac{1}{2}\rho v^2$ is constant in a streamline, the pressure P_a above the wing must be less than the pressure P_b below. (We can neglect the $\rho g y$ terms, since the wing thickness is small.) Then the upward lift force F_L on a wing area A is

$$F_L = (P_b - P_a)A = A\frac{\rho}{2}(v_a{}^2 - v_b{}^2)$$

This expression is not yet useful, since we do not know the velocities v_a and v_b, and there is no simple way to predict their values. However, we would expect that both v_a and v_b are proportional to the initial air velocity v. Hence $(v_a{}^2 - v_b{}^2)$ is proportional to v^2, and the lift force can be rewritten as

$$F_L = AC_L\frac{\rho}{2}v^2 \qquad (14.17)$$

The proportionality factor C_L is called the *lift coefficient.* In a few idealized cases, it can be predicted theoretically using advanced mathematical

Figure 14.15.   (*a*) Streamline flow about a wing. (*b*) Turbulence, loss of lift, and stalling result from large angles of attack. (F. Homann, *Forschung auf dem Gebiete des Ingenieurwesens,* 1936, courtesy VDI Verlag.)

Figure 14.16.   The air above the wing moves at a speed v_a, which is greater than the speed below the wing v_b. Both v_a and v_b are proportional to v.

techniques, but it is usually measured experimentally. The lift coefficient depends in a complicated way on the shape of the wing and on its *angle of attack* α, the angle between the wing and the direction of the airflow. When the angle of attack is small, the lift is approximately proportional to the angle of attack. However, if α is increased sufficiently, turbulence sets in, the lift diminishes, and a stall may result (Fig. 14.15*b*).

Equation 14.17 for the lift force is all we need to discuss flight qualitatively. For flight at constant altitude the lift force must equal the weight. Jet aircraft have high flight speeds, so the area A and the angle α can be made small, thereby reducing dissipative drag forces. However, this also has the effect of making the takeoff velocities quite high. In order to take off at moderate speeds, jets must make an abrupt increase in the angle of attack after reaching a critical speed (Fig. 14.17). This increases the lift coefficient and permits takeoff. Propeller planes, which fly at lower speeds, have proportionately larger wing areas as well as larger angles of attack. This makes the lift coefficient large enough to permit takeoff at lower speeds.

Figure 14.17.   The angle of attack is increased just before the jet takes off. (Courtesy Boeing.)

Figure 14.18.   During level flight the downstroke provides the greatest propulsive and lift force. The front of the wing is lower than the trailing edge during the sequence (*a*) through (*d*). During the upstroke the wing is pulled forward and up, nearly parallel to the plane of the wing. Little propulsive or lift force is obtained, but little flying speed is lost, and little energy is required to execute the motion. (Adapted from Alexander, *Animal Mechanics*.)

Because the wings of flying animals provide both propulsion and lift, their design and motion are complex. The lift force is still determined by $F_L = AC_L\rho v^2/2$ but A, C_L, and v vary during different phases of the wing motion (Fig. 14.18).

Scaling and Flight | Small birds, such as sparrows, and larger ones, such as ducks, take off and land quite differently. We can understand this by once again using the scaling model. We use the simplest form of scaling, which assumes that the volume or weight of a bird varies as the cube of a characteristic length l. (The more complicated scaling procedure of Chapter Nine, which employs the buckling strength criterion $l \propto r^{2/3}$, leads to results very similar to those we obtain here.)

The cross-sectional area A of a wing varies as l^2, so the lift force $F_L = AC_L\rho\, v^2/2$ varies as Av^2 or l^2v^2. In level flight, F_L must equal the weight of the bird, $w \propto l^3$. Equating these we have $l^2v^2 \propto l^3$, or

$$v \propto l^{1/2} \tag{14.18}$$

This result means that a large bird has a greater minimum flying speed than a small bird. A small bird can jump into the air and achieve flying speed with a beat or two of its wings. A large bird first builds up speed by taxiing along the ground or water (Fig. 14.19) or by descending from a high perch.

Figure 14.19.   A large bird builds up speed before taking off. (Horst Schafer/Photo Trends.)

246

Figure 14.20. The pterosaur. (*a*) The size scale. (*b*) An artist's reconstruction of the pterosaur hunting its prey.

It is interesting to compare two birds of greatly different sizes. The minimum flying speed for a swift is about 13 mi h^{-1}. An ostrich has a characteristic length about 25 times that of the swift, so from $v \propto l^{1/2}$, its minimum flying speed is $\sqrt{25} = 5$ times that of the swift or 65 mi h^{-1}. It is small wonder that the ostrich cannot fly.

The largest present-day flying animal is the albatross, which has a wingspan of 11 ft. Larger flying reptiles (pterosaurs) existed during the age of dinosaurs. The largest discovered to date had a wingspan of 51 feet! It is speculated that such large creatures could not take off except by scaling cliffs and soaring off them like gliders. Their continued stay aloft would then depend on their ability to locate rising air currents or columns. (Fig. 14.20).

EXERCISES ON SUPPLEMENTARY TOPICS
Section 14.8 | The Flight of Animals and Airplanes

14-43 An airplane with a weight of 9000 lb must reach 120 mi h^{-1} to take off. If the plane carries an additional 7000 lb, what is its minimum takeoff speed?

14-44 Why do very large airplanes usually require long runways?

14-45 The minimum flying speed of a 0.05 kg swift is 6 m s^{-1}. What is the minimum flying speed of a 3.2-kg goose?

14-46 The swift has a wingspread of about 0.25 m, while the pterosaur had a wingspread of about 16 m. The swift has a minimum flying speed of 6 m s^{-1}. (a) Using scaling, estimate the minimum flying speed of the pterosaur. (b) Comment on the validity of the scaling model for this case.

PROBLEMS ON SUPPLEMENTARY TOPICS

14-47 The power required for a hummingbird (or a helicopter) to hover is

$$\mathcal{P}_H = \sqrt{\frac{w^3}{2\rho A}}$$

where w is the weight of the bird, A is the area swept out by its wings as they swing back and forth, and ρ is the density of air. (a) Show that the right-hand side has the dimensions of power. (b) Can another expression of the form $w^a \rho^b A^c$ be found that has dimensions of power?

14-48 Using the formula for the power expended in hovering flight in Problem 14-47, show how this power scales with the characteristic length of a bird.

14-49 A hovering hummingbird of mass 3×10^{-3} kg has wings that sweep out an area of 3×10^{-3} m². (a) Using the formula in Problem 14-47, find the power expended at 20°C. (b) The bird's muscles have a mass of 0.75×10^{-3} kg. Compare their power output with the 80 W kg^{-1} maximum output of human muscle.

14-50 Apply the scaling law $l^3 \propto r^2$ to the flight of birds. How is Eq. 14.18 modified?

Figure 14.21.   Hancock Building in Boston, Massachusetts. Aerodynamic forces resulting from the asymmetric flow of air around an object are not restricted to wings. In 1973, a prominent feature of the Boston skyline was this unoccupied skyscraper with its windows covered with plywood (white rectangles) because of their unfortunate tendency to blow out during high winds. The pressure differences predicted by Bernoulli's equation provide a partial explanation of this phenomenon. (The Boston Globe.)

Additional Reading

R. McNeill Alexander, *Animal Mechanics,* University of Washington Press, Seattle, 1968. Chapters 5 and 6.

R.A.R. Tricker and B.J.K. Tricker, *The Science of Movement,* Mills and Boon Limited, London, 1967.

Arthur C. Guyton, *Circulatory Physiology: Cardiac Output and Its Regulation,* W.B. Saunders Co., Philadelphia and London, 1963. Chapter 6 discusses flowmeters and their use.

Scientific American articles:

Wallace O. Fenn, The Mechanism of Breathing, January 1960, p. 138.

J.V. Warren, The Physiology of the Giraffe, November 1974, p. 96.

Stanley J. Dudrick and Jonathan E. Rhoads, Total Intravenous Feeding, May 1972, p. 73.

D. James Baker, Jr., Models of Oceanic Circulation, January 1970, p. 114.

Suk Ki Hong and Hermann Rahn, The Diving Women of Korea and Japan, May 1967, p. 34.

Eric Denton, The Buoyancy of Marine Animals, July 1960, p. 118.

John P. Campbell, Vertical-Takeoff Aircraft, August 1960, p. 41.

John H. Storer, Bird Aerodynamics, April 1952, p. 24.

Carl Welty, Birds as Flying Machines, March 1955, p. 88.

Clarence D. Cone, Jr., The Soaring Flight of Birds, April 1962, p. 130.

David S. Smith, The Flight Muscles of Insects, June 1965, p. 76.

Alfred Gessow, The Changing Helicopter, April 1967, p. 38.

Felix Hess, The Aerodynamics of Boomerangs, November 1968, p. 124.

Vance A. Tucker, The Energetics of Bird Flight, May 1969, p. 70.

C.J. Pennynick, The Soaring Flight of Vultures, December 1973, p. 102.

Torkel Weis-Fogh, Universal Mechanisms for the Generation of Lift in Flying Animals, November 1975, p. 80.

CHAPTER FIFTEEN

VISCOUS FLUID FLOW

Real fluids in motion always exhibit some effects of frictional or viscous forces. Whenever the work done against these dissipative forces is comparable to the total work done on the fluid or to its mechanical energy change, Bernoulli's equation cannot be used. Bernoulli's equation always applies for fluids at rest, since viscous forces then have no effect, but for moving fluids an estimate of the viscous forces must be made. For example, Bernoulli's equation can adequately describe the flow of blood in the large main arteries of a mammal but not in the narrower blood vessels.

We begin this chapter by defining the *viscosity* of a fluid. We then examine the effects of viscous forces on the flow of a fluid in a tube. Viscosity is also responsible for the drag force experienced by a small object moving slowly through a fluid. Consequently viscous forces determine the velocities of molecules and small particles in solution in a centrifuge.

15.1 | VISCOSITY

Viscosity is readily defined by considering a simple experiment. Figure 15.1 shows two flat plates separated by a thin fluid layer. If the lower plate is held fixed, a force is required to move the upper plate at a constant speed. This force is needed to overcome the viscous forces due to the liquid and is greater for a highly viscous fluid, such as molasses, than for a less viscous fluid, such as water.

The force F is observed to be proportional to the area of the plates A and to the velocity of the upper plate Δv, and inversely proportional to the plate separation Δy:

$$F = \eta A \frac{\Delta v}{\Delta y} \tag{15.1}$$

The proportionality constant η (eta) is called the *viscosity*. From this equation the dimensions, denoted with brackets [ ], of the viscosity are

$$[\eta] = \left[\frac{F/A}{\Delta v/\Delta y}\right] = \left[\frac{MLT^{-2}/L^2}{LT^{-1}/L}\right] = [ML^{-1}T^{-1}] \tag{15.2}$$

Here M, L, and T stand for mass, length and time, respectively. The S.I. unit of viscosity is $1\ \text{kg m}^{-1}\,\text{s}^{-1} = 1\ \text{Pa s}$. The c.g.s. unit is called the *poise* (P); $1\ \text{P} = 0.1\ \text{Pa s}$.

Table 15.1 lists some typical viscosities. Usually as the temperature decreases, liquids become more viscous, as is readily observed in engine oil, honey, and other viscous fluids. By contrast, gases usually become less viscous as the temperature is lowered.

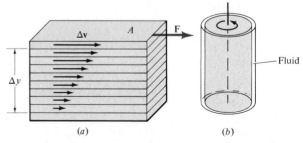

Figure 15.1. (*a*) Apparatus for measuring viscosity. (*b*) To prevent fluid leaking out, an actual measurement of viscosity uses rotating and fixed concentric cylinders.

TABLE 15.1

Typical values of viscosity in units of pascal-seconds (Pa s)

Temperature, °C	Castor Oil	Water	Air	Normal Blood*	Plasma*
0	5.3	1.792×10^{-3}	1.71×10^{-5}		
20	0.986	1.005×10^{-3}	1.81×10^{-5}	3.015×10^{-3}	1.810×10^{-3}
37	—	0.6947×10^{-3}	1.87×10^{-5}	2.084×10^{-3}	1.257×10^{-3}
40	0.231	0.656×10^{-3}	1.90×10^{-5}		
60	0.080	0.469×10^{-3}	2.00×10^{-5}		
80	0.030	0.357×10^{-3}	2.09×10^{-5}		
100	0.017	0.284×10^{-3}	2.18×10^{-5}		

*The relative viscosities (η/η_{water}) of blood and of plasma remain nearly constant for temperatures between 0°C and 37°C.

Because viscous forces are usually small, fluids are often used as lubricants to reduce friction. This is illustrated in the following example.

Example 15.1

An air track used in physics lecture demonstrations supports a cart that rides on a thin cushion of air 1 mm = 10^{-3} m thick and 0.04 m² in area. If the viscosity of the air is 1.8×10^{-5} Pa s, find the force required to move the cart at a constant speed of 0.2 m s⁻¹.

The force required is

$$F = \eta A \frac{\Delta v}{\Delta y}$$

$$= (1.8 \times 10^{-5} \text{ Pa s})(0.04 \text{ m}^2) \frac{(0.2 \text{ m s}^{-1})}{(10^{-3} \text{ m})}$$

$$= 1.44 \times 10^{-4} \text{ N}$$

This is a very small force and is consistent with the observation that an airtrack is nearly frictionless.

Laminar Flow

The fluid in contact with the moving plate in Fig. 15.1 has the same velocity as the plate, as indicated by the arrows. The fluid layer just below moves slightly more slowly, and each successive layer lags a bit more. The layer next to the stationary plate is at rest. This layered structure or *laminar flow* is the kind of streamline flow characteristic of viscous fluids at low velocities. When the fluid velocity is increased sufficiently, the flow changes its character and becomes turbulent.

Turbulent flow is frequently undesirable because it dissipates more mechanical energy than does laminar flow. Airplanes and cars are often designed so that the flow of air around them is as streamlined as possible. Also, nature has arranged that the flow in blood vessels is normally laminar rather than turbulent.

These ideas are discussed further in later sections.

15.2 | LAMINAR FLOW IN A TUBE; DIMENSIONAL ANALYSIS

Many interesting applications of the physics of fluids involve laminar flow in cylindrical tubes such as copper pipes or human arteries. In this section, we consider a formula for the flow rate called *Poiseuille's law*. It was first discovered experimentally by a physician, Jean Louis Marie Poiseuille (1799–1869), who was investigating the flow in blood vessels. Poiseuille's law relates the flow rate to the viscosity of the fluid, the pressure drop, and the radius and length of the tube. We use it extensively in Chapter Seventeen in discussing the flow in blood vessels.

Consider a fluid moving through a tube slowly enough so that there is no turbulence, and the flow is laminar. Just as in the case of the two flat surfaces discussed in the preceding section, the fluid in contact with the wall of the tube clings to it and is at rest. The thin, cylindrical layer of fluid adjacent to this stationary layer moves very slowly, and successive thin layers move at increasing velocities (Fig. 15.2). Hence

 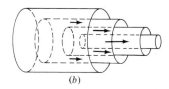

(a)	(b)

Figure 15.2. Laminar flow in a tube. The fluid in contact with the wall is at rest, and successive thin cylindrical layers move with increasing speeds. The fluid at the center has the maximum velocity. If initially the fluid in the tube is as in (a), then a short time later the layers will have moved as in (b).

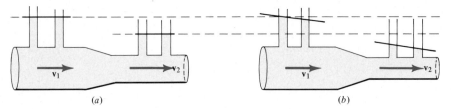

Figure 15.3. Pressures in a horizontal tube containing a moving fluid. (*a*) Ideal nonviscous fluid. In accordance with Bernoulli's equation, the pressure drops when the tube narrows and the average velocity increases. (*b*) Viscous fluid. Additional pressure drops occur due to the work done against viscous or frictional forces.

the fluid at the center has the maximum velocity, v_{max}. It is found that the average velocity $\bar{v}$ is half this, so $\bar{v} = \frac{1}{2}v_{max}$, and from the continuity equation the flow rate is $Q = A\bar{v} = \frac{1}{2}Av_{max}$.

In a horizontal tube with a constant cross section, the equation of continuity implies that the average velocity $\bar{v}$ remains the same, since $A\bar{v}$ must be constant. Nevertheless, the pressure drops as the fluid moves along the tube. This is because work is done against the viscous forces. If the cross section varies or if the tube is not horizontal, additional pressure changes arise in accordance with Bernoulli's equation (Fig. 15.3).

The pressure drop $\Delta P = P_1 - P_2$ along a horizontal tube of constant cross section (Fig. 15.4) is proportional to the viscous forces and hence to the average velocity of the fluid. Also, the pressure drop is proportional to the length l of the tube, since the work done against the viscous forces is proportional to the displacement. Thus $\Delta P \propto \bar{v}l$ or $\bar{v} \propto \Delta P/l$. The average velocity $\bar{v}$ and the flow rate $Q = A\bar{v}$ of the fluid are proportional to the pressure gradient, $\Delta P/l$. Note the analogy with the case of heat flow discussed in Chapter Twelve: the heat flow rate is proportional to the temperature gradient, $\Delta T/l$.

The average velocity depends on other factors in addition to the pressure gradient. It is easier to pump fluid through a wide tube than through a narrow one, and it is easier to pump a relatively nonviscous fluid than a highly viscous fluid. Thus, $\bar{v}$ must also depend on the tube radius R and the viscosity η. An exact

expression for $\bar{v}$ can be found by considering the forces acting on each thin cylindrical fluid layer. However, this approach requires mathematics beyond the level of this textbook. Accordingly, we obtain the main features of the result using a technique called *dimensional analysis*.

We have seen that the average velocity is proportional to the pressure gradient $\Delta P/l$ and also depends on the radius R and viscosity η. Apart from dimensionless numerical coefficients, these are the only quantities that can enter the formula for $\bar{v}$. Thus, we assume that the formula involves a factor of $\Delta P/l$ and unknown powers of R and η:

$$\bar{v} = \beta \frac{\Delta P}{l} R^a \eta^b \tag{15.3}$$

Here β (beta) is a dimensionless numerical factor, and the powers a and b must be chosen so that $\bar{v}$ has the correct dimensions.

The dimensions of η were given in Eq. 15.2; the dimensions of ΔP are those of $[F/A]$ or $[ma/A]$. Thus the physical quantities in Eq. 15.3 have the following dimensions:

$$[\bar{v}] = [LT^{-1}], \qquad [R] = [l] = [L]$$
$$[\Delta P] = [ML^{-1}T^{-2}], \qquad [\eta] = [ML^{-1}T^{-1}]$$

From Eq. 15.3, we require

$$[LT^{-1}] = \frac{[ML^{-1}T^{-2}]}{[L]}[L]^a[ML^{-1}T^{-1}]^b$$

To form the factor T^{-1}, we must choose $b = -1$. Then

$$[LT^{-1}] = [L^{-1}T^{-1}][L]^a$$

Note that the powers of M have cancelled. Now the choice $a = 2$ gives us the same dimensions on each

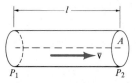

Figure 15.4. The pressure drop $P_1 - P_2$ is proportional to $\bar{v}l$.

side. With these choices for a and b, Eq. 15.3 for the average velocity becomes

$$\bar{v} = \beta \frac{\Delta P}{l} \frac{R^2}{\eta}$$

As we noted above, the exact analysis of this fluid flow problem can be carried out. It leads to an equation for $\bar{v}$ identical to this result, with β replaced by $\frac{1}{8}$. The analysis also shows that the average and maximum velocities are related by $\bar{v} = \frac{1}{2}v_{max}$, as we mentioned earlier. Hence, a flow meter placed at the center of a tube measures twice the average velocity. The average velocity and the flow rate $Q = A\bar{v} = (\pi R^2)\bar{v}$ can be written as

$$\bar{v} = \frac{\Delta P \, R^2}{8\eta l} \tag{15.4}$$

$$Q = \frac{\Delta P \, \pi R^4}{8\eta l} \tag{15.5}$$

The formula for Q is called Poiseuille's law. It indicates that high viscosity leads to low flow rates, which is reasonable. It also shows that the flow rate is proportional to R^4, which is somewhat surprising; the rate is extremely dependent on the radius of the tube. For example, this implies that in blood vessels, moderate adjustments in the radius can produce large changes in the flow rates. Thus, increasing R to $1.19R$ means a $(1.19)^4 =$ twofold increase in the flow rate. We can illustrate the results of this section by considering the blood flow in an artery.

Example 15.2

A large artery in a dog has an inner radius of 4×10^{-3} m. Blood flows through the artery at the rate of $1 \text{ cm}^3 \text{ s}^{-1} = 10^{-6} \text{ m}^3 \text{ s}^{-1}$. Find (a) the average and maximum velocities of the blood; (b) the pressure drop in a 0.1-m long segment of the artery; (c) the power required to maintain this flow in the segment.

(a) The average velocity is

$$\bar{v} = \frac{Q}{A} = \frac{Q}{\pi R^2}$$

$$= \frac{10^{-6} \text{ m}^3 \text{ s}^{-1}}{\pi (4 \times 10^{-3} \text{ m})^2}$$

$$= 1.99 \times 10^{-2} \text{ m s}^{-1}$$

The maximum velocity occurs at the center of the artery and is

$$v_{max} = 2\bar{v} = 2(1.99 \times 10^{-2} \text{ m s}^{-1})$$
$$= 3.98 \times 10^{-2} \text{ m s}^{-1}$$

(b) From Table 15.1, $\eta = 2.084 \times 10^{-3}$ Pa s. Thus the pressure drop is found from $\bar{v} = \Delta P \, R^2/8\eta l$, or

$$\Delta P = \frac{8\eta l \bar{v}}{R^2}$$

$$= \frac{8(2.084 \times 10^{-3} \text{ Pa s})}{(4 \times 10^{-3} \text{ m})^2}(0.1 \text{ m})(1.99 \times 10^{-2} \text{ m s}^{-1})$$

$$= 2.07 \text{ Pa}$$

(c) The power required is the net force $F = \Delta P \, A = \Delta P(\pi R^2)$ on the fluid segment times its average velocity $\bar{v}$. Thus the power is

$$\mathcal{P} = F\bar{v} = \Delta P(\pi R^2)\bar{v}$$
$$= (2.07 \text{ Pa})\pi(4 \times 10^{-3} \text{ m})^2(1.99 \times 10^{-2} \text{ m s}^{-1})$$
$$= 2.07 \times 10^{-6} \text{ W}$$

The metabolic rate of a dog is 10 W or greater, so the power expended in pumping blood through the large arteries is negligible. We see in Chapter Seventeen that most of the pressure drop and loss of energy due to viscous forces occurs in the small arterial subbranches and in the capillaries.

In the discussion above, we obtained the average velocity of the fluid in a tube using dimensional analysis. This method is very useful in many scientific problems where the exact theory is difficult to apply or is not even known, and we use it again in this chapter. However, it has limitations. For example, if several lengths enter a problem, it may not be possible to determine what combinations enter the formula, since $(l_1/l_2)^3$, $(l_1/l_2)^{1/2}$, and e^{-l_1/l_2} are all dimensionless quantities. In general, we must have only three dimensionally independent variables so that unique combinations can be formed which determine the units of mass, time, and length. Another drawback is that dimensionless constant factors such as β occasionally turn out to be numerically very large or small.

15.3 | TURBULENT FLOW

Poiseuille's law holds true only for laminar flow. However, often the flow is not laminar but turbulent; it is similar to the wake of a fast boat, with swirls and eddies. The ink lines we used to describe streamline flow in Chapter Fourteen would become mixed or blurred. Some examples of turbulent flow are shown in Fig. 15.5.

We noted earlier that the mechanical energy dissipated is typically much larger in turbulent flow than

Figure 15.5. (*a*) Laminar flow and (*b*) turbulent flow of water. (*c*) First laminar, then turbulent flow of cigarette smoke. (From Sears, Zemansky and Young, *University Physics*, fifth edition, Copyright © 1976, Addison Wesley, Reading, Mass.)

in laminar flow. Hence, it is often desirable to ensure that the flow does not become turbulent.

It is much harder to analyze turbulent flow than laminar flow. For example, Poiseuille's law for the laminar flow rate in a tube has no analog for turbulent flow. In practice turbulent flow is treated using a variety of empirical rules and relationships developed from extensive experimental studies.

In order to determine whether the flow is laminar and thus whether Poiseuille's law can be applied, we can make use of one of these empirical rules. It states that the value of a dimensionless quantity called the *Reynolds number* N_R determines whether the flow is turbulent or laminar. Consider a fluid of viscosity η and density ρ. If it is flowing in a tube of radius R and has an average velocity $\bar{v}$, then the Reynolds number is defined by

$$N_R = \frac{2\rho\bar{v}R}{\eta} \qquad \text{(flow in a tube of radius } R) \quad (15.6)$$

In *tubes*, it is found experimentally that if

$$
\begin{array}{ll}
N_R < 2000, & \text{flow is laminar} \\
N_R > 3000, & \text{flow is turbulent} \qquad (15.7) \\
2000 < N_R < 3000, & \text{flow is unstable (may change} \\
& \text{from laminar to turbulent or} \\
& \text{vice versa)}
\end{array}
$$

Thus whether the flow is laminar or turbulent is determined by a particular combination of variables; doubling the tube radius and halving the average velocity will leave the character of the flow unchanged.

The Reynolds number also indicates whether the flow around an obstacle, such as a ship hull or an airplane wing, will be turbulent or laminar. In general, the Reynolds number at which turbulence sets in depends significantly on the shape of the obstacle and is *not* given by Eqs. 15.6 or 15.7.

In the preceding section, we computed the flow rate in an artery using Poiseuille's law, which is valid only if the flow is laminar. We now compute the Reynolds number for that case to see whether the flow is laminar or turbulent.

Example 15.3

In Example 15.2, the radius of the artery is 4×10^{-3} m, the average velocity of the blood is 1.99×10^{-2} m s^{-1}, and the viscosity is 2.084×10^{-3} Pa s. Also, the density of the blood from Table 14.1 is 1.06×10^3 kg m^{-3}. Find the Reynolds number and determine whether the flow is laminar.

The Reynolds number is

$$N_R = \frac{2(1.06 \times 10^3 \text{ kg m}^{-3})}{2.084 \times 10^{-3} \text{ Pa s}}$$
$$\times (1.99 \times 10^{-2} \text{ m s}^{-1})(4 \times 10^{-3} \text{ m})$$
$$= 81$$

This is much less than 2000, so the flow is laminar.

In turbulent flow, some energy is dissipated as sound and some as heat. The noise associated with turbulent flow in arteries facilitates blood pressure measurements and makes possible the detection of some heart abnormalities. These topics are discussed further in Chapter Seventeen.

SUMMARY

Viscous or frictional forces in fluids must be considered whenever the work done against them is appreciable. Viscous forces are proportional to the viscosity of the fluid and to the velocity difference between parts of a fluid. If the velocity is low enough to avoid turbulence, the flow of a viscous fluid is laminar in structure, with a smooth variation of the velocity from layer to layer.

In a tube of radius R, the laminar flow rate is proportional to the pressure gradient and to R^4:

$$Q = \frac{\Delta P \, \pi R^4}{8 \eta l}$$

The velocity of the fluid is a maximum at the center of the tube and decreases gradually in successive lamina to zero at the walls.

Turbulence sets in when the fluid velocity is increased sufficiently. The Reynolds number, a dimensionless combination of variables, can be used to predict whether the flow will be laminar or turbulent.

Checklist

Define or explain:

viscous force
viscosity
poise
laminar flow
pressure gradient
dimensional analysis
Poiseuille's law
Reynolds number

EXERCISES

Use 1.2 kg m^{-3} for the density of air and 1000 kg m^{-3} for the density of water.

Section 15.1 | Viscosity

15-1 In Fig. 15.1, at what point in the fluid is its velocity half that of the upper plate? Explain.

15-2 Why do automobile manufacturers recommend the use of "multiviscosity" engine oil in cold weather?

15-3 An experimental car rides on an air cushion 0.06 m thick with an area of 15 m^2. How much power is expended against viscous forces at a speed of 20 m s^{-1} if the viscosity of the air is 1.8×10^{-5} Pa s?

15-4 The air track cart in Example 15.1 has a mass of 0.5 kg. If its initial speed is 1 m s^{-1}, estimate the change in its speed after it has traveled 2 m.

Section 15.2 | Laminar Flow in a Tube; Dimensional Analysis

15-5 A blood vessel of radius 10^{-3} m has a pressure gradient $\Delta P/l$ of 4.5 torr cm^{-1} = 600 Pa m^{-1}. (Assume laminar flow.) (a) What is the flow rate of blood at 37°C in the vessel? (b) What is the maximum velocity of the blood in the vessel?

15-6 An artery has an inner radius of 2×10^{-3} m. If the temperature is 37°C, the average velocity of the blood is 0.03 m s^{-1} and the flow is laminar, find (a) the maximum velocity; (b) the flow rate; and (c) the pressure drop in 0.05 m, if the artery is horizontal.

15-7 The pressure drop along a length of a horizontal artery is 100 Pa. The radius of the artery is 0.01 m, and the flow is laminar. (a) What is the net force on the blood in this portion of the artery? (b) If the average velocity of the blood is 1.5×10^{-2} m s^{-1}, find the power expended in maintaining the flow.

15-8 The radius of an artery is increased by a factor of 1.5. (a) If the pressure drop remains the same, what happens to the flow rate? (b) If the flow rate stays the same, what happens to the pressure drop? (Assume laminar flow.)

15-9   A hypodermic needle of length 0.02 m and inner radius 3×10^{-4} m is used to force water at 20°C into air at a flow rate of 10^{-7} m³ s⁻¹. (a) What is the average velocity of the water in the needle? (Assume laminar flow.) (b) What is the pressure drop necessary to achieve the flow rate?

Section 15.3 | Turbulent Flow

15-10   The average velocity of water at 20°C in a tube of radius 0.1 m is 0.2 m s⁻¹. (a) Is the flow laminar or turbulent? (b) What is the flow rate?

15-11   A water pipe of radius 0.02 m delivers water at a flow rate of 0.01 m³ s⁻¹ at 20°C. (a) What is the average velocity of the water? (b) Is the flow laminar or turbulent? (c) Is there enough information to determine the maximum velocity of the water in the pipe?

15-12   Calculate the Reynolds number for Exercise 15-9. Is the flow laminar as assumed?

15-13   (a) What is the greatest average velocity of blood flow at 37°C in an artery of radius 2×10^{-3} m if the flow is to remain laminar? (b) What is the corresponding flow rate Q?

15-14   (a) Draw an imaginary plane at right angles to the laminar flow in a cylinder. Is the pressure the same everywhere on this plane? Explain. (b) Does the pressure vary on this plane if the flow becomes turbulent? Why?

SUPPLEMENTARY TOPICS

15.4 | VISCOUS DRAG FORCES

Anyone who has held a hand out of a car window has discovered that an object moving through a fluid experiences a force that increases rapidly with the velocity. At very low velocities, this *drag force* results primarily from viscous forces that are proportional to the velocity v. At slightly higher but still relatively low speeds, the object accelerates the fluid moving around it, causing the force to vary approximately as v^2. In this section, we consider *low-speed* or *viscous drag forces*. "High-speed" drag is discussed in the next section.

Viscous drag forces result because the layer of fluid adjacent to an object is at rest with respect to the object. When the object moves through the fluid, this layer experiences a frictional force from the more rapidly moving layer next to it. Successive adjacent

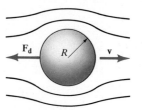

Figure 15.6. $\mathbf{v}$ is the velocity of the spherical object, and $\mathbf{F_d}$ is the viscous drag force. Also shown are the laminar flow lines.

layers near the object produce frictional forces on each other, and the net result is to retard the motion of the object through the fluid.

An exact derivation of the viscous drag force is very difficult and has been carried out only for the simplest cases. However, given the observed direct proportionality at low speeds of the force and the velocity, we can obtain the general low velocity form of the force law by again using dimensional analysis.

The basic condition under which the drag force has the form we will obtain is that the object moves very slowly through the fluid and the Reynolds number is very small. Figure 15.6 shows a spherical object of radius R moving with a small velocity v through a fluid of viscosity η and density ρ_0. To construct from these variables a force proportional to v, we assume that the drag force has the form $F_d = \phi R^a \eta^b \rho_0^c v$. Here, ϕ is a dimensionless numerical factor, and a, b, and c must be chosen so that F_d has the correct dimensions. The calculation from this point is very similar to that in Section 15.2. The result is that the viscous drag force on an object must have the form

$$F_d = \phi R v \eta \qquad (15.8)$$

Note that the density ρ_0 does not enter the result. This is not too surprising, since the fundamental viscous force law $F = \eta A \, \Delta v / \Delta y$ does not depend upon the fluid density.

This result is valid whenever the velocity is small enough. In terms of the Reynolds number for a sphere of radius R, the requirement is

$$N_R = \frac{\rho_0 v R}{\eta} < 1 \qquad \text{(sphere of radius } R) \qquad (15.9)$$

If $N_R \gtrsim 1$, the drag force becomes proportional to v^2. ($\gtrsim$ means "greater than or approximately equal to.") This occurs at velocities well below the onset of turbulence and is due, as noted earlier, to the kinetic energy imparted to the fluid.

For a sphere, ϕ is known to be exactly 6π, so

$$F_d = 6\pi R v \eta \qquad \text{(sphere)} \qquad (15.10)$$

This is called *Stokes' law.* For complex shapes, Eq. 15.8 can still be used if ϕ is determined by experiment. R must be interpreted as some characteristic length of the object, such as the average radius of a red blood cell.

It is sometimes possible to get information about the size or shape of a small object if the viscous drag force on the object is determined from its motion. This idea underlies some of the applications of the centrifuge, which are discussed in Section 15.6.

To illustrate the use of Stokes' law, we obtain the maximum or terminal velocity v_T for a small sphere of radius R and density ρ falling through a fluid of viscosity η and density ρ_0. Terminal velocity is reached when the drag force $\mathbf{F}_d$ exactly balances the weight $\mathbf{w}$ and the buoyant force $\mathbf{B}$ (Fig. 15.7).

The volume of the sphere is $V = \frac{4}{3}\pi R^3$, and its weight is $w = \rho g V$. The upward buoyant force is the weight of the fluid displaced $B = \rho_0 g V$. From Stokes' law, the viscous drag force at the terminal velocity v_T is $F_d = 6\pi R v_T \eta$. Thus terminal velocity is reached when $F_d = w - B$, or

$$6\pi R v_T \eta = \tfrac{4}{3}\pi R^3 \rho g - \tfrac{4}{3}\pi R^3 \rho_0 g$$

Solving for v_T, the terminal velocity of the sphere is

$$v_T = \frac{2}{9}\frac{R^2}{\eta}g(\rho - \rho_0) \qquad (15.11)$$

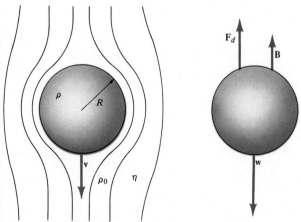

Figure 15.7. A sphere falling through a viscous fluid. Terminal velocity is reached when $F_d = w - B$.

If v_T and R are measured, this equation provides a way of finding the viscosity of the fluid. The terminal velocity of a dust particle is found in the next example.

Example 15.4

(a) What is the terminal velocity in air at $20°C$ of a spherical dust particle of radius 10^{-5} m and density 2×10^3 kg m^{-3}? (b) What is the Reynolds number at terminal velocity? (c) Find the drag force at terminal velocity.

(a) At $20°C$, the density of air is 1.22 kg m^{-3}, which is negligible compared to that of the particle. Using $\eta = 1.81 \times 10^{-5}$ Pa s from Table 15.1 and assuming the velocity is low enough that Stokes' law applies,

$$v_T = \frac{2}{9}\frac{R^2}{\eta}g(\rho - \rho_0)$$

$$= \frac{2}{9}\frac{(10^{-5}\text{ m})^2}{(1.81 \times 10^{-5}\text{ Pa s})}$$

$$\times (9.8\text{ m s}^{-2})(2 \times 10^3\text{ kg m}^{-3})$$

$$= 2.41 \times 10^{-2}\text{ m s}^{-1}$$

(b) The Reynolds number is

$$N_R = \frac{\rho_0 v R}{\eta}$$

$$= \frac{(1.22\text{ kg m}^{-3})(2.41 \times 10^{-2}\text{ m s}^{-1})(10^{-5}\text{ m})}{(1.81 \times 10^{-5}\text{ Pa s})}$$

$$= 0.0162$$

This is much less than 1, so Stokes' law and Eq. 15.11 for v_T apply.

(c) The drag force is

$$F_d = 6\pi R v \eta$$
$$= 6\pi(10^{-5}\text{ m})(2.41 \times 10^{-2}\text{ m s}^{-1})(1.81 \times 10^{-5}\text{ Pa s})$$
$$= 8.23 \times 10^{-11}\text{ N}$$

Eq. 15.11 for the terminal velocity is valid only for very small objects such as dust particles in air or macromolecules in solution. For larger objects the terminal velocity predicted by this equation corresponds to a Reynolds number much larger than 1, so that Stokes' law is not applicable. However, Eq. 15.11 is useful for finding the motion of molecules or small particles in a centrifuge, as is discussed in Section 15.6.

15.5 | "HIGH-SPEED" DRAG FORCES

At the beginning of the preceding section, we noted

that as the speed of an object moving in a fluid increases so that $N_R = \rho_0 vR/\eta \gtrsim 1$, the drag force is no longer proportional to v. Instead, it is approximately proportional to v^2. The term "high speed" is somewhat misleading, since the v^2 dependence is observed whenever N_R is appreciably more than 1, and this happens at surprisingly low speeds. For example, for a sphere of radius 1 cm in air, N_R is 1 at $v = 1.5$ mm s^{-1}. Thus, high-speed drag applies to essentially all problems involving the motion of macroscopic objects.

The formula for the high-speed drag force on a sphere in a fluid can again be found using dimensional analysis. The force is assumed to be proportional to v^2 and to unknown powers of the radius of the sphere R, the viscosity of the fluid η, and the density of the fluid ρ_0. The calculation is left as a problem. The result is that F_d is proportional to $R^2 \rho_0 v^2$, or that

$$F_d = C_D A \frac{\rho_0 v^2}{2} \qquad (15.12)$$

Here factors are grouped for convenience: $A = \pi R^2$ is the cross-sectional area of the moving sphere, $\rho_0 v^2/2$ is the kinetic energy per unit volume of fluid with velocity v, and C_D is the *drag coefficient* that must be obtained from measurements. The viscosity η is absent, which is consistent with the idea that the drag results from accelerating the fluid moving around the sphere and not from viscous effects.

The result given above holds true for objects of any shape. Figure 15.8 gives experimental values of C_D versus the Reynolds number for a long cylinder moving perpendicular to its axis. We see that C_D is nearly constant; it varies between 0.3 and 3.0 as N_R varies from 10 to 10^6. This occurs despite the fact that as N_R increases, the motion of the fluid around the cylinder changes from laminar to turbulent when $N_R \simeq 100$.

Equation 15.12 can be used to obtain the terminal velocity of a macroscopic falling object. The result for an object of cross-sectional area A, length L, and mass ρAL, is

$$v_T = \sqrt{\frac{(\rho - \rho_0)}{\rho_0} \frac{2gL}{C_D}} \qquad (15.13)$$

Note that the cross-sectional area does not appear in the result. This derivation and numerical examples are left for the exercises and problems.

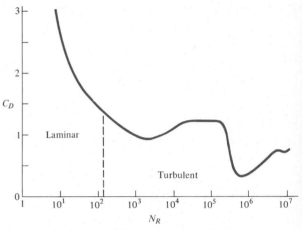

Figure 15.8. Drag coefficients versus Reynolds number for a long cylinder moving perpendicular to its axis.

15.6 | CENTRIFUGATION

Centrifuges use the very large accelerations experienced by rapidly rotating objects to perform many tasks in biological and medical laboratories. For example, centrifuges can separate large and small particles and molecules. The speed of a molecule in a centrifuge is determined by the viscous drag force and the molecular mass. Accordingly, measurement of this speed, together with other data, can be used to find the mass of the molecule.

A centrifuge consists basically of a sample holder that can be rotated (Fig. 15.9). If the sample rotates at a radius r and an angular velocity ω, its centripetal acceleration is $a_r = \omega^2 r$. We saw in Chapter Five that the effective weight of a rotating object of mass m is $\mathbf{w} = m(\mathbf{g} - \mathbf{a}_r)$. Since a_r may be up to $300,000g$ and is always much greater than g, the effective weight is given to a good approximation by $\mathbf{w}^e = -m\mathbf{a}_r$. Thus the sample will behave as though it were on a planet where $w^e = mg^e$ and the gravitational acceleration has a magnitude $g^e = \omega^2 r$. For example, molecules denser than a solvent will *sediment* or settle to the bottom of the holder at a much greater terminal or *sedimentation* velocity than in a stationary solution.

The sedimentation velocity v_s of particles experiencing a "gravitational acceleration" g^e is determined by their effective weight w^e, the buoyant force B due to the fluid, and the low speed or viscous drag force F_d (Fig. 15.10). If the particles are spherical,

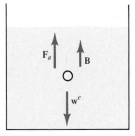

Figure 15.10. Forces on a particle in a centrifuge. The particle moves downward with a sedimentation velocity v_s.

$$F_d = fv \qquad (15.14)$$

Here f is called the *frictional coefficient*. We will see shortly that f can be related to the diffusion constant introduced in Chapter Eleven. If the particles have mass m and volume V, their density is $\rho = m/V$, and their effective weight is $w^e = mg^e$. If the density of the fluid is ρ_0, the buoyant force is $B = \rho_0 g^e V = (\rho_0/\rho)mg^e$. The forces are balanced at a velocity v_s if $F_d = w_e - B$. Thus

$$fv_s = mg^e - \frac{\rho_0}{\rho} mg^e$$

or

$$v_s = \frac{mg^e}{f}\left(1 - \frac{\rho_0}{\rho}\right) \qquad (15.15)$$

This is the sedimentation velocity for molecules in a solution if the acceleration is $g^e = \omega^2 r$.

This result can be used in various ways. If the quantities on the right side are known, the sedimentation velocity can be calculated, and one can determine how long it will take for the material to collect at the bottom of the sample holder (Fig. 15.11). Alternatively, the sedimentation velocity can be measured by shining light through the spinning sample and exploiting the fact that proteins and many other materials absorb light at specific wavelengths. This velocity measurement then provides information about the other variables in Eq. 15.15. Also, the frictional coefficient can be inferred from diffusion experiments, and the densities are readily measured, so sedimentation velocities can provide a determination of molecular masses. If several molecules of different sizes are present, v_s will be different for each. This permits various components of a mixture to be identified.

It is possible to find a relationship among the fric-

Coupling to drive motor

Sample cell

Counter weight

Figure 15.9. A refrigerated centrifuge. Rotational speeds of up to 6000 revolutions per minute corresponding to 6835 g are achieved with this unit. Other types of centrifuges achieve speeds of 60,000 revolutions per minute and 500,000 g. (Photo courtesy of Beckman Instruments, Inc.)

Eq. 15.11, which was obtained from Stokes' law, can be used with g replaced by g^e to find this velocity. However, another expression for v_s is often more useful.

To obtain this expression, we rewrite the low speed drag force on a particle with a velocity v as

$t = 0$ $t = 1000 \, s$ $t = 2000 \, s$

(a)   (b)   (c)   (d)

Figure 15.11. (a), (b), and (c). Molecules sedimenting in a centrifuge at a velocity $v_s = 10^{-3}$ cm s^{-1}. A sharp boundary between the solution and pure solvent moves downwards with velocity v_s. (d) In a mixture of two kinds of solutes, the larger molecules (black dots) will sediment faster than the smaller ones (colored dots).

tional coefficient, the diffusion constant, and the temperature. This relation can only be derived using mathematics beyond the scope of the textbook. However, we can give a rough idea of the argument as follows. From Chapter Eleven, molecules travel randomly because of their thermal energy. In a time t, the distance they travel on the average is $x^2 = 2Dt$, where D is the diffusion constant. If the velocity of the molecules is v, then $t = x/v$ and

$$D = \frac{xv}{2} \qquad (15.16)$$

We can relate x to the frictional coefficient as follows. Suppose that a large molecule in a solution has a velocity v due to collisions with the molecules of the solvent. Its motion is opposed by the drag force $F_d = fv$, and it will come to rest in a distance x. The average velocity is approximately $v/2$, so the work done against the drag force by the molecule is the average force times the distance, $(fv/2)x$. Equating this to the original kinetic energy $\frac{1}{2}mv^2$, we find $x = mv/f$. Thus, Eq. 15.16 becomes

$$D = \frac{mv^2}{2f}.$$

Now, if the temperature of the solution is T, the average value of the kinetic energy $\frac{1}{2}mv^2$ is approximately equal to $k_B T$, so we finally obtain

$$D = \frac{k_B T}{f} \qquad (15.17)$$

This is the same result as obtained by more rigorous treatments. We see in the following examples how it can be used with the expression for the sedimentation velocity.

Example 15.5

Hemoglobin has a diffusion constant in water of 6.3×10^{-11} m^2 s^{-1}, a density of 1.35×10^3 kg m^{-3}, and a molecular mass of 68,000 amu. (a) Find its frictional coefficient at $27°C = 300$ K. (b) If it is in a centrifuge with a centripetal acceleration of 10^5 g, find its sedimentation velocity.

(a) The frictional coefficient is

$$f = \frac{k_B T}{D} = \frac{(1.38 \times 10^{-23} \, \text{J K}^{-1})(300 \, \text{K})}{6.3 \times 10^{-11} \, \text{m}^2 \, \text{s}^{-1}}$$

$$= 6.57 \times 10^{-11} \, \text{N s m}^{-1}$$

(b) In kilograms, the mass of the molecule is $(68,000 \, \text{amu})(1.66 \times 10^{-27} \, \text{amu kg}^{-1}) = 1.129 \times 10^{-22}$ kg. The density of water is 10^3 kg m^{-3}. Thus the sedimentation velocity is

$$v_s = \frac{mg^e}{f}\left(1 - \frac{\rho_0}{\rho}\right)$$

$$= \frac{(1.129 \times 10^{-22} \, \text{kg})(10^5)(9.8 \, \text{m s}^{-2})}{6.57 \times 10^{-11} \, \text{N s m}^{-1}}\left(1 - \frac{1}{1.35}\right)$$

$$= 4.37 \times 10^{-7} \, \text{m s}^{-1}$$

At this velocity the molecules will travel 3.7 cm in a 24-hour period. We can see why it is quite common to run a centrifuge for many hours.

Example 15.6

A protein has a density of 1.35×10^3 kg m^{-3} and a diffusion constant in water at 300 K of 2×10^{-11} m^2 s^{-1}. When the acceleration in the centrifuge is 10^6 m s^{-2}, the sedimentation velocity of the protein is measured as 10^{-6} m s^{-1}. Find the molecular mass of the protein.

Solving Eq. 15.15 for the molecular mass gives

$$m = \frac{v_s f}{g^e}\frac{1}{1 - \rho_0/\rho}$$

Using $D = k_B T/f$ or $f = k_B T/D$, this becomes

$$m = \frac{v_s k_B T}{g^e D} \frac{1}{1 - \rho_0/\rho}.$$

With $\rho_0 = 10^3$ kg m^{-3} for water and the values given,

$$m = \frac{(10^{-6} \text{ m s}^{-1})(1.38 \times 10^{-23} \text{ J K}^{-1})(300 \text{ K})}{(10^6 \text{ m s}^{-2})(2 \times 10^{-11} \text{ m}^2 \text{ s}^{-1})(1 - 1/1.35)}$$

$$= 7.98 \times 10^{-22} \text{ kg} \left(\frac{1 \text{ amu}}{1.66 \times 10^{-27} \text{ kg}}\right)$$

$$= 481,000 \text{ amu}$$

The preceding example illustrates how sedimentation velocity measurements can be used to find the molecular mass if the diffusion constant is known. Diffusion constants are measured by optically observing the concentration changes that occur when a sample is introduced into a fluid. Thus, diffusion and centrifugation experiments taken together are a powerful tool for studying biological molecules.

EXERCISES ON SUPPLEMENTARY TOPICS

Section 15.5 | Viscous Drag Forces

15-15 A spherical blood cell of radius 5×10^{-6} m and density 1.3×10^3 kg m^{-3} is in water at 37°C. What is its terminal velocity? (Assume Stokes' law holds true.)

15-16 A large spherical molecule has a radius of 2×10^{-8} m and a density of 1.5×10^3 kg m^{-3}. (a) What is its terminal velocity when falling in water at 20°C? (b) What is the maximum velocity at which Stokes' law holds true?

15-17 For spherical dust particles of density 3×10^3 kg m^{-3}, find the maximum radius for which Stokes' law can be used to find the terminal velocity (a) in air at 20°C, and (b) in water at 20°C.

15-18 The terminal velocity of a spherical oil droplet falling in air at 20°C is 2×10^{-7} m s^{-1}. What is the radius of the spherical droplet if its density is 930 kg m^{-3}? (Assume that Stokes' law holds true.)

Section 15.5 | "High-Speed" Drag Forces

15-19 A spherical pebble of radius 0.04 m and density 3×10^3 kg m^{-3} is dropped into a pond. Estimate its terminal velocity assuming the drag coefficient is 1.

15-20 Estimate the force a person experiences standing in a wind blowing at 20 m s^{-1}, assuming the drag coefficient is 1.

15-21 A fish swimming in water at a velocity of 0.3 m s^{-1} experiences a drag force proportional to the velocity squared. The cross-sectional area presented to the water is 2.2×10^{-3} m^2. (a) What is the drag force on the fish if the drag coefficient C_D is 1? (b) What is the power expended by the fish against the drag force?

15-22 A baseball has a mass of 0.149 kg and a Ping Pong ball has a mass of 3×10^{-3} kg. Their radii are 0.037 m and 0.018 m respectively. Both are thrown in air with initial horizontal velocities of 15 m s^{-1}. (a) What is the ratio of drag forces on the two balls? (b) What is the initial acceleration of each ball due to the drag force, assuming $C_D = 1$?

15-23 A baseball has a mass of 0.149 kg and a radius of 0.037 m. (a) Calculate its terminal velocity in air at 0°C using Eq. 15.11. (b) Calculate the terminal velocity using Eq. 15.13 assuming $C_D = 1$. (c) Explain the significance of the difference in the results in parts (a) and (b).

Section 15.6 | Centrifugation

15-24 A protein molecule has a mass of 10^5 amu and a density of 1.35×10^3 kg m^{-3}. It is in a centrifuge with an acceleration of 2×10^5 g. If it is in water, find (a) its effective weight, and (b) the buoyant force.

15-25 The sample in a centrifuge is 0.1 m from the axis of rotation. If its centripetal acceleration is 200,000 g, how many times per minute does the centrifuge rotate?

15-26 Bushy stunt virus has a molecular mass of 1.06×10^7 amu, a diffusion constant in water of 1.15×10^{-11} m^2 s^{-1}, and a density of 1.35×10^3 kg m^{-3}. (a) Find its frictional coefficient in water at 300 K. (b) How much time is required for it to seciment 10^{-2} m if the acceleration in the centrifuge is $10^5 g$?

15-27 A red blood cell can be approximated by a sphere of radius 2×10^{-6} m and density 1.3×10^3 kg m^{-3}. How long will it take for it to sediment 1 cm $= 10^{-2}$ m in blood at 37°C (a) under the earth's gravitational acceleration; (b) in a centrifuge with an acceleration of $10^5 g$? (The density of blood is 1.056×10^3 kg m^{-3}.)

15-28 Tobacco mosaic virus has a diffusion constant in water of 0.53×10^{-11} m^2 s^{-1} and a density of 1370 kg m^{-3}. If the acceleration in a centrifuge is $2 \times 10^5 g$, the sedimentation velocity at 300 K is observed to be 3.7×10^{-5} m s^{-1}. What is the molecular mass of the tobacco mosaic virus?

15-29 Myoglobin has a diffusion constant in water at 300 K of 1.13×10^{-10} m^2 s^{-1}. Its density is 1350 kg m^{-3}. (a) What is its frictional coefficient? (b) Its sedimentation velocity is 2.04×10^{-7} m s^{-1} when the acceleration in the centrifuge is 10^6 m s^{-2}. What is the molecular mass of myoglobin?

PROBLEMS ON SUPPLEMENTARY TOPICS

15-30   Red blood cells are usually not spherical in shape. What is the product ϕR in Eq. 15.8 for a cell of mass 10^{-12} kg if the terminal velocity in water at 37°C is 10^{-5} m s^{-1}? (The density of the cell is 1.3×10^3 kg m^{-3}.)

15-31   Estimate the maximum velocity at which Stokes' law applies for a small fish in water.

15-32   Estimate the terminal velocity of an 800-N man falling through air (a) if he falls in a tuck position; (b) if he spreads his body and arms out. (Assume the drag coefficient C_D is 1.)

15-33   Derive Eq. 15.8 using dimensional analysis.

15-34   Derive Eq. 15.12 using dimensional analysis.

15-35   When the drag force is proportional to the velocity squared, show that the terminal velocity of an object of density ρ falling through a fluid of density ρ_0 is given by Eq. 15.13.

15-36   For a spherical molecule of radius R in a fluid with viscosity η, the drag force is given both by $F_d = 6\pi\eta Rv$ (Stokes' law) and by $F_d = fv$. (a) What is the frictional coefficient f in water at 20°C of a spherical molecule with radius 10^{-8} m?

(b) What is the diffusion constant for this molecule?

15-37   Nucleohistone has a mass of 2.1×10^6 amu and a density of 1520 kg m^{-3}. (a) What is its volume? (b) If this molecule is spherical, what is its radius? (c) Assuming the molecule is spherical, use Stokes' law to find the frictional coefficient in water at 20°C = 293 K. (d) The diffusion coefficient in water at 293 K is $D = 9.3 \times 10^{-12}$ m^2 s^{-1}. Find the frictional coefficient from D. (e) Is this molecule spherical? Explain.

Additional Reading

Richard B. Setlow and Ernest C. Pollard, *Molecular Biophysics*, Addison-Wesley, Reading, 1962. Chapter 4 discusses centrifugation.

R. McNeill Alexander, *Animal Mechanics*, University of Washington Press, Seattle, 1968. Chapters 5 and 6.

E. M. Purcell, Life at Low Reynolds Numbers. *American Journal of Physics*, vol. 45, 1977, p. 3.

Scientific American articles:

Jesse W. Beams, Ultrahigh-Speed Rotation, April 1961, p. 134.

F.H. Harlow and J.E. Fromm, Computer Experiments in Fluid Dynamics, March 1965, p. 104.

R.W. Stewart, The Atmosphere and the Ocean, September 1969, p. 76.

D. James Baker, Jr., Models of Oceanic Circulation, January 1970, p. 114.

R.C. Chanaud, Aerodynamic Whistles, January 1970, p. 40.

Victor P. Starr and Norman E. Gant, Negative Viscosity, July 1970, p. 72.

Norman C. Chigier, Vortexes in Aircraft Wakes, March 1974, p. 76.

CHAPTER SIXTEEN

COHESIVE FORCES IN LIQUIDS

When we consider pulling on a substance until it breaks, we usually think of solids. However, liquids also have a strong tendency to remain intact. For example, if pure water with no air dissolved in it is pressed between two smooth plates, enormous forces are necessary to separate the plates.

As in solids, the cohesiveness of liquids results from the attractions among the molecules. Because of these attractions, liquids have well-defined surfaces which, like stretched membranes or rubber sheets, tend to have a minimum surface area. Ripples in a quiet, smooth pool are suppressed because they require an increase in surface area. Aquatic insects are able to move on water surfaces because their weight is opposed by the resistance of the surface to deformation.

In addition to the attractive forces among themselves, the molecules of a liquid experience attractive or repulsive interactions with the molecules of other substances. Thus, water rises near a vertical glass surface, while the opposite is true for mercury.

The cohesive properties of liquids can be changed by the addition of small amounts of other substances. For example, molecules in oils are *hydrophobic* (water-hating), and oils will not dissolve in pure water. Molecules in soaps and detergents have both hydrophobic and *hydrophilic* (water-loving) portions. The hydrophilic part attaches itself to the water surface, and the hydrophobic portion surrounds oil or grease. This aids in the solution and removal of oil and grease.

In this chapter, we consider some effects of the cohesiveness of liquids. Topics discussed include the rise of liquids in narrow tubes or *capillaries*, the formation of bubbles, and the rise of sap in trees.

16.1 | SURFACE TENSION

One way to observe the effects of surface tension is to dip the apparatus shown in Fig. 16.1 into a liquid. This consists of a U-shaped wire, a slide wire of weight w_1, and a suspended weight w_2. A thin film of liquid fills the enclosed area between the wires. If the total weight $w = w_1 + w_2$ is properly chosen, the two surfaces of the film exert a force equal and opposite to the weight and the slide wire remains stationary. Thus, in this situation the force F due to the surface tension is equal in magnitude to w.

The *surface tension* γ is defined as the force per unit length exerted by *one* surface. Then, if the straight wire of Fig. 16.1 has a length l, the net force upward

Figure 16.1. (*a*) A liquid film filling the area A exerts a force F, which balances $w = w_1 + w_2$. (*b*) A cross-sectional view of the sliding straight wire and the film attached to it. The film has two surfaces.

due to the two surfaces is $F = 2\gamma l$. Thus the surface tension is

$$\gamma = \frac{F}{2l} \qquad (16.1)$$

Representative surface tensions are given in Table 16.1.

One may ask why the forces in Fig. 16.1 are attributed to the surfaces. The point is that the liquid film acts somewhat differently than a rubber sheet. If the apparatus is in equilibrium and the straight wire is moved up or down, it is in equilibrium again at each position, even though the film thickness has changed. Thus the thickness of the film is not important. If the film is stretched, molecules from the bulk of the fluid move to the surface, and the surface area is increased. Since the force remains the same while the bulk thickness varies, the force must be attributed to the surface of the film.

We can understand the reason for this behavior with the aid of a microscopic picture of the boundary between a liquid and its vapor (Fig. 16.2). A molecule in the interior of the liquid experiences attractive forces due to all its neighbors which lowers its potential energy. A molecule at the surface does not have as many neighbors, and its potential energy is not so low. Consequently, the molecules at the surface arrange themselves so that they have as many neighbors as possible. In this process, they minimize the

TABLE 16.1

The surface tension for representative liquids in contact with air

Liquid	Surface Tension ($N\,m^{-1}$)	Temperature ($0°C$)
Ethyl alcohol	2.23×10^{-2}	20
Olive oil	3.20×10^{-2}	20
Glycerine	6.31×10^{-2}	20
Water	7.56×10^{-2}	0
	7.28×10^{-2}	20
	6.62×10^{-2}	60
	5.89×10^{-2}	100
Mercury	0.465	20
Silver	0.800	970
Gold	1.000	1070
Copper	1.100	1130
Oxygen	1.57×10^{-2}	-193
Neon	5.15×10^{-3}	-247

Figure 16.2. The molecules of a substance in the liquid and the coexistant vapor phase. The attractive forces on molecules near the surface and in the interior are denoted by arrows.

surface area and the potential energy and produce the observed surface tension. The following example shows that these surface forces are relatively small.

Example 16.1

The U-shaped loop in Fig. 16.1 is dipped into water at 20°C. The slide wire is 0.1 m long and has a mass $m_1 = 1\,g = 10^{-3}\,kg$. (a) How large is the surface tension force? (b) If the wire is in equilibrium, how large is the mass m_2 suspended from the wire?

(a) According to Table 16.1, $\gamma = 7.28 \times 10^{-2}\,N\,m^{-1}$. Thus the force due to the two surfaces of the film of water is

$$F = 2\gamma l = 2(7.28 \times 10^{-2}\,N\,m^{-1})(0.1\,m)$$
$$= 1.46 \times 10^{-2}\,N$$

(b) In equilibrium the force F exerted by the film must equal the total weight $m_1 g + m_2 g$. Solving for m_2,

$$m_2 = \frac{F}{g} - m_1 = \frac{1.46 \times 10^{-2}\,N}{9.8\,m\,s^{-2}} - 10^{-3}\,kg$$
$$= 0.49 \times 10^{-3}\,kg = 0.49\,g$$

Thus the surface forces support the weight of a total mass, $m_1 + m_2$, which is approximately 1.5 g.

16.2 | CONTACT ANGLES AND CAPILLARITY

The surface of a liquid in contact with a solid surface forms an angle with respect to the solid surface (Fig. 16.3). The *contact angle* θ is determined by the competition between the liquid-liquid molecular forces and the liquid-solid forces and depends on the particular solid and liquid involved (Table 16.2). It also depends on how smooth and clean the solid surface is.

If θ is less than 90°, the liquid surface appears as in Fig. 16.3a and will rise in a narrow tube. If θ is greater than 90°, the liquid appears as in Fig. 16.3b and will

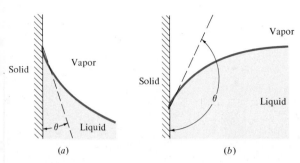

Figure 16.3. The contact angle θ for two liquids in contact with a solid surface. (a) $\theta < 90°$; (b) $\theta > 90°$.

Figure 16.4. Liquid in a tube of radius r with a contact angle θ rises to a height h. The force on a small segment of the liquid in contact with the tube wall is $\mathbf{f}$.

be depressed in a tube. If $\theta = 90°$, the liquid will neither rise nor fall. The rise or fall of a liquid in a narrow tube or capillary is referred to as *capillarity* or *capillary action*.

We will now see how the contact angle determines the height of a liquid in a column. Figure 16.4 shows a liquid of density ρ in a tube of radius r. The contact angle is less than 90°. The net vertical force F_{up} is the vertical component of the surface tension times the length l of the liquid surface in contact with the tube; l equals the circumference $2\pi r$. Thus

$$F_{up} = 2\pi r \gamma \cos \theta \qquad (16.2)$$

The volume of the column of liquid up to the bottom of the curved liquid surface or *meniscus* is $V = \pi r^2 h$, and its weight is $w = \rho g V = \rho g \pi r^2 h$. The liquid rises until $F_{up} = w$, or

$$2\pi r \gamma \cos \theta = \rho g \pi r^2 h$$

This gives, for the height of the liquid in a capillary,

$$h = \frac{2\gamma \cos \theta}{\rho g r} \qquad (16.3)$$

This result has several notable features. If $\theta = 90°$, then $h = 0$, and the fluid neither rises nor is depressed. If θ is greater than 90°, $\cos \theta$ is negative and

TABLE 16.2

The contact angles for several liquid-solid interfaces

Interface	Contact Angle
Water—clean glass	0°
Ethyl alcohol—clean glass	0°
Mercury—glass	140°
Water—silver	90°
Water—paraffin	107°
Methylene iodide—Pyrex glass	30°

so is h. This means that the liquid is depressed. The height h is proportional to γ; the greater the surface tension, the greater the capillary effect. Conversely, h depends on $1/r$, so the effect is most dramatic when r is small.

Although Eq. 16.3 was derived for a circular cross section, capillary rise occurs in tubelike structures or connected channels of any geometry. The height still depends on $\gamma/\rho g r$, where r is some typical dimension, although the overall numerical coefficient changes. For example, water will be easily absorbed by fine, porous fabric if the contact angle is less than 90°. We apply this result to trees in the next example.

Example 16.2

The sap in trees, which consists mainly of water in summer, rises in a system of capillaries of radius $r = 2.5 \times 10^{-5}$ m. The contact angle is 0°. The density of water is 10^3 kg m^{-3}. What is the maximum height to which water can rise in a tree at 20°C?

From Table 16.1, $\gamma = 7.28 \times 10^{-2}$ N m^{-1}. With $\cos \theta = 1$, the height the sap rises is

$$h = \frac{2\gamma \cos \theta}{\rho g r}$$

$$= \frac{2(7.28 \times 10^{-2} \text{ N m}^{-1})}{(10^3 \text{ kg m}^{-3})(9.8 \text{ m s}^{-2})(2.5 \times 10^{-5} \text{ m})}$$

$$= 0.594 \text{ m}$$

Since trees grow to heights of many meters, capillary action cannot account for the supply of water to the top of a tree.

Figure 16.5a shows the shape of a drop of water on paraffin. A wetting agent, a molecule with hydrophobic and hydrophilic portions, would cause the droplet on the paraffin to appear as shown in Fig. 16.5b. The hydrophilic part of the molecule attaches itself to the

Figure 16.5. Water droplets on paraffin. (*a*) Pure water. (*b*) Wetting agent added.

water surface, and the hydrophobic part stays away from the water surface but is attracted to the paraffin surface. Waterproofing agents have the opposite effect. They cause the angle of contact to be increased, and droplets are less likely to penetrate narrow fabric pores.

16.3 | LAPLACE'S LAW

Laplace's law relates the pressure difference across a closed elastic membrane or liquid film to the tension in the membrane or film. Its specific form depends on the shape of the closed surface.

We begin by considering a spherical membrane or rubber balloon filled with fluid at a pressure P_i. The pressure outside the wall is P_0 and the wall can exert a force per unit length or wall tension γ (Fig. 16.6). The lungs and the ventricles of the heart may be approximated by such a model.

The total force to the left on the hemisphere of Fig. 16.6*b* due to the wall tension is the product of γ and the circumference $2\pi r$ of the hemisphere, or $2\pi r\gamma$. The forces due to the pressures are perpendicular to the surface at each point. All the components cancel except those to the right, which add up to the pressure difference $P_i - P_0$ times the *projected area* πr^2. The two forces must balance in equilibrium, so

$$2\pi r\gamma = (P_i - P_0)\pi r^2$$

or

$$P_i - P_0 = \frac{2\gamma}{r} \qquad \text{(spherical membrane)} \quad (16.4)$$

This is called *Laplace's law for a spherical membrane.* It is named after Marquis Pierre Simon de Laplace (1747–1827), a French physicist and mathematician.

Laplace's law implies that it requires a greater pressure difference to sustain a small sphere than a larger one. It holds true not only for the model discussed, a membrane or a balloon, but also for a

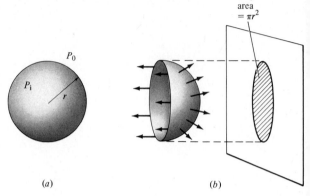

Figure 16.6. (*a*) A spherical membrane or balloon with inner and outer pressures P_i and P_o. (*b*) The sphere is cut in half by an imaginary plane. The arrows to the left represent the forces exerted by the wall. The arrows normal to the surface represent the forces due to the pressure difference.

spherical liquid drop. Both situations are illustrated by the next two examples.

Example 16.3

A rubber balloon is inflated to a radius of 0.1 m. The pressure inside is 1.001×10^5 Pa, and the pressure outside is 10^5 Pa. What is the wall tension γ?

Solving Laplace's law for γ,

$$\gamma = \frac{r}{2}(P_i - P_0) = \left(\frac{0.1 \text{ m}}{2}\right)(1.001 \times 10^5 \text{ Pa} - 10^5 \text{ Pa})$$

$$= (0.05 \text{ m})(10^2 \text{ Pa}) = 5 \text{ N m}^{-1}$$

This is the tension in the wall when the balloon is inflated to a radius of 0.1 m. Since the balloon is elastic, the wall tension will increase if the radius is increased.

When a liquid is in equilibrium with its own vapor, the pressure of the gas phase is called the *vapor pressure*. It may also be thought of as the pressure necessary to keep more of the liquid from evaporating; it balances a pressure difference $P_i - P_0$ across the liquid-vapor surface. Thus, a liquid drop in equilibrium with its vapor will have a pressure difference $P_i - P_0$ equal to the vapor pressure. The droplet size arising from this condition is found in the next example.

Example 16.4

From Table 16.1 the surface tension of water at 20°C is 7.28×10^{-2} N m^{-1}. The vapor pressure of water at 20°C is 2.33×10^3 Pa. What is the radius of the smallest spherical water droplet which can form without evaporating?

The pressure difference $P_i - P_0$ must be no more than 2.33×10^3 Pa. Solving Laplace's law for r, we have

$$r = \frac{2\gamma}{P_i - P_0} = \frac{2(7.28 \times 10^{-2}\ \text{N m}^{-1})}{2.33 \times 10^3\ \text{Pa}}$$
$$= 6.25 \times 10^{-5}\ \text{m}$$

For a soap bubble the film has two surfaces, and a similar derivation leads to the result

$$P_i - P_0 = \frac{4\gamma}{r} \qquad \text{(spherical bubble)} \quad (16.5)$$

Also, for a cylindrical tube of radius r, Laplace's law is

$$P_i - P_0 = \frac{\gamma}{r} \qquad \text{(cylindrical tube)} \quad (16.6)$$

Again we observe the inverse dependence on the radius. In physiology, $P_i - P_0$ is called the *transmural* pressure, the pressure across the walls of a blood vessel. This is quite different from the driving pressure that forces fluid through the vessel.

Equations 16.4, 16.5, and 16.6 can be deceptive in their apparent simplicity. Implicit in each result is the fact that the pressure and wall tension may depend on the radius. Suppose that we use a soap bubble as an example. Then the surface tension γ is constant as the radius changes and, because there are two surfaces, Eq. 16.5 applies. Suppose the outside pressure is reduced gradually to zero. Then $P_i - P_0$ is replaced by P_i, and Eq. 16.5 becomes

$$P_i = \frac{4\gamma}{r}$$

Now P_i is larger than $P_i - P_0$, so the left-hand side of this equation is larger than before. Since γ is constant, the right-hand side can only increase if r decreases! But this is contrary to our intuition; we know the soap bubble will become larger. The solution is that the inside pressure P_i does not remain constant as P_0 decreases. The following example illustrates this.

Example 16.5

The pressure outside a soap bubble of radius r is initially half the pressure inside, so $P_0 = \frac{1}{2}P_i$. The outside pressure is then reduced to $P_0' = 0$. Find the new pressure inside P_i' and the new radius r', assuming the surface tension γ and temperature remain constant.

Initially $P_0 = \frac{1}{2}P_i$ and $P_i - P_0 = \frac{1}{2}P_i$. Thus $P_i - P_0 = 4\gamma/r$ becomes

$$\frac{1}{2}P_i = \frac{4\gamma}{r} \qquad (i)$$

When the outside pressure is reduced to zero, $P_i' - P_0' = P_i'$ and

$$P_i' = \frac{4\gamma}{r'} \qquad (ii)$$

Dividing Eq. (ii) by (i), Laplace's law requires for equilibrium

$$\frac{P_i'}{P_i} = \frac{r}{2r'}$$

Since the temperature remains constant, the ideal gas law gives $P_i V_i = P_i' V_i'$. The volume of a sphere is $4\pi r^3/3$, so we find a second relationship between P_i and P_i',

$$\frac{P_i'}{P_i} = \frac{V}{V'} = \frac{r^3}{r'^3}$$

Equating the two expressions for P_i'/P_i, we find $r' = 2r$. Thus the radius increases as expected.

16.4 | SURFACTANT IN THE LUNGS

The small air sacs of the lungs, the *alveoli*, expand and contract an average of 15,000 times per day in a normal adult. It is across the membrane of the alveoli that oxygen and carbon dioxide are transported. The tension in the walls results from both the membrane tissue and a liquid on the walls containing a long lipoprotein, *surfactant*. We will see that the surfactant gives the membrane the elasticity needed to make the required adjustments in the wall tension.

Figure 16.7. A cross section exposing alveoli at the end of a bronchiole, the smallest branch of a bronchial tube. The pressure inside the alveoli is P_i; the pressure in the fluid of the pleural cavity is P_0. The pleural cavity encloses the entire lung.

267

The alveoli may be approximated by small spheres with tiny entrances. Thus we may use Laplace's law, Eq. 16.4, rewritten as

$$r(P_i - P_0) = r\,\Delta P = 2\gamma \qquad (16.7)$$

For a spherical membrane the product of the radius and the pressure difference must be equal to 2γ for the membrane to be in equilibrium. In the alveoli, there is a special problem. During exhalation the pleural pressure P_0 increases, so ΔP decreases. At the same time, muscle contraction reduces the radii of the alveoli. If both r and ΔP decrease and γ is constant, the equilibrium condition, Eq. 16.7, cannot be satisfied, and the alveoli will collapse, because the inward force due to the walls exceeds the outward force due to the pressure difference. During inhalation the pleural pressure P_0 decreases and the radius increases. Again, if γ is constant, Eq. 16.7 is not satisfied, and the alveoli would tend to increase in size and rupture; the force due to the pressure difference would exceed the force due to the wall.

Nature solves this problem by having the surfactant present. Its long molecules prefer to lie nearly beside each other, making the membrane highly elastic. During inhalation, as the radius increases the molecules are pulled apart and the wall tension increases. Thus as $r\,\Delta P$ increases, so does γ, and Eq. 16.7 is satisfied. During exhalation the molecules slide back together, and the wall tension decreases along with $r\,\Delta P$. Thus the surfactant serves to change the wall tension so that equilibrium is maintained. Insufficient surfactant in the lungs is the cause of death of many newborn infants.

16.5 | THE RISE OF SAP IN TREES; NEGATIVE PRESSURES

Sap is water plus the products of photosynthesis including sugar. The fact that sap can reach the top of Douglas fir trees, 60 m or more high, has long been a puzzle. We have already seen that osmotic pressure (Chapter Eleven) and capillary action (Example 16.2) cannot explain this phenomenon.

Negative pressures arising from the cohesive forces in water appear to provide the answer to this puzzle. These cohesive forces of water can be measured with an apparatus like that of Fig. 16.8.

Figure 16.8. (*a*) Pure water, containing no dissolved gases, fills the volume beneath a piston in a chamber. The piston is pulled upward; the water pulls downward. (*b*) The water finally separates from the piston leaving water and its vapor.

When the piston is pulled upward the water expands very slightly, exerting a downward force on the piston. The force per unit area on the piston is found experimentally to be between 25 and 300 atm just before the column separates from the piston. This pressure is called negative pressure because the water is pulling *inward* on the piston instead of pushing *outward*.

In trees the sap moves through the *xylem,* which forms channels with radii from 2.5×10^{-5} to 2.5×10^{-4} m. The xylem channels are filled with water up to the leaves. As water is evaporated from the leaf, the water column moves upward to keep itself intact. We can determine if the experimentally observed maximum values of the negative pressure are consistent with the rise of water in tall trees.

If we suppose that the pressure at the base of a tree trunk is atmospheric pressure P_A, then the pressure at a height h above the ground is

$$P_h = P_A - \rho g h \qquad (16.8)$$

If h is large enough, P_h can become negative. Expressing pressures in atmospheres, Eq. 16.8 can be written as

$$P_h = 1 \text{ atm} - 9.67 \times 10^{-2}\, h \text{ atm} \qquad (16.9)$$

where h is measured in meters. For a tree 60 m in height, $P_h = -4.8$ atm. This is well within the range of the observed negative pressures in water. Thus, it is believed that the negative pressure arising from the cohesive forces among the water molecules is responsible for the rise of sap to the tops of trees.

SUMMARY

Molecules at the surface of a liquid experience forces attracting them to the bulk of the liquid. These forces are responsible for the surface tension of a liquid.

When a liquid is in contact with a solid, the competition between the liquid-liquid molecular forces and the liquid-solid forces determines the contact angle. This contact angle can be used to calculate the capillary rise or fall of the liquid in a thin tube.

The definition of surface or wall tension can be used to obtain Laplace's law for the pressure drop across a membrane or liquid film. The forces among the molecules of a liquid are also responsible for the negative pressure present in the ascent of sap in trees.

Checklist
Define or explain:

surface tension	meniscus
contact angle	Laplace's law
capillarity	negative pressure

EXERCISES

Section 16.1 | Surface Tension

16-1   The surface tension of a liquid is measured using the apparatus of Fig. 16.1. If the straight wire is 0.05 m long and the weight of the wire plus attached weights is 2×10^{-3} N, what is the surface tension of the liquid?

16-2   If the surface tension of a liquid is 3×10^{-2} N m^{-1} and the length of the wire is 0.2 m, what total weight will the liquid support in the apparatus of Fig. 16.1?

Figure 16.9.   Exercise 16-3.

16-3   Figure 16.9 shows a loop of thread attached to a wire ring that has been dipped into a soap solution. Why does the loop assume a circular shape when the film inside is punctured?

Section 16.2 | Contact Angles and Capillarity

16-4   How high will water at 20°C rise in a tube of radius 10^{-3} m if it is made (a) of glass; (b) of paraffin?

16-5   How far will the meniscus of mercury be depressed in a glass tube of radius 10^{-4} m?

16-6   A glass tube of radius 1 mm $= 10^{-3}$ m is used to construct a mercury barometer (Fig. 16.10). (a) How far will the surface be shifted by capillary effects? Is it raised or lowered? (b) If the height of the mercury column is 0.76 m, what fractional error is introduced if the capillary effects are ignored?

Figure 16.10.   Exercises 16-6 and 16-7.

16-7   Normal atmospheric pressure supports a column of mercury 0.76 m high. A mercury barometer consists of a glass tube filled with mercury (Fig. 16.10). If capillary effects are to be less than 0.01 percent of the height of the column, what is the minimum radius of the tube?

Section 16.3 | Laplace's Law, and
Section 16.4 | Surfactant in the Lungs

16-8   A soap bubble has a radius of 0.05 m. If the pressure difference between the inside and outside is 2 Pa, what is the surface tension of the soap film?

16-9   A typical alveolus has a radius of 10^{-4} m. The wall tension is 0.05 N m^{-1}. (a) What is the pressure difference between the inside and outside of the alveolus? (b) There is an outward normal force everywhere on the alveolar surface. Find the sum of the magnitudes of these forces. (The surface area of a sphere is $4\pi r^2$.)

16-10   At the end of exhalation the radius of the alveoli is 0.5×10^{-4} m. The gauge pressure inside the alveoli is -400 Pa, and in the pleural cavity it is -534 Pa, about -3 torr and -4 torr, respectively. What is the wall tension in the alveoli? How does this compare with the wall tension of 0.05 N m^{-1} in the absence of surfactant?

16-11 Derive Eq. 16.5.

16-12 Derive Eq. 16.6.

16-13 What is the pressure difference between the inside and outside of a droplet of glycerine of radius 10^{-4} m?

Section 16.5 | The Rise of Sap in Trees; Negative Pressures

16-14 If the pressure of sap in a tree is negative, will fluid flow out of a fine tube inserted into the xylem?

16-15 The pressure of sap in the xylem is measured at two points 1 m apart in height. What pressure difference is observed?

16-16 What is the negative pressure at the base of a tree 20 m high?

16-17 Show how Eq. 16.8 leads to Eq. 16.9.

PROBLEMS

16-18 A thin wire loop of weight w and radius r floats on a liquid with surface tension γ (Fig. 16.11). What force F must be exerted to remove the loop?

16-19 Each leg of a six-legged insect standing on water at 20°C produces a depression of radius $r = 10^{-3}$ m (Fig. 16.12). The angle is 30°. (a) What is the surface tension force acting upward on each leg? (b) What is the weight of the insect?

Figure 16.11. Problem 16-18.

Figure 16.12. An insect leg supported by water. Problem 16-19.

16-20 (a) To what height will water rise in a vertical glass capillary tube of radius 3×10^{-5} m? (b) If the tube is only 0.2 m long, will the water overflow the top of the tube? (Assume that the top edges of the tube are smoothly rounded, and that the water is at 0°C.)

16-21 Capillary rise or fall is not limited to cylindrical tubes. Show that when two flat plates separated by a distance, d, are placed vertically in a fluid, the fluid rises a height $h = 2\gamma \cos \theta / \rho g d$ between the plates. (γ is the surface tension of the fluid, ρ is its density, and θ is the contact angle between the fluid and the plates.)

16-22 Use the result of the previous problem to find the height that water at 20°C rises between two flat glass plates separated by 0.3 mm = 3×10^{-4} m.

Additional Reading

Martin H. Zimmerman and C.L. Brown, *Trees: Structure and Function,* Springer-Verlag, New York, 1971.

Alan T. Hayward, Negative Pressure in Liquids: Can It Be Harnessed to Serve Man? *American Scientist,* vol. 59, 1971, p. 434.

P.F. Scholander, Tensile Water, *American Scientist,* vol. 60, 1972, p. 584.

R. McNeill Alexander, *Animal Mechanics,* University of Washington Press, Seattle, 1968. Chapters 5 and 6.

Scientific American articles:

James E. McDonald, The Shape of Raindrops, February 1954, p. 64.

V.A. Greulach, The Rise of Water in Plants, October 1952, p. 78.

John A. Clements, Surface Tension in the Lungs, December 1962, p. 120.

Martin H. Zimmermann, How Sap Moves in Trees, March 1963, p. 132.

Robert E. Apfel, The Tensile Strength of Liquids, December 1972, p. 58.

CHAPTER SEVENTEEN

THE MAMMALIAN CIRCULATORY SYSTEM

The physics of fluids has many applications in biological systems. In this chapter, we study some of the many features of the mammalian circulatory system that may be understood using physical principles introduced earlier in this unit.

The circulatory system contains a number of interacting elements that affect each other. The system as a whole is also affected by outside influences and responds to changing bodily requirements such as those caused by heavy exercise or temperature extremes.

This chapter begins with a brief description of the important elements of the system. We then discuss the effect of gravity on circulation and the changes necessitated by changes in posture. Blood pressure measurements, the flow resistance of various blood vessels, and the work done by the heart are the other topics explored in this chapter.

17.1 | ELEMENTS OF THE CIRCULATORY SYSTEM

The Blood | The circulatory system transports the substances required by the body and the waste products of metabolism. In order to perform a large number of functions, the blood contains many different constituents, including red blood cells, white blood cells, platelets, and proteins. However, for our purposes, it is sufficient to treat the blood as a uniform fluid, with viscosity $\eta = 2.084 \times 10^{-3}$ Pa s and density $\rho = 1.0595 \times 10^3$ kg m^{-3} at normal body temperature. Only more subtle physical properties of the circulatory system require a fuller description.

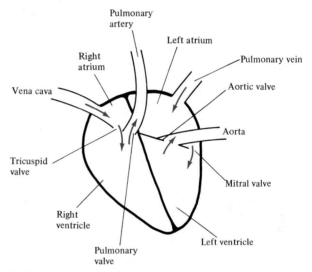

Figure 17.1. The heart in cross section. The small colored arrows show the general direction of blood flow.

The Heart | Figure 17.1 illustrates the heart in cross section. The sequence of events during the heart cycle is as follows:

1 The *right ventricle* relaxes and fills with oxygen-depleted blood returned from the body through the *right atrium*. The *tricuspid valve* is open, and the *pulmonary valve* is closed.
2 The right ventricle contracts; the pressure closes the tricuspid valve, opens the pulmonary valve, and pumps blood toward the lungs.
3 The *left ventricle* relaxes and fills with oxygen-rich blood returned from the lungs through the *left atrium*. The *mitral valve* is open, and the *aortic valve* is closed.

4   The left ventricle contracts; the pressure closes the mitral valve, opens the aortic valve, and pumps blood through the *aorta* to the arterial system.

This is not quite a sequential operation since the right and left portions of the heart act independently, and the ventricular contractions may be nearly simultaneous.

When the heart is studied with a stethoscope, two rather sharp sounds are usually heard. The first is associated with the closing of the tricuspid and mitral valves and results from the vibrations set up in the system when the flow is suddenly cut off. This is similar to the noise associated with the rapid closing of valves in automatic washing machines. The second sound is associated in the same way with the closing of the aortic and pulmonary valves. This second sound may be split into two sounds if these valves do not close simultaneously.

Further sounds, when heard, are associated with high rates of blood flow and turbulence. For example, heart murmurs result from the incomplete opening or closing of the valves and the resulting rapid, turbulent flow through the abnormally small openings.

The Cardiovascular System

The cardiovascular system includes the heart and an extensive system of *arteries, vascular beds* containing *capillaries,* and *veins* (Fig. 17.2). The arteries carry blood to the organs, muscles, and skin, and the veins transport the return flow. Each large artery branches to form several smaller arteries, which in turn branch further. The blood ultimately reaches the vascular beds, where materials are exchanged with the surrounding tissues. The branching process is then reversed in the venous system, culminating in the *vena cava,* which returns blood to the heart. Table 17.1 gives the number and dimensions of several types of blood vessels in one vascular bed. Table 17.2 lists the properties of the human cardiovascular system.

A particularly interesting component of the cardiovascular system is the *arteriovenous anastomosis* (AVA) or *shunt*. Figure 17.2 shows only one of the many AVAs present in the body. These shunts are particularly important, since the surrounding smooth muscle tissue can adjust the vessels' diameter. In the interior of the body, they help to adjust the blood

Figure 17.2.   Schematic diagram of the mammalian circulatory system. The circles represent vascular beds. The small arrows represent the direction of blood flow. The colored shaded areas are those with oxygen-rich blood. The arterial system is on the right at pressure P_2, and the venous system on the left is at pressure P_1.

flow to various organs as conditions change. Smaller shunts in the skin are open if the body needs to release heat or to increase the skin temperature. Under more moderate conditions, they are closed in order to reduce the load on the heart and direct the major portion of the blood supply to other parts of the body.

17.2 | THE ROLE OF GRAVITY IN THE CIRCULATION

When animals evolved to the point where they spent a significant amount of time standing upright, a sur-

TABLE 17.1

Detailed structure of the mesenteric (intestinal) vascular bed of a small dog. This is one of many such vascular beds in the body

Structure	Number, N	Inner Radius, R (cm)	Total Inner Cross-Sectional Area, $N\pi R^2$ (cm²)	Length, l (cm)	Equivalent Flow Resistance R_{fl}/N (torr s cm⁻³)
Mesenteric artery	1	0.15	0.07	6.0	0.05
Main branches	15	0.05	0.12	4.5	1.91
Secondary branches	45	0.03	0.13	3.91	4.27
Tertiary branches	1,900	0.007	0.29	1.42	12.39
Terminal arteries	26,600	0.0025	0.52	0.11	4.21
Terminal branches	328,500	0.0015	2.32	0.15	3.59
Arterioles	1,050,000	0.001	3.30	0.20	7.58
Capillaries	47,300,000	0.0004	23.78	0.10	3.29
Venules	2,100,000	0.0015	14.84	0.10	0.37
Terminal branches	160,000	0.0037	6.73	0.24	0.32
Terminal veins	18,000	0.0065	2.39	0.15	0.19
Tertiary veins	1,900	0.014	1.17	1.42	0.77
Secondary veins	60	0.08	1.47	4.19	0.007
Mesenteric vein	1	0.3	0.28	6.0	0.003

TABLE 17.2

Properties of the human cardiovascular system for a typical adult. All pressures listed are gauge pressures. (1 atm = 760 torr; 1 torr = 133.3 Pa)

Mean pressure in large arteries	96 torr
Mean pressure in large veins	8 torr
Volume of blood (70 kg man)	5.2 liters = 5.2×10^3 cm³
Time required for complete circulation (resting)	54 seconds
Heart flow rate (resting)	97 cm³ s⁻¹
Viscosity of blood (37°C)	2.084×10^{-3} Pa s
Density of blood (37°C)	1.0595×10^3 kg m⁻³

prising number of changes in the circulatory system were required. Of particular importance is the system used to return blood from the lower extremeties to the heart. Humans have adapted to the problems of moving blood upward a large distance against the force of gravity. Animals that have not, such as snakes, eels, and even rabbits, will die if held head upwards; the blood remains in the lower extremeties, and the heart receives no blood from the venous system.

Figure 17.3 shows what is observed if a person's large arteries are cannulated. In the reclining position, the pressures everywhere are almost the same. The small pressure drop between the heart and the feet or brain is due to the viscous forces. However, the pressures at the three points are quite different in the

standing person, reflecting the large difference in their heights.

Figure 17.3. Schematic view of the results of cannulation of arteries in various parts of the human body when standing and reclining. The pressures shown are averaged over the heart cycle.

Since the viscous effects are small, we can use Bernoulli's equation, $P + \rho gh + \frac{1}{2}\rho v^2 = $ constant, to analyze this situation. The velocities in the three arteries are small and roughly equal, so the $\frac{1}{2}\rho v^2$ term can be ignored. Hence the gauge pressures at the heart P_H, at the foot P_F, and at the brain P_B are related by

$$P_F = P_H + \rho gh_H = P_B + \rho gh_B \qquad (17.1)$$

where ρ is the density of blood.

In discussions of the circulatory system, pressures are usually measured in torr, where 1 atm = 760 torr, and

$$1 \text{ torr} = 133.3 \text{ Pa}$$

For a typical adult, $h_H = 1.3$ m, and $h_B = 1.7$ m. With $\rho = 1.0595 \times 10^3$ kg m^{-3}, we find

$$
\begin{aligned}
P_F - P_H &= \rho gh_H \\
&= (1.0595 \times 10^3 \text{ kg m}^{-3})(9.8 \text{ m s}^{-2})(1.3 \text{ m}) \\
&= 1.34 \times 10^4 \text{ Pa} = 101 \text{ torr}
\end{aligned}
$$

P_H is about 100 torr, so $P_F \simeq 200$ torr. In a similar way we find $P_B \simeq 70$ torr. This explains why the pressures in the lower and upper parts of the body are very different when the person is standing, although they are about equal when reclining.

This situation poses several problems. The most important are the tendency for blood to drain out of the venous side of the upper body back to the heart and the difficulty of lifting blood from the lower extremeties up to the heart. To retard drainage from the venous side of the upper body, particularly from the brain where constant volume and flow rate are extremely important, the muscles surrounding the veins contract and cause constriction. In the lower extremeties, because the veins have a much larger capacity for passive expansion and blood storage than do arteries, the problem is to pump the blood "uphill." The veins in the extremeties contain valves that open when blood flows toward the heart and close if the blood moves away from the heart. Blood is returned to the heart, at least partially, by the pumping action associated with breathing and by the flexing of skeletal muscle, as in walking. These muscle contractions squeeze the veins, and the valves ensure that the resultant blood flow is toward the heart. The importance of this is illustrated by the fact that a soldier who is required to stand at strict attention may faint because of insufficient venous return. Once horizontal, the pressures are equalized, and the soldier regains consciousness.

Effects of Acceleration

When a person in an erect position experiences an upward acceleration **a**, the effective weight becomes $m(g + a)$. Applying Bernoulli's equation to the brain and heart with g replaced by $(g + a)$,

$$P_B + \rho(g + a)h_B = P_H + \rho(g + a)h_H$$

or

$$P_B = P_H - \rho(g + a)(h_B - h_H) \qquad (17.2)$$

Thus the blood pressure in the brain will be reduced even further. It has been found that if a is two or three times g, a human will lose consciousness because of the collapse of arteries in the brain. This factor limits the speed with which a pilot can pull out of a dive (Chapter Five). A related experience is the feeling of light-headedness that sometimes occurs when one suddenly stands up. Since muscular movement is required to activate the venous return mechanism, blood will tend to collect in the lower veins until normal activity is resumed.

17.3 | BLOOD PRESSURE MEASUREMENTS USING THE SPHYGMOMANOMETER

Since the upper arm of a human is at about the same level as the heart, blood pressure measurements made there give values close to those near the heart. The sphygmomanometer is the familiar instrument used

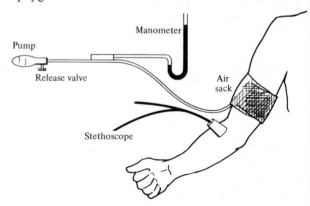

Figure 17.4. The use of the sphygmomanometer for measuring blood pressures.

to make this measurement conveniently and pain-lessly.

In the sphygmomanometer the gauge pressure in an air sack wrapped around the upper arm is meas-ured using a manometer or a dial pressure gauge. The pressure in the sack is first increased until the brachial artery, a major artery of the arm, is closed entirely. The pressure in the sack is then slowly reduced, while a stethoscope is used to listen for noises in the brach-ial artery below the sack. When the pressure is slightly below the *systolic* (peak) pressure produced by the heart, the artery will open briefly. Because it is only partially opened, the flow velocity is high and turbulent and therefore noisy. This resulting noise is heard as a tapping sound.

When the pressure in the sack is lowered further, the artery remains open during longer portions of the heart cycle but is still closed during the *diastolic* (minimum) pressure portion of the cycle. Hence, sounds are heard, but they are interrupted by periods of silence. When the pressure in the sack reaches the diastolic pressure, the artery remains open during the entire heart cycle. At this pressure the flow is still turbulent and noisy (particularly at diastolic pres-sure), but the sound is continuous. Thus, both the systolic and diastolic pressures can be measured without the cannulation technique.

Blood pressures are usually presented as systolic/diastolic ratios. Typical readings for a resting healthy adult are about 120/80. The borderline for high blood pressure (hypertension) is usually defined to be 140/90. Pressures appreciably above that level re-quire medical attention, because prolonged high blood pressure can lead to serious damage of the heart or other organs before a person is aware of any problem. Increasing emphasis has been placed in recent years on mass screenings to discover people with undetected high blood pressure.

17.4 | FLOW RESISTANCE

We now apply some of the ideas developed in our earlier discussion of viscous fluids to the flow of blood in blood vessels. In Chapter Fifteen, we discussed Poiseuille's law for the flow rate Q through a tube, which applies if the flow is laminar. We found that Q is determined by the tube radius R, the pressure

difference across its ends ΔP, the length of the tube l, and the viscosity η:

$$Q = \frac{\pi R^4}{8\eta l} \Delta P \qquad (17.3)$$

The *flow resistance* R_f is defined by

$$R_f = \frac{\Delta P}{Q} \qquad (17.4)$$

A comparison of these two equations shows that for laminar flow,

$$R_f = \frac{8\eta l}{\pi R^4} \qquad \text{(laminar flow)} \qquad (17.5)$$

The flow resistance is seldom computed for physio-logical systems because of the complexity of the sys-tems. Instead, it is usually arrived at by measuring ΔP and Q and hence determining R_f. Note that $R_f = \Delta P/Q$ defines a flow resistance whether the flow is laminar or not. However, Eq. 17.5 is valid only for laminar flow.

From its definition, it follows that the units of flow resistance are those of pressure divided by a volume per unit time. Thus the S.I. unit is the pascal-second per cubic meter (Pa s m^{-3}). As we noted earlier, pres-sures in this field are normally measured in torr. Also, the centimeter is a more convenient length unit than the meter for most blood flow problems. Accordingly, the unit used for the flow resistance is

$$1 \text{ torr s cm}^{-3} = 1.333 \times 10^8 \text{ Pa s m}^{-3}$$

If flow resistances are given in these units, then Q must be in cubic centimeters per second (cm^3 s^{-1}) and ΔP in torr.

From the following example, we see that the flow resistance in a large artery is small. This explains why the pressure drop in such arteries is small.

Example 17.1
The aorta of an average adult human has a radius of 1.3 cm $= 1.3 \times 10^{-2}$ m. What are the resistance and pressure drop over a 20-cm $= 0.2$-m distance, assuming a flow rate of 100 cm^3 s^{-1}?

From Table 17.2, $\eta = 2.084 \times 10^{-3}$ Pa s, so the flow resistance of the aorta is

$$R_f = \frac{8\eta l}{\pi R^4} = \frac{8}{\pi} \frac{(2.084 \times 10^{-3} \text{ Pa s})(0.2 \text{ m})}{(1.3 \times 10^{-2} \text{ m})^4}$$

TABLE 17.3

Approximate flow rates and resistances for the resting, reclining adult. The total flow rate from the aorta is 97 cm³ s⁻¹ and the average pressure drop across the beds is 88 torr. (Refer to Fig. 17.2 for the relationships of the various beds.)

Vascular Bed	Flow Rate (cm³ s⁻¹)	Flow Resistance (torr s cm⁻³)
Brain	12.5	7.0
Arms and shoulders	6.8	12.9
Lungs	97.0	0.07
Bronchial	1.0	88.0
Heart	4.2	21.0
Other organs	10.0	8.8
Liver, L_1	25.0	0.3
Liver, L_2	5.0	16.2
Spleen	8.3	9.8
Small bowel	8.8	9.2
Large bowel	2.9	27.9
Kidneys	18.3	4.8
Skin	5.5	16.0
Legs	13.7	6.4

$$= 3.72 \times 10^4 \text{ Pa s m}^{-3} \frac{(1 \text{ torr s cm}^{-3})}{(1.333 \times 10^8 \text{ Pa s m}^{-3})}$$

$$= 2.79 \times 10^{-4} \text{ torr s cm}^{-3}$$

The pressure drop over the 20-cm distance is then

$$\Delta P = R_f Q$$
$$= (2.79 \times 10^{-4} \text{ torr s cm}^{-3})(100 \text{ cm}^3 \text{ s}^{-1})$$
$$= 0.0279 \text{ torr}$$

This is very small compared to the total pressure drop in the system, which is about 100 torr. Most of the flow resistances and pressure drops occur in the smaller arteries and vascular beds of the body (Table 17.3).

Series and Parallel Flow Resistances

Since blood flows through a succession of vessels of various sizes, the circulatory system is a complicated network of flow resistances. When a given volume of blood flows through several flow resistances in turn, they are said to be in *series* (Fig. 17.5a). The flow rate Q through each is the same, so the total pressure drop is

$$\Delta P = \Delta P_1 + \Delta P_2 + \cdots$$
$$= (R_{f1} + R_{f2} + \cdots + R_{fN})Q \quad (17.6)$$

The N resistances in *series* are equivalent to a single *equivalent resistance* R_s chosen so that $\Delta P = R_s Q$, or

$$R_s = R_{f1} + R_{f2} + \cdots + R_{fN} \quad \text{(series)} \quad (17.7)$$

If a number of resistances are in *parallel*, as in Fig. 17.5b, then the fluid flow splits up, Q_1 through R_{f1}, Q_2 through R_{f2} and so on. The pressure drop ΔP across each resistance is the same, so applying $\Delta P = QR_f$ to each resistance gives

$$Q_1 = \frac{\Delta P}{R_{f1}}, \ Q_2 = \frac{\Delta P}{R_{f2}}, \ \cdots, \ Q_N = \frac{\Delta P}{R_{fN}}$$

Thus

$$Q = Q_1 + Q_2 + \cdots + Q_N$$
$$= \Delta P \left(\frac{1}{R_{f1}} + \frac{1}{R_{f2}} + \cdots + \frac{1}{R_{fN}} \right)$$

If we were to replace this system of parallel resistances by a single resistance R_p, we would have $Q = \Delta P / R_p$. Hence, for N resistances in *parallel*, the *equivalent resistance* is found from

$$\frac{1}{R_p} = \frac{1}{R_{f1}} + \frac{1}{R_{f2}} + \cdots + \frac{1}{R_{fN}} \quad \text{(parallel)} \quad (17.18)$$

The series and parallel resistance formulas can

(a)   (b)

Figure 17.5. N resistances (a) in series and (b) in parallel. Each resistance may represent a single vessel or an array of vessels such as is found in a vascular bed.

Figure 17.6. Example 17.2.

Figure 17.7. The flow resistance diagram equivalent to that of Fig. 17.6 with R_p replacing R_{f2} and R_{f3}.

Figure 17.8 The equivalent flow resistance diagram for the mesenteric vascular bed of the dog of Table 17.1.

often be used together as in the following example.

Example 17.2

In Fig. 17.6, a pattern of flow resistances is shown. Assume $R_{f1} = 1$, $R_{f2} = 4$, $R_{f3} = 4/3$ in units of torr-second per cubic centimeter (torr s cm^{-3}). The flow rate through R_{f1} is $Q_1 = 5$ cm^3 s^{-1}. Find (a) the total pressure drop ΔP; (b) Q_2 and Q_3.

(a) To find ΔP, we find the equivalent resistance R of the entire system. Then we can use $\Delta P = RQ_1$, since Q_1 must also be the flow in from the right-hand end. The resistance R_p equivalent to R_{f2} and R_{f3} is found from

$$\frac{1}{R_p} = \frac{1}{R_{f2}} + \frac{1}{R_{f3}} = \frac{1}{4} + \frac{3}{4} = 1$$

Thus $R_p = 1$ torr s cm^{-3}.

With R_p replacing R_{f2} and R_{f3}, the system appears as in Fig. 17.7. Now R_{f1} and R_p are in series, so $R = R_{f1} + R_p = 1 + 1 = 2$ torr s cm^{-3}, and $\Delta P = R Q_1 = (2$ torr s cm$^{-3})(5$ cm^3 s$^{-1}) = 10$ torr.

(b) To obtain Q_2 and Q_3, we need to know the pressure drops across R_{f2} and R_{f3}. The pressure drop across R_p is $\Delta P' = R_p Q_1 = 5$ torr; this is also the pressure drop across each resistance R_{f2} and R_{f3}. Thus

$$Q_2 = \frac{\Delta P'}{R_{f2}} = \frac{5}{4} = 1.25 \text{ cm}^3 \text{ s}^{-1}$$

$$Q_3 = \frac{\Delta P'}{R_{f3}} = \frac{5}{4/3} = 3.75 \text{ cm}^3 \text{ s}^{-1}$$

As we shall see in Chapter Nineteen, the relationships for series and parallel resistances are exactly the same as for electrical resistances in circuits. The reader familiar with electrical circuits will recognize the correspondence of ΔP with the potential difference, of Q with the electric current, and of R_f with the

electrical resistance. The analog to $R_f = \Delta P/Q$ in electrical circuits is known as *Ohm's law*.

Resistance of a Vascular Bed | To apply

these concepts to the circulatory system, we make the simplifying assumption that all arteries of the same size in a given vascular bed are in parallel, as are all veins and capillaries of the same size. With this assumption, Fig. 17.8 shows the equivalent flow resistance diagram for the data of Table 17.1. We find the flow resistance for each type of vessel using the parallel resistance formula. If one vessel has a resistance R_{f1}, the resistance of N vessels in parallel is

$$R_p = \frac{R_{f1}}{N} \qquad (17.9)$$

To find the overall equivalent resistance of the vessels of various sizes, we then use the series resistance formula. The following example illustrates the determination of the total resistance of the capillaries.

Example 17.3

From Table 17.1, the radius of a single capillary is 0.0004 cm = 4×10^{-6} m, and its length is 0.1 cm = 10^{-3} m. What is the net resistance of the 4.73×10^7 capillaries in the mesenteric vascular bed of a dog if they are assumed to be in parallel?

With $\eta = 2.084 \times 10^{-3}$ Pa s, the resistance of one capillary is

$$R_{f1} = \frac{8\eta l}{\pi R^4}$$

$$= \frac{8(2.084 \times 10^{-3} \text{ Pa s})(10^{-3} \text{ m})}{\pi (4 \times 10^{-6} \text{ m})^4}$$

277

$$= (2.073 \times 10^{16} \text{ Pa s m}^{-3}) \times \frac{(1 \text{ torr s cm}^{-3})}{(1.333 \times 10^8 \text{ Pa s m}^{-3})}$$

$$= 1.56 \times 10^8 \text{ torr s cm}^{-3}$$

There are $N = 4.73 \times 10^7$ capillaries in parallel, so their effective resistance is

$$R_f = \frac{R_{f1}}{N} = \frac{1.56 \times 10^8 \text{ torr s cm}^{-3}}{4.73 \times 10^7}$$

$$= 3.29 \text{ torr s cm}^{-3}$$

The effective resistance of all the capillaries in one vascular bed was calculated in the preceding example by assuming that all the capillaries are in parallel, which is only approximately true. We also assumed that Poiseuille's law holds true for the flow in calculating the resistance of a single capillary. Again, this is an approximation, since the radii of the capillaries are comparable to the sizes of some of the constituents of the blood. Nevertheless, the results of such calculations are in rough agreement with the values found by measuring the pressure drop and flow rate and using $R_f = \Delta P / Q$.

These calculations can also be made for the other portions of the vascular bed. The results are listed in Table 17.1 and summarized again in Fig. 17.9. The arterial portion of this vascular bed has 87 percent of the total resistance and hence of the pressure drop. The capillaries have 9 percent of the resistance, and the venous portion has only 4 percent. Similar results are found in other vascular beds: the arteries have the largest resistance, the capillaries a relatively small resistance, and the veins least of all.

The fact that most of the pressure drop and resistance occur in the arterial systems has important consequences for how the body regulates the circulation. The smallest arterial subdivisions, the arterioles, as

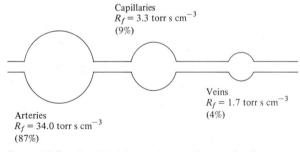

Capillaries
$R_f = 3.3$ torr s cm^{-3}
(9%)

Veins
$R_f = 1.7$ torr s cm^{-3}
(4%)

Arteries
$R_f = 34.0$ torr s cm^{-3}
(87%)

Figure 17.9. Simplified flow resistance diagram for the mesentaric vascular bed of a dog. The total equivalent resistance is $(34.0 + 3.3 + 1.7) = 39.0$ torr s cm^{-3}.

well as some of the larger branches in the vascular beds, have surrounding muscle fibers. These can contract, reducing the radii of the vessels and hence increasing the flow resistance. Because the resistance varies as the fourth power of the radius, the body has a very effective way of adjusting the blood flow and responding to changing requirements.

Since the blood in the arteries is at a high pressure, severing an artery can lead to a serious loss of blood. The risk of this occurring is reduced by the fact that many arteries are deep within the body. When there is a drop in arterial pressure as a result of hemorrhaging, the body responds by constricting the vessels in many of the vascular beds. This temporarily conserves the reduced blood supply for use in the heart and the brain. However, the buildup of metabolic waste products and the shortage of oxygen eventually cause the vessels in the vascular beds to dilate. This lowers the blood pressure further, and the body is then in an unstable and poorly functioning condition referred to as shock.

17.5 | THE HEART AS A PUMP

As our last example of how the physics of fluids can be used in understanding the circulatory system, we examine the mechanical properties of the heart. In particular, we wish to calculate the work done by the heart in pumping blood and compare the result with the total metabolic energy expended.

We can find the mechanical work done by the heart by calculating the work that must be done against the dissipative viscous forces. To do this, we consider the fluid in the segment of a vessel shown in Fig. 17.10. The net force on the fluid is the pressure drop times the area, ΔPA. The power expended against viscous forces is the product of this force and the average velocity, $\Delta PA\bar{v}$. Since $A\bar{v}$ is the flow rate Q, the power expended is

$$\mathcal{P} = Q \Delta P \qquad (17.10)$$

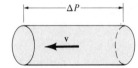

Figure 17.10. The power expended in the segment shown is $\mathcal{P} = Q \Delta P$.

Since the tube is horizontal and has a constant cross section, neither the potential nor the kinetic energy of the fluid can change. Hence $\mathcal{P}$ represents the power dissipated against the viscous forces.

According to Table 17.2, in a normal resting adult the heart pumps 97 cm^3 = 9.7×10^{-5} m^3 of blood per second. The pressure drop from the arterial to the venous systems is 88 torr = 1.17×10^4 Pa. Hence the power expended by the heart in overcoming viscous forces is

$$\begin{aligned} \mathcal{P} &= Q\,\Delta P \\ &= (9.7 \times 10^{-5} \text{ m}^3\text{ s}^{-1})(1.17 \times 10^4 \text{ Pa}) \\ &= 1.1 \text{ W} \end{aligned}$$

The total metabolic energy used by the heart can be estimated by measuring its oxygen consumption. For a 70-kg resting man, it is found that the rate of energy usage by the entire heart is about 5.5 W. Of this, about 2.5 W are available to do useful work; the remainder is converted directly to heat during metabolism. We have just seen that 1.1 W are used to do mechanical work pumping blood. A significant fraction of the remaining 1.4 W is used to maintain the wall tension of the heart.

We can gain some insight into this wall tension by using a model that depicts the left ventricle as a sphere (Fig. 17.11). When the ventrical contracts, its wall must be under active tension to generate the increased blood pressure. The force per unit length or tension γ exerted by the wall was found in the preceding chapter to be related to the pressure difference $\Delta P = P_i - P_0$ and the radius r by Laplace's law, $\Delta P = 2\gamma/r$ or $\gamma = r\,\Delta P/2$. The total force produced along an imaginary cut in the wall (Fig. 17.11b) is then

$$F = 2\pi r\gamma = \pi r^2\,\Delta P$$

The energy required by the heart muscle is related to this force but the exact relationship is complex. However, we can see qualitatively what may happen. If a person suffers from high blood pressure, ΔP is large, and the power, $Q\,\Delta P$, the heart must expend to pump blood is larger than normal. The energy necessary to maintain the wall tension is also greater than normal. After months or even years, this high pressure may lead to enlargement of the heart. Since the wall force increases with r^2, even more energy will be required, and the heart's oxygen requirements will be

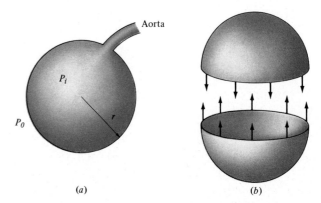

Figure 17.11. (a) A model of the heart pictures the left ventricle as a sphere. (b) Forces on an imaginary cut in the wall.

correspondingly larger. Congestive heart failure occurs when the oxygen supply is insufficient to meet the increased demands of the heart muscle.

SUMMARY

In this chapter we have considered some of the aspects of the circulatory system that can be clarified with the aid of the physics of fluids. Bernoulli's equation can be used in the large arteries to understand the effects of gravity on the system. In the smaller vessels, the pressure drops due to viscosity are large and may be calculated using Poiseuille's equation. This is done most conveniently by introducing the flow resistance and developing rules for combining flow resistances. One finds then that most of the pressure drop occurs in the smaller subdivisions of the arterial system. The work done by the heart against viscous forces in pumping the blood can be calculated from the observed pressure drop and flow rate.

Checklist
Define or explain:

ventricles, atria	systolic, diastolic
vascular bed	flow resistance
AVA	series
sphygmomanometer	parallel

EXERCISES

Section 17.1 | Elements of the Circulatory System

17-1 Which of the flow rates in Table 17.3 will depend most on the environmental temperature?

17-2   Why is it impractical to observe someone's pulse rate using a vein?

17-3   Calculate the volumes of blood in the arterial and venous portions of the mesenteric vascular bed of the dog.

17-4   Describe the condition of the arteriovenous anastomoses in the skin when the environmental temperature is (a) high; (b) low.

17-5   Explain why serious loss of blood is less likely to occur when a vein is damaged than when an artery is damaged.

Section 17.2 | The Role of Gravity in the Circulation

17-6   When a man stands, his brain is 0.4 m above his heart. If he bends so that his brain is 0.5 m below his heart, by how much does the blood pressure in his brain change?

17-7   Explain why we are uncomfortable when we bend with our heads lower than our heart.

17-8   A pilot is accelerating downward at four times the acceleration of gravity during a maneuver. If he is erect, what is the blood pressure in his brain?

17-9   A jet pilot pulls his plane out of a dive in such a way that his upward acceleration is 3 g's. What would you predict as the blood pressure in the brain?

17-10   At what acceleration would you expect the blood pressure in the brain to drop to zero for an erect person? (Assume there are no body mechanisms operating to compensate for such conditions.)

17-11   (a) If an elevator accelerates upwards at 9.8 m s^{-2}, what is the average blood pressure in the brain? What is the average blood pressure in the feet? Assume the person is standing. (b) If the elevator accelerates downwards at 9.8 m s^{-2}, what is the average blood pressure in the brain and feet?

Section 17.3 | Blood Pressure Measurements Using the Sphygmomanometer

17-12   If a sphygmomanometer were used to measure the blood pressure in the leg of a man sitting at rest, would the results give the pressure at the heart? Explain.

17-13   Suppose a man ran into a doctor's office and immediately had his blood pressure measured.

Would a reading above the normal 120/80 indicate he is suffering from high blood pressure? Explain.

Section 17.4 | Flow Resistance

17-14   A blood vessel of radius R branches into several vessels of smaller radius r. If the average fluid velocity in the smaller vessel is half that in the large vessel, how many vessels of radius r must there be?

17-15   Which of the various portions of the arterial system in the vascular bed of Table 17.1 will have (a) the largest pressure drop; (b) the smallest?

17-16   Figure 17.12 shows a portion of the circulatory system from Fig. 17.3. The flow resistances are also shown. Under normal conditions the flow rate through part of the liver, L_2, is $Q_{L2} = 5$ cm^3 s^{-1}, and the resistance is $R_{L2} = 16.2$ torr s cm^{-3}. What is the normal pressure at point A in the portal vein?

17-17   A small artery has a length of 0.11 cm = 1.1×10^{-3} m and a radius of 2.5×10^{-5} m. (a) Calculate its resistance in torr s cm^{-3}. (b) If the pressure drop across the artery is 10 torr, what is the flow rate?

17-18   A glass tube has a radius of 10^{-3} m and a length of 0.1 m. (a) What is its resistance in S.I. units to the flow of a liquid with a viscosity of 10^{-3} Pa s? (b) If the pressure drop along the tube is 10^3 Pa, what is the flow rate?

Section 17.5 | The Heart as a Pump

17-19   Using the assumption that a ventricle may be approximated by a sphere, by what factor must the wall force increase if the ventricle radius increases by 10 percent?

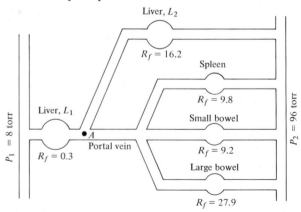

Figure 17.12.   All flow resistances are in torr s cm^{-3}. Exercise 17-16, Problems 17-24 and 17-25.

17-20 A man exercising vigorously is using metabolic energy at a rate 10 times that of the resting human in Table 17.2. His blood pressure rises by 50 percent above the resting level. Assuming the rate of blood flow is proportional to the metabolic rate, estimate (a) the blood flow rate for this person; (b) the work done by the heart against viscous forces in pumping the blood.

PROBLEMS

17-21 In Section 17.2, we assumed that the blood in the various large arteries had the same velocities. Suppose the velocity were 0.2 m s^{-1} in the artery near the heart and 0.01 m s^{-1} in the artery near the foot. How large a percentage error would that introduce in $P_F - P_H$?

17-22 (a) Using the resistances in Table 17.3, compute the pressure drops across the legs and lungs. (b) Explain the differences. What does this imply about the right ventricle?

17-23 Which of the vascular beds in Table 17.3, excluding the lungs, dissipates (a) the most power; (b) the least?

17-24 In certain types of liver damage the resistance R_{L2} is lowered, raising the pressure in the portal vein. This, in turn, decreases the flow rate through the spleen and bowels (Fig. 17.12). (a) If the portal vein pressure increases from a normal value of 15 torr to 25 torr, what is the flow rate through the spleen and bowels? (b) What is the new net flow rate through the main part of the liver, L_1? Compare your result with the normal value of $25 \text{ cm}^3 \text{ s}^{-1}$.

17-25 If the portal vein pressure in Fig. 17.12 is 25 torr, what are the flow rates through the (a) spleen, (b) small bowel, and (c) large bowel?

17-26 The pressure drop along a tube is increased gradually, and the corresponding flow rates are measured. (a) How would one get the flow resistance from this data? (b) What would happen to the flow resistance once the pressure became large enough so the flow became turbulent? Explain.

17-27 In Table 17.1, it is noted that the total cross-sectional area of all the capillaries in a particular vascular bed is 23.78 cm^2, while the corresponding area for the main branches is only 0.12 cm^2. How can this be reconciled with the fact that the main branches have a much smaller resistance?

17-28 A person suffering from high blood pressure has an average difference between the arterial and venous systems of 150 torr instead of the normal 88 torr. Assume that the energy expended in maintaining the tension in the heart wall is proportional to the force it exerts. (a) If the heart is not enlarged, by what factor are its metabolic requirements increased? (b) If the heart is enlarged in radius by 5 percent, estimate the additional metabolic demands this implies.

17-29 The pressure drop along a copper tube is 10^3 Pa, and the flow rate of a liquid in this tube is $0.01 \text{ m}^3 \text{ s}^{-1}$. (a) Find the flow resistance in S.I. units. (b) How much power is required to maintain this flow? (c) Does this calculation assume laminar flow? Explain.

17-30 When a calf is at rest, its heart pumps blood at a rate of $60 \text{ cm}^3 \text{ s}^{-1}$. The pressure drop from the arterial to venous systems is 90 torr. (a) What is the flow resistance of its circulatory system? (b) How much work does the heart do in pumping the blood? (c) Estimate the total energy expenditure of the heart. (d) An experimental artificial heart powered by an electrical pump is implanted in place of the animal's heart. If the pump has an efficiency of 50 percent, how much electrical power does it require?

Additional Reading

Alan C. Burton, *Physiology and Biophysics of the Circulation,* Yearbook Medical Publishers, Inc., Chicago, 1965.

Vernon B. Mountcastle (ed.), *Medical Physiology,* Vol. 1, 12th ed., C.V. Mosby Co., St. Louis, 1968.

Arthur C. Guyton, *Circulatory Physiology: Cardiac Output and Its Regulation,* W.B. Saunders Co., Philadelphia, 1963.

R.L. Whitmore, *Rheology of the Circulation,* Pergamon Press, Oxford and New York, 1968.

Gerald Aiello, Pierre LaFrance, Rogers C. Ritter, and James S. Trefil, The Urinary Drop Spectrometer, *Physics Today,* vol. 27, September 1974, p. 23.

J. Richard Shanebrook, Fluid Mechanics of Coronary Artery Bypass Surgery, *American Journal of Physics,* vol. 45, 1977, p. 677.

Scientific American articles:

J. Edwin Wood, The Venous System, January 1968, p. 86.

James V. Warren, The Physiology of the Giraffe, November 1974, p. 96.

UNIT FIVE

ELECTRICITY AND MAGNETISM

Although electric and magnetic forces have been observed for many centuries, most of the laws governing these forces were first discovered in the nineteenth century. To this day the study of the theoretical implications of these remarkable laws and the development of their practical applications remains a field of very active research.

We have several reasons for studying the properties of electric and magnetic forces in some detail. These forces represent fundamental natural laws, and they are responsible for the structure and indeed the existence of atoms, molecules, and bulk matter. Consequently, any serious attempt to understand the properties of atoms or aggregates of atoms depends on a knowledge of these forces. At the macroscopic level, electric charge flows or *currents* are present in systems as different as the circuits in television sets and the nerve fibers of animals. Since electric currents exert forces, do work, transmit information, and produce electromagnetic waves, they are of major importance.

The first of the five chapters in this unit deals with the basic properties of electric forces, and the next two consider electric currents in circuits and nerves respectively. The fourth chapter explores magnetic forces, and the final chapter discusses the surprising interrelations between electric and magnetic phenomena.

CHAPTER EIGHTEEN

ELECTRIC FORCES, FIELDS, AND POTENTIALS

We have already touched lightly upon electric charges and forces (Chapter Five) and on the electric potential energy (Chapter Six). In this chapter, we explore these and related concepts in more depth, so that we have a better understanding of how charges interact. We frequently employ the ideas developed here in later chapters.

18.1 | ELECTRIC FORCES

Coulomb's law states that the force between two electric charges is proportional to the product of the charges and inversely proportional to their separation squared. If a charge q is at a distance $\mathbf{r}$ from a second charge Q, then the force on q (Fig. 18.1) is

$$\mathbf{F} = \frac{kqQ}{r^2}\hat{\mathbf{r}} \tag{18.1}$$

Here, $\hat{\mathbf{r}}$ is a unit vector directed toward q. In S.I. units, the charge is measured in coulombs, and k has the experimentally determined value

$$k = 9.0 \times 10^9 \, \text{N m}^2 \, \text{C}^{-2} \tag{18.2}$$

The force is attractive if q and Q have opposite signs, and it is repulsive if they have like signs.

When a charge q is near two or more additional charges, the net force on q is the vector sum of the forces due to each of the other charges. The following example illustrates this point.

Example 18.1

A positive charge q is near a positive charge Q and a negative charge $-Q$ (Fig. 18.2). (a) Find the magnitude

Figure 18.1. The force $\mathbf{F}$ between the two charges has a magnitude kqQ/r^2. When the charges have like signs, the force on q is repulsive or away from Q.

Figure 18.2. Both q and Q are positive. The force $\mathbf{F}_+$ on q due to the charge Q is repulsive, so it is directed upward in the direction of $\hat{\mathbf{y}}$. The force $\mathbf{F}_-$ on q due to $-Q$ is attractive and is also directed upward.

and direction of the force on q. (b) If $q = 10^{-6}$ C, $Q = 2 \times 10^{-6}$ C, and $a = 1$ m, find the force on q.

(a) The vector distance from Q to q is $a\hat{\mathbf{y}}$, so the force on q due to this charge is

$$\mathbf{F}_+ = \frac{kqQ}{a^2}\hat{\mathbf{y}}$$

Since both q and Q are positive, the force is parallel to $\hat{\mathbf{y}}$, or upward. Similarly, the distance from $-Q$ to q is $a(-\hat{\mathbf{y}})$, so the force due to this change is

$$\mathbf{F}_- = \frac{kq(-Q)}{a^2}(-\hat{\mathbf{y}}) = \frac{kqQ}{a^2}\hat{\mathbf{y}}$$

Again the force is parallel to $\hat{\mathbf{y}}$. Since $\mathbf{F}_+$ and $\mathbf{F}_-$ are parallel, the net force on the charge q is

$$\mathbf{F} = \mathbf{F}_+ + \mathbf{F}_- = \frac{2\,kqQ}{a^2}\,\hat{\mathbf{y}}$$

(b) Substituting the numerical values given, the net force on q is

$$\mathbf{F} = \frac{2\,kqQ}{a^2}\,\hat{\mathbf{y}}$$

$$= \frac{2(9 \times 10^9 \text{ N m}^2\text{ C}^{-2})(10^{-6}\text{ C})(2 \times 10^{-6}\text{ C})}{(1\text{ m})^2}\,\hat{\mathbf{y}}$$

$$= 3.6 \times 10^{-2}\,\hat{\mathbf{y}}\text{ N}$$

The force is directed upward.

18.2 | THE ELECTRIC FIELD

We have seen that when two or more charges exert forces on a given charge q, the net force is the vector sum of the forces. When there are many charges, it is often more convenient to do this sum indirectly by introducing a quantity called the *electric field*. The electric field is also useful because it characterizes the effects of the other charges without explicit reference to the charge q.

When we have several charges, we may say that these charges produce an electric field in their vicinity. Then if another charge q is present, it experiences a force $\mathbf{F}$ proportional to q and to the electric field $\mathbf{E}$,

$$\mathbf{F} = q\mathbf{E} \qquad (18.3)$$

Suppose q is a unit positive charge, or $+1$ C in S.I. units. Then $\mathbf{F} = q\mathbf{E}$ implies that $\mathbf{F}$ is equal numerically to $\mathbf{E}$.

Thus the electric field is the electric force exerted on a unit positive charge.

The units of the electric field are those of force divided by charge, or newtons per coulomb (N C^{-1}) in S.I. units. Note that since the force on a charge q is $\mathbf{F} = q\mathbf{E}$, *a positive charge will tend to move parallel to $\mathbf{E}$, and a negative charge opposite to $\mathbf{E}$.*

The expression for the field due to a single point charge can be deduced from Coulomb's law. The force on a charge q due to a charge Q at a distance r is

$$\mathbf{F} = \frac{kqQ}{r^2}\,\hat{\mathbf{r}}$$

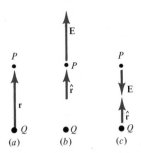

Figure 18.3. (*a*) A point P is at a distance $\mathbf{r}$ from a charge Q. (*b*) If the charge Q is positive, the electric field $\mathbf{E}$ at P is away from Q or parallel to $\hat{\mathbf{r}}$. (*c*) If Q is negative, the field is along $-\hat{\mathbf{r}}$.

Here, $\hat{\mathbf{r}}$ is a unit vector directed from the charge Q toward the point P where q is located (Fig. 18.3). Since the electric field is the force on a unit charge, at point P the field due to Q is $\mathbf{E} = \mathbf{F}/q$ or

$$\mathbf{E} = \frac{kQ}{r^2}\,\hat{\mathbf{r}} \qquad (18.4)$$

If Q is positive, $\mathbf{E}$ points along $\hat{\mathbf{r}}$ or away from Q; if Q is negative, $\mathbf{E}$ points along $-\hat{\mathbf{r}}$ or toward Q.

When there are several charges $Q_1, Q_2, \ldots$ at various positions, the electric field $\mathbf{E}$ at a point P is the vector sum of the individual electric fields $\mathbf{E}_1, \mathbf{E}_2, \ldots$ due to all the charges. If a charge q is placed at P, then the force on it is again given by $\mathbf{F} = q\mathbf{E}$. This is equivalent to the statement that the total force on q is the sum of the forces due to the individual charges.

Once we have measured the electric field at a point, we can immediately find the force on any charge placed there. It is not necessary to know the magnitude or location of the charges producing the field. For example, a Na$^+$ ion has a charge $q = e = 1.6 \times 10^{-19}$ C. If the field in a cell membrane is 10^6 N C^{-1}, the force on the ion has a magnitude $F = qE = (1.6 \times 10^{-19}\text{ C})(10^6\text{ N C}^{-1}) = 1.6 \times 10^{-13}$ N and is directed along the electric field. The following example demonstrates the calculation of the electric field due to two point charges and the force exerted by this field on another charge.

Example 18.2

The charges Q and $-Q$ of the preceding example are shown again in Fig. 18.5. (a) If $Q = 2 \times 10^{-6}$ C and $a = 1$ m, find the electric field at the origin. (b) Find the force on the charge $q = 10^{-6}$ C placed at the origin.

285

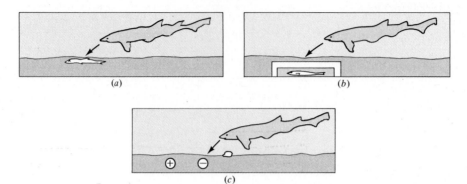

Figure 18.4. Sharks are sensitive to the minute electric fields produced by charges in a body. (*a*) The shark attacks a fish hidden beneath the sand. (*b*) A chamber blocks all but electrical stimuli, and the shark still attacks. (*c*) An artificially produced electric field elicits the same response. Here the shark is ignoring an obvious piece of food to follow the electrical stimulus.

Figure 18.5. The electric field due to a positive charge is directed away from the charge, while the field due to a negative charge is directed toward the charge. Hence, both contributions to the total field at the origin point along the $+y$ direction.

(a) As before, the vector distance from the charge Q to the origin is $a\hat{\mathbf{y}}$. Using Eq. 18.4, the field at the origin due to this charge is

$$\mathbf{E}_+ = \frac{kQ}{a^2}\hat{\mathbf{y}}$$

Also, the distance from the charge $-Q$ to the origin is $a(-\hat{\mathbf{y}})$, so its field there is

$$\mathbf{E}_- = \frac{k(-Q)}{a^2}(-\hat{\mathbf{y}}) = \frac{kQ}{a^2}\hat{\mathbf{y}}$$

The net electric field at the origin is the sum,

$$\mathbf{E} = \mathbf{E}_+ + \mathbf{E}_- = \frac{2\,kQ}{a^2}\hat{\mathbf{y}}$$
$$= \frac{2(9 \times 10^9 \text{ N m}^2 \text{ C}^{-2})(2 \times 10^{-6} \text{ C})}{(1 \text{ m})^2}\hat{\mathbf{y}}$$
$$= 3.6 \times 10^4\,\hat{\mathbf{y}} \text{ N C}^{-1}$$

The net field is directed upward.

(b) The force on a charge $q = 10^{-6}$ C at the origin is

$$\mathbf{F} = q\mathbf{E} = (10^{-6} \text{ C})(3.6 \times 10^4)\hat{\mathbf{y}} \text{ N C}^{-1}$$
$$= 3.6 \times 10^{-2}\,\hat{\mathbf{y}} \text{ N}$$

As expected, this is identical to the force calculated in the previous example by the direct use of Coulomb's law.

Electric Field Diagrams

Figure 18.6 shows the electric field vectors calculated from Eq. 18.4 for

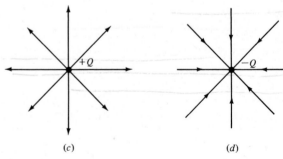

Figure 18.6. (*a*) and (*b*) Electric field vectors at locations near charges that are (*a*) positive, (*b*) negative. The corresponding electric field lines are shown in (*c*) and (*d*).

286

several locations near point charges. The field points away from the positive charge and toward the negative charge. In both cases the field becomes weaker as the distance from the charge increases, since it varies as $1/r^2$.

Another kind of electric field diagram or map can be drawn by joining the vectors in Fig. 18.6 with continuous *lines*. In such a diagram the direction of the lines at any point indicates the field direction there. Also, since the lines become further apart as the field becomes weaker, the spacing of the lines indicates the magnitude of the field. Such diagrams are especially useful in understanding the electric fields due to complex arrangements of charges.

Figure 18.7. Photographs of patterns formed by grass seeds in a liquid near various charge arrangements. (*a*) A point charge. (*b*) Two point charges, opposite signs. (*c*) Two point charges, like signs. (*d*) Two charged metal plates, opposite signs. (*e*) One charged metal plate. (Reproduced by permission of the publisher from *PSSC Physics,* fourth edition, 1976, by D. C. Heath and Company, Lexington, Massachusetts.)

We can actually make an electric field visible by exploiting the fact that oblong objects, such as grass seeds, suspended in a liquid tend to line up along the field lines (Fig. 18.7). The pattern formed near a single point charge is strikingly similar to that in Fig. 18.6. We will see corresponding similarities shortly for more complex charge arrangements.

To summarize this introduction to electric fields, we say that a charge produces an electric field in its vicinity and that the field exerts a force on any other charge that is present. Although this point of view is mathematically equivalent to using Coulomb's law, it is useful since the electric field can often be conveniently measured or calculated. Also, the electric field diagrams convey information about the effects of a charge or set of charges on *any* other charge brought into the field. *Specifically, positive charges will experience forces along the field lines, and negative charges will experience forces opposite to these lines.*

18.3 | THE ELECTRIC FIELD DUE TO ARRANGEMENTS OF CHARGES

The total electric field due to two or more charges is the sum of their individual electric fields. If the individual fields point in different directions, cancellations will occur in the sum. These cancellations may cause the total field to vary with distance in a way that is quite different from the inverse distance squared dependence of the individual point charge fields. We will see, for example, that the field may vary as $1/r^3$ or even be constant.

The Field of an Electric Dipole | A pair of equal but opposite charges, $+q$ and $-q$, is called an electric dipole (Fig. 18.8). At the point P, the fields $\mathbf{E}_+$ due to $+q$ and $\mathbf{E}_-$ due to $-q$ are

$$\mathbf{E}_+ = \frac{kq}{(r-a)^2}\widehat{\mathbf{y}}, \qquad \mathbf{E}_- = \frac{-kq}{(r+a)^2}\widehat{\mathbf{y}}$$

The total field at P is

$$\mathbf{E} = \mathbf{E}_+ + \mathbf{E}_- = kq\left[\frac{1}{(r-a)^2} - \frac{1}{(r+a)^2}\right]\widehat{\mathbf{y}}$$

If r is large compared to a, the total field $\mathbf{E}$ is quite small, since the fields due to the two charges nearly cancel. Rewriting $\mathbf{E}$ by putting both terms over a

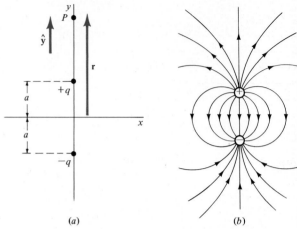

(a)   *(b)*

Figure 18.8. *(a)* A dipole is a pair of equal but opposite charges. *(b)* The field due to a dipole. Notice that the lines go from the positive to the negative charge. Note also the similarity with Fig. 18.7b.

common denominator, some terms cancel in the numerator, and we find

$$\mathbf{E} = \frac{4kqar}{(r^2-a^2)^2}\widehat{\mathbf{y}}$$

When r is much greater than a, we can neglect a in the denominator. Then

$$\mathbf{E} = \frac{4kqa}{r^3}\widehat{\mathbf{y}}$$

Thus the field diminishes as $1/r^3$ at large distances, which is a faster rate of decrease than the $1/r^2$ associated with a single charge. If we repeat this calculation for a distant point on the x axis or somewhere else, the overall numerical factor and the direction of the field will vary, but the field will again diminish as $1/r^3$.

The Field of a Uniformly Charged Plane | Near a plane bearing a large number of uniformly distributed charges, the field is almost constant in direction and magnitude (Fig. 18.9a). The cancellations causing this can be seen with the aid of Fig. 18.9b. The fields at P due to charges at A and B have cancelling vertical components, so their vector sum is horizontal. Similarly, all other pairs of charges located symmetrically about point C have a net horizontal field, and the total field at P is horizontal. If P is not at the midpoint of the plane, there are unpaired charges left over at one edge. However, the fields of

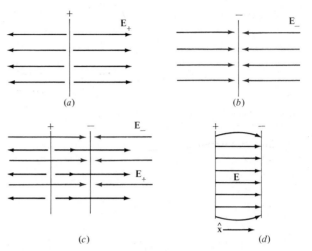

Figure 18.9. (a) The electric field close to a uniformly charged plane is uniform except near the edges. Note the similarity with Fig. 18.7e. (b) The fields at point P due to a pair of charges located symmetrically about point C have cancelling vertical components.

these distant charges are unimportant if P is close to the plane and not near an edge. A detailed calculation shows that the field near a plane of area A with a total charge Q is

$$\boxed{\mathbf{E} = 2\pi k \frac{Q}{A} \hat{\mathbf{x}}} \tag{18.5}$$

Here $\hat{\mathbf{x}}$ is a unit vector directed *away from the plane*.

With two planes of area A bearing equal but opposite charges $+Q$ and $-Q$ (Fig. 18.10), the fields *add* between the planes and *cancel* elsewhere. Hence, between the planes,

$$\mathbf{E} = 4\pi k \frac{Q}{A} \hat{\mathbf{x}} \tag{18.6}$$

This type of uniform field can readily be produced using metal plates. When a metal plate has a net charge, the mutual repulsion of the charges causes them to be almost uniformly distributed. Two such oppositely charged plates produce a nearly uniform field between them. A charged particle, such as an electron, moving through the uniform field experiences a constant force and therefore has a constant acceleration. We see later how this is put to practical use in an oscilloscope. The electric fields in cell membranes are also approximately uniform, as is seen in the next example.

Example 18.3

A thin, flat membrane separates a layer of positive ions outside a cell from a layer of negative ions inside (Fig. 18.11). If the electric field due to these charges is 10^7 N C^{-1}, find the charge per unit area Q/A in the layers on either side of the membrane.

Since the membrane is flat, the charges form uni-

Figure 18.10. (a) The field of a positively charged plane. (b) The field of a negatively charged plane shown in color for clarity. (c) The field lines due to the two planes when they are close together. (d) The total field due to the two charged planes. Note the similarity with Fig. 18.7d. The field is not quite uniform at the edges. The effects of these *fringing fields* at the edges may often be neglected.

Figure 18.11. (a) A portion of a cell membrane seen in perspective. (b) A cross-sectional view showing the charge layers.

formly charged planes on either side. Using $E = 4\pi k Q/A$, we have

$$\frac{Q}{A} = \frac{E}{4\pi k} = \frac{10^7 \text{ N C}^{-1}}{4\pi (9 \times 10^9 \text{ N m}^2 \text{ C}^{-1})}$$

$$= 88.4 \times 10^{-6} \text{ C m}^{-2} = 88.4 \, \mu\text{C m}^{-2}$$

using 1 microcoulomb $= 1 \, \mu\text{C} = 10^{-6}$ C. Thus 1 m^2 of membrane will have a net charge of $+88.4 \, \mu$C along the outside surface and a net charge of $-88.4 \, \mu$C on the inside.

18.4 | THE ELECTRIC POTENTIAL

In Chapter Six, we noted that the electric forces among charges at rest are conservative, so that their effects can be included in the potential energy of a system. We also defined the *electric potential V* as the

Figure 18.12. When a positive charge q is moved opposite to the field from A to B, its potential energy increases by $\Delta \mathcal{U} = q \Delta V = qEl$.

potential energy per unit charge. If at a point P, a charge q has potential energy $\mathcal{U}$, then the electric potential at P is

$$V = \frac{\mathcal{U}}{q} \qquad (18.7)$$

The unit of potential is the volt (V), where from this definition 1 volt = 1 joule per coulomb. Colloquially, potential differences are often referred to as *voltages*. We see in the next chapter that it is the potentials rather than the fields that are most useful in discussing electric circuits.

We can find a relationship between the electric potential and field by considering a positive charge q in a constant electric field **E** (Fig. 18.12). We suppose a force **F** equal but opposite to the electric force $q\mathbf{E}$ is applied so that the charge moves at a constant velocity from A to B. When the charge moves a distance l opposite to the field, the applied force does work $Fl = qEl$. Since the kinetic energy remains constant, this work must equal the change in the potential energy of the charge:

$$\Delta \mathcal{U} = qEl \qquad (18.8)$$

Dividing by q, we obtain the change in the electric potential,

$$\Delta V = El \qquad (18.9)$$

The potential energy of a *positive* charge increases when it is moved *opposite* to the field, just as the potential energy of a mass increases when it is moved opposite to the gravitational force or to a greater height. Since the electric force on a *negative* charge is opposite to the field, its potential energy increases when it moves *parallel* to the field. Note that according to Eq. 18.9, the units of the electric field are those of $\Delta V/l$, or volts per meter. Thus the field can be measured either in newtons per coulomb or volts per meter.

We saw in the preceding section that the field be-

Figure 18.13. The potential difference between the parallel plates is $\Delta V = El$.

tween two oppositely charged plates is uniform. If the plate area is A and the charges are $+Q$ and $-Q$, the magnitude of the field is $4\pi kQ/A$. Thus when the plates are separated by a distance l (Fig. 18.13), the potential difference between them is $\Delta V = El$, or

$$\Delta V = \frac{4\pi kQ}{A} l \qquad (18.10)$$

This result is illustrated by the following example.

Example 18.4

Two oppositely charged parallel plates have an area of 1 m² and are separated by 0.01 m. The potential difference between the plates is 100 V. Find (a) the field between the plates, and (b) the magnitude of the charge on a plate.

(a) Since the field is uniform, $\Delta V = El$ can be used, and

$$E = \frac{\Delta V}{l} = \frac{100 \text{ V}}{0.01 \text{ m}} = 10^4 \text{ V m}^{-1}$$

The field is directed from the plate at the higher potential toward the other plate.

(b) Using Eq. 18.10, the charge on a plate has a magnitude

$$Q = \frac{A \,\Delta V}{4\pi kl} = \frac{(1 \text{ m}^2)(100 \text{ V})}{4\pi(9 \times 10^9 \text{ N m}^2 \text{ C}^{-2})(10^{-2} \text{ m})}$$
$$= 8.84 \times 10^{-8} \text{ C}$$

Notice that a small charge is sufficient to produce a 100-V potential difference.

Potential energy of any kind depends only on the position of an object and not on how it arrived there. Stated somewhat differently, the work done by the

Figure 18.14. The work done by the conservative electric force is the same for the paths ABC and AC.

conservative electric force is independent of the path followed between the initial and final positions. Therefore, we may obtain the difference in potential between two points using any convenient path to find the work per unit charge done against the electric field. For example, if we wish to know the difference in potential between points A and C in Fig. 18.14, we can choose the path ABC. AB is a displacement opposite to the field, so $\Delta V = El$. BC is perpendicular to the field, so no work is done, and the potential is the same at B as at C. Using the dashed path AC, we would find the same potential difference. However, the calculation would be slightly more complicated.

Finding the potential difference between two points is more complicated when the field is not uniform; the path must be broken up so that the field is nearly constant in each part. For a single point charge Q, we saw in Chapter Six that the potential at a distance r from the charge is

$$V = \frac{kQ}{r} \qquad (18.11)$$

This expression implies that the electric potential has been chosen to be zero at $r = \infty$. We may do this since only potential differences can be measured, and we may define the potential to be zero at a convenient reference point, which in this case is at $r = \infty$.

If there are several charges, the sum of their potentials at a point is the net potential. Just as in the calculation of electric fields, cancellations among the individual potentials can lead to various dependences on the distance. This cancellation is illustrated by the following example.

Example 18.5

For the dipole in Fig. 18.15, find the electric potential at (a) a distant point, P_1, on the y axis, and (b) a point, P_2, on the x axis.

(a) At P_1, the potentials due to the two charges are

$$V_+ = \frac{kq}{r - a}, \qquad V_- = \frac{-kq}{r + a}$$

Figure 18.15. Example 13.5.

The total electric potential is then

$$V = V_+ + V_- = kq\left[\frac{1}{r - a} - \frac{1}{r + a}\right] = \frac{2kqa}{r^2 - a^2}$$

If r is large compared to a, then a can be neglected in the denominator, so

$$V = \frac{2kqa}{r^2}$$

Thus even though the potential due to a single point charge varies as $1/r$, this combination of two point charges has a potential that diminishes as $1/r^2$.

(b) The two charges are equally distant from *any* point on the x axis. Therefore, at P_2 the two potentials will have the same magnitude but opposite signs, and their sum will be zero. The net potential of the dipole is zero everywhere on the x axis.

18.5 | EQUIPOTENTIAL SURFACES

At any point on the surface of an imaginary sphere of radius r centered on a point charge, Q, the potential has the same value, $V = kQ/r$. A surface on which the potential is the same everywhere is called an *equipotential surface*. Thus, for a point charge the equipotential surfaces are concentric spheres (Fig. 18.16). The equipotential surfaces for a uniform electric field are planes normal to the field.

When a charge moves at right angles to the electric field, no work is done against electrical forces, so its potential energy remains constant. For this reason, *the equipotential surfaces are always perpendicular to the electric field lines*. Charges may move along an equipotential surface with no change in potential energy.

Conductors and Insulators | Most materials may be considered either electrical *conductors* or *insulators*. A conductor is a material, such as a metal or

291

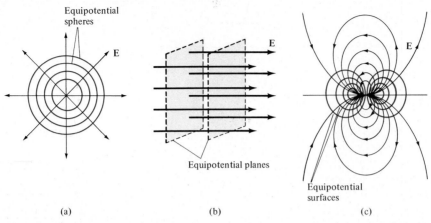

Figure 18.16. Equipotential surfaces for (*a*) a point charge; (*b*) a uniform electric field; (*c*) a dipole. Note that the equipotential surfaces and the electric field lines are always mutually perpendicular.

an ionic solution, in which some charges are relatively free to move about. In an insulator, such as paper or glass, all the charges are relatively immobile.

Conductors have the important property of being equipotential objects when there are no charges in motion. To see this, we note that if all the charges that can move are at rest, the electric field must be zero everywhere in the conductor. Alternatively, only if the field is zero everywhere will the charges stop moving. Consequently the potential is the same at all

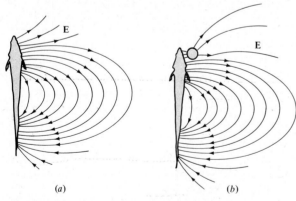

Figure 18.17. A number of fish use electric fields for detection and communication. (*a*) A charge separation in the body of the fish produces electric fields of a dipole type. The strength of the field is monitored by receptors along the body of the fish. (*b*) The field is distorted by the presence of a conducting object. The field lines intersect the surface of the object perpendicular to the surface, which is an equipotential. The resulting distortion of the field is apparent at the fish's body and the receptors sense the change. Thus the fish affectively "sees" using electric fields.

points.

The effect of a conductor on an electric field is illustrated in Fig. 18.17.

18.6 | ELECTRIC DIPOLES

Earlier we defined an electric dipole as a pair of equal and opposite charges, and we briefly discussed the field and potential due to a dipole. Here we consider a dipole placed in a uniform electric field due to other charges.

Atoms and molecules provide many illustrations of electric dipoles. For example, a water molecule has an excess of negative charge near its oxygen atom and a similar positive excess near the hydrogen atoms. Accordingly, it behaves like a small electric dipole, and many of its physical and chemical properties are related to its dipole character.

A dipole is characterized by its *electric dipole moment* (Fig. 18.18). If *l* is the vector distance from $-q$ to $+q$, then the electric dipole moment is defined by

$$\mathbf{p} = q\mathbf{l} \qquad (18.12)$$

This is the definition usually given in physics books. Chemistry books often define **p** so that it points from $+q$ to $-q$. The calculation of the electric dipole moment is illustrated by the following example.

Figure 18.18. An electric dipole consists of equal and opposite charges, $+q$ and $-q$. The electric dipole moment is $\mathbf{p} = q\mathbf{l}$.

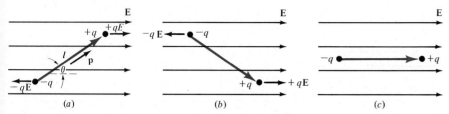

Figure 18.19. The torque on the dipole tend to produce rotations which are (a) clockwise, (b) counterclockwise. (c) The dipole is in equilibrium.

Example 18.6

In a hydrogen atom the electron and proton are separated by 5.29×10^{-11} m. (a) Find the electric dipole moment at one instant in time. (b) The electron moves in a circular path about the proton. Find the average of the electric dipole moment vector over one full orbit.

(a) The proton charge is $e = 1.60 \times 10^{-19}$ C, and the electronic charge is $-e$. The dipole moment points toward the positively charged proton and has a magnitude of

$$p = ql = (1.60 \times 10^{-19} \text{ C})(5.29 \times 10^{-11} \text{ m})$$
$$= 8.45 \times 10^{-30} \text{ C m}$$

(b) If the electric dipole initially points in one direction, half an orbit later it will point in the opposite direction. The average of two vectors equal in magnitude but opposite in direction is zero. Thus the average of the electric dipole moment vector over a complete circular orbit must be zero. For this reason the hydrogen atom has no permanent electric dipole moment.

Figure 18.19 shows an electric dipole in a uniform electric field **E**. The force on the positive charge is $q\mathbf{E}$, and the force on the negative charge is $-q\mathbf{E}$. Their sum is zero, so *the net force on an electric dipole in a uniform electric field is zero.*

However, the net torque on the dipole is not zero, because the equal but opposite forces have different lines of action and form a couple. The torque due to a couple is the same relative to any point, so we may choose to compute torques about the charge $-q$. The force on $-q$ then has no lever arm, and it produces no torque. The charge $+q$ is subjected to a force $q\mathbf{E}$ and acts at a distance l. Thus the torque $\mathbf{\Gamma} = \mathbf{r} \times \mathbf{F}$ becomes

$$\mathbf{\Gamma} = \mathbf{l} \times (q\mathbf{E}) = \mathbf{p} \times \mathbf{E} \qquad (18.13)$$

The magnitude of the torque is $pE \sin \theta$, and it is directed so that the dipole tends to line up with the field. When the dipole is oriented as in Fig. 18.19a, the torque is directed into the page, tending to rotate

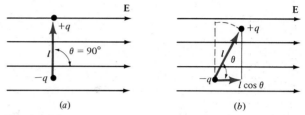

Figure 18.20. The energy of the dipole $\mathcal{U}$ is zero in (a), negative in (b).

the dipole clockwise. The torque is directed out of the page in Fig. 18.19b and tends to rotate the dipole counterclockwise. The torque is zero, and the dipole is in stable equilibrium when it is directed along the field (Fig. 18.19c).

Conventionally the potential energy $\mathcal{U}$ of a dipole in a uniform field is taken to be zero when $\theta = 90°$; the dipole is then perpendicular to the field (Fig. 18.20a). The energy is a minimum when the dipole is parallel to the field. Thus, when the dipole is in the position shown in Fig. 18.20b, its potential energy must be negative. We can find an expression for $\mathcal{U}$ by imagining that the dipole is rotated from $\theta = 90°$, where $\mathcal{U} = 0$, to θ. If the rotation is performed with $-q$ held in place, the energy of the negative charge remains unchanged. The displacement of the positive charge along the field is $l \cos \theta$, so the work done against the electrical force $q\mathbf{E}$ is $-qE(l \cos \theta)$. Equating this work to the change in potential energy, we find

$$\mathcal{U} = -qEl \cos \theta = -pE \cos \theta \qquad (18.14)$$

These ideas are illustrated by the following example.

Example 18.7

An atom with an electric dipole moment of 8.45×10^{-30} C m is in a uniform electric field of 10^4 N C^{-1}. If the angle between **p** and **E** is 30°, find (a) the magnitude of the torque, and (b) the potential energy.

(a) When $\theta = 30°$, $\sin\theta = 0.5$, and the torque has a magnitude

$$\Gamma = pE\sin\theta$$
$$= (8.45 \times 10^{-30}\text{ C m})(10^4\text{ N C}^{-1})(0.5)$$
$$= 4.23 \times 10^{-26}\text{ N m}$$

(b) At $\theta = 30°$, $\cos\theta = 0.866$, and the potential energy is

$$\mathcal{U} = -pE\cos\theta$$
$$= -(8.45 \times 10^{-30}\text{ C m})(10^4\text{ N C}^{-1})(0.866)$$
$$= -7.33 \times 10^{-26}\text{ J}$$

18.7 | THE OSCILLOSCOPE

Except for some simple meters, probably no scientific instrument is used as widely as the *oscilloscope*. The electrical circuitry contained in an oscilloscope is quite elaborate, and a sophisticated model has so many control knobs and switches that it takes some time to become acquainted with its operation. Nevertheless, the basic principles of its major component, the *cathode-ray tube,* can be understood fully with the ideas developed in this chapter.

In the first part of the tube, which is the electron gun, a constant potential difference accelerates electrons emitted from a hot filament (Fig. 18.21). If there are no other forces on the electron, they travel along a straight line and strike the center of the fluorescent screen, producing a bright spot. Horizontal deflections can be produced with horizontal electric fields

Figure 18.21. A cathode-ray tube. Electrons are accelerated from the negative cathode toward the positive accelerating anode and then pass between the two pairs of deflection plates. When they strike the fluorescent screen, light is emitted.

between the parallel plates marked 1 in Fig. 18.21. These plates have a separation l. If they are at a potential difference V_1, there is a uniform horizontal electric field between them $E_1 = V_1/l$. This field accelerates the electrons, which then strike the screen at a horizontal distance from the center, which is proportional to V_1. If V_1 is gradually increased the spot gradually moves or *sweeps* across the screen; when V_1 returns to its original value the spot returns to its starting point. If this sweep is repeated at a high enough frequency, the persistence of the image on the screen and in the eye conceals the motion, and a straight line is seen (Fig. 18.22b).

In the same way, a potential difference V_2 applied to the plates marked 2 will cause vertical deflections. For example, a constant V_2 will shift the line as in

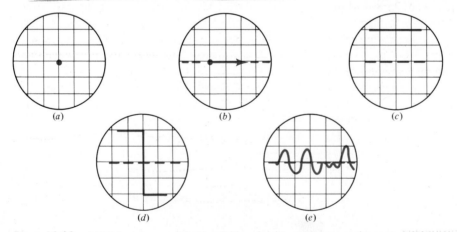

Figure 18.22. (*a*) If there are no deflecting fields, a bright spot is seen at the center of the oscilloscope screen. (*b*) Gradually changing the horizontal potential difference causes the beam spot to move steadily across the screen. If this sweep is repeated rapidly enough, a continuous line is perceived. (*c*) A constant vertical field displaces the line. (*d*) If the vertical field quickly reverses after each half sweep, this pattern is seen. (*e*) Complex variations of the vertical field with time can be displayed if they repeat at the sweep frequency.

294

Fig. 18.22*c*. Usually an unknown potential difference V_2, varying at some frequency f, produces the vertical deflection. The horizontal sweep frequency is adjusted until it is equal to f (or f divided by some integer), so every time the vertical signal V_2 repeats itself the horizontal sweep is at the same point in its cycle. Thus the beam repeatedly hits the same points on the screen, and a stable pattern is seen (Fig. 18.22*d,e*). Accurate measurements of both the frequency and the magnitude of the unknown potential difference can readily be made in this way.

Since almost any type of information can be converted into electric potential differences, oscilloscopes are used in laboratories of virtually every kind. They are extremely valuable in qualitative and quantitative studies, not only of electrical variables but also of mechanical, acoustical, and other quantities. Microphones and television cameras are examples of devices that convert energy from other forms into varying electrical potentials. In general, devices that convert energy from one form to another are called *transducers*.

18.8 | CAPACITANCE

Suppose that two conductors are initially electrically neutral and that we then take small amounts of charge from one and place them on the other. As this process continues, a potential difference develops. The ratio of the amount of charge transferred to the potential difference resulting is defined to be the *capacitance* of the two conductors and turns out to be independent of the charge transferred. An arrangement of two conductors separated by a vacuum or an insulator is called a *capacitor*.

As our description suggests, the capacitance is a measure of the amount of charge separation that can be maintained at a given potential difference. The energy required to separate the charges is stored in the capacitor. Hence the capacitance is also a measure of the ability to store energy.

Examples of capacitance can be found in nature. For instance, cell membranes separate thin layers of ions in the fluids inside and outside a cell. Hence the membrane and the adjacent fluids are considered to have capacitance.

Capacitors are widely used in electrical circuits. They are used in radio and television tuners, in auto-

Figure 18.23. A parallel plate capacitor.

mobile ignition systems, and in the starting circuits of electric motors. Capacitors also influence the way currents change in time. In nerve cells the rate of transmission of a nerve pulse depends on the membrane capacitance. Some of these applications are discussed in later chapters.

If two conductors have equal and opposite charges $\pm Q$ and a corresponding potential difference V, the ratio Q/V is usually found to be a constant independent of Q. The ratio is the capacitance,

$$C = \frac{Q}{V} \qquad (18.15)$$

Note that V is defined to be the potential of the positive plate less that of the negative plate, so C is positive.

The unit of capacitance is the *farad* (F); $1\ \mathrm{F} = 1\ \mathrm{C\ V^{-1}}$. Because the coulomb is a large unit, the farad is also large, and most capacitors have small values in terms of the farad. Hence we define the microfarad and picofarad by

$$1\ \mu\mathrm{F} = 10^{-6}\ \mathrm{F}, \qquad 1\ \mathrm{pF} = 10^{-12}\ \mathrm{F}$$

The Parallel Plate Capacitor | The simplest capacitor is composed of two parallel plates in a vacuum. The plates have surface area A, charges $\pm Q$, and a separation l (Fig. 18.23). From Eq. 18.10, $V = 4\pi k Q l / A$. Thus the capacitance $C = Q/V$ is

$$C = \frac{A}{4\pi k l} \qquad (18.16)$$

It is conventional to define a new constant, ε_0, as

$$\varepsilon_0 = \frac{1}{4\pi k} = 8.85 \times 10^{-12}\ \mathrm{C^2\ N^{-1}\ m^{-2}}$$

295

Our final expression for the capacitance of a parallel plate capacitor is then

$$C = \frac{\varepsilon_0 A}{l} \qquad (18.17)$$

Note that C depends only on the geometry of the arrangement and not on the charge Q. The capacitance increases with the area of the plates and decreases with their separation. This is also true for capacitors with more complicated shapes. Numerical examples of parallel plate capacitors are given in the next section.

18.9 | EFFECTS OF DIELECTRICS

If an insulator or *dielectric* is introduced between the plates of a parallel plate capacitor with fixed charges, the capacitance is increased. This occurs because the electric field due to the charges on the plates distorts the charge distributions of the molecules in the dielectric, giving each molecule a small *induced* electric dipole moment. These dipoles, in turn, reduce the overall electric field and hence the potential difference between the plates.

To study this in detail, consider a neutral molecule that normally has its centers of positive and negative charge coincident so that it has no permanent dipole moment. When an external electric field is applied, it separates the charge centers by a distance proportional to the strength of the field (Fig. 18.24). Thus, each molecule acquires an *induced* electric dipole moment.

Inside the dielectric the effects of the displaced

Figure 18.24. The molecules in a dielectric are distorted by the electric field **E** between the parallel plates. The charge on the plates is the same before and after the dielectric is introduced.

positive and negative charges cancel. However, at the left surface, there is an excess of positive charge, and at the right surface, there is an excess of negative charge. This produces an electric field **E′** directed opposite to **E**. Thus the actual electric field between the plates is reduced to an effective value,

$$E_{\text{eff}} = E - E'$$

Since the separation of the charges increases with **E**, the field **E′** they produce is also proportional to **E**. Thus we can write

$$\mathbf{E}_{\text{eff}} = \frac{1}{K}\mathbf{E} \qquad (18.18)$$

The *dielectric constant K* is a dimensionless number that indicates the reduction of the field due to the dielectric. K is one for a vacuum, and greater than one for a dielectric.

Table 18.1 lists representative dielectric constants. In some substances, such as water, the molecules have permanent electric dipole moments that tend to line up with the field. The degree of alignment increases with the field, but decreases as the temperature rises due to thermal disordering. Thus K depends on temperature.

The potential difference between the plates is $V = E_{\text{eff}} l = El/K$. Since K is greater than one for a dielectric, V is reduced when a dielectric is inserted. Consequently, the *capacitance $C = Q/V$ increases by the factor K.* An increase in C occurs for any capacitor, not just a parallel plate arrangement. In a parallel

TABLE 18.1

Dielectric constants of several common insulators. Note that for air we can usually use $K = 1$, the value for a vacuum

Material	Temperature, °C	Dielectric Constant, K
Air (dry, at 1 atm)	20	1.00059
Glass	25	5–10
Water	25	78
	80	61
Plastics	20	3–20
Titanium dioxide	20	100
Axon membrane (unmyelinated)	37	8
Paper	20	3.5

Figure 18.25. A small section of a cell membrane.

plate capacitor, the capacitance changes from $\varepsilon_0 A / l$ to

$$C = \frac{K\varepsilon_0 A}{l}. \qquad (18.19)$$

The following examples illustrate the effects of dielectrics on the capacitance.

Example 18.8

A capacitor is made of two foils, each of surface area 1 m², separated by paper 0.05 mm $= 5 \times 10^{-5}$ m thick. What is its capacitance?

(a) According to Table 18.1, $K = 3.5$ for paper. Thus the capacitance is

$$C = \frac{K\varepsilon_0 A}{l}$$

$$= \frac{(3.5)(8.85 \times 10^{-12} \text{ C}^2 \text{ N}^{-1} \text{ m}^{-2})(1 \text{ m}^2)}{(5 \times 10^{-5} \text{ m})}$$

$$= 6.19 \times 10^{-7} \text{ F} = 0.619 \, \mu\text{F}$$

Example 18.9

The ions inside and outside a cell are separated by a flat membrane 10^{-8} m thick with a dielectric constant $K = 8$. Find the capacitance of 1 cm² of membrane (Fig. 18.25).

For an area $A = 1$ cm² $= 10^{-4}$ m², the capacitance is

$$C = \frac{K\varepsilon_0 A}{l} = \frac{8(8.85 \times 10^{-12} \text{ C}^2 \text{ N}^{-1} \text{ m}^{-2})(10^{-4} \text{ m}^2)}{10^{-8} \text{ m}}$$

$$= 7.08 \times 10^{-7} \text{ F} = 0.708 \, \mu\text{F}$$

When the electric field in a dielectric becomes sufficiently strong, large numbers of free electrons and ions are produced, and the material becomes an excellent conductor. This *breakdown* occurs at a critical electric field called the *dielectric strength*. Lightning is a spectacular example of this phenomenon in which the air becomes a conductor (Fig. 18.26). Commercial capacitors are marked with the maximum voltage that can be applied without risk of exceeding the dielectric strength and damaging the capacitor or other parts of a circuit.

Figure 18.26 When low-lying clouds accumulate a charge, there is a potential difference between the cloud and the ground. If this potential difference is large enough, the electric field exceeds the dielectric strength of the air. The air then ionizes and becomes a good electrical conductor. (Wide World Photos.)

18.10 | ENERGY STORED IN A CAPACITOR

A charged capacitor stores electrical energy. If its plates are connected by a conducting wire, electrons will move in the wire from the negative plate to the positive plate. This charge flow or *current* continues until the plates are neutralized and can be used to operate an electronic flash gun or trigger an artificial heart pacemaker. Thus, stored electrical energy can be transformed into other forms of energy.

The energy stored in a capacitor may initially be supplied by a battery, which maintains a potential difference between its two terminals. When the plates of an uncharged capacitor are connected by two wires to the terminals, electrons flow in the wires from one plate through the battery to the other until the potential difference across the capacitor reaches a maximum value. We can find the energy stored in the capacitor by calculating the work that must be done by the battery in building up this charge from zero to the final value Q.

To avoid minus signs, we suppose that positive charges are gradually transferred from one plate to the other. (The final result is independent of the sign

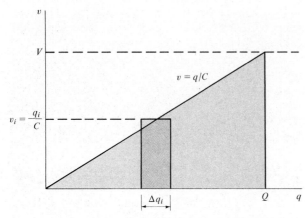

Figure 18.27. The energy stored in a capacitor is equal to the area of the triangle, $\frac{1}{2}$(base)(height) $= \frac{1}{2}QV$. This result follows from summing the areas of small rectangles such as the one shown, which has an area $v_i \Delta q_i = q_i \Delta q_i / C$.

of the moving charges.) As one plate acquires an increasingly large positive charge and the other a corresponding negative charge, the potential difference v between them increases. If at some instant the charges are $+q_i$ and $-q_i$, then from the definition of capacitance, $C = q_i / v_i$, we have $v_i = q_i / C$. Transferring a small additional amount of charge Δq_i therefore requires that work ΔW be done, where

$$\Delta W = v_i \, \Delta q_i = \frac{q_i}{C} \Delta q_i$$

Since the electric force is conservative, the work done by the battery must equal the increase $\Delta \mathcal{U}_i$ in the stored energy of the capacitor, so

$$\Delta \mathcal{U}_i = \frac{q_i}{C} \Delta q_i$$

To find the total energy of the capacitor, we must sum the small increases in its energy $\Delta \mathcal{U}_i$ associated with each charge transfer Δq_i (Fig. 18.27). This sum is mathematically identical to several others we have done and is equal to the triangular area under the straight line graph of $v = q/C$ versus q (Fig. 18.27) as q increases from 0 to Q. Thus the total energy stored in the capacitor is

$$\mathcal{U} = \tfrac{1}{2}(\text{base})(\text{height}) = \tfrac{1}{2}QV$$

Using $C = Q/V$, the energy can be expressed in three equivalent forms,

$$\mathcal{U} = \frac{1}{2}QV = \frac{1}{2}\frac{Q^2}{C} = \frac{1}{2}CV^2 \qquad (18.20)$$

The energy stored in a cell membrane is calculated in the following example.

Example 18.10

One square centimeter of membrane has a capacitance of 7.08×10^{-7} F. If the potential difference across the membrane is 0.1 V, find the electrical energy stored in 1 cm² of membrane.

Since we know C and V, we write the energy as $\frac{1}{2}CV^2$. Then

$$\begin{aligned}\mathcal{U} &= \tfrac{1}{2}CV^2 = \tfrac{1}{2}(7.08 \times 10^{-7} \text{ F})(0.1 \text{ V})^2 \\ &= 3.54 \times 10^{-9} \text{ J}\end{aligned}$$

SUMMARY

A charge produces an electric field in its vicinity. The field, in turn, exerts an electric force on any other charges present. The field due to two or more charges is the vector sum of the fields due to each of the charges. The field of a single charge varies as $1/r^2$, but the net field due to an arrangement of charges can vary in quite different ways. This results from the cancellations that occur when the fields are added. Similarly, the electric potential due to several charges is the algebraic sum of the potentials due to the individual charges and can have a dependence on the distance that is different from the $1/r$ variation due to a single point charge.

A capacitor is a pair of conductors separated by a vacuum or an insulator. If the conductors are given equal but opposite charges, the ratio $C = Q/V$ of the charge to the resulting potential difference is the capacitance. Other arrangements of conductors and insulators, such as those in living cells, can also be regarded as having capacitance.

When an insulator is placed between the plates of a charged capacitor, electric dipoles are induced in the material. The dipole fields oppose the applied field and reduce the potential difference between the plates, thereby increasing the capacitance.

Checklist

Define or explain:

Coulomb's law   electric dipole
electric field   electric potential
electric field lines   voltage

equipotential surface
conductor
insulator
cathode-ray tube
capacitance

dielectric
induced dipole moment
dielectric constant
dielectric strength

EXERCISES

Several of the exercises below involve atomic quantities. The magnitude of the charge on an electron or a proton is $e = 1.60 \times 10^{-19}$ C, and the electron mass is 9.11×10^{-31} kg.

Section 18.1 | Electric Forces

18-1   Find the magnitude and direction of the force on the charge Q in Fig. 18.28.

Figure 18.28   Exercises 18-1, 18-2, 18-3, 18-11 and 18-12.

Figure 18.29.   Exercise 18-4.

18-2   If $Q = 10^{-6}$ C and $b = 0.1$ m, what is the magnitude and direction of the force on the charge $2Q$ in Fig. 18.28?

18-3   An additional charge Q is placed at the origin in Fig. 18.28. What is the force on this charge?

18-4   Find the magnitude and direction of the force on the charge $-Q$ in Fig. 18.29.

18-5   In a NaCl molecule, a Na^+ ion with charge e is 2.3×10^{-10} m from a Cl^- ion with charge $-e$. What is the magnitude of the force between them?

18-6   A cell membrane 10^{-8} m thick has positive ions on one side and negative ions on the other. What is the force between two ions with charges $+e$ and $-e$ at this separation?

Section 18.2 | The Electric Field and
Section 18.3 | The Electric Field Due to
Arrangements of Charges

18-7   A uranium nucleus has a charge of $92e$. (a) What is the direction and magnitude of the electric field due to the nucleus at a distance of 10^{-10} m from the nucleus? (b) What is the direction and magnitude of the force on an electron at this distance?

18-8   What is the force on an electron in a field of 10^5 N C^{-1}?

18-9   An electron is accelerated at 10^8 m s^{-2} by an electric field. What is the direction and magnitude of the field?

18-10   Find the magnitude and direction of the electric field 0.1 m from a charge of -10^{-4} C.

18-11   In Fig. 18.28, find the electric field at the origin.

18-12   In Fig. 18.28, find the electric field at $x = 0$, $y = -b$.

18-13   The electric field near a uniformly charged circular plate of area 0.1 m^2 is directed toward the plate and has a magnitude of 10^4 N C^{-1}. Find the charge on the plate.

18-14   Two square plates of side 0.1 m have equal and opposite charges of $\pm 10^{-6}$ C, which are uniformly distributed. The plates are separated by 10^{-2} m. (a) What is the magnitude and direction of the electric field? (b) What is the magnitude and direction of the force on an electron placed in this field? (c) How much work must be done against the field to move an electron from the positive plate to the negative plate?

Section 18.4 | The Electric Potential

18-15   A carbon nucleus has a charge of $+6e$. At a distance of 10^{-10} m from a carbon nucleus, find (a) the electric potential; (b) the potential energy of an electron.

18-16   In Fig. 18.30, what is the potential at (a) the origin; (b) $x = 3a$, $y = 0$?

18-17   Two uniformly charged metal plates separated by 0.04 m produce a uniform field between them of 10^4 N C^{-1}. Find (a) the charge per unit

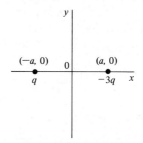

Figure 18.30.   Exercise 18-16.

area Q/A on the plates, and (b) the potential difference between the plates.

18-18   At the center of the square in Fig. 18.31, find (a) the electric field, and (b) the potential.

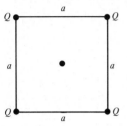

Figure 18.31.   Exercise 18-18.

Section 18.6 | Electric Dipoles

18-19   The NH_3 molecule has a permanent electric dipole moment of 5.0×10^{-30} C m. If this arises from net charges of $+e$ and $-e$ in two regions of the molecule, what is their separation?

18-20   Although a neutron has no net charge, it has a charge distribution consisting of spherical layers of positive and negative charge. Its electric dipole moment has not been detected; if it has one, the moment must be less than 10^{-45} C m according to recent experiments. The diameter of a neutron is about 10^{-15} m. How large must equal and opposite charges at a distance of 10^{-15} m be to produce an electrical dipole moment of 10^{-45} C m? Express your result in multiples of e.

18-21   What energy is required to "flip" an electric dipole **p** from its position parallel to an electric field **E** to the antiparallel position?

Section 18.8 | Capacitance

18-22   A capacitor has a potential difference of 100 V when its plates have charges of magnitude 10^{-5} C. What is its capacitance?

18-23   Two square metal plates with sides of length 0.1 m are separated in vacuum by 10^{-3} m. Find their capacitance.

Section 18-9 | Effects of Dielectrics

18-24   A capacitor made of thin foils of aluminum separated by paper 10^{-4} m thick is found to have a capacitance of 1 μF. What is the area of the foil?

18-25   Metal plates are placed on either side of a sheet of glass, and their capacitance is measured to be 10^{-5} F. If they are kept at the same separation in air, their capacitance is found to be 2×10^{-6} F. What is the dielectric constant of the glass?

18-26   A lightning bolt travels 500 m from a cloud to a mountain top. What is the potential difference between the cloud and the peak? (Assume that the electric field is uniform and that the air breaks down and becomes a conductor when the field reaches 8×10^5 V m^{-1}.)

Section 18.10 | Energy Stored in a Capacitor

18-27   A 50-μF capacitor in an electronic flash gun supplies an average power of 10^4 W for 2×10^{-3} seconds. (a) To what potential difference must the capacitor initially be charged? (b) What is its initial charge?

18-28   An electronic flash gun uses energy stored in a capacitor. (a) How large a capacitor is needed to provide 30 J of energy if it is charged to 1000 V? (b) How much charge is on its plates?

18-29   Can you run a car on energy stored in a capacitor? Make a rough estimate to support your answer.

18-30   Two parallel metal plates of area 0.1 m^2 are separated in air by 0.01 m. Their potential difference is 1000 V. Find (a) the charge on the plates, and (b) the stored energy. (c) A slab of dielectric with $K = 10$ is inserted so that it fills the space between the plates. The charge on the plates remains the same. Find the new potential difference and stored energy. (d) Explain the change, if any, in the stored energy.

PROBLEMS

18-31   An electron is accelerated by a constant electric field from rest to a velocity of 10^6 m s^{-1}. If the accelerating region is 0.2 m long, what is the magnitude of the electric field?

18-32   The charge on the electron was first determined by Millikan in 1909 by measuring the electric field necessary to balance the gravitational force on a tiny oil droplet. Suppose the net charge of the droplet is due to one excess electron and the droplet mass is 10^{-18} kg. Find the magnitude and direction of the vertical electric field required to support the droplet.

18-33   A charge of -10^{-6} C is placed at the origin of an x-y plane. Find the magnitude and direction of the electric field at (a) $x = 1$ m, $y = 0$;

(b) $x = 0$, $y = -2$ m; (c) $x = 2$ m, $y = 2$ m.

18-34 A circular metal plate of radius 0.2 m has 10^{10} excess electrons uniformly distributed over its surfaces. What is the magnitude and direction of the field just outside the plate near its center?

18-35 Estimate the magnitude and direction of the field at a distance of 10 m from the center of the plate in the preceding problem.

Figure 18.32. Problem 18-36.

18-36 An electron is projected with a velocity v_0 into a uniform electric field (Fig. 18.32). (Neglect the gravitational force on the electron.) (a) Find the direction and magnitude of its acceleration. (b) How long will it be in the field? (c) How far will it be deflected vertically as it leaves the field? (d) Find the angle between its velocity as it leaves the field and its original direction.

18-37 A charge q is placed at the origin and a charge $2q$ is placed at $x = a$, $y = 0$. Find the potential at $x = a$, $y = a$.

18-38 A particle of mass m and positive charge q is projected with a velocity $\mathbf{v}$ into a region where a uniform electric field $\mathbf{E}$ is opposite to $\mathbf{v}$. How far will the particle travel before it comes momentarily to rest?

18-39 If an atom is placed in an electric field $\mathbf{E}$, its charge distribution will be distorted so that an induced electric dipole moment $\mathbf{p} = \alpha\mathbf{E}$ is produced, where α is the *polarizability* of the atom. (a) An atom of polarizability α is a distance r from an ion with charge $+e$, where r is large compared to the size of the atom. What is the induced dipole moment? (b) What is the potential energy of the atom and ion?

* **18-40** Two dipoles $\mathbf{p}_1$ and $\mathbf{p}_2$ are a distance R apart, which is large compared to their charge separation. Find the total electric energy of the dipoles when they are oriented as in Fig. 18.33a and b. (A dipole at the origin directed along the y axis has an electric field at a distant point on the x axis that is $\mathbf{E} = -kp\,\hat{\mathbf{y}}/x^3$, and on the y axis the field is $\mathbf{E} = 2kp\,\hat{\mathbf{y}}/y^3$.)

Figure 18.33. Problem 18-40.

18-41 The electric dipole moment of a water molecule is 6.13×10^{-30} C m. (a) If this dipole moment is due to a pair of point charges $\pm e$, how far apart must they be? Find the ratio of this distance to the radius of a hydrogen atom, 5.29×10^{-11} m. (b) If the dipole is parallel to an electric field of 10^6 V m^{-1}, how much energy is needed in joules and in electron volts to "flip" the dipole so that it is opposite to the field? (c) At room temperature, the average kinetic energy of a molecule is about 0.04 eV. What implication does this have for the orientation of water molecules in the relatively strong field of 10^6 V m^{-1}?

Figure 18.34. Problem 18-42.

* **18-42** Figure 18.34 shows a model of a water molecule. Each hydrogen atom has a net positive charge q, and the oxygen atom has a net charge $-2q$. The distance l is 9.65×10^{-11} m. The positive charge on a hydrogen atom and half the negative charge on the oxygen atom form a dipole. The total molecular electric dipole moment $\mathbf{p}$ is the vector sum of the two H—O dipole moments. (a) What is the direction of $\mathbf{p}$? (b) If $p = 6.0 \times 10^{-30}$ C m, what is the charge q in multiples of the proton charge e?

18-43 A certain nerve fiber (axon) is a cylinder 10^{-4} m in diameter and 0.1 m long. Its interior has a potential 0.09 V below that of the surrounding fluid; it is separated from that fluid by a thin membrane. Na$^+$ ions are transported by a chemical reaction out of the fiber at the rate of 3×10^{-11}

moles per second per cm² of membrane. (a) How many coulombs of charge per hour are transported out of the fiber? (b) How much work per hour must be done against the electrical forces?

Figure 18.35 Problems 18-44 and 18-45.

Figure 18.36 Problems 18-47 and 18-48.

* **18-44**   The two capacitors in Fig. 18.35 are said to be connected in *parallel*. They are connected to a battery so the potential difference across each is V. Show that a single capacitor C_p will store the same amount of charge if $C_p = C_1 + C_2$. (C_p is called the *equivalent capacitance*.)

18-45   In Fig. 18.35, $C_1 = 2\,\mu\text{F}$ and $C_2 = 4\,\mu\text{F}$. Using the result of the preceding problem, find the equivalent single capacitance that would have the same charge when connected to the battery.

18-46   Using the result of Problem 18-44, show how an equivalent capacitance of $10\,\mu\text{F}$ can be assembled from a supply of 2-μF capacitors.

* **18-47**   The capacitors in Fig. 18.36 are said to be in *series*. When they are connected to the battery as shown, the charge Q on each is the same. Show that a single capacitor C_s with charge Q will have a potential difference $V = V_1 + V_2$ if

$$\frac{1}{C_s} = \frac{1}{C_1} + \frac{1}{C_2}$$

(C_s is called the equivalent capacitance.)

18-48   In Fig. 18.36, $C_1 = 2\,\mu\text{F}$ and $C_2 = 4\,\mu\text{F}$. Using the result of the preceding problem, find the equivalent single capacitance which would maintain the same potential difference $V = V_1 + V_2$ with the same charge Q.

Additional Reading:

R.E. Orville, The Lightning Discharge, *Physics Teacher*, vol. 14, 1976, p. 7.

R.M. Alexander, *Functional Design in Fishes*, Hutchinson University Library, London, 1967. Chapter 6 describes the electrical senses of fish.

Robert Burton, *Animal Senses*, Taplinger Publishing Company, Inc., New York, 1970.

R.H. Bullock, Seeing the World Through a New Sense: Electroreception in Fish, *American Scientist*, May–June 1973, p. 316.

Carl D. Hopkins, Electric Communication in Fish, *American Scientist*, July–August 1974, p. 426.

R.T. Cox, Electric Fish, *American Journal of Physics*, vol. 11, 1943, p. 13.

Russell K. Hobbie, The Electrocardiogram as an Example of Electrostatics, *American Journal of Physics*, vol. 41, 1973, p. 824.

Scientific American articles:

George Shiers, Ferdinand Brown and the Cathode Ray Tube, March 1974, p. 92.

H. Kondo, Michael Faraday, October 1953, p. 90.

A.D. Moore, Electrostatics, March 1972, p. 46.

Herbert A. Pohl, Nonuniform Electric Fields, December 1960, p. 106.

C. Andrave Bassett, Electrical Effects in Bone, October 1965, p. 18.

L.B. Loeb, The Mechanism of Lightning, January 1949, p. 22.

A.A. Few, Thunder, July 1975, p. 80.

J.E. McDonald, The Earth's Electricity, April 1953, p.32.

Harry Grundfest, Electric Fishes, October 1960, p. 115.

H.W. Lissman, Electric Location by Fishes, March 1963, p. 50.

H.B. Steinback, Animal Electricity, February 1950, p. 40.

CHAPTER NINETEEN

DIRECT CURRENTS

Most applications of electricity and magnetism involve moving charges or *electrical currents* in conductors. Direct currents (dc) are produced when a conducting path exists between the terminals of a battery or a dc generator. These devices tend to maintain a constant potential difference between their terminals and convert other kinds of energy, such as chemical or mechanical energy, into electrical energy. An alternating current (ac) is produced by an ac generator, which has a terminal potential difference that alternates in sign at some characteristic frequency. Many of the ideas we consider in this chapter can be applied immediately or with minor changes to alternating currents as well as to direct currents.

We begin this chapter by defining *electric current* and *resistance*. We then consider sources of energy and the transformation of energy in circuits. In the remaining sections, we discuss methods of analyzing complex circuits, electrical meters, the charging of a capacitor through a resistor, and electrical safety.

19.1 | ELECTRIC CURRENT

The *electric current* in a wire is the rate at which charge moves in the wire. For example, in Fig. 19.1, charges move through a conducting wire under the influence of an applied electric field. If a net charge ΔQ crosses the shaded cross-sectional area in a time Δt, the *average current* is

$$\bar{I} = \frac{\Delta Q}{\Delta t} \qquad (19.1)$$

Figure 19.1. A segment of a conducting wire. Charges enter through the left end and leave through the right end.

and the *instantaneous current* is

$$I = \frac{dQ}{dt} \qquad (19.2)$$

The S.I. current unit is the *ampere* (A). Often it is convenient to use the *milliampere* (mA); $1 \text{ mA} = 10^{-3}$ A. From the definition, it follows that an ampere is a coulomb per second. (The ampere is defined by the magnetic force between two currents under specified conditions. This, in turn, defines the coulomb. See Section 21.8.) The definition of current is used in the following example taken from electrochemistry.

Example 19.1
An electrochemical cell consists of two silver electrodes placed in an aqueous solution of silver nitrate. A constant 0.5-A current is passed through the cell for 1 hour. (a) Find the total charge transported through the cell in coulombs and in multiples of the electronic charge. (b) Each electron reaching the cell discharges one positively charged silver ion, which is then deposited on the negative electrode (cathode). What is the total mass of the deposited silver? (The atomic mass of silver is 107.9 amu.)

(a) Since the current is constant,

$$\Delta Q = I\,\Delta t = (0.5\ \text{A})(1\ \text{h})$$
$$= (0.5\ \text{C s}^{-1})(3600\ \text{s}) = 1800\ \text{C}$$

The ratio of ΔQ to the electronic charge is

$$N = \frac{\Delta Q}{e} = \frac{1800\ \text{C}}{1.60 \times 10^{-19}\ \text{C}} = 1.13 \times 10^{22}$$

This is the number of silver ions transported through the cell and deposited in one hour.

(b) The mass of the deposited silver is the number of atoms N times the mass of an atom. Using 1 amu $= 1.66 \times 10^{-27}$ kg,

$$m = (1.13 \times 10^{22})(107.9\ \text{amu})(1.66 \times 10^{-27}\ \text{kg amu}^{-1})$$
$$= 2.02 \times 10^{-3}\ \text{kg}$$

Conventionally the current in a conductor is assumed to be in the direction of motion of positive charges. However, in metallic conductors the moving charges are electrons. In metals, some of the electrons become detached, leaving behind positively charged ions. The heavy ions form a regularly spaced crystalline lattice and vibrate about their equilibrium positions with an energy and an amplitude that increases with the temperature. The detached *conduction electrons* move randomly among the ions. In the absence of an applied electric field, the average charge flow in any direction is zero. When an electric field is applied, the electrons acquire an average *drift velocity* opposite to the field, and there is a net current.

We can relate the current in a wire to the density of conduction electrons and their drift velocity v. If there are n electrons per unit volume, then the total number of electrons in a volume V is nV. The segment of wire in Fig. 19.1 has length l and cross-sectional area A, so its volume is $V = lA$. Hence there are nlA electrons in the wire, with a total charge of magnitude $enlA$. The time needed for all of them to pass through the end of the segment shown is $\Delta t = l/v$, so the magnitude of the current is

$$I = \frac{\Delta Q}{\Delta t} = \frac{enlA}{l/v} = enAv \qquad (19.3)$$

The current is the product of the electronic charge, the density of conduction electrons, the area, and the average drift velocity. Note that a positive charge flow in one direction is equivalent to a negative flow in the opposite direction. The following example shows that the drift velocity in a typical metal is surprisingly small.

Example 19.2

Number 12 copper wire is often used to wire household electrical outlets. Its radius is 1 mm $= 10^{-3}$ m. If it carries a current of 10 A, what is the drift velocity of the electrons? (Metallic copper has one conduction electron per atom, the atomic mass of copper is 64 amu, and the density of copper is 8900 kg m^{-3}.)

From Eq. 19.3, the drift velocity is $v = I/neA$. We are given the current I, and we can immediately find the area A from the radius of the wire, leaving only the electron density to be found. Since copper has one conduction electron per atom, n equals the number of atoms per unit volume. The number of atoms per unit volume times the mass of one atom M equals the mass of a unit volume of copper, which is its density d. (We will use d here for density rather than ρ as in Chapter Fourteen to avoid confusion with the resistivity ρ defined below.) Thus $nM = d$, or

$$n = \frac{d}{M} = \frac{8900\ \text{kg m}^{-3}}{(64\ \text{amu})(1.66 \times 10^{-27}\ \text{kg amu}^{-1})}$$
$$= 8.38 \times 10^{28}\ \text{m}^{-3}$$

The drift velocity is then

$$v = \frac{I}{neA} = \frac{I}{ne\pi r^2}$$
$$= \frac{(10\ \text{A})}{(8.38 \times 10^{28}\ \text{m}^{-3})(1.6 \times 10^{-19}\ \text{C})\pi(10^{-3}\ \text{m})^2}$$
$$= 2.37 \times 10^{-4}\ \text{m s}^{-1}$$

Thus the electrons move very slowly, contrary to what one might have supposed; it takes about 4200 seconds or over an hour for them to move a meter! The average electron thermal velocity at room temperature is about 10^5 m s^{-1}, which is about 10^{10} times the drift velocity. On the other hand, changes in the electric field travel in a wire at nearly the speed of light. This is analogous to the situation in a fluid such as air or water, where the effects of pressure changes travel much faster than the fluid itself.

19.2 | RESISTANCE

In Chapter Seventeen, we defined the resistance of a section of pipe to the flow of a fluid as the pressure difference divided by the flow rate. Similarly, the *electrical resistance* R of a conductor is the potential difference V between its ends divided by the current I.

$$R = \frac{V}{I} \qquad (19.4)$$

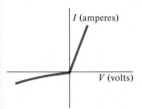

(a)

Figure 19.2. (a) A copper wire has a variable potential difference V between its ends. (b) The current varies linearly with the potential difference in the wire, so it is an ohmic conductor.

Figure 19.3. The current versus potential difference graph for a rectifying transistor. An applied voltage, which produces a large current in one direction, produces only a small current in the opposite direction. This is an example of a conductor that does not satisfy Ohm's law.

The S.I. unit for resistance is the *ohm;* an ohm is a volt per ampere.

For many materials, the potential difference and the current are directly proportional, so the resistance is a constant independent of the current. This is illustrated in Fig. 19.2 for a length of copper wire with its ends at an adjustable potential difference V. The current increases linearly with V and reverses direction when V is reversed, so the ratio $R = V/I$ is constant. Materials with a constant resistance are said to obey *Ohm's law* and are called *ohmic conductors.*

The resistance of some conductors varies with the magnitude or direction of the applied potential difference. The operation of many electronic devices, such as vacuum tubes and transistors, is based on their *nonohmic* character (Fig. 19.3). The calculation of the resistance from the definition $R = V/I$ for an ohmic conductor is illustrated by the following example.

Example 19.3
Find the resistance of the wire in Fig. 19.2.
The potential difference and current are proportional,

so the resistance is a constant. When the current is 10 A, the potential difference is 1 V, and

$$R = \frac{V}{I} = \frac{1\ \text{V}}{10\ \text{A}} = 0.1\ \text{ohm}$$

The resistance of a conductor depends on its size, shape, and composition. We can understand the size and shape dependence with the following argument. If we put two identical wires side by side, the current doubles and hence the resistance is halved. Thus, R must vary inversely with the cross-sectional area A. Now if we halve the length l, the potential change and hence the resistance are also halved, so R must be proportional to l. Thus, we can write the resistance in terms of geometric factors and a constant as

$$R = \frac{\rho l}{A} \qquad (19.5)$$

The proportionality constant ρ depends only on the properties of the material and is called the *resistivity.* The S.I. unit for resistivity is the ohm meter. Typical resistivities are listed in Table 19.1.

The *conductivity* σ, defined by

$$\sigma = \frac{1}{\rho} \qquad (19.6)$$

is sometimes used instead of the resistivity to characterize conductors. The S.I. unit of conductivity is the $\text{ohm}^{-1}\ \text{m}^{-1}$. The calculation of the resistance of a wire is illustrated by the following example.

Example 19.4
Find the room temperature resistance of a copper wire 100 m long with a radius of 1 mm = 10^{-3} m.

TABLE 19.1

Resistivities at 20°C in ohm meters

	Substance	Resistivity (ohm meters)
Conductors	Silver	1.47×10^{-8}
	Copper	1.72×10^{-8}
	Aluminum	2.63×10^{-8}
Semiconductors	Germanium	0.60
	Silicon	2300
Insulators	Sulfur	10^{15}
	Glass	10^{10}–10^{14}
Ionic conductors	Body fluids	approx. 0.15

305

According to Table 19.1, the resistivity of copper at room temperature (20°C) is 1.72×10^{-8} ohm m. Hence the resistance of this wire is

$$R = \frac{\rho l}{A} = \frac{\rho l}{\pi r^2}$$

$$= \frac{(1.72 \times 10^{-8} \text{ ohm m})(100 \text{ m})}{\pi (10^{-3} \text{ m})^2}$$

$$= 0.547 \text{ ohm}$$

19.3 | ENERGY SOURCES IN CIRCUITS

A nonelectrical source of energy, such as a battery or a generator, is needed to maintain a continuous current in a closed conducting path or *circuit*. Such an energy source is called a *source of EMF*, which is short for the archaic term electromotive force. EMF, as we define it below, has the dimensions of potential, not force.

To clarify the role of an EMF in a circuit, consider the analogous situation of water moving under the influence of gravitational forces. Figure 19.4 shows an artificial stream and waterfall constructed for decorative purposes in a garden. Water drops over the falls a vertical distance h, so that the gravitational potential energy of a mass m decreases by mgh. This energy is converted into kinetic energy as the water accelerates and is then transformed into heat $Q = mgh$ as the water splashes among the rocks in the falls. If no water is supplied to the beginning of the stream, the

Figure 19.4. The work done by the pump equals the energy dissipated as heat in the falls. The conservative gravitational force does no net work on water that goes completely around the system.

Figure 19.5. The electrical analog to a falls without a pump.

flow stops almost immediately. However, water is returned to the stream by a pump, which does work $W = mgh$ on the water against the gravitational force. The conservative gravitational force does positive work equal to mgh on the water as it falls and negative work $-mgh$ as it rises. These exactly cancel, so the conservative gravitational force does no net work when the water goes completely around the system. The energy W supplied by the pump exactly equals the heat energy Q generated in the falls. Note that if we introduced a waterwheel connected to a machine into the falling stream, then some of the energy supplied by the pump would be used to do mechanical work rather than being converted into heat.

Figure 19.5 shows an electrical system that is analogous to a waterfall *without* a pump. Two conductors have equal and opposite charges, so there is a potential difference between them. When they are connected by a metal wire, there is a brief charge flow, which stops when the charges are neutralized and the potential difference is reduced to zero.

To maintain a steady current, a *pump* or source of EMF must supply energy. In Fig. 19.6, a battery is connected to a resistor by wires that are perfect conductors. As a charge goes through the battery, it is *pumped* by nonelectrical forces to a position of higher potential energy. The energy required for this is supplied by chemical reactions occurring in the battery. In the resistor, the charge moves in the direction of the electric force, and the kinetic energy it acquires is transformed into heat. The conservative electric field does positive work in the resistor, negative work in the battery, and zero net work when a charge goes completely around the circuit. The net effect of the various energy transfers is that stored chemical energy in the battery is transformed into heat in the resistor. This is completely analogous to the energy

Figure 19.6. (*a*) A battery connected by perfectly conducting wires to a resistor. (*b*) The symbol for an EMF used in circuit diagrams. The longer line and + symbol indicate the higher potential terminal. (*c*) The symbol for a resistance. (*d*) The circuit of (*a*) redrawn using the conventional symbols.

supplied by the pump, which is converted into heat in the waterfall.

Two other aspects of the analogy should be noted. The resistor can be replaced in the circuit by some other *load*, such as an electric motor. Then the chemical energy of the battery is converted into mechanical energy, just as the energy supplied by the pump is used to do mechanical work if a waterwheel is introduced. Also, just as the pump supplied energy but did not alter the original supply of water, the battery cannot produce or destroy charge. Whatever current enters the battery in Fig. 19.6 must leave it, and the current in the resistor and in the connecting wires must also be the same. *A battery or generator converts some other kind of energy into electrical energy, but it is not a source of charge.*

The EMF $\mathcal{E}$ of a battery or generator is defined as the *work done per unit charge by the nonelectrical forces.* Its S.I. unit is a joule per coulomb or volt, which is also the unit of the electric potential.

For the circuit of Fig. 19.6, we can relate the current I to the battery EMF $\mathcal{E}$ and to the resistance R by using the potential changes observed as we go completely around the circuit. This is equivalent to finding the potential energy changes for a unit positive charge. Proceeding clockwise, these changes are:

From *a* to *b*: *V increases* by $\mathcal{E}(\Delta V = \mathcal{E})$; the potential increase in the battery is equal to the EMF.
From *b* to *c*: *V* is constant ($\Delta V = 0$); the potential difference is IR, the current times the resistance, and the resistance is zero for a perfect conductor.
From *c* to *d*: *V decreases* by IR ($\Delta V = -IR$).
From *d* to *a*: *V* is constant ($\Delta V = 0$); again, no potential change occurs in a perfect conductor.

When a charge goes around a closed path and returns to the starting point, its potential energy must return to its original value, since the conservative electrical forces do no net work. Thus the sum of the potential changes must be zero, or $\mathcal{E} - IR = 0$. Equivalently, the potential increase in the battery must equal the decrease in the resistor:

$$\mathcal{E} = IR \qquad (19.7)$$

This result is illustrated by the following example.

Example 19.5

A dry cell with an EMF of 6 V is connected to a light bulb with resistance 4 ohms. Find the current.

With $\mathcal{E} = IR$, the current is

$$I = \frac{\mathcal{E}}{R} = \frac{6\text{ V}}{4\text{ ohms}} = 1.5\text{ A}$$

Real batteries and generators usually have various dissipative effects associated with them which can be thought of as *internal resistances*. These cause the terminal voltages to be different from the EMFs and can result in substantial heat being produced. For example, the electrolytic fluid in storage batteries will boil if the current is large enough. These effects are illustrated in the next two examples.

Example 19.6

A battery with an EMF of 6 V and an internal resistance $r = 2$ ohms is connected to an $R = 4$ ohm lightbulb (Fig. 19.7). Find (a) the current, and (b) the terminal potential difference of the battery.

(a) The potential increase due to the EMF must equal the total of the potential decreases in the resistances, so

$$\mathcal{E} = IR + Ir$$

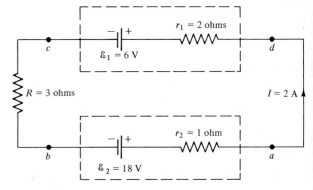

Figure 19.7. The terminal voltage of the battery is less than the EMF here because of the internal resistance.

and

$$I = \frac{\mathcal{E}}{R + r} = \frac{6 \text{ V}}{(2 + 4) \text{ ohms}} = 1 \text{ A}$$

(b) The terminal voltage is the potential difference V_{ab} between points a and b. It equals the EMF minus the potential drop associated with the internal resistance, so

$$V_{ab} = \mathcal{E} - Ir = 6 \text{ V} - (1\text{A})(2 \text{ ohms}) = 4 \text{ V}$$

Alternatively, we can obtain the same result from

$$V_{ab} = IR = (1 \text{ A})(4 \text{ ohms}) = 4 \text{ V}$$

We see that the terminal potential difference of a real battery is different from its EMF when there is a current.

Example 19.7

Two batteries are connected to a resistor as shown in Fig. 19.8. Find (a) the current, and (b) the terminal potential difference of each battery.

(a) Since the batteries are connected so that their polarities are opposite, the direction of the current is determined by the larger EMF and is counterclockwise in Fig. 19.8. As we follow the current through the larger EMF, the potential *increases* by $\mathcal{E}_2$, and in the other EMF the potential *decreases* by $\mathcal{E}_1$. The current is found from

$$\mathcal{E}_2 - \mathcal{E}_1 = I(r_1 + r_2 + R)$$

Thus

$$I = \frac{\mathcal{E}_2 - \mathcal{E}_1}{r_1 + r_2 + R} = \frac{(18 - 6)\text{V}}{(2 + 1 + 3) \text{ ohms}} = 2 \text{ A}$$

(b) The potential differences across the batteries are

$$V_{ab} = \mathcal{E}_2 - Ir_2 = 18 \text{ V} - (2\text{A})(1 \text{ ohm}) = 16 \text{ V},$$
$$V_{dc} = \mathcal{E}_1 + Ir_1 = 6 \text{ V} + (2\text{A})(2 \text{ ohms}) = 10 \text{ V}$$

Note that the terminal potential difference V_{dc} is greater than the EMF in this case. In calculating the potential difference between points c and d, we are proceeding opposite to the current. Thus, we see a potential *rise* in the resistance as well as in the EMF. The terminal

Figure 19.8. Examples 19.7 and 19.8.

potential difference exceeds the EMF only when the current is driven "backwards" through the battery by another larger EMF.

19.4 | POWER IN ELECTRICAL CIRCUITS

In a circuit, energy is initially converted from some other form by a battery or a generator into electrical potential energy. It is then transformed in the load into heat, mechanical work, or some other kind of energy. In this section, we calculate the power or rate of energy conversion for various parts of a circuit.

Figure 19.9. A circuit element is one part, such as a battery or resistor, of a complete circuit.

Figure 19.9 shows a *circuit element* (one part of a circuit) with a potential difference V across its terminals and a current I through it. If the element is a battery, V is the terminal voltage; if it is a resistance R, then V is equal to IR in magnitude. In a time Δt, a charge $\Delta Q = I \Delta t$ passes through the element. Its change in potential energy is $V \Delta Q = VI \Delta t$, which must equal the work ΔW done by the element on the charge. Dividing this work by the time Δt, the power in any circuit element is

$$\boxed{\mathcal{P} = \frac{\Delta W}{\Delta t} = IV} \quad \text{(any circuit element)} \quad (19.8)$$

If the current is in the direction of increasing potential, the circuit element supplies energy. When the current is in the direction of decreasing V, as in a resistor or a motor, the element receives or dissipates energy. For an EMF, the potential difference is $\mathcal{E}$, so

$$\boxed{\mathcal{P} = I\mathcal{E}} \quad \text{(EMF)} \tag{19.9}$$

For a resistor, $V = IR$, and

$$\boxed{\mathcal{P} = I^2 R} \quad \text{(resistor)} \tag{19.10}$$

In any electrical circuit, the power supplied is always equal to the power dissipated. These ideas are illustrated by the following example.

Example 19.8

Calculate the power supplied to or by each element in the circuit of Fig. 19.8.

In the larger battery, the EMF supplies power $I\mathcal{E}_2$, and the internal resistance dissipates heat at the rate $I^2 r_2$. Thus the net power *supplied* by this battery is

$$\mathcal{P}_2 = I\mathcal{E}_2 - I^2 r_2$$
$$= (2 \text{ A})(18 \text{ V}) - (2 \text{ A})^2 (1 \text{ ohm})$$
$$= 32 \text{ W}$$

The other battery has *work done on it* by the current, since the charge is moving from the high potential terminal to the low potential side. Thus the power *absorbed* by this battery is

$$\mathcal{P}_1 = I\mathcal{E}_1 + I^2 r_1$$
$$= (2 \text{ A})(6 \text{ V}) + (2 \text{ A})^2 (2 \text{ ohms})$$
$$= 20 \text{ W}$$

Finally the resistor R generates heat at the rate

$$\mathcal{P}_R = I^2 R = (2 \text{ A})^2 (3 \text{ ohms}) = 12 \text{ W}$$

Notice that the power supplied by the larger battery equals the power dissipated in R plus the power absorbed by the smaller battery. This is just the energy conservation law in another form.

Electric utilities charge their customers for the electric energy used, which is the product of the electric power and the time. Electric energy is usually sold by the kilowatt-hour (kW h). This is a kilowatt of power for one hour, so

$$1 \text{ kW h} = (10^3 \text{ W})(3600 \text{ s})$$
$$= 3.6 \times 10^6 \text{ J} \tag{19.11}$$

Most household electric power in North America is equivalent to 120 V dc but is actually alternating current at a frequency of 60 Hz. As we see in Chapter Twenty-two, for resistive current and power calculations, an *effective* voltage of 120 V ac is the same as 120 V dc. The calculation of the cost of operating a typical electrical device is illustrated by the following example.

Example 19.9

For a household 60-W light bulb operated at 120 V, find (a) the current; (b) the resistance; and (c) the 24-hour operating cost, if energy costs 6 cents per kilowatt-hour.

(a) Since the power $\mathcal{P}$ in any circuit element equals VI,

$$I = \frac{\mathcal{P}}{V} = \frac{60 \text{ W}}{120 \text{ V}} = 0.5 \text{ A}$$

(b) The resistance is

$$R = \frac{V}{I} = \frac{120 \text{ V}}{0.5 \text{ A}} = 240 \text{ ohms}$$

(c) The bulb uses power at the rate of 60 W or 0.060 kW. Thus the cost of operating it for 24 hours is

$$(0.06 \text{ kW})(24 \text{ h})[6\cent \text{ (kW h)}^{-1}] = 8.6\cent$$

19.5 | SERIES AND PARALLEL RESISTORS

Looking inside a television set or stereo one quickly discovers that circuits are often quite complex. Nevertheless, the most complex dc circuit can be analyzed to find the currents and voltages everywhere using two basic principles. These same principles can also be generalized to deal with ac circuits. Here we show how these principles can be used to simplify the discussion of circuits containing certain combinations of resistors.

The first of these principles is already familiar:

> 1 *The sum of the potential changes around any closed path is zero.* This follows from the conservative nature of electrical forces.

The second principle is a restatement of another basic idea, charge conservation. Suppose a current divides into two or more currents at a point where a number of wires come together. No charge can be lost or created nor can it collect at that point. Consequently,

> 2 *The current entering any point must equal the current leaving.*

We can use these principles to discuss series and parallel combinations of resistors. Two or more resis-

Figure 19.10. (*a*) Three resistors in series. (*b*) The equivalent resistance R_s leads to the same current *I*.

tors are in *series* if the same current goes through each (Fig. 19.10). According to the first principle, the sum of the potential charges around the circuit is zero, or

$$\mathcal{E} - IR_1 - IR_2 - IR_3 = 0$$

Thus

$$I = \frac{\mathcal{E}}{R_1 + R_2 + R_3}$$

A single *equivalent resistance* R_s connected to the battery will produce the same current if $I = \mathcal{E}/R_s$, or

$$R_s = R_1 + R_2 + R_3 + \cdots \qquad (19.12)$$

Two or more resistors are in *parallel* if they have a common potential difference (Fig. 19.11.) According to the second principle, the current entering point *a* must equal that leaving, or

$$I = I_1 + I_2 + I_3$$

Applying $I = \mathcal{E}/R$ to each resistor,

$$I_1 = \mathcal{E}/R_1, \qquad I_2 = \mathcal{E}/R_2, \qquad \text{and } I_3 = \mathcal{E}/R_3$$

Hence

$$I = \frac{\mathcal{E}}{R_1} + \frac{\mathcal{E}}{R_2} + \frac{\mathcal{E}}{R_3}$$

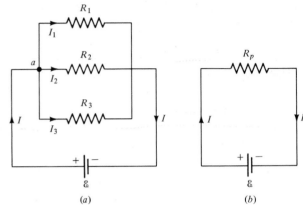

Figure 19.11. (*a*) Three resistors in parallel. (*b*) The equivalent single resistance R_p produces the same current *I*.

A single equivalent resistance R_p will produce the same current *I* if

$$I = \frac{\mathcal{E}}{R_p}$$

Comparing, we see that

$$\frac{1}{R_p} = \frac{1}{R_1} + \frac{1}{R_2} + \frac{1}{R_3} + \cdots \qquad (19.13)$$

The following example illustrates how the series and parallel formulas may sometimes be used together in analyzing a complex arrangement of resistors.

Example 19.10

(a) Find the equivalent resistance of the resistors in Fig. 19.12*a*. (b) Find the current in each resistor.

(a) The current *I* splits at point *a* into two currents, I_1 and I_2. I_1 goes through each of the 1-ohm resistors, so they are in series. Their equivalent resistance is then

$$R_s = 1 \text{ ohm} + 1 \text{ ohm} + 1 \text{ ohm} = 3 \text{ ohms}$$

This reduces the circuit to two parallel 3-ohm resistors

Figure 19.12. (*a*) The complete network. (*b*) The simplified equivalent network obtained using the series resistance formula. (*c*) The final equivalent network obtained with the parallel resistance formula.

(Fig. 19.12b). The effective resistance of the entire system is found from

$$\frac{1}{R_p} = \frac{1}{3 \text{ ohms}} + \frac{1}{3 \text{ ohms}} = \frac{2}{3 \text{ ohms}}$$

Thus

$$R_p = 1.5 \text{ ohms}$$

as shown in Fig. 19.12c.

(b) The current I in the battery is found using the equivalent resistance of the system, 1.5 ohms. Thus

$$I = \frac{\mathcal{E}}{R} = \frac{6 \text{ V}}{1.5 \text{ ohms}} = 4 \text{ A}$$

In Fig. 19.12b, the voltage drop across each resistor must be the same. Since their resistances are equal, the current in each must also be the same. Hence, each current is half the total, or

$$I_1 = I_2 = \tfrac{1}{2}I = \tfrac{1}{2}(4 \text{ A}) = 2 \text{ A}$$

The current in each resistor is 2 A.

19.6 | VOLTMETERS AND AMMETERS

The most basic dc measuring instruments are the *voltmeter* and *ammeter*, which measure potential differences and currents, respectively. Each of these contains a *galvanometer*, which consists of a coil of wire suspended near a magnet. The coil is attached to a spring, which opposes rotational motion. When there is a current in the coil, the magnetic force on the moving charges causes the coil to rotate by an amount proportional to the current. Typically, a few milliamperes will cause full-scale deflection, and the resistance of the coil is about 10 to 100 ohms.

Figure 19.13 illustrates how a voltmeter and an ammeter are used. To measure the potential difference across a circuit element, the *voltmeter is con-*

Figure 19.14. An ammeter is a galvanometer coil in parallel with a small resistance r. The two points b and c label points in the circuit of Fig. 19.13. I is the current being measured.

nected in parallel with that element. According to the parallel resistance formula, the voltmeter resistance must be large compared to that of the element to avoid large changes in the current in the circuit. To measure the current in an element, the *ammeter is inserted in series* with the element. Consequently, from the series resistance formula, the ammeter resistance must be small to minimize current changes.

To construct an ammeter, we connect a small resistance r in parallel with a galvanometer coil (Fig. 19.14). If the circuit has a current I, only a small part of the current, I_g, goes through the relatively high resistance galvanometer coil. The remaining current, $I - I_g$, goes through r. Since the potential drop across the galvanometer and the parallel resistor must be the same, it follows that

$$I_g R_g = (I - I_g)r \qquad (19.14)$$

The choice of r is determined by the desired range of the ammeter. This is illustrated in the following example.

Example 19.11

A galvanometer with a resistance of 100 ohms gives a full-scale deflection for a current of 1 mA. (a) How large a parallel resistor is needed to convert it into an ammeter with a 20-A range? (b) What is the ammeter resistance?

(a) Using Eq. 19.14,

$$r = \frac{R_g I_g}{(I - I_g)} = \frac{(100 \text{ ohms})(0.001 \text{ A})}{(20 \text{ A} - 0.001 \text{ A})}$$

Since we are working to three significant figures, the 0.001 A in the denominator can be neglected. Thus

$$r \simeq \frac{(100 \text{ ohms})(0.001 \text{ A})}{20 \text{ A}} = 0.005 \text{ ohm}$$

($\simeq$ means approximately equal.)

Figure 19.13. The voltmeter (V) is placed in parallel with the resistor R and indicates the potential difference across R. The ammeter (A) is in series with R and gives the current through R.

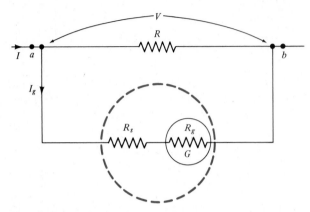

Figure 19.15. A voltmeter is a galvanometer in series with a large resistance R_s. Points a and b label points in the circuit of Fig. 19.13. V is the potential difference between a and b which is being measured.

(b) Since r is much smaller than R_g, the resistance of the ammeter is very close to r:

$$\frac{1}{R_A} = \frac{1}{r} + \frac{1}{R_g} = \frac{1}{0.005 \text{ ohm}} + \frac{1}{100 \text{ ohms}} \simeq \frac{1}{0.005 \text{ ohm}}$$

and $R_A = 0.005$ ohm.

To construct a voltmeter, a large resistance R_s is placed in series with the galvanometer coil (Fig. 19.15). The potential difference V across the voltmeter is

$$V = I_g(R_g + R_s)$$

so

$$R_s = \frac{V}{I_g} - R_g \qquad (19.15)$$

Again the desired range of the meter determines R_s, as is seen in the next example.

Example 19.12

The galvanometer of Example 19.11 has a resistance of 100 ohms and gives a full-scale deflection for a current of 1 mA. (a) How can it be converted into a voltme-

ter with a 100-V range? (b) What is the voltmeter resistance?

(a) The series resistance needed is

$$R_s = \frac{V}{I_g} - R_g = \frac{100 \text{ V}}{0.001 \text{ A}} - 100 \text{ ohms}$$

$$= (100,000 - 100) \text{ ohms} = 99,900 \text{ ohms}$$

(b) The total voltmeter resistance is

$$R_V = R_s + R_g = 99,900 \text{ ohms} + 100 \text{ ohms}$$

$$= 100,000 \text{ ohms}$$

19.7 | CIRCUITS CONTAINING RESISTANCE AND CAPACITANCE

If we connect an uncharged capacitor to a battery, charge moves from one capacitor plate to the other through the battery and the connecting wires. This current stops once the potential difference across the capacitor equals the EMF of the battery. Similarly, if the plates of a charged capacitor are connected by a wire, there will be a current in the wire until the capacitor is fully discharged. These short-lived currents are referred to as *transients*. The constant currents we have been discussing until now are referred to as *steady* currents.

We first consider the transient currents associated with charging a capacitor. Fig. 19.16a shows a circuit containing an initially uncharged capacitance C, a resistance R, and an EMF $\mathcal{E}$. If at time $t = 0$ the switch is closed, the charge q on the capacitor and the current i in the circuit vary with time as in Fig. 19.16b and c. As we show in Section 19.12, the time required for either q or i to reach a specific fraction of its final value is proportional to the time constant τ (tau) given by

$$\tau = RC \qquad (19.16)$$

(a) (b) (c)

Figure 19.16. (a) A circuit containing a resistance, a capacitance, and an EMF. (b) and (c) If the capacitor is initially uncharged and the switch is closed at time $t = 0$, the charge q and current i vary with the time t as shown. The time constant τ is equal to RC.

The greater the resistance or the capacitance, the longer it takes to charge the capacitor. This is physically reasonably, since a larger capacitance requires a larger final charge, and a larger resistance leads to smaller charging currents.

To study this in more detail, we calculate the potential differences across the circuit elements during the charging process. Suppose at time t the charge on the capacitor is q and the current in the circuit is i. Proceeding counterclockwise, there is a potential rise, $\mathcal{E}$, in the EMF, and a drop, iR, in the resistor. For the capacitor, $C = q/V$, so the potential drop is q/C. The sum of these changes is zero. Thus

$$\mathcal{E} - iR - \frac{q}{C} = 0$$

or

$$i = \frac{\mathcal{E}}{R} - \frac{q}{RC} \qquad (19.17)$$

This equation directly gives the current i at $t = 0$ and at $t = \infty$. At $t = 0$, when the switch has just been closed, the charge q on the capacitor is its initial value, zero. Thus the current i_0 at $t = 0$ is

$$i_0 = \frac{\mathcal{E}}{R} \qquad (t = 0) \qquad (19.18)$$

This equals the constant current that would occur if only the resistance and the EMF were connected in series. If we wait a long time, the capacitor will be fully charged, and its potential difference must equal the EMF $\mathcal{E}$. The charge q is then

$$q_f = \mathcal{E}C \qquad (t = \infty) \qquad (19.19)$$

If we substitute q_f for q in Eq. 19.17, we find that the current is zero. As expected, once the capacitor is fully charged, the current ceases.

The charge and current at any time t can be found by solving Eq. 19.17, which will be done in Section 19.12. The results are shown graphically in Fig. 19.16 and are given algebraically by

$$q = q_f(1 - e^{-t/\tau}) \qquad (19.20)$$
$$i = i_0 e^{-t/\tau} \qquad (19.21)$$

Here i_0, q_f, and τ have all been defined above. e is the base of natural logarithms and has the value $2.718\ldots$. Note that when $t = \tau$, $e^{-t/\tau} = e^{-1} = 1/2.718\ldots = 0.37\ldots$. Thus, after one time constant

has elapsed, the current has dropped to about 37 percent of its initial value, i_0. Also, since

$$(1 - e^{-1}) = 1 - 0.37\ldots = 0.63\ldots$$

after a time τ has elapsed, $q = 0.63 q_f$. That means the capacitor charge is equal to approximately 63 percent of its final charge q_f. Alternatively, we can say it is within 37 percent of its final value. When the elapsed time is equal to twice the time constant, $e^{-t/\tau} = e^{-2} = 0.14\ldots$. At this time the charge is $(1 - 0.14)q_f = 0.86q_f$, and the current is $0.14i_0$. After a time equal to several time constants, the charge q and current i are very close to their final values, as is seen in Fig. 19.16. The calculation of some of these quantities is illustrated by the following example.

Example 19.13

In Fig. 19.16, $C = 2\,\mu\text{F}$, $R = 1000$ ohms, and $\mathcal{E} = 6$ V. Find (a) the final charge on the capacitor, (b) the initial current, and (c) the time constant.

(a) The final charge is

$$q_f = \mathcal{E}C = (6\text{ V})(2 \times 10^{-6}\text{ F}) = 1.2 \times 10^{-5}\text{ C}$$

(b) The initial current is

$$i_0 = \frac{\mathcal{E}}{R} = \frac{6\text{ V}}{1000\text{ ohms}} = 6 \times 10^{-3}\text{ A}$$

(c) The time constant is

$$\tau = RC = (1000\text{ ohms})(2 \times 10^{-6}\text{ C}) = 2 \times 10^{-3}\text{ s}$$

We have seen that the time constant $\tau = RC$ determines the rate at which the charge on a capacitor increases. The time constant also determines the rate at which the charge *decreases*. For example, in Fig. 19.17 a capacitor discharges through a resistance R when the switch is closed. If initially it has a charge q_0, after a time t the charge q is reduced to

$$q = q_0 e^{-t/\tau} \qquad (19.22)$$

The initial potential difference across the capacitor is $V_0 = C/q_0$, and the initial current is directed away from the positive plate and has a magnitude of $i_0 = V_0/R$. The current after a time t is

$$i = i_0 e^{-t/\tau} \qquad (19.23)$$

Thus, both the charge on the capacitor and the current in the circuit steadily diminish, reaching very small values after a few time constants.

The artificial pacemaker provides an example of a

Figure 19.17. (a) An initially charged capacitor discharging through a resistor. (b) The charge on the capacitor versus time. (c) The current in the circuit versus time.

circuit in which a capacitor is repeatedly charged and discharged.

Example 19.14

Each cycle in the human heart begins with an electrical *pacemaker* pulse from a group of nerve fibers. Some heart patients are now being helped by surgically implanted *artificial pacemakers,* which are battery-powered circuits that pulse if the person's pacemaker fails to do so. One model has pulses triggered 75 times per minute by a 0.4-μF capacitor, which rapidly charges through a very small resistance r and then slowly discharges through a large resistance R (Fig. 19.18). When the charge drops to $e^{-1} = 0.37$ times its initial value, transistors deliver a short pulse to the heart and then recharge the capacitor almost immediately through r. (a) Find the time constant of the discharging RC circuit, neglecting the small time needed to recharge the capacitor through r. (b) Find the resistance R.

(a) There are 75 pulses per minute, $75/60 = 1.25$ per second, or one pulse each $1/1.25 = 0.8$ seconds. This equals the time for the capacitor charge to drop to e^{-1} times its initial value, since the recharging time is negligible. Such a decrease in the charge requires one time constant to elapse, so $\tau = 0.8$ seconds.

(b) From $\tau = RC$, using $C = 0.4\,\mu$F,

$$R = \frac{\tau}{C} = \frac{0.8 \text{ s}}{0.4 \times 10^{-6} \text{ F}} = 2 \times 10^6 \text{ ohms}$$

The artificial pacemaker in this example stimulates the heart at a fixed frequency. More advanced demand pacemakers trigger only if the patient's natural pacemaker fails to operate within a specified time.

19.8 | ELECTRICAL SAFETY

Electrical equipment must be designed and used with care to avoid possible fire and electrocution hazards. In this section, we consider these hazards and the means of reducing them.

Electric power in American homes and offices is usually supplied as 120-V, 60-Hz alternating current. The outlets in each building are connected in parallel to two power lines from a neighborhood substation (Fig. 19.19). One line is connected to the earth, which is a good conductor. This line is said to be *grounded*. The other line is *hot* or *live*; its potential alternates relative to that of the grounded line, which is considered to be at zero potential.

Each circuit in a building is protected by a *fuse* or a *circuit breaker*. If the current exceeds a conservatively chosen maximum, the fuse melts or the breaker trips, opening the circuit. This reduces the danger of a fire starting because of the heat generated by excessive currents. One cause of excess currents is an *overload*: too many electrical devices in operation on one circuit at the same time. Another cause is a *short circuit:* a very low resistance path between the two power lines leading to a very large current. This sometimes occurs when electrical insulation becomes defective as a result of wear or gradual deterioration.

Electrocution is a serious hazard, since relatively small currents through the human torso can cause injury or death. The average adult can detect a current as small as $1 \text{ mA} = 10^{-3}$ A; a direct current produces slight warmth, and a 60-Hz alternating current causes a tingling sensation. Someone reacting involuntarily to a small, unexpected current might drop a hot pan or fall from a ladder, so the maximum allowable leakage currents are set at less than 1 mA. Considerable muscular reaction and pain occur at a few milliamperes. Between 10 and 20 mA will paralyze some muscles and prevent a person from releasing a conductor; about 18 mA contracts the chest muscles and causes breathing to stop. Unconsciousness and death will occur in minutes unless the current stops. One hundred milliamperes for a few seconds causes ventricular fibrillation, rapid uncoordinated movements of the heart muscle that prevent it from

Figure 19.18. (*a*) The basic pacemaker timing circuit. The capacitor *C* charges quickly through the small resistance *r*. The "switch," which is actually a transistor, then changes position, and the capacitor slowly discharges through the large resistance *R*. When the voltage across *R* reaches a preset level, the triggering circuit *T* sends a pulse to the heart. (*b*) If a patient's heart fails to initiate a pulse, this artificial pacemaker provides both atrial and ventricular stimulation. Its height is 6 cm, its mass is 150 grams, and it contains four lithium-silver chromate cells which may supply power for over four years. (*c*) The implanted pacemaker and leads. ((*b*) and (*c*) courtesy of American Pacemaker Corporation.)

Figure 19.19. Each circuit in every house in a neighborhood is connected in parallel to the power lines. Each circuit contains a fuse or circuit breaker. The meter is normally placed in series with the lines which enter the building.

pumping blood. Fibrillation rarely stops spontaneously, so death usually results unless prompt medical treatment is available.

Grounding | In electrical safety, it is important to understand the role of grounding. Consider an electric fence on a farm. Here, a source of EMF has either one of its terminals attached to a wire that is totally insulated from electrical contact with anything else. The other terminal is attached to a metal rod driven into the earth and is said to be grounded (Fig. 19.20). If an animal standing on the ground touches the wire, it will complete a conducting path from the EMF through the wire and the ground back to the EMF. Thus, a current will pass through the animal, and it

315

Figure 19.20. A person or animal touching the fence completes a path to ground and receives a mild shock.

Figure 19.21 An electric clothes dryer. (*a*) In a two-wire circuit the cabinet is not grounded. A defect in the wiring can therefore make the cabinet "hot." (*b*) In a three-wire circuit the cabinet is connected directly to ground.

will experience a shock. On the other hand, if a bird sits on the wire, its body is at the potential of the wire. It does not receive a shock, since it does not form part of a closed conducting path.

The point of this example is that electric power lines usually have one wire grounded. If anything is simultaneously in contact with the ground and with a part of the circuit not at ground potential, charge will flow through that object.

Electrical accidents usually occur when a person comes into contact with both a live power line and a good path to ground. The current through the body is determined by its electrical resistance, which is highly variable. Dry skin has a resistance of 10^5 ohms per square centimeter or greater, but the resistance may decrease by a factor of 100 when the skin is wet. A person standing in water in a bath tub is in excellent electrical contact with the ground via the water pipes. If this person grasps a defective electrical appliance, the resistance of the body can be as small as 500 ohms. At 120 V the resulting current is $I = V/R = 120$ V/500 ohms $= 0.240$ A $= 240$ mA, which is lethal. The resistance between the two hands of a person perspiring slightly is roughly 1500 ohms, corresponding to an 80-mA current at 120 V. Again this current may be fatal if it persists for a few seconds.

The way shock hazards can arise is illustrated in Fig. 19.21*a*. The heating coil in a clothes dryer, represented by the resistor *R*, is connected to the power line with a common two-prong plug. Consequently the ground and hot lines are determined by which way the plug is inserted into the outlet. Since the plug can usually be inserted in either of two ways, no wire in the dryer may be permitted to touch the metal cabinet. If a wire does come into contact with the

cabinet, it will be either at ground or at the hot line voltage, depending on how the plug is inserted into the outlet. A person simultaneously touching a live cabinet and a water faucet or a concrete floor could receive a lethal shock.

To minimize this hazard, electrical devices such as laundry appliances or power tools should be grounded. This is sometimes done by connecting their metal housings or cabinets to water pipes or other grounded conductors. A more reliable way to ground devices is provided by modern household wiring, which has a third, separate wire. This wire is grounded and is connected to the third prong of the electrical outlets. Thus, when an electrical device with a three-prong plug is plugged into such an outlet, its housing is automatically grounded (Fig. 19.21*b*). When the housing is grounded, accidental contact between the live wire and the housing will result in a short circuit, and the fuse or circuit breaker will open the circuit. Once this happens, there is no danger of a shock.

Ground-Fault-Interrupters | If the failure of

some component in an electrical appliance or instrument leads to large enough currents leaking to ground, the fuse or circuit breaker opens the circuit, eliminating any shock hazard. However, if an electrical device fails in such a way that smaller currents leak to ground, then a person touching a component not quite at ground potential may receive a severe shock.

A device called a *ground-fault-interrupter* is now sometimes installed in the ground wire of three-wire circuits. If the current in the ground wire exceeds some very small preset limit, say 5 mA, this device opens the circuit. The majority of the electrocutions that occur in American homes could probably be prevented if ground-fault-interrupters were in general use.

All-Insulated Construction | In recent years,
many power tools and appliances have appeared with
all-insulated or *double-insulated* construction. They
are designed so that, in the event of an insulation
failure, no exposed metal part will become electrically
live. These devices do not require grounding and represent a useful advance in electrical safety.

Hospitals | The electrical shock hazards in hospi-
tals are generally similar to those in homes and offices, although the use of electrical devices near
grounded metal beds and in the presence of water or
other liquids does pose potential problems. However,
a patient with catheters, probes, and needles inserted
in the body is highly vulnerable to very small electrical currents, since 0.02 mA applied directly to the
heart is sufficient to produce ventricular fibrillation.
Since deaths due to such small currents are indistinguishable from those due to natural causes, there is
some dispute over how often they occur. The special
hazards associated with patients connected directly to
electrical circuits have led to recommendations that
their beds be specially insulated so that they are not
grounded and that no other objects within reach of
the patient be grounded. Then, like the bird on the
fence wire, these patients cannot complete a conducting path to ground. Chapter Twenty-two discusses
how devices called transformers can also be used to
reduce shock hazards.

SUMMARY

Electric current is the rate at which charge is transported. In a metallic conductor, conduction electrons
can move readily among the positive lattice ions.
When there is an electric field in a wire, the electrons
acquire a small net drift velocity in the direction of
the electric force.

The ratio of the potential drop to the current in a
conductor is its resistance. Materials in which the
resistance is independent of the current are said to
obey Ohm's law. The resistance of a conductor is
proportional to its length and inversely proportional
to its cross-sectional area. The proportionality constant is called the resistivity and depends only on the
properties of the material.

In an electric circuit a continuous current can exist
only if energy is supplied by an EMF. The power
supplied by or to any circuit element is VI, where I is

the current in the element, and V is the potential
change across the element. This power is I^2R for a
resistor and $I\mathcal{E}$ for an EMF.

In any circuit, the total current entering a point
where wires come together equals the current leaving,
and the sum of the potential changes around a closed
circuit is zero. Circuits with several resistors can sometimes be analyzed using the series and parallel resistor rules.

When the plates of a capacitance C are connected
to an EMF through a resistance R, the rate at which
its charge changes is determined by the time constant
$\tau = RC$. When a time equal to τ has elapsed, an
initially uncharged capacitor is within 37 percent of
its final charge. Similarly, an initially charged capacitor being discharged through a resistance will lose all
but 37 percent of its charge in one time constant.

Checklist

Define or explain:

electric current	ammeter
conduction electrons	galvanometer
drift velocity	transient, steady
resistance	currents
Ohm's law	time constant
resistivity	overload
EMF	short circuit
series, parallel resistors	ground
voltmeter	

EXERCISES

Resistivities needed in some of the exercises are given
in Table 19.1.

Section 19.1 | Electric Current

19-1 A proton accelerator has a beam with a
current of 10^{-6} A. How many protons are accelerated per second?

19-2 In an electrochemical experiment, a current
of 0.5 A passes through a cell for 1 hour. If two
electrons are needed to discharge an ion, how
many ions are discharged?

19-3 Silver has 5.8×10^{28} free electrons per
cubic meter. If the current in a silver wire is 10 A,
and the wire has a radius of 10^{-3} m, what is the
drift velocity of the electrons?

19-4 In a hydrogen discharge tube positive
hydrogen ions (protons) and negative electrons

travel in opposite directions under the influence of an electric field. Find the direction and magnitude in amperes of the current if in 1 second 8×10^{18} electrons and 3×10^{18} protons reach the electrodes.

19-5   Two copper plates are immersed in a copper sulfate solution and connected to a battery. If there is a 0.4-A current for 1 hour, how much copper is deposited on the plates? (The atomic mass of copper is 63.6 amu, and two electrons are needed to discharge one copper ion.)

19-6   The conduction electrons in a metal can be considered for some purposes as an ideal gas. (a) What is the average thermal kinetic energy $\frac{3}{2}k_B T$ of an electron at 300 K? (b) What is the corresponding velocity? (c) Calculate the ratio of this velocity to the drift velocity found in Example 19.2.

Section 19.2 | Resistance

19-7   A wire has a 10-V potential difference across its ends when it carries a 4-A current. What is its resistance?

19-8   A 10-A current in a wire results in a potential difference of 2 V across its ends. If it is an ohmic conductor, what current will produce a 6-V potential difference?

19-9   The thinnest copper wire normally manufactured has a radius of 4×10^{-5} m. Find the resistance of a 10-m long segment.

19-10   A nerve fiber (axon) may be approximated as a long cylinder. If its diameter is 10^{-5} m and its resistivity is 2 ohm m, what is the resistance of a 0.3-m long fiber?

19-11   A 2-m long copper wire has a resistance of 0.01 ohms. Find its radius.

Section 19.3 | Energy Sources in Circuits

19-12   An automobile lamp draws 1.2 A when connected to a 12-V battery. What is its resistance?

19-13   An electric iron draws 6 A when plugged into a 120-V outlet. What is its resistance?

19-14   A 120-V power line contains a fuse that will open when the current exceeds 15 A. What is the minimum resistance of an appliance operated by this circuit?

19-15   A 12-V battery is connected to a 2-ohm resistor. (a) What is the current? (b) How much charge is transported through the circuit in 10 sec-

Figure 19.22. Exercises 19-16 and 19-18.   **Figure 19.23.** Exercise 19-17.

onds? (c) How much work is done on the charge by electric fields in the battery? (d) How much work is done on the charge by electric fields in the resistor? (e) What is the total work done by the electric fields on the charge? (f) How much energy is converted into heat? (g) What is the source of that energy?

19-16   For the circuit in Fig. 19.22, find (a) the current, and (b) the potential difference across each resistance.

19-17   (a) Find the current in the circuit of Fig. 19-23. (b) Find the potential difference across each circuit element.

Section 19.4 | Power in Electrical Circuits

19-18   Find the power supplied to or by each circuit element in Fig. 19.22.

19-19   (a) A 12-V automobile storage battery has an internal resistance of 0.004 ohms. Find the current and power dissipation when the terminals are connected by a conductor of negligible resistance. (b) The battery delivers 80 A to the starter motor. What resistance would draw the same current? (c) How much power is supplied to the motor? What happens to that power? (d) How much power is dissipated in the battery?

19-20   A toaster uses 1500 W when plugged into a 120-V line. It takes 1 minute to toast a slice of bread. If electrical energy costs 6 cents per kilowatt-hour, how much does it cost to do this?

19-21   (a) What is the resistance of a 100-W light bulb designed for use in 120-V circuits? (b) What current will it draw?

19-22   A 20-A circuit in a home is wired with No. 12 copper wire. A single No. 12 wire 1 m long has a resistance of 5.2×10^{-3} ohms. If a two-wire cable carries 20 A for 30 m, find (a) the potential drop along each wire, and (b) the total power dissipated.

19-23   For the circuit in Fig. 19.24, find (a) the current, (b) the power supplied by each battery, and (c) the power dissipated by each resistor.

Figure 19.24. Exercise 19-23.   Figure 19.25. Exercise 19-24.

19-24 (a) In Fig. 19.25, what is the EMF $\mathcal{E}$? (b) Find the power supplied to each of the two resistors. (c) Find the power supplied to or by each EMF. Compare the results with those in part (b).

19-25 A room air conditioner uses 900 W of electrical power. If it operates an average of 12 hours per day, find the daily operating cost at 6 cents per kilowatt-hour.

19-26 A heater uses 1400 W of power when connected to a 120-V line. (a) What is its resistance? (b) What current does it draw? (c) If the line voltage is reduced to 112 V, how much power does the heater use? (Assume that the resistance remains constant.)

Section 19.5 | Series and Parallel Resistors

19-27 A string of 25 Christmas tree lights wired in series uses 500 W when connected to a 120-V line. (a) What is the current in the lights? (b) What is the resistance of a single bulb?

19-28 Some strings of Christmas lights are connected in series, and others are connected in parallel. Which is less affected by the burning out of one light bulb? Explain your reasoning.

19-29 An electrically heated home is supplied with 200 A at 230 V. How many 100-W appliances could be operated simultaneously in this home?

19-30 An electric heater uses 1800 W when connected to a 120-V source. (a) What is the resistance of the heater? (b) The heater is plugged into a 10-m long extension cord made of No. 18 wire, which is intended only for lamps and small wattage appliances. One meter of a single No. 18 wire has a resistance of 0.021 ohms. How much power is dissipated in the cord?

19-31 Six 60-W light bulbs are used in parallel in a lighting fixture connected to a 120-V power line. (a) What is the resistance of one bulb? (b) What is the effective resistance of the six bulbs?

19-32 Two light bulbs use 100 W each when

separately connected to a 120-V line. (a) If they are connected in series to this line, how much power will they draw? (Assume that the resistance of a bulb does not change when it is used in this way.) (b) Will the bulbs be brighter or dimmer? Explain.

Figure 19.26. Exercise 19-33.

Figure 19.27. Exercise 19-34.

19-33 Find the equivalent resistance for the network in Fig. 19.26.

19-34 Find the equivalent resistance for the network in Fig. 19.27.

Section 19.6 | Voltmeters and Ammeters

19-35 A galvanometer has a resistance of 10 ohms and gives full-scale deflection with a current of 10^{-3} A. How can one convert it into a voltmeter with a 0.1-V range?

19-36 A galvanometer has an internal resistance of 100 ohms and has full-scale deflection for a current of 10^{-5} A. (a) Design a voltmeter with full-scale deflection for 10 V. (b) Design an ammeter with full-scale deflection for 10 A.

Section 19.7 | Circuits Containing Resistance and Capacitance

19-37 A 1-μF capacitor is connected with copper wires to a 12-V battery. The initial current is 120 A. (a) What is the total resistance of the wires and the battery? (b) What is the time constant for the circuit?

19-38 A 1000-ohm resistor and a 10^{-5}-F capacitor are connected in series to a 100-V EMF. (a) Find the time constant. (b) What is the final charge on the capacitor?

19-39 A 10^{-4}-F capacitor is discharged through a 100-ohm resistor. How long does it take for the charge on the capacitor to drop to $1/e^2$ times its original value?

19-40 Verify that the product RC has the dimensions of a time. (*Hint:* Use the defining relations for R and C.)

19-41 A 500-ohm resistor is connected in series with a 6-V battery and a 10-μF capacitor. (a) What is the initial current? (b) How long does it take for the current to drop to $1/e$ times it initial value?

Section 19.8 | Electrical Safety

19-42 A bird stands on one foot on a high-voltage power line. What happens when it puts the other foot down on the same line? What would happen if the bird put this foot instead on a grounded conductor?

19-43 (a) If a person with wet hands grasps two conductors and has a resistance of 1000 ohms, how large a potential difference is needed to produce a 10 mA current that may freeze the hands to the conductors? (b) How large a potential difference is needed to produce a 100 mA current that will cause ventricular fibrillation in a second or so?

19-44 (a) Is one likely to get a dangerous electrical shock from a 12-V automobile battery? (b) What other possible hazards are associated with such batteries?

19-45 An electric shaver plugged into an electric outlet is accidentally dropped into a sink full of water. Is it safe to reach into the water and remove the shaver? Does it matter whether it is turned on? Explain.

19-46 Why is it especially important to properly ground appliances when used outdoors or in basements?

PROBLEMS

19-47 In a determination of Avogadro's number, a constant 4-A current is passed for 30 minutes through a cell consisting of two silver electrodes in a silver nitrate solution. (a) How many coulombs of charge are transported through the system? (b) How many electrons are transported? (c) If 7.84 g of silver are found to be deposited on the plates, what value of Avogadro's number is ob-

tained? (Each electron discharges one ion, and the atomic mass of silver is 107.9 amu).

19-48 The resistivity of body fluids is about 0.15 ohm m. Estimate the resistance of a finger end to end, ignoring the resistance of the skin.

19-49 (a) What is the room temperature resistance of a 1-m long aluminum wire of radius 0.002 m? (b) What is the radius of a 1-m long copper wire with the same resistance? (c) Compare the weights of the two wires. (The density of copper is 8900 kg m^{-3}, and the density of aluminum is 2700 kg m^{-3}.)

19-50 When the terminals of a dry cell are connected by a wire, the current is 2.2 A and the terminal voltage is 1.4 V. When the circuit is opened, the terminal voltage is 1.52 V. Find the internal resistance and the EMF, neglecting any errors due to the meters.

19-51 An electric dryer connected to a 230-V power line draws 20 A. (a) How much power does it use? (b) If it takes 2600 J to evaporate 1 g of water, how long will it take to dry a load of wet laundry containing 4 kg of water?

19-52 A motor is driven by a 12-V battery. This load is equivalent to a 0.2-ohm resistor. (a) What is the current? (b) What is the power supplied to the motor? (c) If the motor operates at 80 percent efficiency, at what rate can it lift a 100-N weight?

19-53 In a student laboratory experiment to measure the mechanical equivalent of heat, a resistor immersed in 0.7 kg of water carries a current of 4.2 A and a potential difference of 12 V. (a) How many joules of heat energy are supplied in 5 minutes? (b) If no heat is lost to the surroundings, what is the temperature change of the water?

19-54 A calorimeter is used to measure the latent heat of fusion of a material. The calorimeter has a 50-ohm resistor connected to a 120-V dc power line. Once the sample reaches the melting point, it takes 2 minutes to melt 0.3 kg. What is the latent heat of fusion of the material?

19-55 A 12-V storage battery with an internal resistance of 0.003 ohms is charged by a generator at a rate of 20 A. (a) Which terminal of the battery is connected to the high potential side of the generator? (b) How much power is supplied by the generator? (c) What is the terminal voltage of the battery while it is being charged?

Figure 19.28. Problem 19-56.

19-56 Find the equivalent resistance for the network in Fig. 19.28.

19-57 The capacitor in an electronic flash gun has a capacitance of $100\,\mu\text{F}$ and is charged so that its potential difference is 1000 V. (a) What is the charge on the capacitor plates? (b) The capacitor is discharged through the flashbulb, and 0.001 seconds after the switch is closed, the charge remaining is 0.37 times the initial charge. What is the resistance of the circuit? (c) What is the current after 0.001 seconds?

* 19-58 If an uncharged capacitance C is connected in series with a resistance R to an EMF, how long does it take for the charge to reach 99 percent of its final value?

* 19-59 A capacitance C and a resistance R are connected in series to a *square-wave generator*. This instrument has a constant terminal potential difference $\mathcal{E}_0$ for a time $T/2$ and then quickly reverses polarity so its terminal potential difference is $-\mathcal{E}_0$ for an equal time. (a) Sketch a graph of the terminal potential difference of the square-wave generator versus time for a time interval equal to $3T$. (b) Sketch graphs of the potential differences across the capacitor and the resistor if $T \gg RC$. (c) Repeat part (b) assuming $T = RC$.

19-60 Figure 19.29 shows a *Wheatstone bridge*,

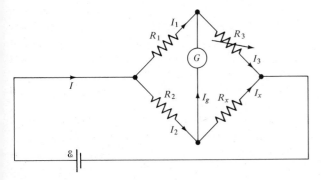

Figure 19.29. A Wheatstone bridge. Problem 19-60.

an instrument used to measure resistance. R_1 and R_2 are 10 ohms, and R_3 is adjusted until the galvanometer current I_g is zero. If R_3 is then 98.2 ohms, find the unknown resistance R_x.

Figure 19.30. A potentiometer. Problem 19-61.

* 19-61 Figure 19.30 shows a *potentiometer*, which is used to measure the EMF of a battery without drawing any current from it. The contact c is moved along the resistive wire ab until the galvanometer current is zero. When this happens, how is the unknown EMF $\mathcal{E}$ related to $\mathcal{E}_1$, R, and R_1?

SUPPLEMENTARY TOPICS

19.9 | ATOMIC THEORY OF RESISTANCE

We now discuss a simple atomic model that gives some insight into the physical origin of electrical resistance. The model assumes that the conduction electrons in a metal move freely among the positive ions with an average *thermal velocity u* determined by the temperature. Their directions are frequently changed by collisions with the ions. The average distance between collisions is called the *mean free path* λ (lambda) and plays a key role in determining the resistance.

In the absence of an applied electric field, the electrons move randomly, and there is no net charge flow in any direction. When an electric field **E** is maintained in the metal, the negatively charged electrons experience a force opposite to **E**, and they acquire an average drift velocity along the force. Although this drift velocity v is very small compared to the thermal velocity u, it is responsible for the current (Fig. 19.31).

Suppose an electric potential difference V is maintained between the ends of a wire of length l. If the

321

Figure 19.31. A typical random electron path with and without an applied electric field. An electric force produces a small net drift velocity in the direction of the electric force. Note how the field steadily deflects the electron toward the right.

field is uniform, then $E = V/l$. The force on an electron has a magnitude $eE = eV/l$, and the acceleration is

$$a = \frac{eV}{ml}$$

Every time an electron collides with an ion, it is deflected randomly and loses its tendency to drift with the electric force. Its next collision will occur after a time t that satisfies $ut = \lambda$ or $t = \lambda/u$. Between the collisions, it will acquire a velocity at. The average drift velocity v is half this, so

$$v = \frac{1}{2} \frac{eV}{ml} \frac{\lambda}{u}$$

If the wire has a cross-sectional area A and there are n conduction electrons per unit volume, then the net current is

$$I = envA = \frac{ne^2\lambda}{2mu} \frac{A}{l} V$$

This result states that the current and the potential difference are proportional. Thus the model has led to Ohm's law, which is found experimentally to correctly describe the behavior of most metals. Now using $R = \rho l/A$, the current $I = V/R$ is

$$I = \frac{V}{\rho l/A} = \frac{VA}{\rho l}$$

Comparing the two equations above for I, we find that in this atomic model the resistivity is given by

$$\rho = \frac{2\,mu}{ne^2\lambda} \qquad (19.24)$$

Only the factors n and λ in this formula for the resistivity depend on the choice of the material, since the electron charge e and mass m are constants, and the thermal speed u is determined by the temperature. For a metal, the number of conduction electrons is typically one or two per atom. Estimating the mean free path λ is a little more difficult. Classical mechanics suggests that λ is comparable to the interatomic spacing, but this value leads to overestimating the room temperature resistivity by a factor of 100 or more. Modern atomic theory leads to the remarkable but experimentally correct result that the electrons travel completely unimpeded in the crystal lattice until they encounter a deviation from the regular lattice structure (Fig. 19.32). Such an imperfection

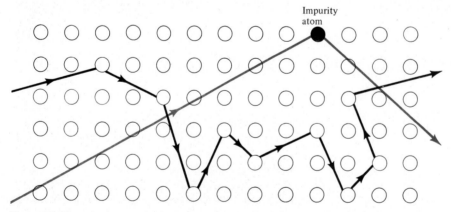

Figure 19.32. Classical mechanics suggests that the mean free path for an electron in a crystal lattice is comparable to the interatomic spacing (black path). According to modern atomic theory, collisions occur only when an electron encounters an imperfection in the lattice, such as the impurity atom in the top row (colored path). The modern theory is in agreement with measured resistivities.

can be an impurity or a lattice ion located some distance from its equilibrium position because of its thermal vibrational motion. At very low temperatures, the amplitude of the lattice ion vibrations is very small, so in very pure metals the mean free path of an electron may be thousands of interatomic spacings.

The resistivities of electrical insulators are larger than those of conductors by a factor of up to 10^{22}. This is a huge number, much larger than the variation of 10^4 between good thermal conductors and thermal insulators. Materials that are good electrical conductors are usually good thermal conductors, because the free electrons transport both charge and thermal energy. Insulators have very few free electrons, so they do not readily conduct either electricity or heat. Some materials called *semiconductors* have resistivities that are intermediate between those of conductors and insulators. They have only a few charge carriers, which can be either electrons or missing electrons (holes). These materials are employed in the manufacture of transistors and other solid-state electronic components.

As the temperature rises, the average electron velocity increases. Also, the mean free path decreases, since the lattice ion vibrational amplitude increases. Thus, Eq. 19.24 predicts that the resistivity will also increase, as is observed experimentally in metals (Fig. 19.33a). However, the resistivity of a semiconductor decreases as the temperature rises (Fig. 19.33b), because the number of charge carriers rises rapidly with the temperature, more than offsetting the changes in the path length λ and velocity u in Eq. 19.24.

At temperatures a few degrees above absolute zero, some materials become superconducting: their resistances apparently drop to zero (Fig. 19.33c). Once currents are started in a superconducting loop, they persist for years; no battery is needed to maintain the currents. Discovered by Heike Kamerlingh Onnes (1853–1926) in 1911, this spectacular effect has been studied extensively by physicists and has a number of useful applications. Powerful magnets built with superconducting coils are used in many laboratories, and superconducting electric power transmission lines with no power loss due to heating may soon be practical. The major difficulty in all applications is that the materials must be kept at very low temperatures, since no material has yet been found that is superconducting much above 15 K.

19.10 | APPLICATIONS OF RESISTANCE MEASUREMENTS

Resistance measurements often provide very useful information. The resistance of some semiconductors varies as much as 5 percent if the temperature changes by 1°C. A small semiconducting bead or *thermistor* can rapidly detect temperature changes as small as 10^{-3} degrees over large temperature ranges (Fig. 19.34).

A *strain gauge* exploits the dependence of the resistance of a wire on its geometry. A small wire grid is bonded to a flexible backing material, which is attached to a flexible diaphragm; the entire unit is covered with rubber. If the diaphragm flexes, the wire

Figure 19.34. The resistance of the thermistor tape to the abdomen of the infant changes rapidly with the temperature. This provides information to an electronic circuit that controls the incubator temperature, compensating for the baby's erratic temperature control system. (Courtesy of Air Shields, Inc.)

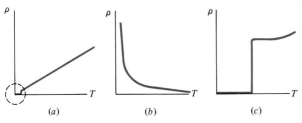

| (a) | (b) | (c) |

Figure 19.33. Resistivity versus temperature for (a) a conductor; (b) a semiconductor. (c) At low enough temperatures a conductor will become superconducting. This graph is an expanded view of the circled part of (a).

Figure 19.35. The conductivity of a solution is smallest at the point of exact acid-base neutralization. (The conductivity σ is the inverse of the resistivity ρ.)

grid is distorted, and its resistance changes. This strain gauge and its attached wires can be swallowed by a patient to measure pressures within the digestive tract or inserted with a catheter into veins or arteries to measure blood pressures.

Resistance measurements are useful in chemistry and biochemistry laboratories. The resistivity of a conducting or *electrolytic* solution is measured with a *conductivity cell,* which consists of two metal plates immersed in the solution. The cell is calibrated by using it to measure the resistance of a solution of known resistivity. For example, strongly acidic or basic solutions have many free ions and therefore have a low resistivity or high conductivity. If one gradually adds measured amounts of a basic solution to an acidic solution of unknown concentration, the conductivity of the sample drops sharply as the point of exact acid-base neutralization is approached, and the number of free ions becomes very small. This procedure is called *titration* and can be used to find the concentration of the acid (Fig. 19.35).

Resistance measurements are also used to study changes in living cells. A decrease in the cell resistance means that ions can more easily pass through the cell membrane. Therefore the changes in the resistance that occur when various materials are added to the cellular environment give clues to their effects on the membrane.

19.11 | ELECTROPHORESIS

Electrophoresis is an efficient technique for separating and analyzing the mixtures of proteins found in human blood and in other biological materials. It is based on the fact that the drift velocities of molecules in an electric field depend on their masses. When a solution is placed in an electric field, large protein molecules with a net charge of a few times the elec-

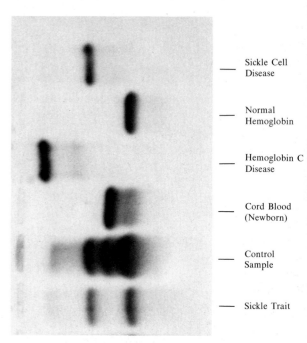

Figure 19.36. Electrophoretic separation of hemoglobin samples can be used to help diagnose many disorders. (Courtesy Helena Laboratories.)

tronic charge but with masses of thousands of amu experience small accelerations. Thus, they drift much more slowly than do small ions such as Na^+ or Cl^-.

In the electrophoretic technique most commonly used for medical diagnostic studies, one end of a moistened filter paper strip is placed in the protein solution. A potential difference is applied across the ends of the strip, and the protein molecules of various sizes migrate at different rates along the strip. If the process is stopped after a while, the various proteins have traveled different distances and are separated into several components (Fig. 19.36). Comparison with standard electrophoresis patterns then indicates whether certain abnormalities are present. More elaborate electrophoretic techniques can separate as many as 40 proteins in human blood plasma.

19.12 | DERIVATION OF THE CHARGE AND CURRENT IN AN *RC* CIRCUIT

In Section 19.7, we stated that the charge and current in an *RC* circuit connected at $t = 0$ to an EMF are

$$q = q_f(1 - e^{-t/\tau}), \qquad i = i_0 e^{-t/\tau}$$

Here $q_f = \mathcal{E}C$, $\tau = RC$, and $i_0 = \mathcal{E}/R$. To show that these two expressions are properly related to each other and satisfy $i = dq/dt$, we use Eq. B.25 from Appendix B,

$$\frac{d}{dt}e^{-at} = -ae^{-at}$$

Differentiating q with the aid of this formula,

$$\frac{dq}{dt} = \frac{q_f e^{-t/\tau}}{\tau}$$

Now $q_f/\tau = \mathcal{E}C/RC = \mathcal{E}/R = i_0$, so $dq/dt = i$ as required.

We must now demonstrate that these expressions satisfy Eq. 19.17, which was obtained by setting the sum of the potential changes equal to zero. Rewriting this equation in a slightly different form, it is

$$i + \frac{q}{RC} = \frac{\mathcal{E}}{R} \qquad (19.25)$$

Using the expressions above for q and i, the left side is

$$\frac{\mathcal{E}}{R}e^{-t/\tau} + \frac{q_f}{RC}(1 - e^{-t/\tau})$$

With $q_f = \mathcal{E}C$, this reduces to

$$\frac{\mathcal{E}}{R}e^{-t/\tau} + \frac{\mathcal{E}C}{RC}(1 - e^{-t/\tau}) = \frac{\mathcal{E}}{R}$$

which is the right side of Eq. 19.25. Thus, Eq. 19.25 is satisfied. Also, q has the correct initial value, $q = 0$, at time $t = 0$. These results show that we have found the correct solutions for q and i.

The equations for a discharging capacitor can be verified in the same way, except that $\mathcal{E}$ in Eq. 19.25 is set equal to zero.

EXERCISES ON SUPPLEMENTARY TOPICS

Section 19.9 | Atomic Theory of Resistance

19-62 At room temperature the average random thermal speed of a conduction electron in copper is about 10^5 m s^{-1}, and there are approximately 10^{29} conduction electrons per cubic meter. (a) Using the measured resistivity, what is the mean free path of an electron in the copper? (b) Find the ratio of this distance to the interatomic spacing, 4.2×10^{-10} m.

Section 19.10 | Applications of Resistance Measurements

19-63 The two parallel electrodes in a conductivity cell each have an area of 1.4×10^{-4} m^2, and they are separated by 4×10^{-2} m. The cell has a resistance of 370 ohms when filled with a particular electrolytic solution. Find the resistivity of the solution.

19-64 A conductivity cell is calibrated by filling it with a solution of known conductivity σ and measuring its resistance R. The _cell constant_ k is defined as $R\sigma$. (a) When a cell is filled with an aqueous solution of potassium chloride with conductivity 0.277 ohm^{-1} m^{-1}, the resistance is 190 ohms. What is the cell constant? (b) If the area of each plate is 1.2×10^{-4} m^2, what is their average separation?

19-65 The resistance R and resistivity ρ of a sample in a conductivity cell are related by $\rho = R/k$, where k is the _cell constant_. The cell constant for a particular cell is 42 m^{-1}. (a) When the cell is filled with a potassium sulfate solution, the resistance is 570 ohms. What is the resistivity of the solution? (b) What is the conductivity of the solution?

19-66 A conductivity cell has a resistance of 156 ohms when filled with a potassium chloride solution with conductivity 0.277 ohm^{-1} m^{-1}. Its resistance is 755 ohms when it is filled with a sodium chloride solution. Find the conductivity of the sodium chloride solution.

Additional Reading:

O.H. Smith, An Inexpensive High Resistance Voltmeter, _American Journal of Physics,_ vol. 19, 1951, p. 224.

Charles F. Dalziel, Electric Shock Hazard, _IEEE Spectrum,_ vol. 7, February 1972, p. 41.

G.D. Friedlander, Electricity in Hospitals, Elimination of Lethal Hazards, _IEEE Spectrum,_ vol. 8, September 1971, p. 40.

Brian B. Schwartz and Simon Foner, Large Scale Applications of Superconductivity, _Physics Today,_ vol. 30, July 1977, p. 34.

Scientific American articles:

Giorgio de Sautillana, Alessandro Volta, January 1965, p. 82.

Henry Ehrenreich, The Electrical Properties of Materials, September 1967, p. 194.

Werner Rieder, Circuit Breakers, January 1971, p. 76.

L.O. Barthold and H.G. Pfeiffer, High Voltage Transmission, May 1964, p. 3.

Curtis A. Williams, Jr., Immunoelectrophoresis, March 1960, p. 130.

R.T. Mattias, Superconductivity, November 1957, p. 92.

W.A. Little, Superconductivity at Room Temperature, February 1965, p. 21.

N.B. Brandt and N.I. Ginzberg, Superconductivity at High Pressure, April 1971, p. 83.

T.H. Geballe, New Superconductors, November 1971, p. 22.

Donald P. Snowden, Superconductors for Power Transmission, April 1972, p. 84.

CHAPTER TWENTY

NERVE CONDUCTION

In the preceding chapter, we applied the concepts of electric potential, current, resistance, and capacitance to electric circuits and instruments. These concepts can also be applied to the biological phenomenon of nerve conduction.

Information is transmitted in the human body by electrical pulses in nerve fibers called *axons*. These pulses differ greatly from pulses in copper telephone wires because an axon is a complex structure in which biochemical processes play an important role. An axon has a very high resistance and is poorly insulated from its surroundings, so after a very short distance nerve pulses become very weak and must be amplified. By comparison, telephone messages need to be amplified only after traveling many miles. Also, the velocity of a nerve pulse is only about one-millionth that of a pulse in a wire; the latter travels with nearly the speed of light.

Our discussion in this chapter centers on understanding two electrical potentials observed in nerves. In its undisturbed or *resting state,* the interior of an axon is at a lower potential than the surrounding *interstitial fluid.* This is the *resting potential.* When a nerve is suitably stimulated, a current pulse travels along the axon. The associated transient potential change is termed the *action potential.*

20.1 | THE STRUCTURE OF NERVE CELLS

Like other cells, a nerve cell is separated from its surroundings by a membrane that restricts the flow of materials. However, it is atypical in shape (Fig. 20.1).

Protuberances called *dendrites,* as well as a long thin structure, the *axon,* are attached to the central core of the cell, the *cell body.* Axons are typically 1 to 20 microns in diameter (1 micron = $1\,\mu = 10^{-6}$ m) and

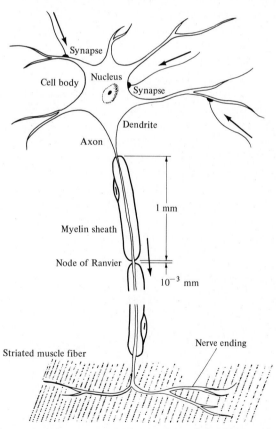

Figure 20.1. A typical nerve. Arrows indicate the direction of nerve impulse conduction.

Figure 20.2. (a) The conducting axoplasm has a resistance R to a current i_{axon} along the axon. (b) The resistance of a unit area of membrane to a leakage current i_{leak} is R_m. (c) Charges can accumulate on the two sides of the membrane. The capacitance of a unit area is C_m.

may be quite long. For example, nerves controlling the muscles have their cell bodies in the spinal column. Since some axons reach as far as the foot, a human axon may be 1 m long. The dendrites are typically shorter and thinner, but like the axon, they may have several branches. One nerve cell can influence another at points called *synapses* where dendrites make functional contact.

Wrapped around some axons of higher animals are *Schwannn cells,* which form a multilayered *myelin sheath,* reducing the membrane capacitance and increasing its electrical resistance. This sheath allows a nerve pulse to travel further without amplification, reducing the metabolic energy required by the nerve cell. Each Schwann cell is about $1\text{ mm} = 10^{-3}\text{ m}$ long, but the distance between successive Schwann cells is only about $1\,\mu = 10^{-6}\text{ m}$. In these short spaces between successive cells, called the *nodes of Ranvier,* the axon is in close contact with the surrounding interstitial fluid. It is at the nodes that the amplification of nerve pulses occurs in a myelinated nerve.

Thus, a myelinated axon resembles an intercontinental submarine cable, with periodic amplifiers used to prevent the signal from becoming too weak. By contrast, signals in unmyelinated axons become weak in a very short distance and require virtually continuous amplification.

20.2 | THE RESISTANCE AND CAPACITANCE OF AN AXON

We can understand many of the electrical properties of an axon with the aid of a model that resembles an electrical cable covered with defective insulation so that current leaks to the surroundings in many places. More specifically, we assume the axon consists of a cylindrical membrane containing a conducting fluid, the *axoplasm* (Fig. 20.2). The current can travel along the axon in this fluid and can also leak out through the membrane.

The electrical properties of the axon are determined by several quantities. The resistance R of a length of the axon to a current i_{axon} along the axon is proportional to the axoplasm resistivity, ρ_a. The resistance of a unit area of membrane to a leakage current i_{leak} is labeled R_m. The membrane also has capacitance, since charges of opposite signs accumulate on the two sides of the membrane. The charge per unit area divided by the resulting potential difference is the capacitance of a unit area, C_m.

From Eq. 19.5, the resistance of a wire of length l, cross-sectional area $A = \pi r^2$, and resistivity ρ_a is $R = \rho_a l/A$. Using the typical axon parameters given in Table 20.1, a $1\text{-cm} = 10^{-2}\text{-m}$ length of axon has a resistance

$$R = \frac{\rho_a l}{\pi r^2} = \frac{(2\text{ ohm m})(0.01\text{ m})}{\pi(5\times 10^{-6}\text{ m})^2}$$

TABLE 20.1

Values of axon parameters used in examples. Measured values vary somewhat with the type of axon. (The resistivity of the interstitial fluid is small and can be neglected)

Quantity	Myelinated Axon	Unmyelinated Axon
Axoplasm resistivity, ρ_a	2 ohm m	2 ohm m
Capacitance per unit area of membrane, C_m	5×10^{-5} F m^{-2}	10^{-2} F m^{-2}
Resistance of a unit area of membrane, R_m	40 ohm m^2	0.2 ohm m^2
Radius, r	$5\,\mu = 5\times 10^{-6}$ m	$5\,\mu = 5\times 10^{-6}$ m

$$= 2.5 \times 10^8 \text{ ohms} \qquad (20.1)$$

This is a huge resistance! It equals that of 44,000 miles of No. 40 copper wire, the thinnest wire normally manufactured, with a diameter of 0.08 mm. Nature has paradoxically devised an efficient communication system using a "wire" that would usually be considered a good insulator.

The membrane is quite thin, so a small section appears nearly flat. This permits us to use the parallel plate capacitance formula, Eq. 18.19, which states that the capacitance is proportional to the plate area, A. A length l of membrane has a surface area $A = 2\pi r l$. Since the capacitance per unit area is C_m, the capacitance of the length l of axon is

$$C = C_m(2\pi r l) \qquad (20.2)$$

Using the data of Table 20.1, a 1-cm length of an unmyelinated axon of radius 5 μ has a capacitance of 3.1×10^{-9} F. A similar myelinated axon segment has a much larger distance between the interstitial fluid and the axoplasm. Its capacitance is therefore smaller by a factor of about 200, because the capacitance of any capacitor diminishes as its two conductors are separated.

Since a membrane is not a perfect insulator, charge leaks from the axoplasm through the membrane into the interstitial fluid. The resistance of a conductor is inversely proportional to its cross-sectional area. If the resistance to leakage currents through a unit surface area of membrane is R_m, then a portion of the membrane with surface area A has a resistance $R' = R_m/A$. For a length l of axon, the surface area of the membrane is $2\pi r l$, and the membrane leakage resistance is

$$R' = \frac{R_m}{2\pi r l} \qquad (20.3)$$

Using the values in Table 20.1, a 1-cm length of unmyelinated axon has a leakage resistance $R' = 6.4 \times 10^5$ ohms, which is less than 1 percent of the axoplasm resistance R. Hence, most of a current entering an axon segment leaks out through the walls in much less than 1 cm.

According to our model, the axoplasm resistance R is proportional to the length l of the axon segment, and the leakage resistance R' is proportional to $1/l$. Thus, there is some distance λ (lambda) for which the resistances R and R' are equal. Using Eqs. 20.1 and 20.3 for R and R' respectively, λ must satisfy

$$\frac{\rho_a \lambda}{\pi r^2} = \frac{R_m}{2\pi r \lambda}$$

or

$$\lambda = \sqrt{\frac{R_m r}{2\rho_a}} \qquad (20.4)$$

The distance λ, called the *space parameter,* indicates how far a current travels before most of it has leaked out through the membrane. The values in Table 20.1 give $\lambda = 0.05$ cm for a typical unmyelinated axon, and 0.7 cm for a myelinated axon. Thus, a current pulse can travel much farther without amplification in a myelinated nerve.

20.3 | IONIC CONCENTRATIONS AND THE RESTING POTENTIAL

So far, we have considered the capacitance and the resistances associated with an axon. In this section, we consider some additional information that is needed to understand the behavior of an axon in its undisturbed or resting state: the differences in the ionic concentrations and potentials inside and outside the axon.

Figure 20.3 lists the concentrations of various ions inside, c_i, and outside, c_o, a resting axon. Of the ions that can pass through the membrane, sodium (Na^+) and chlorine (Cl^-) are much more numerous outside, while potassium (K^+) has a larger concentration inside. Conventionally, the electric potential V_o in the fluid outside the cell is taken to be zero. The potential

Figure 20.3. Concentrations and potentials inside and outside a typical mammalian axon in the resting state.

inside the axon is found to be 90 mV lower, so $V_i = -90$ mV. A number of important consequences follow from these observations, as we will see shortly.

According to Fig. 20.3, there are equal numbers of positive and negative ions inside and outside the cell. However, there is a potential difference across the membrane, so there must be small net charges $\pm Q$ on either side of the membrane. These can be calculated from $Q = CV$, using the capacitance found earlier and the measured resting potential. The excess of negative ions over positive ions inside the cell turns out to be only about 1/100,000 times the number of negative ions in the cell. An equal number of excess positive ions is present in the interstitial fluid. The excess ions form thin charge layers on either side of the membrane.

Let us now consider the effects arising from the fact that the concentration of Na$^+$ is much higher outside the cell than inside. We saw in Chapter Eleven that because of their random thermal motion, particles tend to diffuse from regions where their concentrations are high to regions where they are lower. The permeability of a cell membrane is a measure of the ease with which a given molecule or ion can pass through the membrane. Sodium ions will diffuse *into* the cell at a rate proportional to the Na$^+$ concentration difference $(c_o - c_i)$ and to the permeability of the membrane to Na$^+$ (Fig. 20.4). The resting potential is negative, so the electric field is directed into the cell, and it drives additional positive Na$^+$ ions through the membrane *into* the cell. Since the Na$^+$ concentration remains much larger outside, *sodium must be continually brought back out against both the electrical forces and the effects of the concentration*

differences. This return process requires an ongoing expenditure of metabolic energy.

The discussion of the Cl$^-$ and K$^+$ flows is more complicated. There are more Cl$^-$ ions outside, so diffusion produces a net flow into the cell. However, the electrical force on negative ions is opposite to the field, so the electric field draws some Cl$^-$ ions outward. Conversely, K$^+$ ions diffuse outward because of the concentration difference, but they also flow inward under electrical forces. Thus Cl$^-$ and K$^+$ each have flows in both directions across the membrane (Fig. 20.4). In a resting axon, the effect of the K$^+$ concentration difference exceeds that of the potential difference, and there is a net outward movement of K$^+$ ions. As in the case of sodium, there must be a mechanism that returns potassium into the cell and maintains the imbalanced concentrations. There is no net movement of Cl$^-$ ions, since the effect of the Cl$^-$ concentration difference is exactly balanced by the effect of the resting potential difference.

Nernst Equation | We can determine whether an ion is in equilibrium by calculating a theoretical resting potential for which there would be no net flow of that ion across the cell membrane. At this *equilibrium potential difference* across the cell membrane, the flows due to the concentration and potential differences are exactly balanced.

The equilibrium potential difference for an ion can be found from the *Nernst equation*. Although its derivation involves mathematics beyond the level of this textbook, it is based on a model that treats ions in a dilute solution as an ideal gas. A potential difference across a membrane will produce a concentration im-

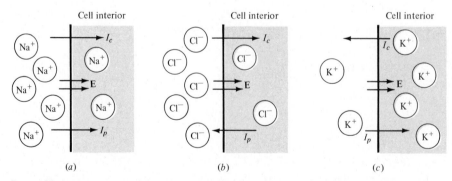

Figure 20.4.   Ionic currents I_c due to concentration differences and I_p due to potential differences. (*a*) Both Na$^+$ flows are into the cell. (*b*) The flows exactly balance for Cl$^-$. (*c*) For K$^+$, I_c is slightly larger than I_p, and there is a slight outward flow.

balance on the two sides of the membrane. Equilibrium occurs when the potential energy of one ion with charge q, $q(V_i - V_o)$, is equal to the work necessary to transfer it to the region of higher concentration. According to the model, this work is $k_B T \ln(c_o/c_i)$, where k_B is the Boltzmann constant, T is the Kelvin temperature, and $\ln(c_o/c_i)$ is the natural logarithm (to the base e) of the concentration ratio. Hence the ion is in equilibrium if the concentrations satisfy the Nernst equation,

$$q(V_i - V_o) = k_B T \ln \frac{c_o}{c_i} \qquad (20.5)$$

Example 20.1 illustrates the use of the Nernst equation by considering the K^+ concentrations.

Example 20.1

Compare the equilibrium potential difference for K^+ with the observed resting potential, -90 mV. (Use the data in Fig. 20.3 and assume a temperature of $37°C = 310$ K.)

The charge on a K^+ ion is $q = e = 1.60 \times 10^{-19}$ C, and $k_B = 1.38 \times 10^{-23}$ J K^{-1}. From Fig. 20.3, $\ln c_o/c_i = \ln 4/155 = -\ln 155/4$. Hence the Nernst equation gives the equilibrium potential difference

$$V_i - V_o = \frac{k_B T}{q} \ln \frac{c_o}{c_i}$$

$$= \frac{(1.38 \times 10^{-23} \text{ J K}^{-1})(310 \text{ K})}{(1.60 \times 10^{-19} \text{ C})}\left(-\ln \frac{155}{4}\right)$$

$$= -98 \text{ mV}$$

This is slightly larger in magnitude than the resting potential, -90 mV. Therefore, in the resting axon the inward flow due to the potential difference is not quite as large as the outward flow due to the concentration difference. If $(V_i - V_o)$ were -98 mV, the two flows would balance exactly.

If we apply the Nernst equation to Cl^-, we find that the equilibrium potential difference is -90 mV. This is equal to the resting potential, so the Cl^- flows are balanced as noted before. The equilibrium potential difference for Na^+ is $+66$ mV, which is opposite in sign to the resting potential. In other words, if enough positive charge entered the axon to reverse its polarity and change the axon potential to $+66$ mV, the outward flow of Na^+ due to the potential difference would balance the inward flow due to the concentration difference. One way to visualize this situation is to think of the concentration difference as producing the same inward flow as a 66-mV battery. When the charge inside the axon builds up enough so that it produces a 66-mV axon potential, there will be no net flow of Na^+ ions.

The Sodium-Potassium Pump

The net flows of Na^+ into the cell and of K^+ out of the cell due to diffusion and to the electrical forces are referred to as *passive flows*, because energy does not have to be supplied for them to occur. Some as yet undetermined process returns the Na^+ and K^+ across the membrane and maintains the nonequilibrium ionic concentrations. This energy expending process is called the *active Na-K transport* or the *Na-K pump*.

In the resting state, the axon membrane is about 100 times more permeable to K^+ than to Na^+. This means that equal concentration or potential differences would produce much larger flows of potassium ions than of sodium ions. However, the net passive sodium and potassium flows are about equal, because the resting potential is much closer to the equilibrium potential for K^+ than for Na^+. It is thought that the pump transports one K^+ ion into the cell for each Na^+ ion it removes.

The Na-K pump is responsible for the establishment and maintenance of the resting potential and the concentration imbalances. We can see this explicitly if we suppose that initially the potentials V_i inside the cell and V_o outside are both zero and that the concentrations of Na^+, K^+, and Cl^- are the same on both sides of the membrane. In a short time the pump transports a few K^+ ions into the cell and as many Na^+ ions out of the cell. Since the permeability of the membrane to K^+ ions is relatively large, some K^+ ions diffuse back out of the cell. A smaller number of Na^+ ions diffuse into the cell, since the membrane is less permeable to Na^+ ions. Thus a net positive charge has developed outside the cell, even though the pump has not transferred any net charge across the membrane. This positive charge attracts negative ions from inside the cell. The only negative ion inside the cell that can readily pass through the membrane is Cl^-, so some Cl^- ions move out of the cell, reducing but not quite eliminating the excess of positive ions outside the cell. Since the total system is electrically neutral, there is then an equal slight excess of negative ions inside the cell. The excess charges on either side of the membrane produce a potential difference across it, and now V_i is negative if V_o is defined to be zero.

As the pumping continues, the outside Na^+ and Cl^- concentrations, the inside K^+ concentration, and the magnitude of V_i all gradually increase. Eventually, equilibrium is reached and the concentrations stop changing when the passive Na^+ and K^+ flows due to diffusion and to the electrical forces exactly balance the active transport due to the pump. Because the membrane is much more permeable to K^+ ions than to Na^+ ions, the resting potential $V_i = -90$ mV, at which equilibrium occurs, is much closer to the -98-mV K^+ equilibrium than to the $+66$-mV Na^+ equilibrium potential.

It is important to note that if the membrane were suddenly to become much more permeable to Na^+, the balance would be upset briefly by an increased flow of Na^+ ions into the axon. A small increase in the number of sodium ions entering the cell would substantially modify V_i without significantly altering the ionic concentrations. We will see in Section 20.5 that such a change of permeability is believed to occur when the nerve is stimulated and an action potential is produced.

In summary the observed resting state potential and ionic concentrations imply a passive flow of Na^+ ions into the axon and of K^+ ions out of the axon due to diffusion and electrical forces. The continued ex-istence of these imbalances is due to the Na-K pump, which actively transports Na^+ out of the cell and K^+ into the cell, expending metabolic energy in the process. The magnitude and sign of the potential of the cell, in turn, is largely determined by the ratio of the permeabilities of the membrane for K^+ and Na^+.

20.4 | THE RESPONSE TO WEAK STIMULI

Having considered the resting state of the axon, we now examine the response of an axon to a weak stimulus. In most experiments, the stimuli are electrical, since these are easily controlled and do not injure the cell if they are sufficiently mild. For an electrical stimulus smaller than a critical threshold value, the response of the axon is similar to that of an analog network of resistors and capacitors. Specifically, if a weak stimulus is applied at some point on an axon, no significant axon potential changes occur beyond a few millimeters from that point. By contrast, a stimulus *above* the threshold level produces a current pulse that travels the length of the axon without attenuation. This current pulse and the associated action potential are discussed in the next section.

We can develop the analog circuit for the axon by dividing the axon into many short segments. The interstitial fluid surrounding the axon has very little

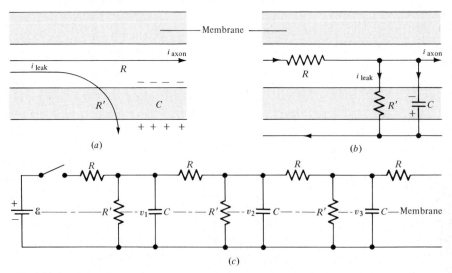

Figure 20.5. (*a*) A short segment of an axon. (*b*) The equivalent analog circuit for the segment. (*c*) The analog circuit for several axon segments. The EMF $\mathcal{E}$ represents the stimulus.

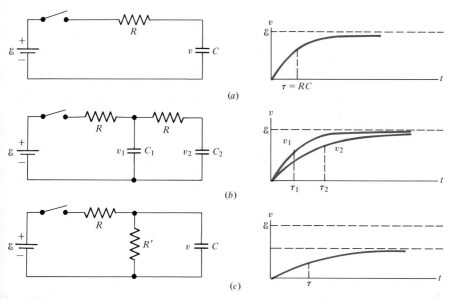

Figure 20.6. Circuit diagrams and capacitor potential differences versus time. The capacitors are initially uncharged, and the switch is closed at $t = 0$. (*a*) Simplest RC circuit. (*b*) v_2 rises more slowly than v_1, but ultimately reaches the same final value, the battery EMF $\mathcal{E}$. (*c*) If a leakage resistor is placed in parallel with the capacitor, the final potential difference v is less than $\mathcal{E}$.

resistance and may be represented by a perfect conductor. Each axon segment has a resistance R to a current i_{axon} along its length. The membrane has a resistance R' to a leakage current i_{leak} plus a capacitance C (Fig. 20.5*a,b*). A series of several segments is then analogous to the complex network of resistors and capacitors in Fig. 20.5*c*. The EMF shown there represents an applied stimulus.

The behavior of this complex axon analog circuit is more readily understood if we first consider the simplest RC network in Fig. 20.6*a*. Suppose that the initial charge q on C is zero, and the switch is closed at $t = 0$. The charge q and the potential difference $v = q/C$ will gradually increase. As we saw in Chapter Nineteen, the time needed to reach a characteristic fraction of the final values of q and v is determined by the time constant, $\tau = RC$.

When there are two resistors and two capacitors, as in Fig. 20.6*b*, the charging process is more complicated. The potential difference v_2 across C_2 must increase more slowly than the potential difference v_1 across C_1, since the path from the battery to C_2 and back has resistance $2R$. As more and more RC pairs are added, the potential difference across each added capacitor rises ever more slowly. Consequently, in the

axon analog circuit (Fig. 20.5*c*), v_2 will change more slowly than v_1, v_3 still more slowly, and so on.

The effect of a leakage resistor can be seen in Fig. 20.6*c*. There is always some current in the conducting path through R and R'. Thus, there is a corresponding potential drop across R, and the final potential difference across the capacitor is less than the EMF. In the axon analog circuit, the final potential differences steadily diminish as we move to the right because of the current lost through the leakage resistors R'.

To summarize, when the switch is closed or a "stimulus" is applied in the axon analog circuit, the potential differences across the capacitors gradually change. As one goes further from the stimulus, the changes occur more slowly, and their final magnitudes diminish (Fig. 20.7).

Very similar behavior is observed when an unmyelinated axon is stimulated weakly, as in Fig. 20.8. A probe connected to a battery is inserted at $x = 0$, and gradually the axon potential V_i at that point changes from -90 to -60 mV. The time required for this change to occur is determined by the membrane capacitance and the external series resistance r. At other values of x, the potentials change more slowly, reaching a final potential between -90 and -60 mV.

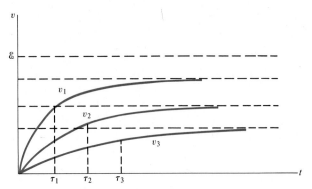

Figure 20.7. The capacitor potential differences for the axon analog circuit of Fig. 20.5c, assuming the switch is closed at $t = 0$. Each potential difference rises more slowly than the preceding one and approaches a successively smaller final value.

As in the axon analog circuit, the time needed to change the potential appreciably increases with the distance x from the stimulus, reflecting the time needed to alter the charges on the membrane. The final magnitude of the potential changes diminishes as x increases, because of the leakage of current through the membrane. Thus the effects of a weak stimulus propagate rather slowly and become negligible after a few millimeters.

We have used the axon analog circuit to make qualitative predictions of the response of an axon to a weak stimulus. The analog circuit can also yield quantitative predictions relating the final axon potential at a distance x from the stimulus to the space parameter λ defined in Section 20.2. This is done by applying Ohm's law to the currents leaking through the membrane and traveling along the axon.

If the difference between the resting and final potentials at $x = 0$ is V_d, then the difference at x is found to be

$$V(x) = V_d e^{-x/\lambda} \qquad (20.6)$$

If λ is 0.05 cm, a typical value for an unmyelinated axon, then $V(x)$ diminishes to $V_d e^{-1} = 0.37 V_d$ after 0.05 cm, to $V_d e^{-2} = 0.135 V_d$ after 0.1 cm, and so on. This predicted dependence on x agrees with the experimental data.

20.5 | THE ACTION POTENTIAL

We have seen that when a weak electrical stimulus is applied to an axon, the potential changes are proportional to the stimulus. The situation is very different if a battery such as that shown in Fig. 20.8 briefly increases the potential at $x = 0$ to a value just above the action potential threshold, which is typically at -50 mV. Shortly after this stimulus is applied, the axon potential at x suddenly increases and becomes positive, reaching a value as high as $+50$ mV for some axons (Fig. 20.9a). The potential then gradually returns to its resting value. For a particular axon, the shape and peak size of the action potential curve are *independent of the strength of the initial above threshold stimulus* or the distance x from the stimulus except very near $x = 0$. Thus the action potential is not proportional to the stimulus. Instead, it is a transient *all-or-nothing* response.

In an unmyelinated axon, the action potential is accompanied by dramatic changes in the permeabil-

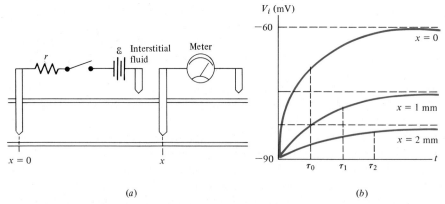

Figure 20.8. (a) Apparatus for changing the axon potential at $x = 0$ and observing the resulting change at position x. The "meter" is usually a cathode-ray oscilloscope. (b) The axon potential versus time at $x = 0$, 1, and 2 mm. Note the similarities with Fig. 20.7.

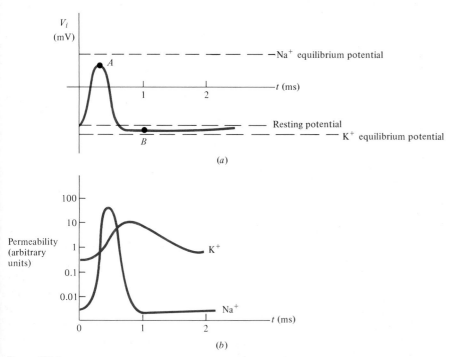

Figure 20.9.   (*a*) The axon potential at a point some distance from the stimulus. (*b*) The associated changes in the membrane permeabilities. Note that the vertical scale is logarithmic.

ity of the membrane to Na^+ and K^+ (Fig. 20.9*b*). When the axon potential V_i in an unmyelinated axon rises above the action potential threshold at some point, the Na^+ permeability there suddenly increases by a factor of more than 1000. This causes a rapid influx of positive sodium ions, which changes the sign of V_i from negative to positive. After about 0.3 ms, the potential approaches the Na^+ equilibrium potential (point *A* in Fig. 20.9*a*) as determined from the Nernst equation, and the sodium influx diminishes. Also the sodium permeability begins to decrease toward its normal low level. Meanwhile the potassium permeability has gradually risen by about a factor of 30. Consequently, potassium ions now begin to flow rapidly out of the cell, and V_i again becomes negative. V_i actually reaches a value below the resting potential (point *B* in Fig. 20.9*a*), close to the K^+ equilibrium potential which, as we saw before, is slightly more negative than the resting potential. This return over about 1 ms to a potential close to the resting potential is due to the changes in the K^+ permeability and *not* to the effects of the Na-K pump, which acts much more slowly. The pump does

gradually reestablish the resting Na^+ and K^+ concentrations that were altered slightly during the action potential pulse; this process takes about 50 ms.

We now see how this mechanism amplifies a pulse and permits an action potential to travel the length of an axon without attenuation. Figure 20.10 shows an unmyelinated axon segment that has been excited at one end, so that V_i is positive at this point. Positive ions move toward this end on the outside of the membrane and away from it inside. This decreases the charge on the adjacent portion of the membrane, so that the axoplasm potential there becomes less negative and rises to the action potential threshold. This triggers an increased sodium permeability, leading to a sodium influx and an action potential at the adjacent portion of the membrane. In this way the action potential proceeds from point to point along the entire length of the axon.

The term "amplification" is applied to the process that occurs each time an action potential is generated at some point along an axon, because it is here that energy is expended. The Na-K pump is continually maintaining both the resting potential and the ionic

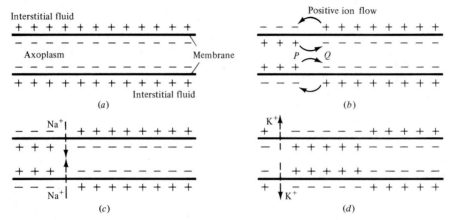

Figure 20.10.   Propagation of an action potential in an *unmyelinated* axon. (*a*) The resting state. (*b*) An action potential pulse has traveled through the axon segment to point *P*. Positive ions move away from *P* toward the right inside the axon and toward *P* from the right outside. The charges on either side of the adjacent portion of the membrane at *Q* gradually diminish in magnitude, and the potential there increases toward the action potential threshold. (*c*) Once that threshold is reached, the membrane allows an influx of sodium ions (dotted arrows). The axon potential at this point now rises rapidly and becomes positive. (*d*) The action potential pulse has traveled further along the axon, and the left end of the segment has returned to a negative potential close to the resting potential. This quick return to a negative axon potential is due to an outward flow of potassium ions (dotted arrows) caused by an increase in the K^+ permeability. The resting state potential and ionic concentrations are restored over a much longer interval by the active Na–K pump.

imbalances of Na^+ and K^+ ions across the membrane. In so doing, considerable electric potential energy is stored in the membrane, much like water is stored above a dam. An action potential occurs because of the increases that occur in the permeability of the membrane to Na^+ and K^+ ions. The increases of permeability are analogous to opening a floodgate in the dam. There is a sudden flow of ions through the cell membrane due to the large concentration differences, and it is this flow of ions that provides the current of an action potential. The Na–K pump then acts analogously to a pump at the dam, which redeposits the water above the dam for the next flow. Like the pump at the dam, the Na–K pump requires much more time to restore the ions than elapses during an action potential.

Myelinated Axon |
In a myelinated nerve, relatively few ions pass through the myelin sheath except at the nodes of Ranvier, which are about 1 mm apart (Fig. 20.11). At the nodes the membrane responds to an above-threshold stimulus much as in an unmyelinated axon: the Na^+ permeability increases rapidly, producing an influx of Na^+ and the charac-

teristic action potential pulse. This, in turn, produces flows of positive ions away from the node inside the axon and toward the node outside. Some of the axon current leaks through the membrane, but most of it reaches the next node, since the 1-mm internodal distance is small compared to the space parameter λ, which is equal to several millimeters in a typical myelinated axon. The axon current reduces the charges on the membrane and increases the potential at the next node to the action potential threshold level. Consequently, this next node is triggered, and the action potential travels along the axon.

Because the amplification and the associated ionic transfers occur only at the nodes, less metabolic en-

Figure 20.11.   Propagation of an action potential in a *myelinated* axon. An above-threshold axon potential at one node triggers an influx of sodium and an action potential pulse, which in turn triggers the next node.

ergy is needed to restore a myelinated axon to its resting state after an action potential pulse than is required for an unmyelinated axon. The propagation of an action potential is also faster in a myelinated axon. Thus, such axons are better suited to transmit rapidly the large amounts of information required by more advanced animals and represent an important evolutionary step.

The velocity of propagation of an action potential between two nodes in a myelinated axon can be predicted with the simple axon model because amplification occurs only at the nodes. Although a rigorous derivation involves advanced mathematics, we can quickly obtain the correct formula for the velocity using plausible physical arguments. This formula is interesting because it indicates some compromises that have been made in the process of evolution.

The action potential velocity v is the distance X between two nodes divided by the time τ needed to reduce the charge on the membrane and increase the potential at the second node to the threshold level. This time must be comparable to the time constant RC of the series circuit containing the axoplasm resistance R and membrane capacitance C. R is the resistance from the first node to $X/2$, or to the midpoint of the "capacitor" being charged. Using $l = X/2$ in Eq. 20.1 for R and $l = X$ in Eq. 20.2 for C gives, for an axon of radius r,

$$\tau = RC = \left(\frac{\rho_a X}{2\pi r^2}\right)(2\pi r X) = \rho_a C_m \frac{X^2}{r}$$

and

$$v = \frac{X}{\tau} = \frac{r}{\rho_a C_m X} = 10r \text{ m s}^{-1} \mu^{-1} \quad (20.7)$$

This last expression is obtained using $X = 1$ mm and the values of ρ_a and C_m in Fig. 20.3. Measured values of v in meters per second are usually between $12r$ and $17r$, so the simple axon model gives reasonable results.

Equation 20.7 shows that nerve pulses travel rather slowly. For example, if $r = 5\,\mu$, v is 50 m s^{-1}, and it takes 0.08 s to send messages from a toe stepping on a sharp object to the brain and back again. Increasing the radius of an axon increases the velocity, but it also increases the metabolic energy requirements, so that a compromise must be made. Clearly, long axons that transmit information requiring rapid responses must have large radii, but other axons need not. Accord-

ingly, many, but by no means all, long axons have large radii. For example, about 60 percent of the axons going to muscles are 6 to 10.5 μ in radius. By contrast, in parts of the brain where the nerves are very short, 90 percent of the axons are 2 μ or less in radius.

A subtle process is also involved in choosing the optimum internode spacing, since the velocity varies as $1/X$. If X is increased, the velocity decreases, which is undesirable but the metabolic energy requirements are also reduced. Thus again, a compromise between high velocity and low energy usage must be made. In nature, X varies over the relatively small range of 1 to 2 mm, suggesting that deviations from this range render the axon less useful in carrying out its function.

20.6 | ELECTROENCEPHALOGRAPH AND ELECTROCARDIOGRAPH

Direct observations of nerve pulses require inserting a probe into a nerve and are impractical for routine medical diagnostic studies. Fortunately, electrodes placed on the skin can pick up small signals related to large-scale electrical activity within the body. This makes possible the *electroencephalograph* (EEG) and *electrocardiograph* (EKG), instruments that are useful aids in studying brain and heart disorders, respectively.

Electrical signals arising within the body reach the surface because the resistance of the interstitial fluid is not quite zero. If many axons are simultaneously active, then the potential of the body nearby changes relative to the potential elsewhere in the body. Although the changes observed at the surface are at most 50 μV or about 0.1 percent of the full action potential changes, they can be amplified and measured or recorded.

An *electroencephalograph* is shown in Fig. 20.12, along with EEGs typical of various states of awareness. Electrodes are placed at standardized locations on the head, and the signals are recorded. Deviations from the norm can aid in the diagnosis of epilepsy, tumors, accidental brain damage, and other disorders.

An *electrocardiograph* records electrical activity associated with the heart. The usual positions for the

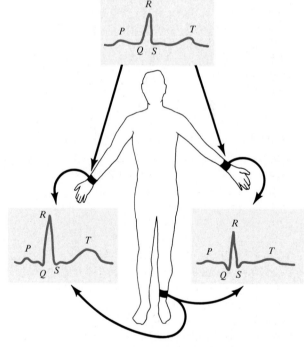

Excited

Relaxed

Drowsy

Asleep

Deep sleep

1 s

(a)   (b)

Figure 20.12.   Electroencephalograph. (*a*) An electrode attached to the ear provides the reference potential. The two recordings are typical of the response from different areas. (*b*) EEG patterns typical of different states of awareness of a patient.

electrodes are shown in Fig. 20.13 along with typical EKG patterns.

SUMMARY

An axon resembles a poorly insulated electrical cable. Many of its electrical properties can be understood using this model or an equivalent *RC* network.

A resting axon has a potential below that of the surrounding interstitial fluid. The resting potential and nonequilibrium concentrations of Na$^+$ and K$^+$ ions are maintained by the active Na-K pump mechanism, which offsets passive ionic flows due to diffusion and to electric forces.

Weak electrical stimuli produce proportional axon responses that are similar to those of the analog *RC* circuit and diminish rapidly with distance. A stimulus that raises the axon potential above a critical threshold triggers a current pulse and an associated action potential. These travel without attenuation along the entire length of the axon, amplified by processes associated with changes in the ionic permeabilities of the membrane.

The velocity of nerve pulses in a myelinated axon is proportional to the axon radius and inversely proportional to the distance between the nodes of Ranvier where amplification occurs. Thus there is a delicate evolutionary compromise between speed and economy of energy usage.

Figure 20.13.   The standard limb leads of the electrocardiogram and normal patterns recorded from these leads. The *P* peak corresponds to the initial electrical pulse which triggers a heart cycle. The interval from *P* to *Q* lasts about 0.12 to 0.20 seconds in a normal person and corresponds to the spread of the pulse over the atria (the upper heart chambers.) The *QRS* interval lasts about 0.06 to 0.10 seconds and corresponds to the spread of the pulse over the ventricles (lower chambers). The *T* peak is associated with the return of the heart to its normal unstimulated state after a pulse.

338

Checklist

Define or explain:

axon	Nernst equation
myelin sheath	Na-K pump
nodes of Ranvier	axon analog circuit
leakage resistance	action potential
space parameter	amplification
equilibrium potential	EEG
difference	EKG

EXERCISES

When numerical values are required that are not given in the exercise, the data in Table 20.1 and Fig. 20.3 should be used.

Section 20.1 | The Structure of Nerve Cells

20-1   Estimate the number of axons in the human spinal column if the average axon is 10μ in diameter and the spinal column is 1 cm in diameter.

20-2   An axon 1 m long has nodes of Ranvier every 10^{-3} m. How many times is a nerve pulse amplified when it is transmitted along this axon?

Section 20.2 | The Resistance and Capacitance of an Axon

20-3   Calculate the axoplasm resistance of a 1-cm long segment of unmyelinated axon of radius 2μ.

20-4   Calculate the axoplasm resistance of a 1-cm long segment of myelinated axon of radius 2μ.

20-5   An unmyelinated segment of axon has a radius of 2μ and a length of 1 cm. Find its (a) membrane capacitance, and (b) membrane leakage resistance.

20-6   A myelinated segment of axon has a radius of 2μ and a length of 1 cm. Find its (a) membrane capacitance; (b) membrane leakage resistance.

20-7   An axon membrane is 7.5×10^{-9} m thick. (a) In the resting state the axon potential is -90 mV. What is the direction and magnitude of the electric field in the membrane? (b) If the membrane has a capacitance of 0.01 F m^{-2} what is its dielectric constant?

20-8   Find the space parameter for an axon of radius 0.2μ if it is (a) myelinated; (b) unmyelinated.

20-9   A membrane has a potential difference of 90 mV across it. How much charge per square meter is located on either side if it is (a) myelinated; (b) unmyelinated.

20-10   What is the space parameter for an axon of radius 3μ if it is (a) unmyelinated; (b) myelinated?

20-11   A myelinated axon has a space parameter of 1 cm. Find its radius.

20-12   A myelinated axon and an unmyelinated axon have the same space parameter. Find the ratio of their radii.

20-13   Find the radius of an unmyelinated axon with a space parameter of 2×10^{-4} m.

Section 20.3 | Ionic Concentrations and the Resting Potential

20-14   How many potassium ions are there inside a 1-cm long segment of an axon of radius 2μ?

20-15   How many sodium ions are there in a 1-cm length of an axon of radius 6μ?

20-16   The concentration of an ion is 100 moles m^{-3}. How many ions are there is 1 m^3?

20-17   Find the equilibrium potential at 37°C for an ion with charge $+e$ if its concentration outside the axon is 160 moles m^{-3} and inside is 10 moles m^{-3}.

20-18   The equilibrium potential at 37°C for Cl$^-$ ions in a particular axon is -80 mV. If the Cl$^-$ concentration outside the cell is 110 moles m^{-3}, what is the concentration inside?

20-19   The active sodium transport due to the Na-K pump occurs at the rate of 3×10^{-7} moles m^{-2} s^{-1}. For 1 m^2 of membrane, find (a) the current in amperes due to the sodium ions; (b) the power expended against electrical forces if the resting potential is -90 mV.

20-20   The K$^+$ concentration inside an axon is 165 moles m^{-3} and outside it is 8 moles m^{-3}. (a) What is the equilibrium potential at 37°C? (b) In which directions are the potassium flows due to diffusion and to the electric field if the axon potential is -90 mV? Which flow is greater?

Section 20.4 | The Response to Weak Stimuli

20-21   Estimate the space parameter for the axon in Fig. 20.8*b*.

20-22   A myelinated axon of radius 3μ is stimulated so that the potential at $x = 0$ is -100 mV, which is 10 mV below the resting potential. (a) What is the space parameter? (b) Sketch the axon potential-versus-distance graph when the potential has achieved its final steady state value.

20-23 A myelinated nerve with a space parameter of 0.5 cm is disturbed at a point so that its potential is raised from the resting value of -90 mV to -80 mV. Find the steady state potential at distances from this point of (a) 0.5 cm; (b) 1 cm.

Section 20.5 | The Action Potential

20-24 A probe with a resistance of 10^7 ohms connects an electrical circuit to an axon. Estimate the time constant for charging a 1-cm long axon segment assuming $C_m = 10^{-2}$ F m^{-2}, $\rho_a = 0$, $R_m = \infty$, and a radius of 5 μ.

20-25 Find the velocity of propagation for an action potential and the time required for it to travel 2 m in a myelinated nerve (a) with a radius of 1 μ; (b) with a radius of 20 μ.

20-26 A nerve pulse can travel the length of a 0.5-m long myelinated axon in 0.05 s. What is its radius?

PROBLEMS

Use the data in Table 20.1 and Fig. 20.3 when numerical values are required that are not given in the problem.

20-27 A square meter of axon membrane has a resistance of 0.2 ohms. The membrane is 7.5×10^{-9} m thick. (a) What is the resistivity of the membrane? (b) Suppose the membrane resistance is due to fluid-filled cylindrical pores through the membrane. The pores have a radius of 3.5×10^{-10} m and a length equal to the membrane thickness, 7.5×10^{-9} m. The fluid in the pores has a resistivity of 0.15 ohm m, and the remainder of the membrane is assumed to be a perfect insulator. How many pores must there be in 1 m^2 to account for the observed resistance? (c) If the pores are in a square pattern, how far apart are they?

20-28 If an ion in an electric field E has a drift velocity v, then the *mobility* of the ion is $\mu = v/E$. (a) Show that the resistivity of a fluid containing a single type of ion is $(qn\mu)^{-1}$, where q is the magnitude of the charge on an ion, and n is the number of ions per unit volume. (b) Show that if there are m

different kinds of ions in a solution, the resistivity is $(q_1 n_1 \mu_1 + q_2 n_2 \mu_2 + \cdots + q_m n_m \mu_m)^{-1}$.

20-29 Using the relation between mobility and resistivity given in the preceding problem and the ionic concentrations in Fig. 20.3, estimate the resistivities of axoplasm and interstitial fluid. Explain any difference between the calculated result for ρ_a and the value in Table 20.1. (The mobilities of Na$^+$, K$^+$, and Cl$^-$ are 5.20×10^{-8}, 7.64×10^{-8}, and 7.91×10^{-8} m^2 V^{-1} s^{-1}, respectively. Neglect the effects of the "other" ions.)

20-30 A 1-cm long segment of myelinated axon has a radius of 10^{-5} m. Its capacitance is 6×10^{-9} F, and the resting potential is -90 mV. (a) The resting potential is due to an excess of positive ions along the membrane outside the axon and an equal negative excess inside. How much excess charge is there on either side? (b) If these excesses are due to singly charged ions, how many excess ions are there on either side? (c) Find the ratio of this number of ions to the total number of negative ions inside the segment.

20-31 A 1-cm long segment of unmyelinated axon has a capacitance of 3×10^{-9} F. (a) If the axoplasm potential changes from -90 to $+40$ mV, by how much do the excess charges on either side of the membrane change? (b) If this change is due to an influx of sodium ions, how many sodium ions enter the axon?

20-32 A 1-cm long segment of unmyelinated axon has a radius of 5×10^{-6} m and a capacitance of 3×10^{-9} F. (a) If the axoplasm potential changes from $+40$ to -96 mV, by how much do the excess charges on either side of the membrane change? (b) If this change is due to an outflow of potassium ions, how many ions must leave the axon segment? (c) If the original potassium concentration inside the axon is 155 moles m^{-3}, what fraction of the potassium ions leave the axon?

20-33 The axon analog network in Fig. 20.5c can be made more complete by adding EMFs that simulate the resting potential. Show how this may be done.

20-34 The *Goldman equation*

$$V_i - V_o = \frac{k_B T}{e} \ln \left[\frac{c_{oNa} + c_{oK}(P_K/P_{Na})}{c_{iNa} + c_{iK}(P_K/P_{Na})} \right]$$

gives the net equilibrium potential difference for an

axon, where P_K/P_{Na} is the ratio of the membrane permeabilities for potassium and sodium, the c's are the ionic concentrations, and e is the proton charge, Assume $T = 310$ K. (a) If the resting potential is -90 mV, what is the permeability ratio? (b) An above threshold stimulus alters the permeabilities so the potential jumps to $+40$ mV. What is the new ratio? (c) Explain to what extent it is correct to use the same concentrations in parts (a) and (b).

*20-35 Draw a rough graph of the equilibrium potential versus the permeability ratio using the Goldman equation given in Problem 20-34.

*20-36 Using the Goldman equation in Problem 20-34, find the resting potential of an axon if the concentrations are as in Fig. 20.3 and the potassium to sodium permeability ratio is 100.

20-37 The total influx of Na^+ during the transmission of a single action potential pulse is 4×10^{-8} moles m^{-2} of active axon surface. (a) How many pulses may be transmitted in an unmyelinated nerve of radius $5\,\mu$ before the Na^+ concentration within the axon is increased by 10 percent? (Neglect the effects of the Na-K pump.) (b) How many pulses in a myelinated axon of radius $5\,\mu$ are required to increase the Na^+ concentration by 10 percent? (Assume the active surface is only at the nodes, which are $1\,\mu$ long and 1 mm apart. Again neglect the effects of the Na-K pump.)

Addition Reading:

Bernard Katz, *Nerve, Muscle, and Synapse,* McGraw-Hill Book Company, New York, 1966. Paperback.

Arthur C. Guyton, *Textbook of Medical Physiology,* 4th ed., W.B. Saunders Co., Philadelphia, 1971, chapters 4 and 5. More advanced but quite readable.

Theodore C. Ruch, Harry D. Patton, J. Walter Woodbury, and Arnold L. Towe, *Neurophysiology,* W.B. Saunders Company, Philadelphia, 1965. A very complete discussion is given in Chapters 1 and 2.

D.V. Aidley, *Physiology of Excitable Cells,* Cambridge University Press, Cambridge, 1971.

Scientific American articles:

C.L. Stong, Amateur Scientist, January 1962, p. 145. Description of a demonstration apparatus for nerve and muscle potentials in aquatic animals.

Arthur K. Solomon, Pumps in the Living Cell, August 1962, p. 100.

Peter F. Baker, The Nerve Axon, March 1966, p. 74.

Sir John Eccles, The Synapse, January 1965, p. 56.

Bernard Katz, The Nerve Impulse, November 1952, p. 55.

Werner R. Lowenstein, Biological Transducers, August 1960, p. 98.

Graham Hoyle, How is Muscle Turned On and Off?, April 1970, p. 85.

H.E. Huxley, The Contraction of Muscle, November 1958, p. 66.

Allen M. Scher, The Electrocardiogram, November 1961, p. 132.

Bruce I.H. Scott, Electricity in Plants, October 1962, p. 107.

Keir Pearson, The Control of Walking, December 1976, p. 72.

CHAPTER TWENTY-ONE

MAGNETISM

Most of us have had the opportunity to observe some of the fascinating properties of permanent magnets. The north pole of one magnet attracts the south pole and repels the north pole of another magnet. Either pole of a magnet attracts unmagnetized iron or steel objects, and when such objects are in contact with a magnet, they can attract other iron or steel objects.

It has been known since antiquity that *lodestones,* pieces of the mineral *magnetite* (iron oxide), are able to attract iron objects. Also, the fact that the earth is a large magnet was long known from its orienting effect on compass needles. However, it was only in 1820 that Hans Christian Oersted (1777–1851) discovered that an electric current in a wire can deflect a compass needle. Electric currents in wires, as well as charges moving in a vacuum, produce magnetic effects indistinguishable from those due to permanent magnets. Since currents exert forces on magnets, we would expect from Newton's third law relating action and reaction forces that magnets also exert forces on currents. Such forces are indeed observed.

In discussing electric forces, we found it advantageous to avoid the direct use of Coulomb's law for the force between two charges. Instead, we said that one charge produces an electric field **E**, which in turn exerts a force on a second charge. Similarly, we will consider that a magnet or moving charge produces a magnetic field **B**. The magnetic field exerts a force on another magnet or moving charge.

We begin this chapter by describing magnetic fields. We then discuss the forces on moving charges and currents, along with some applications. Finally

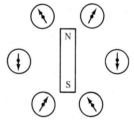

Figure 21.1. The equilibrium orientation of a compass needle indicates the direction of the magnetic field.

Figure 21.2. If the pole faces of a magnet are closely spaced the field between them is nearly uniform.

we show how the fields due to currents can be calculated.

21.1 | MAGNETIC FIELDS

The direction and magnitude of the magnetic field **B** produced by a magnet can be determined with the aid of a compass, a permanently magnetized steel needle pivoted at its center. The equilibrium orientation of the needle gives the direction of the field, and the torque tending to align the needle is proportional to its magnitude (Fig. 21.1).

Magnetic fields are represented by diagrams or maps similar to those used for electric fields. Again

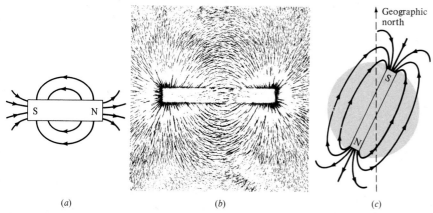

(a)   (b)   (c)

Figure 21.3. (a) Magnetic field of a bar magnet. (b) The pattern formed by iron filings near a bar magnet. (c) The magnetic field of the earth resembles that of a bar magnet. The *magnetic* south pole is located near the *geographic* north pole; it attracts the north pole of a compass needle. [Part (b) reproduced by permission of the publisher from *PSSC Physics*, fourth edition, 1976, by D. C. Heath and Company, Lexington, Massachusetts.]

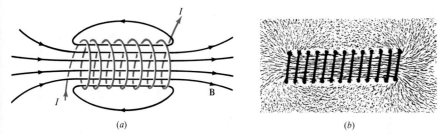

(a)   (b)

Figure 21.4. (a) Magnetic field lines and (b) iron filing patterns due to a current in a long coil of wire. Note the similarities to the field lines and patterns for the long bar magnet in Fig. 21.3. [Part (b) reproduced by permission of the publisher from *PSSC Physics*, fourth edition, 1976, by D. C. Heath and Company, Lexington, Massachusetts.]

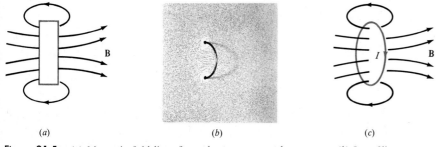

(a)   (b)   (c)

Figure 21.5. (a) Magnetic field lines for a short permanent bar magnet. (b) Iron filing patterns, and (c) magnetic field lines due to a single current turn. [Part (b) reproduced by permission of the publisher from *PSSC Physics*, fourth edition, 1976, by D. C. Heath and Company, Lexington, Massachusetts.]

the lines indicate the field direction, and the field is strongest where the lines are closest together (Figs. 21.2 and 21.3). *Notice that by convention the magnetic field lines are always directed from the north pole to the south pole outside a magnet.* The field lines around a magnet can be made visible by sprinkling iron filings on a sheet of paper. The filings become magnetized and tend to align themselves with the field (Fig. 21.3).

As we noted earlier, electric currents also produce magnetic fields (Figs. 21.4, 21.5, and 21.6). The field due to a current in a long coil of wire looks much like that of a long bar magnet, and the field of a single current turn is similar to that of a short bar magnet. A current in a long straight wire produces circular magnetic field lines.

A convenient rule aids in remembering the ob-

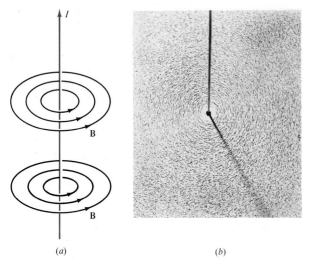

(a)   (b)

Figure 21.6. (a) Magnetic field lines and (b) iron filing patterns due to a current in a long straight wire. [(b) Reproduced by permission of the publisher from *PSSC Physics*, fourth edition. 1976, by D. C. Heath and Company, Lexington, Massachusetts.]

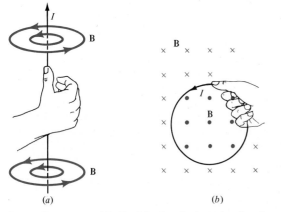

(a)   (b)

Figure 21.7. (a) With the right thumb along the direction of the current, the fingers curl in the direction of the magnetic field. (b) Inside the loop, **B** points out of the page; outside the loop, **B** points into the page.

served relationship between the directions of the magnetic field and the current. We may use the long straight wire as an example. If the right thumb is placed along the wire in the direction of the current, the fingers curl in the direction of the field (Fig. 21.7a). Similarly, for the circular current loop in Fig. 21.7b, this rule indicates that inside the loop, **B** points out of the page; outside the loop, **B** points into the page.

The close similarity between the magnetic fields of currents and permanent magnets led André Marie

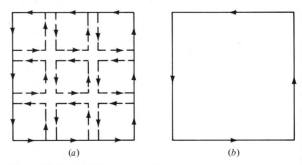

(a)   (b)

Figure 21.8. (a) The microscopic currents in a permanent magnet. (b) Since the effects of the opposing pairs of currents cancel in the interior, the net field produced is the same as for a current around the edge.

Ampere (1775–1836) to realize that microscopic currents in atoms or molecules are responsible for the fields of permanent magnets. Within a permanent magnet, the atoms are aligned so that their currents are in the same direction (Fig. 21.8a). In the interior of the material the currents occur in opposing pairs, so their fields exactly cancel. The net field is then due to the currents around the edge of the material and is identical to that produced by an electrical current in a wire with the same shape (Fig. 21.8b).

The S.I. magnetic field unit is the *tesla* (T). An older c.g.s. unit still in frequent use is the *gauss* (G), where

$$1 \text{ gauss} = 10^{-4} \text{ teslas} \qquad (21.1)$$

The tesla is a rather large unit. The largest fields produced in the laboratory for an extended time are about 10 T; fields up to about 100 T can be produced very briefly. The earth's magnetic field at the surface of the earth is about 10^{-4} T. One centimeter away from a long straight wire carrying the relatively large current of 100 A, the field is only 2×10^{-3} T. The small electrical currents associated with the human nervous system produce very weak magnetic fields. These fields are about 10^{-11} T near the chest (Fig. 21.9).

21.2 | THE MAGNETIC FORCE ON A MOVING CHARGE

Electric charges in motion near a magnet experience forces that can readily be demonstrated. For example, a magnet brought near a cathode-ray tube will deflect a beam of electrons and alter the places where they strike a fluorescent screen (Fig. 21.10).

5×10^{-11} T

Magnetocardiogram

1 mV

Electrocardiogram

0.2 sec

Figure 21.9. Magnetic fields near the human chest vary in time much like the electrical potential differences measured with an electrocardiagraph. Possible diagnostic uses of magnetic field measurements near the heart and brain are under investigation. (From Alexander Kolin, *Physics Today*, November 1968, p. 39 © American Institute of Physics.)

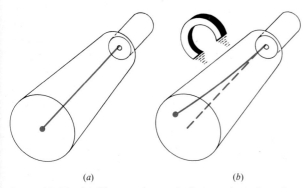

(a)   (b)

Figure 21.10. (a) Electrons in a cathode ray tube strike a fluorescent screen. (b) The electrons are deflected by a magnet.

The magnetic force law is more complicated in form than is the electric force law, $\mathbf{F} = q\mathbf{E}$. A charge q moving with velocity $\mathbf{v}$ in a magnetic field $\mathbf{B}$ experiences a force perpendicular to $\mathbf{v}$ and to $\mathbf{B}$. The magnitude of this force is the product of q, v, and the component of $\mathbf{B}$ perpendicular to $\mathbf{v}$. Using the cross product notation introduced in Chapter Four, the magnetic force $\mathbf{F}$ is given (Fig. 21.11) by

$$\mathbf{F} = q\mathbf{v} \times \mathbf{B} \tag{21.2}$$

The magnitude of this force is

$$F = |q|vB \sin \theta \tag{21.3}$$

where θ is the angle between $\mathbf{v}$ and $\mathbf{B}$. The force is greatest when $\sin \theta = 1$ or when the velocity is at right angles to the field. It is zero when $\sin \theta = 0$ or when the velocity is parallel or opposite to $\mathbf{B}$ (Fig. 21.12). Note that a charge at rest experiences no force; magnetic fields exert forces only on moving

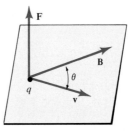

Figure 21.11. The magnetic force $\mathbf{F}$ on a positive charge q has a magnitude $qvB \sin \theta$ and is directed as shown.

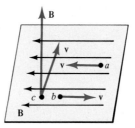

Figure 21.12. Positively charged particles in a magnetic field experience no force if the velocity is parallel (a) or opposite (b) to the field. The force has its maximum magnitude qvB when v is perpendicular to the field (c).

charges. The following example illustrates the calculation of the magnetic force on a moving charge.

Example 21.1

At Boston, Massachusetts, the magnetic field due to the earth is at 17° to the vertical direction and has a magnitude of 5.8×10^{-5} T (Fig. 21.13). (a) Find the force $\mathbf{F}$ on an electron moving straight down at 10^5 m s^{-1}. (b) Find the ratio of F to the weight mg.

(a) With $q = -e = -1.6 \times 10^{-19}$ C and $\sin 17° = 0.292$, the magnitude of the magnetic force is

$$F = evB \sin \theta$$
$$= (1.6 \times 10^{-19} \text{ C})(10^5 \text{ m s}^{-1})(5.8 \times 10^{-5} \text{ T})(0.292)$$
$$= 2.7 \times 10^{-19} \text{ N}$$

Since $\mathbf{F} = -e\mathbf{v} \times \mathbf{B}$, the force is opposite to $\mathbf{v} \times \mathbf{B}$ or into the page in Fig. 21.13.

(b) The mass of an electron is 9.1×10^{-31} kg, so

$$\frac{F}{mg} = \frac{2.7 \times 10^{-19} \text{ N}}{(9.1 \times 10^{-31} \text{ kg})(9.8 \text{ m s}^{-2})}$$
$$= 3.0 \times 10^{10}$$

The force on the electron due to the relatively weak magnetic field of the earth is larger than its weight by a factor of more than 10^{10}. For this reason the motion of charged cosmic ray particles entering the upper atmosphere is determined primarily by the earth's magnetic field and not by gravitational forces.

345

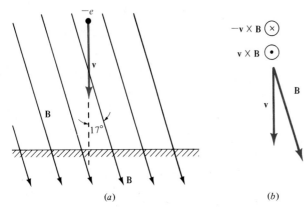

Figure 21.13. (*a*) The magnetic field of the earth varies in strength and direction. At Boston, it is at 17° to the vertical. (*b*) **v** × **B** is out of the page. The force on an electron is **F** = −*e***v** × **B** and is opposite to **v** × **B** or into the page.

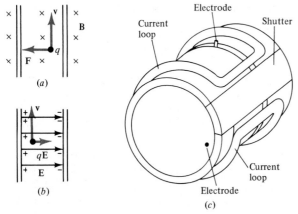

Figure 21.14. The electromagnetic flow meter as used for blood flow measurements. (*a*) When the magnetic field is into the page, the magnetic force on a positive ion is toward the left. (*b*) Charges accumulate at the edges, producing an electric field, *E*, which exerts a force toward the right on a positive ion. In equilibrium the electric force balances the magnetic force. (*c*) In one apparatus the artery is inserted into a metal sleeve by opening a shutter. The magnetic field is provided by a single current loop. The voltage between the electrodes is proportional to the flow rate.

21.3 | ELECTROMAGNETIC FLOW METERS

Electromagnetic flow meters, which employ the force on charges moving in a magnetic field, provide an excellent way to measure blood flow rates in patients undergoing heart or arterial surgery. Unlike the flow meters discussed in Chapter Fourteen, they do not require the insertion of a probe into a vessel and may be used even if the flow is not laminar but turbulent.

Figure 21.14*a* illustrates the principle of this device. A magnetic field is applied perpendicular to the motion of the blood. This produces magnetic forces to the left on positive ions in the blood and to the right on negative ions. Thus, charges accumulate at the edges of the vessel. Equilibrium occurs when the electric field due to these accumulations produces a force *qE* on an ion of charge *q*, which exactly balances the magnetic force *qvB*, that is, when *E* = *vB* (Fig. 21.14*b*). The potential difference associated with the electric field is then proportional to the average velocity *v* of the blood and can be measured with a sensitive voltmeter.

In a version of this instrument used for arteries 4 mm or larger in diameter, a closely fitting metal sleeve is placed around the artery to assure that its diameter remains constant (Fig. 21.14*c*). A current in a single loop of wire provides a small magnetic field that is rapidly alternated in direction to avoid *polarization,* the formation of an insulating gas layer. Voltmeter leads are attached to electrodes on opposite sides of the sleeve. The meter is calibrated by making measurements on an artery with a known flow rate.

21.4 | THE MAGNETIC FORCE ON A CURRENT CARRYING WIRE

We can find the expression for the magnetic force on a current carrying wire directly from the force **F** = *q***v** × **B** on a moving charge. Consider a straight segment of wire of length ℓ, which is part of a closed circuit carrying a current *I* (Fig. 21.15). If the moving charges in the wire have a velocity *v*, they travel the length of the segment in a time $t = \ell/v$. The total moving charge *q* is the product of the current and the time,

$$q = It = \frac{I\ell}{v}$$

If the magnetic field **B** is perpendicular to the wire, then the magnetic force on the wire is

$$F = qvB = \left(\frac{I\ell}{v}\right)vB$$

or

$$F = I\ell B \qquad (\ell \text{ perpendicular to } \mathbf{B}) \quad (21.4)$$

Figure 21.15. When the segment of wire carrying the current is perpendicular to the field, the force has a magnitude $F = I\ell B$.

Figure 21.16. If the segment is at an angle θ to the field, the magnitude of the force is $F = I\ell B \sin \theta$.

If the wire and the field are not perpendicular, $\mathbf{F} = q\mathbf{v} \times \mathbf{B}$ leads to

$$\mathbf{F} = I\boldsymbol{\ell} \times \mathbf{B} \qquad (21.5)$$

where ℓ is directed along the wire in the direction of the current (Fig. 21.16). If the wire is not straight, or if $\mathbf{B}$ varies with position, the wire may be regarded as having many small segments so that Eq. 21.5 holds true for each. The total force on the wire is then the vector sum of the forces on each segment.

In the next example, we see that the force due to the earth's magnetic field on a current carrying wire is relatively small.

Example 21.2

The magnetic field of the earth at Boston, Massachusetts, is at 17° to the vertical and has a magnitude of 5.8×10^{-5} T (Fig. 21.17). A vertical wire carries a current of 10 A. Find the force on a 2-m length of the wire.

The magnitude of the force is

$$\begin{aligned} F &= I\ell B \sin \theta \\ &= (10 \text{ A})(2 \text{ m})(5.8 \times 10^{-5} \text{ T})(\sin 17°) \\ &= 3.40 \times 10^{-4} \text{ N} \end{aligned}$$

With the right-hand rule for cross products, the force is out of the page in Fig. 21.17.

21.5 | MAGNETIC DIPOLES

In Chapter Eighteen, we discussed the electric dipole, two neighboring electric charges $+q$ and $-q$. A loop

Figure 21.17. The force on the current-carrying wire is directed out of the page.

of wire carrying a current behaves in a magnetic field much as an electric dipole does in an electric field and is called a *magnetic dipole*. Magnetic dipoles are important in discussions of the magnetic properties of atoms and molecules, including many molecules of importance in biochemistry.

We saw in Chapter Eighteen that an electric dipole in a uniform electric field experiences no net force, but it does experience a torque. We now show that a magnetic dipole in a uniform magnetic field also experiences no net force but experiences a torque. The rectangular current loop in Fig. 21.18 is an example of a magnetic dipole. The uniform magnetic field is parallel to sides 1 and 3, so no forces are exerted on them. Using $\mathbf{F} = I\boldsymbol{\ell} \times \mathbf{B}$, the forces on sides 2 and 4 are

$$F_2 = IbB \text{ (downward)}, \qquad F_4 = IbB \text{ (upward)}$$

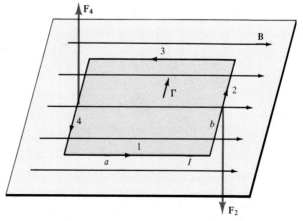

Figure 21.18. A magnetic dipole in a uniform magnetic field experiences no net force, but it does experience a torque.

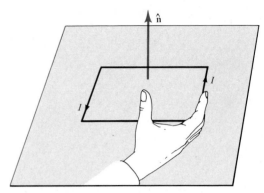

Figure 21.20. The normal direction $\hat{\mathbf{n}}$ indicated by the right thumb is upward when the fingers point along the current direction.

Figure 21.19. The torque on the magnetic dipole is $\mathbf{\Gamma} = I A \hat{\mathbf{n}} \times \mathbf{B} = \mathbf{\mu} \times \mathbf{B}$.

These are equal but opposite, so the net force on the loop is zero. However, the net torque on the loop is not zero, since $\mathbf{F}_2$ and $\mathbf{F}_4$ have different lines of action and form a couple.

We found in Chapter Four that the torque due to a couple is independent of the position of the axis of rotation. With the convenient choice of an axis along side 4, the torque due to $\mathbf{F}_4$ is zero. The lever arm for $\mathbf{F}_2$ is a, so its torque, $\mathbf{\Gamma} = \mathbf{r} \times \mathbf{F}$ is equal in magnitude to $\Gamma = a(IbB)$. Since $ab = A$ is the area of the loop, $\Gamma = IAB$ and is directed as shown.

If the loop is oriented as in Fig. 21.19, then the torque has a magnitude of $aF \sin\theta = IAB \sin\theta$. This can be rewritten in vector form as

$$\mathbf{\Gamma} = IA\hat{\mathbf{n}} \times \mathbf{B}$$

Here $\hat{\mathbf{n}}$ is a unit vector perpendicular or *normal* to the plane of the loop; its direction can be found using the right-hand rule as in Fig. 21.20. The *magnetic dipole moment* $\mathbf{\mu}$ is defined by

$$\mathbf{\mu} = IA\hat{\mathbf{n}} \qquad (21.6)$$

With this definition, the torque on a magnetic dipole in a uniform magnetic field is

$$\mathbf{\Gamma} = \mathbf{\mu} \times \mathbf{B} \qquad (21.7)$$

This result is correct for a current loop of *any* shape. It is identical in form to the torque $\mathbf{p} \times \mathbf{E}$ on an electric dipole in a uniform electric field.

The potential energy $\mathcal{U}$ of an electric dipole mo-

ment $\mathbf{p}$ at an angle θ to a uniform electric field $\mathbf{E}$ is $-pE \cos\theta$. Similarly, a magnetic dipole moment $\mathbf{\mu}$ at an angle θ to a uniform magnetic field has a potential energy

$$\mathcal{U} = -\mu B \cos\theta \qquad (21.8)$$

These properties of magnetic dipoles are illustrated in the following example.

Example 21.3

A circular turn of wire of radius 0.5 m carries a current of 4 A. (a) What is its magnetic moment? (b) If the normal to the loop is at 90° to a field of 2 T, what is the magnitude of the torque? (c) What is its potential energy at this angle?

(a) The area of the loop is πr^2, so its magnetic moment is

$$\mu = IA = I\pi r^2 = (4 \text{ A})\pi(0.5 \text{ m})^2 = 3.14 \text{ A m}^2$$

(b) With $\sin 90° = 1$,

$$\Gamma = \mu B \sin\theta = (3.14 \text{ A m}^2)(2 \text{ T})(1) = 6.28 \text{ N m}$$

The direction of the torque is shown in Fig. 21.21.

(c) Since $\cos 90° = 0$, the potential energy of the loop is $\mathcal{U} = -\mu B \cos\theta = 0$.

Figure 21.21. Example 21.3.

Figure 21.22. The magnetic moment due to an orbiting positive charge is parallel to its angular momentum.

Magnetic Moment of an Orbiting Charge |

When an electron moves around an atomic nucleus, there is a flow of charge or a current around the nucleus. The corresponding magnetic dipole moment determines how an electron interacts with an applied magnetic field. We now obtain a formula for the dipole moment of an orbiting charge, which we will use in later chapters.

Suppose a particle of mass m and charge q (Fig. 21.22) moves in a circular orbit. To find the magnetic moment $\mu = IA$, we need the current I and the area A. The time t required for the charge to complete one orbit of radius r at a speed v satisfies $vt = 2\pi r$ or $t = 2\pi r/v$. The magnitude of I is then

$$I = \frac{q}{t} = \frac{q}{2\pi r/v} = \frac{qv}{2\pi r}$$

Since the area A of the orbit is πr^2, the magnetic moment is

$$\mu = IA = \frac{qv}{2\pi r}\pi r^2 = \frac{qrv}{2}$$

This result can be put into another form involving the angular momentum of the particle. Its moment of inertia relative to the center of its orbit is mr^2 and its angular velocity is $\omega = v/r$. The angular momentum is the product of these two factors, $L = (mr^2)(v/r) = mvr$. Hence $rv = L/m$, and the magnetic dipole moment can be rewritten in vector form as

$$\mu = \frac{q}{2m}\mathbf{L} \qquad (21.9)$$

The magnetic moment of a charged particle is proportional to its angular momentum; it is parallel to the angular momentum for a positive charge and opposite for a negative charge (Fig. 21.22). This result is illustrated by the next example.

Example 21.4

In the Bohr model of the hydrogen atom, the electron moves in circular orbits for which the angular momentum $L = nh/2\pi$, where n is an integer, and h is Planck's constant; $h = 6.63 \times 10^{-34}$ J s. (a) Find the magnetic moment due to the orbital motion in the lowest ($n = 1$) orbit. (b) Suppose an electron in the $n = 1$ orbit has its moment parallel to a magnetic field of magnitude 10 T. How much energy must be supplied to reverse the orientation of the moment?

(a) With $n = 1$, the angular momentum is $L = h/2\pi$. Since $q = -e$, the magnetic moment has a magnitude

$$\mu = \frac{e}{2m}L = \frac{eh}{4\pi m}$$

Substituting $e = 1.6 \times 10^{-19}$ C and $m = 9.11 \times 10^{-31}$ kg,

$$\mu = \frac{(1.60 \times 10^{-19}\,\text{C})(6.63 \times 10^{-34}\,\text{J s})}{4\pi(9.11 \times 10^{-31}\,\text{kg})}$$

$$= 9.26 \times 10^{-24}\,\text{A m}^2$$

(b) Since the potential energy of the magnetic dipole is $\mathcal{U} = -\mu B \cos\theta$, the potential energy is $-\mu B$ when μ is along the field and $+\mu B$ when μ is opposite to the field. Hence the energy needed to reverse the direction of the dipole is

$$2\mu B = 2(9.26 \times 10^{-24}\,\text{A m}^2)(10\,\text{T})$$

$$= 1.85 \times 10^{-22}\,\text{J}\left(\frac{1\,\text{eV}}{1.60 \times 10^{-19}\,\text{J}}\right)$$

$$= 1.16 \times 10^{-3}\,\text{eV}$$

We will use the relation between the magnetic dipole moment and the angular momentum of a particle in atomic and molecular applications in Chapters Thirty and Thirty two. There we will see that the energy needed to reorient the dipole moment as in the example above can be supplied by a time varying magnetic field.

21.6 | MOTORS; GALVANOMETERS

The magnetic force on a current carrying wire has many useful applications. Two of the most important ones are electrical motors and galvanometers.

Figure 21.23 illustrates the principle of the dc motor. A loop connected to an EMF is mounted so that it can rotate in a magnetic field that is produced either by a permanent magnet or by an electromagnet. When the normal to the loop becomes parallel to the field, the *split ring* or *commutator* reverses the current direction. As a result the torque on the loop is

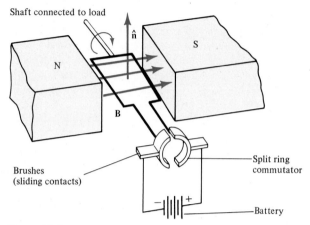

Shaft connected to load

$\hat{n}$

N

S

B

Brushes
(sliding contacts)

Split ring
commutator

Battery

Figure 21.23. Elements of a dc motor.

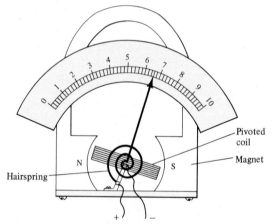

Pivoted
coil

Magnet

N   S

Hairspring

Figure 21.24. A galvanometer.

always in the same direction. This causes the loop to rotate and enables it to do work on a load. Practical dc motors have many turns with each successive one rotated slightly, so that the net torque is nearly constant as the motor turns. Alternating current motors of various types are also widely used. In these no commutator is needed, since an alternating current is supplied to the electromagnet in such a way that the field reverses along with the current in the coil.

The pivoted coil galvanometer used in voltmeters and ammeters has a coil suspended in a radial magnetic field (Fig. 21.24). Hairsprings provide restoring torques proportional to the angular displacement. When there is a current in the coil, the torque due to the magnetic field causes it to rotate until equilibrium is established. A pointer is used to indicate the deflection of the coil, and a calibrated scale gives the corresponding current.

$\Delta\mathbf{B}$

P

r

I

$\Delta\boldsymbol{\ell}$

$\Delta\mathbf{B}$

Q

Figure 21.25. The field $\Delta\mathbf{B}$ due to the current through $\Delta\boldsymbol{\ell}$ is out of the page at P and into the page at Q.

21.7 | MAGNETIC FIELDS PRODUCED BY CURRENTS

We saw in Chapter Eighteen how to calculate the electric field due to an arrangement of charges. We now discuss the analogous procedure for finding the magnetic field due to a current.

The basic law for the magnetic field is an expression for the field due to a short current segment. It is named the *Biot-Savart law,* after the two physicists who deduced it from experimental studies soon after Oersted discovered that currents produce magnetic fields. Consider a short segment of length $\Delta\ell$ in a closed circuit with a current I (Fig. 21.25). At point P, a distance $\mathbf{r}$ from the segment, the magnetic field due to the segment is

$$\Delta\mathbf{B} = k'I\frac{\Delta\boldsymbol{\ell} \times \hat{\mathbf{r}}}{r^2} \qquad (21.10)$$

Here $\hat{\mathbf{r}}$ is a unit vector directed along $\mathbf{r}$. The net magnetic field $\mathbf{B}$ at P is the sum of the fields due to all the elements in the complete circuit. In S.I. units, the proportionality constant k' has the value

$$k' = 10^{-7} \, \text{T m A}^{-1} \qquad (21.11)$$

Like Coulomb's law for the electric field of a point charge, the Biot-Savart law is an inverse square law. However, it is more complex, since it involves a vector cross product. Also, we can never directly observe the field $\Delta\mathbf{B}$ at P due to the single element in Fig. 21.25, because only the net field $\mathbf{B}$ due to the complete circuit can be measured.

Magnetic Field of a Circular Current Loop | To illustrate the use of the Biot-Savart law, we calculate the field at the center of a circular wire loop of radius a carrying a current I (Fig. 21.26). We

Figure 21.26. A circular current loop is divided into many small segments to calculate the magnetic field. The field is directed out of the page at the center of the circle.

divide the loop into many small segments. For any one segment $\Delta\ell$, the distance r to the center is a. Since $\Delta\ell$ is perpendicular to the unit vector $\hat{r}$, $|\Delta\ell \times \hat{r}| = \Delta\ell$; $\Delta\ell \times \hat{r}$ is directed out of the page. Thus the field due to the one segment is directed out of the page and has a magnitude

$$\Delta B = \frac{k'I\,\Delta\ell}{a^2}$$

Because $\Delta\mathbf{B}$ is out of the page for all the $\Delta\ell$'s, summing over the segments just replaces $\Delta\ell$ by the circumference, $2\pi a$, and the net field has a magnitude

$$B = \frac{2\pi k'I}{a} \qquad (21.12)$$

Thus at the center of the circle $\mathbf{B}$ varies inversely with the first power of the radius, even though the fundamental Biot-Savart law varies as $1/r^2$. (The complete field was shown in Fig. 21.5.)

A single loop carrying a relatively large current produces a fairly weak field. For this reason, many turns of wire are needed to produce even modest sized fields. This is illustrated by the following example.

Example 21.5

A coil consists of 1000 circular turns of thin wire with an average radius of 0.1 m (Fig. 21.27). If the current in the coil is 10 A, find the magnetic field at its center due to (a) one turn; (b) the entire coil.

(a) For any one turn, the field has a magnitude

$$B = \frac{2\pi k'I}{a} = \frac{2\pi(10^{-7}\text{ T m A}^{-1})(10\text{ A})}{0.1\text{ m}}$$

$$= 6.28 \times 10^{-5}\text{ T}$$

(b) Since the field due to each turn is in the same

Figure 21.27. Example 21.5.

direction, the net field is 1000 times the field of one turn or $(1000) \times (6.28 \times 10^{-5}\text{ T}) = 0.0628$ T.

We will see in the next chapter that larger fields can be obtained if materials such as iron are present in the coil.

In most situations the use of the Biot-Savart law requires mathematics beyond the level of this textbook. Nevertheless, the resulting formulas are sometimes reasonably simple. Two illustrations of this are the solenoid and the long straight wire.

Magnetic Field of a Solenoid

A long coil with many circular turns placed next to each other is called a *solenoid*. It is a particularly useful device because within the solenoid the magnetic field is nearly uniform (Fig. 21.28). A solenoid with length ℓ and N turns has $n = N/\ell$ turns per unit length. The field inside the solenoid is

$$B = 4\pi k'In \qquad \text{(solenoid)} \qquad (21.13)$$

Magnetic Field of a Long Straight Wire

At a distance r from a long straight wire carrying current I, the magnitude of the magnetic field is approximately

$$B = \frac{2k'I}{r} \qquad \text{(long straight wire)} \qquad (21.14)$$

This formula is accurate if r is small compared to the

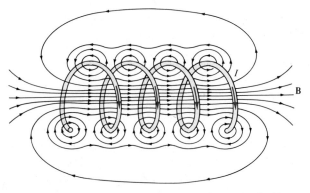

Figure 21.28. The field of a loosely wound solenoid. Inside the solenoid the field is nearly uniform.

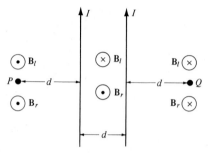

Figure 21.29. The fields due to the two currents are opposite in the region between the two wires and in the same directions elsewhere, in the plane of the wires.

Figure 21.30. Two currents in the same direction attract each other. **B** is the field due to I'.

length of the wire. The lines of **B** form circles about the wire as shown before in Fig. 21.6. The following example illustrates how the field due to two wires is found.

Example 21.6

Two parallel long straight wires a distance d apart each carry a current I in the same direction (Fig. 21.29). Find the net magnetic field (a) midway between the wires; (b) at points P and Q in Fig. 21.29.

(a) According to the right-hand rule, the field $\mathbf{B}_l$ due to the left-hand wire points into the page in the region between the wires. Similarly, the field $\mathbf{B}_r$ due to the right-hand wire points out of the page in this region. At the midpoint, the distance to either wire is the same, so $\mathbf{B}_l$ and $\mathbf{B}_r$ are equal in magnitude. Since they are oppositely directed, their vector sum is zero.

(b) At point P, the fields due to both wires point out of the page. Using $B = 2k'I/r$, we have

$$B_l = \frac{2k'I}{d}, \qquad B_r = \frac{2k'I}{2d}$$

Hence

$$B = B_l + B_r = \frac{2k'I}{d} + \frac{2k'I}{2d} = \frac{3k'I}{d}$$

At point Q, the field has the same magnitude, but it points into the page.

21.8 | THE FORCE BETWEEN TWO PARALLEL CURRENT-CARRYING WIRES

Two long, parallel wires attract each other when their currents are in the same direction and repel when the currents are opposite. The force between them can be found using $B = 2k'I/r$ for the field of a straight wire. In Fig. 21.30, I and I' are in the same direction, and their separation is d. The field **B** due to I' at the

segment $\Delta \ell$ is into the page, and its magnitude is $2k'I'/d$. The force **F** on the segment is $I \Delta \ell \times \mathbf{B}$. By the right-hand rule for cross products, **F** is directed to the left or toward the other wire, so the force is attractive as anticipated. Since $\Delta \ell$ is perpendicular to **B**,

$$F = I \Delta \ell B = I \Delta \ell \frac{2k'I'}{d}$$

The *force per unit length* of wire is

$$\frac{F}{\Delta \ell} = \frac{2k'II'}{d} \qquad (21.15)$$

This result is illustrated by the following example.

Example 21.7

Two parallel wires carry currents of 10 A and are separated by 1 mm $= 10^{-3}$ m. What is the force on a 2-m long portion of a wire?

The force per unit length is

$$\frac{F}{\Delta \ell} = \frac{2k'II'}{d} = \frac{2(10^{-7} \text{ T m A}^{-1})(10 \text{ A})(10 \text{ A})}{10^{-3} \text{ m}}$$

$$= 0.02 \text{ N m}^{-1}$$

Thus the force on a 1-m segment of the wire is 0.02 N. The force on a 2-m segment is twice this force, or 0.04 N.

A standard arrangement of this type is used to define the ampere as the current that gives rise to a stated force between the two wires. The unit of charge, the coulomb, is defined as the charge carried by an ampere of current in 1 second.

SUMMARY

Magnetic fields are established by permanent magnets and by moving charges. These fields, in turn, exert forces on other permanent magnets and on moving charges. The magnetic force on a charge q moving with velocity $\mathbf{v}$ in a magnetic field $\mathbf{B}$ is $q\mathbf{v} \times \mathbf{B}$.

From the expression for the magnetic force on a moving charge it follows that the magnetic force on a wire segment of length ℓ carrying a current I is $I\ell \times \mathbf{B}$.

In a uniform magnetic field, a current loop experiences a torque but no net force. Such a loop is called a magnetic dipole, and its dipole moment is $\boldsymbol{\mu} = IA\hat{\mathbf{n}}$. Its behavior in a magnetic field is very similar to that of an electric dipole in an electric field.

Given a current, the magnetic field it produces can be found by applying the Biot-Savart law. Like Coulomb's law, this is an inverse square law, although its form is more complex.

Checklist

Define or explain:

magnetic field	magnetic dipole
magnetic force on a	motor
moving charge	galvanometer
magnetic force on a	Biot-Savart law
current-carrying wire	solenoid

EXERCISES

Section 21.1 | Magnetic Fields

21-1   Consider the magnetic field in Fig. 21.31 at points P_1, P_2, P_3, P_4. (a) At which point is the field largest? (b) At which point is it smallest?

21-2   In the magnetic field shown in Fig. 21.31, near which point is the field most nearly uniform?

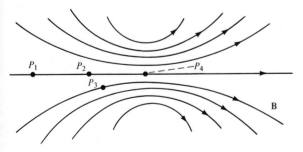

Figure 21.31.   Exercises 21-1 and 21-2.

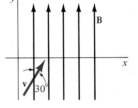

Figure 21.32.   Exercise 21-4.   Figure 21.33.   Exercise 21-5.

Section 21.2 | The Magnetic Force on a Moving Charge

21-3   An electron passing through an apparatus is undeflected. Does that imply that the apparatus has no magnetic field? Explain.

21-4   A particle with charge 10^{-6} C is moving in the $+y$ direction at 10^4 m s^{-1} at right angles to a magnetic field of 2 T (Fig. 21.32). (a) Find the magnitude and direction of the force on the particle. (b) What would the force be if the velocity were in the $-y$ direction?

21-5   An object of mass 0.01 kg moves at 100 m s^{-1} at an angle of 30° to a magnetic field of 10^{-2} T (Fig. 21.33). If its charge is -10^{-3} C, find the magnitude and direction of the magnetic force and of the acceleration.

21-6   The velocity of a typical hydrogen ion (proton) at room temperature is about 3000 m s^{-1}. (a) If the magnetic field of the earth is 10^{-4} T, what is the maximum magnetic force on the hydrogen ion? (b) Find the ratio of this force to the electric force between two protons 10^{-10} m apart, a typical interatomic distance. (c) Is the earth's magnetic field likely to affect most biochemical processes? Explain.

Section 21.4 | The Magnetic Force on a Current-Carrying Wire

21-7   A straight, 2-m long segment of a circuit is at 30° to a magnetic field. (a) If the field is 3 T and the current is 10 A, find the magnitude of the force on the segment. (b) Indicate the force direction with the aid of a sketch.

21-8   A power line carries a current of 1000 A. The earth's magnetic field makes an angle of 73° with the line and has a magnitude of 7×10^{-5} T. Find the magnitude of the magnetic force on a 100-m long section of the line.

353

Figure 21.34. Exercises 21-9 and 21-10.

21-9 (a) Find the magnitude and direction of the force on each of the three straight segments in Fig. 21.34. (b) What is the net force on the circuit?

Section 21.5 | Magnetic Dipoles

21-10 What is the torque on the circuit in Fig. 21.34?

21-11 A square wire loop with sides 0.1 m and a current of 10 A is placed in a magnetic field of magnitude 0.1 T. (a) What is magnetic dipole moment of the loop? (b) What is the largest torque the loop can experience in this field? (c) For what orientation of the loop will this torque occur?

21-12 What orientation of a magnetic dipole in a magnetic field will result in (a) stable equilibrium; (b) unstable equilibrium; (c) maximum torque?

21-13 A square current loop is placed in a magnetic field directed into the page (Fig. 21.35). The magnitude of **B** increases as x increases. Is there a net force on the loop? If there is, what is its direction? Explain.

21-14 The z component of the magnetic dipole moment due to the spin of an electron about its axis is $\mu_z = \pm eh/4\pi m$, where $h = 6.63 \times 10^{-34}$ J s is Plank's constant. (a) If μ_z is parallel to a magnetic field of 10 T, how much energy in electron volts must be supplied to reverse its direction so that it is opposite to the field? (b) Find the ratio of this energy to the 13.6 eV needed to remove an electron from a normal hydrogen atom.

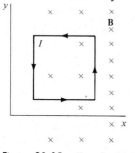

Figure 21.35. Exercise 21-13.

Section 21.7 | Magnetic Fields Produced by Currents

21-15 A circular current loop of radius 0.2 m has a resistance of 100 ohms and is connected to a 12-V battery. What is the magnetic field at the center of the loop?

21-16 The magnetic field at the center of a circular current loop is 0.05 T. If the radius of the loop is 1.2 m, find the current.

21-17 A thousand turns of wire are wound on a thin tube 0.4 m long. If the current in the wire is 2 A, find the field in the tube.

21-18 The magnetic field 0.1 m from a long, straight wire is 10^{-4} T. How large is the current in the wire?

21-19 A long, straight wire carries a current of 4 A. Find the magnitude and direction of the field at a distance of (a) 0.1 m, and (b) 1 m.

Section 21.8 | The Force Between Two Parallel Current-Carrying Wires

21-20 Two long, straight parallel wires 0.1 m apart each carry a current of 10 A. Find the direction and magnitude of the force on a 0.5-m long segment of one of the wires if the currents are (a) in the same direction; (b) in opposite directions.

21-21 A 10-m long segment of wire carries a 5-A current and has a weight of 0.5 N. It is constrained so that it can move vertically but not horizontally above another wire carrying a current of 10 A in the opposite direction. (a) At what separation would the weight of the wire be supported by the magnetic forces? (a) If the wires each have a radius of 1 mm = 10^{-3} m, will they be in contact?

PROBLEMS

21-22 An electron is moving at 10^5 m s^{-1} directly toward a long straight wire which carries a 50-A current. (a) Find the force on the electron when it is 0.5 m from the wire. (Indicate the direction of the force using a sketch.) (b) Find the resulting acceleration.

21-23 In Fig. 21.36, electrons in a conductor are moving into the page with a drift velocity $\mathbf{v}_d$. A magnetic field **B** is perpendicular to $\mathbf{v}_d$. (a) Find the magnitude and the direction of the magnetic force on the electrons. (b) Find the direction and magni-

Figure 21.36. Problem 21-23.

Figure 21.37. Problem 21-27.

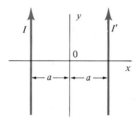

Figure 21.38. Problems 21-28 and 21-29.

tude of the electric field **E** that would exert an equal but opposite force on the electrons. (c) What potential difference would have to be applied across the conductor to produce this electric field? (d) If no external electric field is applied, then electrons will tend to move toward one side of the conductor until a constant potential difference is established across the conductor. How large is this potential difference? (The establishment of a potential difference across a conductor in a magnetic field is called the *Hall effect*.) (e) Find the Hall potential difference if $v_d = 10^{-3}$ m s^{-1}, $B = 2$ T, and $a = 10^{-2}$ m. (f) What is the direction of the current? (g) In some conductors, the charge carriers are missing electrons, or "holes," which behave as positive charges. Will the Hall potential difference be the same as when the carriers are negative? Explain.

21-24 It has been proposed to build huge magnets using superconducting wires to store energy for periods of peak electrical power demands. In one design, turns of radius 100 m carry 150,000 A. The average field due to this current is 5 T. The magnet is placed in a tunnel cut into bedrock to obtain the necessary structural support, since each portion of the wire is subjected to large forces due to the field. (a) If the magnetic field is parallel to the coil axis and perpendicular to the wires, what is the force on 1 m of wire? (b) Show that the force is radially outward. (c) If there are 10 turns per meter of length in the coil, what is the average outward pressure on the bedrock?

21-25 A wire with a resistance of 10 ohms is bent into a rectangle 0.5 m by 0.8 m and connected to a 6-V battery. (a) What is the dipole moment of the loop? (b) The loop is placed in a magnetic field of 0.5 T. What is the maximum torque on the loop? (c) How is it oriented when the torque is a minimum?

21-26 Suppose that the wire in the preceding problem is bent to obtain the maximum magnetic dipole moment. (a) What is the shape of the wire? (b) What is the dipole moment?

21-27 Find the net force on the loop in Fig. 21.37.

*****21-28*** In Fig. 21.38, $I = I'$. Show that the field between the wires at a distance x from the origin is $4k'Ix/(a^2 - x^2)$ if $|x| < a$. (Positive B values are out of the page, and negative values are into the page.)

21-29 In Fig. 21.38, $I = 10$ A, $I' = 5$ A, and $a = 0.1$ m. Find the magnitude and direction of **B** at (a) $x = 0$; (b) $x = 0.2$ m; (c) $x = -0.2$ m. (d) Where is the field zero?

21-30 The coil of a galvanometer has 100 turns, each with an area of 10^{-3} m^2. The magnetic field is 0.02 T and is arranged so that it is always in the plane of the coil. A current of 1 mA $= 10^{-3}$ A produces a rotation of $\theta = 10°$. If the torque due to the hairspring equals $k\theta$, find the spring constant k in newton meters per radian.

21-31 Five hundred meters of wire are wound on a tube 0.5 m long and 0.2 m in circumference. How large a current is needed to produce a field inside the tube of 0.1 T?

*****21-32*** (a) Show that the magnetic field along the axis of a circular loop of radius a carrying a current I at a distance y from its center is equal in magni-

Figure 21.39.   Problem 21-32.

Figure 21.40.   Problem 21-34.   **Figure 21.41.**   Problem 21-35.

tude to $2\pi Ik'a^2/(a^2 + y^2)^{3/2}$ (Fig. 21.39). (b) Find an approximate expression for the field in terms of the dipole moment and y when y is large compared to a. (c) Compare this result to the corresponding formula for the electric field due to an electric dipole (Chapter Eighteen).

*__21-33__   According to the Bohr model, in the normal hydrogen atom the electron orbits the proton in a circle of radius 5.1×10^{-11} m at a frequency of 6.8×10^{15} Hz. (a) What is the current due to the orbital motion of the electron? (b) What is the magnetic field at the proton due to this current?

*__21-34__   A current splits into two equal currents in Fig. 21.40. What is the magnetic field at the center c of the circle?

*__21-35__   What is the magnetic field at the center c of the semicircle in Fig. 21.41?

SUPPLEMENTARY TOPICS

21.9 | MEASUREMENT OF CHARGE TO MASS RATIOS

In this and the following two sections, we consider some applications of the magnetic force on a moving charge. Here we discuss an arrangement of electric and magnetic fields that can be used to measure the charge to mass ratio of a charged particle.

Figure 21.42 shows a particle of mass m and charge q accelerated from rest by a known potential difference V. Since the sum of the kinetic and potential

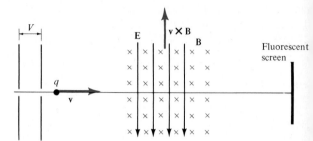

Figure 21.42.   The charged particle is accelerated from rest to a velocity **v** by a potential difference V. The magnetic field **B** is directed into the page. The vector **v** $\times$ **B** is opposite to **E** so the net force **F** $= q$**E** $+ q$**v** $\times$ **B** is zero for some value of E/B.

energies must remain constant, the resulting velocity **v** satisfies $\frac{1}{2}mv^2 = qV$, and the charge to mass ratio is

$$\frac{q}{m} = \frac{v^2}{2V} \qquad (21.16)$$

Thus the charge to mass ratio q/m of the particle can be found if the velocity is measured. This is accomplished with the aid of the *crossed* or perpendicular electric and magnetic fields shown in the center of Fig. 21.42. The magnetic force on the particle is q**v** $\times$ **B**, and the electric force in q**E**. Since **v** $\times$ **B** points upward or opposite to **E**, these forces exactly balance when $qvB = qE$ or

$$v = \frac{E}{B} \qquad (21.17)$$

This means the velocity may be determined by adjusting the crossed fields until the beam spot on the fluorescent screen is undeflected from its position when there are no crossed fields.

J.J. Thomson (1856–1940) used an arrangement of this kind in 1897 to measure the charge to mass ratio for the negative cathode rays in a cathode-ray tube. He concluded that these cathode rays were negatively charged particles (electrons). He also found that their velocity in his apparatus was about one-tenth of the velocity of light, which was much greater than any velocity previously observed.

One other feature of this apparatus is significant. The crossed **E** and **B** fields permit charged particles to travel through them without deflection whenever the condition $v = E/B$ is satisfied, no matter what the charge or mass of the particles. If a beam of several kinds of charged particles with various velocities passes through such a crossed field region, the particles emerging undeflected all have the same velocity. Thus, perpendicular electric and magnetic fields act as a *velocity selector*. One use of such an arrangement is seen in the next section.

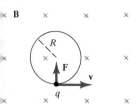

Figure 21.43. A charged particle moves with velocity **v** perpendicular to a uniform magnetic field **B** directed into the page. Since the force and the acceleration are always perpendicular to **v** and are constant in magnitude, the particle moves in a circular orbit with constant speed.

21.10 | MASS SPECTROMETERS

The *mass spectrometer* was originally developed as a nuclear physics research tool. Today, mass spectrometers are widely used in many kinds of laboratories to measure and identify minute quantities of various substances.

The principle of the mass spectrometer can be understood by considering a particle with positive charge q and mass m moving perpendicular to a uniform magnetic field **B** (Fig. 21.43). The magnetic force $\mathbf{F} = q\mathbf{v} \times \mathbf{B}$ is perpendicular to the velocity **v**, so it changes direction but not magnitude. The magnitude of the force, qvB, does not change, because **v** and **B** remain constant in magnitude and perpendicular to each other.

Since the acceleration $\mathbf{a} = \mathbf{F}/m$ is constant in magnitude and always perpendicular to **v**, the particle moves in a circular orbit with constant speed. The centripetal acceleration $a_r = v^2/R$ is produced by the magnetic force. Thus the radius R of the orbit must satisfy $qvB = mv^2/R$, and

$$R = \frac{mv}{qB} \qquad (21.18)$$

The use of this result is illustrated by the following example.

Example 21.8

How large a magnetic field is needed to cause an O_2^+ ion to move in a circular orbit of radius 2 m at 10^6 m s^{-1}? (The mass of the O_2^+ ion is approximately 32 amu, where 1 amu = 1.66×10^{-27} kg.)

The charge on the ion is $e = 1.60 \times 10^{-19}$ C. Using $R = mv/qB$, the magnetic field required is

$$B = \frac{mv}{qR}$$

$$= \frac{(32 \times 1.66 \times 10^{-27} \text{ kg})(10^6 \text{ m s}^{-1})}{(1.60 \times 10^{-19} \text{ C})(2 \text{ m})}$$

$$= 0.167 \text{ T}$$

Figure 21.44 shows the major parts of the mass spectrometer. In the ion source, molecules are ionized by bombarding them with electrons, and the ions are extracted by an electric field. Crossed electric and magnetic fields in the velocity selector only permit ions with a velocity $v = E_1/B_1$ to proceed without deflection through the slits into the uniform magnetic field **B**. There the ions move in circular paths of radius $R = mv/qB$ until they strike the photographic plate at a distance proportional to their mass, since the radius of curvature increases with the mass. This allows one to separate ions with the same charges but different masses.

Historically the mass spectrometer made possible the systematic study of *isotopes*. Isotopes are forms of an element with very nearly the same chemical properties but different numbers of neutrons in their nuclei and hence different atomic masses. Later the mass spectrometer was also used to separate the fissionable uranium 235 isotopes from the more abundant ^{238}U in the World War Two development of nuclear weapons.

The ability of the mass spectrometer to distinguish among nuclear isotopes makes it invaluable in situations involving *stable* rather than *radioactive* isotopes.

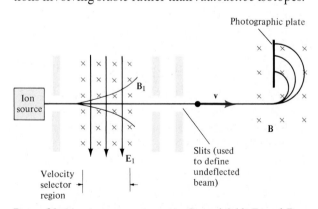

Figure 21.44. A mass spectrometer. Crossed fields E_1 and B_1 select only those particles that have the same velocity. The radius of the circular path in the **B** field then determines m/q.

The latter spontaneously transform into other nuclear species, emitting ionizing radiation in the process. If a radioactive material is administered to a plant or animal, its motion within the organism may be observed by detecting the ionization due to the radiation. Such *radioactive tracers* are widely used in biological research and medical diagnosis. However, some biologically important elements such as nitrogen and oxygen lack suitable radioactive isotopes, although they do have stable isotopes. For example, oxygen normally has 99.756 percent oxygen 16, which is a nucleus containing 8 protons and 8 neutrons; 0.039 percent oxygen 17, which has one more neutron; and 0.205 percent oxygen 18, which has yet another neutron. If the oxygen in material administered to an organism contains extra amounts of the rare isotopes, samples can be taken from the organism at various places or times and analyzed with a mass spectrometer. The presence of the rare oxygen isotopes signals the arrival of the administered substance or its metabolic derivatives. Thus the mass spectrometer makes possible the use of stable isotopes as tracers. It is also used to identify the ratios of the abundances of stable isotopes in geological samples to aid in determining their source or age.

Mass spectrometers are sometimes employed in situations where one is not concerned about the isotopic composition. They have been used to analyze the respiratory gases of patients with various disorders and to study the composition of the gas surrounding plants during photosynthesis. Also, they are used in the petroleum industry to distinguish among complex compounds having identical chemical compositions but different molecular configurations.

21.11 | CYCLOTRONS

The *cyclotron,* invented by Ernest O. Lawrence (1901–1958) in 1930, was the first machine developed to accelerate charged particles to high velocities by causing them to pass repeatedly through the same accelerating region. Its operation depends on the remarkable fact that the period or the time required for a charged particle to complete one circular orbit in a uniform magnetic field **B** is independent of the speed of the particle v.

For a particle of charge q and mass m, the radius R of the orbit was shown in the preceding section to be $R = mv/qB$. The period T satisfies $vT = 2\pi R$, or

Figure 21.45. The principle of the cyclotron. A magnetic field perpendicular to the plane of the dees maintains the ions in circular orbits. The orbital radius increases as the ions are accelerated by the potential difference between the dees, which reverses at the orbital frequency of the ions.

$$T = \frac{2\pi R}{v} = \frac{2\pi m}{qB} \qquad (21.19)$$

Thus, increasing the velocity increases the radius of the orbit but has no effect on the period T or the orbital frequency $f = 1/T$.

A cyclotron consists of two evacuated hollow metal *dees* in a uniform magnetic field perpendicular to their plane (Fig. 21.45). Protons or other positive ions are injected near the center. An electric generator reverses the potential difference between the dees at the orbital frequency of the ions, so they are accelerated each time they pass through the gap between the dees. This increases their velocity and consequently their orbital radius, $R = mv/qB$, but does not alter their period.

The operation of the cyclotron depends on the fact that the period is independent of the velocity. However, it is found that the inertial mass of a particle increases rapidly as its velocity approaches the speed of light. This was originally predicted by Einstein's special theory of relativity in 1905 and has been well confirmed. This mass increase causes the period to increase and sets a limit to the velocity and kinetic energy attainable with a cyclotron. Higher energies can be achieved in more complex accelerators that gradually vary the magnetic field or the generator frequency as the ions accelerate.

Originally developed for nuclear physics research, cyclotrons have largely been replaced in that area by newer machines. However, they are sometimes used today in hospitals to bombard various targets, inducing nuclear reactions that produce medically useful radioactive materials. Most radioactive materials are supplied to hospitals by nuclear reactor facilities, but some are so short-lived that they cannot be transported, and others can only be made in cyclotrons.

EXERCISES ON SUPPLEMENTARY TOPICS

Section 21.10 | Mass Spectrometers

21-36 A beam of particles with charge $+e$ moves in a circle of radius 3 m in a magnetic field of magnitude 0.2 T. (a) What is the momentum of the particles? (b) If the particles are protons, what is their velocity? (The proton mass is 1.67×10^{-27} kg.)

21-37 An electron moves at a speed of 10^7 m s^{-1} in a circle of radius 2 m in a magnetic field. (a) How large is the field? (b) How large is the acceleration? (c) Draw a sketch showing the directions of the electron path and the field.

21-38 Protons with a kinetic energy of 5 MeV = 5×10^6 eV move along a circular path in a magnetic field. Alpha particles, which are helium nuclei, have twice the charge and about four times the mass of a proton. How much energy must they have to move along the same path?

21-39 A beam of ions passes through a velocity selector in a mass spectrometer that has an electric field of 1.4×10^5 N C^{-1}. (a) If the ions emerging from the selector have a velocity of 2×10^5 m s^{-1}, what is the magnetic field in the selector? (b) The magnetic field in the bending region is 1 T. What is the radius of the path followed by a He$^+$ ion with charge e and mass 6.68×10^{-27} kg?

21-40 Charged particles passing through a bubble chamber leave visible tracks that consist of very small hydrogen gas bubbles in liquid hydrogen that is almost at the boiling point. In Fig. 21.46, the magnetic field is directed into the page, and the tracks are in the plane of the page and moving in

Figure 21.46. Exercise 21–40.

the directions indicated by the arrows. (a) Which of the tracks C, D, and E correspond to positively charged particles? (b) If all three particles have the same mass and their charges are equal in magnitude, which is moving the fastest? (c) If all three particles are moving with the same speed, which has the greatest mass?

Section 21.11 | Cyclotrons

21-41 Helium nuclei with mass 6.68×10^{-27} kg and charge $2e$ are accelerated in a cyclotron. The period of their orbit is 10^{-7} s. (a) What is the magnetic field? (b) If the maximum orbital radius is 2 m, what is the maximum velocity attained?

PROBLEMS ON SUPPLEMENTARY TOPICS

21-42 A mass spectrometer has an electric field of 10^5 N C^{-1} and a magnetic field of 0.6 T in its velocity selector, and a magnetic field of 0.8 T in its bending region. (a) What is the velocity of the ions passing through the velocity selector? (b) Find the spatial separation of singly ionized neon 20 and neon 22 isotopes with charges $+e$ after they have been bent through a half circle. (Neon 20 has a mass of approximately 20 amu, and neon 22 has a mass of approximately 22 amu. 1 amu = 1.66×10^{-27} kg.)

21-43 A cyclotron accelerates protons of mass 1.67×10^{-27} kg to a velocity 3×10^7 m s^{-1}, a tenth that of light. The magnetic field is 1.5 T. Find the maximum orbital radius and the orbital frequency.

*21-44 An electron is projected with velocity **v** at an angle of 30° into a uniform magnetic field **B**. (a) Find the magnitude and direction of the acceleration. (b) Find the component of the velocity along the magnetic field. What is its rate of change? (c) Find the component of the velocity perpendicular to the field. Show that this component rotates

359

about the field and find the associated radius. (d) How far does the electron move along the field direction during one full rotation? (e) Sketch the trajectory of the electron.

*21-45   (a) Electrons in a television picture tube are accelerated from rest by a potential difference of 2×10^4 V. What is their speed? (b) If the earth's magnetic field is at right angles to the electron beam, and its magnitude is 6×10^{-5} T, what is the magnetic force on an electron? (c) Estimate the displacement resulting from this force when the beam travels 0.15 m through the tube.

Additional Reading:

Francis Bitter, *Magnets: The Education of a Physicist,* Science Study Series, Doubleday and Co., Garden City, N.Y., 1959. Paperback.

Louis W. McKeehan, *Magnets,* Momentum Series, D. Van Nostrand and Co., Princeton, N.J., 1967. Paperback.

Alexander Kolin, Magnetic Fields in Biology, *Physics Today,* November 1968, p. 39.

David Cohen, Magnetic Fields of the Human Body, *Physics Today,* vol. 29, August 1975, p. 34.

Richard Blakemore, Magnetostatic Bacteria, *Science,* vol. 190, 1975, p. 377.

D. Brenner, S.J. Williamson, and L. Kaufman, Visually Evoked Magnetic Fields of the Human Brain, *Science,* vol. 190, 1975, p. 480.

Wolfgang Wiltschko and Roswitha Wiltschko, Magnetic Compass of European Robins, *Science,* vol. 176, 1972, p. 62.

Robert P. Green, Orientation of Homing Pigeons Altered by a Change in the Direction of an Applied Magnetic Field, *Science,* vol. 184, 1974, p. 180.

Charles Walcott, The Homing of Pigeons, *American Scientist,* vol. 62, 1974, p. 542.

Scientific American articles:

M. Wilson, Joseph Henry, July 1954, p. 72.

Joseph J. Becker, Permanent Magnets, December 1970, p. 92.

H.P. Furth, Strong Magnetic Fields, February 1958, p. 28.

Henry H. Kolm and Arthur J. Freeman, Intense Magnetic Fields, April 1965, p. 66.

Francis Bitter, Ultrastrong Magnetic Fields, July 1965, p. 64.

Alfred O.C. Nier, The Mass Spectrometer, March 1953, p. 68.

A.B. Benfield, The Earth's Magnetism, June 1950, p. 20.

S.K. Runcorn, The Earth's Magnetism, September 1955, p. 152.

W.M. Elsasser, The Earth on a Dynamo, May 1958, p. 44.

S.W. Angrist, Galvanomagnetism and Thermomagnetism, December 1961, p. 124.

Raymond Wolfe, Magnetothermoelectricity, June 1964, p. 70.

Laurence L. Cahill, Jr., The Magnetosphere, March 1965, p. 58.

Allen Cox *et al.,* Reversals of the Earth's Magnetic Field, February 1967, p. 44.

Palmer Dyal and Curtis W. Parkin, The Magnetism of the Moon, August 1971, p. 62.

J.E. Kurszler and M. Tanenbaum, Superconducting Magnets, June 1962, p. 60.

W.B. Sampson *et al.,* Advances in Superconducting Magnets, March 1967, p. 114.

John H. Reynolds, The Age of the Solar System, November 1960, p. 171.

Kenneth W. Ford, Magnetic Monopoles, December 1963, p. 122.

Henry Koln, John Oberteuffer, and David Kelland, High-Gradient Magnetic Separation, November 1975, p. 46.

R.R. Wilson, The Batavia Accelerator, February 1974, p. 72.

William T. Keeton, The Mystery of Pigeon Homing, December 1974, p. 96.

CHAPTER TWENTY-TWO

INDUCED CURRENTS AND FIELDS

We have seen that charged objects at rest may experience electrical forces and that moving charged objects may experience electrical and also magnetic forces. However, we have not yet encountered any situations in which the electrical and magnetic forces are related; electricity and magnetism have appeared to be essentially distinct phenomena. We see in this chapter that when electric and magnetic fields are changing in time, they are, in fact, related to each other in a remarkable fashion. A changing magnetic field produces an *induced* electric field, and a changing electric field produces an *induced* magnetic field. One major consequence of these effects is the existence of *electromagnetic waves,* which travel with the speed of light.

22.1 | FARADAY'S LAW

As we saw in the preceding chapter, Oersted discovered in 1820 that electric currents produce magnetic fields. Michael Faraday (1791–1867), a self-educated English physicist and chemist, guessed that magnetic fields might also produce electric currents. He found that this did not happen if a wire loop and a magnet were both stationary. However, he discovered in 1831 that when the magnetic field through a loop is changing, a current is *induced* in the loop.

This phenomenon, called Faraday's law, can be illustrated with a coil of wire connected to a battery and a second separate loop of wire attached to an ammeter (Fig. 22.1). When the switch is closed, the ammeter indicates that a current occurs momentarily in the second loop; when it is reopened, a current in the opposite direction is briefly observed. Moving the

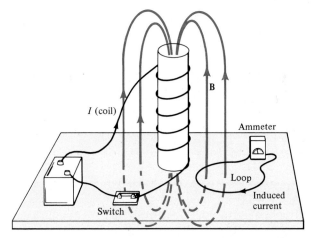

Figure 22.1. When the switch is closed, a current is temporarily produced in the loop in the indicated direction; a current in the opposite direction occurs when the switch is opened. If the switch is left closed so that a constant magnetic field is present, moving the loop or changing its size or orientation will also produce a current.

loop further away from the coil or reducing its area by bending the wires produces a current in the same direction, as does opening the switch. Rotating the loop so that its plane is not horizontal also induces a current.

Faraday's studies led him to observe that the single common feature in these and related phenomena was the occurrence of a change in the "amount" of the magnetic field passing through the loop in which the current is induced. Suppose, for example, that we count the number of magnetic field lines that pass through the loop of Fig. 22.1. Then, a current will be induced in the loop when any change occurs in the number of lines intersecting the loop area. Such a

Figure 22.2.   The magnetic flux Φ through the loop is $BA \cos \theta = B_n A$.

Figure 22.3. $\hat{n}$ is chosen to be directed into the page. (With this choice, clockwise currents are positive, counterclockwise currents are negative.) Hence **B** is parallel to $\hat{n}$ and $B_n = B$.

change occurs in each of the situations we described above.

To make this idea more precise, we define the *magnetic flux*. We will see shortly that the magnetic flux is proportional to the number of lines that pass through an area. Suppose a loop of area A has its normal vector $\hat{n}$ at an angle θ to a uniform magnetic field **B** (Fig. 22.2). The magnetic flux Φ (phi) through the loop is a scalar quantity defined by

$$\Phi = BA \cos \theta = B_n A \qquad (22.1)$$

Thus the magnetic flux is the product of the area and the component B_n of the magnetic field normal to the loop. Notice that the flux is proportional to the area, since a larger area has more lines passing through it. The $\cos \theta$ factor indicates that the maximum number of lines of **B** will pass through the area when $\hat{n}$ and **B** are parallel, and this number decreases until no lines pass through the loop when $\theta = 90°$. We will see shortly that it is the changes in the magnetic flux that are most directly related to the induced currents.

The definition we have given above for the magnetic flux holds true only when the magnetic field is constant over the loop. If **B** varies, the loop must be subdivided into areas small enough so that the field is nearly constant in each. The total flux is then the sum of the fluxes through the small areas.

In S.I. units, magnetic flux is measured in *webers* (Wb). Since the field **B** is measured in teslas and the area A in square meters, $\Phi = B_n A$ implies

$$1 \text{ weber} = 1 \text{ tesla(meter)}^2 = 1 \text{ T m}^2 \quad (22.2)$$

Magnetic fields are sometimes given in webers per square meter (Wb m^{-2}) instead of in teslas.

The calculation of the flux is illustrated by the following example.

Example 22.1
The loop in Fig. 22.3 has an area of 0.1 m^2. The

magnetic field is perpendicular to the plane of the loop and has a constant magnitude of 0.2 T. Find the magnetic flux through the loop.

The normal vector $\hat{n}$ must be perpendicular to the loop, but it may be chosen to be either into or out of the page in Fig. 22.3. If we choose $\hat{n}$ to be into the page, then by the convention discussed in the preceding chapter, clockwise currents will be positive, and counterclockwise currents will be negative. (See Section 21.5.) Since **B** is parallel to $\hat{n}$, $B_n = B$, and

$$\Phi = B_n A = BA$$
$$= (0.2 \text{ T})(0.1 \text{ m}^2) = 0.02 \text{ Wb}$$

Faraday's law can now be written in terms of the magnetic flux. Suppose that the magnetic flux through a loop is changed by an amount $\Delta\Phi$ in a time Δt.

Faraday's law states that the average EMF induced in the circuit equals the average rate of change of the flux:

$$\bar{\mathcal{E}} = -\frac{\Delta\Phi}{\Delta t} \qquad (22.3a)$$

The instantaneous EMF is

$$\mathcal{E} = -\frac{d\Phi}{dt} \qquad (22.3b)$$

The minus signs indicate the direction of the induced EMF and current and will be discussed below. *Note that Faraday's law contains only the rate of change of the flux and does not depend on how that rate is achieved.* This is consistent with the observation that moving the loop away in Fig. 22.1, reducing the loop area, or rotating the loop, all produce the same qualitative effect as reducing the field; all of these result in a reduction in the flux. Calculation of the induced EMF and the resulting current is illustrated by the next example.

Example 22.2
A wire loop of area 0.1 m^2 has a resistance of 10 ohms

Figure 22.4. Example 22.2.

(Fig. 22.4). A magnetic field **B** normal to the loop initially has a magnitude of 0.2 T and is reduced to zero at a uniform rate in 10^{-4} s. Find the induced EMF and the resulting current.

Let us take $\hat{n}$ parallel to the field or into the page, so that clockwise currents are positive. The initial flux is $\Phi_i = BA$ and the final flux Φ_f is zero, so $\Delta\Phi = \Phi_f - \Phi_i = -BA$. Thus

$$\mathcal{E} = -\frac{\Delta\Phi}{\Delta t} = -\frac{(-BA)}{\Delta t} = \frac{BA}{\Delta t}$$

$$= \frac{(0.1 \text{ m}^2)(0.2 \text{ T})}{10^{-4} \text{ s}} = 200 \text{ V}$$

and

$$I = \frac{\mathcal{E}}{R} = \frac{200 \text{ V}}{10 \text{ ohms}} = 20 \text{ A}$$

Since I is positive, the current is clockwise.

Note that $\hat{n}$ can be chosen to be out of the page, so that counterclockwise currents are positive. With this choice of $\hat{n}$, $\Delta\Phi$ is positive, and the induced EMF and the current are negative. Since with this convention a negative current is clockwise, we have the same result as before.

It is important to notice that the induced current will also produce a magnetic field. In the example above, the induced current is clockwise. The direction of the field can be found with the right-hand rule given in the last chapter (see Fig. 21.7). Applying the right-hand rule, we place our thumb along the current; our fingers show that the current produces a field into the page within the loop. This means that the field inside the loop due to the induced current is in the same direction as the initial field and *tends to slow the reduction in the field and in the flux*. Similarly, if the original field is increasing, the field produced by the induced current is opposite to the original field, *tending to retard its increase*. For example, if in Fig. 22.4, **B** is into the page and increasing, then the induced current will produce a field out of the page, so the current must be counterclockwise. *The observation that the field due to the induced current always*

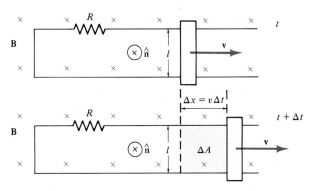

Figure 22.5. Example 22.3.

opposes the change in flux is called Lenz's law. Note that it is the *change* in flux, not the flux, which is opposed. This is seen again in the next example.

Example 22.3

According to Faraday's law, an EMF is induced when the flux through a loop is changed by varying its area. To illustrate this, consider a metal bar of width l sliding with a velocity **v** along two conducting rails that form a closed circuit (Fig. 22.5). A uniform magnetic field is directed into the page. What is the induced EMF?

We choose the normal vector $\hat{n}$ into the page or parallel to **B**; with this choice, clockwise currents are positive. If the area of the circuit is A, then the flux is $\Phi = BA$. A short time Δt later, the bar will have moved a distance $\Delta x = v\,\Delta t$. Hence the area will have increased by $\Delta A = l\,\Delta x = lv\,\Delta t$, and the flux will have increased by $\Delta\Phi = B\,\Delta A = Blv\,\Delta t$. The induced EMF is therefore

$$\mathcal{E} = -\frac{\Delta\Phi}{\Delta t} = -Blv$$

The minus sign indicates that the induced current, $I = \mathcal{E}/R$, will be negative, or counterclockwise. Using the right-hand rule, we see that the field within the loop due to this current will be out of the page, or opposite to **B**. This opposes the increase in the flux that occurs as the areas increases, and again our result is in accordance with Lenz's law.

22.2 | EDDY CURRENTS

So far, we have considered examples of currents induced in circuits with a single current path. However, whenever the magnetic flux through a conducting object changes, EMFs and currents are induced in that object. Such *eddy currents* occur, for example, in electrical machinery. As we will see in Section 22.6, iron has the property of greatly increasing an applied

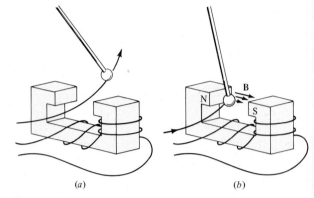

Figure 22.6. (a) An increasing magnetic field will induce eddy currents in a conducting object. (b) If the object is made of thin conducting sheets separated by insulating layers, the induced currents are confined to the individual sheets. The average current is smaller than in (a) and less power is dissipated as heat.

Figure 22.7. (a) If the magnet is turned off, the metal pendulum swings freely between the poles. (b) When the magnet power is on, the pendulum slows abruptly as it enters the magnetic field due to the magnetic forces on the eddy currents induced in the pendulum.

magnetic field. For this reason, motors and generators usually have large amounts of iron in magnetic fields. Since the flux through the iron is continually changing, eddy currents are produced that generate heat and reduce the efficiency of the machine. This effect is minimized by using thin iron sheets separated by insulating coatings rather than large solid pieces. This increases the resistance of the iron and decreases the eddy currents and hence the power loss (Fig. 22.6).

Eddy currents also arise when the flux through a conductor changes because of its motion. For example, if a metal pendulum swings between the poles of a strong magnet, it rapidly comes to rest (Fig. 22.7). This happens because the changing flux induces a current, and the magnetic force on that current is opposite to the motion of the pendulum. The kinetic energy lost by the pendulum is dissipated as heat by the eddy currents.

Eddy currents can also be induced in nonmetallic conductors, such as biological tissues. For example, a field of about 1 T alternating at 60 Hz will induce a large enough current in the retina of a human eye to produce a sensation of intense brightness.

22.3 | ELECTRIC GENERATORS

Faraday's law is the basic physical principle underlying electric power generators. To understand their operation, consider a wire loop of area A rotated by some power source at an angular velocity ω in a magnetic field (Fig. 22.8a). Its ends slide along two fixed rings. If at some instant the angle between the normal vector $\hat{\mathbf{n}}$ and the field $\mathbf{B}$ is $\theta = \omega t$, then the

flux through the loop is

$$\Phi = B_n A = BA \cos \omega t \qquad (21.4)$$

Since this flux is continually changing, there is an EMF induced in the loop, which is the generator voltage. We show in Section 22.15 that this EMF varies with time as

$$\mathcal{E} = \mathcal{E}_0 \sin \omega t \qquad (22.5)$$

where the *amplitude* $\mathcal{E}_0$ is

$$\mathcal{E}_0 = \omega BA \qquad (22.6)$$

Since $\sin \omega t$ varies between $+1$ and -1, the potential difference between the two ends of the loop varies between $+\mathcal{E}_0$ and $-\mathcal{E}_0$ with a frequency $f = \omega/2\pi$ (Fig. 22.8b). Hence this rotating loop is an alternating current (ac) generator.

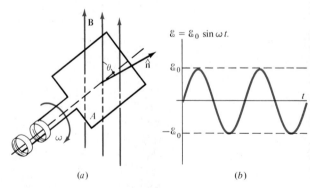

Figure 22.8. (a) A simplified ac generator consists of a coil rotating in a magnetic field. (b) The potential difference between the rings.

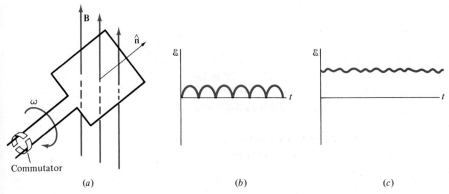

Figure 22.9. (*a*) A split ring or commutator reverses the connections each half turn. (*b*) The terminal voltage measured at the split ring. (*c*) The total EMF due to a large number of turns each at slightly different angles.

Direct current or dc generators supply an EMF that does not reverse polarity. The two rings of the ac generator are replaced by a single *split ring* or *commutator* (Fig. 22.9*a*). This reverses the connectors every half-turn and produces the EMF plotted in Fig. 22.9*b*. A practical dc generator has many loops each at slightly different angles, so the resulting total EMF is nearly constant (Fig. 22.9*c*).

22.4 | TRANSFORMERS

The fact that an ac voltage can efficiently be increased or decreased with a *transformer* is of great practical importance to the transport and distribution of electrical power. To see why, we recall that the rate at which electrical energy is lost as heat in a power line of resistance R carrying a current i is i^2R. Since the power supplied by a pair of lines with a potential difference v is iv, choosing the largest feasible voltage and a correspondingly small current minimizes the heat losses. For example, power is sent hundreds of miles from hydroelectric plants to cities at some hundreds of thousands of volts. The power lines used are of moderate size. By contrast, if power were transported over such distances at a low voltage, huge amounts of copper would be needed to build lines with acceptable losses. However, safety and insulation limitations require that electricity be supplied to consumers at low voltages. This requirement is satisfied by using transformers to reduce the voltage in several steps in the local transmission system.

Figure 22.10 illustrates the principle of a transformer. *Primary* and *secondary* coils with N_1 and N_2

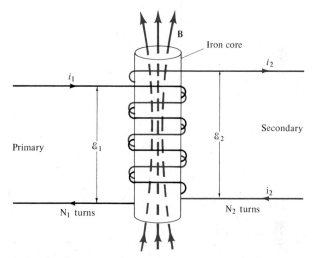

Figure 22.10. A transformer consists of a primary coil and a secondary coil wound around a core which is usually made of iron to increase the field and the flux.

turns, respectively, are wound around an iron core. A variable current i_1 in the primary produces a magnetic field through both coils. The iron serves to increase the field and the flux. The induced EMF in any one turn of either coil is $-\Delta\Phi/\Delta t$; the total induced EMF in a coil is this times the number of turns. Since $\mathcal{E}_1 = -N_1(\Delta\Phi/\Delta t)$ and $\mathcal{E}_2 = -N_2(\Delta\Phi/\Delta t)$, it follows that

$$\frac{\mathcal{E}_2}{\mathcal{E}_1} = \frac{N_2}{N_1} \qquad (22.7)$$

In the next example we see that if the secondary coil has more turns than the primary, the EMF will be increased by the transformer.

365

(The Royal Institution, London.)

MICHAEL FARADAY
(1791 – 1867)

Michael Faraday discovered many of the fundamental laws of physics and chemistry, despite the fact that he had virtually no formal education. The son of an English blacksmith, he was apprenticed at the age of 14 to a bookseller and bookbinder. He read every book on science in the bookshop and attended lectures given at the Royal Institute by various scientists, including Sir Humphrey Davy, the discoverer of 12 chemical elements. In 1812, he applied to Davy for a job, citing his interest in science and showing Davy the extensive lecture notes he had taken. Davy hired Faraday to assist with his research and lecture demonstrations.

Within a few years, Faraday began to do original research on his own, submitting two papers on chemistry to the Royal Society in 1820. In that same year, Oersted discovered that a current in a wire will deflect a compass needle. Faraday repeated Oersted's experiments and found that a magnet also exerts a force on a wire carrying an electric current. Soon afterward, he also showed how to liquefy chlorine, and he isolated benzene, a compound now widely used in chemical products.

Faraday's important discoveries brought him considerable fame, much to the discomfort of Davy, whe felt he should have shared the credit for some of the advances. Davy preferred to regard Faraday as a technical assistant and even forced him to serve as a valet on an extended tour of European research centers. Despite Davy's objections, Faraday was elected to the Royal Society in 1824 and was made director of the laboratory at the Royal Institute in 1825.

After Faraday discovered, in 1831, that a changing magnetic field can induce a current, he performed a series of experiments that showed clearly that the induced EMF is equal to the rate of change of magnetic flux. Also, generalizing from the patterns formed by iron filings around magnets, he invented the concepts of magnetic and electric field lines. Faraday knew little mathematics and found this concrete approach to electricity and magnetism much more useful than equations giving the forces between charges or currents. He also suggested that the propagation of light through space consisted of vibrations of these lines. His concepts of electric and magnetic fields were put into a mathematical form a generation later by Maxwell, who showed that light is, in fact, an oscillatory electromagnetic disturbance.

Faraday made many other notable contributions. He devised the first electrical generator, which consisted of a copper disk rotating between the poles of a magnet. He discovered the correct laws of electrochemistry after proving that earlier theories disagreed with experiments. He studied optical phenomena and found that when light passes through a medium, a magnetic field will rotate the direction of the oscillating electric field. Ignoring scorn from his contemporaries, he attempted unsuccessfully in laboratory experiments to find a link between gravitation and electromagnetism. Such a link was observed 70 years later in a test of Einstein's general theory of relativity, when light rays passing near the sun were found to be deflected.

Despite his achievements, Faraday remained a modest and humble person. He declined to be knighted or to receive honorary degrees and only reluctantly accepted a small pension on his retirement in 1858.

Example 22.4

The primary coil of a transformer has 100 turns and is connected to a 120-V ac power line. If the secondary coil has 1000 turns, what is its terminal voltage?

According to Eq. 22.7,

$$\varepsilon_2 = \frac{N_2}{N_1}\varepsilon_1 = \left(\frac{1000}{100}\right)120 \text{ V}$$
$$= 1200 \text{ V}$$

Transformers are useful in increasing electrical safety. Since the primary and secondary coils are not directly connected to each other, the secondary is electrically *"isolated,"* and its potential relative to ground is independent of that in the primary. If a person simultaneously touches a wire in the secondary circuit and a ground, this will ground the wire, but no current will pass through the person. This is especially useful in reducing the risk associated with high voltages. Also, the extreme vulnerability of patients wired into EKGs, pacemakers, and other devices has led to the proposal that hospital units serving such people should have all their instrumentation supplied with power via carefully designed isolation transformers.

22.5 | INDUCED FIELDS AND ELECTROMAGNETIC WAVES

We have seen that a varying magnetic field induces an EMF in a wire loop. This EMF, in turn, does work on the charges in the conductor and sustains a current. Alternatively, we may say that the *changing magnetic field* **B** *produces an induced electric field* **E** in the loop, and this electric field exerts an electric force on the charges.

Since an induced electric field *does net work* on a charge as it goes around a closed path, *an induced electric field is not conservative.* In this respect, an induced electric field due to a changing magnetic field is quite different from the conservative electrostatic field produced by electrical charges, which does no net work around a closed path.

It is not necessary to have a conductor present in the changing magnetic field in order to observe the effects of the induced electric fields. One example of this provided by the *betatron*, which accelerates electrons to energies of about 100 MeV = 10^8 eV. If the electrons are allowed to strike a target, they produce highly penetrating x-rays that can be used for cancer therapy. In a betatron, a magnetic field maintains the electrons in circular orbits, just as in the cyclotron (Chapter Twenty-one). However, the magnetic field is not kept constant. Instead, it is gradually increased with time, and the resulting induced electric field accelerates the electrons. With proper shaping of the magnetic field, the radius of the electronic orbits remains constant as the electronic speed and the field increase. Thus the electrons can be confined to a narrow doughnut-shaped tube instead of requiring a large chamber as in a cyclotron.

With the introduction of the idea of an induced electric field resulting from a changing magnetic field, there is a fundamental connection between electrical and magnetic phenomena. However, the situation is not symmetric unless we suppose that a changing electric field can also induce a magnetic field.

Maxwell's Hypothesis and Electromagnetic Waves | James Clerk Maxwell (1831–1879) made the brilliant hypothesis in 1864 that

changing electric fields do indeed induce magnetic fields. He was led to this idea by considering the relationships among the basic laws of electromagnetism, which had been discovered decades before. These were:

1. Coulomb's law for the force between two charges, or equivalently, for the electric field due to a point charge.
2. The Biot-Savart law for the magnetic field due to a current.
3. Faraday's law, which states that a changing magnetic field induces an electric field.
4. The conservation of electric charge.

Maxwell showed that these laws were not mathematically consistent when the electric fields were changing in time. However, if one hypothesized that a changing electric field could induce a magnetic field, the inconsistency was removed.

Although it was based on purely theoretical reasoning, Maxwell's hypothesis led immediately to the prediction that *electromagnetic waves* can be produced by oscillating charges or currents. For example, suppose a current oscillates (reverses direction) at some frequency. The current produces a magnetic field nearby that also oscillates at this frequency. According to Faraday's law, the magnetic field induces an oscillatory electric field in its vicinity. Maxwell's hypothesis provides the critical next step: the changing electric field induces a magnetic field. This magnetic field in turn induces an electric field, and so on (Fig. 22.11).

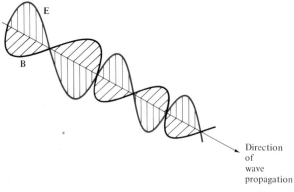

E

B

Direction
of
wave
propagation

Figure 22.11. An electromagnetic wave. An oscillating magnetic field induces an oscillating electric field, which in turn induces an oscillating magnetic field, and so on. This disturbance travels at the speed of light.

Maxwell showed that this electromagnetic disturbance would travel away from the oscillating current with a velocity equal to $(k/k')^{1/2} = 3.00 \times 10^8 \text{ m s}^{-1}$, where k and k' are the constants in the electric and magnetic force laws, respectively. Since this velocity was equal to the velocity of light to the accuracy then known, he inferred that light is an electromagnetic wave.

The first experimental verification of Maxwell's prediction of electromagnetic waves came nearly a quarter-century after he had published his work and several years after his death. In 1887, Heinrich Hertz (1857–1894) constructed two circuits that tended to oscillate at the same frequency. He found that if one circuit was connected to an EMF so that it had a current, the other circuit located some distance away would also have a current.

The "technological fallout" from this demonstration of electromagnetic waves quickly followed. In 1890, Guglielmo Marconi (1874–1937) became interested in wireless telegraphy, and by 1901 he succeeded in transmitting signals across the Atlantic. Today, we enjoy the benefits of radio, television, and other communications systems based on electromagnetic waves produced by oscillating currents.

The wavelength of the radiation produced by electric circuits ranges from millimeters to kilometers or longer, while visible light has a wavelength of 4×10^{-7} to 7×10^{-7} m. It has been verified to a very high accuracy that all the waves produced by oscillating currents have the same velocity as that of visible light. Thus, Maxwell's remarkable theoretical predictions have been fully verified.

22.6 | MAGNETIC MATERIALS

We saw in Chapter Eighteen that the introduction of a dielectric into an applied electric field $\mathbf{E}$ reduces the field to $\mathbf{E}/K$, where K is the dielectric constant. Similarly, the introduction of materials into an applied magnetic field $\mathbf{B}$ changes that field to $K_m\mathbf{B}$, where K_m is the *magnetic constant.*

There are three major kinds of magnetic materials. In *diamagnetic* materials, K_m is slightly less than 1. In *paramagnetic* materials, K_m is slightly greater than 1. Finally, in *ferromagnetic* materials, K_m varies with the applied field and with the way the material has been treated, but it is typically very large compared to 1.

When a diamagnetic material is placed in a changing applied magnetic field, changes are induced in the electron currents in the atoms. When the applied field is increasing, the field due to the induced currents opposes the applied field, in accordance with Lenz's law, so the net field is smaller than the applied field. Diamagnetism is a weak effect; typically, K_m is less than 1 by 0.001 to 0.01 percent in diamagnetic materials. Diamagnetism is present in all materials, although it is sometimes masked by other effects.

In many materials, when there is no applied field, the magnetic effects associated with the orbital and spin motions of the atomic electrons cancel exactly. However, in some materials, which are called paramagnetic, there is a permanent residual magnetic dipole moment associated with the individual atoms. In the absence of an applied field, thermal agitation will cause these dipoles to be randomly aligned, and no macroscopic magnetic field will be observed. However, when such a paramagnetic substance is placed in a magnetic field, the dipoles tend to align themselves with their moments along the field, increasing the total field inside the material. Since the effects of the permanent dipoles are usually larger than those of induced currents, paramagnetism, when present, will mask diamagnetic effects. Paramagnetic materials typically have values of K_m which are larger than 1 by about 0.01 percent.

Five elements (Fe, Co, Ni, Gd, and Dy) are ferromagnetic, as are many alloys. Ferromagnets are characterized by strong interactions between neighboring atomic dipoles that are sufficiently strong to cause a spontaneous alignment of the dipoles, even in the absence of an applied magnetic field.

Although the dipoles align themselves without an applied field, the fields produced outside the material are often very small. This is associated with the behavior of *magnetic domains*, regions in which all of the dipoles have the same alignment. If the directions of the domains are sufficiently varied, their net field will be small. These domains are rather stable, but when an external magnetic field is applied, the domains most nearly parallel with the field grow in size at the expense of others. The domains may also rotate into alignment with the field. In both cases the ordering of the dipoles may enhance the field by a factor of 1000 or more. For this reason, ferromagnetic materials play an important role in transformers and other

(a)   (b)

Figure 22.12. Ferromagnetic particles in the body can be detected by magnetizing them with a strong magnetic field and observing the fields they produce afterwards. (a) This man has eaten beans from a can, and his stomach contains about 10^{-4} grams of iron oxide. The areas of the squares are proportional to the field at their centers; the largest and smallest nonzero values are indicated in units of 10^{-11} T. Solid squares indicate fields in one direction, and open squares show fields in the opposite direction. (b) A similar plot for a man with 5×10^{-4} g of iron oxide in his lungs inhaled while doing arc welding. Since asbestos has small amounts of iron oxide adhering to its fibers, measurements of this type can also be used to determine the amount of the dangerous asbestos dust inhaled by workers. (D. Cohen, *Physics Today,* August, 1975, p. 41. © American Institute of Physics.)

devices requiring large magnetic fields. When the applied field is removed the domains in a ferromagnet may retain some order. In this case the sample is a permanent magnet (Fig. 22.12).

22.7 | INDUCTANCE

A current in a circuit always produces a magnetic field and a magnetic flux through that circuit. If this current changes, so does the flux, and this causes a *self-induced* EMF $\mathscr{E}$ in the circuit. According to Lenz's law, the induced current opposes the change in the flux and therefore also opposes the change in the current. *Thus self-induced EMFs tend to prevent rapid current changes in circuits.*

One familiar example of the effect of self-induced EMFs in limiting current changes is the flash seen when a high current appliance, such as a toaster, is on and its plug is pulled out of the electric outlet. This flash or arc occurs because the sudden current change induces a large EMF, which produces an electric field strong enough to ionize the air and permit current to pass through it briefly.

The self-induced EMF in a circuit can be related as follows to the rate of change of the current in that circuit. The EMF is proportional to the rate of change

369

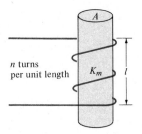

Figure 22.13. A solenoid wound on a core with magnetic constant K_m.

of flux. The flux, in turn, is proportional to the field and therefore to the current. Thus the EMF is proportional to the rate of change of current, and

$$\bar{\mathcal{E}} = -L \frac{\Delta i}{\Delta t} \qquad (22.8)$$

The constant L is called the *inductance* of the circuit. The minus sign indicates that the induced EMF opposes the current change.

The inductance depends on the geometry of the circuit and on the magnetic materials present in its vicinity. In S.I. units, L is measured in *henries* (H). This unit is named after Joseph Henry (1799–1878), an American contemporary of Faraday, who independently discovered induced EMFs.

We can calculate L if the flux Φ through the circuit is known. In a coil with a flux Φ through each of N turns, the total average induced EMF is $-N \Delta\Phi/\Delta t$. Comparing this with Eq. 22.8, we have

$$Li = N\Phi \qquad (22.9)$$

Consider, for example, a coil with n turns per unit length, length l, and cross-sectional area A wound on a core with magnetic constant K_m (Fig. 22.13). From Chapter Twenty-one, the field in the coil is $B = 4\pi K_m k' in$, where we have added a factor K_m to the earlier formula to account for the core. The inductance is then $L = N\Phi/i = (nl)(BA)/i$ or

$$L = 4\pi K_m k' n^2 A l \qquad (22.10)$$

Note that the inductance is proportional to K_m and to the coil volume Al. Such a coil is referred to as an *inductor*. Its effect is seen in the next example.

Example 22.5

A coil with 1000 turns has a cross section of $1 \text{ cm}^2 = 10^{-4} \text{ m}^2$ and a length of 0.1 m. (a) If K_m is 1, find the inductance. (b) The current increases from 0 to 1 A in 10^{-3} s. Find the average EMF in the coil.

(a) Using Eq. 22.10, the inductance is

$$L = 4\pi K_m k' n^2 A l$$
$$= 4\pi (1)(10^{-7} \text{ T m A}^{-1}) \left(\frac{1000}{0.1 \text{ m}}\right)^2 (10^{-4} \text{ m}^2)(0.1 \text{ m})$$
$$= 1.26 \times 10^{-3} \text{ H} = 1.26 \text{ mH}$$

Note that this relatively large coil has an inductance of only a small fraction of a henry. The henry is a comparatively large unit.

(b) The average EMF induced in the coil is

$$\bar{\mathcal{E}} = -L \frac{\Delta i}{\Delta t} = -(1.26 \times 10^{-3} \text{ H}) \frac{1 \text{ A}}{10^{-3} \text{ s}} = -1.26 \text{ V}$$

The minus sign indicates that the EMF opposes the current change.

22.8 | ENERGY STORED IN AN INDUCTOR

Just as a capacitor can store electrical energy, so also can an inductor store magnetic energy. To calculate this energy, we note that the power that must be *supplied* to change a current i in an inductor at a rate $\Delta i/\Delta t$ is $\mathcal{P} = -\mathcal{E}i = L(\Delta i/\Delta t)i$. In a time Δt, the work done *against* the induced EMF $\mathcal{E}$ is

$$\Delta W = \mathcal{P} \, \Delta t = \left(Li \frac{\Delta i}{\Delta t}\right)\Delta t = Li \, \Delta i$$

Now the energy stored in an inductor with current I equals the work done in increasing the current from 0 to I. In the derivation of the energy stored in a capacitor, we found that $\Delta W = q \, \Delta q / C$ and that $\mathcal{U} = \frac{1}{2}Q^2/C$ (Eq. 18.20). We can use that result here with the replacements $q \to i, \Delta q \to \Delta i, 1/C \to L$, and $Q \to I$. This gives, for the energy stored in an inductor,

$$\mathcal{U} = \frac{1}{2}LI^2 \qquad (22.11)$$

For example, if a 10^{-2}-H inductor carries a current of 1 A, the energy stored is $\frac{1}{2}LI^2 = \frac{1}{2}(10^{-2} \text{ H})(1 \text{ A})^2 = 5 \times 10^{-3}$ J.

The quantitative effects of inductances on circuits are considered in the supplementary topics at the end of this chapter.

SUMMARY

In a uniform magnetic field, the magnetic flux through a loop of area A is $\Phi = B_n A$, where B_n is the normal component of the field. Whenever the flux is changing, Faraday's law states that there is an induced EMF in the loop equal to the time rate of change of the flux:

$$\mathcal{E} = -\frac{d\Phi}{dt}$$

Lenz's law states that the current due to an induced EMF always has a magnetic field that tends to retard the change in the flux. The induced EMF is determined only by the flux change; it does not matter whether it is the magnetic field or the area of the loop in the field that is changing.

A changing magnetic field will induce an EMF even if no conducting loop is present, as is illustrated by the betatron. This is equivalent to stating that the changing magnetic field induces a nonconservative electric field. Since changing electric fields also induce magnetic fields, oscillating charges or currents produce electromagnetic waves that travel with the speed of light.

The self-induced EMF in a circuit is proportional to its inductance, and has the effect of slowing current changes.

Checklist

Define or explain:

induced current   primary, secondary coils
Faraday's law   induced electric field
magnetic flux   induced magnetic field
Lenz's law   electromagnetic waves
eddy currents   diamagnetism
amplitude   paramagnetism
generator   ferromagnetism
transformer   inductance

EXERCISES

Section 22.1 | Faraday's Law

22-1 In Fig. 22.14 the magnetic field is uniform in the region shown and is into the page. A wire loop is in the plane of the page. State whether there is an induced current when the loop is moved from (a) P_1 to P_2; (b) P_2 to P_3; (c) P_3 to P_1. Give reasons for your answers.

22-2 The wire loop in Fig. 22.14 is moved in the plane of the page from position P_1 to P_2. Find the direction of the induced current, if any.

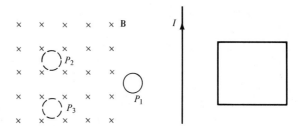

Figure 22.14. Exercises 22-1 and 22-2.   **Figure 22.15.** Exercise 22-4.

22-3 A copper loop and a rubber loop of the same size and shape are placed in a uniform magnetic field perpendicular to the plane of the loops, which is gradually increased. Discuss (a) the EMFs induced in the two loops; (b) the currents induced in the two loops.

22-4 In Fig. 22.15, (a) what is the direction of the field inside the rectangular wire loop due to the current I in the long straight wire? (b) If I is increasing, what is the direction of the current induced in the loop? (c) If I is constant and the loop is moved toward the right, what is the direction of the induced current?

22-5 A wire loop of radius 0.1 m has a 2-ohm resistance. A uniform magnetic field perpendicular to the plane of the loop increases from zero to 2 T in 0.01 s. Find the average induced current.

22-6 A circular wire loop has a resistance of 10 ohms and a radius of 0.1 m. Its normal is parallel to a uniform magnetic field of magnitude 3 T. If the loop is rotated so that in 0.3 s its normal is perpendicular to the field, find (a) the average induced EMF; (b) the average induced current.

22-7 A loop of area 0.1 m² is at right angles to a uniform magnetic field. The field alternates direction at 60 Hz and has a peak magnitude of 2 T. (a) What is the average induced EMF during one-half cycle when the field varies from normal to the loop to zero and to normal again but in the opposite direction? (b) What is the average EMF over a full cycle?

22-8 In Fig. 22.5, the resistance of the loop is 100 ohms, the velocity of the moving conductor is 0.7 m s⁻¹, its length is 0.05 m, and B is 0.5 T. Find the induced current.

Section 22.2 | Eddy Currents

22-9 In Fig. 22.16, the field through a conductor

371

Figure 22.16. Exercises 22-9 and 22-10.

Figure 22.17. Exercise 2-12.

Figure 22.18. Exercise 2-16.

Figure 22.19. Exercise 22-18.

is upward and increasing in magnitude. What is the direction of the eddy currents?

22-10 Suppose the field in Fig. 22.16 is constant and the conductor is gradually removed to the left. What is the direction of the eddy currents induced?

Section 22.3 | Electric Generators

22-11 Find the average EMF induced during the one-quarter cycle it takes for the normal vector in Fig. 22.8 to rotate from parallel to the field to perpendicular to the field. (Express the result in terms of A, B, and ω.)

22-12 A square wire loop of resistance 10 ohms with sides 0.2 m rotates 100 times per second about a horizontal axis (Fig. 22.17). The magnetic field is vertical and has a magnitude of 0.5 T. Find the amplitude of the induced current.

Section 22.4 | Transformers

22-13 The tube of an x-ray machine requires an ac voltage of 50,000 V. If this is obtained from a 120-V line, what is the ratio N_2/N_1 of the number of turns in the secondary and primary coils of the transformer?

Section 22.7 | Inductance

22-14 Why is it important that the materials used in or near electrical switches and outlets be fireproof?

22-15 How can one make a variable inductance?

22-16 The variable resistance in the left-hand loop of Fig. 22.18 is increased. Find the direction of the current induced in the (a) left-hand loop, and (b) right-hand loop.

22-17 The current in a 0.2-H inductor is increased from zero to 10 A in 0.1 s. Find the induced EMF.

22-18 Figure 22.19 shows the current in a solenoid during some time interval. Sketch how the voltage drop across the solenoid varies during that interval.

22-19 A solenoid with 1000 turns is wound on a core of radius 0.01 m and length 0.2 m. The magnetic constant of the core is 1000. Find the inductance of the coil.

Section 22.8 | Energy Stored in an Inductor

22-20 When the current in a coil is increased uniformly from zero to 10 A in 2 s, the self-induced EMF is 8 V. (a) How large is the inductance of the coil? (b) How much energy does the coil store when the current is 10 A?

22-21 An inductance stores 4 J of energy when the current is 5 A. How large is it?

PROBLEMS

22-22 Suppose the opposite to Lenz's law were true, so that induced currents would tend to increase changes in flux instead of opposing them. Would this be consistent with the energy conservation principle? Explain.

22-23 (a) In Fig. 22.5, what is the magnetic force on a charge q in the moving conductor? (b) If this charge moves the length l of the moving conductor, how much work W is done on it? (c) The EMF is defined as the work done per unit charge, W/q. Compare this result with the EMF obtained in Example 22.3.

22-24 A wire loop of resistance R and area A has its normal along the direction of a uniform mag-

Figure 22.20. Problem 22-25. **Figure 22.21.** Problem 22-26.

netic field, **B**. The loop is then flipped over in a time Δt so that its normal is opposite to the field. (a) Calculate the average induced EMF. (b) Find the average induced current. (c) If the field is out of the page, what is the direction of the induced current? (d) What is the total charge transported through the circuit in the time interval Δt? (This charge turns out to be independent of Δt and proportional to the field. Therefore, it is possible to determine magnetic fields by using such a coil and measuring the total charge.)

22-25 A bar magnet is dropped through a wire loop (Fig. 22.20). Find the direction of the induced current when the magnet is (a) above the loop; (b) halfway through the loop; (c) below the loop. Draw a rough graph of the induced current versus time assuming the velocity of the magnet is constant.

22-26 In Fig. 22.21, a metal pendulum is moving into a uniform magnetic field. (a) Are the induced eddy currents generally clockwise or counterclockwise? (b) What is the direction of the magnetic force on the pendulum? (Explain your answers in both parts.)

22-27 When a motor coil turns, the magnetic field in the motor induces a *back EMF* in its coil which opposes the current. (a) How does the back EMF depend upon the speed of the motor? (b) How does the EMF affect the current? (c) When a motor is connected to a large load, its speed decreases. What effect will this have on the current in the motor? (d) Why is there a danger that an excessive load will burn out a motor?

22-28 The ratio of the power transmitted into the secondary of a transformer to that supplied to the primary is usually over 90 percent, but it is never 100 percent. Suggest some reasons for the power losses.

22-29 Explain why the magnetic constant of a diamagnetic material does not change as the temperature is increased. Would a paramagnetic material have the same temperature dependence? Explain.

22-30 It has been proposed to build very large magnets using superconducting wires that have no resistance and therefore dissipate no heat. In one proposed design, 1000 turns of radius 100 m form a coil of height 100 m in rock ($K_m = 1$). (a) Using the formula for the inductance of a solenoid, estimate the inductance of this coil. (This is only a rough estimate since this coil is not long in relation to its width). (b) Estimate the energy stored if the current is 1.5×10^5 A. (c) A very large nuclear reactor or fossil fuel electric power plant produces 10^9 W of power. For how long can the superconducting magnet supply energy at this rate if its current is initially 1.5×10^5 A?

22-31 A solenoid carrying a current I has N turns, area A, length l, and magnetic constant K_m. (a) What is the magnetic field B inside the solenoid? (b) How much energy is stored per unit volume of the solenoid? (c) The energy stored per unit volume in a magnetic field is a constant times B^2/K_m. Use your result in part (b) to find that constant.

* **22-32** A change in the current i_1 in a loop (1) will induce an EMF $\mathcal{E}_2$ in a second loop (2) nearby. The *mutual inductance* M is defined by $\overline{\mathcal{E}_2} = -M \Delta i_1/\Delta t$. Two solenoids with N_1 and N_2 turns, respectively, are wound on a common core of area A, length l, and magnetic constant K_m. Show that the mutual inductance is $M = 4\pi K_m k' \dfrac{N_1 N_2}{l} A$.

* **22-33** Show that two inductances L_1 and L_2 connected in series have an equivalent inductance equal to $L_1 + L_2$.

* **22-34** Show that two inductances L_1 and L_2 in parallel are equivalent to a single inductance L, where

$$\frac{1}{L} = \frac{1}{L_1} + \frac{1}{L_2}$$

SUPPLEMENTARY TOPICS

22.9 | *RL* CIRCUITS

We saw in Section 22.7 that any circuit always has

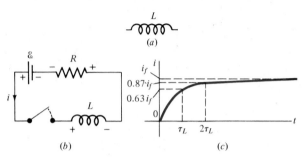

Figure 22.22. (*a*) The symbol for an inductor. (*b*) An *RL* series circuit. The switch is closed at $t = 0$. (*c*) The current in the circuit versus time.

some inductance. The inductance retards current changes, but it has no effect on a current that is constant. We now examine this in more detail for a series circuit consisting of a resistance R, an EMF $\mathcal{E}$ provided by a battery, and an inductance L (Fig. 22.22).

When the switch is closed in the circuit, the inductance prevents a sudden change in the current. Instead, the current gradually increases from zero toward the current predicted by Ohm's law,

$$i_f = \frac{\mathcal{E}}{R} \qquad (22.12)$$

The *time constant* τ_L, which characterizes this buildup, is proportional to L, since the greater the inductance the longer it takes to achieve the final current. If the resistance R is large, then the magnitude of the overall current change is small, and the final current may be achieved more rapidly. Thus, we also expect τ_L to be proportional to $1/R$, and the time constant is

$$\tau_L = \frac{L}{R} \qquad (22.13)$$

This time constant plays a role in the *RL* series circuit that closely parallels that of the time constant $\tau_C = RC$ in an *RC* series circuit. Suppose that the switch is closed at $t = 0$. When $t = \tau_L$, the current differs from its final value i_f by $i_f e^{-1} = 0.37\,i_f$; when $t = 2\tau_L$, the current differs from i_f by $0.14\,i_f$, and so on. At any time t, the current i is given (Fig. 22.22*c*) by

$$i = i_f(1 - e^{-t/\tau_L}) \qquad (22.14)$$

These results are illustrated by the following example and are derived below.

Example 22.6

A coil with a resistance of 10 ohms and an inductance of 0.1 H is connected to a 12-V battery. (a) Find the time constant. (b) What is the final current? (c) Find the current after 0.03 seconds.

(a) The time constant is

$$\tau_L = \frac{L}{R} = \frac{0.1\text{ H}}{10\text{ ohms}} = 0.01\text{ s}$$

(b) From Ohm's law, the final current is

$$i_f = \frac{\mathcal{E}}{R} = \frac{12\text{ V}}{10\text{ ohms}} = 1.2\text{ A}$$

(c) Since 0.03 s is equal to $3\tau_L$, the current at that time is

$$i = i_f(1 - e^{-t/\tau_L}) = (1.2\text{ A})(1 - e^{-3}) = 1.14\text{ A}$$

The current is within a few percent of its final value.

*Derivation of Eq. 22.14 |

Consider again the *RL* circuit in Fig. 22.22. At $t = 0$, the switch is closed and the current is zero, because the inductor prevents a sudden current change. At a later time, there will be a current i, so the potential drop across the resistor is iR. Since the current is increasing, there will be an induced EMF in the inductance equal in magnitude to $L\,di/dt$. The potential *drops* by this amount as we go through the inductance *with* the current. The algebraic sum of the potential changes around a closed path must be zero, so

$$-iR - L\frac{di}{dt} + \mathcal{E} = 0 \qquad (22.15)$$

If we wait long enough, so that di/dt is zero, then this equation implies that the final current is $i_f = \mathcal{E}/R$, as expected. By differentiating, it can be verified that Eq. 22.14 is a solution. It also satisfies the condition that the current is zero at $t = 0$, so it is the correct solution to our problem.

22.10 | EFFECTIVE OR ROOT MEAN SQUARE ALTERNATING CURRENTS AND VOLTAGES

In the remainder of this chapter, we discuss some basic features of alternating currents, beginning here with the concept of *effective* or *root mean square* (*rms*) *currents* and *voltages*. Alternating currents are important in many applications. As we noted earlier,

*The remainder of this section may be omitted without loss of continuity.

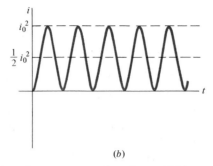

Figure 22.23. (a) An alternating current $i = i_0 \sin \omega t$ has an average value of zero. (b) The average of $\sin^2 \omega t$ over a full cycle is $\frac{1}{2}$, so the average value of i^2 is $i_0^2/2$.

electrical power is normally supplied as 60 Hz ac. Higher frequency currents are present in the circuits used to produce radio waves with an antenna or sound waves with a loudspeaker. Combinations of currents with many frequencies can also occur. For example, to reproduce a musical sound or a human voice, a loudspeaker is supplied with a complex current that is a sum of many small currents with various frequencies. Such complex currents can be discussed by analyzing each small current separately.

In discussing ac circuits, it is customary to use effective or rms currents and voltages, since these are convenient for power calculations. Alternating current ammeters and voltmeters normally are calibrated to read these quantities, rather than the amplitudes or peak values.

To define the effective current, consider a circuit connected to an ac generator. The current varies with the generator frequency $f = \omega/2\pi$ (Section 22.3) and has the form $i = i_0 \sin \omega t$, where i_0 is the amplitude of the current. Since $\sin \omega t$ is positive for one half cycle and negative for the other, the current averages to zero over a full cycle. Nevertheless, the current still has some effects; for example, it heats a resistance R at an instantaneous rate $\mathcal{P} = i^2 R$ (Fig. 22.23). Since i^2 is always positive, its average cannot be zero. According to Eq. B.13 in Appendix B, the average of $\sin^2 \omega t$ over a full cycle is $1/2$. Hence the average value of $i^2 = i_0^2 \sin^2 \omega t$ is $i_0^2/2$, and the average power dissipated in a resistor is

$$\overline{\mathcal{P}} = \frac{i_0^2}{2} R = i_e^2 R \qquad (22.16)$$

Here the *effective* or *root mean square (rms) current* i_e is defined by

$$i_e = \frac{i_0}{\sqrt{2}} \qquad (22.17)$$

Thus, an effective current of 1 ampere in an ac circuit produces exactly as much heat in a resistor as a 1 ampere dc current. Notice that both i_0 and i_e are always positive quantities.

Effective ac voltages are defined in an analogous fashion. If the voltage amplitude is v_0, the *effective or rms voltage* is

$$v_e = \frac{v_0}{\sqrt{2}} \qquad (22.18)$$

Again v_0 and v_e are always positive.

The effective voltage across a resistor v_e^R can be related to the current using Ohm's law, $v = iR$. At any instant, the voltage and current are related by

$$v^R = (i_0 \sin \omega t)R \qquad (22.19)$$

Squaring and averaging, we find $(v_e^R)^2 = \frac{1}{2}i_0^2 R^2$, or

$$v_e^R = i_e R \qquad (22.20)$$

The power dissipated in a resistor can then be expressed as

$$\overline{\mathcal{P}} = \frac{(v_e^R)^2}{R} = i_e v_e^R \qquad (22.21)$$

Thus with effective quantities, we have exactly the same formulas as for the power dissipation in a dc circuit. These quantities are used in the following example.

Example 22.7

A broiler draws 1000 W from a 120-V rms power line. Find (a) the amplitude of the power line voltage; (b) the rms current; (c) the amplitude of the current.

(a) Since the rms voltage is 120 V, $v_e = v_0/\sqrt{2}$ gives for the amplitude or peak voltage

$$v_0 = \sqrt{2}\, v_e = 1.414(120 \text{ V}) = 170 \text{ V}$$

(b) Using $\overline{\mathcal{P}} = i_e v_e{}^R$, with $v_e{}^R$ equal to the rms line voltage, the rms current is

$$i_e = \frac{\overline{\mathcal{P}}}{v_e{}^R} = \frac{1000 \text{ W}}{120 \text{ V}} = 8.33 \text{ A}$$

(c) From $i_e = i_0/\sqrt{2}$, the current amplitude is

$$i_0 = \sqrt{2}\, i_e = 1.414(8.33 \text{ A}) = 11.8 \text{ A}$$

22.11 | REACTANCE

In an ac circuit, the rms current and the potential drop across a resistor are related by the resistance R. The analogous quantities for inductances and capacitances are called *reactances*. They can be used to determine the effective and instantaneous potential differences in an ac circuit.

The induced EMF in an inductance L is proportional to L and to the rate of change of the current, which increases with the frequency. As we show below, if the generator frequency of $f = \omega/2\pi$, then the *inductive reactance* is

$$X_L = \omega L \qquad (22.22)$$

This means that the rms voltage drop $v_e{}^L$ across an inductance carrying an effective current i_e is

$$v_e{}^L = i_e X_L \qquad (22.23)$$

This is similar in form to the voltage drop across a resistor, $v_e{}^R = i_e R$.

The potential difference across a capacitor with charge q is q/C, so the *capacitive reactance* X_C is proportional to $1/C$. X_C also varies as $1/\omega$, since at high frequencies very little charge accumulates on the capacitor before the current direction reverses, and the rms voltage drop $v_e{}^C$ is then small. Thus

$$X_C = \frac{1}{\omega C} \qquad (22.24)$$

and the effective potential drop across a capacitor is

$$v_e{}^C = i_e X_C \qquad (22.25)$$

The reactance of an inductor is calculated and used in the following example.

Example 22.8
    A 5 mH $= 5 \times 10^{-3}$ H inductor is connected to a

120-V rms 60-Hz power line. Find (a) the inductive reactance, and (b) the rms current.

(a) Since the frequency f is 60 Hz, the inductive reactance is

$$X_L = \omega L = 2\pi f L = 2\pi(60 \text{ s}^{-1})(5 \times 10^{-3} \text{ H})$$
$$= 1.88 \text{ ohms}$$

(b) Since the rms voltage across the inductor is 120 V, $v_e{}^L = i_e X_L$ gives

$$i_e = \frac{v_e{}^L}{X_L} = \frac{120 \text{ V}}{1.88 \text{ ohms}} = 63.8 \text{ A}$$

Reactances can also be used to determine the instantaneous potential differences in a circuit. Suppose a circuit consisting of a resistance R, an inductance L, and a capacitance C connected in series to a generator has a current $i = i_0 \sin \omega t$ (Fig. 22.24). Then the potential drops across the three elements vary with time as

$$v^R = i_0 R \sin \omega t \qquad (22.19)$$

(a)

(b)

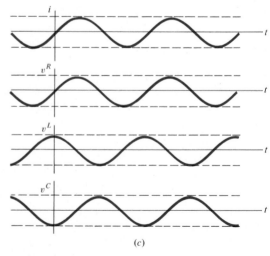

(c)

Figure 22.24. (*a*) Symbol for an ac generator. (*b*) An *RLC* series circuit. (*c*) The potential drops across each of the elements and the current versus time.

$$v^L = i_0 X_L \cos \omega t \qquad (22.26)$$
$$v^C = -i_0 X_C \cos \omega t \qquad (22.27)$$

These potential differences and the current are plotted in Fig. 22.24c. The drop v^R across the resistor is always proportional to the instantaneous current, but this is not true for the other two voltages. For example, at $t = 0$, the current is zero, but its rate of change is a maximum. Since the induced EMF is proportional to $\Delta i/\Delta t$, v^L has its maximum positive value. Also, at $t = 0$, the current has just completed a negative half cycle. Hence the capacitor charge q and its potential drop $v^C = q/C$ have their most negative values. These concepts are illustrated by the next example.

Example 22.9

The current in Fig. 22.25 has an amplitude of 0.5 A and a frequency $f = \omega/2\pi = 159$ Hz, so $\omega = 1000 \text{ s}^{-1}$. (a) Find the instantaneous voltage drop across each circuit element at $t = 0$. (b) Find the potential difference v between points a and b at an arbitrary time t, and the effective voltage between these points. (c) Find the effective voltages across the reactances, X_L and X_C.

(a) At $t = 0$, $\sin \omega t = \sin 0 = 0$, and $\cos \omega t = \cos 0 = 1$. With $X_L = X_C = 20$ ohms, the voltage drops are

$$v^R = i_0 R \sin \omega t = 0,$$
$$v^L = i_0 X_L \cos \omega t = (0.5 \text{ A})(20 \text{ ohms})(1) = 10 \text{ V}$$
$$v^C = -i_0 X_C \cos \omega t = -(0.5 \text{ A})(20 \text{ ohms})(1) = -10 \text{ V}$$

Thus at this instant, if we proceed along the circuit in the direction of positive currents, we see no voltage drop across the resistor, a 10-V drop across the inductor, and a 10-V *rise* across the capacitor.

(b) Since X_L and X_C are equal, $v^L = -v^C$ at all times. Thus the sum $v^R + v^L + v^C$ of the voltage drop across the three circuit elements equals v^R, and with $R = 10$ ohms, the total drop from point a to point b is

$$v = v^R = i_0 R \sin \omega t$$
$$= (0.5 \text{ A})(10 \text{ ohms}) \sin (1000 \, t)$$
$$= 5 \sin (1000 \, t) \text{ V}$$

Figure 22.25.   Example 22.2.

Figure 22.26.   Effective voltages are *not* additive.

Since the voltage amplitude is 5 V, the effective voltage is

$$v_e = v_e{}^R = \frac{v_0}{\sqrt{2}} = \frac{(5 \text{ V})}{1.414} = 3.54 \text{ V}$$

(c) The potential differences across the inductor and the capacitor each have an amplitude $v_0 = i_0 X_L = i_0 X_C = (0.5 \text{ A})(20 \text{ ohms}) = 10 \text{ V}$, so

$$v_e{}^L = v_e{}^C = \frac{v_0}{\sqrt{2}} = \frac{(10 \text{ V})}{1.414} = 7.07 \text{ V}$$

From this example, we can see that it is *not* correct to proceed in analogy with dc circuits and simply add rms ac voltages. Alternating current voltmeters connected to the terminals of the resistor, inductor, and capacitor will give rms voltages of 3.54, 7.07, and 7.07 V, respectively. These add up to 17.68 V, but the rms voltage across the three elements according to part (b) is 3.54 V (Fig. 22.26). The procedure for combining ac voltages is considered in the next section.

*Derivation of the Reactance Formulas

Suppose the current in an inductance is $i = i_0 \sin \omega t$. Then the induced EMF is $\mathcal{E} = -L \, di/dt$, and the voltage *drop* is $-\mathcal{E}$ or

$$v^L = L \frac{di}{dt} = L i_0 \frac{d}{dt} \sin \omega t$$

According to Eq. B.26 in Appendix B,

$$\frac{d}{dt} \sin \omega t = \omega \cos \omega t$$

Thus, with $X_L = \omega L$, we have

$$v^L = L i_0 (\omega \cos \omega t) = i_0 X_L \cos \omega t$$

*The remainder of this section may be omitted without loss of continuity.

which is Eq. 22.26. Now $\cos^2 \omega t$, like $\sin^2 \omega t$, averages to 1/2 over a cycle. Hence, if this equation is squared and averaged, we obtain $v_e^L = i_e X_L$ for the rms voltage.

The charge q on a capacitor is related to the current entering it by $i = dq/dt$. With Eq. B.27 in Appendix B,

$$\frac{d}{dt}\cos \omega t = -\omega \sin \omega t$$

Thus, in order that $dq/dt = i = i_0 \sin \omega t$, q must be given by

$$q = -\frac{i_0}{\omega}\cos \omega t$$

Hence, with $X_C = 1/\omega C$, the voltage drop is

$$v^C = \frac{q}{C} = -\frac{i_0}{\omega C}\cos \omega t = -i_0 X_C \cos \omega t$$

which is Eq. 22.27. Again, squaring and averaging gives the rms voltage formula, $v_e^C = i_e X_C$.

22.12 | IMPEDANCE

The current in a network of resistors connected to a dc EMF can be found by calculating the equivalent resistance of the network and using Ohm's law. Similarly, the current in a network of resistors, inductors, and capacitors can be found by calculating the equivalent *impedance* Z of the network and using an analog to Ohm's law.

As we saw in the preceding section, the instantaneous potential drop v from point a to point b in Fig. 22.27 is the sum of the individual voltage drops.

$$v = v^R + v^L + v^C$$

To find the rms voltage drop v_e, we must substitute the explicit formulas for the individual voltage drops, square, and average. We show below that the result is

$$v_e = i_e Z \qquad (22.28)$$

This is similar in form to Ohm's law with the resistance replaced by the *impedance* Z. However, Z is frequency dependent, since it is given by

$$Z = [R^2 + (X_L - X_C)^2]^{1/2} \qquad (22.29)$$

where $X_L = \omega L$ and $X_C = 1/\omega C$. If the *net reactance* X is defined by

$$X = X_L - X_C = \omega L - \frac{1}{\omega C} \qquad (22.30)$$

then Z has a somewhat simpler-looking form,

$$Z = (R^2 + X^2)^{1/2} \qquad (22.31)$$

When the circuit contains an inductor but no capacitor, $X = X_L$; similarly if there is only a capacitor, $X = -X_C$. These results are used in the next example, which has a resistor, capacitor, and inductor in series with an ac generator.

Example 22.10

An *RLC* series circuit is connected to an ac generator (Fig. 22.28). If $\omega = 1000$ rad s^{-1}, find (a) the impedance, and (b) the effective current.

(a) The net reactance is

$$X = X_L - X_C = \omega L - \frac{1}{\omega C}$$

$$= (1000 \text{ rad s}^{-1})(0.4 \text{ H}) - \frac{1}{(1000 \text{ rad s}^{-1})(10^{-5} \text{ C})}$$

$$= 400 \text{ ohms} - 100 \text{ ohms} = 300 \text{ ohms}$$

Hence the impedance is

$$\begin{aligned} Z &= (R^2 + X^2)^{1/2} \\ &= [(400 \text{ ohms})^2 + (300 \text{ ohms})^2]^{1/2} \\ &= 500 \text{ ohms} \end{aligned}$$

(b) The effective voltage across the *RLC* series combination must equal the effective generator voltage $\mathcal{E}_e$, so $v_e = i_e Z$ gives

$$i_e = \frac{\mathcal{E}_e}{Z} = \frac{100 \text{ V}}{500 \text{ ohms}} = 0.2 \text{ A}$$

Figure 22.27. Reactances and voltages in an *RLC* circuit.

Figure 22.28. Examples 22.10, 22.11, and 22.12.

In the preceding section, we also saw that the sum of the rms ac voltages is *not* equal to the rms potential difference v_e across two or more circuit elements. The correct procedure for combining these voltages can be found from Eq. 22.29 for the impedance. If we multiply this by i_e, then we have

$$i_e Z = [(i_e R)^2 + (i_e X_L - i_e X_C)^2]^{1/2}$$

Now, $i_e Z = v_e$, $i_e R = v_e^R$, $i_e X_L = v_e^L$, and $i_e X_C = v_e^C$, so

$$v_e = [(v^R)^2 + (v^L - v^C)^2]^{1/2} \qquad (22.32)$$

These ideas are illustrated by the next two examples.

Example 22.11

(a) Find the rms voltage drops across each of the elements in the preceding example, where $i_e = 0.2$ A, $X_L = 400$ ohms, $X_C = 100$ ohms, and $R = 400$ ohms. (b) Calculate the rms voltage drop across the three elements, and show it is equal to the generator rms EMF, 100 V.

(a) The rms drops across the three elements are

$$v_e^R = i_e R = (0.2 \text{ A})(400 \text{ ohms}) = 80 \text{ V}$$
$$v_e^L = i_e X_L = (0.2 \text{ A})(400 \text{ ohms}) = 80 \text{ V}$$
$$v_e^C = i_e X_C = (0.2 \text{ A})(100 \text{ ohms}) = 20 \text{ V}$$

(b) Using Eq. 22.32,

$$v_e = [(80 \text{ V})^2 + (80 \text{ V} - 20 \text{ V})^2]^{1/2}$$
$$= [10{,}000 \text{ V}^2]^{1/2} = 100 \text{ V}$$

Example 22.12

Suppose the frequency of the generator in Fig. 22.28 is varied. (a) At what value of $\omega = 2\pi f$ will the impedance be smallest? (b) Find the effective current at that value of ω.

(a) Since $Z = [R^2 + (X_L - X_C)^2]^{1/2}$, its minimum value occurs when $X_L - X_C = \omega L - 1/\omega C = 0$, or when $\omega L = 1/\omega C$. Thus $\omega^2 = 1/LC$, or

$$\omega = \sqrt{\frac{1}{LC}}$$
$$= \sqrt{\frac{1}{(0.4 \text{ H})(10^{-5} \text{ F})}} = 500 \text{ rad s}^{-1}$$

(b) Since X is zero, $Z = R$, and the effective current is

$$i_e = \frac{v_e}{Z} = \frac{v_e}{R} = \frac{100 \text{ V}}{400 \text{ ohms}} = 0.25 \text{ A}$$

This last example illustrates the phenomenon

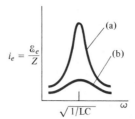

Figure 22.29 Effective current versus frequency for two circuits with the same EMF and resonant frequency. The narrow curve (a) corresponds to a circuit with a small resistance. The broad flat curve (b) corresponds to a circuit with a large resistance.

$$i_e = \frac{\mathscr{E}_e}{Z}$$

called *resonance*. At a frequency below or above the resonance frequency, the impedance is greater, so the current for a given applied generator EMF is smaller (Fig. 22.29). This is very similar to the behavior of a mechanical oscillator. We found in Chapter Ten that an oscillating object will have its greatest amplitude if the driving force is applied at the characteristic frequency of the oscillator. The similarity in the behavior of the *RLC* circuit and the mechanical oscillator reflects the fact that the two systems satisfy the same equation of motion, except for a relabeling of the variables.

Resonant ac circuits have many applications. For example, in a television or radio antenna, currents are produced by the electromagnetic waves broadcast at different frequencies by many stations. A tuner in the receiver contains an inductor and a variable capacitor adjusted so that the resonant frequency corresponds to the desired station. Since the resistance of the tuner is small, the impedance rises rapidly as the frequency varies from the resonant frequency, and therefore only one station will normally be received.

*Derivation of the Alternating Current Analog to Ohm's Law, $v_e = i_e Z$ |
We saw in the last section that the instantaneous voltage drops across R, L, and C in Fig. 22.27 are given by

$$v^R = i_0 R \sin \omega t \qquad (22.19)$$
$$v^L = i_0 X_L \cos \omega t \qquad (22.26)$$
$$v^C = -i_0 X_C \cos \omega t \qquad (22.27)$$

Hence the total voltage drop across the three elements is

$$v = v^R + v^L + v^C = i_0(R \sin \omega t + X \cos \omega t)$$

*The remainder of this section may be omitted without loss of continuity.

where $X = X_L - X_C$. This must be squared and averaged over a full cycle to find $\overline{v_e^2}$. Now $\sin^2 \omega t$ and $\cos^2 \omega t$ each average to 1/2, while $(\sin \omega t)(\cos \omega t)$ averages to zero. (See Appendix B, Eqs. B.13 and B.14.) Thus we find

$$\overline{v^2} = \frac{i_0^2}{2}(R^2 + X^2)$$

Since $\overline{v^2} = v_e^2$ and $(i_0)^2/2 = i_e^2$, this can be rewritten as

$$v_e^2 = i_e^2(R^2 + X^2)$$

Hence

$$v_e = i_e \sqrt{R^2 + X^2} = i_e Z$$

which is the desired result.

22.13 | POWER IN ALTERNATING CURRENT CIRCUITS

The equation for the power dissipated in an ac circuit differs somewhat from the dc formula. During half of each cycle, the capacitor charge is increasing, so the capacitor is absorbing energy. However, this stored electrical energy is released when the capacitor discharges during the other half cycle. Similarly, the inductor temporarily stores and releases magnetic energy. Neither of these circuit elements absorbs any *net* energy over a full cycle. However, as we have seen in Section 22.10, a resistor dissipates energy as heat at the average rate $\bar{\mathcal{P}} = i_e^2 R$. Since $v_e = i_e Z$, this can be rewritten as $\bar{\mathcal{P}} = i_e(v_e/Z)R$, or

$$\bar{\mathcal{P}} = i_e v_e \left(\frac{R}{Z}\right) \qquad (22.33)$$

The power dissipation in a dc circuit is iv, so this result has an extra factor (R/Z) called the *power factor*. It is always less than 1 if there is any reactance. If R is small compared to Z, then very little power is dissipated by the circuit, even though i_e is large; energy is absorbed and released by the inductor and the capacitor, but little heating occurs in the resistor. The power factor is less than 1 for ac motors, as in the next example.

Example 22.13

An ac motor connected to a 120-V rms power line is equivalent in the circuit to a 80-ohm resistor and a 60-ohm inductive reactance in series. Find (a) the effective current, and (b) the power supplied to the motor.

(a) To find the effective current, we first calculate the impedance:

$$\begin{aligned} Z &= [R^2 + X^2]^{1/2} \\ &= [(80 \text{ ohms})^2 + (60 \text{ ohms})^2]^{1/2} \\ &= 100 \text{ ohms} \end{aligned}$$

Then

$$i_e = \frac{v_e}{Z} = \frac{120 \text{ V}}{100 \text{ ohms}} = 1.2 \text{ A}$$

(b) The power supplied to the motor is

$$\bar{\mathcal{P}} = i_e v_e \left(\frac{R}{Z}\right) = (1.2 \text{ A})(120 \text{ V})\left(\frac{80 \text{ ohms}}{100 \text{ ohms}}\right)$$
$$= 115 \text{ W}$$

22.14 | IMPEDANCE MATCHING

It is often desirable to arrange that the maximum possible power transfer occurs between a power source and a load. These could be a battery and a resistor, respectively, as in the example we discuss shortly. They could also be an amplifier and a loudspeaker or probes connected to a person and an EKG recorder. *In general, the maximum power transfer occurs when the impedances of a source and a load are matched.*

We begin by considering the dc case. Figure 22.30 shows a battery with EMF $\mathcal{E}$ and internal resistance R_s connected to a load that is a resistance R_l. With Ohm's law, the current is

$$I = \frac{\mathcal{E}}{(R_s + R_l)}$$

The power supplied to the load is then

$$\mathcal{P} = I^2 R_l = \frac{R_l}{(R_l + R_s)^2}\mathcal{E}^2$$

To see how this varies with the ratio $x = R_l/R_s$ of the

Figure 22.30. The power transfer to R_l is greatest if the resistances are matched ($R_l = R_s$).

$$\frac{x}{(1 + x)^2}$$

Figure 22.31. The peak power transfer to the load is at $x = R_l/R_s = 1$.

load and source resistances, we divide out a factor of $R_s{}^2$ in the denominator and rewrite $\mathcal{P}$ as

$$\mathcal{P} = \frac{\mathcal{E}^2}{R_s} \frac{x}{(1 + x)^2}$$

The factor $\mathcal{E}^2/R_s$ is the power that would be dissipated in the battery if the load resistance R_l were equal to zero, or if the battery were short-circuited. The second factor is proportional to x, when x is small compared to 1 (Fig. 22.31). Its maximum occurs when $x = 1$, or $R_l = R_s$.

The maximum dc power transfer occurs when the load and source resistances are equal.

Note that when $x = 1$, the power supplied to R_l is $(1/4)(\mathcal{E}^2/R_s)$, or one-fourth the power dissipated in a short-circuited source.

Now let us consider, instead, an ac generator with effective EMF $\mathcal{E}_e$, internal resistance R_s, and internal reactance X_s connected to a load with resistance R_l and net reactance X_l (Fig. 22.32). Again we want to know what load will receive the maximum power transfer. If the above analysis is repeated using the formulas appropriate to alternating currents, the re-

sult is that the maximum power transfer occurs if

$$R_l = R_s \quad \text{and} \quad X_l = -X_s \quad (22.34)$$

This means that the maximum power transfer in an ac circuit occurs when the source and load resistances are equal and the reactances cancel, so that the total reactance $X_l + X_s$ of the circuit is zero.

We recall that the net reactance is $X = X_L - X_C$, which may be positive or negative. Thus, for example, if the source has a resistance of 50 ohms and an inductive reactance of 100 ohms, then to be matched the load must have a resistance of 50 ohms and a capacitive reactance of 100 ohms.

Impedance matching is important whenever "black boxes" are assembled in a laboratory. Even though we can often ignore the internal details of complex electronics instruments, we must be sure that the input and output electrical impedances of two interconnected instruments are not badly mismatched. Impedance matching also is important in applications other than electrical circuits. For example, if sound is to be transferred efficiently without reflections from one medium to another, then the *acoustical impedances* must be matched. Thus, impedance matching is a concept with wide applicability.

22.15 | DERIVATION OF THE EMF OF A GENERATOR

Equation 22.5 for the EMF of a generator can be derived by differentiating the expression for the flux through the loop. According to Eq. B.27 in Appendix B

$$\frac{d}{dt} \cos \omega t = -\omega \sin \omega t$$

Hence the instantaneous rate of change of $\Phi = BA \cos \omega t$ is

$$\frac{d\Phi}{dt} = BA(-\omega \sin \omega t)$$

and we have the desired result

$$\mathcal{E} = -\frac{d\Phi}{dt} = \omega BA \sin \omega t$$

EXERCISES ON SUPPLEMENTARY TOPICS
Section 22.9 | RL Circuits
22-35 Using the definitions of inductance and

Figure 22.32. The power transfer to the load is greatest when the impedances are matched. ($R_l = R_s$, $X_l = X_s$.)

resistance, show that L/R has the units of time.

22-36 A 10-ohm resistor and a 0.2-H inductor are connected in series to a 12-V battery. (a) What is the final current? (b) What is the time constant? (c) What is the current after one time constant? (d) What is the current after 4 seconds?

22-37 A coil of wire connected to a 6-V battery has a current of 0.063 A after 0.01 seconds and a current of 0.1 A after a few seconds. (a) What is its resistance? (b) What is its time constant? (c) What is its inductance?

Section 22.10 | Effective or Root Mean Square Alternating Currents and Voltages

22-38 A fuse is designed to open if the current exceeds 20 A rms. What is the peak current allowed if this fuse is used in an ac circuit?

22-39 A light bulb draws 60 W from a 120-V rms 60-Hz line. (a) What is the rms current? (b) What is the peak current? (c) How many times per second does this peak current occur?

22-40 Homes are often supplied with 240-V rms alternating current to provide the power for high-wattage appliances or electric heat. The lines coming into the home include a ground and two "hot" lines, which have voltages equal in magnitude but opposite in sign at all times. Explain how 120-V rms alternating current can be obtained from these lines.

22-41 Number 14 wire is rated to safely carry 15 A rms. How much power can this wire supply at (a) 120-V rms alternating current; (b) 240-V rms alternating current?

22-42 The EMF of a generator is given by $\mathcal{E} = 100 \sin(1000t)$V. (a) What is the frequency f of the generator? (b) What is its rms voltage?

Section 22.11 | Reactance

22-43 A capacitor has a reactance of 10 ohms at 400 Hz. (a) Find the capacitance. (b) What is the reactance at 60 Hz? (c) What will be the rms voltage drop across the capacitor if it is connected to a 240-V 60-Hz line?

22-44 At 60 Hz, find the reactance of a (a) 0.05-H inductor; (b) a 1-μF = 10^{-6}-F capacitor. (c) At what frequency will the inductive and capacitive reactances be equal?

22-45 A 2-H inductor is connected to a 120-V rms 60-Hz power line. (a) Find its reactance. (b) Find the rms potential difference across the inductor. (c) What is the peak voltage drop across the inductor?

Section 22.12 | Impedance

22-46 A 100-ohm resistance and a 0.2-H inductance are connected in series to a 240-V rms 60-Hz generator. (a) Find the reactance of the inductor. (b) Find the impedance of the circuit. (c) Find the rms current.

22-47 For the circuit in the preceding exercise, draw sketches showing how the following quantities vary with time: (a) the current; (b) the potential difference across the resistor; (c) the potential difference across the inductor.

22-48 A 100-ohm resistor, a 10^{-4}-F capacitor, and a 0.1-H inductor are connected in series to a 120-V rms generator. If the generator frequency is 60 Hz, find (a) the impedance, and (b) the effective current.

22-49 A tuning circuit in a radio has a coil of inductance 2×10^{-5} H in series with a variable capacitor. What is the capacitance if the radio is tuned to receive a station at $f = \omega/2\pi = 10^6$ Hz?

22-50 Alternating current voltmeters are connected across the terminals of an inductance, a capacitance, and a resistance connected in series. Their readings are 100 V, 300 V, and 150 V, respectively. What is the rms voltage across the three circuit elements?

Section 22.13 | Power in Alternating Current Circuits

22-51 An air conditioner connected to a 120-V rms ac line is equivalent to a 10-ohm resistance and a 1-ohm inductive reactance in series. (a) What is its impedance? (b) What is the power supplied to the air conditioner?

22-52 A 5-ohm resistor, a 10-ohm inductive reactance, and a 22-ohm capacitive reactance are connected in series to a 120-V rms ac generator. (a) Find the impedance. (b) How much power is dissipated in the resistor?

22-53 A 5-ohm resistor, a 0.01-H inductor, and a 10^{-4}-F capacitor are connected in series to a power supply with a variable frequency. (a) At what value of ω is the current the greatest? (b) What is the corresponding frequency f? (c) How much power is dissipated at this frequency if the effective EMF of the power supply is 10 V?

22-54 For the circuit in the preceding exercise, find (a) the impedance at $\omega = 2000$ rad s^{-1}; (b) the power supplied to the resistor if the effective EMF is 10 V.

Section 22.14 | Impedance Matching

22-55 It is more economical to extract energy from a battery by using a load resistance that is larger than the internal source resistance rather than a load resistance equal to the source resistance. Explain why.

22-56 Why is impedance matching *not* desirable when using a voltmeter to measure a potential difference?

PROBLEMS ON SUPPLEMENTARY TOPICS

22-57 Suppose a light bulb and a variable inductance are connected in series to an ac power line. (a) How would the brightness of the bulb vary as the inductance is increased? (b) One can also vary the brightness by replacing the inductor with a variable resistance. Why is that less satisfactory?

*22-58 A tuning circuit in a radio receiver consists of a coil with an inductance of 10^{-5} H and a variable capacitor. (a) What is the capacitance if the receiver is tuned to $f_0 = 1.4 \times 10^6$ Hz? (b) Find the reactance at a frequency 1 percent above f_0. (c) At this higher frequency the impedance is larger than

at f_0 by a factor of 4. Find the resistance of the circuit.

*22-59 Verify that if the resistance and reactance of a load are adjusted as in Eq. 22.34 then the maximum power is transferred to the load.

22-60 A battery with EMF $\mathcal{E}$ and internal resistance R_s is connected to a resistance R_l. Compare the power dissipated in the battery and in the load if (a) $R_l = R_s$; (b) R_l is much larger than R_s.

Additional Reading:

E. Schlömann, Recovery of Nonmagnetic Metals from Municipal Wastes, *Physics Teacher*, vol. 14, 1976, p. 116. The process uses eddy currents.

Kurt S. Lion, *Elements of Electrical and Electronic Instrumentation*, McGraw-Hill Book Co., New York, 1975.

Scientific American articles:

Herbert Kondo, Michael Faraday, October 1953, p. 90.

J.R. Newman, James Clerk Maxwell, June 1955, p. 58.

P. Morrison and E. Morrison, Heinrich Hertz, December 1957, p. 98.

H.L. Sharlin, From Faraday to the Dynamo, May 1961, p. 107.

G. Shiers, The Induction Coil, May 1971, p. 80.

James R. Heirtzler, The Longest Electromagnetic Waves, March 1962, p. 128.

UNIT SIX

WAVE MOTION

Most of the wave motions with which we are familiar involve a large-scale coordinated disturbance of many particles or objects. While the individual particles do not move far, the disturbance may travel great distances, carrying with it energy and momentum. The motions of the particles vary with the type of wave. For example, in a water wave, the water molecules move in small, approximately circular paths; in a sound wave, molecules move back and forth; and in a wave on a string, the parts of the string move up and down.

Light waves are also coordinated disturbances involving changing electric and magnetic fields. Here, no particles move, but the waves nevertheless carry energy and momentum, and the mathematical description of these waves is nearly identical to the mechanical examples given above.

All waves have a number of characteristics in common. Many of these are described in the first chapter of this unit (Chapter Twenty-three) where we mainly use strings and springs for illustration. The special features of sound waves are described in Chapter Twenty-four and the properties of light are covered in Chapters Twenty-five and Twenty-six.

The importance of waves in physics derives from the transmission of energy and momentum from one place to another or from a source to a detector. These wave motions are described using the ideas of mechanics or electromagnetic theory. Their properties follow from the ideas presented earlier in this book and merely represent cooperative or coordinated motions in systems we have already studied.

A severe jolt to physics in the early twentieth century was the realization that the sharp distinction between wave and individual particle motions was not valid when studying atoms and molecules. In the years since, a dramatic revolution in our description of nature has taken place. Modern quantum physics, which was developed to describe the microscopic properties of molecules, atoms, and nuclei, has forced us to regard particles as having some wave properties and to regard waves as though they are composed of particles or *quanta*. This wave-particle duality of our present description of nature may be due to our lack of a true understanding of nature or of our inability to express our understanding clearly. Nevertheless, with the language we do have, we must, for example, regard an electron as an entity with both particle and wavelike characteristics.

This revolution of modern physics does not mean that all of our studies have been for naught. It simply means that when very small dimensions are involved, we must further refine our understanding. The remaining units of this book are devoted to those refinements.

CHAPTER TWENTY-THREE

THE DESCRIPTION OF WAVE MOTION

In this chapter, concepts common to all kinds of wave phenomena are developed. We primarily use the readily visualized waves along strings and springs as examples and analogies for other types of wave motion that are more difficult to observe.

When a stone is dropped into a lake or when a string is wiggled briefly at an end, a single wave *pulse* travels away from the disturbance. However, waves also occur in a regular continuing series; they are then said to be *periodic*. For example, a vibrating tuning fork produces alternate compressions and rarefactions of the air nearby. These disturbances, which are perceived as sound, occur at the frequency of the tuning fork.

When a string is disturbed, its particles move at right angles to the string. Waves with displacements perpendicular to the direction of the wave are called *transverse waves*. Light is a transverse wave. Waves that have displacements along the wave direction are termed *longitudinal*. Examples of this latter type are sound waves and compressional waves in coiled springs.

All the waves we consider have the important property of *linearity*. Linearity means that when two or more waves pass the same point, the resulting wave is the sum of the individual waves, and after passing, the waves continue along their paths as if no encounter had occurred. If one observes the circular waves traveling outward from separate points where two pebbles are dropped in still water, the peaks add up when they meet. Afterward, they travel along undisturbed. The description of such linear behavior is called the *principle of superposition*.

23.1 | THE REPRESENTATION OF WAVES

Figure 23.1 shows single wave *pulses* on a string and in a spring produced by a disturbance at their left ends and traveling toward the right. Figure 23.2 shows *periodic waves* produced by oscillatory motion of the left ends. The waves on the string are transverse, and the waves in the spring are longitudinal. Nevertheless, both types of waves can be represented symbolically by similar graphs.

Many kinds of wave phenomena occur in nature. Light is a transverse electromagnetic wave, with changing electric and magnetic fields at right angles to each other and to the direction of the light wave. Sound is an alternate compression and rarefaction of the medium along the direction of motion, so it is a longitudinal wave. Water waves are a mixture of longitudinal and transverse waves, with water molecules moving in roughly circular paths. All these wave phenomena can be represented by graphs similar to those for strings and springs and share many common properties (Fig. 23.3).

Periodic waves of any type are characterized by several quantities. The *frequency f* is the number of waves passing a point per second and is determined by the source of the waves. For example, in the string and spring of Fig. 23.2, the frequency is the rate at which the oscillations occur at the left end. The *period T* is the time between successive wave crests, or the inverse of the frequency; $T = 1/f$. The *velocity c* of a wave is the speed at which a wave peak travels. The *wavelength* λ of a periodic wave is the distance be-

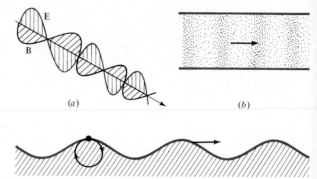

Figure 23.3. (*a*) An electromagnetic wave. (*b*) A sound wave. (*c*) A water wave. All these waves can be represented by the same kind of graphs as used for springs and strings.

tween successive wave peaks. Finally, the *amplitude A* is the maximum magnitude of the displacement; the displacement of a periodic wave varies back and forth between A and $-A$. Most of the periodic waves we consider are *sinusoidal* waves, and their graphs look like sine or cosine graphs.

The relationship among the frequency, wavelength, and velocity of a periodic wave can be found from Fig. 23.4. In one period T, the time required for one complete wave oscillation, the wave travels one wavelength λ. The velocity of the wave c is the distance traveled divided by the time, so $c = \lambda/T$. Since $T = 1/f$, this implies

$$f\lambda = c \qquad (23.1)$$

The following examples illustrate these relationships.

Example 23.1

A wave pulse on a string moves a distance of 10 m in 0.05 s. (a) What is the velocity of the pulse? (b) What is the frequency of a periodic wave on the same string if its wavelength is 0.8 m?

(a) The velocity of the pulse is $c = x/t$, where $x = 10$ m and $t = 0.05$ s, so

$$c = \frac{10 \text{ m}}{0.05 \text{ s}} = 200 \text{ m s}^{-1}$$

(b) The periodic wave has the same velocity, 200 m s^{-1}. Using $f\lambda = c$, the frequency of a 0.8-m wave is

$$f = \frac{c}{\lambda} = \frac{200 \text{ m s}^{-1}}{(0.8 \text{ m})} = 250 \text{ s}^{-1} = 250 \text{ Hz}$$

Figure 23.1. (*a*) A single wave pulse is produced at the left end of a string. The appearance of the string is shown at three successive times. Note that the motion of the particles is transverse to the motion of the wave. (*b*) A similar wave pulse on a coiled spring. The spring is alternately compressed and extended along the direction of motion, so the wave is longitudinal. (*c*) Either wave pulse can be represented symbolically by the same graph. For the string, *y* is the displacement of the string from its undisturbed position; displacements above the equilibrium position are positive. For the spring, *y* is a measure of the compression or extension of the spring; a compression is regarded as a positive displacement.

Figure 23.2. (*a*) A periodic disturbance produced in a string by an oscillating lever travels toward the right. The dashed line indicates the undisturbed position of the string. (*b*) A spring is alternately compressed and extended. (*c*) The same graph can represent either wave.

388

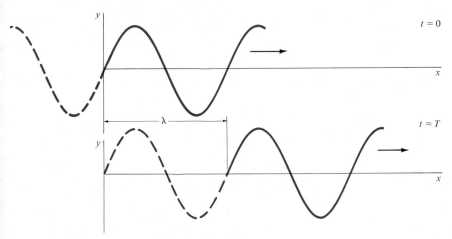

Figure 23.4. In the period T required for a complete wave oscillation, the wave moves a wavelength λ.

Example 23.2

A typical sound wave associated with human speech has a frequency of 500 Hz, while the frequency of yellow light is about 5×10^{14} Hz. In air, sound travels at 344 m s^{-1} and light at 3×10^8 m s^{-1}. Find the wavelengths of the two waves.

For both waves, $f\lambda = c$, so for the sound wave

$$\lambda = \frac{c}{f} = \frac{344 \text{ m s}^{-1}}{500 \text{ Hz}} = 0.688 \text{ m}$$

For the light wave,

$$\lambda = \frac{c}{f} = \frac{3 \times 10^8 \text{ m s}^{-1}}{5 \times 10^{14} \text{ Hz}} = 6 \times 10^{-7} \text{ m}$$

The fact that these characteristic wavelengths are so different is responsible for some of the apparent differences between the two types of waves. Waves tend to bend around obstacles comparable in size to their wavelength. Such effects are often apparent with sound or water waves but are seldom noticed for light because the wavelength of visible light is so short. Light waves will bend around objects that are small enough.

23.2 | THE VELOCITY OF WAVES

While waves are common phenomena, each type of wave disturbance has its own specific physical origin. Accordingly, each type of wave has a characteristic velocity.

Electromagnetic waves are unique in that they require no medium in which to propagate. As we discussed in Chapter Twenty-two, these waves are due to mutually induced time varying electric and magnetic fields. In a vacuum, they have a velocity of 3×10^8 m s^{-1}; in matter, their velocity is always smaller.

Sound is a coordinated mechanical disturbance involving large numbers of molecules. These molecules move and collide as a wave disturbance passes, but on the average they suffer no net change in position. As we show in the next chapter, the velocity of sound in a substance depends on how the pressure changes when the density changes. The velocity of sound in air at 30°C is 344 m s^{-1}, but it is much higher in solids; in aluminum for example, $c = 6400$ m s^{-1}. In solids there may also be transverse as well as longitudinal sound waves. The two types of sound waves have different velocities.

The velocity of a wave can be predicted from the physical laws that describe the specific wave phenomenon. For example, we saw in Chapter Twenty-two that Maxwell was able to derive an expression for the speed of an electromagnetic wave using the fundamental properties of electric and magnetic fields. Similarly, the velocities of the various kinds of mechanical waves, such as sound waves and waves in strings, springs, and water, can be predicted using Newton's laws of motion. Such calculations are mathematically complex and will not be carried out here. However, it is important to remember that the velocity of a wave depends on the type of wave, on the properties of the system in which the wave travels, and sometimes on the frequency. As an example, we describe the result for a wave on a string.

When a string is displaced, the restoring force is proportional to the tension in the string. Also, a thick string will respond more slowly to this restoring force than will a thin string. Hence, we expect that the wave velocity on a string will depend on the tension T and the mass per unit length μ of the string. This is the case and the detailed analysis gives for the wave velocity c on a string,

$$c = \sqrt{\frac{T}{\mu}} \qquad (23.2)$$

This result has the features described above. The following example leads to a typical value for the wave velocity on a string.

Example 23.3

The tension on the longest string of a grand piano is 1098 N, and the mass per unit length is 0.065 kg m^{-1}. What is the velocity of a wave on this string?

Using the above result,

$$c = \sqrt{\frac{T}{\mu}} = \sqrt{\frac{1098\ \text{N}}{0.065\ \text{kg m}^{-1}}} = 130\ \text{m s}^{-1}$$

23.3 | WAVE INTERFERENCE AND STANDING WAVES

When two or more waves travel in a medium, the resulting wave is the sum of the displacements associated with the individual waves. This property is referred to either as *linearity* or as the *principle of superposition* and is applicable to all the waves we consider here. As we have noted, the resulting wave may have a very complex shape when the waves overlap, but each individual wave is unchanged and displays its original form when the waves separate. Since waves add algebraically, the resultant of overlapping waves can be larger or smaller than the individual waves, depending on their relative sign. This characteristic interaction of waves is termed *interference* and leads to many interesting and curious effects. In this section, we see that waves traveling in opposite directions can combine to form a wave disturbance that does not appear to travel at all and is therefore called a *standing wave*. Many other examples of interference are considered later.

To illustrate these ideas, suppose first that a string is held at each end by individuals who simultaneously produce wave pulses. The only difference between the pulses is that they are inverted mirror images of each

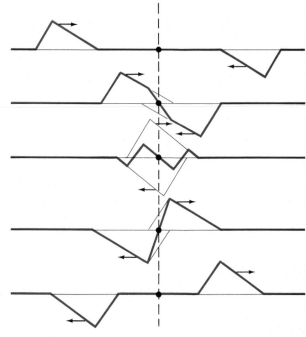

Figure 23.5. Two wave pulses that are inverted mirror images of each other are produced at the ends of a string. As the waves pass each other, the string has a complex shape but the one point on the dashed line is always at rest. The string shape is found by adding the displacements of the two pulses at every point.

other (Fig. 23.5). As the waves pass by the same points on the string, the resulting wave is found by adding the displacements of the individual waves.

An important feature of this particular experiment is that despite the fact that the wave shape becomes quite complex, there is one point on the string that is never displaced. This point is called a *node*. The string could actually be held fixed at this point without affecting the results. The pulses are said to be interfering with each other *destructively* at this point.

The same experiment can be performed with two periodic waves traveling in opposite directions but having the same amplitude and wavelength (Fig. 23.6). Now there are a number of nodes spaced one half wavelength apart. Midway between two successive nodes are points called *antinodes* where the string can experience a maximum displacement. At an antinode the waves add *constructively*.

Figure 23.7 shows a multiple exposure series of pictures of a string vibrating in this way. Notice that the traveling waves are no longer observed. What is

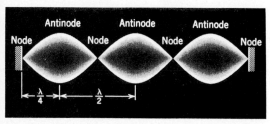

(a)

(b)

Figure 23.6. (a) Two oppositely directed periodic waves of the same wavelength. The waves are shown separately but should be considered as propagating on one string. Views are shown at four times a quarter cycle apart. (b) The shape of the string changes with time but the displacement is always zero at the nodes.

Figure 23.7. The pattern formed by the string of Fig. 23.6 if viewed at many different times.

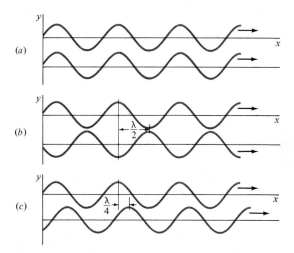

Figure 23.8. (a) Two waves exactly in phase. (b) Two waves exactly out of phase. Addition of these two waves results in complete destructive interference. (c) These two waves are out of phase by $\lambda/4$.

seen is a pattern described as a *standing wave:* a wave with nodes and antinodes at fixed points.

The interference effects produced by waves depend on their *phases*. If two waves reaching a point have their maxima at the same time, they are *in phase* and add constructively. If a maximum of one wave coincides with a minimum of the other, they are a half wavelength out of phase and interfere destructively. Under these conditions the waves are said to be exactly out of phase. In general, there may be an arbitrary phase difference between two waves (Fig. 23.8).

23.4 | THE EFFECTS OF BOUNDARIES

Usually when a wave encounters a *boundary,* a point where the medium changes, part of the wave is reflected and part is absorbed or transmitted. The de-

tails depend on the nature of the boundary. In this section, we consider two special types of boundaries. The end of a string will either be held firmly fixed or will be free to move transversely. In both cases the wave is nearly completely reflected because almost no energy is lost from the system.

Figure 23.9 shows what happens to a wave pulse which reaches a fixed or a free end of a string. If the string end is fixed, the reflected wave is the *inverted*

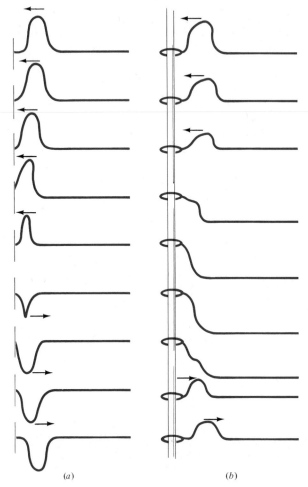

Figure 23.9.   The reflection of a wave pulse on a string at (a) a fixed end boundary and (b) a free end boundary.

mirror image of the incident pulse; the wavelength and shape are unchanged. If the string end is free, the reflected wave is the mirror image of the incident wave but is *not inverted.*

The results can be understood from the principles of mechanics. When a pulse arrives at a fixed end of a string, it exerts an upward force on the support. The support will exert an *equal and opposite* force on the string, producing a reflected pulse opposite to the incident pulse or reversed in phase. If a pulse reaches the free end of a string, the particles of the string there acquire momentum in the upward direction. When the end of the string reaches the maximum height of the wave pulse, this momentum is not zero, and the string overshoots or continues upward. Now the free end of the string exerts a force on the remainder of the string and produces a reflected wave of exactly the same shape as the incident wave. No inversion takes place, and the phase is unchanged.

Once we know how a pulse behaves at a boundary, we can consider the effects of boundaries on periodic waves. If a sinusoidal wave reaches the fixed end of a string, the reflected wave will be the inverted mirror image of the incident wave. This means that two waves are present, traveling in opposite directions (Fig. 23.10). The incident and reflected waves interfere to produce a standing wave with a node at the end. The reflection of sinusoidal waves at a free end is shown in Fig. 23.11. It has an antinode at the end.

In summary, the reflection of sinusoidal waves at a boundary results in standing waves. On a string, a node appears at a fixed end and an antinode at a free end. Analogous standing waves are formed for all kinds of waves when reflections occur.

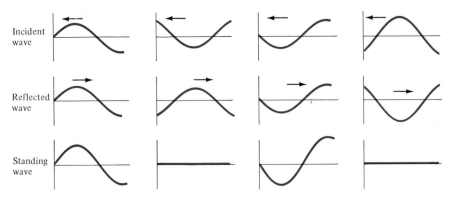

Figure 23.10.   Four views a quarter cycle apart of the incident, reflected, and resulting standing waves for a fixed end string.

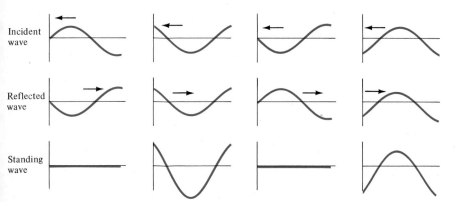

Figure 23.11. The analogous quantities for a string with a free end. Note that there is an antinode at the boundary of a free end string.

23.5 | RESONANT STANDING WAVES

In Chapter Ten on harmonic motion, we found that many structures have specific resonant frequencies at which large-amplitude vibrations are readily produced. Musical instruments and the voice involve systems that combine vibrating strings or air columns with structures that have certain resonant frequencies.

The fact that strings have specific resonant frequencies can be seen by tying one end of a rope to a post and shaking the other end. After a few tries, one finds standing waves are produced only at specific frequencies. At other frequencies the rope vibrates erratically and with a small amplitude.

To understand why only certain frequencies result in standing waves, consider a string of length l. With both ends rigidly fixed, only standing waves with nodes at each end can be produced. Figure 23.12 shows the five longest standing waves for a fixed end string, that is, the longest waves with nodes at both ends. The waves that fit on the string are called the *harmonics* of the string. The longest is called the *fundamental* or first harmonic; the first five harmonics of the fixed end string are shown in Fig. 23.12.

In Fig. 23.12 the wavelengths can be labeled with a subscript n, where n is a positive integer. For the nth harmonic, we see that

$$\lambda_n = \frac{2l}{n},$$
$$n = 1, 2, 3, \ldots \text{ (fixed end string)} \qquad (23.3)$$

The corresponding frequencies are found from

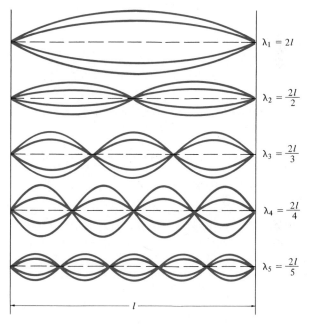

Figure 23.12. The five longest wavelength standing waves on a string of length l with fixed ends.

$f_n\lambda_n = c$, where c is the wave velocity on the string. Thus on a string of length l

$$f_n = \frac{n}{2l} c,$$
$$n = 1, 2, 3, \ldots \text{ (fixed end string)} \qquad (23.4)$$

Since $c = \sqrt{T/\mu}$ for a string, we also have

$$f_n = \frac{n}{2l}\sqrt{\frac{T}{\mu}}$$
$$n = 1, 2, 3, \ldots \text{ (fixed end string)} \qquad (23.5)$$

393

The following examples illustrate the use of these results.

Example 23.4

What are the frequencies of the first three harmonics of the longest string in a grand piano? The length is 1.98 m, and the velocity of the wave on the string is $c = 130$ m s^{-1}.

Using the result for the fixed end string with $n = 1$,

$$f_1 = \frac{c}{2l} = \frac{130 \text{ m s}^{-1}}{2(1.98 \text{ m})} = 32.8 \text{ Hz}$$

The second harmonic is at twice this frequency, or $2(32.8) = 65.6$ Hz, and the third harmonic is at $3(32.8) = 98.4$ Hz.

Example 23.5

The wave velocity on the highest frequency violin string is 435 m s^{-1}, and its length l is 0.33 m. If a violin player lightly touches the string at a point a distance $l/3$ from an end, a node is produced there. What is the lowest frequency that can now be produced by the string?

Referring to Fig. 23.12, the third harmonic is the longest wavelength and lowest frequency standing wave with a node one-third of the distance from one end. Hence with $\lambda_3 = 2l/3$,

$$f_3 = \frac{c}{\lambda_3} = \frac{c}{\left(\dfrac{2l}{3}\right)} = \frac{3c}{2l} = \frac{3(435 \text{ m s}^{-1})}{2(0.33 \text{ m})} = 1977 \text{ Hz}$$

A violinist can, by lightly touching a string at some point along its length, produce a node on the string at that point so that only harmonics with nodes at that point are excited. However, if the string is pushed firmly against the finger board, its effective length is reduced, and all the harmonic frequencies are shifted upward.

The relationship between the frequency and mass per unit length of a string is discussed in the following example.

Example 23.6

The highest and lowest frequency strings of a piano are tuned to fundamentals of $f_H = 4186$ Hz and $f_L = 32.8$ Hz. Their lengths are 0.051 m and 1.98 m, respectively. If the tension in these two strings is the same, what is the ratio of the masses per unit length of the two strings?

Using Eq. 23.5 and solving for μ, we find for the fundamental, $n = 1$,

$$\mu = \frac{T}{(2lf_1)^2}$$

The ratio of μ_L for the low frequency string to μ_H for the high frequency string is

$$\frac{\mu_L}{\mu_H} = \frac{\dfrac{T}{(2l_L f_L)^2}}{\dfrac{T}{(2l_H f_H)^2}} = \frac{(l_H f_H)^2}{(l_L f_L)^2}$$

$$= \left[\frac{(0.051 \text{ m})(4186 \text{ Hz})}{(1.98 \text{ m})(32.8 \text{ Hz})}\right]^2 = 10.8$$

This is a large difference. It was found empirically by piano makers that a thick solid string does not produce a pleasing tone. This is apparently due to the great stiffness of such strings. The solution developed was to wind a wire in a tight coil about a straight wire (here of length 1.98 m). This increases the mass per unit length without changing the stiffness of the wire very much.

23.6 | COMPLEX WAVES AND BEATS

Often when a string is disturbed or standing waves are set up in air columns, not only is the fundamental frequency present but so also are some of the higher harmonics. The higher harmonics that are actually present are called *overtones*. If all the harmonics are present, the second harmonic is the first overtone, the third harmonic is the second overtone, and so on. In the next chapter, we discuss situations where every other harmonic is present. In this case, the third and fifth harmonics are the first and second overtones, respectively. The higher overtones are numbered similarly.

The higher harmonics are not all excited with the same amplitude; instead the higher harmonics usually have progressively smaller amplitudes. If the fundamental and several harmonics are simultaneously present, the actual shape of the string at any instant may be quite complex (Fig. 23.13). Any com-

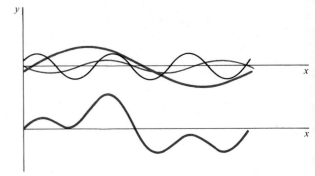

Figure 23.13. The addition of three harmonics of a fixed end string results in the string having the complex shape shown.

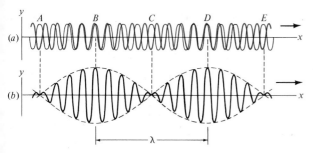

Figure 23.14. (a) Two waves of slightly different frequencies, f_1 (shown colored) and f_2, add to form the resultant wave in (b). The dashed lines show how the amplitude changes.

plex wave can be regarded as a sum of many sinusoidal waves.

In Fig. 23.14, we show two traveling waves with slightly different wavelengths and frequencies. At points B and D the two waves add constructively. At the intermediate point C the waves interfere destructively. The resultant wave is a rapid oscillation that changes amplitude with time. At points A, C, and E the amplitude of the total wave is zero, and at points B and D the amplitude is a maximum. This phenomena is called *beating*. The frequency with which the nodes pass a given point on the x axis is called the *beat frequency*.

We can relate the beat frequency f to the frequencies f_1 and f_2 of the two waves. Since the two waves in Fig. 23.14a and the resultant wave in Fig. 23.14b all travel with a velocity c, we have, from $f\lambda = c$,

$$f\lambda = f_1\lambda_1 = f_2\lambda_2 \qquad (23.6)$$

Here, λ is the wavelength associated with the overall change in amplitude of the total wave; λ_1 and λ_2 are the wavelengths of the waves of frequency f_1 and f_2. Now for the particular case shown in Fig. 23.14a, we see that nine wavelengths of the colored wave fit into the distance λ. Thus $\lambda = 9\lambda_1$. Similarly, $\lambda = 8\lambda_2$. Using these in Eq. 23.6, we have

$$f(9\lambda_1) = f_1\lambda_1$$

and

$$f(8\lambda_2) = f_2\lambda_2$$

λ_1 and λ_2 cancel out of these equations. We then subtract them and find $9f - 8f = f_1 - f_2$, or

$$f = f_1 - f_2 \qquad (23.7)$$

This result holds true quite generally and shows that the beat frequency is just equal to the difference in frequencies of the two waves producing the effect.

The human ear is sensitive to beats from nearly 0 to about 20 Hz, and the waves need not have exactly equal amplitudes. The following example illustrates how beats may be put to practical use.

Example 23.7

A tuning fork with a known frequency of 520 Hz is sounded at the same time that a violin string is plucked. Beats are heard with a frequency of 5 Hz. What is the frequency of the violin string?

From our discussion, we note that the difference between the frequencies of the tuning fork and violin string is 5 Hz. However, we cannot tell immediately which frequency is higher and which is lower. Since the fork has a frequency of 520 Hz, the violin string vibrates at either 525 Hz or 515 Hz.

We could determine which is the case by adjusting the tension in the violin string. From Eq. 23.5, the string frequency is proportional to the square root of the tension. If reducing the tension reduces the beat frequency, we know the violin was at 525 Hz. If the string were at 515 Hz, the tension would have to be increased to reduce the beat frequency.

Use is made of beats in tuning a piano. Most piano keys strike three strings that are supposed to produce identical notes. A modern piano tuner stops two of the three strings from vibrating and adjusts the tension of the third until it is synchronized with an electronic oscillator set at the desired frequency. Then the other two strings are released one at a time and adjusted until no beats are heard. A single note played on a poorly tuned piano has beats that are too rapid to be heard separately, but the sound is discordant and may be unpleasant. This is the sound produced by a honky-tonk piano.

Modern methods for studying large molecules with light use beats in a sophisticated way. A laser beam is split into two parts. One portion passes through a sample chamber, and its frequency is modified by the interaction with the molecules. Since the two portions of the beam now have different frequencies, they produce beats when recombined. Information about the molecules is obtained from the frequency changes.

23.7 | ENERGY AND MOMENTUM IN WAVES

Although it is not always apparent, waves of every kind carry energy and momentum obtained originally

395

from their sources. This may be illustrated by some familiar examples. Light from the sun provides the energy that makes life possible on our planet. Ocean waves gradually transform the coastlines, and they can exert large forces on someone standing in shallow water. Intense sound waves can crack windows and cause other kinds of damage to mechanical structures including the human ear.

The specific expressions for the energy and momentum of a wave depend on the nature of the particular wave. However, it is always true that the energy and momentum stored in a sinusoidal wave are proportional to the square of its amplitude.

This fact can be understood if we make an analogy with the simple harmonic oscillator (Chapter Ten). If we observe a sinusoidal wave passing a given point, the displacement at that point varies with time as a sine or cosine. A simple harmonic oscillator, such as a mass on a spring, also has a displacement that varies sinusoidally with time. Hence, each point in a sinusoidal wave is undergoing simple harmonic motion. In an oscillator, the potential energy $\frac{1}{2}kx^2$ is proportional to the displacement squared and hence to the amplitude squared. The kinetic energy, $\frac{1}{2}mv^2$, and the total energy are also proportional to the amplitude squared. Accordingly, at any point in a sinusoidal wave, the energy stored is proportional to the amplitude squared. If the wave is a standing wave, no

energy will be transported; the stored energy will be transformed from potential energy to kinetic energy and back again repeatedly, but it will remain in the same location. However, if the wave is moving in some direction, it will transport energy in that direction.

Discussions of sound and light waves frequently use the *intensity I* of the wave. This is the power transported across a unit cross sectional area. In S.I. units, the intensity is measured in watts per square meter. From the discussion above it follows that *the intensity of a wave is proportional to its amplitude squared,* or

$$I \propto A^2 \tag{23.8}$$

The momentum carried by the wave also varies as the amplitude squared.

23.8 | THE POLARIZATION OF TRANSVERSE WAVES

Transverse waves have a property called *polarization* that is not shared by longitudinal waves. For example, in Fig. 23.15, a wave on a string is traveling along the x direction. The transverse wave disturbance can be along any line perpendicular to this direction of motion. If the disturbance is always along the same line, the wave is said to be *polarized* along that line. A

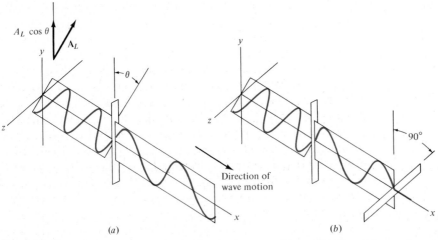

Figure 23.15. (*a*) A wave on a string is vibrating in some arbitrary direction. After passing through the slot, the wave is polarized in the *y* direction and has a smaller amplitude.
(*b*) With the second slot at right angles to the first, the wave is completely suppressed. If the two slots were not perpendicular, the final wave would be polarized in the direction of the second slot.

wave whose direction of oscillation changes randomly from time to time or which is composed of many waves of random polarization is said to be *unpolarized*.

An unpolarized wave can be polarized by a *polarizer*. For the string shown in Fig. 23.15a, the polarizer is a slot. Beyond the slot the string can only vibrate in the y direction. In Fig. 23.15b, there is also a second slot. When the two slots are perpendicular the wave is totally suppressed.

The amplitude of the wave after passing through a polarizing structure is reduced, since only the *component* of the original wave parallel to the polarizer passes through. In Fig. 23.15a the amplitude of the wave on the left side of the slot is A_L and the intensity is $I_L = A_L^2$. The angle between the planes of polarization on the left side and right side of the slot is θ. The amplitude of the wave on the right side of the slot is $A_R = A_L \cos \theta$. The intensity is then

$$I_R = I_L \cos^2 \theta$$

In Fig. 23.15b the angle between the two slots is $90°$, and since $\cos 90° = 0$, the amplitude becomes zero.

Electromagnetic waves are transverse waves. We refer to the direction of the electric vector as the direction of polarization. Polarized sunglasses use the transverse property of light to reduce glare. Light that is reflected from smooth horizontal surfaces tends to be polarized in the horizontal direction. Polaroid sunglass lenses almost totally suppress the component of light along one direction and are oriented so that only the vertically polarized component is transmitted. If one rotates a Polaroid lens relative to a second, the transmitted light is seen to decrease nearly to zero when the polarizing axes are perpendicular. A more extensive discussion of light polarization is given in Chapter Twenty-five.

SUMMARY

Waves of all kinds share many common features. Periodic waves are characterized by their frequency f, wavelength λ, and velocity c, which are related by $f\lambda = c$. The velocity depends on the properties of the medium and in some cases on the frequency. The amplitude of a wave is the maximum magnitude of its displacement.

Waves can interfere with one another. When two waves are present at a point, the resulting wave is found by algebraically adding the displacements of the individual waves. Two waves in phase add constructively, while two waves a half wavelength out of phase interfere destructively. The property of combining waves by the addition of displacements is called the principle of superposition or linearity.

Waves incident on a boundary can be transmitted, absorbed, or reflected. Waves that reach a fixed end of a string are reflected as inverted mirror images. At a free end the reflected wave is not inverted.

Standing waves are formed when two waves of equal amplitude but opposite direction interfere. Standing waves on a string can persist only at certain special frequencies, called harmonics, which are determined by the length of the string and the velocity of the wave.

The energy and momentum in a sinusoidal wave are proportional to the square of its amplitude. The intensity of a wave is the power transported per unit area and also varies as the amplitude squared.

Checklist

Define or explain:

wave pulse	standing waves
periodic wave	phase
longitudinal waves	boundary
wavelength	fundamental frequency
amplitude	harmonics
sinusoidal waves	overtones
wave velocity	beat frequency
superposition	intensity
destructive interference	polarized wave
node	

EXERCISES

Section 23.1 | The Representation of Waves

23-1 What is the wavelength of a sound wave with a frequency of 1000 Hz and a velocity 344 m s^{-1}?

23-2 What is the frequency of a wave of velocity 200 m s^{-1} and wavelength 0.5 m?

23-3 A radar antenna emits electromagnetic radiation ($c = 3 \times 10^8$ m s^{-1}) with a wavelength of 0.03 m for 0.5 s. (a) What is the frequency of radiation? (b) How many complete waves are emitted in this time interval? (c) After 0.5 s, how far is the front of the wave from the antenna?

23-4   A TV station broadcasts using 2-m waves. What is the frequency of the broadcast wave if the speed of the wave is 3×10^8 m s^{-1}?

Section 23.2 | The Velocity of Waves

23-5   A string on a steel guitar has a mass per unit length of 3×10^{-3} kg m^{-1}. If the tension in the string is 90 N, what is the velocity of a wave on the string?

23-6   The tension in a string is four times that in a second identical string. What is the ratio of the wave velocities of the strings?

Section 23.3 | Wave Interference and Standing Waves

23-7   Two wave pulses on a string (Fig. 23.16) travel toward each other with speeds of 1 m s^{-1}. Sketch the shape of the string at $t = 1$, 1.25, and 1.5 s after the instant shown.

23-8   Two wave pulses travel toward each other on a string (Fig. 23.17). If the wave velocity is 2 m s^{-1}, sketch the string 1, 1.25, and 1.5 s after the instant shown.

Figure 23.16.   Exercise 23-7.

Figure 23.17.   Exercise 23-8.

Figure 23.18.   Exercise 23-9.

Figure 23.19.   Exercises 23-10 and 23-11.

23-9   Sketch the string of Fig. 23.18, 1, 1.25 and 1.5 seconds after the instant shown. The wave velocity is 1 m s^{-1}.

Section 23.4 | The Effects of Boundaries

23-10   Sketch the shape of a square-wave pulse on a string (Fig. 23.19) if it is totally reflected from a fixed boundary.

23-11   Sketch the shape of the square-wave pulse on a string of Fig. 23.19 if it is totally reflected from a free end boundary.

Section 23.5 | Resonant Standing Waves

23-12   The lowest frequency string of a violin is 0.33 m long and is under a tension of 55 N. The fundamental frequency is 196 Hz. What is the mass per unit length of the string?

23-13   The E string of a violin has a length of 0.33 m and a fundamental frequency of 659 Hz. The tension of the string is 55 N. (a) What is the wave velocity of the string? (b) What is the mass per unit length of the string?

23-14   If the heaviest and lightest strings of a violin have masses per unit length of 3×10^{-3} kg m^{-1} and 2.9×10^{-4} kg m^{-1}, what is the ratio of the radii of these strings?

23-15   The fundamental frequency of a string with fixed ends is 100 Hz and the wave velocity is 350 m s^{-1}. (a) What is the wavelength of the fundamental? (b) What is the length of the string?

23-16 If the wave velocity of a 0.5-m long guitar string is 170 m s^{-1}, what is its fundamental frequency?

22-17 The highest C string of a piano has a frequency of 4186 Hz and a length of 0.051 m. The lowest C string has a frequency of 32.8 Hz. If these strings had the same tension and mass per unit length, what would be the length of the lowest C string?

23-18 The A string on a violin is 0.33 m long and tuned to a fundamental frequency of 440 Hz. How far from the end of the string should one press it against the finger board to obtain the same fundamental frequency, 659 Hz, as the E string?

23-19 A harp string of length 0.5 m is tuned to a fundamental frequency of 650 Hz. (a) What is the wavelength of the fourth harmonic of the string? (b) What is the wavelength of the sound produced in the air if the fourth harmonic is excited? (Use $c = 344 \text{ m s}^{-1}$ in air.)

23-20 If a string of an instrument is touched lightly at a point one-third of its length from one end, a node is formed there. Which harmonics of the string can be excited?

Section 23.6 | Complex Waves and Beats

23-21 The first and second harmonics of a string are shown at a certain instant in Fig. 23.20. Sketch the actual shape of the string at this instant.

23-22 The fundamental frequency of the heaviest string on a cello is 65.4 Hz. What is the beat frequency of the third harmonic of this string with the 196 Hz fundamental of the heaviest string on a violin?

23-23 What beat frequency is heard if two tuning forks vibrate with frequencies $f_1 = 200$ Hz and $f_2 = 205$ Hz?

23-24 What are the possible frequencies of a tuning fork that produces a beat frequency of $f = 4$ Hz with a standardized tuning fork which has a frequency of 300 Hz?

Section 23.7 | Energy and Momentum in Waves, and Section 23.8 | The Polarization of Transverse Waves

23-25 A traveling wave on a rope has a vertical amplitude of 0.1 m. It passes through a slot which is tilted at 30° with the vertical. What is the amplitude of the wave after passing through the slot?

23-26 A traveling wave on a rope has a vertical amplitude of 0.2 m. It passes through two successive slots; the first is tilted at 45° with the vertical and the second is vertical. (a) What is the amplitude of the wave after it has passed through the first slot? (b) What is the amplitude of the wave after it has passed through both slots?

23-27 What is the ratio of the energy of the wave of Exercise 23-26 after it has passed through both slots to its original energy?

PROBLEMS

23-28 Using dimensional arguments, show that the velocity of a wave on a string with tension T and mass per unit length μ is proportional to $\sqrt{T/\mu}$.

*23-29 The restoring force on long wavelength surface water waves is due to gravity as long as the water is deep and the bottom does not affect the wave. Using dimensional arguments and explaining your reasoning, show that the velocity of long water waves of wavelength λ is of the form

$$c = A\sqrt{g\lambda}$$

where g is the gravitational acceleration, and A is a numerical constant. (Exact calculation shows that $A = \sqrt{1/8\pi}$.)

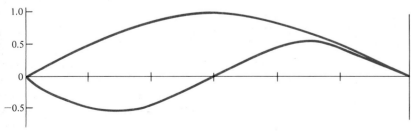

Figure 23.20. Exercise 23-21.

*23-30 Using dimensional reasoning, show that for very short surface waves of wavelength λ on water the wave velocity depends on the surface tension and has the form

$$c = B \sqrt{\frac{\gamma}{\rho\lambda}}$$

Here γ is the surface tension, ρ is the density, and B is a numerical constant. (Exact calculations yield $B = \sqrt{9\pi/2}$.)

23-31 Using the results of Problem 23-29, what is the velocity of a water wave of wavelength 30 m?

23-32 The general expression for the velocity of surface water waves of wavelength λ is

$$c = \frac{1}{2} \sqrt{\frac{g\lambda}{2\pi} + \frac{18\pi\gamma}{\rho\lambda}}$$

where ρ is the density, g the acceleration of gravity, and γ is the surface tension. At what wavelength λ does the transition from long to short waves take place? (This result is only valid if the water is very deep compared to the wavelength and crest height.)

23-33 How long will it take a 1-m bow wave from a boat to reach the shore of a river 25 m away? (See Problem 23-29.)

23-34 The 0.33-m long A string of a violin is tuned to a fundamental frequency of 440 Hz. The 0.69-m long A string of a cello is tuned to a fundamental of 220 Hz. If both strings have the same tension, what is the ratio of the masses per unit length of these strings?

23-35 The E string of a violin is 0.33 m long, and the wave velocity of the string is 435 m s^{-1}. (a) What is the time required for the wave produced by plucking the string to make one complete circuit along the string and return to its original position? (b) Does this result have any bearing on the proper frequency of plucking or bowing the string?

23-36 If one of the three strings corresponding to middle C on a piano is adjusted to the proper frequency of 261.6 Hz and is struck simultaneously with an untuned string so that a beat frequency of 10 Hz is heard, (a) what two frequencies might the untuned string have? (b) If the tension in the tuned string is 1160 N, what are the possible tensions in the untuned string?

23-37 The amplitude of the horizontally polarized light on a beach is twice that of the vertical component. If a woman standing erect puts on Polaroid sunglasses, only a negligible portion of the horizontally polarized light reaches her eyes. (a) What is the percentage reduction of the light energy that reaches her eyes when she puts on the sunglasses? (b) If the woman is lying on her side, what is the percentage reduction in energy that reaches her eyes when she puts on the sunglasses? (*Hint:* The total amplitude of a wave in the vector sum of its horizontal and vertical components.)

SUPPLEMENTARY TOPICS

23.9 | THE DOPPLER EFFECT

When a train passes by, the frequency at which its whistle is heard drops suddenly. This *Doppler effect* arises because as the source approaches a listener, more waves arrive each second than are emitted; conversely, as the source recedes, fewer waves arrive each second than are emitted. A similar effect is heard when the listener moves and the source is fixed.

All types of waves exhibit a Doppler effect, although the results for light waves are somewhat different than those for other types of waves. In this section, we describe the Doppler effect for sound and other mechanical waves in detail and merely state the results for light waves.

Consider first a sound source stationary relative to the air with frequency f and wavelength $\lambda = c/f$, and a listener moving away from the source at a velocity v_l (Fig. 23.21). The speed of the sound waves will appear to the listener to be $c' = c - v_l$. The distance between two successive wave maxima in the listener's moving frame of reference is the same as for the stationary frame of the source, λ. The apparent frequency f' heard by the listener then satisfies $f'\lambda = c'$, or $f' = c'/\lambda$. With $\lambda = c/f$ and $c' = c - v_l$, this becomes

$$f' = f\left(\frac{c - v_l}{c}\right) \qquad (23.9)$$

Thus the frequency heard is lower than the source frequency when the listener is moving away from the source. When the listener is approaching the source, v_l in Eq. 23.9 must be regarded as negative, and the frequency heard is greater than the source frequency.

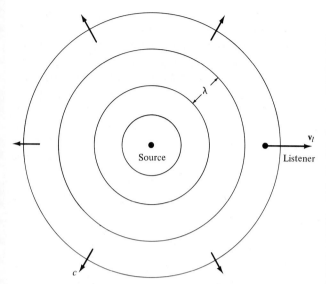

Figure 23.21. Waves leave the stationary source with a speed c but appear to the moving listener to have a speed of $c - v_l$.

(a)

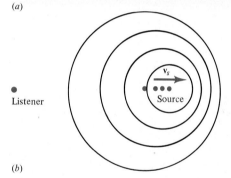

(b)

Figure 23.22. *(a)* A moving source of waves in a water-filled ripple tank produces waves that are bunched together in the direction of motion and spread out in the direction opposite the motion. *(b)* With a sound source moving with velocity v_s, a listener hears sound waves that have a longer wavelength than those from a source at rest. ((a) From Education Development Center, Inc., Newton, Massachusetts.)

The following example illustrates this type of Doppler effect.

Example 23.8

A stationary civil defense siren has a frequency of 1000 Hz. What frequency will be heard by drivers of cars moving at 15 m s⁻¹ (a) away from the siren; (b) toward the siren? The velocity of sound in air is 344 m s⁻¹.

(a) With $v_l = 15$ m s⁻¹, drivers moving away from the siren hear a frequency

$$f' = f\frac{(c - v_l)}{c} = (1000 \text{ Hz}) \left(\frac{344 \text{ m s}^{-1} - 15 \text{ m s}^{-1}}{344 \text{ m s}^{-1}}\right)$$

$$= 956 \text{ Hz}$$

(b) Using $v_l = -15$ m s⁻¹ for drivers moving towards the source

$$f' = (1000 \text{ Hz}) \left(\frac{344 \text{ m s}^{-1} + 15 \text{ m s}^{-1}}{344 \text{ m s}^{-1}}\right)$$

$$= 1044 \text{ Hz}$$

A different expression is obtained for a source moving at a velocity v_s away from a listener who is at rest relative to the air (Fig. 23.22). The sound waves are bunched together in the direction of motion and spread out in the opposite direction. If f_0 is the frequency at which the waves are emitted, then in the direction opposite to the motion, the distance between successive crests is increased from $\lambda = c/f$ to

$$\lambda' = \frac{c + v_s}{f}$$

The frequency heard by the stationary listener is determined by $f'\lambda' = c$, so

$$f' = \frac{c}{\lambda'} = f\left(\frac{c}{c + v_s}\right). \qquad (23.10)$$

Thus, when the source is moving away from the listener, the frequency heard is smaller than the source frequency. If the source is approaching the listener, v_s is negative, and the observed frequency is greater than f. The following example illustrates this effect.

Example 23.9

A police car with a 1000-Hz siren is moving at 15 m s^{-1}. What frequency is heard by a stationary listener when the police car is (a) receding from and (b) approaching the listener?

(a) With $c = 344 \text{ m s}^{-1}$ and $v_s = 15 \text{ m s}^{-1}$ for the receding source,

$$f' = f\left(\frac{c}{c + v_s}\right)$$

$$= (1000 \text{ Hz})\frac{(344 \text{ m s}^{-1})}{(344 \text{ m s}^{-1} + 15 \text{ m s}^{-1})}$$

$$= 958 \text{ Hz}$$

Note this is a slightly higher frequency than our earlier result of 956 Hz for a listener receding from a stationary siren with the same frequency and speed.

(b) When the source is approaching the listener, $v_s = -15 \text{ m s}^{-1}$, and

$$f' = f\left(\frac{c}{c + v_s}\right)$$

$$= (1000 \text{ Hz})\frac{(344 \text{ m s}^{-1})}{(344 \text{ m s}^{-1} - 15 \text{ m s}^{-1})}$$

$$= 1046 \text{ Hz}$$

Again this frequency is slightly greater than the frequency of 1044 Hz heard by a listener approaching a stationary siren.

Doppler Flow Meter | The Doppler frequency shift can be used to measure blood flow rates by using a high frequency sound source on one side of a vessel and a detector on the other side (Fig. 23.23). The sound from the transmitter is reflected from red cells moving away from the source with a velocity v_c and

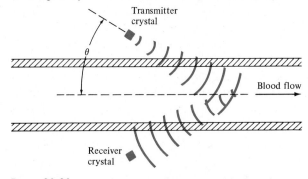

Figure 23.23. A schematic view of the use of the Doppler flow meter.

detected at the receiver. If θ is small, the frequency of the sound incident on the red cells moving away from the source is given by Eq. 23.9,

$$f_1 = f\left(\frac{c - v_c}{c}\right)$$

c is the speed of sound in the bloodstream. The cells then act as a moving source with a frequency f_1. From Eq. 23.10 the frequency f_r at the stationary receiver is then

$$f_r = f_1\left(\frac{c}{c + v_c}\right) = f\left(\frac{c - v_c}{c}\right)\left(\frac{c}{c + v_c}\right)$$

$$f_r = f\left(\frac{c - v_c}{c + v_c}\right) \qquad (23.11)$$

Doppler Shift for Light | A Doppler shift is also observed with light waves, but it differs somewhat from that for sound waves. This arises from the fact that no medium is necessary for light wave propagation. Therefore, there is no way to distinguish between motion of an observer relative to a stationary source or motion of a source relative to a stationary observer. One can only say that they are moving with respect to each other. This is called the *principle of relativity* and is discussed further in Chapter Twenty-seven.

When the analysis is carried out it is found that if a light source of frequency f and an observer move away from each other with a relative velocity u, then the observed frequency is

$$f' = f\frac{c - u}{\sqrt{c^2 - u^2}} \qquad (23.12)$$

The velocity u should be interpreted as negative when the observer and source approach each other. This result is equivalent to our earlier result, Eq. 23.9, when u^2/c^2 is very small.

EXERCISES ON SUPPLEMENTARY TOPICS

Section 23.9 | **The Doppler Effect**

23-38 An automobile moves at 20 m s^{-1} toward a factory whistle that has a frequency of 1000 Hz. What is the frequency heard by a passenger?

23-39 A locomotive is moving at 40 m s^{-1}. Its whistle has a frequency of 2000 Hz. What is the frequency heard by a stationary observer when the locomotive is (a) approaching, and (b) moving away from the observer?

23-40   A source moves away from a stationary listener. If the frequency heard is 8 percent lower than the source frequency, what is the speed of the source?

23-41   The average velocity of blood flow in the aorta during systole is 1.5×10^{-2} m s^{-2}. What is the frequency shift in a Doppler flow meter with a source frequency of 10^5 Hz? (The speed of sound in blood is 1570 m s^{-1}.)

23-42   A bat emits squeaks of short duration at a frequency of 80,000 Hz. If the bat flies towards an obstacle at a speed of 20 m s^{-1}, what is the frequency of the reflected wave detected by the bat?

23-43   A bat travels towards a stationary obstacle. It emits sounds at a frequency of 50,000 Hz and detects a reflected sound of 51,000 Hz. How fast is the bat flying?

23-44   One of two identical 500-Hz tuning forks is at rest and the other is moving. These produce a beat frequency of 5 Hz. (a) What is the speed of the moving fork? (b) Do you have enough information to determine whether the moving fork moves toward or away from the stationary fork? What additional information, if any, might be useful in this determination?

23-45   A galaxy is moving away from us with a speed of 3×10^7 m s^{-1}. What is the frequency of the light we would observe if it is emitted at 6×10^{14} Hz?

23-46   Light from distant galaxies that are moving away from us is said to be red-shifted; the wavelength observed is at longer wavelengths than that emitted by the galaxy. What is the observed wavelength shift for the galaxy of the preceding exercise?

PROBLEMS ON SUPPLEMENTARY TOPICS

23-47   A typical red blood cell has a radius of 5×10^{-6} m. Doppler shift flow meters depend on reflection from red cells and utilize ultrasonic frequencies. (a) If the source frequency is 10^7 Hz, how many red cells will "fit" into one wavelength of the sound? (b) Why are high frequencies necessary?

23-48   A tuning fork of frequency 500 Hz moves away from a stationary listener and toward a stationary wall at a speed of 2 m s^{-1}. (a) What is the frequency of the direct sound heard by the listener? (b) What is the detected frequency of the reflected wave? (c) How many beats per second are heard by the listener?

** **23-49**   The frequency of sound reflected from an object moving away from a detector was found in the text to be (Eq. 23.10)

$$f_r = f_0 \left(\frac{c - v_c}{c + v_c} \right)$$

(a) Show that if $v_c \ll c$, then

$$\Delta f = f_r - f_0 \simeq -\frac{2 f_0 v_c}{c}$$

(b) If the object moves toward the detector, show that

$$\Delta f = f_r - f_0 \simeq \frac{2 f_0 v_c}{c}$$

23-50   Show that the beat frequency detected when the reflected wave from an object moving toward or away from the source is combined with the unshifted wave of frequency f_0 is

$$f_B = \frac{2 f_0 v}{c}$$

(Assume v, the speed of the reflecting object, is small compared to c.)

23-51   The average velocity of blood flow in a dog's artery with an inner radius of 4.0×10^{-3} m is 2.3×10^{-2} m s^{-2}. (a) What is the average frequency of the sound detected in a Doppler shift flow meter if the source frequency is 10^5 Hz? (b) What is the blood flow rate? (The speed of sound in blood is 1570 m s^{-1}.)

23-52   An average frequency shift of 100 Hz is detected in a Doppler shift flow meter with a source frequency of 5×10^6 Hz. What is the average blood flow velocity in the vessel being studied? (The speed of sound in blood is 1570 m s^{-1}.)

25-53   If one approaches a red light of wavelength 6.5×10^{-7} m, at a speed of 10^8 m s^{-1}, what is the observed wavelength?

23-54   Show that if $u^2/c^2 \ll 1$ that the formula for the Doppler effect for light, Eq. 23.12, is the same as Eq. 23.9 for sound.

Additional Reading:

Sources marked with asterisks (*) are appropriate references for this and the next chapter.

R.A. Waldron, *Waves and Oscillations,* Momentum Series, D. Van Nostrand and Co., Princeton, N.J., 1964.

Willard Boscom, *Waves and Beaches: The Dynamics of the Ocean Surface,* Science Study Series, Doubleday and Co., Garden City, N.Y., 1964.

* Arthur H. Benade, *Horns, Strings and Harmony,* Anchor Books, Doubleday and Company, Inc., Garden City, N.Y., 1960.

* Alexander Wood, *The Physics of Music,* University Paperbacks, Chapman and Hall, London, 1975.

* Charles A. Culver, *Musical Acoustics,* McGraw-Hill Book Co., New York, 1957.

* C.A. Taylor, *The Physics of Musical Sounds,* American Elsevier Publishing Co., New York, 1965.

* John Backus, *The Acoustical Foundations of Music,* W.W. Norton and Co., Inc., New York, 1969.

Scientific American articles:

J. Bernstein, Tsunamis, August 1954, p. 60.

Frank Press, Resonant Vibrations of the Earth, November 1965, p. 28.

John R. Percy, Pulsating Stars, June 1975, p. 66.

Eugene Helm, The Vibrating String of the Pythagoreans, December 1967, p. 92.

* Carleen Maley Hutchings, The Physics of Violins, November 1962, p. 79.

* Arthur H. Benade, The Physics of Brasses, July 1973, p. 24.

J.T. Gosling and A.J. Huudhausen, Waves in the Solar Wind, March 1977, p. 36.

CHAPTER TWENTY-FOUR

SOUND

When a gas, liquid, or solid is mechanically disturbed, sound waves are often produced. In these waves the molecules of the substance vibrate and collide with one another but maintain the same average position. However, since their motions are coordinated, a wave results and energy is transmitted, even though no net particle displacement occurs.

The speed of sound depends on the physical properties of the substance in which it travels. When sound encounters a boundary between substances in which the sound speeds differ, some energy is transmitted and some is reflected. Thus a study of the characteristics of the production, propagation, detection, and uses of sound is necessarily a study of the transfer of mechanical energy.

Animals use sound for information exchange and for the detection and location of objects. Some bats and porpoises use sound for navigation and to locate food where inadequate light is present for vision. Humans also use sound as a substitute for light and even for x-rays. *Sonar* is used for underwater navigation and observation, and *ultrasonic* or high frequency sound is now commonly used for medical diagnosis and therapy. Very low frequency sound is also used in geophysical studies.

The first part of this chapter deals with some basic properties of sound, including the speed of sound and the energy carried by sound waves. This is followed by a description of sound sources and detectors, including the human voice and ear. These are examples of *transducers*, devices that convert energy from one form to another. The remainder of the chapter is devoted to some special examples of man's use of sound.

24.1 | THE NATURE AND SPEED OF SOUND

In our discussion of the physics of fluids, we found that it is impractical to apply Newton's laws of motion directly to small segments of fluid or individual molecules. Instead, we introduced the density and pressure and explored their relationships. These same variables are useful in characterizing a sound wave, which has millions of molecules in a single wavelength.

Figure 24.1 illustrates the production of a sound

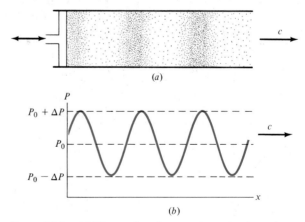

Figure 24.1. (*a*) The moving piston produces density and pressure changes which move to the right with a velocity *c*. (*b*) The graph of pressure versus position. P_0 is the equilibrium pressure.

wave in a medium by a piston that oscillates back and forth at a frequency f. When it moves forward, it compresses the medium, and a compressional wave moves outward. When the piston moves backward, there is a region of reduced pressure, a *rarefaction*. This disturbance also travels outward. Note that the individual molecules do not travel any appreciable distance but only oscillate about their average positions.

The speed c with which a sound wave travels in a medium is determined by the strength of the forces among the molecules. At the macroscopic level, these forces are characterized by the *bulk modulus K*, which was defined in Chapter Fourteen. This quantity relates the pressure change ΔP and the fractional density change $\Delta \rho / \rho$:

$$\Delta P = K \frac{\Delta \rho}{\rho} \qquad (24.1)$$

The velocity of sound c is found to depend only on the bulk modulus and the density ρ of the medium. Using dimensional analysis, one finds that c must be proportional to $\sqrt{K/\rho}$. A detailed derivation leads to the conclusion that the proportionality constant is 1, so

$$c = \sqrt{\frac{K}{\rho}} \qquad (24.2)$$

TABLE 24.1

Representative densities and sound velocities
(Temperature is 20°C unless otherwise noted)

Material	Density (kg m^{-3})	Sound Velocity (m s^{-1})
Air	1.05	344
Carbon dioxide (0°C)	1.98	258
Hydrogen (H$_2$) (0°C)	8.99×10^{-2}	1270
Alcohol (ethyl)	791	1210
Benzine	810	1166
Water (pure)	1000	1480
Aluminum	2700	5100
Copper	8900	3560
Glass	2500	5600
Iron	7900	5130
Oak	800	3850
Blood (37°C)	1056	1570
Body tissue (37°C) (average)	1047	1570

K is the adiabatic bulk modulus, not the isothermal bulk modulus that is tabulated in Table 14.2, because, while there are temperature changes that accompany the compressions and rarefactions, these are so rapid that little heat flow can occur, and hence the motion is adiabatic. Representative values of ρ and c are given in Table 24.1.

The following example uses the difference in sound velocities in air and in iron.

Example 24.1

Two children are at opposite ends of an iron pipe. One strikes an end of the pipe with a stone. What is the ratio of times it takes the sound waves in air and in iron to reach the second child?

The time necessary for a sound wave of velocity c to travel the length of the pipe d is found from $d = ct$ or $t = d/c$. Using Table 24.1, the ratio of the times for air and iron is

$$\frac{t_{air}}{t_{iron}} = \frac{\dfrac{d}{c_{air}}}{\dfrac{d}{c_{iron}}} = \frac{c_{iron}}{c_{air}} = \frac{5130 \text{ m s}^{-1}}{344 \text{ m s}^{-1}} = 14.9$$

Note that the sound travels faster in the denser material. Sound velocities are generally larger in solids and liquids than in gases.

As in all other waves, the velocity of sound, the frequency, and the wavelength are related by $f\lambda = c$. Typical wavelengths for sound waves in air are calculated in the next two examples.

Example 24.2

A typical young adult has a hearing range from 20 Hz to 20,000 Hz. What are the wavelengths of sound waves in air corresponding to these two frequencies?

For air, $c = 344$ m s^{-1}, so with $f\lambda = c$ and $f = 20$ Hz,

$$\lambda = \frac{c}{f} = \frac{344 \text{ m s}^{-1}}{20 \text{ s}^{-1}} = 17.2 \text{ m}$$

At 20,000 Hz,

$$\lambda = \frac{344 \text{ m s}^{-1}}{20{,}000 \text{ s}^{-1}} = 0.0172 \text{ m} = 1.72 \text{ cm}$$

Example 24.3

A bat can hear sound at frequencies up to 120,000 Hz. What is the wavelength of sound in air at this frequency?

Again we use

$$\lambda = \frac{c}{f} = \frac{344 \text{ m s}^{-1}}{120{,}000 \text{ s}^{-1}} = 2.87 \times 10^{-3} \text{ m}$$

$$= 0.287 \text{ cm}$$

One may ask why a bat utilizes sound waves of such high frequencies and short wavelengths. This is because a wave will be disturbed only by objects comparable to or larger than a wavelength; it will pass by smaller objects with little effect. A bat is nearly blind and avoids obstacles and finds food using sound waves. The bat emits a series of high-frequency squeaks and senses the time it takes for the waves to return after being reflected by an object. The wavelength must be short enough so that reflections can occur from small objects.

Porpoises are sensitive to frequencies up to 2×10^5 Hz and use a similar system for underwater navigation and location.

24.2 | STANDING SOUND WAVES

In the preceding chapter, we discussed standing waves on strings. However, standing waves can occur for any type of wave. In particular, standing waves are an important feature of sound-producing instruments.

We first consider sound waves in long narrow tubes or air columns. The standing waves for sound are set up exactly as are those on a string: periodic waves are reflected at a boundary, and the incident and reflected waves combine to form a standing wave.

Sound waves can be reflected at either open or closed ends of air columns. At the closed end of an air column, the molecules cannot oscillate normally; those adjacent to the end do not move at all, and a node appears at a closed end. At an open end, part of the wave is transmitted and part reflected. Since molecules at the open end can move freely, the open end is an antinode of the standing wave. The longest wavelength standing waves for two types of air columns are shown in Fig. 24.2.

For a column with two open ends, there is an antinode at each end. The wavelengths λ_n are related to the length of the column by $\lambda_n = 2l/n$, where n is an integer (Fig. 24.2a). From $f_n \lambda_n = c$, we find that the standing waves have frequencies

$$f_n = \frac{c}{\lambda_n} = \frac{nc}{2l},$$

$$n = 1, 2, 3, \ldots \quad \text{(two open ends)} \tag{24.3}$$

These are the same as the standing wave frequencies on a string fixed at both ends. For the string, however,

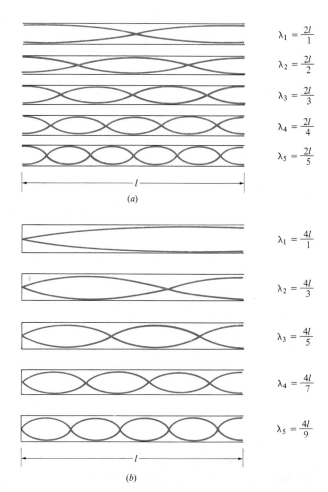

Figure 24.2. The first five resonant frequencies of an air column (a) with both ends open; (b) one end open.

the ends are nodes rather than antinodes.

For a column with one end closed and one end open $\lambda_n = 4l/(2n - 1)$, where n is an integer. (Note that $2n - 1$ is always an odd integer). The frequencies for this column are

$$f_n = \frac{c}{\lambda_n} = \frac{(2n - 1)c}{4l},$$

$$n = 1, 2, 3, \ldots \quad \text{(one open end)} \tag{24.4}$$

Only odd harmonics, odd multiples of the fundamental $f_1 = c/4l$, are possible for this type of column. The lowest frequencies for the two types of air column are illustrated in Fig. 24.3. These are the fundamentals and overtones for these columns.

The clarinet is an example of a standing wave instrument.

407

Figure 24.3. The lowest possible frequencies for air columns with one or both ends open.

Example 24.4

Instruments such as the clarinet employ air columns with one open end. What is the effective length of a clarinet which has a fundamental frequency of 147 Hz?

The fundamental frequency is found from Eq. 24.4 with $n = 1$,

$$l = \frac{c}{4f_1} = \frac{344 \text{ m s}^{-1}}{4(147 \text{ Hz})} = 0.585 \text{ m}$$

Because the end of a clarinet is flared, its actual length is closer to 0.67 m. Similar differences occur in many instruments.

24.3 | THE INTENSITY OF SOUND WAVES

For most purposes the power per unit area or intensity of a sound wave is more important than the total energy carried over a period of time. For example, the noise of a jet engine will have very different effects on people and objects nearby than will the sound of a violin played for a much longer interval, even though the total sound energy produced might be the same in the two cases. However, a jet engine far away may be no louder than a nearby violin, since a sound wave spreads out as it travels (Fig. 24.4). Also, some sound energy is dissipated by the internal friction or viscosity of the medium in which it is moving. Since this energy loss is usually small, we neglect it.

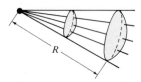

Figure 24.4. Sound waves spread out as they leave a source. Since the surface area of a sphere of radius R is $4\pi R^2$, the area the wave spreads over varies as R^2, and the intensity or power per unit area varies as $1/R^2$.

As noted in the preceding chapter, the intensity of any wave is proportional to its amplitude squared. The intensity of a sound wave with pressure amplitude ΔP is proportional to $(\Delta P)^2$ and is given exactly by

$$I = \frac{(\Delta P)^2}{2\rho c} \tag{24.5}$$

The following example involves typical values of the pressure and intensity of audible sound waves.

Example 24.5

The maximum amplitude ΔP of a sound wave that is tolerable to a human ear is about 28 Pa. (a) What fraction is ΔP of normal atmospheric pressure? (b) What intensity of sound does ΔP correspond to in air at room temperature?

(a) Since normal atmospheric pressure is 1.01×10^5 Pa,

$$\frac{\Delta P}{P} = \frac{28 \text{ Pa}}{1.01 \times 10^5 \text{ Pa}} = 2.77 \times 10^{-4}$$

Thus even very loud sounds correspond to pressure fluctuations that are only a small fraction of a percent of atmospheric pressure.

(b) From Table 24.1, $\rho = 1.05$ kg m^{-3} and $c = 344$ m s^{-1}, so

$$I = \frac{(\Delta P)^2}{2\rho c} = \frac{(28 \text{ Pa})^2}{2(1.05 \text{ kg m}^{-3})(344 \text{ m s}^{-1})}$$
$$= 1.09 \text{ W m}^{-2}$$

Because the sound intensity decreases with the distance squared, the intensity of a sound source decreases rapidly as the distance from it increases. This is illustrated in the next example.

Example 24.6

The low frequency speaker of a powerful stereo set has a surface area of 0.05 m^2 and produces 1 W of acoustical power. (a) What is the intensity at the speaker? (b) If the speaker projects sound uniformly into

Figure 24.5. Example 24.6.

the forward hemisphere, at what distance from the speaker is the intensity 0.1 W m^{-2}?

(a) At the speaker, the intensity is

$$I = \frac{\mathcal{P}}{A} = \frac{1 \text{ W}}{0.05 \text{ m}^{-2}} = 20 \text{ W m}^{-2}$$

(b) At a distance R from the speaker, the wave has spread out over a hemisphere whose area is half the surface area of a sphere, or $\frac{1}{2}(4\pi R^2)$ (Fig. 24.5). Thus $I = \mathcal{P}/2\pi R^2$ or

$$R = \sqrt{\frac{\mathcal{P}}{2\pi I}} = \sqrt{\frac{1 \text{ W}}{2\pi(0.1 \text{ W m}^{-2})}} = 1.26 \text{ m}$$

24.4 | SOUND SOURCES

Many phenomena produce sound in an incidental but unavoidable fashion. For example, the combustion of fuel in an engine always produces some sound as a by-product. This sound is both annoying and wasteful of energy. However, there are many man-made and natural sources for which sound is the desired output. These usually have two primary components: a mechanism for producing a vibration and a *resonant structure*.

Musical instruments present a variety of arrangements for the production of sound. In a violin the strings vibrate, and their vibrations are efficiently transmitted to the air by the resonant hollow body of the instrument. In woodwinds and brasses the vibrations are produced by causing the air in the mouthpiece to puff, swirl, and eddy. This causes the reeds in woodwinds to vibrate. In brasses the lips themselves vibrate as air is blown into the mouthpiece. In both cases the oscillatory flow of air results in standing waves in the extended hollow body of the instrument, and the energy is then efficiently transmitted to the air outside. Similarly, the oral and nasal cavities in humans serve as resonant structures for vibrations produced by the vocal cords.

Some insight into these somewhat complicated systems can be obtained by considering the experiment illustrated in Fig. 24.6. We know that a cylindrical tube of length l that is open at each end will sustain resonant standing waves of frequency $f_n = nc/2l$, where n is an integer. The experiment is performed by using a number of tuning forks with different but closely spaced frequencies and monitoring the sound with a microphone on the opposite side of the tube. When tuning forks with frequencies equal to the

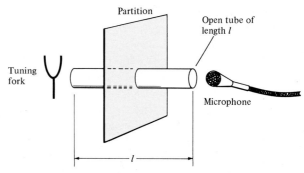

Figure 24.6. A tuning fork on the left of the partition produces sound at a single frequency. Sound can only reach the microphone by passing through the tube.

resonant frequencies of the tube produce sound, standing waves are produced in the tube, and sound is transmitted efficiently to the microphone. Forks with other frequencies produce sound that does not result in standing waves. Much less sound reaches the

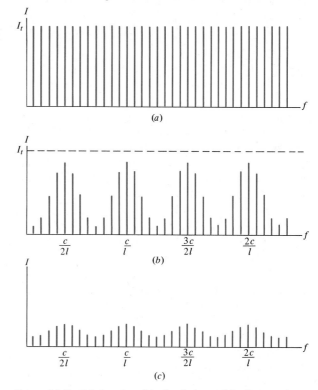

Figure 24.7. (*a*) A series of tuning forks used in the experiment of Fig. 24.6 produces sound of intensity I_t at the tube entrance at closely spaced frequencies. (*b*) The intensity reaching the microphone is greatest at the resonant frequencies of the tube. (*c*) The intensity received by the microphone through a narrower tube than that used in (*b*) is less sharply peaked about the resonant frequencies.

microphone; the actual amount depends on how close one is to a resonant frequency (Fig. 24.7).

When sound enters the tube, the air inside vibrates, and the air near the tube walls loses some energy because of viscous forces. As we have seen with the harmonic oscillator and the *LRC* circuit, whenever the energy dissipation in a resonant system is increased, the intensity versus frequency curve or *spectrum* is lowered and proportionately broadened. If a narrower tube of the same length is used, the broadening is more pronounced because the relative amount of air near the walls is larger, and the viscous effects are proportionately greater (Fig. 24.7c). The sound energy not transmitted is partially reflected at the tube entrance and partially dissipated in the walls of the tube.

Thus, we conclude that the tube serves as a resonant structure that selectively transmits sound at and near specific resonant frequencies and suppresses all other frequencies. Similar selective transmission occurs in instruments such as the violin and in the human speech mechanism.

The Violin |
The body of a violin is a more complex resonant structure than the tube. When the strings of a violin are plucked or bowed, their vibrations are transferred to the body through the bridge (Fig. 24.8). Although the strings may vibrate with many different frequency components, the body resonates at and amplifies only certain frequencies.

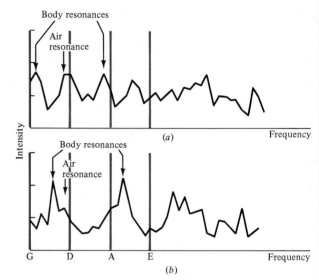

Figure 24.9. (*a*) The intensity-versus-frequency graph for a good violin. The vertical colored lines represent the characteristic frequencies of the four violin strings. Note that the resonances occur at nearly the same frequencies as those characteristic of the strings. (*b*) The spectrum of a poor violin.

The violin body vibrates so that its volume varies, and air is forced in and out through the f-holes. This is called the air resonance. The front and back plates of the body can also vibrate at characteristic frequencies called body resonances. Just as a string can vibrate at more than one frequency, so also can the violin plates, and several body resonances exist (Fig.

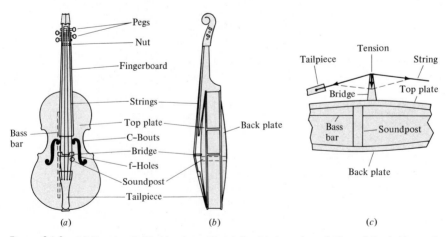

Figure 24.8. (*a*) Front and (*b*) side views of a violin. (*c*) An enlarged view of the bridge and part of the violin body.

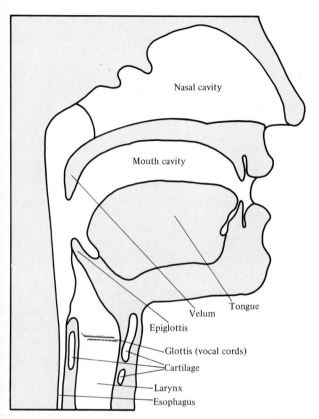

Figure 24.10. Essential components of the human speech mechanism.

24.9). The frequencies of the air and body resonances should be at or near the fundamental string frequencies. If this is not achieved some notes will be muted or distorted.

The Human Voice

In human speech, the vocal cords initiate vibrations of the air; the throat and the nasal and oral cavities serve as resonant structures (Fig. 24.10). The remarkable ability of humans to produce so many different sounds stems from two facts. First, the vocal cord tension can be varied; hence the frequencies produced and the proportions of the harmonics present can be changed. Second, the resonant structures, particularly the oral cavity, can be changed in shape and dimensions to modify the frequency content of the amplified sounds. (These considerations apply to voiced sounds as opposed to nonvoiced sounds such as "sh" in "shut" and "f" in "feet," which do not require the use of the vocal cords.) Normally the vocal cords are relaxed and present no obstruction to air passing through the larynx. During preparation for speech the tension in the vocal cords increases and the larynx is closed (Fig. 24.11). The air pressure below the vocal cords increases until the cords are forced open. The air then rushes through the opening causing the vocal cords to vibrate.

The vibrations of the vocal cords can be understood from Bernoulli's equation (Chapter Fourteen). As air rushes through the opening, its velocity is large, and hence the pressure is lowered, allowing the vocal cords to start to close. Once the opening becomes sufficiently small, the pressure below the vocal cords increases and forces the cords to spread apart again. This process is continuous as long as exhalation persists and the fundamental as well as many harmonics of the vocal cords are excited (Figs. 24.12 and 24.13).

As illustrated in Fig. 24.13, the spectrum of sound produced by the vocal cords is quite uniform up to about 3000 Hz, the range of frequencies used in speech. Without modification this would result simply in noise. The modification takes place in the resonant structures.

The human oral cavity is a complex structure, but its properties are surprisingly close to those of a simple model. This model is a pipe 0.17 m long with one end open at the mouth and nose and the other end nearly closed at the vocal cords. The resonant fre-

Figure 24.11. (a) The vocal cords are relaxed during normal breathing (a). When speech is begun the vocal cord tension increases (b), (c), and the larynx is closed (d).

411

| 124 hertz | 174 hertz | 248 hertz | 330 hertz |

Figure 24.12. The vocal cords are shown for several fundamental frequencies during singing. Higher frequencies occur when the tension increases. Notice the stretching of the cords just as in a string.

Figure 24.13. An example of the fundamental and harmonics produced by the vocal cords during speech. This spectrum will change with the tension in the vocal cords.

Figure 24.14. The intensity versus frequency spectrum of the sound "ae" in "at." The formants are wide resonances labeled F_1, F_2, and F_3. The vertical lines show how the vocal cord frequencies shown in Figure 24.13 are selectively transmitted by the resonance structure.

quencies of a pipe open at one end were found to be

$$f_n = \frac{(2n-1)c}{4l}, \qquad n = 1, 2, 3 \ldots \quad (24.4)$$

In speech the frequencies of importance are between 300 and 3000 Hz. Using $c = 344 \text{ m s}^{-1}$ and $l = 0.17$ m, we find that the fundamental and first two overtones of the pipe, which are near 500, 1500, and 2500 Hz, respectively, are within this range. If one analyzes a typical voiced sound, one finds that there usually are three resonant peaks that are close to the resonant frequencies of the model. These resonant peaks are referred to as *formants*. For example, the sound "ae" in "at" has formants at about 600, 1700, and 2300 Hz.

From Fig. 24.14, we see that there are many possible ways in which the formant composition may be varied to produce the different sounds characteristic of a language. The formants may be broad or narrow and may be shifted in frequency. In Fig. 24.15 the peak formant intensities of some English vowel sounds are shown.

The Loudspeaker | The reproduction and amplification of sound using electronic equipment has revolutionized human communication. Sound that is modified by electronic systems is reconverted to sound by a speaker, the most common of which utilizes a permanent magnet surrounded by a current carrying coil (Fig. 24.16). The coil carries a time-varying current from the amplifier that produces a variable magnetic field inside the coil. This results in varying forces on the permanent magnet. As the magnet moves, so does the diaphragm, and its motion produces sound waves in the air.

24.5 | SOUND DETECTORS

Sound detection requires the conversion of the mechanical vibrations of sound waves into a form that permits the analysis of their frequency and intensity. Here, we discuss how this is done by the condenser microphone and the human ear.

The condenser microphone contains a flexible membrane that includes an electrically conducting layer. This membrane is close to a second rigid conductor, so together they form a capacitance that varies as the membrane vibrates in response to a sound wave. In a circuit, this variation of the capacitance causes a time variation in the current. Once this varying component of the current has been amplified electronically, it can drive a speaker or record information on magnetic tape.

Any useful sound detector must respond accurately to variations in the sound frequency and intensity.

Figure 24.15. The position and relative intensity of the formants for some English vowel sounds as voiced by males.

Figure 24.16. A speaker has a permanent magnet that is free to move along its length. One end of the magnet is attached to the speaker diaphragm and the magnet is surrounded by a current-carrying coil.

The human ear is remarkable, since it can do this well and is nevertheless virtually unaffected by motion and vibration of the body or by the sounds produced by the blood flow and in the internal organs. It also permits the listener to locate sound sources and to concentrate on specific sounds in a confused sound environment.

Figure 24.17*a* shows the anatomy of the human ear. The outer ear collects sound waves and transmits them to the eardrum. Bones called *ossicles* in the middle ear transmit the eardrum vibrations through the *oval window* to the *perilymph* fluid of the inner ear canals. The ossicles function as a hydraulic press and amplify the force on the oval window to nearly 15 times that of the force on the eardrum. In addition, muscles connected to the ossicles control the amplitude of their motion, so that loud sounds do not damage the sensitive inner ear.

The inner ear (Fig. 24.17*b*) has two canals filled with fluid. The *cochlear duct,* containing nerve endings in the *organ of Corti,* divides the two chambers except at the end farthest from the windows. The *oval window* is driven by one of the ossicles. A flexible *round window* in the other chamber flexes as the perilymph moves, so the volume of the inner ear stays constant. Because the cochlear duct is thicker near the narrow end of the cochlea, different frequencies of vibration of the perilymph cause the partition to flex at different points along its length. The flexing is sensed by the nerve hairs in the region of excitation, and nerve impulses travel toward the brain.

The mechanical operation of the ear is well understood; however, the perception of sound involves complex neurological processing of information that is not yet fully explained. An example is the use of earphones. Any sound with a wavelength much larger than the diameter of a speaker cannot be efficiently produced by that speaker. Earphone speakers are too small to reproduce the low-frequency fundamentals of many musical sounds, although they may accurately reproduce the overtones. Nevertheless, music heard on earphones sounds as though the fundamental were present because the brain, recognizing the harmonics, effectively supplies the fundamental.

24.6 | AUDITORY RESPONSE

The ears of animals are remarkable mechanical

413

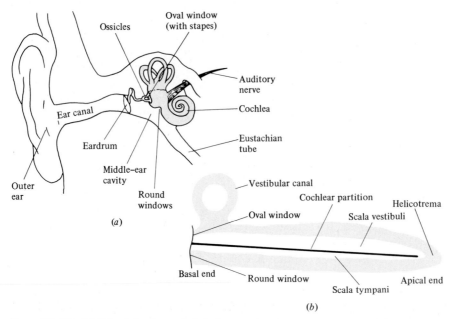

Figure 24.17. (*a*) Schematic diagram of the human ear. (*b*) The uncoiled inner ear.

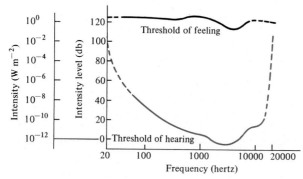

Figure 24.18. A logarithmic graph showing the range of audible intensities versus frequency. The human ear is most sensitive to frequencies slightly below 3000 Hz.

this high intensity, a tickling sensation is experienced when the ossicles vibrate so strongly that they strike the middle ear wall. The normal hearing range lies between these two curves.

The ear is sensitive to such an enormous intensity range in part because muscles around the eardrum and the ossicles respond to neural feedback and modify the tension in these parts. Thus the eardrum is somewhat like a flexible drum head with an adjustable tension.

Because of the enormous intensity range of the ear, it is common to measure intensities in logarithmic units or *decibels* (dB). The *intensity level* is related to the intensity by

$$\beta = 10 \log \frac{I}{I_0} \qquad (24.6)$$

β is measured in decibels, I is the sound intensity, $I_0 = 10^{-12}$ W m^{-2} is an arbitrary reference level roughly equal to the lowest intensity normally audible, and log denotes the base 10 or common logarithm. The decibel range of hearing at 1000 Hz is from near 0 dB up to about 120 dB (Fig. 24.18). Table 24.2 lists the intensity levels of common sound sources.

The following is an example of the relationship between intensity level and perception.

structures. For example the human ear can comfortably respond to intensities from as low as 10^{-12} W m^{-2} up to 1 W m^{-2}, 12 orders of magnitude. Without this range, many sounds in our natural environment would either be inaudible or unbearable.

Measurements of auditory response are somewhat subjective, but two objective characteristics have been fairly well established. One is the *threshold of hearing;* the minimum intensity that is just audible at a given frequency. This is represented by the lower curve of Fig. 24.18. The second is the *threshold of feeling.* At

TABLE 24.2

The intensity level in decibels of common sound sources

Source	Intensity Level (in decibels)
Large orchestra (maximum)	98
Riveter	95
Bass drum (maximum)	94
Trumpet (maximum)	75
Busy street traffic	70
Clarinet (maximum)	67
Ordinary conversation	65
Quiet radio	40
Whisper	20

Example 24.7

If a 20-dB sound varies in frequency from 20 to 30,000 Hz, which frequencies will the normal ear hear?

From Fig. 24.18, we note that the horizontal line at 20 dB intersects the threshold of hearing curve at about 200 and 14000 Hz. Thus, only frequencies within this range will be audible.

When more than one sound source is present, the intensity level does not usually increase markedly. This is shown in the next example.

Example 24.8

If the average intensity level of each of two radios is 45 dB, what is the average intensity level when both radios are on and tuned to different stations?

The intensity of one radio is I_r, where the corresponding intensity level is

$$\beta_1 = 45 \text{ dB} = 10 \log \frac{I_r}{I_0}$$

With both radios on, the intensity is $I_r + I_r = 2I_r$, so the intensity level is

$$\beta_2 = 10 \log \frac{2I_r}{I_0} = 10 \log 2 + 10 \log \frac{I_r}{I_0}$$
$$= 10 \log 2 + \beta_1$$

Since $10 \log 2 = 3$,

$$\beta_2 = 3 \text{ dB} + 45 \text{ dB} = 48 \text{ dB}$$

Notice that although the intensity I_1 doubles β increases by only 3 dB.

Subjective Factors |
Properties such as *loudness, pitch,* and *quality* are often used to describe sound. These concepts are difficult to deal with because they are subjective. For example, a 20-dB sound at 1000 Hz will sound louder than a 20-dB sound at 600 Hz. Thus loudness depends both on intensity and frequency. The pitch of a sound is closely related to the frequency. However, above about 3000 Hz, the pitch increases with intensity even if the frequency is constant. Below 2000 Hz the pitch decreases with increasing intensity.

The quality of a sound is even more ambiguous. We know that some types of sounds that have certain harmonics present with specific intensities sound pleasing. However, again we cannot characterize quality with a precise statement about measurable physical parameters.

The combination of a mechanical ear structure, neurological processing, and the psychological aspects of perception has so far allowed us to classify but not to understand the subjective qualities of sound. These aspects are under intensive study by those concerned with hearing defects and the science of musical sounds.

Auditory Localization |
We can sometimes locate a sound source by turning our heads, since the ear nearest the source will often hear the loudest sound. However, people can often accurately locate sound sources without turning their heads, unless the source is directly in front of or behind them. There are different explanations for this at low and high frequencies.

The human head is approximately a sphere of diameter 0.2 m, which is the wavelength of a 1700 Hz sound wave in air. Sound waves of longer wavelengths or lower frequencies will pass around the head largely unaffected. If at some time a peak in the wave is at one ear, the wave has a different pressure at the other ear. The nerve impulses from the two ears then contain information about the relative pressures at the two ears which the brain uses to locate the source. This mechanism is most effective below 1000 Hz.

Sound waves with wavelengths much shorter than 0.2 m tend to be reflected when they strike the head, leaving a sound shadow with little or no sound on the far side. Thus, above 5000 Hz the source is usually located because the sound is distinctly louder in the ear nearest it. Between 1000 and 5000 Hz, both mechanisms are used but the localization is less accurate.

SUMMARY

Sound waves are the result of mechanical disturbances in materials. The speed of sound depends on the ratio of the pressure changes to the accompanying density changes. This ratio, called the bulk modulus, depends on the molecular properties of the substance.

Standing sound waves are often produced and play an important role in the production and detection of sound. In sound sources, a resonant structure produces the standing waves, amplifying certain frequencies and suppressing others.

The intensity of a sound wave is proportional to the square of the amplitude of the pressure change in the wave and is the transported power per unit area. As a sound wave spreads out from a source, its intensity decreases with the square of the distance from the source.

The intensity level of sound in decibels is often used when one is dealing with auditory response. This decibel or logarithmic scale is used because of the enormous range of ear sensitivities of humans and other animals.

Checklist

Define or explain:

pressure variations	loudspeaker
bulk modulus	middle ear
standing waves	inner ear
sound intensity	threshold of hearing
source of vibration	threshold of feeling
resonant structure	loudness
sound spectrum	pitch
formant	quality

EXERCISES

Section 24.1 | The Nature and Speed of Sound

24-1   A hifi speaker has a diameter of 0.3 m. What is the frequency of the sound produced if the wavelength is equal to the circumference of the speaker? (This is comparable to the resonant frequency of the speaker, and special enclosures must be designed so that most of the power is not emitted by the speaker at this frequency.)

24-2   A lightening bolt is seen by an observer. The accompanying thunder is heard 5 seconds later. What is the approximate distance of the observer from the storm cloud?

24-3   A stone is dropped into a well. The sound of the splash is heard 3 seconds after the stone is dropped. What is the depth of the well?

24-4   When one end of a copper pipe is struck, the time difference between the sounds heard in the copper and in air at the other end is 1 second. How long is the pipe?

24-5   The depth of a body of water can be found by emitting sound pulses at the surface detecting the pulses reflected from the bottom. If the time interval from emission to detection is 2 seconds, what is the depth of the water?

24-6   A marching band takes a step each 0.8 seconds. At what distance will a marching band appear to be one half step out of time with the music being played?

24-7   When approaching an object, a bat decreases the duration of its chirps and also the time interval between chirps. When the chirps last 3×10^{-4} seconds, what is the minimum distance at which the first part of the echo overlaps with the end of the chirp?

24-8   The telephone is designed to efficiently transmit frequencies from about 50 to 3000 Hz. What are the wavelengths of sound in air corresponding to these two frequencies?

24-9   Bats utilize sound up to frequencies of about 1.2×10^5 Hz, while porpoises are sensitive to frequencies up to 2×10^5 Hz. Why might porpoises be expected to have a greater frequency range?

Section 24.2 | Standing Sound Waves

24-10   The fundamental frequency of the longest pipe on an organ is 16.35 Hz. If the pipe is open at both ends, how long is the pipe?

24-11   A bugle has the same characteristic frequencies as a cylindrical tube 1.3 m long with both ends open. What are the frequencies of the first four harmonics?

24-12   A clarinet has a fundamental frequency of 147 Hz and, when played, has one end closed. (a) How many harmonics appear below 1350 Hz? (b) If an open end tube has the same fundamental frequency, how many harmonics appear below 1350 Hz?

24-13 A pipe organ has open end pipes and spans a frequency range from 65 to 2090 Hz. What are the lengths of the longest and shortest pipes of this organ?

24-14 The outer ear can be thought of as a pipe 2.7×10^{-2} m long with one closed end. (a) Using this model predict what frequency sound would be most effectively detected by the ear. (b) How does this compare to the frequency of the minimum of the threshold of hearing curve (Fig. 24.18)?

Section 24.3 | The Intensity of Sound Waves

24-15 What is the pressure amplitude of thunder with an intensity of 0.1 W m^{-2}?

24-16 What is the total power output of a loud-speaker for which the intensity over the surface of a hemisphere 10 m away is 10^{-4} W m^{-2}?

24-17 Two sound waves of the same intensity travel in air and water. What is the ratio of the pressure amplitudes of the sounds in the two materials?

24-18 If the pressure amplitudes of sound in air and water are the same, what is the ratio of the two intensities?

24-19 The amplitude of a sound wave is usually given in terms of the maximum pressure change from the equilibrium value. If one sound wave has twice the pressure amplitude of another, what is the ratio of the intensities of the two waves?

24-20 The ratio of the maximum intensities of a piano and a flute is 8. What is the ratio of the pressure amplitude produced by a piano to that of the flute?

24-21 The intensity of a large orchestra is equal to that of 216 trumpets. What is the ratio of the pressure amplitude of the orchestra to that of 1 trumpet?

24-22 What is the intensity of a sound wave in air with a pressure amplitude of 1 Pa?

24-23 What is the pressure amplitude of a sound wave in water of intensity $I = 10^{-12}$ W m^{-2}?

24-24 What is the pressure amplitude in air of a sound at the threshold of feeling, $I = 1$ W m^{-2}?

Section 24.4 | Sound Sources, and
Section 24.5 | Sound Detectors

24-25 A bat needs two ears for direction finding,

just as humans do. If the distance between the ears is 0.01 m, what is the minimum frequency for which the ears are separated by at least one half wavelength? (Assume that the sound approaches directly from one side of the head.)

24.26 In hearing tests, a tone of a given frequency is gradually reduced in intensity until it becomes inaudible. Why are tones of different frequencies used during the test?

24-27 Sound can be heard by humans when the vibrations are transmitted to the inner ear via bone. The outer and middle ear play no role in such hearing. (a) Suggest a method of testing hearing such that conclusions may be drawn about whether damage is present in the inner ear only or if middle ear damage is also present. (b) For what types of hearing losses would hearing aids which transmit sounds to the skull bones be useful?

24-28 Bone is a better conductor of low-frequency vibrations than is air. Why does a person feel that his or her voice is richer and lower pitched than do listeners?

24-29 Why do sinus infections in which the eustachian tubes are infected often impair hearing?

Section 24.6 | Auditory Response

24-30 Two sound waves have intensities of 10^{-9} W m^{-2} and 5×10^{-8} W m^{-2}. What is the difference in the intensity levels of the two sounds?

24-31 Find the ratio of intensities of two sounds, one of which is 10 dB louder than the other.

24-32 What is the intensity level of ultrasound of intensity 25×10^4 W m^{-2}?

24-33 If the intensity level of one person speaking is 50 dB, what is the intensity level when 10 such people are speaking?

24-34 What is the intensity of a sound that is 5 dB louder than that of a sound of intensity 10^{-9} W m^{-2}?

24-35 The surface area of the eardrum is about 8×10^{-5} m^2. What is the power transmitted to the eardrum by a sound wave of 40 dB if no sound is reflected?

PROBLEMS

24-36 Bats, which use echo location for navigation, are observed to emit chirps lasting 2×10^{-3} seconds with 7×10^{-2} seconds of silence between.

How close to an object can a bat be so that the reflected sound from the first part of the chirp is not masked by the emission of the first part of the next chirp?

24-37 Equation 24.2 for the sound velocity, $c = \sqrt{K/\rho}$, hides an understanding of the physics involved in sound propagation. In fact, c is larger for dense substances than for less dense materials (see Table 24.1). If dominoes are placed on end beside one another and the end one is toppled, the entire row falls. By noticing that the dominoes fall fastest when they are close together, discuss the dependence of the sound velocity on density.

24-38 During cruising flight the interval between chirps of a bat is 7×10^{-2} seconds. What is the maximum distance a bat can be from an object so that the complete reflected wave returns before the next chirp begins?

24-39 For an ideal gas the adiabatic bulk modulus is $K = \gamma P$, where P is the pressure, and γ is the ratio of the specific heat capacities at constant pressure and volume, $\gamma = c_p/c_v$. (a) Show that the ideal gas law can be written as $P/\rho = RT/M$, where ρ is the mass density, and M is the molecular weight of the gas. (b) Show that the sound velocity in an ideal gas is $c = \sqrt{\gamma RT/M}$.

24-40 In Problem 24-39 the velocity of sound in an ideal gas is found to be $c = \sqrt{\gamma RT/M}$. Using this expression, can one account for the difference in sound velocities in air and hydrogen, H_2?

24-41 A bat using echo location emits chirps at intervals of 5×10^{-3} seconds, each having a duration of 0.3×10^{-3} seconds. (a) How far from an object will the bat be so that the first part of the chirp is being reflected at the time the chirp ends? (b) How much time is there between the return of the complete reflected wave of part (a) and the beginning of the succeeding chirp?

24-42 A sonar source used for underwater detection emits pulses of duration 0.1 second once every T seconds. What should T be if the echo and pulse do not overlap from objects (a) 50 m away; (b) 1 km away?

24-43 A vertical tube of length 2 m can be filled to any level with water. Sound entering the open end of the tube is reflected at the water surface and this is the position of a node. (a) If the tube is filled with water to a depth of 1 m, what is the lowest

frequency at which resonance will occur? (b) What is the depth of the water if the lowest frequency resonance occurs at 500 Hz?

24-44 The maximum tolerable pressure amplitude for the human ear in air is $\Delta P = 28$ Pa. What is the ratio of the corresponding density change to the mean density for sound waves in (a) air, and (b) water?

24-45 The acoustic intensity of an antiaircraft shell exploding at an altitude of 1000 m is 10^{-3} W m^{-2} at ground level. (a) What is the total acoustic power released? (b) What total acoustic energy is released during the explosion if it lasts 0.1 second?

24-46 If the air and body resonances of a violin had a very narrow frequency width or range the instrument might not be effective. Explain why.

24-47 A jet plane flying at an altitude of 3000 m produces a sound of 40 dB at ground level. What would the intensity level be if the altitude were 1000 m? (Assume that the total sound energy produced does not change.)

24-48 An outdoor public address system is adjusted to a level of 70 dB for listeners 10 m distant. What intensity level is heard at 50 m?

24-49 The average intensity level of a radio is adjusted to 40 dB at a distance of 10 m. (a) What is the intensity in watts per square meter at this distance? (b) What is the intensity level in decibels 3 m from the radio? (c) What is the output power of the radio in watts if the sound is spread uniformly over the forward hemisphere?

24-50 The normal human ear can distinguish a difference in intensities of about 0.6 dB at a given frequency. What percentage of power increase is required to raise the intensity level 0.6 dB.

SUPPLEMENTARY TOPICS

24.7 | ULTRASOUND

In this section, we describe some of the properties and medical applications of ultrasound, which is sound with frequencies above 20,000 Hz. Ultrasound can presently be produced at frequencies up to 10^9 Hz. It is widely used as a diagnostic, therapeutic, and surgical tool in medicine.

Physical Principles and Limitations | In most

applications, ultrasound is sent out in pulses that are partially reflected. Between pulses the transmitter acts as a receiver, detecting the reflected wave or echo. Accurately locating small objects at substantial distances requires narrow, short wavelength beams with sufficient range. These criteria tend to conflict, since short wavelengths (high frequencies) are strongly absorbed. For example, at a frequency of 1 megahertz (1 megahertz = 1 MHz = 10^6 Hz), the intensity drops 50 percent in 7 cm of soft animal tissue. Although in some situations the intensity is limited by the capacity of the apparatus, in medical applications the intensity limit is determined mainly by the destructive effects of intense ultrasound on animal tissues.

Ultrasound is useful because it is reflected from the boundaries between materials of nearly the same density and can be used without apparent harmful effects where x-rays cannot. For example, ultrasonic scanning of the uterus during pregnancy is considered safe, while x-rays must be avoided.

The fraction of the sound intensity reflected at a boundary can be found from a formula derived using advanced mathematical methods. Consider a sound wave passing from one medium of density ρ_1 and sound velocity c_1 to a second medium of density ρ_2 and sound velocity c_2. The ratio of the reflected to incident intensities when the wave travels perpendicular to the interface is

$$\frac{I_r}{I_i} = \left(\frac{\rho_1 c_1 - \rho_2 c_2}{\rho_1 c_1 + \rho_2 c_2}\right)^2 \tag{24.7}$$

Since the sound velocity does not vary much, the amplitude of the reflected wave depends primarily on the density difference. It is this dependence that causes detectable reflections from boundaries even if the density change is small. This is illustrated by the next example.

Example 24.9

The densities of two types of muscle tissue are 1026 and 1068 kg m^{-3}. What is the ratio of the intensities of the reflected and incident waves if the wave passes from the more dense to the less dense medium? (Assume that the sound velocities are the same.)

Using the densities given and Eq. 24.7, the sound velocities factor out, and we have

$$\frac{I_r}{I_i} = \left(\frac{1068 \text{ kg m}^{-3} - 1026 \text{ kg m}^{-3}}{1068 \text{ kg m}^{-3} + 1026 \text{ kg m}^{-3}}\right)^2 = 0.00040$$

While this ratio is small the reflected sound is still measurable. The unreflected part of the wave continues on and further reflections at other boundaries can occur.

Equation 24.7 becomes simpler if we assume that the sound velocities c_1 and c_2 are equal, so we have

$$\frac{I_r}{I_i} = \frac{(\rho_1 - \rho_2)^2}{(\rho_1 + \rho_2)^2}$$

We see that the reflection is minimal when ρ_1 and ρ_2 are nearly equal. Conversely, if one of the densities is much larger than the other the reflection is comparatively large.

This result can be partially understood with an anology. If a moving billiard ball collides head-on with another one initially at rest, the second ball starts moving and the first comes to rest. The energy is transmitted perfectly if the masses are equal and no reflection (motion of the first ball) occurs. However, if we observe a collision between a moving billiard ball and a ping-pong ball at rest, the billiard ball continues to move afterward. Less energy is transmitted now and more is reflected. The analogous result holds when the ping pong ball collides with a billiard ball at rest. The reflected energy (the energy of the ping-pong ball) is large.

An analogous case for ultrasonic waves is an air-tissue boundary, where virtually all the sound is reflected because the density change is large. In most medical work this necessitates using a liquid to provide good transmission between the ultrasound source and the tissue. Nearly total reflection occurs at air-tissue boundaries in the body.

Destructive Effects | Intense ultrasound produces large density and pressure changes within each small wavelength. This results in large stresses, and molecules are forced to move rapidly. It also produces heat in most materials and *cavitation* in liquids and perhaps in tissue. Cavitation is the formation of bubbles of vapor caused by the mechanical fracture of the liquid in a region where the pressure is decreasing. These bubbles may then collapse violently. This catastrophic collapse can be used for cleaning, since the implosion scatters the liquid along with impurities. Cavitation is also used to produce minute droplets of liquid. Medication used in inhalation therapy

Figure 24.19. Electron photomicrographs of Baker's Yeast cells (*a*) before and (*b*) after exposure to ultrasound. (Courtesy Heat Systems-Ultrasonics, Inc.)

is broken with ultrasound into droplets fine enough to enter the alveoli of the lungs.

Another violent result of ultrasound treatment is the mechanical rupture of cell membranes and breakage of cell constituents such as chromosomes (Fig. 24.19). A major fraction of the energy lost in tissue is absorbed by proteins.

All these effects limit the therapeutic use of ultrasound to intensities below 3×10^4 W m^{-2}. This may be compared with surgical intensities of 25×10^4 W m^{-2}.

Instrumentation and Display |

Ultrasound sources are usually crystals that are either *piezoelectric* or *magnetostrictive*. A piezoelectric crystal is one in which an applied electric field alters the molecular positions, producing stresses in the crystal. If the applied field is periodic the crystal vibrates, producing sound. Magnetostrictive materials exhibit similar behavior in an applied magnetic field. Both types of transducers also act as receivers, since the mechanical vibrations produce electric and magnetic fields that can be detected and used to monitor incoming sound waves.

In diagnostic medical work, two techniques of scanning and display are common. In an A-scan the vertical deflection plates of an oscilloscope display voltages associated with the emission or reception of sound pulses. The horizontal motion of the beam is adjusted so that the sweep time is long enough to display the original pulse and its echo (Figs. 24.20 and 24.21).

A second technique, the B-scan, uses a movable transducer. In this case, the voltage due to the pulse or echo is added to the accelerating voltage in the electron gun of the cathode-ray tube. The oscilloscope is adjusted so that only when a pulse or echo voltage is present are the electrons fast enough to produce a bright spot on the screen (Fig. 24.22).

The echos then appear on the screen as bright spots whose brightness depends on the strength of the echo. Thus if the transducer were not moved the peaks would appear only as bright spots along a line. When the transducer is moved parallel to the body surface the vertical deflection plates are adjusted so that the traces are displaced along horizontal lines one above another (Figs. 24.23 and 24.24).

Ultrasound is very useful for detecting motion. For example, the motion of the heart wall or even of the mitral valve can be detected (Fig. 24.25). Special source probes can even be used to do ultrasonic scanning from inside the heart. The Doppler effect with ultrasound can be used to detect fetal heart beats and the pulsation of arterial walls.

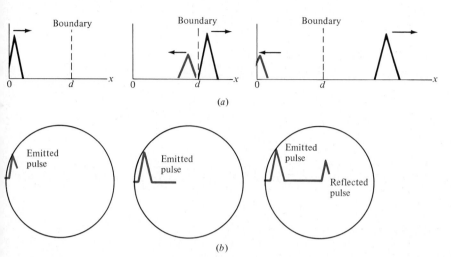

(a)

(b)

Figure 24.20. (a) A pulse is emitted and travels toward a tissue boundary a distance d away. Part of the pulse (shown colored) is reflected at the interface and returns to the transducer. (b) The oscilloscope trace monitors the emitted pulse and the reflected pulse a time $t = 2d/c$ later. Because these events are so rapid, the eye only sees the completed trace.

Figure 24.21. (a) An echoencephalogram to locate the midline between the hemispheres of the brain of a normal person. (b) Right to left and left to right A-scans showing a displaced midline due to a pathological enlargement of the right side of the brain. (From J. R. Frederick, *Ultrasonic Engineering*. Copyright © 1965, John Wiley & Sons, New York. Used with permission.)

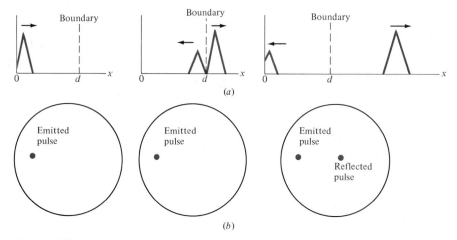

(a)

(b)

Figure 24.22. The sequence of events in (a) appears on a single scan of the oscilloscope trace (b) as bright spots.

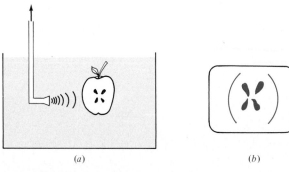

(a)   (b)

Figure 24.23. A B-scan performed on an apple (shown in cross section). (a) The apple and transmitter-receiver are immersed in water to improve transmission. The ultrasonic beam is directed to the right, and partial reflection occurs at the apple surface and from the seeds. The receiver is moved upward during the scan producing the oscilloscope trace shown in (b).

Figure 24.24. B-scan of a fetus during the last trimester of gestation. The transducer is directed downward and is moved back and forth mechanically to obtain this result. (Photo courtesy of Picker Corporation.)

EXERCISES ON SUPPLEMENTARY TOPICS
Section 24.7 | Ultrasound

24-51 (a) If sound travels from air to water perpendicular to the interface, what percentage of the intensity is reflected? (b) If sound travels from water to air, what percentage of the intensity is transmitted?

PROBLEMS ON SUPPLEMENTARY TOPICS

24-52 The echo from the midline of the brain is detected on an oscilloscope 10^{-4} seconds after the source pulse. How far from the source is the midline? (Assume that the velocity of sound in brain tissue is 1540 m s^{-1}. Neglect corrections due to the skull bone.)

24-53 What would you expect the approximate limit of accuracy of distance measurements in human tissue to be when using sound of frequency 10^6 Hz?

24-54 In a brain scan the echos from the right side of the skull, the midline of the brain, and the left side of the skull are observed after times of 0.10×10^{-4}, 1.26×10^{-4}, and 2.40×10^{-4} seconds, respectively. If the sound velocity is 1540 m s^{-1}, (a) how far is the midline of the brain displaced? (b) Which hemisphere of the brain is enlarged?

Figure 24.25. A display of (a) normal and (b) abnormal mitral valve activity. Here the transducer is held fixed and the horizontal axis is a time axis. A scale in centimeters appears at s. The reflections in region c are from the skin, those in region m are from the mitral valve, and the corresponding EKG trace is shown at e. (From J. R. Frederick, *Ultrasonic Engineering*. Copyright © 1965, John Wiley & Sons, New York. Used with permission.)

Additional Reading:

See also the references marked with an asterisk (*) at the end of Chapter Twenty-three.

Winston E. Kock, *Sound Waves and Light Waves,* Science Study Series, Doubleday and Co., Garden City, N.Y., 1965.

William A. Van Bergeigk, John R. Pierce, and Edward B. David, Jr., *Waves and the Ear,* Science Study Series, Doubleday and Co., Garden City, N.Y., 1960.

Jess J. Josephs, *The Physics of Musical Sound,* D. Van Nostrand Company, Inc., Princeton, N.J., 1967.

D.R. Griffin, How Bats Guide Their Flight by Supersonic Echoes, *American Journal of Physics,* vol. 12, 1944, p. 343.

F.S. Crawford, Singing Corrugated Pipes, *American Journal of Physics,* vol. 42, 1974, p. 278.

T.D. Rossing, Musical Acoustics, (Resource Letter MA-1), *American Journal of Physics,* vol. 43, 1975, p. 944.

Arthur H. Benade, *Fundamentals of Musical Acoustics,* Oxford University Press, New York, 1976.

J.W. Coltman, Acoustics of the Flute, *Physics Today,* vol. 21, November 1968, p. 25.

H. Fletcher, The Pitch, Loudness, and Quality of Musical Tones, *American Journal of Physics,* vol. 14, 1946, p. 215.

J.L. Flanagan, *Speech Analysis: Synthesis and Perception,* 2nd ed., Springer-Verlag, Berlin, 1972.

Peter B. Denes and Elliot N. Pruson, *The Speech Chain,* Bell Telephone Laboratories, Murray Hill, N.J., 1963.

Dale Esnminger, *Ultrasonics; The Low and High Energy Applications,* Marcel Dekker, Inc., New York, 1973. Chapter 14 is a summary of the medical applications of ultrasound.

Julian R. Frederick, *Ultrasonic Engineering,* John Wiley and Sons, Inc., New York, 1965, chapters 7, 8, and 9.

P.N.T. Wells, *Physical Principles of Ultrasonic Diagnosis,* Academic Press, London and New York, 1969.

Edgar A.G. Shaw, Noise Pollution—What Can be Done?, *Physics Today,* January, 1975, p. 46.

Scientific American articles:

Maurice Ewing and Leonard Engel, Seismic Shooting at Sea, May 1962, p. 116.

Herbert A. Wilson, Jr., Sonic Boom, January 1962, p. 36.

Mark R. Rosenzweig, Auditory Localization, October 1961, p. 132.

Richard M. Warren and Roslyn P. Warren, Auditory Illusions and Confusions, December 1970, p. 30.

Robert C. Chanard, Aerodynamic Whistles, January 1970, p. 40.

Edward E. David, Jr., The Reproduction of Sound, August 1961, p. 72.

Victor E. Ragosine, Magnetic Recording, November 1969, p. 70.

Leo L. Beranck, Noise, December 1966, p. 66.

Arthur A. Few, Thunder, July 1975, p. 80.

Veru O. Knudsen, Architectural Acoustics, November 1963, p. 78.

Gordon S. Kiuo and John Shaw, Acoustic Surface Waves, October 1972, p. 50.

Georg von Békésy, The Ear, August 1957, p. 66.

James L. Flanagan, The Synthesis of Speech, February 1972, p. 48.

Adrian M. Wenner, Sound Communication in Honeybees, April 1964, p. 116.

Crawford Greenewalt, How Birds Sing, November 1969, p. 126.

Peter F. Ostwald, Acoustic Methods in Psychiatry, March 1965, p. 82.

Gerald Oster, Auditory Beats in the Brain, October 1973, p. 94.

Peter F. Ostwald and Phillip Peltzman, The Cry of the Human Infant, March 1974, p. 84.

F.A. Saunders, Physics and Music, July 1948, p. 33.

Arthur H. Benade, The Physics of Woodwinds, October 1960, p. 144.

E. Donnell Blackham, The Physics of the Piano, December 1965, p. 88.

John C. Schelleng, The Physics of the Bowed String, January 1974, p. 87.

B. Patterson, Musical Dynamics, November 1974, p. 78.

Diana Deutsch, Musical Illusions, October 1975, p. 92.

Johan Sundberg, The Acoustics of the Singing Voice, March 1977, p. 82.

G.E. Henry, Ultrasonics, May 1954, p. 54.

Klaus Dransfeld, Kilomegacycle Ultrasonics, June 1963, p. 60.

Kenneth D. Roeder, Moths and Ultrasound, April 1965, p. 94.

CHAPTER TWENTY-FIVE

WAVE PROPERTIES OF LIGHT

Historically, the wave theory of light was not readily accepted because many observations seemed easier to explain with a model that treated a light beam as a stream of particles. Both particle and wave theories could adequately describe the *reflection* of light, as well as *refraction,* the bending of light as it crosses a boundary between two media. Waves tend to bend around obstacles, as is readily observed with water or sound waves. Since we cannot see around obstacles, this was taken as evidence against wave theories until experiments in the early part of the nineteenth century demonstrated interference effects. These experiments also showed that the bending of light around ordinary objects is difficult to observe because the wavelength of visible light is very short. Later in the century, Maxwell's work on electromagnetic radiation gave the wave model of light a firm theoretical foundation.

Despite the great success of Maxwell's electromagnetic wave theory in predicting many phenomena involving light, it failed to correctly describe some processes in which light is absorbed or emitted by matter. The modern *quantum theory,* developed early in the twentieth century, states that light is composed of *quanta* or *photons*. These are little packets or bundles of light waves, each with an energy proportional to the frequency. Photons have a particlelike aspect associated with the discreteness of their energy in addition to their wave attributes. The quantum theory of light will be discussed in Chapter Twenty-eight, but for the situations considered in this unit, the classical electromagnetic wave theory is adequate.

In the first part of this chapter, we discuss the speed of light, reflection, and refraction. The next few sections cover some of the interference properties of light waves. The chapter concludes with a brief discussion of the polarization of light. Applications to optical instruments will be treated in the next chapter.

Although we concentrate mainly on visible light in this chapter, it should be noted that much of what we discuss applies also to other electromagnetic waves and to other wave phenomena such as sound and water waves.

25.1 | THE INDEX OF REFRACTION

Electromagnetic waves of any frequency travel in a vacuum with the same velocity, $c = 3.00 \times 10^8$ m s^{-1}. In a material medium, the velocity v depends on the frequency of the wave, but it is never greater than the velocity c in a vacuum. The ratio of these velocities is the *index of refraction* of the medium,

$$n = \frac{c}{v} \qquad (25.1)$$

Since v is never greater than c, n is never less than 1. In comparing two media, the one with the larger refractive index is said to be *optically denser*. Representative indices of refraction are given in Table 25.1 for yellow light. The refractive index varies somewhat with the frequency. In a typical material the variation is 1 or 2 percent within the visible spectrum.

The frequency of a light wave is determined by its source and is unaffected by the medium. Since $f\lambda = v = c/n$, the wavelength changes when the

TABLE 25.1

Indices of refraction of representative materials for yellow sodium light (λ = 589 nanometers) We use the approximate values $n = 1$ for air and $n = 4/3$ for water

Material	Index
Air	1.00029
Carbon dioxide	1.00045
Water	1.333
Ethyl alcohol	1.362
Benzene	1.501
Carbon disulfide	1.628
Glass, light crown	1.517
Glass, heavy flint	1.647
Fluorite	1.434
Diamond	2.417

index of refraction changes. If a beam of light goes from a medium with index n_1 to a medium with index n_2, then $f\lambda_1 = c/n_1$, and $f\lambda_2 = c/n_2$. Dividing these two equations,

$$\frac{\lambda_2}{\lambda_1} = \frac{1/n_2}{1/n_1} = \frac{n_1}{n_2} \qquad (25.2)$$

This means that the wavelength is smaller in the optically denser medium, as is illustrated by the following example.

Example 25.1

Yellow light with a wavelength of 5×10^{-7} m in a vacuum enters a glass plate with refractive index 1.5. (a) What is the velocity of light in the glass? (b) What is the wavelength of the light in the glass?

(a) From the definition $n = c/v$, the velocity in the glass is

$$v = \frac{c}{n} = \frac{3.00 \times 10^8 \text{ m s}^{-1}}{1.5}$$

$$= 2.00 \times 10^8 \text{ m s}^{-1}$$

(b) In the vacuum, $v = c$ and $n = 1$. Thus

$$\lambda_2 = \lambda_1 \frac{n_1}{n_2} = (5 \times 10^{-7} \text{ m})\left(\frac{1}{1.5}\right)$$

$$= 3.33 \times 10^{-7} \text{ m}$$

Several units of length are commonly employed in discussions of electromagnetic waves. Meters, centimeters, and millimeters are convenient for radio and microwaves. Discussions of optical instruments and biological applications of visible light often use the *nanometer* (nm), where

$$1 \text{ nanometer} = 1 \text{ nm} = 10^{-9} \text{ m}$$

The visible spectrum for humans extends approximately from 400 nm (violet) to 700 nm (red) (Fig. 25.1). *X-rays* are very high frequency electromagnetic waves. They have many diagnostic and therapeutic uses in medicine and also serve as a powerful probe of the structure of complex molecules. A typical x-ray wavelength is about an *angstrom* (Å), where

$$1 \text{ angstrom} = 1 \text{ Å} = 10^{-10} \text{ m}$$

25.2 | HUYGENS' PRINCIPLE

In 1678, almost two centuries before Maxwell's work on electromagnetic waves, Christian Huygens (1629–1695) proposed a wave theory of light. It is still very useful for understanding many properties of light and other waves, since it makes no reference to the physical nature of the wave phenomenon.

To discuss Huygens' idea, now called *Huygens' principle,* it is useful to introduce the concept of a *wave front*. We recall from Chapter Twenty-three that two waves which have their maximum amplitude at

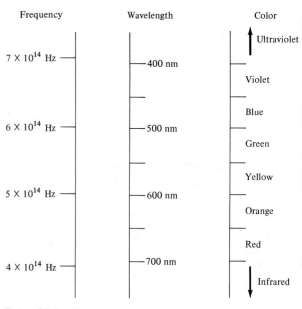

Figure 25.1. The spectrum of visible light. Color ranges are approximate.

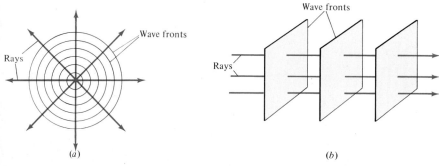

Figure 25.2. (*a*) The wave fronts corresponding to successive maxima in a spherical wave. (*b*) Wave fronts for a plane wave.

Figure 25.3. Water waves in a shallow ripple tank encounter a narrow slit in an obstacle. Circular wave fronts are produced to the left of the slit. (From Educational Development Center, Inc., Newton, Massachusetts.)

Huygens' principle enables us to find the future shape and location of a wave front from its present shape and location. It states that *each point on a wave front can be considered as a source of small secondary spherical wavelets* (Fig. 25.3). The wave front at a later time is the surface tangent to the secondary wavelets, their *envelope*.

We can illustrate the use of Huygens' principle by considering a plane wave (Fig. 25.4). At several points on a wave front, we draw spheres of radius $r = ct$, representing the distance traveled by the secondary wavelets in time t. The tangent to these spheres is a plane displaced a distance ct from the original wave front. Thus the wave front has moved the expected distance in the time interval.

Huygens' principle is applied extensively in this chapter.

the same time are said to be in phase. Also, if one wave has a maximum amplitude when the other has a minimum, the two waves are exactly out of phase. *Wave fronts are surfaces on which the waves at every point are in phase.* They travel outward from the source with the speed of the wave. A useful picture of the wave is obtained by drawing several wave fronts (Fig. 25.2). A wave with spherical wave fronts is called a *spherical wave*. A wave that travels in a single direction has wave fronts that are planes, so it is called a *plane wave*. *A line perpendicular to the wave fronts is called a* ray *and indicates the direction of motion of the wave.* Often it is easier to draw only the rays and not the wave fronts.

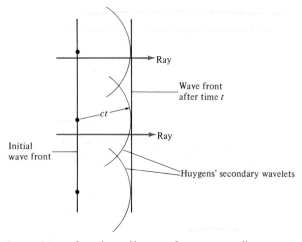

Figure 25.4. In a time t, the wave front moves a distance ct.

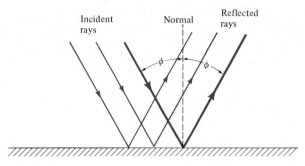

Figure 25.5. The reflected light rays are in the same plane as the incident rays and the normal and make the same angle with the normal.

25.3 | REFLECTION OF LIGHT

When a beam of light reaches the boundary between two media, some light is transmitted, some is absorbed, and the remainder is reflected. The smooth surface of a piece of glass or polished metal reflects light in a particular direction. This is called *specular reflection*. Sometimes the reflection is *diffuse, and the reflected light travels in all directions*. This happens when light strikes a surface such as a sheet of paper or a painted wall with random irregularities that are large compared to a wavelength. Each smooth section of such a surface produces specular reflection, but because of the varying orientations of the sections, the total reflected beam has no unique direction.

In specular reflection the directions of the incident and reflected light are related very simply:

The incident rays and the reflected rays make the same angle with the normal to the reflecting surface (Fig. 25.5).

This equality of the angles of incidence and reflection is a general property of waves and follows from Huygens' principle (Fig. 25.6). Note that the incident and reflected rays and the normal are all in the same plane.

Two additional properties of reflected light are important. First, we saw in Chapter Twenty-three that when waves on a string reach a fixed end, the reflected waves are *inverted* or *reversed in sign*. In this case, one says that the *phase is reversed*. This also happens to a light wave if it is incident on a boundary

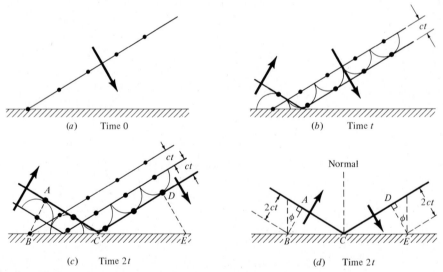

Figure 25.6. Huygens' principle applied to reflection. (*a*) A wave front reaches a reflecting surface. (*b*) After a short time *t* the secondary wavelets from five equally spaced points on the original wave front have spread a distance *ct*. The new wave front is found by drawing lines tangent to the wavelets. (*c*) The wavelets from points on a wave front in part (*b*) have spread a distance *ct*. (*d*) Two right triangles redrawn from (*c*) for clarity. The reflected wave has traveled from *B* to *A* in time 2*t*, and the incident wave will travel from *D* to *E* in an equal time. Hence *AB* = *DE*. Also, since the points shown on the wave front are equally spaced, *AC* = *CD*. This means that the triangles *ABC* and *CDE* have two sides and an included angle equal, so they are congruent and all their corresponding sides and angles are equal. Consequently the two wave fronts are at the same angle with the surface. This means that the two rays also make the same angle with the normal.

from an optically less dense medium moving toward a more dense medium. For example, a phase reversal occurs at an air-glass boundary when light is incident from the air. No phase reversal occurs when light is incident in the glass, which is the denser medium.

Second, the intensity I_r of the reflected wave is determined by the refractive indices n_1 and n_2 in the two media. If the incident intensity is I_0, then for normal incidence ($\phi = 0°$) in either direction

$$\frac{I_r}{I_0} = \left(\frac{n_2 - n_1}{n_2 + n_1}\right)^2 \quad \text{(normal incidence)} \quad (25.3)$$

This formula is also approximately correct if the angle of incidence is small but not zero. As the following example shows, significant amounts of light may be reflected at the air-glass surfaces of optical instruments.

Example 25.2

A lens is made of glass with $n = 1.5$. What fraction of the light is reflected at normal incidence?

With $n_1 = 1$ and $n_2 = 1.5$,

$$\frac{I_r}{I_0} = \left(\frac{n_2 - n_1}{n_2 + n_1}\right)^2 = \left(\frac{1.5 - 1}{1.5 + 1}\right)^2 = 0.040$$

Thus, 4 percent of the light is reflected. In a microscope or camera with several lenses, about 4 percent of the intensity is lost at each lens surface. Methods used to diminish these losses are discussed later in the chapter.

25.4 | REFRACTION OF LIGHT

When light rays go from one medium to another with a different index of refraction, they are bent or *refracted* (Fig. 25.7). As in the case of reflection, it is possible to find a relationship between the two directions with the aid of Huygens' principle. This relationship is called *Snell's law,* after Willebrod Snell (1591–1626) who discovered it experimentally. If the indices of refraction of the media are n_1 and n_2, and the *angle of incidence,* ϕ_1, and the *angle of refraction,* ϕ_2, are measured relative to the normal direction, then Snell's law states

$$\boxed{n_1 \sin \phi_1 = n_2 \sin \phi_2} \quad (25.4)$$

Note that if n increases, then $\sin \phi$ and consequently ϕ decrease. Thus, a ray bends toward the normal when it enters an optically denser medium ($n_2 > n_1$) and away from the normal when it enters a rarer medium ($n_2 < n_1$). Illustrations of Snell's law are given in the following two examples.

Example 25.3

A beam of light is incident from air on water at an angle of 30°. Part of the light is reflected and part is refracted (Fig. 25.8). Find the angles of the two beams.

The incident and reflected rays form equal angles with

Figure 25.8.   Example 25.3.

(a)   (b)

Figure 25.7.   (a) Light bends toward the normal going from an optically rare to an optically dense medium. (b) Light bends away from the normal going from the denser to the rarer medium.

429

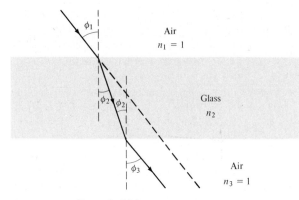

Figure 25.9.   Example 25.4.

the normal, so the angle of reflection is

$$\phi_1' = \phi_1 = 30°$$

By Snell's law, with $n_1 = 1$, $n_2 = 4/3$, and $\sin \phi_1 = \sin 30° = 0.5$, the angle of refraction is found from

$$\sin \phi_2 = \frac{n_1}{n_2} \sin \phi_1 = \frac{1}{4/3}(0.5) = 0.375$$

or $\phi_2 = 22°$.

Example 25.4

Light is incident in air at an angle ϕ_1 on a flat glass plate with index n_2 (Fig. 25.9). At what angle does the transmitted beam emerge from the other side?

At the first surface, with $n_1 = 1$, Snell's law gives

$$n_2 \sin \phi_2 = n_1 \sin \phi_1 = \sin \phi_1$$

Applying Snell's law at the second surface, with $n_3 = 1$,

$$n_2 \sin \phi_2 = n_3 \sin \phi_3 = \sin \phi_3$$

Comparing these equations, we have $\sin \phi_3 = \sin \phi_1$, or $\phi_3 = \phi_1$.

From the above example we see that a flat plate does not change the direction of a ray, although it does displace or shift the ray by a distance proportional to the thickness of the plate. This result will be useful in our discussion of lenses in the next chapter.

Dispersion | Since the index of refraction of materials varies with the wavelength of the light, the amount of bending at the boundary will vary with the wavelength. This phenomenon is called *dispersion*. *White light*, such as the light from an incandescent lamp, contains a mixture of wavelengths extending over the visible range, so its various wavelengths or colors are separated at an air-glass boundary except

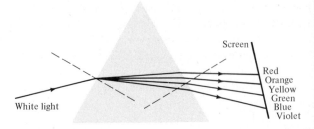

Figure 25.10.   The variation of the index of refraction with wavelength causes different colors in a beam of white light to be separated by a glass prism.

at normal incidence. This effect occurs twice in a triangular glass *prism*, producing a fan-shaped beam of light and a colorful pattern or *spectrum* on a screen behind the prism (Fig. 25.10).

25.5 | TOTAL INTERNAL REFLECTION

Figure 25.11 shows what happens to rays from a light source inside a tank of water. When the rays reach the surface, some of the light is reflected and some is refracted. As the angle of incidence increases, the intensity of the reflected beam increases. The transmitted beam gradually becomes weaker, and its *intensity diminishes to zero as the angle of refraction reaches 90°*. The corresponding angle of incidence is called the *critical angle,* ϕ_c. If the angle of incidence exceeds ϕ_c, no refracted beam is observed, and all the light is reflected. This is called *total internal reflection*.

The critical angle can be found from Snell's law by setting $\phi_2 = 90°$ or $\sin \phi_2 = 1$, corresponding to the

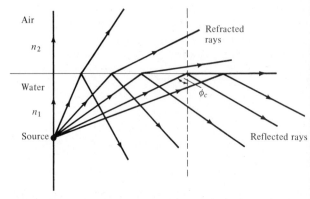

Figure 25.11.   Total internal reflection occurs if the angle of incidence exceeds the critical angle ϕ_c.

maximum possible angle of refraction. This gives $n_1 \sin \phi_c = n_2$, or

$$\sin \phi_c = \frac{n_2}{n_1} \qquad (25.5)$$

If a value of ϕ_1 larger than ϕ_c is substituted in Snell's law, $\sin \phi_2$ turns out to be greater than 1. Since no angle has a sine greater than 1, this implies that there is no refracted beam. This is in agreement with the observations that at any angle equal to or greater than ϕ_c, the beam is totally reflected.

Example 25.5

What is the critical angle for light going from glass with refractive index 1.5 into air?

Using $n_1 = 1.5$ and $n_2 = 1$,

$$\sin \phi_c = \frac{n_2}{n_1} = \frac{1}{1.5} = 0.667$$

and $\phi_c = 42°$. Thus, light incident from the glass on a glass-air boundary will be completely reflected if the angle of incidence exceeds 42°. By contrast, light going from air into glass can enter at any angle of incidence.

In ordinary specular reflection the reflected beam is always weaker than the incident beam, even if the surface is highly polished. By contrast, *no loss of intensity occurs in total internal reflection.* For this reason, totally reflecting prisms are used rather than mirrors in binoculars, periscopes, and reflex cameras (Fig. 25.12).

The fact that no intensity loss occurs in total internal reflection is also the basis for *fiber optics,* a new and rapidly growing branch of optics. The basic principle of fiber optics is illustrated by a long transparent rod or *light pipe* (Fig. 25.13a). A ray of light entering the pipe is totally reflected if the angle of incidence is large enough when the ray reaches the surface. Even if the pipe curves gradually, light will travel its entire length and arrive unattenuated after many reflec-

tions. Consequently, a single light pipe can transmit light energy quite efficiently. However, the rays from different parts of an object are completely scrambled by the multiple reflections, so a single pipe cannot transmit an image. Images can be transmitted using bundles of fine glass or plastic fibers, since each fiber transmits rays from a very small region of the object. The quality of the image is largely determined by the diameter of the fibers, which can be as small as 10^{-6} m (Fig. 25.13b).

Medical applications of fiber optics include instruments used in urinary bladder examinations and in studying the bronchi. Light pipes are also used to transmit intense light beams for surgery. Fiber optics bundles have recently been developed which can transmit light a distance of several kilometers. These have made possible the construction of optical communications systems, which are expected to have widespread uses in the near future. Since visible light has a very high frequency compared to radio or microwaves, and since the information-handling capacity of a wave is proportional to its frequency, optical communication systems have a much greater capacity than those now in general use (Fig. 25.14).

25.6 | YOUNG'S DOUBLE-SLIT INTERFERENCE EXPERIMENT

We have seen that when two waves are present in a string or in an acoustical medium, their interference results in a wave that is a superposition of the two waves. This can lead to standing waves and to beats. The existence of interference effects for light was first demonstrated in 1801 by Thomas Young (1773–1829), an English physician, physicist, and Egyptologist. In Young's experiment, single frequency or *monochromatic light* passes through two narrow slits spaced less than 1 mm apart. It then falls on a distant screen, forming a series of light and dark bands or

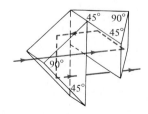

Figure 25.12. Totally reflecting prisms.

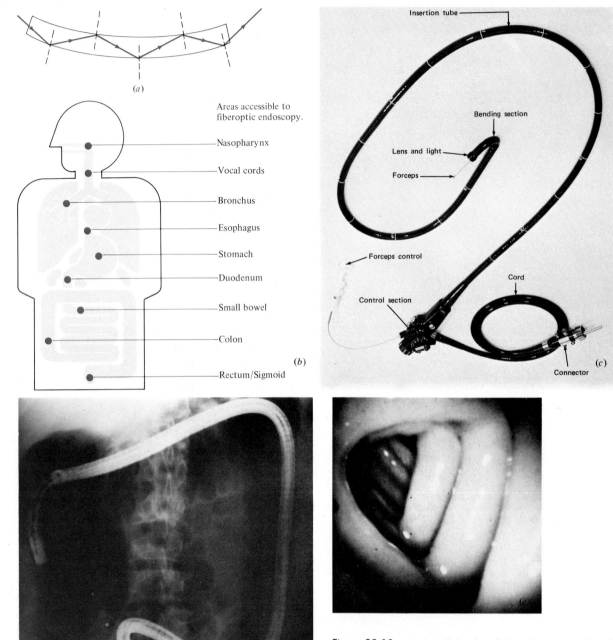

(a)

Areas accessible to fiberoptic endoscopy.

Nasopharynx

Vocal cords

Bronchus

Esophagus

Stomach

Duodenum

Small bowel

Colon

Rectum/Sigmoid

(b)

Insertion tube

Bending section

Lens and light

Forceps

Forceps control

Control section

Cord

Connector

(c)

(d)

(e)

Figure 25.13.   (a) In a light pipe, light suffers repeated total internal reflection, so it is transmitted without any loss of intensity. (b) Bundles of very fine glass fibers can transmit a sharp image even if the bundle is gradually bent. This permits a doctor to visually inspect a variety of internal organs without resorting to surgery. (c) A colonofiberscope. This instrument is fitted with a light source and can be used to remove abnormal growths. (d) An X-ray picture showing the colonofiberscope in use. (e) A photograph of a normal colon as seen with the colonofiberscope located as in (d). (Courtesy Olympus Corporation. Photos (d) and (e) by Dr. Hiromi Shinya, Beth Israel Hospital, New York.)

Figure 25.14. An experimental optical telephone system. Each thin glass fiber can carry up to 672 conversations simultaneously.

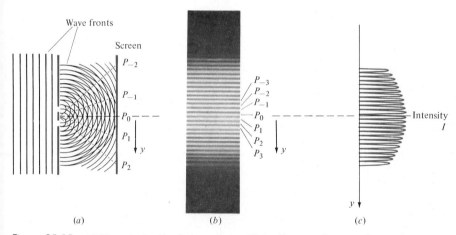

Figure 25.15. (a) Young's double slit experiment. If the distances from a point on the screen to the two slits differ by an integer number of wavelengths, maximum constructive interference occurs there. (b) A photograph of the interference fringes produced on the screen. (c) A graph of the intensity I versus position y. ((b) From Sears, Zemansky and Young, fifth edition, 1976, Addison-Wesley, Reading, Mass.)

fringes. These arise from the interference between the Huygens' wavelets originating at the two slits (Fig. 25.15).

Similar double-slit interference patterns can be observed with many kinds of waves, including water waves in a ripple tank (Fig. 25.16). The nodes correspond to points where the distances to the wavelet sources differ by a half wavelength, one and a half wavelengths, and so on. Exact cancellation or *destructive interference* occurs between the two out-of-phase waves. The wave maxima or antinodes occur when these two distances differ by an exact integer multiple of a wavelength. Then the waves are in phase, the amplitudes add, and maximum *constructive interference* occurs.

Huygens' principle can be used to locate the maxima and minima in Young's experiment. At an arbitrary point P on the screen in Fig. 25.17a the distances to the two slits are r_1 and r_2. The slit separation d is very small compared to the distance D to the screen. Consequently, if we draw a circle of radius r_1 with P as the origin, then the arc AB is approximately a straight line perpendicular to both AP and BP. From Fig. 25.17b we see that the difference in path lengths $x = r_2 - r_1$ is then equal to $d \sin \theta$. Maximum constructive interference occurs when this difference is an exact integer number of wavelengths, or when

Figure 25.16. Interference of water waves in a ripple tank. The two waves are produced by identical synchronized vibrations. (Reproduced by permission of the publisher from *PSSC Physics*, fourth edition, © 1976, by D. C. Heath and Company, Lexington, Massachusetts.)

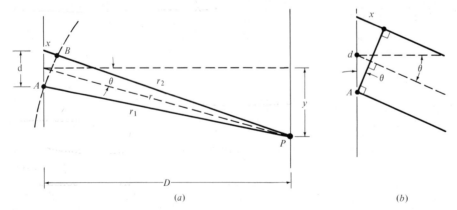

Figure 25.17. (*a*) The geometry of Young's double slit apparatus. The scale is distorted for clarity; the screen distance D is actually very large compared to the slit spacing d. (*b*) Enlarged view of the region near the slits $x = d \sin \theta$.

$$d \sin \theta = m\lambda, \qquad m = 0, \pm 1, \pm 2, \ldots \quad (25.6)$$

The maxima, or fringes, are located symmetrically about the central $m = 0$ maximum. The negative integers correspond to maxima above the axis, as indicated by the points P_{-1}, P_{-2}, ... in Fig. 25.15.

The intensity at the maxima is not simply the sum of the intensities of the waves from the two slits. The intensity of a wave is proportional to the square of its amplitude. If the amplitude of each wave is A, when the two waves are in phase the total amplitude of the combined wave is $2A$. The intensity at the maxima is then proportional to $(2A)^2 = 4A^2$, or four times the intensity from a single slit. When the double-slit interference pattern is averaged over both the maxima and the minima, it turns out that the resulting average intensity is twice that of a single slit. This is the result required by energy conservation, since the net energy reaching the screen is the sum of the energies from the two slits. The position of a maximum is found in the next example.

Example 25.6

Find the angle of the $m = 3$ fringe if two slits 0.4 mm $= 4 \times 10^{-4}$ m apart are illuminated by yellow light of wavelength 600 nm.

Substituting $m = 3$ in $d \sin \theta = m\lambda$,

$$\sin \theta = \frac{m\lambda}{d} = \frac{3(600 \times 10^{-9} \text{ m})}{4 \times 10^{-4} \text{ m}}$$

$$= 4.50 \times 10^{-3}$$

When the sine is very small, we can use the approximation $\sin \theta = \theta$, where θ is in radians. Thus

$$\theta = 4.50 \times 10^{-3} \text{ rad} \left(\frac{180°}{\pi \text{ rad}} \right)$$

$$= 0.258°$$

This is a small angle, but the fringes can be seen readily if the screen is a meter or more from the slits.

When the slit separation is known, the interference pattern can be used to determine the wavelength of the light used.

Example 25.7

Two slits 4×10^{-4} m apart are 1 m from a screen. If the distance y between the central fringe and the $m = 1$ fringe is 1 mm $= 10^{-3}$ m, what is the wavelength of the light?

In Fig. 25.17a, the distance r from the center of the slits to the fringe is approximately equal to the distance D to the screen, since θ is small. Thus

$$\sin \theta = \frac{y}{r} \simeq \frac{y}{D} = \frac{10^{-3} \text{ m}}{1 \text{ m}} = 10^{-3}$$

Consequently, using $m = 1$ in $m\lambda = d \sin \theta$,

$$\lambda = \frac{d}{m} \sin \theta = \frac{(4 \times 10^{-4} \text{ m})}{(1)} (10^{-3})$$

$$= 4 \times 10^{-7} \text{ m} = 400 \text{ nm}$$

The interference fringes in the double-slit pattern (Fig. 25.15) diminish appreciably in intensity beyond the first few fringes on either side of the central maximum. This is due to the interference among wavelets produced in different parts of a *single* slit. This phenomenon, called *diffraction*, will be discussed later in this chapter.

25.7 | COHERENCE

In order to form fringes in a double-slit experiment, the light reaching the slits must originate from a *single* source. By contrast, if the slits are illuminated by light from separate lamps, only a relatively uniform level of brightness will appear on the screen, because the light waves from an ordinary source are not emitted continuously, but rather as short pulses at random intervals. Consequently the phases of the waves from two lamps frequently change, as do the positions of the maxima and minima of the interference pattern; no single interference pattern persists long enough to be detectable. The two lamps are said to be *incoherent* light sources. Only *coherent* waves, those that have a stable phase relationship, can produce interference effects.

Coherent and incoherent sources can be illustrated by two vibrators in a ripple tank. If the vibrators oscillate in unison, they produce coherent waves, and the pattern in Fig. 25.16 results. But suppose that once every few seconds one of the vibrators pauses for a brief but random period and then begins to vibrate again. The two sources will no longer be in phase. There will still be nodes and antinodes at the locations of maximum destructive and constructive wave interference. However, every time a vibrator pauses and restarts, these positions will shift. If these shifts occur frequently enough, no interference pattern will be discernable.

The electromagnetic waves produced by ordinary light sources such as heated wires or flames are emitted at random by single atoms or molecules. These act independently, much like the two vibrators in the ripple tank illustration of incoherent sources. Typically the atoms or molecules produce wave pulses of about 10^{-8} seconds duration, or a few meters in length. Such a light source can produce interference effects if a colored filter and a small hole are placed directly in front of it. The filtered light is nearly monochromatic and has originated in a small portion of the source. This light is at least approximately coherent, and it will produce fringes in a double-slit apparatus.

In recent years, intense, coherent light sources called *lasers* have become available. In a laser the atoms emit light in phase, instead of randomly or incoherently. The light produced from a laser has an extremely small range of frequencies, so it is almost exactly monochromatic, and it is highly coherent. As a result, light emerging from the laser at two different points or at two slightly different times can be used to produce interference patterns.

25.8 | THE DIFFRACTION GRATING

The wavelength of a light beam can be found by measuring the spacing of the interference fringes in a double-slit arrangement. However, this is hard to do with accuracy, since the fringes are relatively wide and weak in intensity. A *diffraction grating,* which consists of many closely spaced slits, permits much more precise wavelength determinations.

The principle of the diffraction grating is illustrated by considering the case of six slits (Fig. 25.18). If the difference $x = d \sin \theta$ of the distances from a point on the screen to any two adjacent slits is exactly an integral multiple of a wavelength, all the waves reaching the screen are in phase. Thus a maximum in the intensity occurs when

$$d \sin \theta = m\lambda, \quad m = 0, \pm 1, \pm 2, \dots \quad (25.7)$$

m is called the *order* of the maximum or spectrum produced by the grating.

This is the same formula as was found for the angular positions of the maxima in the two slit apparatus. However, here the fringes are much more intense, since the amplitudes from each of the six slits

add (Fig. 25.19). Equally important is the fact that the fringes are now much narrower. This happens because the waves from the six slits cancel almost completely as soon as the angle differs slightly from a value at which a maximum intensity occurs. Both the intensity increase and the line narrowing are accentuated as the number of slits is increased further.

Gratings are manufactured by cutting or ruling equally spaced parallel grooves or lines with an automatic ruling engine. Those used to measure wavelengths in the visible region usually have about 4000 to 12,000 lines per centimeter. Spacings not too much greater than a wavelength are employed so that the angular spacing of the maxima will be reasonably large. This makes it possible to measure the angles accurately. The large angles obtainable are illustrated by the next example.

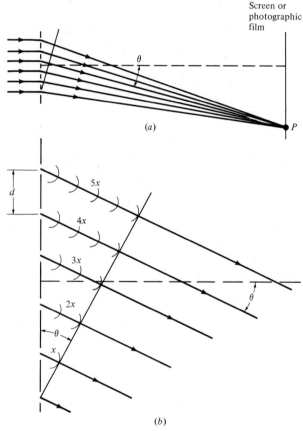

Figure 25.18. (*a*) Light rays reaching a point P on a screen from six closely spaced slits. (Light rays reaching other points are not shown.) (*b*) Enlarged view of the region near the slits.

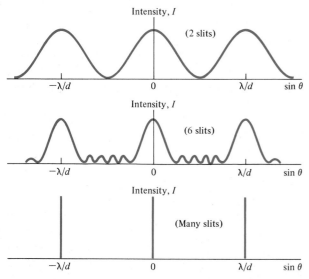

Figure 25.19. Central and $m = \pm 1$ maxima for 2, 6 and a very large number of slits. The intensity units in the three graphs are not to scale; the heights of the maxima are proportional to the square of the number of slits. Note that the fringes become more intense and narrower as the number of slits is increased. In the six-slit case, there is significant but not total cancellation among the waves from the various slits except at the peaks given by $d \sin \theta = m\lambda$. When there are a great many slits, this cancellation is almost exact.

Example 25.8

A grating has 4000 lines per centimeter. At what angles are maxima formed if it is illuminated with yellow light at 600 nm?

The slit spacing is

$$d = \frac{1}{4000} \text{ cm} = 2.5 \times 10^{-4} \text{ cm}$$

$$= 2.5 \times 10^{3} \text{ nm}$$

Using $\lambda = 600$ nm, $d \sin \theta = m\lambda$ gives

$$\sin \theta = \frac{m\lambda}{d} = \frac{m(600 \text{ nm})}{(2.5 \times 10^{3} \text{ nm})}$$

$$= m(0.240)$$

Substituting $m = 1$ gives $\sin \theta = 0.24$, or $\theta = 14°$; maxima are formed at 14° to either side of the central $m = 0$ maximum. Similarly, $m = 2, 3$, and 4 give maxima at 29°, 46°, and 74°, respectively. There are no other maxima, since $m = 5$ requires $\sin \theta = 1.2$, and no angle has a sine greater than 1.

When a diffraction grating is used to measure wavelengths, a narrow slit is placed in front of the

Figure 25.20. The line spectrum produced by iron atoms. This is a photographic negative; the dark lines seen here correspond to bright lines formed by the diffraction grating. (Courtesy of Lick Observatories.)

light source. If the source is monochromatic, the grating forms several bright lines on the screen, as in the above example. If the light contains several specific frequencies, lines are formed for each frequency, and the pattern is called a *line spectrum* (Fig. 25.20). Such a spectrum results, for example, when atoms or molecules in a gas are disturbed by an electrical current. Measurements of the wavelengths of line spectra provide important atomic and molecular structure information. Other sources, such as incandescent lamps, produce white light, which is a continuous distribution of all frequencies in the visible region, and give rise with a grating to a *continuous spectrum*. The following example shows how rainbowlike patterns are formed when a grating is illuminated by white light.

Example 25.9

A grating with a slit spacing of 2.5×10^{3} nm is illuminated with white light containing wavelengths from 400 to 700 nm. Describe the spectrum formed by the grating.

Using $m = 1$ and $\lambda = 400$ nm,

$$\sin \theta = \frac{m\lambda}{d} = \frac{(1)(400 \text{ nm})}{2.5 \times 10^{3} \text{ nm}} = 0.16$$

and $\theta = 9°$. For $m = 1$ and $\lambda = 700$ nm, $\theta = 16°$. Thus the first-order ($m = 1$) spectrum extends from 9° to 16° on each side of a central white maximum, with the shortest wavelength (violet) at the smallest angle. In the same way, one finds that the second-order ($m = 2$) spectrum extends from 19° to 34°, and the third-order spectrum, from 29° to 57°. Note that the second- and third-order spectra overlap. In general, the first-order visible spectrum of a grating is isolated, but all the higher orders overlap.

(Radio Times Hulton Picture Library.)

THOMAS YOUNG
(1773–1829)

Thomas Young was an unusually talented English physician and physicist. At the age of 2, he could read fluently; at 4, he had read the Bible twice, and at 14, he knew eight languages. Throughout his adult life he was active in both medicine and physics, and he also found the time to be the first person to make major progress on the problem of deciphering Egyptian hieroglyphics.

Young worked on various topics in the physics of fluids and made significant contributions relating to surface tension, capillarity, and the tides. He was the first to use the concept of energy as the ability of a system to do work, and he argued against the caloric theory of heat. He also studied the elastic properties of materials; as we have seen in Chapter Nine, his name is still associated with the parameter characterizing elastic deformations. However, Young's most important research related to the properties of light.

While he was still a medical student, Young discovered how the lens of the eye changes shape in order to focus on objects at different distances. He received his medical degree from Göttingen in Germany in 1796 and went into practice in London in 1799. In 1801, he discovered that irregularities in the cornea are responsible for astigmatism. Some years later, he proposed a theory of color vision based on the idea that the perception of three basic colors would, in combination, generate all the many shades we see. This theory was later refined by Helmholtz and is referred to as the Young-Helmholtz theory.

Turning his attention from the eye to the nature of light itself, Young performed some experiments that were crucial in demonstrating its wave properties. In 1803, he showed that when light went through a narrow slit, bright bands appeared in a region that would be totally dark if light did not bend. Even more striking was his double-slit experiment, which showed that two light beams can combine to produce alternating light and dark bands. The dark bands indicated a cancellation of the two beams, which was easily understood with waves but rather hard to explain with a particle model.

Young's demonstration of the wave nature of light was not well received in England, since Newton and his spiritual descendents had argued for a particle model. As a result the work needed to extend his ideas was carried out by two French physicists, Fresnel and Arago. On the other hand, French physicists found it hard to abandon Lavoisier's "French" caloric theory. Several decades passed before this theory of heat was permanently put to rest.

Angular Width of the Maxima; Resolution

We saw above that as the number of lines is increased, the maxima formed by a grating become progressively narrower. The widths of the maxima determine how accurately a wavelength can be measured with the grating and set a limit on the ability of the grating to separate or *resolve* two closely spaced wavelengths present in a beam of light.

The amplitude at a maximum is proportional to N, the number of slits, and the intensity varies therefore as N^2. Since the total power reaching the screen varies as the number of slits N, the widths of the maxima must vary as $1/N$.

Suppose now that we wish to use a grating to separate two light waves of nearly equal wavelength, λ and $\lambda' = \lambda + \Delta\lambda$. Clearly, this is possible only if the two maxima do not overlap too much. A conventional criterion for setting a limit on this overlap was introduced by Lord Rayleigh (1842–1919). *Rayleigh's criterion* is that one can just resolve two adjacent maxima if their intensities are each half their respective peak values at the point where the overlap is greatest (Fig. 25.21). Since the width of each peak is proportional to $1/N$, it is found that the peaks of the mth order spectrum can just be distinguished when

$$\frac{\Delta\lambda}{\lambda} = \frac{1}{Nm} \tag{25.8}$$

The resolution obtainable with typical gratings is quite impressive, as can be seen in the following example.

Example 25.10

Trace amounts of sodium in a flame result in a characteristic bright yellow color due to a *doublet*, light at two nearby wavelengths, 589.59 and 589.00 nm. Can a grating with 10,000 lines resolve the two wavelengths in the first-order spectrum?

For the two wavelengths in the yellow sodium doublet,

$$\frac{\Delta\lambda}{\lambda} = \frac{(589.59 - 589.00)\text{nm}}{589 \text{ nm}} = 10^{-3}$$

The smallest wavelength difference that can be resolved by this grating in first order ($m = 1$) satisfies

$$\frac{\Delta\lambda}{\lambda} = \frac{1}{Nm} = \frac{1}{(10^4)(1)} = 10^{-4}$$

Since this is only one-tenth of the ratio $\Delta\lambda/\lambda$ for the sodium doublet, the grating readily resolves the two wavelengths.

Rayleigh's criterion for the minimum resolvable separation of two maxima is a good estimate of what can be achieved if a spectrum is examined visually. However, if an intensity pattern such as the one in Fig. 25.21b is carefully measured and analyzed, the minimum distinguishable wavelength separation can be greatly reduced.

25.9 | DIFFRACTION

Thus far, we have discussed examples of interference between waves from two discrete sources, as in a double-slit experiment, or from many discrete

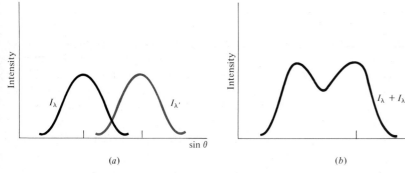

Figure 25.21. Intensity distributions in a very small angular region of a diffraction grating spectrum. (*a*) Light of wavelength λ produces the maximum shown by the black curve, while light with wavelength λ' produces the maximum shown by the colored curve. (*b*) When both waves are present, the intensity observed is the sum of the two curves in (*a*). In the case illustrated the two peaks are just barely discernable. According to Rayleigh's criterion, when the wavelengths λ and λ' are significantly closer, the two peaks cannot be resolved.

Figure 25.22. The shadow produced when a razor blade is illuminated by a monochromatic point source. (*From Sears, Zemansky and Young, fifth edition, 1976, Addison-Wesley, Reading, Mass.*)

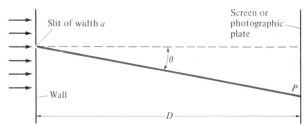

Figure 25.23. Geometry of the single slit different experiment. The pattern seen on the screen is shown in Fig. 25.24.

sources, as in the diffraction grating. However, parts of a wave from a single source will also interfere with each other. This causes effects such as the bending of waves around an obstacle. Single-source interference is referred to as *diffraction* and occurs for all kinds of waves.

Figure 25.22 shows the shadow of a razor blade placed between a point source of monochromatic light and a photographic plate. Some light is bent inside the *geometrical shadow,* the region that would be totally dark in the absence of any bending. Near the edge of the shadow, a diffraction pattern of alternate light and dark bands appears. Thus, even in this fairly simple situation, the diffraction pattern is quite complex.

Under ordinary conditions, we seldom notice the diffraction of light. Light sources such as incandescent lamps or the sun are not monochromatic point sources, and the diffraction patterns due to different parts of the source and to different wavelengths usually overlap and obscure each other. Nevertheless, diffraction patterns are visible when we look at a distant light source, such as a street lamp, through a crack between two fingers or through a cloth umbrella. By contrast, the diffraction of sound waves is hard to avoid. Sound readily bends around obstacles of ordinary size, such as furniture, and fills a room rather uniformly. This difference between the diffraction of visible light and of sound is due to their wavelengths, which are typically 5×10^{-7} m and

1 meter, respectively. *Diffraction effects are large only when we deal with obstacles or apertures comparable in size to the wavelength.*

Diffraction by a Narrow Slit | Calculations of

diffraction patterns are usually complex. However, for the case of light diffracted by a narrow slit, some of the main features can be obtained without elaborate mathematics. Qualitatively similar results are found for other situations.

In Fig. 25.23, monochromatic light passes through a slit whose width a is comparable to the wavelength λ and falls on a screen or photographic plate at a large distance. The Huygens' wavelets from different parts of the slit interfere and produce the diffraction pattern shown in Fig. 25.24.

To locate the minima in this diffraction pattern, we start by dividing the slit into two halves (AB and BC in Fig. 25.25). The distance to the screen for the wavelet originating at A is greater than that for the wavelet from B by $x = \frac{1}{2}a \sin \theta$. Wavelets originating at any point on AB and at the corresponding point on BC will have the same path difference. Consequently, if x is exactly $\frac{1}{2}\lambda$, all the pairs of wavelets will be out of phase and cancel exactly, producing a minimum. Hence a minimum occurs when $x = \frac{1}{2}a \sin \theta = \frac{1}{2}\lambda$, or

$$a \sin \theta = \lambda$$

At this point one might suppose that a maximum occurs when $x = \lambda$, since the wavelets from A and B will then be exactly in phase. However, this is not the case, as can be seen by dividing the slit into four equal parts (Fig. 25.25). Wavelets from A and D will have a path difference which is $\frac{1}{2}x = \frac{1}{2}\lambda$, so they will be exactly out of phase and cancel completely. The same is also true for wavelets from any two points $a/4$ apart, so $x = \lambda$ also implies a minimum. Similarly, we can divide the slit into 6, 8, . . . parts and find

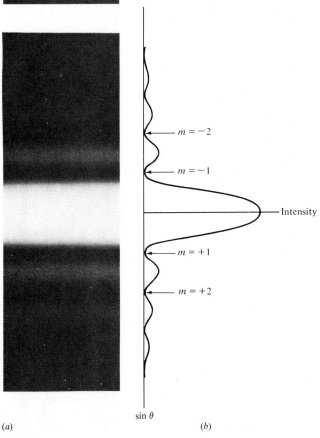

(a)   *(b)*

Figure 25.24. (*a*) Photograph of the diffraction pattern produced by a single slit illuminated by monochromatic light as in Fig. 25.23. (*b*) Intensity graph for this diffraction pattern. (From Sears, Zemansky and Young, fifth edition, 1976, Addison-Wesley, Reading, Mass.)

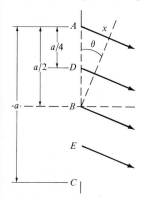

Figure 25.25. A diffraction minimum occurs when wavelets from different parts of the slit interfere destructively.

that total destructive interference occurs when $x = \frac{1}{2}a \sin \theta = \frac{1}{2}\lambda,\ \lambda,\ \frac{3}{2}\lambda,\ 2\lambda,\ \dots$. Thus, in general, diffraction minima occur at angles satisfying

$$a \sin \theta = m\lambda, \qquad m = \pm 1, \pm 2, \dots$$
$$\text{(diffraction minima)} \quad (25.9)$$

The negative integers correspond to minima above the axis in Fig. 25.24.

The maxima are located approximately (but not exactly) midway between the minima, or at $m = 0$, $\pm\frac{3}{2}\lambda, \pm\frac{5}{2}\lambda, \dots$. The central $m = 0$ maximum is very bright, since all the wavelets have nearly the same path length and are in phase. At the other maxima the intensity is much smaller, and it diminishes rapidly as m increases (Fig. 25.24). This happens because, except at $\theta = 0$, the wavelets from some parts of the slit cancel even at the maxima. This partial cancellation becomes more and more complete as the angle increases.

The angular width of the diffraction pattern depends on the size of the slit relative to the wavelength of light, as is seen in the next example.

Example 25.10

A slit is illuminated by light of wavelength λ. Find the angular position of the first diffraction minimum as the slit width expands from λ to 5λ and finally to 10λ.

Substituting $m = 1$ and $a = \lambda$ in $a \sin \theta = m\lambda$, we find

$$\sin \theta = m \frac{\lambda}{a} = (1)\frac{\lambda}{\lambda} = 1$$

so $\theta = 90°$. Similarly, $a = 5\lambda$ gives $\sin \theta = 0.2$ and $\theta = 12°$; $a = 10\lambda$ gives $\sin \theta = 0.1$ and $\theta = 6°$. Thus, as the slit becomes wider, the central diffraction peak becomes narrower. A similar narrowing of the diffraction pattern would be observed if the slit size were held fixed and the wavelength decreased.

Diffraction explains some of the features of the double-slit interference experiment that we neglected in Section 25.6. Earlier we supposed that all the light coming from a single slit had the same phase. This is equivalent to assuming that the slit is very narrow compared to a wavelength and that its diffraction pattern is very wide. However, if the slit widths are comparable to the wavelength, the waves reaching the screen from each slit will have an intensity that varies with position in accordance with the single-slit diffraction formulas. Thus the complete pattern for a double slit is actually formed from two superimposed diffraction patterns, and the double-slit interference maxima do not all have the same intensity. The single-slit diffraction pattern forms an *envelope* for the interference pattern so that the intensity of the peaks becomes smaller further away from the center of the pattern (Fig. 25.26).

Diffraction by a Circular Aperture | When

a light wave enters an optical instrument with a circular opening, a diffraction pattern is produced by the interference of the wavelets originating at different points in the aperture. The diffraction pattern due to a distant point source of light has a bright central circular region, surrounded by concentric dark and light rings (Fig. 25.27). A detailed analysis shows that the first minimum for a circle with diameter d occurs when

$$\sin \theta = 1.22 \frac{\lambda}{d} \qquad (25.10)$$

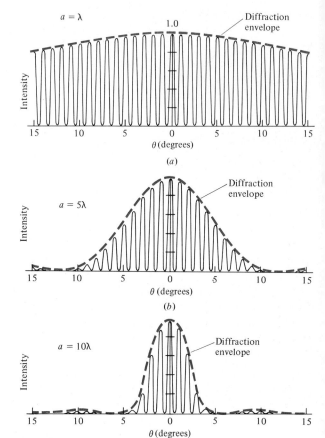

Figure 25.26. Interference patterns for two slits separated by $d = 50\lambda$. The fringes seen are centered at the peaks in the solid curve. Note how the pattern changes as the slit width a is increased and the single slit diffraction envelope narrows. (From Halliday and Resnick, *Physics*, Part II, third edition, Copyright © 1978, John Wiley and Sons, New York.)

Figure 25.27. The image of a star formed on a photographic plate by a telescope. A central diffraction maximum and a circular secondary maximum are visible. Additional maxima also occur but their intensities are too small to be seen here. (From Halliday and Resnick, *Physics*, Part II, third edition, Copyright © 1978, John Wiley and Sons, New York.)

This is similar to the formula for the first minimum of a slit, $\sin \theta = \lambda/a$, but the circular geometry results in the factor of 1.22.

We see in the next chapter how diffraction limits the sharpness of images formed by optical instruments and the human eye.

25.10 | POLARIZATION OF LIGHT

In Chapter Twenty-three, we noted that light, like any other transverse wave, can be polarized. We now examine this idea in more detail.

Electromagnetic waves have electric and magnetic fields oscillating at right angles to the direction of motion of the waves. If the electric field vector is always along a certain direction, the wave is said to be *linearly polarized* along this direction. The radiation from a single atom or molecule is polarized, but since the many individual atoms or molecules usually act randomly, the resultant light beam from most sources is unpolarized.

Polarized light can be produced from an unpolarized beam in several ways, including *absorption, reflection,* and *scattering*. The most familiar way, mentioned in Chapter Twenty-three, is by *absorption* in Polaroid filters. These filters contain long molecules that are aligned. When light has its electric field vector along the molecules, electric currents are set up, and the light is absorbed. However, little happens to a light beam polarized at right angles to the molecules. A similar effect can be demonstrated nicely with microwaves, which are electromagnetic waves with wavelengths of the order of centimeters (Fig. 25.28). Microwaves with vertical electric fields pass readily through horizontal wires. However, when the wires are vertical, the electric field sets up currents in the wires, and the waves are absorbed.

Light can also be polarized by *reflection*. Except at normal incidence, the fraction of the light reflected by a surface depends on its polarization. It was discovered in 1812 by Sir David Brewster (1781–1868) that the reflected beam is completely polarized at an angle of incidence, ϕ_p, which is determined by the indices of refraction n_1 and n_2 of the two media (Fig. 25.29). This angle is called *Brewster's angle* and satisfies

$$\tan \phi_p = \frac{n_2}{n_1} \qquad (25.11)$$

Figure 25.28.   Microwaves polarized so that the electric field is vertical are beamed toward a detector. (*a*) Horizontal metal wires permit the waves to pass unimpeded. (*b*) Vertical wires absorb the waves, as shown by the reduced detector reading. (Courtesy of Worth Publishers, from P. A. Tipler, *Physics*, 1976. Photos by Larry Langrill.)

If an unpolarized beam is incident at this angle, the relatively weak reflected beam will be fully polarized perpendicular to the *plane of incidence* defined by the incident ray and the normal. The stronger transmitted beam will be slightly polarized. The use of Brewster's angle is illustrated by the following common situation.

Example 25.11
Sunlight reflected off a still lake is fully polarized. At what angle is the light incident on the lake?

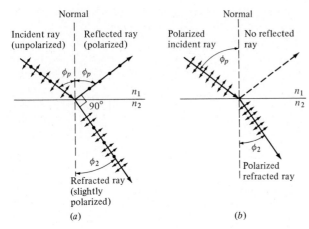

Figure 25.29. Polarization by reflection, (*a*) On the average the unpolarized incident beam has equally large electric fields perpendicular to the plane of incidence (dots) and in the plane (arrows). When the wave is incident at Brewster's angle, the reflected beam is fully polarized. The transmitted beam is only slightly polarized, since only a small fraction of the light is reflected. (*b*) A wave polarized with its electric field in the plane of incidence is fully transmitted at Brewster's angle.

Since $n_1 = 1$ for air and $n_2 = 4/3$ for water, the reflected beam is polarized horizontally when

$$\tan \phi_p = \frac{n_2}{n_1} = \frac{4/3}{1} = 1.333$$

or $\phi_p = 53°$. Light reflected off the lake at angles near 53° will be mostly horizontally polarized.

Polarized light is also produced by *scattering,* which is the absorption and reradiation of light. An interesting demonstration of the polarization of scattered light can be done using a tank of water contain-

ing some powdered skim milk or soap. Using a Polaroid filter, one finds that the light scattered at right angles to the incident beam is polarized perpendicular to the plane of the incident and scattered rays (Fig. 25.30). If a Polaroid filter is used to polarize the incident beam vertically, no change is observed in the light scattered at right angles in the horizontal plane. However, the vertical beam now disappears.

These observations are explained by the fact that the incident light sets charges in the atoms into oscillatory motion along the direction of the electric field. These oscillating charges then emit radiation with an electric field that has a component along the direction of oscillation. Since the waves are transverse, this means that no radiation is emitted along the direction of oscillation. Accordingly the horizontal beam originates from vertical oscillators and is polarized vertically.

The daytime sky appears blue because short wavelength light is most readily scattered. The sunlight scattered at right angles is fully polarized, and all the scattered sunlight is at least partially polarized. Bees are able to sense this polarization and can use the information to determine the direction of the sun from a view of only a small part of the sky.

25.11 | X-RAY DIFFRACTION AND THE STRUCTURE OF BIOLOGICAL MOLECULES

Because waves are unaffected by objects that are small compared to a wavelength, wavelengths com-

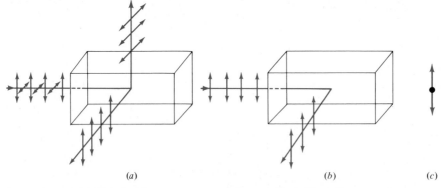

Figure 25.30. (*a*) Light incident on water containing powdered skim milk or soap is scattered in all directions. Light scattered at right angles is fully polarized. (*b*) If the incident light is vertically polarized, the vertically scattered beam disappears. (*c*) A vertically polarized incident beam causes charges to oscillate vertically. These oscillating charges produce radiation with an electric field component along the direction of oscillation.

444

Figure 25.31. (a) An x-ray tube. Electrons from the heated filament are accelerated through the potential difference V, and produce x-rays when they strike the metal target at the anode. (b) The spectrum of x-rays produced by a typical x-ray tube.

parable to the interatomic spacing are needed to "see" the arrangement of atoms in a molecule. This spacing is typically a few angstroms, where 1 angstrom = 10^{-10} meters. Visible light has a wavelength of 400 to 700 nm or 4000 to 7000 Å, so it cannot be used to study molecular structure. However, x-rays, which are electromagnetic waves with typically a 1-Å wavelength, are well suited to this task. A crystalline material, which consists of a regular array of atoms, serves for x-rays as a three-dimensional analog to the diffraction grating. Information obtained from such x-ray diffraction studies has contributed greatly to unraveling the structure of complex biological molecules such as proteins and DNA, the carrier of the genetic code.

When electrons from a heated filament are accelerated through a large potential difference and allowed to strike a metal target, x-rays are produced with a continuous distribution of wavelengths (Fig. 25.31). If the x-rays then strike a crystal, the deflected x-rays form intense *Laue spots* on a screen or film (Fig. 25.32). These are due to the constructive interference of wavelets produced by many atoms.

To understand how one Laue spot is formed, consider a crystal made up of identical atoms in a cubic arrangement that is exposed to a beam of x-rays containing many different wavelengths. We can imagine the atoms as forming a series of partially reflecting planes, so that some x-rays are reflected by each plane (Fig. 25.33). If the distance between two successive planes is d, then the path lengths for waves reflected by two successive planes will differ by $2x = 2(d \sin \alpha)$, where α is the angle between the incident x-ray direction and the crystal plane. (Conventionally α is used in x-ray diffraction discussions rather than the angle between the beam and the normal direction.) Constructive interference occurs among the reflected beams from all the parallel planes if the path length difference is an integer number of wavelengths, or if

$$2d \sin \alpha = m\lambda, \qquad m = 1, 2, 3, \ldots \quad (25.12)$$

This equation was first obtained by Sir William Henry Bragg (1862–1942), a pioneer in x-ray research, and is called the *Bragg condition*.

Although the incident beam contains many wavelengths, for a given angle of incidence α, spots will be produced only for those wavelengths satisfying Eq. 25.12. The remainder of the beam, containing all other wavelengths, is either absorbed or transmitted. (Because there are many parallel reflecting planes, nearly total cancellation occurs among reflected beams if the Bragg condition is not satisfied.) Notice that spots, and not lines, are formed, because the incident beam is very narrow and would itself form a single spot if it were not deflected. The selective nature of the crystal diffraction is illustrated by the next example.

Example 25.12

The interatomic spacing d in Fig. 25.33 is 2 Å, and the angle α is 10°. If the shortest wavelength in the x-ray beam is 0.4 Å, which wavelengths will be reflected strongly from the planes shown?

The wavelengths strongly reflected satisfy $m\lambda = 2d \sin \alpha$. Using $m = 1$ and $\sin 10° = 0.174$,

$$\lambda = \frac{2d \sin \alpha}{m} = 2 \frac{(2 \text{ Å})}{(1)} (0.174) = 0.696 \text{ Å}$$

445

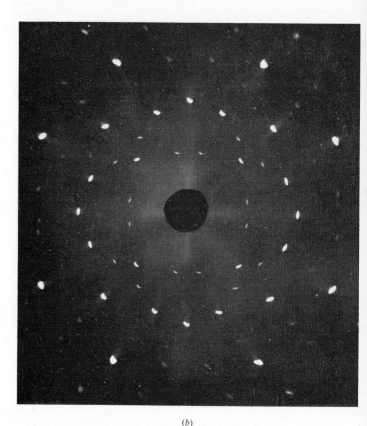

Figure 25.32. (a) A narrow beam of nonmonochromatic x-rays strikes a crystal and forms a pattern of Laue spots on a photographic plate. (b) The Laue spots from a sodium chloride crystal. ((b) From Halliday and Resnick, *Physics*, Part II, third edition, Copyright © 1978, John Wiley and Sons, New York.

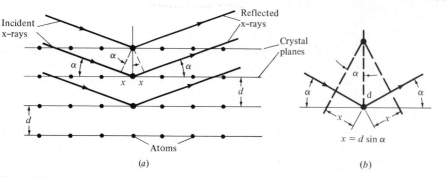

Figure 25.33. (a) X-rays are reflected by parallel planes of atoms in a crystal. (b) An enlarged view of the geometric details.

Using $m = 2$, we get half this wavelength, or 0.348 Å, which is less than the 0.4-Å minimum wavelength of the beam. Hence only the 0.696 Å x-rays will interfere constructively, and the reflected beam will be monochromatic. We see then that reflection from crystals provides a way of making monochromatic x-ray sources.

The Laue pattern in Fig. 25.32 contains a large number of spots. This is because there are many sets of parallel reflecting planes in a crystal, with various orientations (Fig. 25.34). Each set of planes produces one or more spots in accordance with the Bragg condition.

The pattern of spots formed by a mixture of x-ray wavelengths cannot be used to determine the spacing of crystal planes, because there is no way to find the wavelength corresponding to a particular spot. Thus to obtain information about the crystal structure and

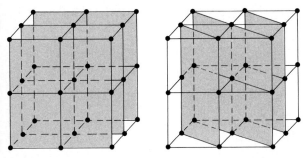

Figure 25.34. Two of the many sets of reflecting planes in a NaCl crystal.

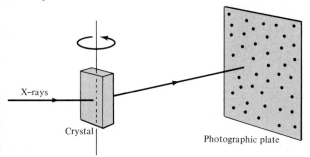

Figure 25.35. Monochromatic x-rays are allowed to strike a crystal that is gradually rotated, forming a series of spots characteristic of the crystal structure.

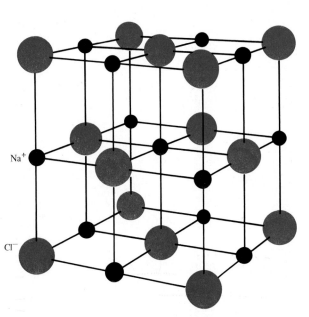

Figure 25.36. The crystal structure of sodium chloride. (From Halliday and Resnick, *Physics*, Part II, third edition, Copyright © 1978, John Wiley and Sons, New York.)

the spacing of its planes, monochromatic x-ray beams are required. However, when monochromatic x-rays are used, for most crystal orientations the Bragg condition is not satisfied for *any* set of planes and *no* spots are formed. To overcome this, the crystal is gradually rotated (Fig. 25.35). Spots are formed on the plate as sets of planes are brought to the required angles.

The locations and intensities of these spots contain remarkably detailed information about the crystal. Since a spot is formed whenever the Bragg condition is satisfied for some set of planes, the spot *positions* depend only on the geometric structure of the crystal and the spacing of its planes. For example, sodium chloride (Fig. 25.36) and all other cubic crystals have the same characteristic x-ray diffraction pattern, with the overall scale of the pattern set by the atomic spacing. However, the *relative intensities* of the spots depend on the chemical composition of the crystal. This is because the x-rays are actually reflected by the electronic clouds of the atoms, and heavy atoms, which have more electrons, are better reflectors. For example, in Fig. 25.36 different sets of planes contain different relative numbers of Na^+ and Cl^- ions, and

this determines the intensity of the associated spots.

Although x-ray diffraction was originally used with relatively simple inorganic crystals, it has also been applied with spectacular success to biological molecules, such as proteins and nucleic acids, that can be put into crystalline form. For example, in the 1950s, Perutz compared the positions and intensities of thousands of spots for hemoglobin with those expected from various models of the molecule. He found that this oxygen-carrying protein of the blood consists of about 10,000 atoms assembled into four chains, each with a helical form and several bends. Combining the x-ray data with chemical determinations of the sequence of the amino acid building blocks in the hemoglobin molecule, he was able to construct a detailed picture of the location and spacing of the atoms. Perutz also showed how the shape of hemoglobin changes when it gains or loses oxygen.

Another major advance in biology based in part on x-ray studies was the discovery by Crick and Watson in 1953 that the structure of the nucleic acid DNA is a double helix. The helical structure of DNA is apparent when its diffraction pattern is compared with that expected for an array of helices (Fig. 25.37).

447

Figure 25.37. (*a*) An optical simulation of x-ray diffraction from an array of helices. (*b*) The x-ray diffraction pattern for a lithium salt of DNA. Helical structures always produce characteristic clear areas similar to those seen in both patterns. ((*a*) After A. R. Stokes, "The Theory of X-ray Fibre Diagrams," Prog. Biophysics vol. 5, 1955, p. 140. (b) Courtesy of Professor M.H.F. Wilkins, Biophysics Department of King's College, London.)

SUMMARY

Electromagnetic waves of any frequency travel in a vacuum at a speed $c = 3.00 \times 10^8 \, \text{m s}^{-1}$. In a medium, the speed is $v = c/n$, where the index of refraction n is never less than 1.

The wave nature of light is demonstrated by many kinds of interference effects. For any wave, Huygens' principle states that each point on a wavefront can be considered as the source of small secondary wavelets. These wavelets can be used to predict many properties of light waves.

When light is reflected or refracted by objects large compared to its wavelength, the wave character of light plays a minor role and can usually be neglected. When a light ray goes from one transparent medium into another, it bends in accordance with Snell's law. A ray bends toward the normal on entering an optically denser medium and away from the normal on entering a rarer medium. Total internal reflection will occur in the latter case if Snell's law gives $\sin \phi > 1$ for the refracted ray.

For a pair of narrow slits illuminated by a coherent beam of monochromatic light, Huygens' principle predicts a pattern of alternating dark and bright fringes on a distant screen. These appear at the angles of maximum destructive and constructive interference, respectively. In a diffraction grating, which has many slits, nearly total destructive interference occurs except in very narrow angular regions.

Secondary Huygens' wavelets from different parts of a single wavefront also interfere with each other. This diffraction property gives rise to a small but detectable bending of light around objects of ordinary size and is a limiting factor in the resolution obtained with optical instruments.

When the electric field vector of an electromagnetic wave is always directed along some line, the wave is said to be linearly polarized in that direction. Light can be polarized by various means, including absorption, reflection, and scattering.

Checklist
Define or explain:

index of refraction	dispersion
angstrom	total internal reflection
nanometer	critical angle
Huygens' principle	coherence
wavefront	diffraction grating
secondary wavelets	line, continuous spectra
diffuse, specular	diffraction
reflection	Brewster's angle
Snell's law	scattering
white light	

EXERCISES

Section 25.1 | The Index of Refraction

25-1 Find the speed of light in water.

25-2 What is the speed of light in diamond?

25-3 A lamp emits yellow light with wavelength 600 nm in air. (a) What is the wavelength of the light in water? (b) What color will the lamp appear to be to a diver who is not wearing a mask? Explain.

25-4 A beam of light has a wavelength of 640 nm in glass with a refractive index of 1.5. What is the frequency of the light?

Section 25.3 | Reflection of Light

25-5 What fraction of the light intensity is reflected when light is normally incident in air on water?

25-6 In which direction must light initially travel to have a phase reversal at (a) an air-water surface; (b) a water-glass surface?

25-7 Light is incident in water along the normal to a glass plate with refractive index 1.5. What fraction of the light intensity is transmitted into the glass?

Section 25.4 | Refraction of Light

25-8 A ray of light is incident on a glass-water surface from the glass at an angle of 45°. Find the angle of refraction if the index of refraction of the glass is 1.5.

25-9 Light is incident in air on water at an angle of 15° to the normal. (a) At what angle is the reflected ray? (b) At what angle is the refracted ray?

25-10 Light is incident in water on an air-water boundary at an angle of 30° to the normal. What are the angles of reflection and refraction?

Section 25.5 | Total Internal Reflection

25-11 What is the critical angle for total internal reflection in diamond? (Assume the diamond is in air.)

25-12 A glass light pipe in air will totally internally reflect a light ray if its angle of incidence is at least 39°. What is the minimum angle for total internal reflection if the pipe is in water?

Section 25.6 | Young's Double-Slit Interference Experiment

25-13 Two narrow slits are illuminated with light of wavelength 500 nm. Adjacent maxima near the center of the interference pattern are separated by 1.5°. How far apart are the slits?

25-14 Two narrow slits $0.2\,mm = 2 \times 10^{-4}\,m$ apart are illuminated with red light of wavelength 700 nm. At what angles are the five maxima closest to the center of the pattern formed?

25-15 A double-slit apparatus is illuminated with yellow sodium light ($\lambda = 589$ nm). The maxima on a screen 1 m away are 1 cm apart. Find the slit separation.

Section 25.7 | Coherence

26-16 A double-slit apparatus is illuminated with a narrow beam of white light from a small portion of a lamp. A filter that transmits only red light is placed just before one slit and a similar green filter before the other. Will a double-slit pattern be observed on the screen? Explain.

Section 25.8 | The Diffraction Grating

25-17 A diffraction grating with 5000 lines per centimeter is illuminated with yellow sodium light of wavelength 589 nm. (a) What is the angular position of the $m = 1$ line? (b) How many lines can be seen?

25-18 A grating has 4000 lines per centimeter. What is the longest wavelength for which the fourth order line can be observed?

25-19 A source emits a red doublet near 656 nm with a separation of 0.2 nm. (a) At what angles will lines be seen if the diffraction grating has 8000 lines per centimeter? (b) If the two first order lines are just resolved, how many lines does the grating have?

25-20 What is the minimum number of lines needed in a diffraction grating to resolve in second order the sodium doublet at 589.59 and 589.00 nm?

25-21 A diffraction grating with 8000 lines per centimeter is illuminated with light from a hydrogen lamp. In the first-order spectrum, what is the angular separation between the 656 and 410 nm lines emitted by atomic hydrogen in the lamp?

25-22 Explain why diffraction gratings have (a) closely spaced slits; (b) a large number of slits.

Section 25.9 | Diffraction

25-23 A narrow slit is illuminated with white light. What will be seen on a screen beyond the slit?

25-24 A narrow slit is illuminated with yellow light of wavelength 589 nm. If the central diffraction maximum extends from 0° to 40°, how wide is the slit?

25-25 A circular aperture of radius 10^{-5} m is illuminated with light of wavelength 500 nm. At what angle is the first diffraction minimum?

Section 25.10 | Polarization of Light

25-26 Light is incident from air on glass with an index of refraction 1.5. At what angle will the reflected light be fully polarized?

Section 25.11 | X-Ray Diffraction and the Structure of Biological Molecules

25-27 It is found that x-rays are reflected strongly from crystal planes 2 Å apart when they are incident at 20° from the planes. What is their longest possible wavelength?

25-28 A set of planes in a crystal has a spacing of 3 Å. If a beam of monochromatic x-rays of wavelength 1 Å is incident on the crystal, what is the smallest angle relative to the planes at which constructive interference occurs?

PROBLEMS

25-29 A monochromatic light ray enters the prism shown in Fig. 25.38 at an angle of 30° to the normal. The refractive index is 1.5. At what angle θ does the ray leave the prism?

Figure 25.38. Problem 25-29.

*25-30 An object 1 m under water is observed from the air. (a) What is the angle of refraction for a light ray from the object reaching the surface at an angle of 10° to the normal? (b) Where does the object appear to be? (*Hint:* Calculate where the

refracted ray appears to intersect a ray normal to the surface as in Fig. 25.39. Q is the actual position and Q' is the apparent position.)

Figure 25.39. Problem 25-30.

*25-31 Snell's law of refraction can be derived from Huygens' principle by using a procedure similar to that used in Fig. 25.6 to derive the law of reflection. (a) Draw a series of diagrams showing how a wavefront changes at it travels from a medium with light velocity v_1 into a medium with velocity v_2. (b) Using similar triangles derive Snell's law.

25-32 Obtain a formula for the angular positions of the minima in a double-slit apparatus.

25-33 One does not normally detect interference effects between the sound coming from two loudspeakers of a stereo system. Explain why.

25-34 Show how one could devise a simple experiment to demonstrate the interference between sound waves from two sources.

25-35 What will be observed in a double-slit experiment if white light is used? Explain.

*25-36 A coherent light beam is produced by placing a filter and a slit of adjustable width in front of an incandescent lamp. The light then falls on a double slit apparatus and forms an interference pattern on a screen. Describe what happens as the source slit is gradually increased in width. (Ignore the effects of source slit diffraction.)

25-37 A grating with 6000 lines per centimeter is illuminated with white light. (a) What is the highest order spectrum that includes the entire visible spectrum? (b) What is the highest order spectrum that includes any portion of the visible spectrum?

25-38 A grating with 4000 lines per centimeter forms a line at 45°. (a) What are the possible wavelengths of the light illuminating the grating? (b) How could one determine the actual wavelength?

25-39 When a small opaque circular disk is illuminated by a coherent plane wave, a bright spot is seen at the center of the shadow formed on a distant screen. Explain why.

*25-40 Suppose that light is incident on a diffraction grating at an angle ϕ to the normal direction (Fig. 25.40). (In the discussion in this chapter, ϕ was always chosen to be zero.) Show that the maxima occur at angles θ such that $d(\sin \phi + \sin \theta) = m\lambda$, where $m = 0, 1, 2, \ldots$, and d is the slit separation.

Figure 25.40. Problem 25-40.

25-41 A slit of width 10^{-5} m is illuminated with light of wavelength 500 nm. How wide is the central diffraction peak on a screen 0.5 m away?

25-42 A lens when properly focused on a distant point light source of wavelength 550 nm forms a circular spot 0.1 mm $= 10^{-4}$ m in diameter on a screen 0.2 m from the lens. If the spot size is entirely due to diffraction by the lens, what is the diameter of the lens?

25-43 Light of wavelength 600 nm is incident on a circular hole of radius 10^{-4} m. How large a spot is formed on a screen 1 m away? (Neglect secondary diffraction maxima.)

25-44 Why are high-frequency loudspeakers (tweeters) more directional than low-frequency speakers (woofers)? (*Hint:* Regard the speaker as a round hole through which the sound waves pass.)

25-45 Using Snell's law, show that when light is incident at Brewster's angle, the reflected and refracted beams are at right angles.

25-46 Suppose a sodium chloride crystal is ground into a powder, so that it becomes a large number of randomly oriented small crystals. If monochromatic x-rays are incident on the powder, what kind of interference pattern will be formed? (*Hint:* Consider what will happen to the spots in Fig. 25.32 if the axis of rotation is changed.)

SUPPLEMENTARY TOPICS

25.12 | HOLOGRAPHY

One of the most intriguing and versatile advances of recent years is *holography*. A *hologram* is a photograph of the interference pattern produced when monochromatic light reflected or transmitted by an object interferes with a coherent *reference* light beam. When a hologram is illuminated by a coherent light source, an image of the original object is formed. Unlike an ordinary two dimensional photograph, this *reconstructed* image is three dimensional. Consequently, a moving viewer observes *parallax:* the relative displacement of nearer and more distant parts of the object (Fig. 25.41).

Holography was invented in 1947 by D. Gabor, who received the 1972 Nobel prize for his work. However, the lack of sufficiently strong coherent light sources limited the usefulness of holography until the advent of the laser in 1960. Since then, increasingly varied and ingenious applications of holography have been developed.

In order to understand how holography works, it is useful to first consider a circular arrangement of concentric transparent and opaque rings called a *zone plate* (Fig. 25.42) illuminated by coherent monochromatic light of wavelength λ. Going outward from the center of the plate, the average distance from a ring to a particular point P on the axis is one wavelength greater than from the preceding ring. Accordingly the Huygens' wavelets from all the transparent rings arrive approximately in phase and interfere constructively. The opaque rings block out the wavelets that would interfere destructively with those from transparent rings.

Because of the opaque rings, about half of the incident light is blocked by the zone plate. Much of the remaining light travels straight ahead. However, the Huygens' wavelets from the transparent rings arrive at P approximately in phase, forming a bright spot due to the constructive interference (Fig. 25.43a). Wavelets spreading outward as though they had originated at point P' a distance d to the left of the zone plate are also in phase. If they are eventually brought together by a lens, as in the eye or a camera, they interfere constructively. Hence, if one looks through a zone plate toward a light source, one sees a bright spot at P', even though no light actually originates

451

Figure 25.41. (*a*) A hologram. (*b*), (*c*), and (*d*) are photographs taken of the reconstructed images taken at slightly different angles. The parallax shows that holography reconstructs the images in three dimensions. (From Howard K. Smith, *Principles of Holography,* second edition, Copyright © 1975, John Wiley, and Sons, New York.)

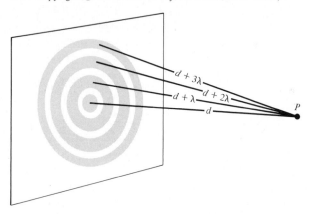

Figure 25.42. A zone plate has alternate opaqe and transparent rings. All the wavelets reaching point *P* are in phase and interfere constructively.

there (Fig. 25.43*b*). This is called a *virtual image,* while the bright spot actually formed at *P* is called a *real image.*

The hologram produced by a single point object is very similar to a zone plate. In the idealized arrangement of Fig. 25.44*a*, a laser beam is incident on a pointlike object, producing spherical Huygens' wavelets. At some places on the photographic plate, the spherical wave from the object and the undeflected reference wave are exactly in phase, and maximum constructive interference occurs. At other places the waves are out of phase, and destructive interference occurs. This produces a pattern that looks like a zone plate, although the abrupt changes from opaque to transparent rings are replaced by a more gradual variation in intensity. If the developed photographic plate is later illuminated by a laser beam, virtual and real images are formed of the point object just as they were for the zone plate (Fig. 25.44*b*).

When a hologram is made of a complex object, each point on the object produces spherical waves that interfere with the reference wave and form a series of zone platelike rings on the photographic plate. These rings are superimposed, so the resulting

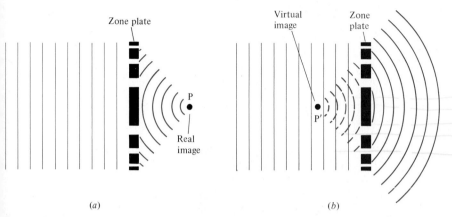

Figure 25.43. A coherent monochromatic wave incident on a zone plate has alternate zones blocked by the plate. Some transmitted light goes straight through, some forms a real image at P as shown in (a), and some forms a virtual image at P' as shown in (b).

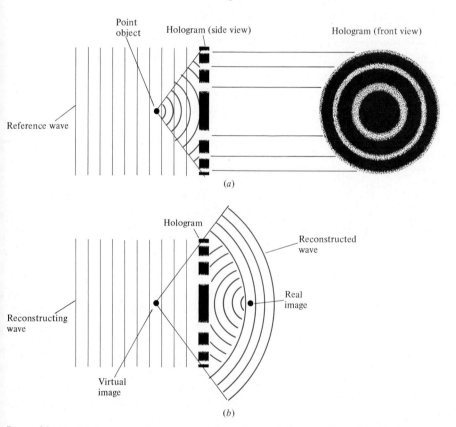

Figure 25.44 (a) An idealized arrangement for making a hologram of a point object. (b) The corresponding reconstruction of the virtual and real images.

hologram is an incredibly complex pattern of lines and swirls. Nevertheless, it represents a coded diffraction pattern containing the information about all the points on the object necessary to reconstruct the wavefronts when the hologram is illuminated by a coherent source (Fig. 25.45).

(a)

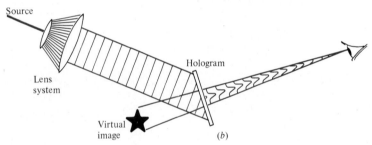

Figure 25.45. (a) Production of a hologram for a complex object. A single highly coherent laser beam is split into two by a partially silvered mirror. Lenses are used to broaden the beams without altering their coherence properties. One beam illuminates the object, while the other acts as a reference beam. When the reference beam and the reflected beam interfere at the photographic plate, a hologram is formed. (b) The reconstruction stage. Only the virtual image is shown, although a real image is also formed.

Applications of Holography | Holography makes it possible to overcome a basic limitation of the microscope. If the magnification is large, the range of depths over which an object is in focus at any microscope setting is very small. Biological specimens often are suspended in a fluid and tend to drift about. Hence they move in and out of focus, unless the motion is restricted by putting the specimen into a solid form or by making it very thin. Either procedure may alter the specimen. However, if one makes a holographic snapshot with the microscope, this "freezes" the motion while preserving all the three-dimensional information. The reconstructed three-dimensional image can then be examined at leisure with a microscope, and the instrument can readily be adjusted to focus on successive layers. Similarly, if a sample changes gradually, a series of holograms can be made which, when reconstructed, allows one to study the sample both at various depths and times.

Holography can be used in various ways to detect minute changes that have occurred in an object. Suppose an object and its hologram made at an earlier

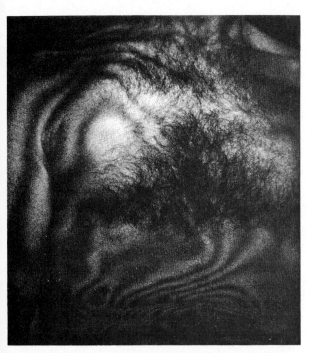

Figure 25.46 Two holograms were made on a single photographic plate by illuminating a subject with two short laser pulses 1.5×10^{-4} seconds apart. This photograph shows the interference fringes produced when the double hologram was illuminated and the wave fronts were reconstructed. Destructive interference occurs where the subject's chest has moved far enough so that the two reconstructed waves are out of phase. The fringes are most closely spaced in regions where the motion is most rapid. This man has had his left lung removed, and is inhaling. The distortion in his muscular action is readily apparent. (Courtesy of The Athletic Institute, from an article by S.M. Zivi in *Biomechanics*, 1971.)

time are returned to precisely the positions they had when the hologram was originally made. If the same laser again is used to illuminate them, then the wave from the object and the reconstructed wave from the hologram will interfere. Any changes that have occurred in the appearance of the object will produce light and dark interference fringes, which directly pinpoint changes not readily found by visual inspection. Differences between two carefully machined parts can be determined in a similar way.

A closely related procedure is to make two successive holograms on a single photographic plate by a double exposure. When the image is reconstructed interference effects will occur if the object has changed in any way. A remarkable photograph made from a double exposed hologram is shown in Fig. 25.46.

So far we have discussed cases in which the original light source and the reconstructing beam have the same wavelength. However, this is not necessary. If a shorter wavelength is used in the reconstruction, this will have the effect of increasing the apparent depth of the object. Also, it is not necessary that the original and reconstructing waves have the same physical nature. Using special photographic techniques, holograms have been made from the interference between reference and reflected or transmitted beams of ultrasonic waves. Holograms are also made with x-rays. In both cases visible light is used for reconstructing the image. The ability of ultrasound and x-rays to penetrate where light may not is thereby combined with the great flexibility of image manipulation afforded by lenses for visible light.

25.13 | INTERFERENCE EFFECTS IN THIN FILMS

The colors produced by light reflected from soap bubbles and oil films, the irridescent eye of a peacock's tail feather, and the purple or amber color of coated camera lenses all are due to the interference of light reflected by the opposite surfaces of thin films. When the two reflected beams are in phase, there is constructive interference, and a maximum reflected intensity results. Conversely, when the two beams are exactly out of phase, destructive interference reduces the intensity of the reflected light. Since the interference conditions depend on the wavelength, the intensity of the light reflected by a given film varies considerably with the wavelength. This causes the colored effects mentioned above when the film is illuminated with white light.

We now investigate reflection by films quantitatively. Figure 25.47 shows monochromatic light incident in air on a thin film of refractive index n and thickness d. At normal incidence, light reflected by the second surface travels a distance $2d$ further than light reflected by the first surface. If the wavelength in vacuum or air is λ, then the wavelength in the film is $\lambda' = \lambda/n$. The light reflected at the first surface is reversed in phase. However, the light reflected at the second surface is not, since the phase reversal occurs only when the light is incident from the less dense medium. Thus, if $2d$ is exactly one wavelength, the waves are exactly out of phase and interfere destructively. In general, destructive interference occurs if $2d$

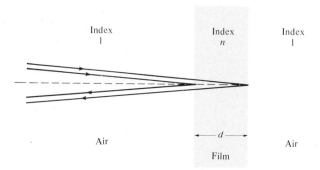

Figure 25.47.   When light is incident on a thin film, some of the light is reflected at each surface, although most of it is transmitted (not shown). At normal incidence the light reflected by the second surface travels a distance $2d$ further than the light reflected by the first surface.

is any integer multiple of λ', or if $2d = m\lambda' = m\lambda/n$. Thus the reflected intensity is a minimum if

$$2nd = m\lambda, \qquad m = 0, 1, 2, \ldots \quad (25.13)$$

Similarly, the constructive interference and the reflected intensity are a maximum when

$$2nd = (m + \tfrac{1}{2})\lambda, \qquad m = 0, 1, 2, \ldots \quad (25.14)$$

These two equations also hold true if the film is less dense than the media on either side. However, the conditions for the maxima and minima are reversed when the film is intermediate in density between the media on either side since then either both reflected beams or neither is reversed in sign. The following example illustrates the reflecting properties of a typical film.

Example 25.13

A film of soap solution ($n = 1.33$) is just thick enough to cause the maximum reflection of red light of wavelength 700 nm at normal incidence. (a) How thick is the film? (b) The film is illuminated with white light at close to normal incidence. If an observer sees the reflected light, what color does the film appear to be?

(a) Using $m = 0$ in Eq. 25.14 for the maximum reflected intensity,

$$d = \frac{(m + \tfrac{1}{2})\lambda}{2n} = \frac{(\tfrac{1}{2})(700 \text{ nm})}{2(1.33)} = 132 \text{ nm}$$

(b) The distance across the film and back is a half wavelength at 700 nm, so it is a full wavelength at 350 nm. This is the longest wavelength at which the reflected intensity is a minimum, and it is just beyond the limit of visible light, 400 nm. This means that partial

destructive interference also occurs at wavelengths in the visible near 400 nm, so relatively little light is reflected at the shorter wavelengths. Conversely, the reflection at wavelengths close to the long wavelength or red end of the spectrum will be enhanced. Thus if the observer sees light reflected at or near normal incidence, the film will appear red. However, if the film is part of a curved surface or a bubble, the angle of incidence and path length will vary over the film, and various colors will be seen in different parts of the film.

Nonreflective Coatings | We saw in Section 25.3 that about 4 percent of the light is reflected at each of the several air-glass surfaces in an optical instrument such as a camera or a microscope. Hence the reduction of these intensity losses made possible by nonreflective coatings is of considerable practical value.

A nonreflective coating is illustrated in Fig. 25.48. A thin film of a material with index n_c is evaporated onto glass with index n_g. Since n_c is chosen to be less than n_g, the waves reflected from *both* sides of the film are reversed in sign. Hence, maximum destructive interference and minimal reflection occurs when the path lengths differ by a half wavelength. For normal incidence this happens if the film thickness d is a quarter wavelength. For light with wavelength λ in air, the wavelength in the coating is λ/n_c, and $d = \lambda/4n_c$. This result is illustrated by the following example.

Example 25.14

A camera lens has a coating of magnesium fluoride, which has a refractive index $n_c = 1.38$. If the reflected intensity at normal incidence of blue light of wavelength 500 nm is a minimum, how thick is the coating?

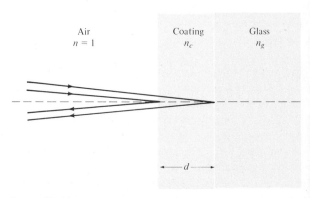

Figure 25.48.   A nonreflecting coating on glass. The refractive index of the coating is less than that of the glass.

From the discussion above, the thickness of the coating is

$$d = \frac{\lambda}{4n_c} = \frac{500 \text{ nm}}{4(1.38)} = 90.6 \text{ nm}$$

Total cancellation or zero reflected light intensity can occur only if the two reflected beams have the same intensity. It follows from Eq. 25.3 that the intensities are equal if $n_c = \sqrt{n_g}$. For glass with index 1.5, this means that the coating should have an index of $\sqrt{1.5} = 1.22$. However, suitable materials with the necessary hardness are not available with this index, and magnesium fluoride, which has an index equal to 1.38, is used as a compromise.

A single coating on a lens reduces the reflected light intensity to varying degrees throughout the visible range. Camera lenses designed for color film are least reflective in the blue, and they appear amber because the reflection is greater at longer wavelengths. Lenses designed primarily for black and white film are least reflective for green light. They reflect more red and violet light, which gives them a purple appearance. In recent years, techniques have been developed to place two or even three layers on lenses. Although difficult and expensive, this process further reduces the reflective intensity (Fig. 25.49).

EXERCISES ON SUPPLEMENTARY TOPICS

Section 25.12 | Holography

25-47 An object is located so that the light diffracted from any one point on the object interferes with the reference beam at every point on a photographic plate. (a) If the resulting hologram is cut in half and one half is illuminated in the usual way with a laser beam, what portion of the original object will be seen? (b) Compare what happens here to the case where a diffraction grating is cut in half, parallel to the lines.

Section 25.13 | Interference Effects in Thin Films

25-48 A lens with refractive index 1.6 is coated with magnesium fluoride with index 1.38. (a) If it is designed to be nonreflecting for green light of wavelength 550 nm, how thick is the film? (b) If this lens is used in water, will it be nonreflecting for green light? Explain.

25-49 A soap film with refractive index 1.33 is just thick enough to produce destructive interference of the shortest wavelength visible light (400 nm) at normal incidence. How thick is the film?

25-50 Room acoustics can be modified by placing slots in the walls. (a) How deep should the slots be to minimize sound reflections at 500 Hz? (Use 344 m s^{-1} as the speed of sound.) (b) What effect will such slots have at 1000 Hz?

PROBLEMS ON SUPPLEMENTARY TOPICS

25-51 The zone plate in Fig. 25.42 is constructed so that the center of the mth dark ring is at a distance $d_m = d + m\lambda$ from point P. (a) Show that the distance r_m from the center of the plate to the center of the mth ring satisfies $r_m^2 = d_m^2 - d^2$. (b) Show that when d is large compared to λ, $r_m^2 = 2md\lambda$ is a good approximation.

*25-52 If a hologram is made with ultrasonic waves and reconstructed with visible light, the depth of the image is highly distorted. (a) Use the result of the preceding problem to explain this observation. (b) Make a numerical estimate of the effect.

*25-53 A zone plate is constructed to produce a bright spot 0.1 m from the plate when illuminated with red light of wavelength 700 nm. Using the result of Problem 25-51 determine what will happen when the plate is illuminated with yellow light of wavelength 600 nm.

*25-54 Using the result of Problem 25-51, describe what will happen when a zone plate is illuminated with white light.

25-55 Explain why a soap film appears black when illuminated with white light if it is much

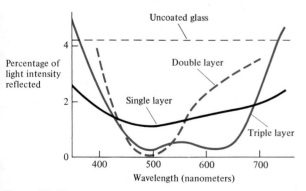

Figure 25.49. Percentage of light intensity reflected by glass with one, two, or three coating layers.

thinner than the average wavelength of visible light.

*25-56 Show that if a lens with refractive index n_g is coated with a film of index n_c, the intensity ratio I_r/I_0 is the same for the light reflected from the two sides of the film when $n_c{}^2 = n_g$. (Assume that the lens is used in a vacuum and that the light is incident normally.)

25-57 Explain why the reflections from thin oil layers are brightly colored but those from thick ones are not.

*25-58 Two glass plates are placed in contact, and a sheet of paper is inserted at one end so that an angle of 0.0005 rad is made by the plates. If the plates are illuminated normally from above with light of wavelength 600 nm, what is the spacing of the observed bright and dark fringes?

Additional Reading

J.B. Cohen, The First Explanation of Interference, *American Journal of Physics,* vol. 8, 1940, p. 99.

R.W. Pohl, Discovery of Interference by Thomas Young, *American Journal of Physics,* vol. 28, 1960, p. 530.

R.B. Setlow and E.C. Pollard, *Molecular Biophysics,* Addison-Wesley, Reading, Mass., 1962, chapter 5. X-ray diffraction.

Eugene Ackerman, *Biophysical Science,* Prentice-Hall, Englewood Cliffs, N.J., 1962, chapter 15. X-ray diffraction.

J.F. Mulligan and D.F. McDonald, Recent Determinations of the Speed of Light, *American Journal of Physics,* vol. 20, 1952, p. 165; vol. 25, 1957, p. 180.

Winston E. Kock, *Radar, Sonar, and Holography,* Academic Press, New York, 1973.

B.G. Ponseggi and B.J. Thompson (eds.), Holography, *Proceedings of the Society of Photo-Optical Instrumentation Engineers,* vol. 15, 1968.

Winston E. Kock, Sound Visualization and Holography, *The Physics Teacher,* January 1975, p. 14.

A.G. Porter and S. George, An Elementary Introduction to Holography, *American Journal of Physics,* vol. 43, 1975, p. 954.

Scientific American articles:

J.H. Rush, The Speed of Light, August 1955, p. 62.

The entire issue of September 1968 is devoted to the subject of light.

J.A. Giordmaine, The Interaction of Light with Light, April 1964, p. 38.

A.G. Ingalls, Ruling Engines, June 1952, p. 45.

Karl H. Drexhage, Monomolecular Layers and Light, March 1970, p. 108.

Phillip Baumeister and Gerald Pincus, Optical Interference Coatings, December 1970, p. 59.

Eric Deuton, Reflectors in Fish, January 1971, p. 64.

H. Moyses Nussensvieg, The Theory of the Rainbow, April 1977, p. 116.

Richard J. Wirrtman, The Effects of Light on the Human Body, July 1975, p. 69.

N.S. Kapany, Fiber Optics, November 1960, p. 72.

J.S. Cook, Communication by Optical Fibers, November 1973, p. 28.

Henri Birsigonies, Communication Channels, September 1972, p. 99.

W.S. Boyle, Light-Wave Communications, August 1977, p. 40.

Arthur Ashkin, The Pressure of Laser Light, February 1972, p. 62.

Victor Vali, Measuring Earth Stresses by Laser, December 1969, p. 88.

Stewart E. Miller, Communication of Laser, January 1966, p. 19.

Glenn L. Berge and George A. Seilestad, The Magnetic Field of the Galaxy, June 1965, p. 46.

Rüdiger Wehner, Polarized Light Navigation by Insects, July 1976, p. 106.

Don R. Sullenger and C.H. Keunard, Boron Crystals, July 1966, p. 96.

M.F. Perutz, The Hemoglobin Molecule, November 1964, p. 64.

J.C. Kendrew, The Three-Dimensional Structure of a Protein Molecule, December 1961, p. 96.

D.C. Phillips, The Three-Dimensional Structure of an Enzyme Molecule, November 1966, p. 78.

Sir Laurence Bragg, X-ray Crystallography, July 1968, p. 58.

Edward A. Stern, The Analysis of Materials by X-ray Absorption, April 1976, p. 96.

Emmett N. Leith and Juris Upatnieks, Photography by Laser, June 1965, p. 24.

S. Henman, in C.L. Stong's "Amateur Scientist" section, How to Make Holograms, February 1967, p. 122.

Keith S. Pennington, Advances in Holography, February 1968, p. 40.

Alexander F. Metherell, Accoustical Holography, October 1969, p. 36.

Emmett N. Leith, White Light Holograms, October 1976, p. 80.

CHAPTER TWENTY-SIX

MIRRORS, LENSES, AND OPTICAL INSTRUMENTS

Cameras, microscopes, telescopes, and the human eye are examples of optical instruments that employ lenses and, in some cases, mirrors. Ordinarily lenses and mirrors are large compared to the wavelengths of visible light, so their principal effects on beams of light can be discussed without reference to interference or diffraction phenomena. However, as expected from the preceding chapter, these wave phenomena do play a role in limiting the resolution and sharpness of the images formed by optical instruments.

The properties of mirrors and lenses are discussed in the first part of this chapter. The later sections cover applications to specific optical instruments.

26.1 | MIRRORS

When we stand a meter away from a plane mirror and look into it, we see someone apparently standing a meter behind the mirror who looks much like us but whose hair is parted on the wrong side. The light reaching our eyes seems to originate at a place behind the mirror called the *image*. The image is *virtual* rather than *real,* since the light never actually passes through that location. It is called an *erect* image, since we do not seem to be standing upside down, or *inverted*.

It is useful to infer these observed properties of the image formed by a mirror from the equality of the angles of incidence and reflection for a light ray, since similar reasoning can be applied to the more complex problem of image formation by a lens. Figure 26.1*a* shows two of the many light rays produced by a point source of light or a point *object* at O, a distance d from the mirror. The ray incident along the normal to the mirror is reflected directly back along the normal, so it appears to have originated behind the mirror on the

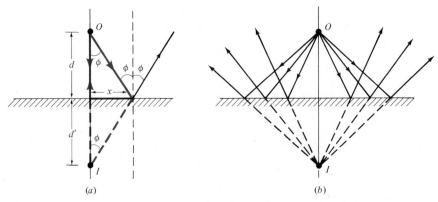

Figure 26.1. (*a*) The two colored triangles are congruent, so $d = d'$. (O is the object, and I is its image.) (*b*) All the light reflected by the mirror appears to originate at the image I.

Figure 26.2. In a plane mirror, the image of each point on an object is directly in front of it. (K. Bendo.)

normal. The light ray that strikes the mirror at an angle φ to the normal is reflected at an equal angle and appears to come from somewhere behind the mirror along the dashed projection of the outgoing ray. The two projected paths cross behind the mirror at point *I*, the image position. In Fig. 26.1a, the two colored right triangles have a common side *x* and an equal angle φ. Hence the triangles are congruent, and all their angles and sides are equal. *This means that the object distance d and the image distance d' are the same.*

When we have a complex illuminated object instead of a point, the image of each point on the object is directly in front of it (Fig. 26.2). This image is sometimes said to be reversed side to side, but not vertically; that is, the image is erect rather than inverted. This remark is not quite correct, as can be seen

if we lie on our sides in front of a mirror. Then we perceive our images as reversed vertically but not horizontally!

26.2 | LENSES

A lens is a piece of transparent material that can focus a transmitted beam of light so that an image is formed. The lenses in manmade optical instruments are usually manufactured from glass or plastic, while the lens in the human eye is formed by a transparent membrane filled with a clear fluid. For our purposes, it is sufficient to consider *thin, spherical lenses*. These have two spherical surfaces or a spherical and a plane surface and a thickness that is small compared to the radii of the surfaces.

We can categorize all lenses as either *converging* or *diverging*. A converging lens is thicker at its center than at the edge, while the opposite is true for a diverging lens (Fig. 26.3). A converging lens bends light rays toward its *axis,* the line through its centers of curvature, so that a beam of parallel rays converges at a point (Fig. 26.4). For example, in bright sunlight, a converging lens may produce a spot of light intense enough to ignite paper. A diverging lens bends rays outward from its axis.

Suppose a very distant object is on the axis of a lens, so that the light rays are nearly parallel to the axis when they reach the lens. A converging lens will refract these parallel rays so that they meet or form an *image* at a *focal point F'* beyond the lens, while a diverging lens will bend the rays outward so that they

Figure 26.3. (*a*) Converging lenses. (*b*) Diverging lenses.

Figure 26.4. Focal points of a converging lens.

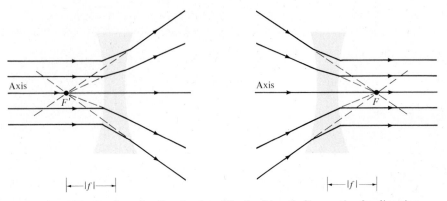

Figure 26.5. Focal points of a diverging lens. The focal length f is negative for diverging lenses.

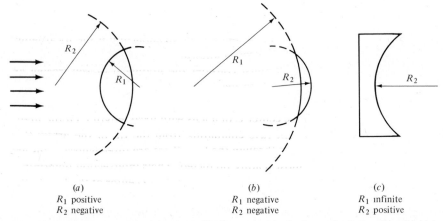

(a)	(b)	(c)
R_1 positive	R_1 negative	R_1 infinite
R_2 negative	R_2 negative	R_2 positive

Figure 26.6. Sign conventions for the radii of curvature in the lensmaker's equation. Light is always taken to be incident from the left.

appear to have come from a focal point F' before the lens (Figs. 26.4 and 26.5). The distance from the center of the lens to the focal point is called the *focal length f.* Conventionally f is taken to be positive for converging lenses and negative for diverging lenses.

Lenses also have a second focal point. If an object is placed at the point F, which is a distance f in front of a converging lens, then light rays from the object will emerge parallel to the axis (Fig. 26.4). Similarly, light rays aimed at the point F a distance $|f|$ beyond a diverging lens emerge parallel to the axis (Fig. 26.5).

The focal length of a lens is related to its index of refraction n and the radii of curvature R_1 and R_2 of its surfaces by a formula that can be derived using Snell's law and the approximation that the angles of incidence are small. Since the derivation is lengthy, we only present the result here. The focal length of a lens with index n in a medium of index 1 is

$$\frac{1}{f} = (n - 1)\left(\frac{1}{R_1} - \frac{1}{R_2}\right) \qquad (26.1)$$

This is called the *lensmaker's equation.*

To use this equation it is necessary to observe certain sign conventions.*

1 We draw diagrams so that light is always incident from the left.

2 The radius of curvature of a lens surface is positive if its center of curvature is to the right of the lens, and negative if the center is to the left.

3 R_1 refers to the first or left surface, and R_2 refers to the second or right surface (Fig. 26.6).

---

*Some books use different sign conventions. Corresponding changes must then be made in the lensmaker's equation and in other equations below.

4 A plane surface can be thought of as part of a sphere of infinite radius.

These conventions are illustrated by the following example.

Example 26.1

A lens made of glass with $n = 1.5$ has two convex surfaces (Fig. 26.6a). If its radii of curvature are 0.1 and 0.2 m, what is the focal length?

According to our sign conventions, R_1 is positive since the center of curvature is to the right of the lens, and R_2 is negative. Taking $R_1 = 0.1$ m and $R_2 = -0.2$ m, the lensmaker's equation gives

$$\frac{1}{f} = (n - 1)\left(\frac{1}{R_1} - \frac{1}{R_2}\right)$$

$$= (1.5 - 1)\left[\frac{1}{0.1 \text{ m}} - \frac{1}{(-0.2 \text{ m})}\right]$$

$$= (0.5)(10 + 5)\text{m}^{-1} = 7.5 \text{ m}^{-1}$$

or $f = 0.133$ m. Note that if we turn the lens around, so that $R_1 = 0.2$ m and $R_2 = -0.1$ m, we obtain the same result for f. This means that a lens can be turned over without changing its properties.

The lensmaker's equation in the form given in Eq. 26.1 assumes that the lens is in a medium with index of refraction one. If the lens is not used in air or vacuum then the symbol n in the equation must be interpreted as the *relative* index of refraction, $n(\text{lens})/n(\text{medium})$. Lenses have longer focal lengths in water than in air, as seen in this example.

Example 26.2

What is the focal length of the lens in the preceding example if it is placed in water?

Since the index of refraction of water is 1.333, we should use $n = 1.5/1.333 = 1.125$ in the lensmaker's equation. Then with $R_1 = 0.1$ m and $R_2 = -0.2$ m, we find

$$\frac{1}{f} = (n - 1)\left(\frac{1}{R_1} - \frac{1}{R_2}\right)$$

$$= (1.125 - 1)\left[\frac{1}{0.1 \text{ m}} - \frac{1}{(-0.2 \text{ m})}\right]$$

and $f = 0.533$ m. This is four times the focal length of the lens in air as determined in the preceding example.

The increase in the focal length of a lens that occurs when it is placed in water explains why we see so poorly under water. Our eyes contain fluids whose indices of refraction are close to that of water. Light bends appreciably when it enters the eye through the curved transparent *cornea*, but it bends very little when it enters from water. Consequently the eye forms a badly focused image when it is in contact with water. When a person wears goggles or a face mask, light passes through the glass into an air layer at or near normal incidence without appreciable deflection. The light is then refracted in the usual fashion as it enters the eye from the air, so that vision is improved.

26.3 | IMAGE FORMATION

We saw above that light rays from a very distant point on a lens axis arrive parallel to the axis and meet to form an image at the focal point. Rays from other points form images whose locations can be found graphically or algebraically if the focal length of the lens is known.

We first consider the graphical or *ray tracing* approach to locating an image. Figure 26.7 shows an illuminated arrow located at an *object distance s* from the center C of a lens with focal length f. Of the many light rays emanating from the tip of the arrow that pass through the lens, there are three whose paths are most readily predicted. Their intersection determines the *image distance s'* from the center of the lens to the image. Two rays are actually sufficient to find the image location; the third serves as a check.

1 The ray leaving the arrowhead parallel to the axis is deflected by the lens so that it passes through the focal point F', in accordance with the definition of the focal point.

2 The ray going through the focal point F emerges from the lens parallel to the axis.

3 The ray directed at the center of the lens is undeflected. This happens because at that point

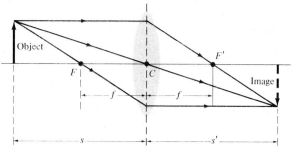

Figure 26.7. Three rays are used to locate an image graphically.

Figure 26.8. A ray passing through a flat plate is displaced but not deflected. Accordingly, a ray passing through or near the center of a thin lens is displaced very slightly.

the two sides of the lens are almost parallel, so a ray is effectively going through a flat plate (Fig. 26.8). Since the lens is thin, the ray is displaced from its original path by an amount that is negligibly small.

This construction is shown in Fig. 26.7 for the case of a *real object* and a *real image*. With the convention that light always goes from left to right,

1   real objects are to the left of the lens, and real images are to the right;
2   virtual images are to the left of the lens, and virtual objects are to the right.

A real image is one that is actually formed; if a screen is placed at that location, the image forms on that screen. A virtual image represents the place from which the outgoing light appears to be coming from if you look into the lens. However, the light never passes through that location; no image is observed if a screen is placed there (Fig. 26.9a). Virtual objects occur in multilens systems when the converging rays from one lens passes through a second lens (Fig. 26.9b).

Although ray tracing provides good qualitative insight into the formation of an image by a lens or system of lenses, numerical work is best done with algebraic formulas. In order to develop and apply these formulas, we adopt the following sign conventions for the quantities shown in Fig. 26.10:

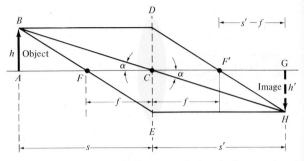

Figure 26.10. The object distance s, the image distance s', and the focal length f are related by the thin lens formula.

1   s is positive for a real object, negative for a virtual object.
2   s' is positive for a real image, negative for a virtual image.
3   The *object height* h is positive if it points above the axis and negative if it points below the axis.
4   The *image height* h' is positive if it points above the axis and negative if it points below the axis.

The *linear magnification m* is the ratio of the image and object heights, h'/h. The linear magnification is negative when the image is inverted, as in Fig. 26.7; it is positive when the image is erect, as in Fig. 26.9a.

We can derive formulas relating s, s', f, and m using similar triangles. Two triangles are similar, and their corresponding sides are therefore proportional, if two of their angles are equal. In Fig. 26.10, the triangles ACB and GCH each have a right angle and the same angle α, and so they are similar. Thus

$$\left| \frac{h'}{h} \right| = \left| \frac{s'}{s} \right|$$

With our sign conventions, h' is negative in Fig. 26.10, so the *linear magnification m* can be written as

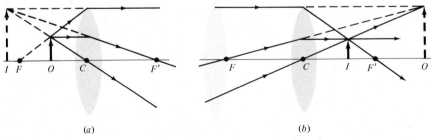

(a)   (b)

Figure 26.9. (a) Real object, virtual image. (b) Virtual object, real image. The virtual object is formed by the lens shown in color.

(a)   (b)

Figure 26.11. (a) When the distance from a real object to a converging lens is greater than the focal length, a real image is formed. (b) When the object is between the focal point and the converging lens, the image is virtual.

$$m = \frac{h'}{h} = -\frac{s'}{s} \tag{26.2}$$

Also, triangles CDF' and GHF' are similar, so in the same way

$$m = \frac{h'}{h} = -\left(\frac{s'-f}{f}\right) \tag{26.3}$$

Comparing these equations, we find

$$\frac{s'}{s} = \frac{s'-f}{f}$$

Dividing by s' and rearranging, we find the *thin lens formula*.

$$\frac{1}{s} + \frac{1}{s'} = \frac{1}{f} \tag{26.4}$$

We can check to see that this equation agrees with our earlier discussion. For a very distant object, s is infinite and $1/s$ is zero, so the thin lens formula gives $1/s' = 1/f$, or $s' = f$, as expected from our definition of the focal point. Similarly, for a very distant image, s' is infinite and $s = f$. Note that for a given focal length, the image distance s' depends only on the object distance s and not on the height h of the object. *This means that all object points at a distance s from the lens have their images on the same plane.* The following examples further illustrate the thin lens formula.

Example 26.3

A lens has a focal length of $+0.1$ m. Find the image distance when the object distance is (a) 0.5 m; (b) 0.08 m.

(a) Using the thin lens formula,

$$\frac{1}{s'} = \frac{1}{f} - \frac{1}{s} = \frac{1}{0.1 \text{ m}} - \frac{1}{0.5 \text{ m}}$$

and $s' = 0.125$ m. Note that s' is positive, corresponding to a real image (Fig. 26.11a).

(b) Again using the thin lens formula,

$$\frac{1}{s'} = \frac{1}{f} - \frac{1}{s} = \frac{1}{0.1 \text{ m}} - \frac{1}{0.08 \text{ m}}$$

and $s' = -0.4$ m. Here s' is negative, so the image is virtual (Fig. 26.11b).

Example 26.4

A camera lens has a focal length of $+0.1$ m. (a) If the camera is focused on a child 2 m from the lens, what is the distance from the lens to the film (Fig. 26.12)? (b) If the child has a height of 1 m, how tall is the image on the film?

(a) If a sharp image is to be formed, the distance from the lens to the film must equal the image distance s'. Using the thin lens formula

$$\frac{1}{s'} = \frac{1}{f} - \frac{1}{s} = \frac{1}{0.1 \text{ m}} - \frac{1}{2 \text{ m}}$$

and $s' = 0.105$ m. Thus the film should be at a distance from the lens slightly greater than the focal length, 0.1 m. Except when extreme closeups are taken, the object distance is always large compared to f, so the image and the correct film location are just beyond the focal point.

(b) The image height can be found from the object height, $h = 1$ m, and the linear magnification, $m = h'/h = -s'/s$. Approximating s' by f in accordance with our discussion above,

$$m = -\frac{s'}{s} \simeq -\frac{f}{s} = \frac{-0.1 \text{ m}}{2 \text{ m}} = -0.05$$

Hence

$$h' = mh = (-0.05)(1 \text{ m}) = -0.05 \text{ m}$$

The height of the image is 0.05 m. The minus sign indicates that the image is inverted.

Since the magnification is approximately proportional

465

Figure 26.12. Principle of the camera.

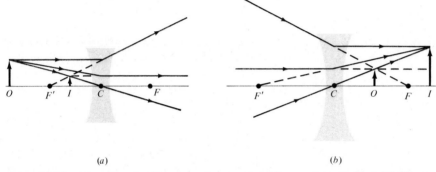

Figure 26.13. (*a*) A diverging lens always forms a virtual image of a real object. (*b*) When there is a virtual object between the lens and *F* formed by another lens (not shown), a diverging lens forms a real image.

to the focal length, cameras are often equipped with interchangeable lenses of varying focal lengths. The film size remains constant, so the field of view decreases as the focal length and magnification increase.

Example 26.5

A diverging lens has a focal length of -0.4 m. (a) Find the image location for an object placed 2 m from the lens. (b) If there is a real image 1 m from the lens, where is the object?

(a) Since the object is real, $s = +2$ m. Thus

$$\frac{1}{s'} = \frac{1}{f} - \frac{1}{s} = \frac{1}{(-0.4 \text{ m})} - \frac{1}{2 \text{ m}}$$

and $s' = -0.333$ m. The image is virtual, and it is located between the focal point F' and the lens (Fig. 26.13a).

(b) With $s' = +1$ m,

$$\frac{1}{s} = \frac{1}{f} - \frac{1}{s'} = \frac{1}{(-0.4 \text{ m})} - \frac{1}{(1 \text{ m})}$$

so $s = -0.286$ m. The object is virtual, and therefore

must be due to another lens. The object is located between the lens and the focal point F (Fig. 26.13b).

26.4 | ABERRATIONS

No matter how perfectly spherical its surfaces, any lens suffers from various kinds of *aberrations*, which limit the sharpness of its images independently of diffraction effects. Since the index of refraction of glass varies with the wavelength of the light, the focal length of a lens also varies with the wavelength. When an object is illuminated with white light, if its image on a screen is in focus for one color component, it will be slightly out of focus for the others. This is *chromatic aberration*. Also, the lensmaker's formula is derived using small angle approximations. Corrections to this formula show that rays parallel to the axis have image locations that vary slightly with their distance from the axis. For this reason, a parallel beam of light actually forms an image of finite size

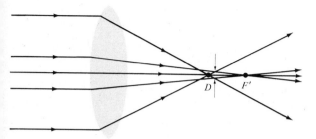

Figure 26.14. Monochromatic light rays near the axis are focused at F', while those near the edge of the lens meet at D. The arrows show where the bundle of rays has the smallest diameter.

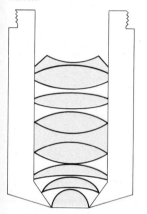

Figure 26.15. A modern multielement microscope lens.

rather than a true point image (Fig. 26.14). Aberrations of this and similar types occur even for light of a single wavelength and are called *monochromatic aberrations*. Complex lens systems with several elements are designed so that their separate aberrations tend to cancel (Fig. 26.15).

26.5 | THE SIMPLE MAGNIFIER

The normal human eye can just barely distinguish two well illuminated point objects with an angular separation $\theta_0 \simeq 5 \times 10^{-4}$ rad $\simeq 0.03°$. This minimum angular separation is called the *visual acuity* and can, in effect, be reduced with a *simple magnifier* or *magnifying glass*.

To see fine details, a person holds an object as close to the eye as possible, or at the *near point:* the closest point at which one can focus comfortably. For a normal young adult the distance x_n to the near point is about 0.25 m (Fig. 26.16a). At the near point, two points a small distance y apart have an angular separation small enough so that $\theta \simeq \tan \theta = y/x_n$ is a good approximation. If θ is $\theta_0 = 5 \times 10^{-4}$ rad, then

$$
\begin{aligned}
y = x_n\theta &= (0.25 \text{ m})(5 \times 10^{-4}) \\
&= 1.25 \times 10^{-4} \text{ m} \\
&= 0.125 \text{ mm}
\end{aligned}
$$

Thus the finest details discernable to the naked eye have a size of about 0.1 mm.

The simple magnifier is a converging lens that allows the object to be brought closer to the eye so that it subtends a larger angle and permits one to see finer details. Usually the object under study is placed just inside the focal point of the lens, which is held close to the eye (Fig. 26.16b). The resulting virtual image is far from the eye and thus can be viewed comfortably. The angle subtended by the image is

$$
\theta' \simeq \tan \theta' = \frac{y}{f}
$$

The *angular magnification M* is the ratio (Fig. 26.16b)

$$
M = \frac{\theta'}{\theta} = \frac{y/f}{y/x_n} = \frac{x_n}{f}
$$

or with $x_n = 0.25$ m,

$$
M = \frac{0.25 \text{ m}}{f} \tag{26.5}
$$

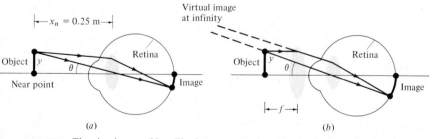

Figure 26.16. The simple magnifier. The lens permits the object to be brought closer to the eye thereby subtending a larger angle.

Figure 26.17. The ordinary bright-field light microscope. (Courtesy of American Optical Corporation.)

The following example illustrates the use of a typical magnifier.

Example 26.5

A collector uses a lens with a focal length of 0.1 m to examine a stamp. What angular magnification is provided by the lens?

The angular magnification is

$$M = \frac{0.25 \text{ m}}{f} = \frac{0.25 \text{ m}}{0.1 \text{ m}} = 2.5$$

Details will appear 2.5 times larger with this magnifier.

Although lenses with very short focal lengths can be made, the effects of aberrations limit simple magnifiers to a useful angular magnification of only two or three. Magnifiers containing two or more lenses can have considerably greater useful magnification, since the aberrations can be partially corrected.

26.6 | THE BRIGHT-FIELD LIGHT MICROSCOPE

Although the microscope is one of the oldest and most widely used physical instruments in biology and medicine, new types of microscopes have been developed in recent decades. These make possible more detailed study of cellular structures and sometimes avoid the need for destructive methods in observing living cells.

Figure 26.17 shows a bright-field light microscope, the ordinary microscope found in every biological laboratory. The lenses in the *condenser* focus the incident light on the specimen, and the *diaphragm* regulates the intensity. The magnification is determined by the focal lengths of the *objective* and *ocular* lenses. In practice, both of these are multielement lenses.

The principle of the microscope is shown in Fig. 26.18. The object under study is placed just beyond

Figure 26.18. The microscope. The ocular and objective in practice are multielement lenses.

the focal point of the objective, so $s_1 \simeq f_1$. Its image is real and inverted and is formed just within the focal point of the ocular. The image of the objective is the object for the ocular. This lens acts as a simple magnifier, producing an enlarged virtual image at a comfortable distance for viewing.

The real image produced by the objective has a linear magnification $m_1 = -s'_1/s_1 \simeq -s'_1/f_1$. The angular magnification of the ocular used to enlarge this real image is $M_2 = 0.25$ m$/f_2$. The overall angular magnification of the microscope is the product of these two magnifications,

$$M = m_1 M_2 = \frac{-s'_1 \times 0.25 \text{ m}}{f_1 f_2} \qquad (26.6)$$

This result is illustrated by the following example.

Example 26.6

The focal length of a microscope objective is 0.4 cm $= 4 \times 10^{-3}$ m, and the focal length of the ocular is 3.2 cm $= 3.2 \times 10^{-2}$ m. The image formed by the objective is 0.2 m from the objective. (a) Where is the object under study? (b) What is the angular magnification? (c) Neglecting any diffraction effects, what is the smallest separation between two points that can be resolved by the eye with this instrument?

(a) From the thin lens formula, the object distance s_1 satisfies

$$\frac{1}{s_1} = \frac{1}{f_1} - \frac{1}{s'_1} = \frac{1}{4 \times 10^{-3} \text{ m}} - \frac{1}{0.2 \text{ m}}$$

so $s_1 = 4.08 \times 10^{-3}$ m. The object is just outside of the focal point of the objective, since $f_1 = 4 \times 10^{-3}$ m.

(b) The angular magnification is

$$\begin{aligned} M &= \frac{-s'_1 \times 0.25 \text{ m}}{f_1 f_2} \\ &= \frac{-(0.2 \text{ m})(0.25 \text{ m})}{(4 \times 10^{-3} \text{ m})(3.2 \times 10^{-2} \text{ m})} \\ &= -391 \end{aligned}$$

The minus sign means that the image is inverted.

(c) The smallest resolvable separation of two points for an unaided eye is about 0.1 mm $= 10^{-4}$ m. If the angular size is increased by a factor of $391 \simeq 400$, then the minimum resolvable separation becomes 10^{-4} m$/400 = 2.5 \times 10^{-7}$ m $= 250$ nm, or about half the average wavelength of visible light. This separation is comparable to the resolution limit of the microscope arising from the diffraction of light by the specimen.

Resolution and contrast in the bright field micro-

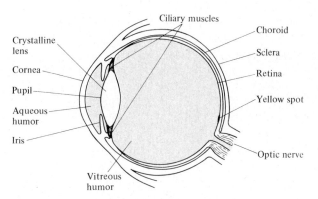

Figure 26.19. The human eye.

scope are discussed in the supplementary topics at the end of this chapter, along with polarizing, interference, and phase contrast microscopes.

26.7 | THE HUMAN EYE

The human eye is a remarkable evolutionary achievement. It has an intensity range of 10^9, covers a field of view of over 180°, can rapidly shift its focus from very short distances to infinity, and has a resolution close to the limit imposed by diffraction. Also, as we see in Chapter Twenty-eight, its threshold sensitivity is comparable to the theoretical limit imposed by the quantum properties of light.

The eye and camera have many similarities. In both, a lens system forms an inverted real image on a light-sensitive surface. The eyeball is approximately spherical in shape with a diameter of about 2.3 cm or nearly 1 in. (Fig. 26.19). Its outer covering is a nearly opaque fibrous layer called the *sclera*. Inside this is a dark membrane, the *choroid* which, like the black interior of a camera, absorbs stray light. The inner surface of the eyeball is the *retina*, a membrane containing numerous nerves and blood vessels. The nerve fibers terminate at *rods* or *cones* in the retina which respond to light by generating electrical nerve pulses. The eye is most sensitive at a small retinal depression, the *yellow spot* or *macula;* its central portion, the *fovea centralis*, is about $\frac{1}{4}$ mm in diameter and contains only densely packed cones. The eye tends to rotate so that the object under examination is imaged on the fovea centralis.

Light enters the eye through a thin membrane called the *cornea,* which covers a transparent bulge on the surface of the eyeball. The *iris* is a colored ring behind the cornea; like a camera diaphragm, it adjusts in size and aids in regulating the amount of light entering the eye through the *pupil*. The *crystalline lens* is composed of a fibrous, jellylike material. Its shape and therefore its focal length are controlled by the *ciliary muscles*. The space between the cornea and the lens contains a watery fluid called the *aqueous humor;* behind the lens is a thin jelly, the *vitreous humor*. Both humors have an index of refraction equal to 1.336, very close to that of water, which is 1.333. The crystalline lens has a slightly larger index, 1.437.

Somewhat surprisingly, most of the bending of light occurs at the cornea. This is because the cornea has a small radius of curvature (0.8 cm), and the change in the refractive index is large when the light goes from the air ($n = 1$) into the aqueous humor ($n = 1.336$). The function of the lens is to provide the fine adjustments needed to focus on objects at different distances.

If the ciliary muscles are relaxed, the front surface of the lens is kept relatively flat, and light from distant objects is focused on the retina. When the ciliary muscles contract, the lens assumes a more rounded shape, and its focal length decreases, bringing light from nearby objects into focus on the retina. The ability of the lens to adjust its focal length is called *accommodation*.

Numerical statements about accommodation are made in terms of a quantity called *power*. The power P of a lens is the reciprocal of its focal length,

$$P = \frac{1}{f} \qquad (26.7)$$

If f is measured in meters, P is measured in *diopters;* 1 diopter $= 1 \text{ m}^{-1}$. For example, a lens with a focal length of -0.4 m has a power $P = 1/(-0.4 \text{ m}) = -2.5$ diopters. It is left as a problem to show that two thin lenses with focal lengths f_1 and f_2 placed next to each other are equivalent to a single lens with a focal length f satisfying

$$\frac{1}{f} = \frac{1}{f_1} + \frac{1}{f_2} \qquad (26.8)$$

Alternatively, with $P_1 = 1/f_1$ and $P_2 = 1/f_2$, the power of the pair of lenses is

$$P = P_1 + P_2 \qquad (26.9)$$

Thus the power of lenses in contact may be added to find the net power. For example, an opthamologist fitting someone for glasses might place a 3-diopter lens and a 0.25-diopter lens in front of the eye. This combination is equivalent to a single 3.25-diopter lens.

The *accommodation* of the eye is the maximum variation of its power for focusing on near and distant objects. Suppose a person sees clearly objects at a distance x_f with the eye relaxed. For someone with normal vision this *far point* is at infinity. The image distance s' is approximately the diameter D of the eye. Therefore, at the far point the power P_f is given by

$$P_f = \frac{1}{f} = \frac{1}{s} + \frac{1}{s'} = \frac{1}{x_f} + \frac{1}{D} \qquad (26.10)$$

When the eye adjusts its focal length so that it focuses on an object at the near point, the object distance is $s = x_n$. Since s' is again equal to D, its power is now

$$P_n = \frac{1}{f} = \frac{1}{s} + \frac{1}{s'} = \frac{1}{x_n} + \frac{1}{D} \qquad (26.11)$$

The difference of these powers is the *power of accommodation A* of the eye,

$$A = P_n - P_f = \frac{1}{x_n} - \frac{1}{x_f} \qquad (26.12)$$

For a normal young adult, $x_n = 0.25$ m and $x_f = \infty$, so A is 4 diopters. The accommodation of young children is considerably larger; they can comfortably read books held quite close to their eyes. The accommodation decreases with aging, and most people find that their near point gradually recedes until they cannot read comfortably without corrective glasses.

Corrective lenses for optical defects of the eye are discussed in the supplementary topics at the end of this chapter.

26.8 | VISUAL ACUITY, SENSITIVITY, AND COLOR PERCEPTION

The subject of vision is a complex one, involving elements of physics, neurophysiology, psychology and biochemistry, and it is an active field of research. We will comment only briefly on visual acuity, sensitivity, and color perception.

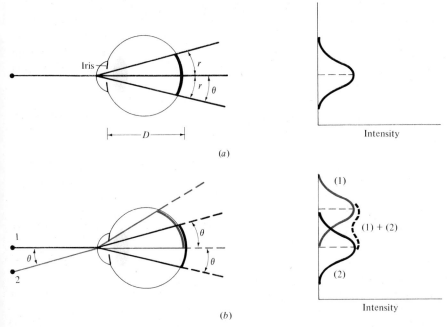

Figure 26.20. (*a*) The minimum image size for a point object due to diffraction at the iris. (*b*) If two point objects are separated by θ, their images will be barely distinguishable.

Acuity |

We noted in our discussion of the magnifier that the visual acuity of a typical person is about 5×10^{-4} radians; objects with a smaller angular separation cannot be distinguished. It is reasonable to inquire whether this limit is due to diffraction effects. According to Chapter Twenty-five, when light from a distant source passes through a small circular aperture of diameter d, the first diffraction minimum is at $\sin \theta = 1.22 \lambda/d$. To estimate what this implies for the eye, let us take the iris diameter to be $5\,\text{mm} = 5 \times 10^{-3}\,\text{m}$, and the wavelength to be $500\,\text{nm} = 5 \times 10^{-7}\,\text{m}$. Since θ is small, to one significant figure

$$\theta \simeq \sin \theta = 1.22 \frac{\lambda}{d}$$
$$= 1.22 \frac{(5 \times 10^{-7}\,\text{m})}{5 \times 10^{-3}\,\text{m}}$$
$$= 10^{-4}\,\text{radians}$$

According to the Rayleigh criterion, two objects will just barely be resolvable if they are separated by this angle (Fig. 26.20). Experiments show that while a few people under optimum conditions have an acuity of twice the diffraction limit, or 2×10^{-4} radians, nobody can reach the limit of 10^{-4} radians.

An explanation for this failure of the eye to quite match the diffraction limit is provided by the structure of the retina. From Fig. 26.20a, the radius of the image circle due to diffraction is

$$r = D \tan \theta \simeq D\theta$$
$$= (2.3 \times 10^{-2}\,\text{m})(10^{-4})$$
$$= 2.3 \times 10^{-6}\,\text{m}$$

This is about equal to the separation of the cones in the fovea, which is the most sensitive region of the retina and contains only cones and not rods. Now the best resolution observed is about 2×10^{-4} radians, corresponding to diffraction centers separated by about 4.6×10^{-6} m, or by two cones. *Thus, it appears that to distinguish two small objects, at least one unexcited cone must intervene between the excited cones.*

The interpretation of the images detected on the retina as meaningful patterns involves complex neurological processing and is a fascinating problem inviting computer simulations and theories of pattern recognition. Given a complex scene, the brain suddenly locks into an interpretation which looks "correct" (Fig. 26.21).

Sensitivity |

The *minimum* or *threshold intensity*

471

Figure 26.21. When our neurological decoding of a visual image contradicts our experience, we find it disturbing. (M.C. Escher, "Waterfall", Escher Foundation—Haags Gemeentemuseum—The Hague.)

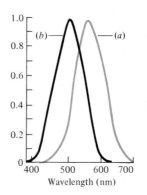

Figure 26.22. Relative sensitivity curves for the (a) light-adapted and (b) dark-adapted eye. Although the dark-adapted eye is more sensitive, both curves are drawn with the same maximum values. (Adapted from Eugene Ackerman, *Biophysical Science*, Prentice-Hall, Englewood Cliffs, N.J., 1962.)

needed to see a flash of light depends on the wavelength. The cornea is opaque to wavelengths shorter than 300 nm, and the crystalline lens to wavelengths below 380 nm, so ultraviolet light does not normally contribute to vision. However, some persons have had their lenses surgically removed because they have become opaque or developed *cataracts*. These people have poor acuity and no accommodation, but to them objects illuminated with ultraviolet light alone are visible and appear violet.

One long wavelength limit on the sensitivity of the eye is set by the strong absorption of light by water in the cornea and the aqueous humor at wavelengths above 1200 nm. However, the sensitivity of the eye goes to zero rapidly above 700 nm. It is thought that the photosensitive molecules in the rods and cones do not respond to the longer wavelengths.

Figure 26.22 shows the variation with wavelength of the sensitivity, which is the inverse of the threshold intensity, for *dark-adapted* and *light-adapted* eyes.

When we go from daylight into a dimly lit room, our eyes gradually adapt and reveal initially invisible details. This adaptation takes over half an hour to be complete. Since the dark-adapted eye is much more sensitive, the two curves have been adjusted to have the same maximum value.

The cones are active only in light-adapted vision, while the rods are always active. Accordingly, the fovea, which lacks rods but has the greatest concentration of cones, is the most sensitive part of the retina of the light-adapted eye, while the edges of the retina, which have more rods, are more sensitive in the dark-adapted eye. The visual acuity of the dark-adapted eye is much poorer than that of the light-adapted eye.

From Fig. 26.22 it can be seen that the sensitivity is greatest near 500 and 550 nm for dark- and light-adapted eyes, respectively. Both wavelengths correspond to green light. Often, green glass is used in sunglasses and in tinted windows. Since green glass absorbs green light less than other wavelengths, it provides the most useful illumination while reducing the total transmitted intensity.

Color Perception | The dark-adapted eye responds to a large range of wavelengths, but the sensation of color is perceived only by the light-adapted eye. This sensation is related to the spectral composition of the light from an object in a complicated and far from fully understood fashion. For example, light

with a wavelength of 600 nm is perceived as yellow, but so is a mixture of green and red light at 500 and 700 nm respectively. Experiments of this type suggest that there are three *primary colors,* usually taken as red, green, and blue. Any perceived color can be stimulated by a suitable mixture of these primaries. It is hypothesized that there are three pigments in the eye that absorb light in overlapping wavelength ranges. By comparing the color sensitivity of people with normal vision and those with forms of color-blindness apparently due to the lack of one pigment, response curves for the three pigments have been obtained. However, these pigments have not yet been isolated chemically. Furthermore, the perceived color of an object depends on the lighting and on the colors of the surroundings. The latter fact suggests rather intricate neurological processing. Color perception remains an old but still intriguing problem.

SUMMARY

A plane mirror forms a virtual image whose distance from the mirror is equal to the object distance. The image of each point on a complex object appears directly in front of that point.

A converging lens will bend rays parallel to its axis so that they meet at the focal point, while a diverging lens will bend such rays outward so that they appear to have come from a focal point before the lens. The focal length is determined by the index of refraction of the lens and the radii of its surfaces. The image of an object can be located graphically using three special rays or algebraically using the thin lens formula.

Multiple lens systems such as microscopes are analyzed by considering the image of one lens as the object of the next. The common bright-field light microscope consists of an objective and an ocular. The objective forms a real image just inside the focal point of the ocular. The ocular then forms an enlarged virtual image, which is seen by the viewer.

In the human eye most of the bending of light occurs at the cornea. The lens of the eye provides the additional adjustments needed to focus on objects at varying distances. The limit of resolution of the eye corresponds to the requirement that the diffraction circles corresponding to the images of two object points must fall on cones in the retina separated by at least one unexcited cone.

Checklist

Define or explain:

virtual, real	linear magnification
erect, inverted	thin lens formula
converging, diverging lenses	aberration
	angular magnification
focal length	visual acuity
lens axis	accommodation
lensmaker's equation	power

EXERCISES

Section 26.1 | Mirrors

26-1   A man stands 2 m from a vertical plane mirror. What is the distance from the man to his image?

26-2   A cat sees its image in a plane mirror. The cat is 1 m from the mirror. (a) Where is the image? (b) The cat lunges at the mirror at 2 m per second. How fast does the cat approach its image?

Section 26.2 | Lenses

26-3   An eyeglass lens has one concave surface with a radius of 0.5 m and a convex surface with a radius of 0.7 m. If the index of refraction of the lens is 1.6, find the focal length.

26-4   A lens is made of glass with $n = 1.5$. One side is convex and has a radius of curvature equal to 0.1 m. Find the radius of curvature of the other surface and draw a sketch of the lens if (a) $f = 0.15$ m; (b) $f = 0.1$ m; (c) $f = -0.15$ m.

26-5   A lens made of glass with refractive index 1.6 has a focal length of 0.5 m in air. What is the focal length in water?

26-6   Crown glass has a refractive index of 1.523 for blue light and an index of 1.517 for red light. If a crown glass lens has a focal length of 1 m for red light, what is its focal length for blue light?

26-7   Flint glass has a refractive index of 1.645 for blue light and an index of 1.629 for red light. A lens made of flint glass has two convex surfaces of radius of curvature 0.1 m. Find its focal lengths for blue and red light.

Section 26.3 | Image Formation

26-8   A lens has a focal length of 0.2 m. A real object is placed 0.08 m from the lens. (a) Locate the image approximately using graphical methods.

473

(b) Locate the image algebraically. (c) What is the magnification?

26-9   An object is placed 1 m from a lens with a focal length of -0.5 m. Locate the image position (a) graphically, and (b) algebraically.

26-10   A lens 0.1 m from a lamp forms a real image of this lamp which is 10 times larger. What is the focal length of the lens?

26-11   A lens of focal length 0.1 m is held 0.08 m from an insect. (a) Where is the image of the insect? (b) What is the magnification of the image? Is it erect or inverted?

26-12   An eyeglass lens has a focal length of -2 m. If it is 4 m from a book, where is the image of the book?

26-13   A camera is focused on a group of people 3 m from the lens, which has a focal length of 50 mm = 0.05 m. (a) What is the lens to film distance? (b) What is the linear magnification? (c) If the film height is 24 mm = 0.024 m, what is the maximum height of a person whose image can completely fit on the film?

26-14   Accessories are available for some cameras which permit the lens-to-film distance to be increased when extreme close-ups are taken of very small objects. Why are these accessories needed?

26-15   A photographer replaces a lens with a focal length of 50 mm by a lens with a focal length of 200 mm. What happens to the image size for distant objects?

Section 26.5 | The Simple Magnifier

26-16   A lens of focal length 0.1 m is used as a simple magnifier. What is its magnification?

26-17   A farsighted woman has a near point 1 m from her eyes. If her visual acuity is 10^{-3} rad, what is the smallest separation between two objects that she can distinguish?

Section 26.6 | The Bright-Field Light Microscope

26-18   A microscope objective of focal length 5 mm forms an image 150 mm from the objective. If the ocular has a focal length of 28 mm, find the magnification of the microscope.

26-19   A microscope has an objective with a 4-mm focal length and an ocular with a 30-mm focal length. The two lenses are separated by 0.16 m, and the final image is formed 0.25 m from the ocular.

(a) Where is the image formed by the objective? (b) Where is the specimen relative to the objective? (c) What is the magnification of the microscope?

26-20   A microscope provides an angular magnification of 150. What is the smallest separation that can be distinguished with this instrument?

Section 26.7 | The Human Eye

26-21   A doctor finds that a nearsighted person needs one lens with a power of -8 diopters and one with a power of -6 diopters. Find the focal lengths of the lenses.

26-22   Find the focal length of a lens with a power of 4 diopters.

26-23   A person wears reading glasses with a focal length of 2 m. What is their power?

26-24   An optometrist fitting a person for eyeglasses places a 0.25-diopter lens directly in front of a 4.5-diopter lens. (a) What is the power of the combination? (b) What is the effective focal length of the combination?

26-25   A lens has a focal length of 2 m. When a second lens is placed in contact the pair has a focal length of 1.5 m. What is the focal length of the second lens?

26-26   Estimate the power of a normal eye focused at infinity.

Section 26.8 | Visual Acuity, Sensitivity, and Color Perception

26-27   Explain why faint stars are best seen by looking "out of the corner of the eye" rather than directly at the stars.

26-28   Why is it believed that color perception is due only to cones and not to the rods?

PROBLEMS

26-29   A girl 1.5 m tall stands in front of a vertical mirror that is just large enough so that she can see her entire body. How tall is the mirror?

26-30   A man holds a plane mirror of height 0.1 m vertically at a distance of 0.25 m from his eyes and observes that the image of a building just fills the height of the mirror. If the building is 200 m from the mirror, what is its height?

26-31 For a lens with a positive focal length f, (a) calculate the image position for $s = 0$, $f/2$, f, $3f/2$, and $3f$. (b) Sketch a graph of the image position as s varies from 0 to $+\infty$.

26-32 A lens has a focal length of -1 m. (a) Calculate the image position when the object is placed at $s = 0$, 0.5, 1, 1.5, and 3 m from the lens. (b) Sketch a graph of the image position as s varies from 0 to $+\infty$.

26-33 Show that if two thin lenses with focal lengths f_1 and f_2 are in contact, they are equivalent to a single lens with focal length f satisfying

$$\frac{1}{f} = \frac{1}{f_1} + \frac{1}{f_2}$$

(*Hint:* The image formed by the first lens is the object for the second, so $s_2 = -s_1'$.)

26-34 An object is 1 m from a screen. At what points may a 0.05-m focal length lens be placed so as to produce a sharp image on the screen? What are the corresponding magnifications?

26-35 A lens of index 1.5 has one plane side and one concave side of radius 0.2 m. The lens is placed horizontally with the concave side up and is filled with water. What is the focal length of the water-glass system? (*Hint:* Treat the system as a pair of thin lenses in contact.)

SUPPLEMENTARY TOPICS

26.9 | THE CAMERA

The essential elements of any camera are a light-tight box, a converging lens, a shutter that can open briefly, and a film that records the image. In order to admit enough light to permit short exposures, a lens with a large opening or aperture must be used. This, in turn, requires that in a high-quality camera, complex multielement lens designs be used in order to reduce aberrations.

Camera lenses are specified by two quantities. One of these is the focal length, f. As we saw in Example 26.4, the image size is approximately proportional to the focal length. The other quantity is the diameter d of the lens. Usually this is expressed in terms of the f-number. For example, an $f/8$ lens has a diameter $d = f/8$ which is one-eighth its focal length. The light-gathering power of a lens is proportional to its area or to its diameter squared. All but the simplest camera lenses have a diaphragm which can be adjusted to vary the aperture or effective diameter of the lens. For example, changing from $f/8$ to $f/16$ means halving the diameter and reducing the area by a factor of $2^2 = 4$. The shutter time must be increased by a factor of 4 to admit the same total light energy.

Because of lens aberrations, a point on the object

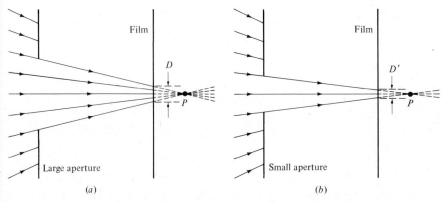

Figure 26.23. The effect of aperture size on the depth of focus. The lens (not shown) of a camera is positioned so that distant objects are imaged sharply on the film. (a) Light rays from a point object closer to the lens converge toward a point P behind the film. An image circle of diameter D is formed on the film. (Diffraction and aberration effects would make the actual diameter larger.) (b) When the aperture is made smaller, the beam of light rays is narrower, and the diameter D' of the image circle on the film is smaller. Consequently the blurring is reduced, and objects located over a larger range of distances appear adequately focused. However, the light intensity reaching the film is reduced, and a longer exposure time is required.

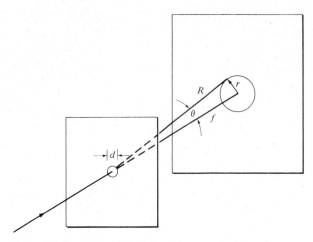

Figure 26.24. The central diffraction circle for light passing through a small circular lens aperture of diameter d.

being photographed will always have an image of finite size no matter how well the lens is positioned or focused. This image size will be reduced if the effective lens diameter or aperture is reduced with a diaphragm. Reducing the aperture also increases the *depth of field*, the range of distances over which object points are imaged with satisfactorily small circles (Fig. 26.23). Thus, a photographer must compromise between the need for short exposure times to stop the motion of a scene and the need for small apertures to reduce aberrations and increase the depth of field.

When the lens aperture is small, diffraction effects may be more important than lens aberrations in limiting the sharpness of photographs. We saw in Chapter Twenty-five that when light passes through a small circular aperture, a diffraction pattern is formed consisting of a bright central circle and weak concentric

rings. If the diameter of the aperture is d, the bright central region extends to the first diffraction minimum at $\sin\theta = r/R \simeq r/f$ (Fig. 26.24). With $\sin\theta = 1.22\lambda/d$, we have

$$r = 1.22\lambda\,\frac{f}{d} \qquad (26.13)$$

This result indicates then when a camera is properly focused on a point object, the image formed on the film will be a circle of radius r even in the absence of aberrations. (The weak concentric rings are not normally noticeable.) Since this radius is proportional to f/d, the effects of diffraction grow as the diaphragm is closed (Fig. 26.25). This sets an upper limit on the useful aperture, as can be seen from the following example.

Example 26.7

If the lens of a 35-mm camera is stopped down to $f/22$, and a photograph is taken, what are the effects due to diffraction? (The film size in a 35-mm camera is 35×24 mm.)

In accordance with our discussion, each point in the scene being photographed will have an image that, at best, is a circle whose radius r is determined by diffraction. If the scene is illuminated with white light, the wavelengths range from 400 to 700 nm, averaging about 550 nm. Using this wavelength and $f/d = 22$,

$$r = 1.22\lambda\,\frac{f}{d} = 1.22(550 \times 10^{-9}\text{ m})(22)$$

$$= 1.5 \times 10^{-5}\text{ m}$$
$$= 0.015\text{ mm}$$

This result means that all details on the film smaller than about 0.015 mm are hopelessly blurred by the overlap-

Figure 26.25. In these enlargements, aberrations limit image quality at large apertures (*a*) and diffraction sets the limit at small apertures (*b*, *c*). (Courtesy of Eastman Kodak.)

ping diffraction patterns. Enlarging the photo 10 times produces a picture 35 by 24 cm, or about 14 by 10 in. It also increases the diffraction circles to 0.15 mm in radius, making them large enough to be discernable to someone looking carefully at the photograph. For this reason, apertures smaller than $f/22$ are not normally provided on 35-mm cameras. Instead, filters are used if the light intensity must be reduced further.

26.10 | RESOLUTION AND CONTRAST IN MICROSCOPES

When the separation between two points in a microscopic specimen is comparable to the wavelength λ of the light, we would expect diffraction effects to be important. A detailed analysis shows that the minimum separation d that can be resolved by a microscope is

$$d = \frac{\lambda}{2n \sin \theta} \qquad (26.14)$$

Here, λ is the wavelength in air, n is the refractive index of the medium between the objective lens and the object under study, and θ is the angle subtended by the objective lens (Fig. 26.26). The product $n \sin \theta$ is called the *numerical aperture* and is sometimes marked on the instrument. If two points in a specimen are separated by less than d, their diffraction patterns overlap so much that their images cannot be distinguished.

A typical value of this minimum separation is found in the next example.

Example 26.8

What is the minimum resolvable separation for objects in air illuminated by green light ($\lambda = 500$ nm) if the angle subtended by the objective is 90°?

With $n = 1$ and $\sin 90° = 1$,

$$d = \frac{\lambda}{2n \sin \theta} = \frac{500 \text{ nm}}{2(1)(1)} = 250 \text{ nm}$$

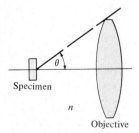

Figure 26.26. The numerical aperture of this lens is $n \sin \theta$, where n is the refractive index of the medium between the lens and the specimen.

We saw in Example 26.6 that a magnification of 400 leads to a minimum resolvable separation of about 250 nm if diffraction is ignored. However, we see now that diffraction cannot be ignored, since it independently limits the resolution of any light microscope to about 250 nm. Consequently, the useful magnification of the ordinary light microscope is limited to roughly 400. Larger magnifications may make viewing more comfortable, but they reveal no additional details. Since a typical bacterial cell has a diameter of about 1000 nm, it is not possible to make very detailed studies of bacterial structures with light microscopes.

Equation 26.14 suggests two ways to improve the resolution of a microscope: use shorter wavelength light and media with larger indices of refraction. With *oil immersion objectives,* some improvement is obtained by immersing the object in a medium such as oil of cedar which has $n = 1.4$. *Ultraviolet microscopes* use light with wavelengths somewhat shorter than visible light. In addition to having a smaller minimum resolvable separation, ultraviolet microscopes are useful because substances such as nucleic acids and proteins strongly absorb ultraviolet light. This leads to very good contrast, which is also necessary to obtain good resolution.

Much greater resolution can be achieved with *electron microscopes,* in which electrons are accelerated by a potential difference and focused by magnetic fields. Although electrons behave in many ways as particles, they also have wave attributes which are discussed in Chapter Twenty-nine. The wavelength associated with electrons accelerated through 50,000 V is 5×10^{-3} nm or about $1/10^5$ times that of visible light. In practice the resolution is limited to about 0.2 nm, which is about 1000 times better than can be achieved with light microscopes.

Contrast | To distinguish an object from its surroundings, there must be sufficient contrast or variation in light intensity. Without good contrast, the actual resolution achieved will be much less than that implied by the design of the microscope. *Staining* with dyes that are absorbed differently in various parts of an object is sometimes used to improve contrast. Other dyes are used to make a sample *fluoresce.* In this case the sample is illuminated with intense ultraviolet light, and the fluorescent constituents emit

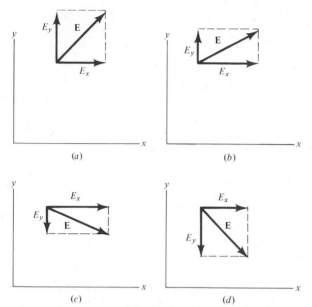

Figure 26.27. The polarization of an electromagnetic wave is parallel to the electric field vector **E**. In a birefringent medium, waves with electric fields along different crystal axes travel with different velocities. (*a*) Traveling into the page, the wave enters the birefringent medium. The initial electric field has x and y components that now travel at different velocities. (*b*) At a point inside the medium, E_x has a maximum at some instant. The wave with an electric field in the y direction travels slower, so E_y is not yet a maximum. Hence the **E** vector is rotated somewhat from its original direction. (*c*) Further into the medium, E_y is still negative when E_x is a maximum. (*d*) After an additional distance, E_y has a maximum negative value when E_x is a maximum. The polarization has now been rotated 90° from its original direction by the birefringent material.

light at a longer wavelength. This light is detected after the ultraviolet light is removed from the beam by a filter.

26.11 | POLARIZING, INTERFERENCE, AND PHASE CONTRAST MICROSCOPES

We now briefly describe *polarizing, interference,* and *phase contrast* microscopes. These instruments ingeniously exploit the wave properties of light in improving the contrast of transparent structures.

A *polarizing microscope* uses polarized light to illuminate an object. When the object has a random or uniform character, the polarization of the light passing through it is unaffected. If this light then passes through an analyzer set at 90° to the original polarization direction, no light is transmitted. However, if

Figure 26.28. Interference microscope. The elements are described in the text.

the sample contains *birefringent* materials, whose indices of refraction depend on the direction of the light beam and on the electric field direction, then the polarization vector of the light is rotated and some light passes through the analyzer (Fig. 26.27). Proteins and nucleic acids are birefringent and can be seen in a polarizing microscope.

In an *interference microscope,* the illuminating light is split into two beams (Fig. 26.28). One beam passes through a sample whose refractive index varies with the position. The phases of the different parts of this beam have corresponding variations after it leaves the sample. The second beam travels through an identical optical path, except that it does not go through the sample. When the two beams are recombined, interference between them produces intensity variations. Thus, many kinds of structures can be made visible, even though the sample is completely transparent.

The *phase contrast microscope,* invented in 1932 by Fritz Zernike, is an ingenious instrument that employs interference to enhance contrast but involves only a single beam of light. Consequently, it is less costly and complicated than the interference microscope.

Since the operation of this microscope depends on diffraction effects, we first consider some aspects of diffraction. We have seen in Chapter Twenty-five that when a parallel beam of monochromatic light passes through a narrow slit, an undeviated bright central line is formed on a screen, with progressively weaker lines on either side. (Similarly, when light goes through a circular aperture, a bright central circular spot and weak concentric rings are seen.) The central

(a)

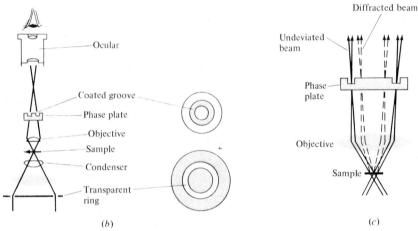

(b)   (c)

Figure 26.29. (a) The undeviated beam from a slit and the adjacent diffracted beam are a quarter wavelength out of phase. If the phase of the diffracted beam is shifted an additional quarter wavelength by a glass plate, it is exactly out of phase with the undeviated beam. (b) A phase contrast microscope. (c) An expanded view near the objective.

line and the lines immediately adjacent on either side can be shown to be exactly a quarter wavelength out of phase. Now suppose the screen is removed, and the bright central beam passes through a partially absorbing coated glass plate so that its amplitude becomes equal to that of an adjacent diffracted beam (Fig. 26.29a). The latter passes through a thicker transparent glass plate. The beam travels slower in the glass than in air, so it emerges an *additional* quarter wavelength out of phase with the central beam. These two beams are now a half wavelength out of phase, or exactly opposite in sign, and equal in amplitude. Hence, they interfere destructively when they are recombined with a lens.

Essentially, this idea is applied to a rather different geometry in the phase contrast microscope (Fig. 26.29b). A transparent ring is placed in front of the light source, producing a hollow cone of light with its point at the specimen. If there is no specimen, this

light forms a bright circular image just beyond the focal point of the objective. The same thing happens if there is a completely uniform transparent specimen. However, suppose the specimen contains a small transparent structure. This structure will have a diffraction pattern consisting of an undeviated bright circle and weak diffracted concentric circles (Fig. 26.29c). The adjacent diffracted circles are again a quarter wavelength out of phase with the indeviated beam. A *phase plate* placed at the location of this image has a circular groove corresponding to the bright circle. The groove is coated so that it absorbs much of the light passing through this part of the plate, reducing its amplitude to that of the adjacent diffracted light. The latter passes through a thicker portion of the phase plate, emerging exactly out of phase with the light passing through the groove, so that destructive interference occurs when the two beams are later recombined. In this way, a dark spot

479

Figure 26.30. (*a*) An unstained cross-section of a grain of wheat as seen with a conventional bright field microscope. (*b*) The same scene as viewed with a phase contrast microscope. (Courtesy of Bausch and Lomb.)

(*a*)   (*b*)

Figure 26.31. (*a*) Parallel light rays from a distant object are focused at a point before the retina of a nearsighted or myopic eye. (*b*) A divergent lens bends the rays so that they appear to come from a closer location and enables the eye to image the rays on the retina.

is formed due to a structure with a slightly different refractive index than the remainder of a transparent object (Fig. 26.30).

26.12 | OPTICAL DEFECTS OF THE EYE

Four common optical defects of the eye can be corrected by the use of eyeglasses. In three of these defects, the glasses are used to shift the apparent position of an object, so that the defective eye is able to focus properly. In the last, *astigmatism*, the glasses correct for a distortion produced by the eye.

In *myopia* or nearsightedness, parallel light from a distant object is focused by the relaxed eye at a point in front of the retina (Fig. 26.31). Consequently a nearsighted person cannot focus clearly on objects further away than the *far point* located at a distance x_f. This condition can be corrected with diverging eyeglass lenses that bend the light from a very distant object outward *so that it seems to originate at the far point*. To find the required focal length, we use the thin lens formula with $s = \infty$ and $s' = -x_f$; the minus sign is needed because the image is virtual. Then

$$\frac{1}{f} = \frac{1}{s} + \frac{1}{s'} = \frac{1}{\infty} - \frac{1}{x_f} = 0 - \frac{1}{x_f}$$

so

$$f = -x_f \qquad (26.15)$$

With a lens of this focal length the relaxed eye can form a sharp image on the retina.

This lens will also shift the near point to a distance x_n', which is greater than x_n, the near point distance without glasses. An object at x_n' has a virtual image at x_n. With $s = x_n'$ and $s' = -x_n$, the thin lens formula

$$\frac{1}{s} = \frac{1}{f} - \frac{1}{s'}$$

gives

$$\frac{1}{x_n'} = \frac{1}{f} + \frac{1}{x_n} \qquad (26.16)$$

A lens with focal length f will change the near point distance from x_n to x_n'. This result is illustrated in the following example.

Example 26.9

A very nearsighted man has his near point at 0.1 m. His power of accommodation is 4 diopters. (a) Where is his far point? (b) What power lenses does he require? (c) What is his near point when wearing the glasses?

(a) From Eq. 26.12, with $A = 4$ diopters and $x_n = 0.1$ m,

$$\frac{1}{x_f} = \frac{1}{x_n} - A = \frac{1}{0.1 \text{ m}} - 4 \text{ m}^{-1}$$

so $x_f = 1/6$ m. This means that without glasses, he cannot focus on objects more distant than $1/6$ m.

(b) In accordance with Eq. 26.15, he requires diverging (negative) lenses with a focal length $f = -x_f = -(1/6)$m. Their power is $P = 1/f = -6$ diopters.

(c) According to Eq. 26.16, his near point with the glasses on is given by

$$\frac{1}{x_n'} = \frac{1}{f} + \frac{1}{x_n} = \frac{1}{(-1/6 \text{ m})} + \frac{1}{0.1 \text{ m}}$$

so $x_n' = 0.25$ m. This is the near point distance for a person with normal vision and average accommodation; the lenses have fully corrected his vision.

Hypermetropia, or farsightedness, is the opposite of myopia. Light from an object close to the eye is focused toward a point behind the retina, even when the lens is adjusted by the ciliary muscles to have its maximum power (Fig. 26.32). Eyeglasses with converging lenses correct this problem by bending the light rays from an object at the desired near point distance x_n' inward, so that it has a virtual image at x_n, the actual near point distance. The focal length or power of the lens needed to shift the near point in this way can be found as in the next example.

Example 26.10

A woman has her near point 1 m from her eyes. What power eyeglasses does she require in order to have her near point at 0.25 m?

Using Eq. 26.16, the power of the lenses required to shift her near point from $x_n = 1$ m to $x_n' = 0.25$ m is

$$P = \frac{1}{f} = \frac{1}{x_n'} - \frac{1}{x_n} = \frac{1}{0.25 \text{ m}} - \frac{1}{1 \text{ m}} = 3 \text{ diopters}$$

The eyeglasses will permit her to read at a normal distance.

Presbyopia, the reduction in accommodation that occurs with age, is the result of a gradual weakening of the ciliary muscles and diminishing flexibility of the lens. The far point of a person who has normal vision as a young adult eventually recedes far enough so that converging lenses are needed for close work or reading, much like the person with hypermetropia. Many people eventually require bifocal lenses. In the most common type of bifocals, the upper portion of the lens is used for distant vision, and the lower part is used for close work. For example, an elderly myopic person will wear bifocals with a diverging lens in the upper part and a weaker diverging lens below.

A person with *astigmatism* cannot simultaneously focus on both horizontal and vertical lines. Usually this is due to a cornea that is not perfectly spherical, so that it has different curvatures in different directions. Occasionally, astigmatism is caused by irregularities elsewhere in the eye. Astigmatism can be corrected by cylindrical lenses oriented to compensate for the distortion (Fig. 26.33a). If a correction for myopia or hypermetropia is also needed, then lenses shaped like the outer surface of a doughnut are used. The maximum and minimum radii of curvature are chosen to correct both defects.

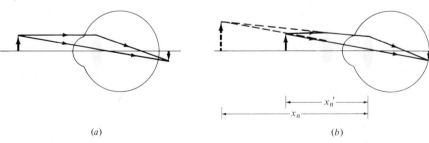

(a)	(b)

Figure 26.32. (a) Rays from a nearby object are bent so they would focus at a point beyond the retina of a farsighted or hypermetropic eye. (b) A converging lens bends the rays so they appear to come from a more distant location. The distance x_n' to the near point with the glasses is less than the distance x_n without the glasses.

(a)

(b)

Figure 26.33. (a) The type of lens used in eyeglasses to correct for astigmatism. (b) A soft contact lens being inserted. ((b) Courtesy Bausch & Lomb, Soflens Division.)

Contact lenses are an alternative to ordinary eyeglasses and have many advantages (Fig. 26.33b). Hard contact lenses are made of rigid plastic and are about 1 mm thick and 1 cm in diameter. Placed directly on the eye, they are separated from the cornea by a layer of tears, *effectively replacing the cornea as the front of the eye.* Hence, they automatically correct for astigmatism due to corneal irregularities. For the same reason, they avoid the aberrations present at the edges of strong eyeglass lenses. Recently soft contact lenses have been introduced. Somewhat larger than hard contact lenses, soft lenses conform to the shape of the cornea and are often more comfortable, but do not correct for astigmatism.

EXERCISES ON SUPPLEMENTARY TOPICS

Section 26.9 | The Camera

26-36 An $f/1.4$ camera lens has a focal length of 50 mm = 0.05 m. What is the lens diameter?

26-37 With a certain set of lighting conditions, a photographer finds that an exposure of 1/25 second is needed at $f/16$. What exposure is needed at $f/8$?

26-38 The diaphragm of a lens is stopped down from $f/2$ to $f/16$. By what factor is the light intensity reduced?

26-39 Why do telephoto (long focal length) lenses usually have larger f-numbers than normal camera lenses of similar cost and quality?

26-40 A camera lens is focused on a distant scene illuminated with white light. Find the radii of the diffraction circles on the film at f-numbers of (a) $f/2$, and (b) $f/32$.

Section 26.10 | Resolution and Contrast in Microscopes

26-41 An oil immersion objective is immersed in oil of cedar with a refractive index 1.4. If the objective subtends an angle of 80°, find the (a) numerical aperture, and (b) minimum resolvable separation at a wavelength of 400 nm.

Section 26.12 | Optical Defects of the Eye

26-42 An elderly man has his near point 2 m from his eyes. What power lenses does he need to read comfortably at 0.25 m?

26-43 A woman has a far point 0.5 m from her eyes. (a) If she is to see distant objects clearly, what focal length lenses does she require? (b) If her power of accommodation is 4 diopters, where is her near point without the glasses? (c) Where is her near point with the glasses?

26-44 A farsighted person with an accommodation of 3 diopters has a near point 2 m from the eyes. (a) What power eyeglasses are needed to move the near point to 0.25 m from the eyes? (b) Where is the far point with these eyeglasses?

PROBLEMS ON SUPPLEMENTARY TOPICS

26-45 A nearsighted middle-aged man has a near point at 0.1 m and a power of accommodation of 2 diopters. Find his far point (a) without glasses, and (b) with the correct glasses to move his near point to 0.25 m.

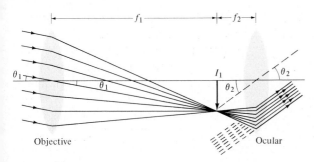

Figure 26.34. A refracting telescope.

*26-46 A refracting telescope (Fig. 26.34) focused on an astronomical object forms an image at infinity. (a) If the focal lengths of the objective and ocular are f_1 and f_2, respectively, show that the separation between the lenses is $f_1 + f_2$. (b) Using small angle approximations, show that the angular magnification $M = \theta_2/\theta_1$ satisfies $M = -f_1/f_2$.

26-47 A telescope is constructed using lenses of focal lengths 2 m and 0.1 m. Using the result of the preceding problem, find its angular magnification.

26-48 The ocular of a refracting telescope has a focal length of 0.1 m. If the length of the telescope is 3 m, what is its angular magnification? (Use the results of Problem 26-46.)

Additional Reading

The following books have chapters on microscopes and the eye:

R.B. Setlow and E.C. Pollard, *Molecular Biophysics,* Addison-Wesley, Reading, Mass., 1962, chapters 12 and 14.

Simon G.G. MacDonald and Desmond M. Burns, *Physics for the Life and Health Sciences,* Addison-Wesley, Reading, Mass. 1975, chapters 20, 21, and 23.

D. Ackerman, *Biophysical Science,* Prentice-Hall, Englewood Cliffs, N.J., 1962, chapters 2, 7, 19, and 29.

Other references on the eye:

S.F. Jacobs and A.B. Stewart, Chromatic Aberration in the Eye, *American Journal of Physics,* vol. 20, 1952, p. 247.

M.H. Pirenne, *Vision and the Eye,* Chapman and Hall, London, 1967.

S.M. Luria, Color Vision, *Physics Today,* March 1968, p. 35.

Joseph J. Sheppard Jr., *Human Color Perception,* American Elsevier Pub. Co., New York, 1968.

Kenneth Neil Ogle, *Optics: An Introduction for Ophthalmologists,* Charles C Thomas, Springfield, Ill., 1968.

Karl Von Frisch, *The Dance Language and Orientation of Bees,* Belknap Press of Harvard University Press, Cambridge, Mass. 1967.

Scientific American articles:

F. Dow Smith, How Images are Formed, September 1968, p. 97.

R. Clark Jones, How Images are Detected, September 1968, p. 111.

P.K. Tien, Integrated Optics, April 1974, p. 28.

Ivo Kohler, Experiments with Goggles, May 1962, p. 62.

P. Connes, How Light in Analyzed, September 1968, p. 72.

W.R.A. Muntz, Vision in Frogs, March 1964, p. 110.

Talbot H. Waterman, Polarized Light and Animal Navigation, July 1955, p. 88.

John F. Harris and Igor Gamow, The Infrared Receptors of Snakes, May 1973, p. 94.

George Wald, Eye and Camera, August 1950, p. 32.

Charles R. Michael, Retinal Processing of Visual Images, May 1969, p. 105.

Derek H. Fender, Control Mechanisms of the Eye, July 1964, p. 24.

Ulric Neisser, The Processes of Vision, September 1968, p. 204.

G. Adrian Horridge, The Compound Eye of Insects, July 1977, p. 108.

Edwin H. Land, Experiments in Color Vision, May 1959, p. 84.

George Wald, Life and Light, October 1959, p. 92.

W.A.H. Rushton, Visual Pigments and Color Blindness, March 1975, p. 64.

Jacob Bech, The Perception of Surface Color, August 1975, p. 62.

Bela Julesz, Experiments in the Visual Perception of Textures, April 1975, p. 34.

Gunnar Johannson, Visual Motion Perception, June 1975, p. 76.

Olga Eizner Favreau and Michael C. Corballis, Negative Aftereffects in Visual Perception, December 1976, p. 42.

Edward F. MacNichol, Jr., Three-Pigment Color Vision, December 1964, p. 48.

Alistair B. Fraser and William H. Mack, Mirages, January 1976, p. 102.

Leonard A. Herzenberg, Richard G. Sweet, and Leonore A. Herzenberg, Fluorescence-Activated Cell Sorting, March 1976, p. 108.

William H. Price, The Photographic Lens, August 1976, p. 72.

483

UNIT SEVEN

MODERN PHYSICS

In the latter part of the nineteenth century, physics was considered by many to be a completed science. The fundamental laws of motion and electromagnetism, including light waves, were well understood, and it seemed that only calculational and experimental difficulties would hinder further progress. However, after only 20 years of the twentieth century, this description of the physical world was severely shaken. The very few unsolved problems that existed in 1900 proved to be explainable only by drastic assumptions that had no historical precedents. The illusion of a complete science proved to be a result of man's lack of experience with atomic-size particles and with objects that move at nearly the speed of light.

By 1912 the work of Max Planck, Niels Bohr, and Albert Einstein had given us a new picture of the world. These three set forth the tentative but crucial ingredients of what is now known as quantum mechanics. During the same period, Einstein also developed the special theory of relativity. Quantum mechanics and relativity are our present basis for understanding nature. Since their predictions agree with many kinds of experiments, they must be considered not as speculative but as a largely correct description of nature.

The three chapters of this unit set forth our fundamental concepts of special relativity, the particle nature of light, and the wave nature of matter. These concepts are the foundations of the remaining chapters in this book, which cover atoms, molecules, and atomic nuclei.

It should be recognized at the outset that while

485

modern physics has forced us to change our philosophy of natural processes, the procedures of the old or classical physics are still essentially correct under certain conditions. Those conditions are that the objects under consideration are large compared to atomic sizes and that their velocities are much less than the speed of light. In nearly all the situations that we have considered up to now, these conditions are well satisfied.

CHAPTER TWENTY-SEVEN

SPECIAL RELATIVITY

At the end of the nineteenth century, several fundamental problems existed in physics. One set of problems eventually led to the development of the revolutionary concepts of quantum mechanics, although it took several decades and many people to develop them. A second set of problems was solved in a single stroke by Einstein when his theory of special relativity was published in 1905.

The problem solved by Einstein was almost obscure enough to be largely ignored, but his solution has had a profound effect on our conception of the physical world. The *special theory of relativity* relates time and length measurements made by inertial observers moving at constant velocities with respect to one another. The *general theory of relativity*, which Einstein began to develop in 1911, deals with accelerated systems and gravitational forces. However, in this book, we only discuss the special theory of relativity.

The problem that Einstein originally addressed was an apparent violation of a fundamental rule of physics. This rule is called the principle of relativity and was developed first by Galileo. It states that the laws of physics should be the same for all observers moving at constant velocities with respect to each other. We describe this idea further in the next section. Here, we need only remark that Newton's laws of mechanics seemed in agreement with this rule, but the laws of electricity and magnetism did not. It appeared that the effects of moving charges on each other depended on which ones were moving. If this were the case, there would be different results measured by different moving observers.

Einstein found that electromagnetism would be a consistent theory if one insisted that the speed of light in the vacuum is the same for all observers and is independent of the motion of the source and the detector. In this way, light waves are different from sound or water waves, whose velocities are determined by their media. Einstein's view, which is now well supported by experiment, is that there is no medium required for the propagation of electromagnetic waves.

If one accepts Einstein's principles that the laws of physics and the speed of light are the same for all inertial observers, one is led directly to some remarkable conclusions. Moving clocks run slow, and moving objects are shortened. Thus, according to an earth observer, a fast-moving astronaut will live longer, and his rocket ship is shorter than before blastoff. Also, events that appear simultaneous according to the astronaut are not simultaneous as viewed by the earth observer.

Because our ideas of space and time are altered for rapidly moving objects, the laws of mechanics are also affected. The definitions of energy and momentum must be changed if these quantities are to obey the usual conservation laws. If an object has a mass m and a velocity v, its relativistic energy turns out to be

$$E = \frac{mc^2}{\sqrt{1 - v^2/c^2}}$$

When the velocity is zero, the energy becomes the *rest energy*

$$E_0 = mc^2$$

This is the most famous equation of twentieth-century physics. It asserts that mass (or matter) and energy are equivalent. Mass is converted into energy in the nuclear reactions that provide the energy of the sun and in nuclear explosives and reactors. If just 1 kg of matter is converted into energy, the energy released is $mc^2 = (1 \text{ kg})(3 \times 10^8 \text{ m s}^{-1})^2 = 9 \times 10^{16}$ J. This is comparable to the total energy used per day to generate electricity in the United States.

27.1 | The Fundamental Principles of Special Relativity

We saw in Chapter Three that Newton's laws of motion apply only to measurements made with respect to inertial reference frames. These are frames in which Newton's first law holds true: an object with no net force on it remains at rest or in motion with a constant velocity. For example, observers in accelerating vehicles or on rotating merry-go-rounds cannot apply Newton's laws directly to their observations. Special relativity also deals only with measurements made by observers in inertial reference frames.

The theory of special relativity is based on two fundamental principles.

1 *All the laws of physics (and nature) have exactly the same form in all inertial reference frames. This is the principle of relativity.*

2 *The speed of light in free space is the same for all observers in inertial reference frames.*

The full implications of these ideas will become gradually clearer in this chapter, but we can immediately see some of the consequences. For example, the first principle implies that two people moving at a constant velocity with respect to each other can never decide which is moving and which is at rest. This is because all physical laws are the same in both systems, and they do not depend on any absolute velocity. Thus the observers can only determine their relative motion. For example, a woman on a train moving with constant velocity relative to the ground might say that she is stationary and the earth is moving. No experiment will decide whether this is true or false. The assumption that the earth is at rest and the train is moving is no more and no less valid.*

*Actually the earth is not exactly an inertial frame; as it rotates about the sun and spins on its axis, objects on the earth experience a centripetal acceleration. In most situations the effects of this acceleration are small, so we assume in this chapter that the earth is an inertial frame.

Turning to the second principle, it is apparent that if the speed of light is the same for all inertial observers, light must be different from all other waves we have studied. For example, a listener moving toward a sound source will observe a higher speed of sound than someone at rest relative to the air. Nothing like this happens for light.

It was long thought that there was some medium, the *ether,* in which light moved at a speed *c*. This ether was the counterpart of the medium in which sound moves. Thus if one were moving relative to the ether, the speed of light would appear different from *c*. However, no experiment ever detected the ether or measured changes in the speed of light due to motion relative to the ether, and as far as we know now, there is no ether. In fact the strongest evidence we have that there is no ether is the success of the theory of relativity. The theory of relativity could not be correct if the ether existed.

Einstein's motivation for the theory of relativity was to remove the contradictions then present in the laws of mechanics and electromagnetism. To the objection that the results were disturbing to our common sense view of things, he answered, "Common sense is that layer of prejudices laid down in the mind prior to the age of eighteen."

27.2 | MOVING CLOCKS AND TIME DILATION

Einstein's theory predicts that if a clock is moving with respect to an observer in an inertial reference frame, he will observe it to be running slower than a clock at rest relative to the observer. The latter clock is said to be in the observer's *rest frame*. In order to show how this follows directly from the principles of special relativity, we first consider a particular type of clock called a *light clock*.

This clock is a stick of length *l* with a mirror *R* and a photodetector *P* at opposite ends (Fig. 27.1). A flash of light emitted at one end will be reflected by the mirror at the other end and return to the photodetector next to the light source. Each time a light flash is detected, the clock "ticks" and emits another flash.

We can easily relate the time *t* between ticks of the clock to the length *l* when the clock is at rest in a laboratory. From Fig. 27.1a the total distance trav-

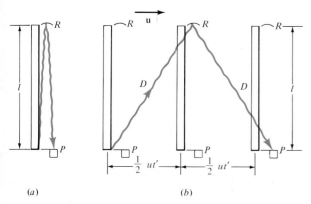

(a)   *(b)*

Figure 27.1. *(a)* Each time a flash of light is reflected from the mirror R and returns to the photodetector P beside the source, the light clock "ticks" and emits a new pulse. *(b)* The same clock moving to the right at a constant velocity **u**. The light pulse must travel the distance $2D$ to return to the photodector P.

eled by a light pulse is $x = 2l$, and the speed of the light is c. Thus $x = vt$ becomes $2l = ct$, or $l = ct/2$, where t is the time between ticks.

When the clock is moving with a speed u relative to the laboratory, the speed of light is again c (principle 2) but the distance it must travel between ticks is $2D$ (Fig. 27.1b). With the Pythagorean theorem, we have

$$D^2 = l^2 + \left(\frac{1}{2}ut'\right)^2$$

where t' is the time for the light to return to the photodetector of the moving clock. But we also know that $2D = ct'$ or $D = ct'/2$ and that $l = ct/2$. Substituting these for D and l, we find

$$\left(\frac{ct'}{2}\right)^2 = \left(\frac{ct}{2}\right)^2 + \left(\frac{ut'}{2}\right)^2$$

Solving for t',

$$t' = \frac{t}{\sqrt{1 - u^2/c^2}} \qquad (27.1)$$

The time t' is greater than t because the factor $\sqrt{1 - u^2/c^2}$ is less than 1. (We see later that u must be less than c.) This result is referred to as *time dilation*. It states that as measured by the observer at rest in the laboratory, time does not pass as rapidly in the moving frame. Note that we have *not* concluded that an observer moving with the clock would see any difference in the interval between ticks of his clock. He would, however, observe that a clock at rest in the

laboratory has slowed down by exactly the same time dilation factor. *Each observer observes that the other's clocks are slowed down.* To illustrate the remarkable implications of time dilation, consider an imaginary world in which the effects of special relativity are very pronounced.

Example 27.1

Assume for this example that the speed of light is 100 km h^{-1}. A woman relates the story of the delivery of her baby. She says that she drove herself at 80 km h^{-1} to the hospital, which was 100 km away according to the highway signs. She also says that according to her watch the baby was born 1 hour after she left. Was the baby born in the hospital?

If we use our usual reasoning, we would say no, since it would take $t = x/v = 100 \text{ km}/80 \text{ km h}^{-1} = 1.25$ hours to get to the hospital. However, because of time dilation, once she is traveling at 80 km h^{-1}, we see that her watch slows down. Thus when her watch indicates 1 hour, our watches indicate that the elapsed time is.

$$t' = \frac{t}{\sqrt{1 - u^2/c^2}} = \frac{1 \text{ h}}{\sqrt{1 - \left(\frac{80 \text{ km h}^{-1}}{100 \text{ km h}^{-1}}\right)^2}}$$

$$= \frac{1 \text{ h}}{0.6} = 1.67 \text{ h}$$

Since all clocks, *including biological ones,* must behave in the same way, in 1 hour of her time she can drive (80 km h^{-1})(1.67 h) = 133.6 km. The hospital is only 100 km away, so she must have arrived with time to spare.

In this example one might object to the idea that the woman's wristwatch and her biological processes obey the same law that we derived for a light clock. However, *all* clocks and processes must obey this law; otherwise, the principle of relativity would be violated. For example, we can synchronize a light clock and several other types of clocks when they are all at rest. Now when they are set in motion with a common velocity, all the clocks must behave the same way. If some but not all apparently slow down, those that do not could be said to operate correctly in only one reference frame.

There is one additional point to clarify about this example. From the woman's point of view, when she is moving the clocks of people at rest appear to her to be running slow. However, when she arrives, her watch will be behind those of the people at the hospital. The difference between her and someone who did

489

not make the trip is that the woman accelerated and decelerated. Since she was not always in an inertial reference frame, her experience is unique, and her watch will appear slow when she arrives. An experiment that represents one of many direct confirmations of special relativity is described in the next example.

Example 27.2

Short-lived subatomic particles called mu mesons are created by cosmic rays in the upper atmosphere, approximately 10,000 m above us. They travel at $0.999 c (c = 3 \times 10^8 \text{ m s}^{-1})$. In laboratory experiments with mu mesons at rest, they have an average lifetime of 2.2×10^{-6} s. (a) How long will the moving particles appear to live to an observer on the earth? (b) Will the average meson reach the earth?

(a) The average lifetime of the moving meson will appear increased to the earth observer. The average lifetime will appear to be

$$t' = \frac{t}{\sqrt{1 - u^2/c^2}} = \frac{2.2 \times 10^{-6} \text{ s}}{\sqrt{1 - (0.999c/c)^2}}$$

$$= 4.92 \times 10^{-5} \text{ s}$$

This lifetime is more than 20 times the lifetime of a stationary mu meson.

(b) Using this as the average lifetime, the earth observer sees the average meson travel a distance of

$$D = vt'$$
$$= (0.999)(3 \times 10^8 \text{ m s}^{-1})(4.92 \times 10^{-5} \text{ s})$$
$$= 14,700 \text{ m}$$

This is more than the 10,000-m height at which they were produced, so the average meson does reach the earth. If the lifetime were 2.2×10^{-6} s, a meson would travel only 660 m, and these particles would not be observed at the surface of the earth. Since mu mesons are observed at the earth's surface, this represents a direct confirmation of time dilation.

27.3 | LENGTH CONTRACTION

An object in motion is shorter than it is at rest. This is called *length contraction* and is an immediate consequence of the time dilation effect. To take a concrete situation, consider the woman in Example 27.1. She and an observer stationary with respect to the road must agree on their relative velocity, in accordance with the principle of relativity. As far as she can tell, her clocks are working normally. Therefore, she will

reason that she was able to reach the hospital safely because the distance she had to travel was less than the posted distance of 100 km. In other words, as measured by this moving observer, the length of the road was reduced.

According to a stationary observer, the woman's velocity was $u = l/t'$, where t' is the time as measured by *a clock at rest relative to the road,* and l is the length of the road. The woman measures the same velocity. As we saw in the preceding section, according to *her clock* the time elapsed, t, is less than t' by the time dilation factor, so

$$t = t' \sqrt{1 - \frac{u^2}{c^2}}.$$

Hence the length l' of the road as observed by the woman is $l' = ut = (l/t')(t' \sqrt{1 - u^2/c^2})$, or

$$l' = l \sqrt{1 - \frac{u^2}{c^2}} \qquad (27.2)$$

This means that the road is measured to be shorter by the factor $\sqrt{(1 - u^2/c^2)}$ by an observer moving relative to the road. *An object is shortened or contracted according to an observer moving along its length.* Let us again examine the mu meson example from this point of view.

Example 27.3

A mu meson approaches the earth from a height of 10,000 m at a speed of $0.999c$. According to an observer moving with the velocity of the meson, what is the height of the atmosphere?

Using the length contraction formula, the height as measured by this observer is

$$l' = l \sqrt{1 - \frac{u^2}{c^2}}$$

$$= (10^4 \text{ m}) \sqrt{1 - (0.999)^2}$$

$$= 447 \text{ m}$$

Since the atmosphere rushes by at $0.999c$, the meson reaches the ground in $(447 \text{ m})/(0.999 \times 3 \times 10^8 \text{ m s}^{-1}) = 1.49 \times 10^{-6}$ s, which is less than its average lifetime of 2.2×10^{-6} s. Thus the average meson reaches the earth's surface, in agreement with our earlier conclusion.

Relativistic effects are very important in particle accelerators where particle speeds approach the speed of light. Effects of relativity are also significant in electron microscopes, as is seen in this example.

ALBERT EINSTEIN
(1879–1955)

(California Institute of Technology Archives.)

By the end of 1905, at the age of 26, Albert Einstein had developed the special theory of relativity and explained the photoelectric effect. He was also well on his way to formulating the general theory of relativity. This precocious activity was not preceded by exhibitions of genius. On the contrary, Einstein had dropped out of secondary school in his native Germany and later returned to school only because he could not pass the university entrance examinations without further preparation. He was graduated from the University of Zurich in 1900 and settled for a position in the Swiss Patent Office in Bern, since his mediocre performance at Zurich precluded his obtaining an academic position.

The Patent Office was quiet and afforded him the time to do research in theoretical physics. Because of his isolation from others working on contemporary problems, some of his work was a duplication of that already done. However, by 1905, Einstein was working on unsolved problems and was having immense success in developing new ideas and concepts. The work on the photoelectric effect was later cited in his Nobel prize award. The special theory of relativity not only explained a number of fundamental problems in physics but also changed the way in which we regard space and time. In this same year, Einstein finished his doctoral dissertation, and in 1907, he joined the faculty at the University of Zurich. By 1913, his work had brought great professional praise, an important position at the University of Berlin, and increasing fame among nonphysicists. Popular descriptions of the theory of relativity captured the imagination of his time.

The general theory of relativity was published in 1916. It predicted the deflection of light in a gravitational field, which was confirmed in England in 1919. The emotions of World War One affected these scientific achievements of a German-born physicist. In Einstein's words, "Today in Germany I am called a German man of science and in England I am represented as a Swiss Jew. If [my theory is overthrown] the description will be reversed, and I shall become a Swiss Jew for the Germans and a German for the English."

While the theory never was overthrown, the situation did change. As the Nazis gained power, his Jewish heritage and his advocacy of world government and disarmament led to vitriolic personal attacks on him. Consequently, he emigrated to the United States in 1933 and never returned to his homeland.

Although we will see that he also helped to initiate the development of quantum mechanics, it evolved in a way that he found unsatisfying. Modern quantum mechanics holds that only the relative likelihood of various outcomes from an experiment may be predicted and not the result of a specific measurement. Einstein did not like this notion of "God playing dice with the Universe." While he played an important role as a consultant and sympathetic protagonist in the development of quantum mechanics, in his later years Einstein's research concentrated on attempts at a unified theory of gravitation and electromagnetism. His philosophy and work took him out of the mainstream of physics, although his early contributions continued to play an important role in the theory of modern physics.

Einstein's reputation was greater and more lasting than that of any other scientist of this century. His name was and is still used as a synonym for the revolutionary developments of modern natural science.

Example 27.4

An electron traveling at $0.3c$ travels through a segment of an electron microscope 0.1 m long. What is the length of this segment as measured in the rest frame of the electron?

In the frame of reference where the electron is at rest, the 0.1-m distance is contracted to

$$l' = l\sqrt{1 - \frac{u^2}{c^2}}$$

$$= (0.1 \text{ m})\sqrt{1 - \left(\frac{0.3c}{c}\right)^2}$$

$$= (0.1 \text{ m})(0.954) = 0.0954 \text{ m}$$

In electron microscopes, the electrons are deflected by magnetic lenses, and the force on the electrons in these lenses is proportional to their velocity. Calculations of the focusing power of the lenses must include relativistic effects to avoid significant errors.

No change in length is seen by an observer moving at right angles to the length of an object. It is left as a problem to show this is a direct consequence of the principle of relativity.

27.4 | MOMENTUM AND ENERGY

In Chapter Seven, we saw that the momentum $p = mv$ is a useful quantity in ordinary nonrelativistic mechanics because the momentum of a system remains constant in a collision. When collisions occur between objects moving at speeds approaching that of light, it is found that momentum is not conserved if the nonrelativistic definition is used. However, the momentum can be redefined so that momentum conservation does hold true in collisions. Specifically, it is found by considering collisions that the correct definition of the relativistic momentum of an object of mass m and velocity v is

$$p = \frac{mv}{\sqrt{1 - v^2/c^2}} \tag{27.3}$$

Note that the square root factor becomes 1 at low velocities, so we recover the nonrelativistic expression for the momentum.

In this equation, as well as everywhere else in this chapter, m is the ordinary mass of the object as measured by an observer in its rest frame. (Some books refer to this quantity as the rest mass and also define a velocity dependent mass. We do not do this.)

In nonrelativistic mechanics, Newton's second law is $F = ma = m\,\Delta v/\Delta t$ or

$$F = \frac{\Delta p}{\Delta t} \tag{27.4}$$

The force is the rate of change of the momentum. In this form, Newton's second law applies in relativistic mechanics as well.

If an object is initially at rest, the work done on the object as it accelerates is equal to its final kinetic energy. The work is the product of the displacement and the force $F = \Delta p/\Delta t$. Using this approach, it is found that the kinetic energy of an object of mass m and velocity **v** is,

$$K = mc^2\left(\frac{1}{\sqrt{1 - v^2/c^2}} - 1\right) \tag{27.5}$$

If the denominator is expanded in powers of v^2/c^2, this reduces at low velocities to the nonrelativistic kinetic energy, $\frac{1}{2}mv^2$. However, the kinetic energy of a rapidly moving object is much larger than $\frac{1}{2}mv^2$.

The following example shows that it takes much more work to accelerate an object to high velocities than predicted by nonrelativistic mechanics.

Example 27.5

An object of mass m has a speed $v = 0.9c$. What is the ratio of the nonrelativistic kinetic energy of this object to its relativistic kinetic energy?

The nonrelativistic kinetic energy is

$$K_n = \frac{1}{2}mv^2 = \frac{1}{2}m(0.9\,c^2) = 0.405\,mc^2$$

The relativistic kinetic energy is

$$K_r = mc^2\left(\frac{1}{\sqrt{1 - v^2/c^2}} - 1\right)$$

$$= mc^2\left(\frac{1}{\sqrt{1 - (0.9)^2}} - 1\right)$$

$$= 1.29\,mc^2$$

Thus the relativistic energy is more than three times the nonrelativistic value. This means that the work necessary to accelerate an object from rest to a velocity $v = 0.9c$ is more than three times greater than we would have predicted nonrelativistically. As the velocity approaches the velocity of light, the kinetic energy of the object increases very rapidly. Hence, it takes proportionately more work to achieve these speeds. In fact, it would require an infinite amount of work to have an object achieve a speed of c. This is one way of seeing that *the velocity of light cannot be attained by objects with a nonzero mass.*

The *rest energy* of an object of mass m is defined to be

$$E_0 = mc^2 \qquad (27.6)$$

This says that the mass m of an object is equivalent to an amount of energy mc^2. The *total energy* of the object is then the sum of the rest energy and the kinetic energy, $E = E_0 + K$, or

$$E = \frac{mc^2}{\sqrt{1 - v^2/c^2}} \qquad (27.7)$$

The equation $E_0 = mc^2$ is perhaps the most famous equation of twentieth-century physics. It is a statement that mass and energy are two forms of the same thing, and that one can be converted into the other. This was a revolutionary idea when Einstein introduced it, but it is now very well supported by experiment.

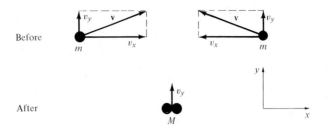

Figure 27.2. Two objects each with mass m collide and stick together.

Before we examine some of the experimental evidence for the equivalence of mass and energy, let us consider an imaginary experiment in which two identical masses m collide and stick together (Fig. 27.2). They form a single object of mass M moving with a velocity v_y, which does not change, since by symmetry neither object can exert a vertical force on the other. We can determine the mass of this object using momentum conservation.

The x component of the total momentum of the system is zero after and before the collision. Equating the y components after and before, we have

$$\frac{Mv_y}{\sqrt{1 - v_y^2/c^2}} = \frac{2mv_y}{\sqrt{1 - v^2/c^2}}$$

Dividing out the factor of v_y and multiplying by c^2,

$$\frac{Mc^2}{\sqrt{1 - v_y^2/c^2}} = \frac{2\,mc^2}{\sqrt{1 - v^2/c^2}}$$

This equation must hold for *any* value of v_y. In particular, if v_y is very small, then this becomes

$$Mc^2 = \frac{2\,mc^2}{\sqrt{1 - v^2/c^2}}$$

This is a remarkable result. It states that the rest energy of the combined object is equal to the total energy—rest plus kinetic—of the original two objects; *kinetic energy has been converted into mass.* Note that this result has followed directly from the conservation of the relativistic momentum. Thus the equivalence of mass and energy is an automatic consequence of momentum conservation.

There are many interesting situations in which the opposite of this experiment occurs, and mass is converted into energy. Every *particle* in nature has its *antiparticle,* which is identical in mass but opposite in charge. Our world is made up of particles, but a few particle-antiparticle pairs may be created by very energetic cosmic rays entering the atmosphere or by an accelerator beam striking a target. If, for example, an *antiproton* with a charge $-e$ encounters a proton with charge $+e$, their entire mass may be converted into gamma rays, very high frequency electromagnetic waves. Both the proton and antiproton cease to exist.

Other examples of the conversion of matter into energy are provided by *nuclear fission* and *fusion*. In nuclear fission, a uranium nucleus splits into two smaller nuclei whose total mass is less than that of the uranium nucleus by about 0.25 amu. This is approximately 0.1 percent of the original mass and is equivalent to an amount of energy

$$E_0 = mc^2 = (0.25 \text{ amu})(1.66 \times 10^{-27} \text{ kg amu}^{-1})$$
$$\times (3 \times 10^8 \text{ m s}^{-1})^2$$
$$= 3.7 \times 10^{-11} \text{ J} = 2.3 \times 10^8 \text{ eV}$$

This is about 10^8 times the energy released in a typical chemical reaction, which is a few electron volts. We can see that nuclear fission reactors and explosives employ huge energy sources.

Nuclear fusion, in which two light nuclei fuse to form a larger nucleus, releases even more energy in proportion to the mass of the constituents. This is the source of energy in the stars and may eventually provide us with almost unlimited amounts of energy. Both fission and fusion will be discussed further in Chapter Thirty-three.

Mass is also converted into energy in chemical reactions. However, because the energies released are very small compared to the rest energies of the atoms, the mass changes are too small to be directly measured.

Relation Between Energy and Momentum |

If one squares the defining equations for the relativistic energy E and momentum p, one finds that

$$E = \sqrt{p^2c^2 + m^2c^4} \qquad (27.8)$$

When v/c is small, pc is small compared to mc^2, and this equation can be expanded to give $E = mc^2 + p^2/2m$. The first term is the rest energy. The second term can be rewritten as $(mv)^2/2m = mv^2/2$, which is the usual nonrelativistic kinetic energy.

If the mass m is zero, which is the case for light, Eq. 27.8 becomes

$$E = pc \qquad (m = 0) \qquad (27.9)$$

The energy equals the momentum times the speed of light. Exactly this result had been obtained decades earlier by Maxwell, when he calculated the energy and momentum carried by an electromagnetic wave. This is another indication that Einstein's theory of relativity is in full agreement with Maxwell's theory of electromagnetism.

SUMMARY

The special theory of relativity is based on two fundamental facts about our natural world. First, there is no way for inertial observers moving with a constant relative velocity to determine which is at rest and which is in motion, since all the laws of nature are the same for both. Second, the speed of a light beam in a vacuum is $c = 3 \times 10^8$ m s^{-1}, no matter which observer makes the measurement. These two facts are our clues to the nature of space and time and lead to a number of interesting results.

If two observers are moving with a constant relative velocity, each will observe that objects used by the other are foreshortened in the direction of motion. Each will also say that the clocks of the other run more slowly.

The modifications of our ideas of space and time require corresponding changes in our identification of momentum and energy. The work necessary to increase the energy of an object by a given amount increases with the speed of the object. This increase is such that an infinite amount of energy is necessary to cause an object with nonzero mass to travel with the speed of light.

Checklist
Define or explain:

inertial reference frames	relativistic momentum
principle of relativity	relativistic kinetic
constancy of the speed	energy
of light	rest energy
time dilation	energy-momentum
length contraction	relationship

EXERCISES

Use $c = 3 \times 10^8 \text{ m s}^{-1}$ unless the problem specifies otherwise.

Section 27.1 | The Fundamental Principles of Special Relativity

27-1   An observer at rest on the ground sees a jet plane fly overhead at a constant speed of 2000 km h^{-1}. At what speed does the pilot see the ground moving?

27-2   A spaceship is traveling away from the earth at a constant speed of $c/2$. A light pulse is emitted by a lamp on the earth and travels toward the rocket. Find the speed of the light pulse according to observers on (a) the earth, and (b) the rocket.

27-3   A spaceship traveling away from the earth at a speed of $c/4$ emits a light pulse when it is a distance of 3×10^{10} m from the earth as measured by earth observers. How long will it take for the light pulse to reach the earth according to earth clocks?

Section 27.2 | Moving Clocks and Time Dilation

27-4   A person on a spaceship travels for 1 year (earth time) at a speed of $0.95c$. (a) How much time will have elapsed for the traveler? (b) How far will the traveler observe he has traveled?

27-5   A beam of pi mesons has a velocity of $0.6c$. Their average lifetime when at rest is 3×10^{-8} s. (a) On the average, how long will the moving pi mesons appear to live to an observer at rest? (b) How far will they travel in this time?

27-6   Two identical twins A and B are each 20 years old when B starts on a round trip to a star 10 light years away. If he travels at an average speed of $u = 0.99c$, what are the ages of the two twins when B returns?

Section 27.3 | Length Contraction

27-7   A pi meson is moving at $0.9c$ relative to a magnet. If the magnet has a length of 2 m in its rest frame, how long is it in the frame of the meson?

27-8   (Assume that the speed of light is $c = 100 \text{ m s}^{-1}$ for this exercise.) Suppose you are driving a racing car at Indianapolis. The homestretch is 1300 m long according to the spectators and you are driving at 90 m s^{-1}. (a) How long does the homestretch appear to you? (b) How long do you think it takes you to drive this far? (c) How long do the spectators think it took you to drive that far?

27-9   At what speed would a moving meter stick appear to be 0.5 m long to an observer at rest?

27-10   Suppose we see a pole passing us, moving parallel to its length at $0.6c$. What length would we measure it to be if its length in its rest-frame is 20 m?

Section 27.4 | Momentum and Energy

27-11   The kinetic energy and rest energy of a particle are equal. What is the speed of the particle?

27-12   The energy obtained by burning one gram of coal is about 3×10^4 J. What is the ratio of this energy to the rest energy of the 1 g of coal?

27-13   The energy released in chemical reactions is on the order of 5 eV per molecule. (a) Estimate the mass change in atomic mass units (1 amu = 1.66×10^{-27} kg). (b) If the mass of the constituent atoms is 30 amu, what is the fractional mass change?

27-14   A nuclear power plant produces 10^9 W of electrical power and simultaneously releases 2×10^9 W to the environment as waste heat. (a) At what rate is mass converted to energy? (b) How much mass is used up on 1 year? (1 year = 3.16×10^7 seconds) (c) What fraction is this of the 10^5 kg of uranium oxide in the reactor core?

27-15   In the decay of a neutron at rest into a proton, an electron e^-, and an antineutrino $\bar{v}$, one observes a total kinetic energy release of 1.25×10^{-13} J. What is the difference between the mass of a proton plus electron and the mass of a neutron? (The antineutrino has zero mass.)

27-16   Although our world is made up of matter, it has been conjectured that some galaxies may be composed of antimatter. If a 10-kg meteor of antimatter struck the earth, (a) how much mass would be converted into energy? (b) How much energy would be released?

27-17   A proton has a rest energy of 938 MeV. (1 MeV = 10^6 eV). When its velocity is $0.9c$, what is its (a) total energy, and (b) kinetic energy?

27-18   An electron has a rest energy of 0.51×10^6 eV. (a) What is its total energy if it is accelerated from rest through a potential difference of 12×10^6 V? (b) What is its velocity?

PROBLEMS

27-19 A person on a rocket notes that he is traveling toward a star 1 light year away from the earth with a speed u. A person on earth sees the rocket traveling with speed u but says the star is 10 light years away. (a) How long will the traveler say the trip takes? (b) How long does the earth observer say the trip takes? (c) Discuss who is correct.

27-20 If the speed of light were $c = 100$ m s^{-1}, what would a sprinter say about his time needed to run the 100-m dash? (The officials say the time is 10 seconds.)

27-21 Consider two islands 10 miles apart on a straight river flowing at 3 mi h^{-1}. Using a boat that moves at 5 mi h^{-1} with respect to the water (a) how long does the upstream trip from island to island take? (b) How long does the downstream trip take? (c) How long does the total round trip take? (d) Compare the duration of the trip with the time it would take in still water. (e) Describe the analogy between the results of this problem and the derivation leading to Eq. 27.1.

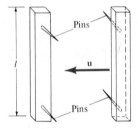

Figure 27.3. Two identical sticks with length l in their rest frames are in relative motion perpendicular to their length (Problem 27-22).

27-22 Two identical sticks both have pins at their ends and will leave scratch marks on each other if they come into contact. Suppose they are set into relative motion perpendicular to their length as in Fig. 27.3. Show that it follows from the principle of relativity that they cannot change in length.

27-23 The energy flux from the sun on the earth averages about 1 kW m^{-2} (24 hours per day). How much of the sun's mass reaches each square meter of the earth each year in the form of energy? (1 year = 3.15×10^7 seconds.)

27-24 (a) Using the relativistic expressions for the energy and momentum, show that the velocity of a particle satisfies $v = pc^2/E$. (b) Show how this goes over into a correct nonrelativistic formula when v/c is small.

27-25 An electron is accelerated from rest through a potential difference of 60,000 V. (a) What is the ratio of the total energy of the moving electron to its rest energy? (b) What is the final speed of the electron? (c) What is the final speed of the electron according to nonrelativistic mechanics?

27-26 Verify that $E = \sqrt{p^2c^2 + m^2c^4}$ follows from the definitions of E and p.

27-27 From the binomial theorem, $(1 - x)^{-1/2} = 1 + \frac{1}{2}x + \frac{3}{8}x^2 + \cdots$. Use this to show that the relativistic expression for the kinetic energy reduces to the nonrelativistic formula when v/c is small.

SUPPLEMENTARY TOPICS

27.5 | THE PROBLEM OF SIMULTANEOUS EVENTS

We have seen that Einstein's postulates led to the conclusion that the time elapsed between events is not an absolute quantity; rather, it depends on the motion of an observer. The mu meson observations described in Example 27.2 represent just one of many experimental verifications of this prediction.

A further consequence of Einstein's postulates is that observers may disagree on the question of whether two spatially separated events occur simultaneously. To see this qualitatively, suppose that an observer S standing at the center of a room sends two light pulses simultaneously toward the front and back of the room. Since the distances to the two walls are the same, the pulses strike the walls simultaneously according to this observer. If an observer M passes S at a velocity u just as the pulses are emitted, he too sees the two pulses traveling at a speed c (principle 2) toward the walls. However, he sees one wall approaching at a speed u and the other wall receding at the same rate. Consequently, M sees the pulse arrive at the approaching wall first, and he disagrees with S's observation that the two pulses arrive simultaneously at the walls.

Figure 27.4. An observer S is stationary with respect to the stick and clocks shown. He will say that the clocks are synchronized if clock A reads $t = 0$ when a light pulse is emitted and clock B reads $t = l/c$ when the pulse arrives.

To consider this question of simultaneity quantitatively, we consider what happens when an observer S synchronizes two clocks at different locations. Suppose S is stationary relative to a stick of length l with a clock at each end. To synchronize the clocks so that they both read $t = 0$ at the same time, S has a light pulse emitted at one clock at $t = 0$. This pulse will travel the distance l to the second clock in a time $t = l/c$. If the second clock reads $t = l/c$ at the instant the pulse arrives, S will conclude that both his clocks read $t = 0$ when the pulse was emitted and that they are synchronized (Fig. 27.4).

We now ask how an observer M moving at a speed u relative to the stick observes these events (Fig. 27.5). He agrees with S that a light pulse is emitted when

S's clocks reads $t = 0$ and that the pulse reaches clock B when that clock reads $t = l/c$. According to S, the time it takes for the light pulse to travel from clock A to clock B is $t = l/c$. M disagrees, since he observes the stick to have a foreshortened length $l' = l\sqrt{1 - u^2/c^2}$ and also sees clock B moving toward the pulse. According to M, the pulse must travel a distance $ct' = l' - ut'$. Thus

$$t' = \frac{l'}{c + u} = \frac{l\sqrt{1 - u^2/c^2}}{c(1 + u/c)}$$

This is the transit time t' for the pulse as observed by M *using his own clocks*. It is less than the time interval l/c measured by S. Now suppose M looks at the *clocks belonging to S*. According to M, these clocks run slower by the time dilation factor. Hence, when a time interval t' has elapsed on M's clocks, M will observe that S's clocks indicate an interval t given by $t = t'\sqrt{1 - u^2/c^2}$. Using our result for t', this becomes

$$t = \frac{l}{c}\frac{1 - u^2/c^2}{1 + u/c} = \frac{l}{c}\left(1 - \frac{u}{c}\right)$$

This is the transit time for the pulse as observed by M using S's clocks. Thus, M concludes that even according to S's clocks, the light pulse does not take as long in transit as the time l/c measured by S. In fact, M reasons that since the pulse arrives at clock B when it reads $t = l/c$, then when the pulse was emitted, clock B read $t = lu/c^2$, not $t = 0$ as S concluded! While S says the clocks read $t = 0$ simul-

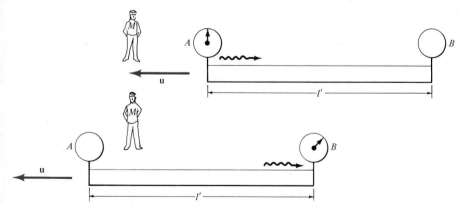

Figure 27.5. An observer M in motion relative to the stick sees the stick and clocks of Figure 27.4. moving past him with a speed u. M sees the pulse emitted when clock A reads $t = 0$ and sees the pulse arrive when clock B reads $t = l/c$. When M measures the length of the moving stick, he finds it has a contracted length $l' = l\sqrt{1 - u^2/c^2}$.

497

taneously, M says that when clock A read $t = 0$, clock B read $t = lu/c^2$. This is a real difference; neither M nor S is wrong.

This result can be restated in more general terms. Suppose two clocks a distance l apart are synchronized according to an observer in their rest frame. An observer who sees them in motion with a speed u will see the *trailing* clock reading fast by

$$\Delta t = \frac{lu}{c^2} \qquad (27.10)$$

The following example illustrates the implications of this result.

Example 27.6

People in a rocket of length $l = 30$ m traveling toward the earth at a speed of $u = 0.6c$ turn on lights in the front and back of the rocket simultaneously (Fig. 27.6). What will observers on earth see?

Observers on earth see the trailing light B turned on Δt before light F, where

$$\Delta t = \frac{lu}{c^2} = \frac{(30\ \text{m})(0.6c)}{c^2}$$

$$= \frac{(30\ \text{m})(0.6)}{3 \times 10^8\ \text{m s}^{-1}} = 6 \times 10^{-8}\ \text{s}$$

This is the time measured according to *rocket clocks*. According to *earth clocks*, the light B will have been turned on earlier than light F by

$$\Delta t' = \frac{\Delta t}{\sqrt{1 - u^2/c^2}} = \frac{6 \times 10^{-8}\ \text{s}}{\sqrt{1 - (0.6\,c/c)^2}}$$

$$= \frac{6 \times 10^{-8}\ \text{s}}{0.8}$$

$$= 7.5 \times 10^{-8}\ \text{s}$$

The results we have obtained in this section can sometimes lead to very intriguing apparent paradoxes. The following example illustrates one of these.

Figure 27.6. Observers on a rocket turn on lights at F and B simultaneously.

Figure 27.7. The barn and board *as measured by the observers F and B who are at rest relative to the barn*. F and B plan to close the doors of the barn simultaneously when the board is inside. When the board is at rest it has a length of 5 m.

Example 27.7

(For the purposes of this example, assume that the speed of light is $c = 5$ m s^{-1}.) Three people own some boards 5 m long and a barn 4 m wide. They know that a moving object is contracted in length. Hence, they decide to move the boards at a speed $u = 3$ m s^{-1}, so that the length of the boards will be contracted to

$$l' = l\sqrt{1 - \frac{u^2}{c^2}} = 5\ m\ \sqrt{1 - \left(\frac{3}{5}\right)^2}$$

$$= (5\ \text{m})(0.8) = 4\ \text{m}$$

Thus the boards will fit into the barn. They station one person (B) at the back door of the barn and another (F) at the front door. The third (M) will run with the board (Fig. 27.7). F and B are to close the doors simultaneously when the board is exactly inside the barn.

The apparent paradox is that M, moving with the board, sees the board as 5 m long and the barn as having a contracted width of only $(4\ \text{m})(0.8) = 3.2$ m. How can the two set of observations be compatible?

We first consider the situation as measured by F and B who are at rest with respect to the barn. F and B observe:

1 The barn and board are the same length.
2 When the board is just inside the barn they both close the doors.
3 They both observe the board crashing into the closed back door just after both doors are closed.

However, M's measurements indicate that (Fig. 27.8):

1 The barn is 3.2 m long and the board is 5 m long.
2 B does shut his door just as the front of the board reaches him, but the back of the board is still $5\ \text{m} - 3.2\ \text{m} = 1.8$ m outside the barn.
3 F does *not* shut his door when B does. According to F and B's clocks, M sees F shut his door

$$\Delta t = \frac{lu}{c^2} = \frac{(4\ \text{m})(3\ \text{m s}^{-1})}{(5\ \text{m s}^{-1})^2} = 0.48\ \text{s}$$

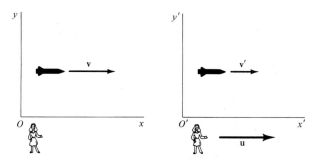

Figure 27.8. (a) The board and barn as viewed by M, the person moving with the board. (b) The back door is shut as the board reaches it. The board crashes through this door an instant later. (c) The front door is shut later, when the trailing end of the board is just inside the barn.

later than B. In M's frame, this is

$$\Delta t' = \frac{0.48 \text{ s}}{\sqrt{1 - (3/5)^2}} = 0.6 \text{ s}$$

Now, in 0.6 s, M sees the barn move toward him a distance of $u \Delta t' = (3 \text{ m s}^{-1})(0.6 \text{ s}) = 1.8 \text{ m}$. This is just the distance that places the back of the board inside the barn when F shuts his door.

We see that all observers agree that the back door was shut just before the board reached it and the front door was shut just after the end of the board was inside the door. What the observers do not agree on was when the board crashed through the back door. F and B say it happened just after they both shut their doors, while M says it happened just after B shut his door but before F shut his. This disagreement is real and is typical of those that can arise when two sets of observers view events that occur at different positions.

27.6 | THE ADDITION OF VELOCITIES

According to special relativity, two observers in relative motion disagree about the properties of their meter sticks and clocks. Accordingly, when they measure the velocity of an object moving relative to both of them, their observations are related in a way that reflects these disagreements.

The derivation of this relationship is similar to those of previous sections but lengthy, so we simply state the results. Consider two observers with a relative velocity u. An object moves along the x direction

Figure 27.9. Observer O' moves relative to O at a velocity **u**. An object has a velocity **v'** according to O' and **v** according to O.

with a velocity v' according to O'. (Fig. 27.9). According to O, the velocity of the object is

$$v = \frac{v' + u}{1 + \frac{uv'}{c^2}} \qquad (27.11)$$

This is called the *velocity addition formula.*

When the velocities are small, we can neglect u/c and v'/c. This gives the usual nonrelativistic result, $v = v' + u$. However, when v' is comparable to or equal to c, the results are quite different. In particular, it turns out that if v' and u are less than c, but $v' + u$ is greater than c, then v is still less than c. This means that nothing can be observed to move faster than the speed of light. The case $v' = c$ is considered in the next example.

Example 27.8

An observer moving away from the earth at $0.5c$ sends a light pulse along his direction of motion and measures its speed to be c. How fast does the pulse travel according to the earth observer?

Here $v' = c$ and $u = 0.5c$. Using the velocity addition formula, the velocity according to an earth observer is

$$v = \frac{v' + u}{1 + \frac{v'u}{c^2}} = \frac{c + 0.5c}{1 + \frac{(c)(0.5c)}{c^2}} = c$$

Thus both observers measure the same speed of light, in agreement with Einstein's second principle.

When the motion of the object is not along the x axis, the addition of velocities involves more complicated expressions. However, these expressions again insure that no object can be observed to travel faster than the speed of light.

Figure **27.10.**   Exercise 27-29.

Figure **27.11.**   Problem 27-35.

EXERCISES ON SUPPLEMENTARY TOPICS

Section 27.5 | The Problem of Simultaneous Events

27-28   An observer on an earth satellite traveling at 8×10^3 m s^{-1} relative to the earth sees rocket launchings occuring in New York and in Los Angeles, which he judges to be simultaneous. (a) The distance between the two cities is 4.5×10^6 m. What will the time interval between these launchings be according to observers on earth? (The satellite is moving directly from Los Angeles toward New York.) (b) In which city does the rocket launching occur first according to earth observers?

27-29   An observer on a train sees two lights flash on the station platform. Light A is seen to flash 0.8 seconds after light B (Fig. 27.10), and this observer measures the lights as being 100 m apart. If the train is moving at 16 m s^{-1} and the speed of light is $c = 20$ m s^{-1}, (a) what is the distance between the lights as measured by an observer at rest on the platform? (b) Which light flashes first according to an observer at rest?

Section 27.6 | The Addition of Velocities

27-30   In a large particle accelerator facility, two beams of protons are directed toward each other, each with a velocity of $0.8c$ as measured in the laboratory. What is the speed of one beam as measured by an observer at rest with respect to the other beam?

27-31   A space ship approaching the earth at $0.6c$ fires a rocket at a speed of $0.6c$ relative to the space ship. According to earth observers, what is the speed of the rocket if it is fired (a) toward the earth; (b) away from the earth?

27-32   A pi meson decays into a mu meson and a neutrino. In the rest frame of the pi meson, the velocity of the mu meson is always found to be $0.507c$. Suppose now the pi meson is moving relative to the laboratory at $0.6c$ when it decays, emitting the mu meson along its direction of motion. What will be the speed of the mu meson relative to the laboratory?

27-33   Particle A decays into particles B and C. When the A particle is at rest, B is always found to have a speed $0.6c$. If the A particle is moving at $0.4c$ relative to the laboratory when it decays, find the velocity of the B particle as measured in the laboratory frame if the B particle moves (a) parallel to the direction of the A particle; (b) opposite to the direction of the A particle.

PROBLEMS ON SUPPLEMENTARY TOPICS

27-34   An observer on earth sees a rocket P traveling away from the earth at 2×10^8 m s^{-1}. P is overtaking a second rocket Q, which is traveling away from the earth at 1.5×10^8 m s^{-1}. Find the velocity of (a) P measured by observers in Q, and (b) Q measured by observers in P.

27-35   A rocket ship moves with a speed $u = 0.6c$ past two observers, A and B, at rest. A and B say they are 5 m apart. They arrange to fire guns simultaneously, striking the rocket at points A' and B' (Fig. 27.11). (a) How far apart are A and B as measured by those on the rocket? (b) If the people at rest and those in the rocket agree that B fired at $t = 0$, what do the observers on the rocket say about when A fired? (c) What is the distance between the points A' and B' as measured by passengers on the rocket?

27-36   Two observers O and O' are in relative motion along their x axes at a speed u. If O' observes an object in motion along the y' axis with a speed v', then O will observe it to have both x and y velocity components, where

$$v_x = u, \qquad v_y = v' \sqrt{1 - \frac{u^2}{c^2}}$$

If $v' = c$, find $v = \sqrt{v_x{}^2 + v_y{}^2}$.

27-37 Observer O' observes a rocket ship traveling along the y direction at a speed $0.9c$. O' is moving at $0.8c$ along the x direction relative to observer O. Using the formulas in the preceding problem, find the components of the rocket ship velocity as observed by O and show that v is less than c.

Additional Reading

N. David Mermin, *Space, Time, and Relativity*, McGraw-Hill Book Co., New York, 1965.

Hermann Bondi, *Relativity and Common Sense*, Science Study Series, Doubleday and Company, Garden City, N.Y., 1964.

Scientific American articles:

R.S. Shankland, The Michelson-Morley Experiment, November 1964, p. 107.

J. Bronowski, The Clock Paradox, February 1963, p. 134.

Milton A. Rothman, Things That Go Faster Than Light, July 1960, p. 142.

V.L. Ginzburg, Artificial Satellites and Relativity, May 1959, p. 149.

Gerard K. O'Neill, Particle Storage Rings, November 1966, p. 107.

Albert B. Stewart, The Discovery of Stellar Aberration, March 1964, p. 100.

CHAPTER TWENTY-EIGHT

PARTICLE PROPERTIES OF LIGHT: THE PHOTON

In retrospect, it is easy to understand why electromagnetic radiation or light was considered to consist of continuous waves. Any particlelike behavior was masked by the enormous number of light quanta usually present. The situation is analogous to that encountered in electrostatics where the amount of charge present is usually very large. The fact that charge comes in discrete amounts is irrelevant under these conditions.

Only a very few of the observations that had been made before 1900 turned out to be inexplicable in terms of quantities that were huge multiples of discrete units. That light could, under certain conditions, have particlelike characteristics was not only not believed, it was not necessary. Since the wave theory of light gave a complete picture of the light-related phenomena observed, it was naturally considered to be correct.

In this chapter, we show how a few crucial experiments have forced us to consider light as having particlelike characteristics sometimes and wavelike properties at other times. This leads to the concept of *photons,* small bundles of light with discrete energies. The last section introduces the role of photons in vision.

28.1 | BLACKBODY RADIATION

In Chapter Twelve we found that the power radiated from a heated object depends on the wavelength of the radiation. The details of how this power varies with wavelength and temperature usually depend on

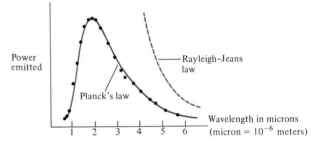

Figure 28.1. Blackbody radiation at 1600 K. The curve predicted by Planck's law passes through the data points. The Rayleigh-Jeans prediction becomes very large at short wavelengths.

the characteristics of the object. However, the radiation emitted through a small opening from a cavity does not depend either on the material of its walls or on the shape of the cavity and the hole. The cavity radiation obeys a universal law of nature and acts as a perfect blackbody: a perfect absorber and emitter with emissivity equal to 1 (Fig. 28.1).

Many attempts were made to derive the blackbody radiation curve (Fig. 28.1) using classical electromagnetic theory, but all of them failed. The most famous of these, called the Rayleigh-Jeans law, was correct at long wavelengths, but at short wavelengths the theoretical power emitted became infinite. This failure was referred to as the ultraviolet (short wavelength) catastrophe.

In 1900, Max Planck (1858–1947) found that two radical assumptions could be combined with classical theory to give an excellent fit to experiment. We now know that these assumptions correctly represent fundamental ideas of quantum physics.

Planck assumed that the radiation in the cavity was emitted and absorbed by atomic oscillators in the walls of the cavity. The oscillators had two unique characteristics:

1   The oscillators could only have energies given by

$$E = nhf, \qquad n = 0, 1, 2, 3, 4, \ldots \quad (28.1)$$

where f is the oscillator frequency and the proportionality factor h is called *Planck's constant.*

2   The oscillators could emit energy only in discrete or quantized amounts corresponding to a change in n of one unit:

$$E_{n+1} - E_n = (n + 1)hf - nhf = hf$$

Once radiated, Planck also assumed that the energy was that of a classical electromagnetic wave. Planck's calculation of the power radiated then led to excellent agreement with experiment (Fig. 28.1) if h was chosen to be

$$h = 6.625 \times 10^{-34}\,\text{J s} = 4.135 \times 10^{-15}\,\text{eV s}$$

The small size of this constant is our first hint that classical physics may fail when extremely small energies are important. This is illustrated by the following example.

Example 28.1

A 1-kg mass is suspended from a spring with constant $k = 16\,\text{N m}^{-1}$. The amplitude of vibration is 0.01 m. (a) If the energy is quantized as Planck suggested, what is the quantum number associated with the energy of the spring? (b) If the quantum number changes by one unit, what is the ratio of the energy change to the total energy of the spring?

(a) The frequency of the spring-mass system was found in Chapter Ten to be

$$f = \frac{1}{2\pi} \sqrt{\frac{k}{m}} = \frac{1}{2\pi} \sqrt{\frac{16\,\text{N m}^{-1}}{1\,\text{kg}}} = 0.637\,\text{Hz}$$

The energy of the spring-mass system is

$$E = \tfrac{1}{2}kx_{max}^2 = \tfrac{1}{2}(16\,\text{N m}^{-1})(0.01\,\text{m})^2$$
$$= 8 \times 10^{-4}\,\text{J}$$

Using Planck's hypothesis, $E = nhf$,

$$n = \frac{E}{hf} = \frac{8 \times 10^{-4}\,\text{J}}{(6.63 \times 10^{-34}\,\text{J s})(0.637\,\text{s}^{-1})}$$
$$= 1.90 \times 10^{30}$$

(b) The energy change when n changes by one unit is $\Delta E = hf$ and the ratio $\Delta E/E$ is

$$\frac{\Delta E}{E} = \frac{hf}{nhf} = \frac{1}{n} = 5.27 \times 10^{-31}$$

Thus the change is so small compared to the energy itself that the discreteness of the energy would ordinarily be undetectable.

28.2 | THE PHOTOELECTRIC EFFECT

It would seem from Planck's analysis that if atomic oscillators emit energy in discrete amounts, then the radiation itself should be composed of "lumps" with distinct energies. However, since the assumptions that Planck had already made were subject to a great deal of skepticism, if not outright disbelief, there seemed no reason to extend them unnecessarily. But nature is not so tentative, and the explanation of the photoelectric effect required the acceptance of *photons* as discrete constituents of electromagnetic radiation.

The photoelectric effect apparatus is shown in Fig. 28.2. Monochromatic light incident on the metal plate yields sufficient energy to allow electrons to escape from the metal. Some of these emitted electrons reach a collection plate, and the ammeter measures the resulting *photoelectric current*. The intensity and frequency of the incident light can be varied, as can the potential difference V between the metal plate and the collector.

The conduction electrons in a metal move in the attractive electric field of the stationary positive ions of the lattice. While electrons move relatively freely

Figure 28.2.   The photoelectric effect apparatus. Light incident on the metal plate causes electrons to be emitted. These can travel to the collector producing a current.

inside, it requires a minimum energy to actually pull them from inside to outside the metal. This minimum energy is called the *work function* W and depends on the properties of the metal and of the surface. If an electron is given an energy E larger than W, it can escape the metal and will have a maximum kinetic energy

$$\tfrac{1}{2}mv_{max}^2 = E - W$$

The results of the photoelectric experiment can be summarized as follows:

1   When $V = 0$, photoelectrons are detected whenever the metal is illuminated by light at a frequency f, which is greater than a critical or threshold frequency f_0. However, no matter what the intensity of the light, no current is observed if the frequency is below f_0.

2   At each light frequency above the threshold, the potential V can be increased until, at some value V_0, the current becomes zero. If the emitter and collector are made of the same material, this occurs when the potential energy difference eV_0 of an electron with charge $-e$ is just equal to the *maximum* kinetic energy of the emitted electrons

$$eV_0 = \tfrac{1}{2}mv_{max}^2$$

A graph of V_0, the stopping potential, versus the frequency of the incident light appears in Fig. 28.3.

3   Above the threshold frequency, an increase in intensity results in an increase in the number of photoelectrons but the maximum kinetic energy of the electrons does not change.

These observations are in direct conflict with predictions based on the wave picture of light. If light is

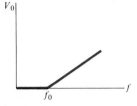

Figure 28.3.   The stopping voltage V_o versus incident light frequency. If f is less than f_o no current is observed, even with zero stopping potential. Above f_o the stopping potential increases linearly with f. This graph is valid for *all* nonzero light intensities.

a classical wave, the electrons should absorb energy continuously, and at any intensity it should be merely a matter of time until an electron has sufficient energy to escape. Hence there would be no threshold frequency, although there would be a delay in production of photoelectrons at low intensities until enough light energy had been absorbed by the material. Furthermore, at high intensities the electrons should receive more energy, so the stopping potential should be greater. No such effects are observed.

Einstein's Explanation | In order to explain the photoelectric effect, Einstein extended the ideas of Planck. If, as Planck suggested, light is produced when atoms give up discrete amounts of energy, then the light itself may be considered to be composed of quanta called *photons*. These light quanta behave like particles that travel at the speed of light. If the light has a frequency f and wavelength $\lambda = c/f$, the photons each have an energy

$$E = hf \quad \text{(photon energy)} \quad (28.2)$$

The intensity of monochromatic light is proportional to the number of photons present. A beam of white light contains photons of many different energies.

The photon theory of light offers a complete explanation of the photoelectric effect. An electron will leave the metal only if it absorbs a photon of energy equal to or greater than the work function W. The threshold corresponds to the frequency at which the photon energy hf_0 equals W or

$$f_0 = \frac{W}{h} \quad (28.3)$$

If the photon frequency is above threshold, the excess energy appears as kinetic energy of the photoelectrons. Since W is the minimum energy necessary to remove an electron, the maximum kinetic energy is

$$\tfrac{1}{2}mv_{max}^2 = hf - hf_0 = hf - W \quad (28.4)$$

Increasing the intensity with f constant will result in more electrons being emitted, but with the same maximum kinetic energy.

Since the stopping voltage V_0 is adjusted so that $eV_0 = \tfrac{1}{2}mv_{max}^2$, we see that $eV_0 = hf - W$, or

$$V_0 = \frac{h}{e}f - \frac{W}{e} \quad (28.5)$$

This is an important result, because it predicts that the slope of the V_0 versus f curve is h/e. Planck's constant h and the electronic charge e were known, and the slope in Fig. 28.3 turned out just as predicted. This convincing demonstration was one of the first in a series of discoveries that led to the acceptance of a particle or photon description of electromagnetic radiation.

28.3 | THE PHOTON

As we saw in the preceding section, a light beam is composed of quanta. For monochromatic light of frequency f, each quantum of light or photon has an energy hf. The number of photons present determines the intensity. The following example shows that under usual conditions the number of photons present is so large that their quantum nature may be overlooked.

Example 28.2

Monochromatic green light of frequency 6×10^{14} Hz is produced by a laser. The power emitted is 2×10^{-3} W. (a) What is the energy of a photon in the beam? (b) How many photons per second pass a point in the beam?

(a) Each photon has an energy

$$E = hf = (6.63 \times 10^{-34} \text{ J s})(6 \times 10^{14} \text{ s}^{-1})$$
$$= 3.98 \times 10^{-19} \text{ J}$$

(b) Let N be the number of photons passing a point in the beam each second. The power $\mathcal{P}$ transmitted in the beam must be equal to N times the energy per photon, $\mathcal{P} = NE$. Thus

$$N = \frac{\mathcal{P}}{E} = \frac{2 \times 10^{-3} \text{ W}}{3.98 \times 10^{-19} \text{ J}}$$
$$= 5.03 \times 10^{15} \text{ photons per second}$$

The Compton Effect | Several further experiments indicated convincingly that the photon picture of light is correct. One, performed in 1923 by A.H. Compton (1892–1962), involved the observation of x-rays scattered from free electrons. Compton found that the scattered x-rays have a reduced frequency. He showed that this implies that a photon-electron collision is much like an elastic collision of two billiard balls (Fig. 28.4). When a photon of energy hf collides with an electron at rest, the electron recoils with an energy E_{el}. The scattered photon has an

Figure 28.4. (*a*) A photon incident on a stationary electron is scattered (*b*) through an angle θ. The energy and momentum lost by the photon are taken up by the electron.

energy hf', where f' must be less than f since energy conservation requires

$$hf' = hf - E_{el}.$$

The analogy of a billiard ball collision implies that the total momentum of the photons and the electron is conserved. Using classical electromagnetic theory, Maxwell showed that the energy E carried by a light wave is related to the momentum of the wave by $E = pc$, where c is the velocity of light. (This is also the result we obtained from relativity for a particle with zero mass.) Thus the photon momentum is given by

$$p = \frac{E}{c} = \frac{hf}{c} \qquad (28.6)$$

When momentum and energy conservation are applied to the collision of Fig. 28.4, the predicted dependence of the frequency change $f - f'$ on the angle θ is found to be in very good agreement with experiment. Thus, Compton's experiment provides further evidence of the particle nature of light.

X-Ray Spectrum | X-rays provided another significant verification of the photon theory. We saw in Chapter Twenty-five that when an electron beam is accelerated in a vacuum and allowed to strike a target, x-rays are emitted with various frequencies up to some maximum frequency, f_{max} (Fig. 28.5). It was found that f_{max} depends only on the accelerating potential difference V and is independent of the number of electrons in the beam. Classically, these observations are difficult to comprehend, but the photon theory gives a ready explanation. If the electrons are accelerated from rest through a potential difference V, their kinetic energy is eV. The electrons

Figure 28.5. X-rays produced by the electron bombardment of a target result in x-ray photons of a maximum frequency $f_{max} = eV/h$ and a minimum wavelength $\lambda_{min} = c/f_{max}$. The sharp peaks appear at different wavelengths for different target materials. They are due to the rearrangement of electrons disturbed from their normal atomic orbits.

lose part or all of their kinetic energy when they interact with an atom in the target. The highest frequency photon will be produced when all the kinetic energy of one electron is converted into a single photon, or when

$$hf_{max} = eV \qquad (28.7)$$

This equation is in excellent agreement with the observed frequency maxima.

Example 28.3

What are the (a) maximum frequency and (b) minimum wavelength of x-rays produced by 40,000-V electrons?

(a) Using Eq. 28.7 for the photon frequency,

$$f_{max} = \frac{eV}{h} = \frac{(1.60 \times 10^{-19}\ \mathrm{C})(4 \times 10^4\ \mathrm{V})}{6.63 \times 10^{-34}\ \mathrm{J\ s}}$$
$$= 9.65 \times 10^{18}\ \mathrm{Hz}$$

(b) The corresponding minimum wavelength is

$$\lambda_{min} = \frac{c}{f_{max}} = \frac{3 \times 10^8\ \mathrm{m\ s^{-1}}}{9.65 \times 10^{18}\ \mathrm{Hz}}$$
$$= 3.11 \times 10^{-11}\ \mathrm{m} = 0.311\ \mathrm{\mathring{A}}$$

The dual wave-particle nature of light is now a basic part of our theory of light and matter. The theory as formulated mathematically contains both aspects of light and the outcome of any experiment can in principle be predicted. On the other hand, a verbal description of the behavior of light is not so unambiguous. As a tentative rule, we may say that light behaves as a wave as long as no absorption or emission of light occurs. Whenever light is absorbed, its quantum or particle nature may be expected to be evident.

28.4 | PHOTONS AND VISION

Light energy is received by the visual receptors of the eye, that is, the rods and cones, in discrete amounts as photons are absorbed. Hence, there is necessarily a minimum number of photons needed for vision. This minimum number is called the *absolute threshold of vision.*

A distinction must be made between the number of photons incident on the eye and the number of photons causing excitation of the receptors. About 90 percent of the photons incident on the eye are absorbed or scattered in the lens and fluid of the eye. Furthermore, less than 40 percent of the photons reaching the retina are actually absorbed by the receptors. Thus the number of photons incident on the eye is far larger than the number of photons used by the receptors for visual perception. Our discussion will involve only photons absorbed by the receptors.

We might guess that a single photon is sufficient for vision. However, theoretical and experimental evidence seems to rule this out. Theoretically we know that molecules in the eye have an average thermal energy that is quite small. However, there is a small probability that an occasional atom or molecule will have an energy large enough to emit a photon that could cause excitation of the receptor. Multiplying this small probability by the large number of molecules present and the large number of receptors available indicates that about one photon per second might be detected. These photons have nothing to do with the external visual scene, and hence their perception would be a constant source of annoyance and confusion. This *background* would be eliminated by the requirement that *coincident* or nearly simultaneous absorption of more than one photon must occur to stimulate the visual process.

Experimentally, coincident absorption of photons apparently must take place in a region where the neurons are interconnected. This region includes about 100 rods in the dark-adapted eye. In this region, if the coincidence does not occur during the eye's "storage time" of 0.2 seconds, the coincidence is

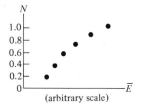

Figure 28.6. The experimental data indicating the fraction of positive responses to light pulses of varying average energy $\bar{E}$.

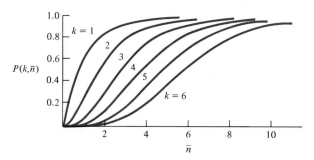

Figure 28.7. Graphs of the predicted value of $P(k,\bar{n})$ versus $\bar{n}$ for $k = 1, \ldots., 6$. One of the curves of this type should correspond to the results shown in Figure 28.6. (Adapted from R.K. Clayton, *Light and Living Matter, Volume II The Biological Part*, McGraw-Hill, New York, 1971. Used with permission.)

not counted. Thus the visual memory of the absorption of one photon lasts about 0.2 seconds.

The actual experiment is performed with pulses that illuminate about 100 rods and last 0.2 seconds. Subjects are asked to say when they see a pulse. The fraction of positive responses is then plotted versus the average pulse energy $\bar{E}$ (Fig. 28.6). The average number of photons $\bar{n}$ absorbed by the rods is proportional to $\bar{E}$,

$$\bar{n} = A\bar{E} \tag{28.8}$$

Here A is a constant that must be determined from the experiment.

A light source emits photons randomly in time, so not every pulse in a group with average energy $\bar{E}$ will result in exactly $\bar{n}$ absorbed photons. For each pulse of the group with an average of $\bar{n}$ absorbed photons, there is some probability, denoted by $P(k, \bar{n})$, that at least k photons are actually absorbed.

This is analogous to the following situation. When rain falls on a sidewalk divided into squares, the average number of drops per square $\bar{n}$ equals the total number of drops divided by the number of squares. However, examination will show that some squares have fewer and some have more than the average number of drops. The number of squares receiving k or more drops divided by the total number of squares is $P(k, \bar{n})$.

The visual threshold experiment effectively measures $P(k, \bar{n})$, where k is the minimum number of photons necessary for vision. That is, there should be a correspondence between the graph of Fig. 28.6 and a graph of the theoretical value of $P(k, \bar{n})$ versus $\bar{n}$ (Fig. 28.7).

The problem is that neither k nor the quantity A in Eq. 28.8 is known. By plotting the number of positive responses versus $\bar{n}$, using A as an adjustable parameter, N of Fig. 28.6 can be compared with $P(k, \bar{n})$. Typical results are shown in Fig. 28.8. It is found that

an approximate fit to the data can be obtained for k between 2 and 6. Hence at least 2 but less than 7 photons are apparently necessary for vision.

SUMMARY

The wave nature of light was well established by the end of the nineteenth century. However, several experiments including blackbody radiation and the photoelectric effect could not be explained in terms of such a picture. Max Planck made the first tentative assertion that vibrating atoms emitted light energy in specific amounts, and Einstein extended this idea to say that light itself is composed of quanta called photons. These carry a specific amount of energy equal to Planck's constant times the light frequency.

Compton's studies of photon-electron collisions provided further evidence that light has a particlelike nature and that photons have a definite energy and

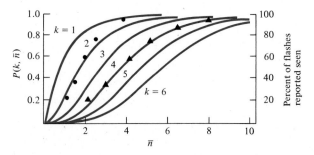

Figure 28.8. The curves showing the fractional probability of absorption of at least k photons versus the average number of photons $\bar{n}$. The symbols ● and ▲ represent two attempts to fit a single set of data to the curves for $k = 2$ and $k = 4$. The fits are achieved by adjusting the constant A in Eq. 28.8. (Adapted from Clayton.)

momentum. X-ray production was also understood using the photon concept.

Present-day theories describe the wave-particle nature of light correctly as far as we know. When experimental results of the interaction of light with matter are discussed in other than a formal mathematical way, the nature of the experiment will determine whether a wave or particle description is best suited for understanding the results. The eye is an excellent example of this. The gathering and focusing of light is well described in terms of waves, but the ultimate absorption of light in rods and cones involves the absorption of photons.

Checklist

Define or explain:

blackbody radiation	photon
Planck's law	Compton effect
Planck's constant	x-ray spectrum
photoelectric effect	absolute threshold of
work function	vision

EXERCISES

Section 28.2 | The Photoelectric Effect

28-1 When light of frequency 7×10^{14} Hz shines on a metal surface, electrons with a maximum speed of 6×10^5 m s^{-1} are emitted. What is the photoelectric threshold frequency of the sample?

28-2 The work function of metallic sodium is 2.3 eV. What is the maximum wavelength of the incident light for which photoemission of electrons will occur?

28-3 An electroscope has two metal foil leaves that normally hang side by side. They are both connected to a metal plate (Fig. 28.9). If light of a sufficiently high frequency is shone on the plate, the leaves are observed to spread apart. Describe what causes this.

Section 28.3 | The Photon

28-4 A radio antenna radiates 10^4 W of power at 9.2×10^5 Hz. How many photons per second are emitted?

28-5 A sodium vapor lamp emits 10 W of light at $\lambda = 590$ nm uniformly in all directions. How far from the lamp will the number of photons per square meter per second be equal to 10^{15}?

Metal plate and rod

Foil leaves

Figure 28.9. When light strikes the plate, the leaves of an electroscope spread apart at the bottom.

28-6 When the sun is directly overhead, the power incident on the earth at ground level is about 10^3 W m^{-2}. Assuming the average wavelength of the light is 550 nm, how many photons per square meter are incident on the earth per second?

28-7 What are the energies in electron volts of photons at the ends of the visible spectrum, (a) $\lambda = 400$ nm, (b) $\lambda = 700$ nm? (1 electron volt $= 1$ eV $= 1.60 \times 10^{-19}$ J.)

28-8 In a Compton scattering experiment, the initial and final photon frequencies are 2.4×10^{20} Hz and 1.6×10^{20} Hz, respectively. What is the recoil energy of the scattering electron in electron volts?

28-9 In order to disintegrate a deuteron (heavy hydrogen nucleus) into a proton and neutron, a γ ray must have an energy of at least 2.2 MeV. (γ rays are very energetic x-rays; 1 MeV $= 10^6$ eV $= 1.60 \times 10^{-13}$ J.) Find the minimum γ-ray frequency for photodisintegration of the deuteron.

28-10 In a monochromatic beam of light of wavelength 500 nm, what is the photon (a) energy, and (b) momentum?

28-11 The maximum frequency of x-ray photons from a certain apparatus is 9.5×10^{18} Hz. What is the accelerating voltage of the x-ray tube?

28-12 The electrons in an x-ray tube are accelerated from rest through a potential difference of 50,000 V. (a) What is the maximum frequency x-ray photon which is produced? (b) What is the wavelength of this photon in angstroms?

Section 28.4 | Photons and Vision

28-13   The center-to-center distance between rods is 5×10^{-6} m in one portion of the eye. Estimate the area on the retina taken up by 100 rods.

28-14   A minimum of 100 photons incident on 2.5×10^{-9} m^2 of the pupil of the eye is necessary for vision. These must be incident during the eye's storage time of 0.2 seconds. (a) What is the minimum intensity of the light if $\lambda = 500$ nm? (b) If the source is 2 m from the eye and radiates uniformly in all directions, what is the power of the source?

28-15   A candle emits energy at a rate of 1 W at an average wavelength of 550 nm. (a) How many photons are emitted per second? (b) How far from the eye can the candle be seen if the threshold intensity is 10^{-7} W m^{-2}? (c) How many photons per square meter per second is this?

28-16   How far from a lamp emitting 5 W of power at a wavelength of 590 nm can one be and still see it with the naked eye? (The threshold intensity is 10^{-7} W m^{-2}.)

PROBLEMS

28-17   The vibrational motions of the atoms of a diatomic molecule are like Planck's oscillators. If we assume that the two atoms are connected by a spring with constant k, the vibrational energy of the molecule is $E = nhf$, where $f = \sqrt{k/\mu}$. μ is called the reduced mass and for two atoms of masses m_1 and m_2,

$$\mu = \frac{m_1 m_2}{m_1 + m_2}$$

For the diatomic molecule H$_2$, $m_1 = m_2 = 1.67 \times 10^{-27}$ kg, and the measured energy level spacing is 0.55 eV $= 8.74 \times 10^{-20}$ J. (a) What is the reduced mass of the molecule? (b) What is the spring constant k for the molecule?

28-18   The minimum frequency of photons emitted in vibrational transitions of the HCl molecule is 8.97×10^{13} Hz. The masses of the two atoms are 1.67×10^{-27} kg and 5.85×10^{-26} kg for hydrogen and chlorine, respectively. (a) What is the reduced mass of the molecule? (b) What is the spring constant of the molecule? (See Problem 28-17.)

28-19   The masses of O and H in the OH group are 2.67×10^{-26} kg and 1.67×10^{-27} kg, respectively. The effective spring constant is 50.5 N m^{-1}. (a) What is the reduced mass of the group? (b) What is the characteristic vibrational frequency of the group? (See Problem 28-17.)

28-20   The work function of tungsten is 4.49 eV. (a) Find the threshold wavelength for photoemission. (b) Ultraviolet light of wavelength 250 nm falls on the surface. What is the maximum kinetic energy of the emitted electrons? (c) What is the stopping potential?

*28-21   Light of intensity 10^{-2} W m^{-2} falls on a metallic surface. The energy required to emit a photoelectron is 3 eV (1 eV $= 1.60 \times 10^{-19}$ J). If a single electron absorbes light from a surrounding area 10 atoms in radius (10^{-9} m), how long will it take the electron to absorb enough energy to be emitted if light is assumed to be a classical wave?

28-22   In a Compton effect experiment, the incident x-ray photons have an energy of 10^5 eV $= 1.6 \times 10^{-14}$ J. (a) What is the frequency of the incident photons? (b) An electron gains 4000 eV of kinetic energy when a photon scatters through a certain angle. What is the frequency of the scattered photons?

28-23   In a Compton effect experiment, the incident photons have an energy of 20,000 eV. (a) What is the momentum of one of these photons? (b) What is the wavelength associated with a photon?

28-24   Nuclear reactions can produce γ rays, very energetic x-rays. When a π° meson decays at rest, 135 MeV is shared equally by two γ rays. Find the wavelength of the γ rays. (1 MeV $= 10^6$ eV $= 1.60 \times 10^{-13}$ J.)

Additional Reading

J. Andrade, E. Silva, and G. Lochak, *Quanta*, World University Library, McGraw-Hill Book Co., Inc., New York, 1969. Paperback.

A.H. Compton. The Scattering of X-rays as Particles, *American Journal of Physics*, December 1961, p. 817.

Roderick K. Clayton, *Light and Living Matter, Volume I: The Physical Part; Volume II: The Biological Part,* McGraw-Hill Book Co., New York, 1971.

Albert Rose, Quantum Effects in Human Vision, in *Advances in Biological and Medical Physics,* vol. V., Academic Press, New York, 1957, p. 211.

Scientific American articles:

Karl K. Darrow, The Quantum Theory, March 1952, p. 47.

Richard Gordon, Image Reconstruction From Projections, October 1975, p. 56.

CHAPTER TWENTY-NINE

WAVE PROPERTIES OF MATTER

The photon theory of radiation did not resolve all the difficulties facing physicists in the early part of the twentieth century. In many experiments, it was noted that matter absorbed and emitted radiation only at certain frequencies. For example, it had long been known that if hydrogen gas is placed in an electrical discharge tube, only a specific set of wavelengths is observed when the light emitted is analyzed with a diffraction grating. Furthermore, every pure gas had its own characteristic frequencies of absorption and emission. By contrast, all classical models of atoms predicted a continuous range of light frequencies.

By 1911, experiments had also shown conclusively that atoms have a positively charged but very small nucleus. The nucleus accounts for most of the mass of the atom, and the negatively charged electrons occupy a region outside the nucleus. This immediately suggests a "planetary" model of atoms with electrons orbiting about the nucleus and the charges held together by the electric forces among them. The electrons must move or they will be pulled into the nucleus, just as the earth would fall into the sun if it stopped moving. However, according to classical physics, accelerating charges produce electromagnetic radiation and hence lose energy. For example, if electrons are in circular orbits in atoms, they have a centripetal acceleration. Consequently, they must radiate energy and very soon fall into the nucleus. Thus the experimental structure of the atom and classical physics led to the unacceptable result that all atoms are unstable.

In discussing the resolution of these and other difficulties, we focus our attention on the work of two physicists. In 1912, Niels Bohr proposed a model of the one-electron atom that was successful in accounting for many observations. In a more refined form, many of the ideas of Bohr's model are contained in our present-day theories. Also, in 1924, Louis de Broglie hypothesized that electrons and all other "particles" possess wave attributes. This was strikingly confirmed when it was found that electrons reflected from a crystal form a pattern similar to that produced when x-rays are diffracted by a crystal.

By about 1930, most of the fundamental puzzles involving the wave-particle behavior of light and matter had been resolved. Although the sequence of events that occurred in the unraveling of these problems is extremely interesting, we do not present them in chronological order here. Rather, we take a retrospective view of how the ideas and experiments were made compatible and how a solid foundation for our modern picture of nature was established. Specifically, we are concerned in this chapter with the wave properties of matter which, along with the photon theory of light, give us a consistent view of nature.

29.1 | FAILURES OF CLASSICAL PHYSICS

In the 1880s, experiments showed that when electrodes of opposite polarity are immersed in a gas, a current may be produced in the gas. In 1897, J.J. Thomson (1856–1940) showed that this current is composed of negatively charged particles now called

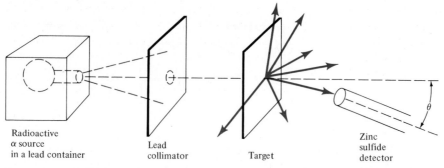

Figure 29.1. In Rutherford's experiments, α particles are emitted by a radioactive source. The lead collimator absorbs all the α particles except those which pass through a tiny hole and form a beam. The α particles scatter in all directions after striking the target and are detected at different angles by a moveable zinc sulfide detector.

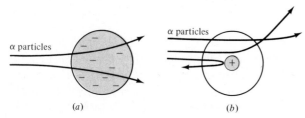

Figure 29.2. (a) In Thomson's model of the atom all the α particles would travel through the foil, deflected slightly by the diffuse positive charge (shown colored) and by the light electrons. (b) Rutherford explained the relatively frequent large angle scatterings as due to the repulsive electric force of a small positively charged nucleus. The light electrons occupying the remainder of the atom would not scatter the α particles significantly.

electrons. These particles can be deflected by a magnetic field and the charge to mass ratio, e/m, measured. The same particles appear no matter what gas or electrode material is used. Hence, Thomson concluded that these electrons are a fundamental constituent of matter.

In 1906, Thomson proposed that atoms consist of electrons imbedded in a positively charged background. The total charge added up to zero because atoms are normally electrically neutral. This model could be tested by bombarding materials with charged particles and observing the way in which they are deflected.

Radioactivity was receiving a great deal of attention at this time, and one of those most prominent in that work was Ernest Rutherford (1871–1937). One of his most important early discoveries was the spontaneous emission by some heavy radioactive elements of *alpha* (α) *particles*. These particles have a positive charge that is twice the magnitude of the electronic charge; they are about 7000 times as massive as an electron. Starting in 1907, Rutherford began an intensive study of scattering α particles by various targets. The scattered α particles were detected by observing the flashes of light they produced upon reaching a zinc sulfide screen (Fig. 29.1).

According to Thomson's atomic model, the α particles should not be disturbed much by the diffuse positive charge, and the light electrons should also deflect them only slightly. Thus, most of the α particles should pass almost straight through the very thin gold foil target (Fig. 29.2a).

When the experiment was performed, it was found that most of the α particles were scattered through small angles, but a significant fraction were scattered through much larger angles. And occasionally an α particle was stopped and sent back out toward the source! This remarkable fact was described by Rutherford: "It was almost as incredible as if you had fired a 15-inch shell at a piece of tissue paper and it came back and hit you."

The only explanation consistent with Rutherford's data is that of a small positively charged nucleus (of radius on the order of 10^{-14} m) with the electrons occupying the remainder of the atom. (Atoms were known to be on the order of 10^{-10} m in radius, a factor of 10,000 times larger than the nucleus.) The α particles experience a repulsive force from the heavy, positively charged nuclei. Occasionally, in a head-on collision, an α particle is stopped close to a nucleus and scattered directly back toward the source (Fig. 29.2b).

Rutherford's nuclear atom was an important step forward, but it presented serious problems. It now appeared that the electrons must be in some sort of orbit in the attractive electric field of the nucleus. The energy of an electron was the sum of its kinetic energy in its orbit plus the potential energy due to the positive nucleus. If the kinetic energy were to decrease, the electron would move closer to the nucleus. The problem was that accelerated charges were known to radiate energy as electromagnetic waves. For example, an electron in a circular orbit has a centripetal acceleration and should radiate energy. Its energy should decrease, and eventually the electron should fall into the nucleus. All classical theories of this nature led to unstable pictures of atoms. It remained for Bohr and those who followed to resolve these problems through the development of quantum mechanics.

Discrete Spectra

A second important body of data existed during the same period which showed that atoms emitted and absorbed light only at specific frequencies. For illustrative purposes, we describe the data for hydrogen.

In examining a flame or an electric discharge tube containing hydrogen through a diffraction grating, only certain characteristic frequencies and wavelengths of emitted light were observed (Fig. 29.3).

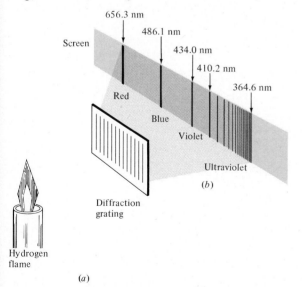

656.3 nm
486.1 nm
434.0 nm
410.2 nm
364.6 nm

Screen

Red

Blue

Violet

Ultraviolet

(b)

Diffraction grating

Hydrogen flame

(a)

Figure 29.3. (a) When the light from a flame containing hydrogen passes through a diffraction grating and is projected on a screen, discrete lines are seen corresponding to the emission of monochromatic light at several different wavelengths. (b) The visible and near ultraviolet lines of hydrogen.

Johann Balmer (1825–1898) discovered in 1884 that the wavelengths of the visible and near ultraviolet lines of hydrogen obey a remarkably simple formula almost exactly,

$$\frac{1}{\lambda} = R_H\left(\frac{1}{2^2} - \frac{1}{n^2}\right), \qquad n = 3, 4, 5, \ldots \quad (29.1)$$

Here $R_H = 1.097 \times 10^7 \text{ m}^{-1}$ is called the *Rydberg constant*.

The longest visible wavelength in the hydrogen spectrum is found in the next example.

Example 29.1

What is the longest wavelength light predicted by the Balmer formula?

The longest wavelength is found using Balmer's formula for $n = 3$,

$$\frac{1}{\lambda} = R_H\left(\frac{1}{2^2} - \frac{1}{3^2}\right) = \frac{5}{36}R_H$$

Then

$$\lambda = \frac{36}{5\,R_H} = \frac{36}{5(1.097 \times 10^7 \text{ m}^{-1})}$$
$$= 6.56 \times 10^{-7} \text{ m} = 656 \text{ nm}$$

which is the longest visible wavelength observed (Fig. 29.3).

Other spectral lines were also observed including some not in the visible spectrum. The wavelengths of *all* the observed lines could be predicted using

$$\frac{1}{\lambda} = R_H\left(\frac{1}{n_f^2} - \frac{1}{n_i^2}\right), \qquad (29.2)$$
$$n_i = n_f + 1,\ n_f + 2, \ldots$$

The visible lines correspond to $n_f = 2$, the ultraviolet spectrum has $n_f = 1$, and the infrared spectrum corresponds to $n_f \geqslant 3$.

Using energy conservation, when a photon of energy hf is emitted by an atom the internal energy of the atom must decrease. Because the observed photons only appear at certain frequencies, the atoms must only change their energies by fixed amounts. This too contradicts a classical picture of electrons orbiting a nucleus at an arbitrary radius and a corresponding arbitrary energy.

From the point of view of classical physics, it seemed impossible for Rutherford's nuclear atom to be able to change its energy in discrete jumps. In the remainder of this chapter we describe the new ideas needed to explain the observations we have described.

29.2 | THE de BROGLIE WAVE HYPOTHESIS

In 1924 Louis de Broglie suggested that matter as well as light might have wave properties. This would give both light and matter a dual wave-particle nature and put them both on the same footing. De Broglie also suggested a formula relating the wavelength of any object to its momentum. This formula was rapidly verified by experimental measurements.

De Broglie noted that photons are zero mass particles and their energy and momentum are related by $E = hf = pc$ (Chapter Twenty-eight). Using $f\lambda = c$, the relation between the photon momentum and wavelength is $hc/\lambda = pc$ or

$$\lambda = \frac{h}{p} \qquad (29.3)$$

Using this as a guide to the possible dual nature of particles, de Broglie suggested that particles of momentum p should also have a wavelength λ associated with them, given by exactly the same relation.

This idea was tested beautifully in 1926 in an experiment performed by C. Davison (1891–1958) and L.H. Germer (1896–1971). They directed a beam of electrons at a crystal and observed that the electrons scattered in various directions for a given crystal orientation (Fig. 29.4). In this experiment the pattern formed by the electrons reflected from the crystal lattice of aluminum is almost identical to that produced by x-rays (Fig. 29.5). This strongly suggests that the electrons have a wavelength λ associated with them and that the Bragg condition for x-ray

(a)

(b)

Figure 29.5. The diffraction pattern in aluminum obtained with (a) x-rays, and (b) electrons. (From Educational Development Center, Inc., Newton, Massachusetts.)

diffraction discussed in Chapter Twenty-five holds true here also,

$$m\lambda = 2d \sin \theta, \qquad m = 1, 2, 3, \ldots$$

The following examples illustrate how the wave character of electrons is predicted by de Broglie's hypothesis.

Example 29.2

What is the wavelength associated with the electrons which scatter as shown in Fig. 29.6?

From the figure $\theta = 90° - \frac{1}{2}(52°) = 64°$, and $\sin 64° = 0.92$. With $d = 0.9$ Å $= 9 \times 10^{-11}$ m, the Bragg condition with $m = 1$ gives

$$\lambda = 2d \sin \theta = 2(9 \times 10^{-11} \text{ m})(0.92)$$
$$= 1.66 \times 10^{-10} \text{ m}$$
$$= 1.66 \text{ Å}$$

Example 29.3

What is the momentum and corresponding de Broglie wavelength of the electrons scattered in Fig. 29.6?

Figure 29.4. The apparatus of Davisson and Germer. Electrons emitted from a hot filament are accelerated through a potential difference V. They strike a crystal, and the scattered electrons are detected at an angle ϕ with the incident beam.

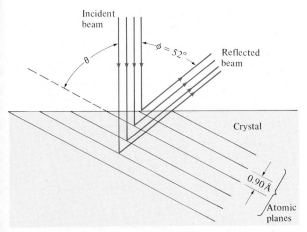

Figure 29.6. When the electron accelerating voltage is 56 V, electrons scattered from atomic planes 0.9 Å apart in the crystal appear mostly at $\phi = 52°$.

The momentum of electrons accelerated through a 56-V potential difference can be found by noting that their kinetic energy is

$$\tfrac{1}{2}mv^2 = eV$$

where m is the electron mass and e is the magnitude of the electronic charge. Using $p = mv$, $\tfrac{1}{2}mv^2 = p^2/2m = eV$, so

$$p = \sqrt{2emV}$$
$$= \sqrt{2(1.6 \times 10^{-19}\ \text{C})(9.1 \times 10^{-31}\ \text{kg})(56\ \text{V})}$$
$$= 4.0 \times 10^{-24}\ \text{kg m s}^{-1}$$

Using de Broglie's hypothesis,

$$\lambda = \frac{h}{p} = \frac{6.63 \times 10^{-34}\ \text{J s}}{4 \times 10^{-24}\ \text{kg m s}^{-1}}$$
$$= 1.66 \times 10^{-10}\ \text{m} = 1.66\ \text{Å}$$

This value agrees with that found from experiment (Example 29.2).

As a result of this and other experiments, we now accept de Broglie's idea as one of the cornerstones of modern physics. The diffraction effects of electrons and neutrons are now important tools to the scientist.

Neutron Diffraction | The wave properties of matter provide us with ways to study microscopic objects on a much smaller scale than is possible with the ordinary microscope. Both neutrons and electrons have been used extensively for this purpose.

No analog to the lens is available for neutrons, so it is not possible to construct a neutron microscope.

However, a crystal serves as a three-dimensional diffraction grating for neutrons, just as it does for electrons and x-rays. In many situations, neutron diffraction provides molecular structure information that cannot be obtained in other ways.

The reactions occurring in a nuclear reactor produce many neutrons with a wide range of energies. If neutrons collide with several atoms before or after leaving the reactor, their average kinetic energy is comparable to the average thermal energy of the atoms. When the temperature is T, this energy is $\tfrac{3}{2}k_BT$. Thus the average energy and de Broglie wavelength of such *thermal neutrons* can be controlled by varying the temperature of the materials.

The wavelength of a thermal neutron is approximately the size of an atom, as is shown in the next example.

Example 29.4

What is the wavelength of a neutron with kinetic energy $\tfrac{3}{2}k_BT$, when $T = 300$ K?

The neutron kinetic energy is $\tfrac{1}{2}mv^2 = \tfrac{3}{2}k_BT$ so the momentum $p = mv$ is found from $p^2 = (mv)^2 = 3k_BT$ or

$$p = \sqrt{3mk_BT}$$

Using $m = 1.67 \times 10^{-27}$ kg and $T = 300$ K,

$$\lambda = \frac{h}{p} = \frac{h}{\sqrt{3mk_BT}}$$
$$= \frac{6.63 \times 10^{-34}\ \text{J s}}{\sqrt{3(1.67 \times 10^{-27}\ \text{kg})(1.38 \times 10^{-23}\ \text{J K}^{-1})(300\ \text{K})}}$$
$$= 1.46 \times 10^{-10}\ \text{m} = 1.46\ \text{Å}$$

This is a typical atomic or molecular distance.

Electron Microscope | The electron microscope is an astonishing improvement over the optical microscope. The shortest useful wavelength and resolution for optical work is about 2000 Å but electrons produced by using accelerating voltages of 50 kV have a de Broglie wavelength of 0.055 Å. Theoretically, this would be the resolution of an electron microscope. In practice, electrons must be focused using electric and magnetic fields as lenses, and these limit the resolution to about 2 Å. However even this is 1000 times better than that obtained with optical instruments and allows the study of minute cellular constituents and molecules.

Figure 29.7. (*a*) Light microscope picture of a plant cell. (*b*) Transmission electron microscope picture of the same type of cell. (*c*) A scanning electron microscope picture of a cross section of the leaf stalk of a fern. The large hollow cells are xylem tracheids and function as water conducting tissues. The three dimensional structure is clearly evident. ((*a*) and (*b*) courtesy William A. Jensen, from CELL ULTRASTRUCTURE by W.A. Jensen and R.B. Park, Wadsworth, 1967; (*c*) courtesy John R. Troughton, from PLANTS by J.R. Troughton and F.B. Sampson, John Wiley & Sons, 1973.)

Two types of electron microscopes are in common use. The *transmission microscope* requires the use of samples at most several hundred angstroms thick. If the sample is too thick the energy lost by off-axis electrons is much greater than that lost by electrons going directly through the sample. The wavelengths are then sufficiently varied so that the electrons cannot all be focused simultaneously. Although the resolution of these microscopes is excellent, sample preparation limits their usage somewhat, and the depth of field is very small. Thus, three-dimensional structures will not be properly focused. These disadvantages are overcome with a *scanning microscope,* but at the price of a reduction of the resolution to about 100 Å. Live samples can be used, and the results show three-dimensional structures clearly (Fig. 29.7).

The transmission microscope is schematically much like an optical microscope. Produced at a hot cathode, the electrons pass through the sample and are focused into a real image (Fig. 29.8*a*). In the scanning microscope the electrons are focused to a very small spot, which is swept across the sample. The image is formed by monitoring secondary electrons knocked out of the sample by the main beam. These secondary electrons are collected and accelerated in a cathode-ray tube. The cathode-ray beam is swept across the

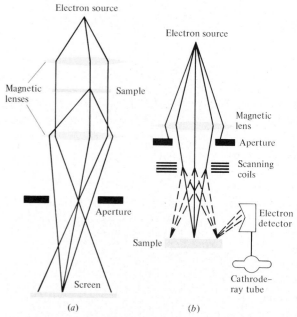

Figure 29.8. (*a*) Transmission and (*b*) scanning electron microscopes.

screen at the same rate as the main beam is swept across the sample. The number of secondary electrons varies with the composition and orientation of the surfaces of the sample, and the brightness of the trace

Figure 29.9. Scanning electron microscope pictures of (*a*) a fruit fly and (*b*) a portion of the eye alone. ((*a*) Photo from Thomas Eisner, Cornell University; (*b*) Photo from H. Hartman and T.L. Hayes, *Journal of Heredity*, Vol. 62, p. 41, 1971. Copyright 1971 by the American Genetic Association.)

on the cathode ray screen varies accordingly (Fig. 29.8*b*).

29.3 | THE BOHR ATOM

In the remaining sections of this chapter, we describe the resolution of the atomic spectra and nuclear atom problems discussed earlier. The wave-particle nature of matter provides the basis for these discussions.

In 1913, well before de Broglie postulated the wave nature of matter, Niels Bohr (1885–1962) proposed a model of one-electron atoms. The immediate purpose of the model was to explain the observed emission spectrum of hydrogen. Although the model was successful in this respect, it is now known to be quite incomplete and oversimplified. However, it does provide a very simple picture that we sometimes use to visualize complex atomic processes.

Bohr adopted Rutherford's nuclear atom model with electrons moving in the electric field of a small massive nucleus. He also suggested that certain orbits are stable; the accelerating electrons do not radiate energy when in these orbits. We can find these orbits by attributing a wavelength $\lambda = h/p$ to the electron, although Bohr used a different approach.

We consider an atom composed of a heavy nucleus containing Z protons and a single electron in a cir-

cular orbit of radius r around the nucleus. Since the nuclear mass is much larger than that of the electron, we assume that the nucleus is at rest. The electric force between the positive nucleus (with charge Ze) and the electron (with charge $-e$) holds the electron in orbit (Fig. 29.10). Its magnitude is kZe^2/r^2.

If the electron is in a circular orbit of radius r the acceleration is $a = v^2/r$. Using Newton's second law, $\mathbf{F} = m\mathbf{a}$, where $\mathbf{F}$ is the electric force,

$$k\frac{Ze^2}{r^2} = \frac{mv^2}{r} \tag{29.4}$$

From this we see that the kinetic energy of the electron is

$$K = \frac{1}{2}mv^2 = k\frac{Ze^2}{2r} \tag{29.5}$$

Figure 29.10. An electron of mass m and charge $-e$ in a circular orbit around a nucleus with charge Ze. The massive nucleus is assumed fixed in place.

The potential energy of the electron is $\mathscr{U} = -kZe^2/r$, so the total energy is

$$E = K + \mathscr{U} = -k\frac{Ze^2}{2r} \qquad (29.6)$$

Since E is negative, the electron does not escape from the region of the nucleus.

In order to proceed further, we must introduce the wave character of the electron. Bohr did this indirectly by postulating that the orbital angular momentum $L = I\omega$ of the electron can only have certain values. For a point mass at radius r, $I = mr^2$, and $\omega = v/r$. Hence $L = (mr^2)(v/r) = rmv$, and Bohr's requirement becomes

$$L = rmv = n\frac{h}{2\pi} = n\hbar, \qquad n = 1, 2, 3, \ldots \quad (29.7)$$

Since the combination $h/2\pi$ appears frequently, it is denoted by

$$\hbar = \frac{h}{2\pi} = 1.055 \times 10^{-34}\,\text{J s}$$

and is read as "h bar."

This postulate does not contain any obvious reference to the wave property of the electron. However, if the electron has a wavelength $\lambda = h/p$ as suggested by de Broglie, then it should "fit" into the orbit. If it does not "fit," it will interfere with itself (Fig. 29.11). Notice that this idea is similar to the one that we used for waves on strings. Resonant standing waves occur when the wavelength "fits" on a string. Here the wavelength must fit into the orbit.

With this assumption, we must fit an integer number of waves of wavelength $\lambda = h/p = h/mv$ into the orbital circumference $2\pi r$. Thus

$$2\pi r = n\lambda = n\frac{h}{mv}$$

or

$$rmv = n\frac{h}{2\pi} = n\hbar, \qquad n = 1, 2, 3, \ldots \quad (29.7)$$

This is exactly Bohr's criterion for the angular momentum! Thus our assumption is equivalent to Bohr's postulate.

We now have two equations relating r and v, Eq. 29.7 and

$$\frac{1}{2}mv^2 = k\frac{Ze^2}{2r} \qquad (29.5)$$

We will solve these equations for r. From Eq. 29.7, $v = n\hbar/mr$, so in Eq. 29.5, we have

$$\frac{1}{2}mv^2 = \frac{1}{2}m\left(\frac{n\hbar}{mr}\right)^2 = k\frac{Ze^2}{2r}$$

This can be solved for r_n, the radius associated with n wavelengths,

$$r_n = \frac{n^2\hbar^2}{kZme^2} = \frac{n^2}{Z}a_0, \qquad n = 1, 2, 3, \ldots \quad (29.8)$$

Here $a_0 = \hbar^2/kme^2 = 5.29 \times 10^{-11}\,\text{m}$ is called the *Bohr radius*. There are an infinite number of allowed values of r_n corresponding to all the integer values of n. The electrons can only be in these orbits. This is quite different from the classical idea that any value of r is possible.

We can now find the energy from $E = -kZe^2/2r$. Using the expression for r_n,

$$E_n = -\frac{kZe^2}{2r_n} = -\frac{kZe^2}{2\left(\dfrac{n^2}{Z}a_0\right)}$$

or

$$E_n = -\frac{kZ^2e^2}{2a_0n^2} = -\frac{Z^2}{n^2}E_0, \quad n = 1, 2, 3, \ldots \quad (29.9)$$

Here

$$E_0 = \frac{ke^2}{2a_0} = 13.6\,\text{eV} = 2.18 \times 10^{-18}\,\text{J}$$

E_0 is the lowest possible energy of the electron. The corresponding orbit is called the *ground state*.

This is the fundamental result that Bohr obtained.

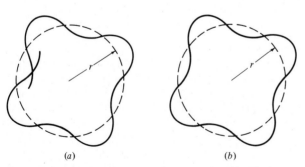

Figure 29.11. (*a*) The wavelength associated with the electron does not fit into the orbit. In (*b*) the wave "fits" correctly. The orbital circumference is $2\pi r$.

The electrons can only occupy orbits with certain energies. The allowed values of the electron energy E_n are termed *energy levels,* and n is called a *quantum number.*

Bohr's result is in close agreement with experiment for *any* atom with one electron, such as hydrogen ($Z = 1$), singly ionized helium, He^+ ($Z = 2$), and doubly ionized lithium, Li^{++} ($Z = 3$). The lowest levels for He^+ are found in the next example.

Example 29.5

(a) What are the lowest three energy levels of singly ionized helium? (b) What are the radii associated with these levels?

(a) Helium has two protons in the nucleus ($Z = 2$), and when it is singly ionized, it has one electron. Using $Z = 2$ and $n = 1$,

$$E_1 = -\frac{Z^2}{n^2} E_0 = -\frac{(2)^2}{(1)^2}(13.6 \text{ eV})$$

$$= -54.4 \text{ eV}$$

If instead we use $n = 2$, the result obtained is smaller by $1/2^2$, and is -13.6 eV; for $n = 3$, the energy is -6.04 eV.

(b) Using Eq. 29.8, the $n = 1$ radius is

$$r_1 = \frac{n^2}{Z} a_0 = \frac{(1)^2}{2}(5.29 \times 10^{-11} \text{ m})$$

$$= 2.65 \times 10^{-11} \text{ m}$$

Similarly, $r_2 = 10.6 \times 10^{-11}$ m,   and $r_3 = 23.8 \times 10^{-11}$ m (Fig. 29.12).

Bohr also postulated that electrons in allowed orbits do not radiate energy, even though they are accelerated. Electrons can only lower their energy by dropping into lower energy orbits and emitting a photon in the process. If the initial and final energies of the electron are $E_{n,i}$ and $E_{n,f}$, from energy conservation the photon energy hf must be

$$hf = E_{n,i} - E_{n,f}$$

Using our result for the energy levels, this gives for hydrogen

$$hf = \frac{hc}{\lambda} = hcR_H\left(\frac{1}{n_f^2} - \frac{1}{n_i^2}\right),$$
$$n_i = n_f + 1, n_f + 2, \dots \tag{29.10}$$

where the theoretical value of R_H is

$$R_H = \frac{mk^2e^4}{2\hbar^2 c} = 1.097 \times 10^7 \text{ m}$$

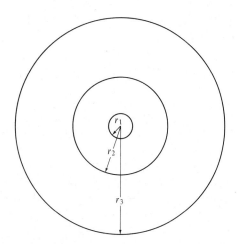

Figure 29.12. The radii predicted by the Bohr model for the three lowest energy orbits in a one electron atom. Here $r_3 = 9r_1$ and $r_2 = 4r_1$.

This is exactly the empirical result found by Balmer and others and described in the preceding section.

Ordinarily the electrons in hydrogen atoms are in the energy level with the lowest or most negative energy, the $n = 1$ level or ground state. In a flame or an electrical discharge tube, many of the atoms are in *excited states,* with the electrons in less tightly bound orbits that have larger values of n. An electron may return to the lowest level in a single step, emitting one photon, or in a series of steps from higher to lower energy levels, emitting photons at each step. Figure 29.13 is an energy level diagram that shows the allowed energy levels for the electron and the transitions that may occur.

An atom will also absorb photons with these same frequencies and wavelengths. In absorption, the energy provided by the photon must be just equal to the energy required to raise an electron from one stable level to a higher one. The following example illustrates photon absorption.

Example 29.6

What energy photon is necessary to raise an electron in hydrogen from the $n = 1$ to the $n = 3$ energy level?

From Fig. 29.13, we see that this is the reverse of the process indicated by the second line in the Lyman series. The electron increases its energy from E_1 to E_3, and the photon energy must be

$$hf = E_3 - E_1 = -\frac{1}{3^2} E_0 - \left(-\frac{1}{1^2} E_0\right)$$

$$= \left(1 - \frac{1}{9}\right) E_0$$

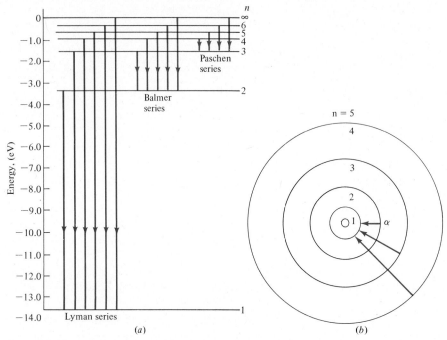

Figure 29.13. (*a*) The energy levels of hydrogen corresponding to the quantum numbers $n = 1$ through 6 and the highest level ($n = \infty$) which has zero energy. The vertical arrows show possible electron transitions which are accompanied by photon emission. For example, the longest wavelength line in the Balmer series corresponds to an electron dropping from the $n = 3$ to the $n = 2$ energy level in its hydrogen atom. The three series shown involve ultraviolet (Lyman), visible (Balmer), and infrared (Paschen) photons. (*b*) The transitions corresponding to the first three lines of the Balmer series.

With $E_0 = 13.6$ eV, the photon energy is

$$hf = \frac{8}{9}(13.6 \text{ eV}) = 12.1 \text{ eV}$$

Bohr's results are correct for any one-electron atom, despite the fact that the model on which they are based is now considered to be a great oversimplification. However, we do see how the two problems described earlier are explained. The fact that the electron is not a particle as we usually think of one classically leads to quantized energy levels. The discrete spectra observed are the result of electrons moving from one orbit to another. The wavelike behavior of the electron also means that it is not an object that radiates energy when accelerated. The orbital radii we have calculated represent the average distances of electrons from the nucleus but are not actually the radii of electronic trajectories.

Despite the success of Bohr's model for one-electron atoms, most attempts to extend these ideas to multielectron atoms and molecules were unsuccessful.

The requirement that the angular momentum is quantized is insufficient to determine the orbits in these complex systems.

In 1925, two detailed theories of *quantum mechanics* were proposed, one by Erwin Schroedinger (1887–1961) and the other by Werner Heisenberg (1901–1975). Although the two theories were very different in appearance, they were soon recognized to be equivalent in content. Unlike Bohr's theory, these are complete theories which, in principle, can be applied to all physical systems, including atoms, molecules, and macroscopic objects. We discuss Schroedinger's form of quantum mechanics in the next chapter.

We have seen that Bohr's model predicts that there is a lowest energy level. It is interesting to ask, why can't the electron go into a deeper energy level, one closer to or even inside the nucleus? In nature, systems tend to stay in the lowest energy state whenever possible, and the electron could reduce its energy enormously by moving much closer to the nucleus.

This question can be partially answered using the *uncertainty principle*.

29.4 | THE UNCERTAINTY PRINCIPLE

Both Schroedinger's and Heisenberg's formulations of quantum mechanics implicitly contain the de Broglie hypothesis of the wave character of matter. They also contain the uncertainty principle, which was first discussed by Heisenberg in 1927.

The uncertainty principle describes limits imposed by nature on the precision of *simultaneous* measurements of the position and momentum of an object. Mathematically stated, if an object is said to be at a position x within an uncertainty of Δx, then any simultaneous measurement of the x component of momentum must have an uncertainty Δp_x consistent with

$$\Delta x\, \Delta p_x \geq \hbar \qquad (29.11a)$$

This means that an increase in the accuracy of a position measurement must be accompanied by a decrease in the accuracy of the momentum measurement. The best that can be done even with idealized experiments is to have the equality hold, $\Delta x\, \Delta p_x = \hbar$. The x direction here is chosen arbitrarily; the uncertainty relation could equally well be stated as

$$\Delta y\, \Delta p_y \geq \hbar \qquad (29.11b)$$

Before pursuing the ramifications of these statements, we may ask just how important are these naturally occurring limitations.

Example 29.7

Suppose the velocities of an electron and of a rifle bullet of mass 0.03 kg are each measured with an uncertainty of $\Delta v = 10^{-3}$ m s^{-1}. What are the minimum uncertainties in their positions according to the uncertainty principle?

Using $\Delta p_x = m\, \Delta v_x$ for each, the minimum position uncertainty satisfies $\Delta x\, m\, \Delta v_x = \hbar$. For the electron $m = 9.11 \times 10^{-31}$ kg, so

$$\Delta x = \frac{\hbar}{m \Delta v_x} = \frac{1.055 \times 10^{-34}\text{ J s}}{(9.11 \times 10^{-31}\text{ m})(10^{-3}\text{ m s}^{-1})}$$

$$= 0.116 \text{ m}$$

For the bullet,

$$\Delta x = \frac{\hbar}{m\, \Delta v_x} = \frac{(1.055 \times 10^{-34}\text{ J s})}{(0.03\text{ kg})(10^{-3}\text{ m s}^{-1})}$$

$$= 3.5 \times 10^{-30} \text{ m}$$

We can see from the preceding example that for normal macroscopic objects such as the bullet the uncertainty principle does not impose any effective limit on experimental measurements because errors in position measurements are always very much larger than 10^{-30} m. However, the opposite is true of objects as small as electrons. For example, since atoms in a solid are about 10^{-9} m apart, a position measurement with an uncertainty of about 0.1 m means that the electron could be anywhere among billions of atoms!

The uncertainty principle leads to the conclusion that, in any experiment, one cannot simultaneously observe the wave and particle properties of light or matter. This can be illustrated with a double-slit interference experiment performed with electrons. Consider a beam of monoenergetic electrons incident on two slits of width $d/4$ a distance d apart (Fig. 29.14). Since the electrons have a wavelength $\lambda = h/p$ associated with them, an interference pattern will be detected at the screen. From Chapter Twenty-five we know that there will be an intensity maximum at $\theta = 0$, and the first minimum occurs when $\sin \theta = \lambda/2d$. As long as $\lambda/2d$ is small, we can use $\sin \theta \simeq \theta$, so the first minimum occurs at

$$\theta_m = \frac{\lambda}{2d}$$

Thus a measurement of θ_m can be used to determine the wavelength λ. This value can be compared with de Broglie's prediction to test his hypothesis.

The interference pattern is a manifestation of the wave properties of electrons. However, electrons also have particle properties. If they are regarded as particles, we should be able to say which slit each electron passes through. This might be done by having a beam of photons passing to the right of the slits, so that each time an electron goes through a slit, it will collide with a photon. The electron proceeds toward the screen, but the scattered photon can be detected and the area of the collision located. For example, if the collision takes place above the centerline of Fig. 29.14, we would say the electron went through the upper slit, and its position uncertainty at the slits would be, at most, $\Delta y = d/4$, the slit width.

We must find out what our measurements mean in the light of the uncertainty principle. The uncertainty

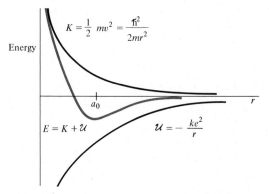

Figure 29.14. (*a*) Electrons of wavelength $\lambda = h/p$ are incident from the left. After passing through the slits, they form an interference pattern at the screen. (*b*) The slit area is shown much enlarged; in practice $D \gg d$ and θ is very small.

principle requires that $\Delta y\, \Delta p_y \geq \hbar$. Since $\hbar = h/2\pi$,

$$\Delta p_y \geq \frac{h}{2\pi\, \Delta y} = \frac{h}{2\pi d/4} = \frac{2h}{\pi d}$$

The angular deflection of the electron is then at least

$$\theta = \frac{\Delta p_y}{p} = \frac{2h/\pi d}{h/\lambda} = \frac{2\lambda}{\pi d}$$

But this deflection is larger than $\theta_m = \lambda/2d$, the angular position of the interference minimum. This means that the interference pattern will be hopelessly blurred if we determine which slit the electron passes through. The act of measurement that pinpoints the electron as a particle going through one slit also destroys the evidence of the wave property of the electron!

Viewed in this way, the uncertainty principle is nature's way of ensuring that the wave and particle properties of an object cannot be simultaneously observed. We also face the extraordinary fact that *the measurement process itself affects the results*. In the case illustrated, the measurement of the electron position deflects the electron enough to destroy the interference pattern.

We conclude this section by considering the question raised in the previous section: Why doesn't an electron radiate energy and drop into the nucleus where its potential energy would be much lower than it is in the first Bohr orbit? Suppose an electron is within a distance r of the nucleus. Then we could say that we know its position within an uncertainty $\Delta x = r$. From the uncertainty principle, the electron momentum cannot be zero, and it must be at least

$p = \Delta p_x = \hbar/r$. The total energy of the electron is then at least (for $Z = 1$)

$$E = \frac{1}{2}mv^2 - \frac{ke^2}{r} = \frac{p^2}{2m} - \frac{ke^2}{r} = \frac{\hbar^2}{2mr^2} - \frac{ke^2}{r}$$

If the electron approaches the nucleus, r becomes small; the potential energy $-ke^2/r$ becomes more negative, but the kinetic energy $\hbar^2/2mr^2$ increases (Fig. 29.15). We would expect the electron to stay at an average radius r such that its total energy is a minimum. This occurs when r equals the Bohr radius, $a_0 = \hbar^2/kme^2$. Thus the lowest energy state of an atom is not one with the electron in the nucleus. The uncertainty principle requires that the kinetic energy becomes large as the position uncertainty decreases, more than offsetting the reduction in the potential energy.

Figure 29.15. The kinetic and potential energy contributions to give the total energy $E = K + \mathcal{U}$. The total energy is a minimum at $r = a_0$.

We see from this discussion that the stability of atoms and of matter depends on the uncertainty principle.

SUMMARY

Matter, as well as light, has both wave and particle characteristics. The wavelength associated with an object is related to the object's momentum by $\lambda = h/p$. This relationship, suggested by de Broglie, is one of the fundamental ideas of quantum theory.

By combining the wave properties of the electrons with the idea that an atom is composed of a massive, positively charged nucleus with electrons outside it, we can find the allowed energy levels for a one-electron atom. Photons with specific energies are emitted or absorbed when an electron goes from one energy level to another. The predictions of this Bohr model are in agreement with experiment.

The uncertainty principle describes a natural limit on the precision of simultaneous measurements of the position and momentum of particles. It indicates that either wave or particle properties, but not both, may be observed in a given experiment.

Checklist

Define or explain:

nuclear atom
discrete spectra
de Broglie wavelength
electron diffraction
neutron diffraction
Bohr atom
quantized angular
  momentum

Bohr radius
ground state
excited state
energy levels
quantum number
uncertainty principle
wave-particle duality

EXERCISES

Section 29.1 | Failures of Classical Physics

29-1   A 4.8 MeV (1 MeV $= 10^6$ eV) α particle makes a head-on collision with a gold nucleus ($Z = 79$ for gold). What is the distance of closest approach of the α particle and the gold nucleus at which the α particle has zero kinetic energy?

29-2   In hydrogen, the proton radius is 0.8×10^{-15} m and the average distance of the electron from the proton center is 5.3×10^{-11} m. If the

proton is represented by an orange 0.06 m in radius, how far away will the electron be?

29-3   What is the wavelength of the visible light of atomic hydrogen corresponding to $n = 4$?

29-4   What is the wavelength of the ultraviolet radiation from hydrogen corresponding to $n_i = 2$ and $n_f = 1$?

Section 29.2 | The de Broglie Wave Hypothesis

29-5   An electron and a photon each have a wavelength of 1 Å. What are their momenta?

29-6   Compare the de Broglie wavelengths of an electron and a neutron each having energy $\frac{3}{2}k_BT$, where $T = 300$ K.

29-7   What is the de Broglie wavelength associated with the earth in its motion about the sun? (The earth's mass is 6×10^{24} kg, and its orbital radius is 1.5×10^{11} m.)

29-8   An oxygen atom has a mass of 16 amu, where 1 amu $= 1.66 \times 10^{-27}$ kg. If the atom has a kinetic energy of 1 eV, what is its de Broglie wavelength?

29-9   A 1-kg mass has a velocity of 1 m s^{-1}. (a) What is its de Broglie wavelength? (b) Should one expect to see interference effects from these waves? Explain.

29-10   The protons and neutrons in the atomic nucleus are typically 10^{-15} m apart, so any probe of nuclear structure must have a de Broglie wavelength of this order of magnitude. (a) What momentum must a probe have if its de Broglie wavelength is 10^{-15} m? (b) If the probe is an alpha particle of mass 6.64×10^{-27} kg, what is its kinetic energy in electron volts?

29-11   Electrons incident on a set of crystal planes at an angle of 20° to the planes interfere constructively and form a spot on a photographic plate. If the crystal planes are 1.2 angstroms apart, find (a) the wavelength of the electrons; (b) the energy of the electrons.

29-12   Neutrons with a de Broglie wavelength of 3 Å are needed to perform a molecular structure experiment. (a) What is the velocity of these neutrons? (b) A monoenergetic neutron pulse can be made by opening a nuclear reactor port and using only those neutrons which reach the target within a certain time interval. How long will it take the neutrons in part (a) to travel 10 m?

NIELS HENRIK DAVID BOHR

(1885-1962)

(American Institute of Physics.)

The years from 1910 to 1930 were among the most exciting in physics. A unique generation of physicists participated in the conception and development of the revolution called quantum physics. Most of this excitement centered on three places, Rutherford's laboratory in Manchester, England, the University of Göttingen in Germany, and the Institute for Theoretical Physics at the University of Copenhagen. The center of attraction in Copenhagen was Niels Bohr. In addition to his abilities as a physicist, Bohr was able to cultivate an atmosphere in which incredible progress and understanding was possible by physicists from all over the world. The flavor of this period is illustrated by the following story told by George Gamow*:

> The evening work in the Institute's library was often interrupted by Bohr, who would say that he was very tired and would like to go to the movies. The only movies he liked were wild Westerns (Hollywood style), and he always needed a couple of students to go with him and explain the complicated plots. . . . But his theoretical mind showed even in these movie expeditions. He developed a theory to explain why although the villain always draws first, the hero is faster and manages to kill him. This Bohr theory was based on psychology. Since the hero never shoots first, the villain has to decide when to draw, which impedes his action. The hero, on the other hand, acts according to a conditional reflex and grabs the gun automatically as soon as he sees the villain's hand move. We disagreed with this theory, and the next day I went to a toy store and bought two guns in Western holsters. We shot it out with Bohr, he playing the hero, and he "killed" all his students.

Bohr was the son of a physiology professor. He received his doctorate in physics in Copenhagen in 1911 and then spent three years in England, two of them in Manchester with Rutherford. It was during this period that Bohr took the revolutionary steps described now as the Bohr model for the hydrogen atom.

Bohr's model combined elements of classical physics with ideas and postulates which, at the time, were not verifiable nor even readily believable. For example, Max Planck, whose work Bohr used as a point of departure, was very reluctant to accept Bohr's ideas. It was the task of physicists during the next two decades to fill in the missing pieces and put the entire picture together in a coherent way. In this process, Bohr represented the generation that had to cast

*George Gamow, *Thirty Years That Shook Physics*, Anchor Books, New York, 1966.

off the old and press forward with the new. The generation he helped educate was not so tied to classical physics and felt less need to look back.

It is important to realize that during this period it was not simply new physics problems that were being solved; rather, an entirely new view of natural law was at issue. During this time, one needed not only new equations but also new philosophies and new pictures of a world that was beyond the direct reach of the senses but not beyond experimental reach. Bohr's role in forging the physics and philosophy of this age was immense.

By 1930 the quantum picture of atoms was fairly firm, and Bohr joined others in trying to understand the atomic nucleus. This led to Bohr's interest in fission. In 1939, in a paper with G.V. Wheeler, Bohr showed how one may understand the role of neutrons in causing the ^{235}U nucleus to break apart. This work was fundamental in the process that ultimately led to nuclear chain reaction.

In 1943, Bohr's ever-present concern for others resulted in his having to flee Denmark to avoid arrest by the Nazis. He was flown to England from Sweden in a British fighter plane. Given an oxygen mask that was too small, he was saved when the pilot noticed that Bohr had lost consciousness, and he reduced the altitude.

From England, Bohr went to the United States and participated in the atomic bomb development program. After the war, he became an early and persistent voice in calling for international reason in the face of the tremendous power he had helped to release. He also played an important role in developing international and Danish facilities for atomic and nuclear research. Having given birth to a model that was awesome in both its simplicity and its novel character, Bohr was an active participant of the scientific and political events that followed for another half-century.

Section 29.3 | The Bohr Atom

29-13   Can a hydrogen atom absorb a photon of energy greater than 13.6 eV? Explain.

29-14   In an electrical discharge tube an electron in a hydrogen atom is excited from the $n = 1$ to the $n = 4$ level. (a) How much energy has the atom absorbed? (b) Sketch an energy level diagram showing all the transitions which the electron may make in returning to the $n = 1$ level. (c) What is the wavelength of the most energetic photon which may be emitted?

29-15   The binding energy of a one-electron atom is the energy necessary to completely remove an electron from the atom. What are the binding energies in electron volts of (a) hydrogen; (b) singly ionized helium; (c) doubly ionized lithium?

Section 29.4 | The Uncertainty Principle

29-16   What is the position uncertainty of a 0.1-kg stone if its velocity is uncertain by 0.003 m s^{-1}?

29-17   What is the minimum kinetic energy of a particle of mass 10^{-29} kg if its position uncertainty is 10^{-10} m?

PROBLEMS

29-18   If the angular momentum of the earth in its motion about the sun satisfies Bohr's angular momentum hypothesis, $L = rmv = n\hbar$, what is the quantum number associated with the earth's orbit? (The earth's mass is 6×10^{24} kg, and the orbital radius is 1.5×10^{11} m.)

* **29-19**   Bohr's theory can be used for positronium, a negative electron and a positron (positive electron) orbiting around one another. (*Note:* Since both particles have the same mass, they circle about the center of mass, halfway between them. The atom must be analyzed as if one particle with half the mass of an electron orbits around the stationary second particle.) (a) What is the binding

525

energy of positronium? (b) What is the radius of the lowest energy orbit?

* **29-20** The dimensionless number $\alpha = ke^2/\hbar c$ is called the *fine structure constant* and is almost exactly $1/137$. (c is the velocity of light.) (a) What is the velocity of an electron in the lowest energy level of a one electron atom with atomic number Z, expressed as a fraction of c, the velocity of light? (b) The model of a one-electron atom given in this chapter does not take relativistic effects into account. Thus, we would expect that for atoms with v/c fairly large that the numerical predictions of the model will fail. For what value of Z would the model fail completely? Explain your reasoning.

29-21 A negative *mu meson* (μ^-) has a mass 207 times that of an electron and charge $-e$. It can orbit a nucleus much like an electron. If a μ^- orbits a sulfur nucleus ($Z = 16$), find (a) the lowest energy level, and (b) the corresponding average radius. (c) Find the ratio of this orbital radius to that of the sulfur nucleus, 4×10^{-15} m.

* **29-22** The conduction electrons in a metal may be considered as free particles in a box the size of the sample. As a start to such a model, consider an electron moving in one dimension with momentum p. The de Broglie wavelength must "fit" between walls a distance D apart with a node at each wall. (a) Draw the longest and next to longest waves which will "fit" between the walls. (b) What is the momentum associated with the longest wave which will "fit" between the walls? (c) What are all of the allowed values of the kinetic energy of an electron? (d) If $D = 0.01$ m, what is the spacing between the lowest two energy levels?

29-23 Hydrogen atoms initially in the ground state are exposed to photons of energies up to 13 eV. What energy photons will be produced as the atoms are excited and return to their ground states? (Neglect the recoil of the atoms during the absorption and emission processes.)

29-24 What are the average velocities of electrons in the lowest energy states of (a) hydrogen, and (b) singly ionized helium? (c) Would you expect relativistic effects to be important in either case?

* **29-25** Bohr's quantum condition states that the angular momentum L must be some integer times $\hbar$. This condition can also be applied to the rotation of a diatomic molecule. If we assume that a dia-

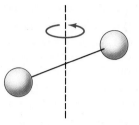

Figure 29.16. A dumbbell model of a diatomic molecule.

tomic molecule is like a dumbbell (Fig. 29.16) with a moment of inertia I, its angular momentum is $L = I\omega$, where ω is the rotational angular velocity of the molecule. The kinetic energy is $K = \frac{1}{2}I\omega^2 = L^2/2I$. (a) Apply Bohr's quantum condition to the rotating molecule and find a general expression for the energy levels. (b) Sketch an energy level diagram for the four lowest levels. (The complete quantum theory gives the result that $L^2 = n(n + 1)\hbar^2$. The energy levels of the rotating diatomic molecule are then correctly given by $n(n + 1)\hbar^2/2I$.)

29-26 The moment of inertia of the O_2 molecule is 1.92×10^{-46} kg m². (a) What is the energy of the photon emitted during a rotational transition of the molecule from the $n = 2$ to the $n = 0$ level? (See Problem 29-25.) (b) What is the wavelength of this photon?

29-27 Hydrogen atoms in their ground states bombarded by an electron beam absorb up to 12.5 eV and emit photons. What photon energies are observed?

29-28 A double-slit interference experiment is performed with electrons. No attempt is made to determine which slit each electron passes through so the position uncertainty of the electrons equals the slit separation d. Show that the uncertainty in the angle through which the electrons are deflected is less than $\theta_m = \lambda/2d$.

* **29-29** Assume that the smallest average distance of an electron from the proton in hydrogen is $a_0 + \varepsilon$, where a_0 is the Bohr radius, and ε is much smaller than a_0. Show, using the uncertainty principle argument of Section 29.4, that the energy of the electron is smallest for $\varepsilon = 0$.

29-30 Before Chadwick discovered the neutron in 1932, it was suggested that atomic nuclei were composed of protons and electrons. For example, a carbon nucleus has a mass of about 12 proton masses, but only 6 units of charge, so it was considered to have 12 positive protons and 6 negative electrons. (a) Show that, according to the uncertainty principle, the minimum kinetic energy for an electron confined to a box of size 10^{-14} m is much larger than the typical 1 MeV $= 10^6$ eV spacing of nuclear energy levels. (b) Is this true also for neutrons? (This problem should actually be solved using the theory of special relativity, but the qualitative conclusion is the same as found using $E = p^2/2m$.)

Additional Reading

C.E. Behrens, Atomic Theory from 1904–1913, *American Journal of Physics,* vol. 11, 1943, p. 60; The Early Development of the Bohr Atom, p. 135; Further Developments of Bohr's Early Atomic Theory, p. 272.

T.H. Osgood and H.S. Hirst, Rutherford and His Alpha Particles, *American Journal of Physics,* vol. 32, 1964, p. 681.

G.P. Thomson, J.J. Thomson and the Discovery of the Electron, *Physics Today,* August 1956, p. 19.

M.A. Medicus, Fifty Years of Matter Waves, *Physics Today,* vol. 27, February 1974, p. 38.

C. Jönsson, Electron Diffraction at Multiple Slits, *American Journal of Physics,* vol. 42, 1974, p. 4.

George Gamow, *Mr. Tomkins in Wonderland: Stories of c, G, and h.* The Macmillan Company, New York, 1940. A world in which h is so large that quantum mechanics is part of daily life.

George Gamow, *Thirty Years That Shook Physics,* Anchor Books, New York, 1966.

John L. Heilbron, J.J. Thomson and the Bohr Atom, *Physics Today,* vol. 30, April 1977, p. 23.

Scientific American articles:

Karl K. Darrow, The Quantum Theory, March 1952, p. 47.

R. Furth, The Limits of Measurement, July 1950, p. 48.

George Gamow, The Principle of Uncertainty, January 1958, p. 51.

Karl K. Darrow, Davisson and Germer, May 1948, p. 50.

Erwin Schroedinger, What Is Matter?, September 1953, p. 52.

E.N. DaCosta Andrade, The Birth of the Nuclear Atom, November 1956, p. 93.

Albert V. Crewe, A High Resolution Scanning Electron Microscope, April 1971, p. 26.

Thomas E. Eberhart and Thomas L. Hayes, The Scanning Electron Microscope, January 1972, p. 54.

Urve Essmann and Hermann Träuble, The Magnetic Structure of Superconductors, March 1971, p. 74.

F. Reif, Quantized Vortex Rings in Superfluid Helium, December 1964, p. 116.

Lester H. Germer, The Structure of Crystal Surfaces, March 1965, p. 32.

Donald M. Engelman and Peter B. Moore, Neutron Scattering Studies of the Ribosome, October 1976, p. 44.

S.W. Hawking, The Quantum Mechanics of Black Holes, January 1977, p. 34.

UNIT EIGHT

ATOMS AND MOLECULES

We have seen that the failure of classical physics to explain several crucial experiments led to fundamental changes in scientists' view of the physical world. Specifically, physicists now believe that wave-particle duality, and in some cases special relativity, must be built into any theory of microscopic phenomena that can hope to be successful. The modern quantum mechanics developed by Schroedinger and others has extended the basic ideas of Planck, Einstein, de Broglie, and Bohr into a more general and complete theory of matter and energy. In this unit, we study the application of quantum mechanics to the structure of atoms and molecules, including topics such as lasers and methods of probing the structure of complex biological molecules.

CHAPTER THIRTY

QUANTUM MECHANICS AND ATOMIC STRUCTURE

In 1925, Erwin Schroedinger proposed an equation whose solutions would represent the matter waves associated with electrons or other "particles." This *wave equation* and some related concepts introduced soon afterward comprise what has come to be called quantum mechanics. Quantum mechanics has been highly successful in predicting and correlating a vast amount of information about atoms. It is also able to deal successfully with aggregates of atoms and with atomic nuclei, as we will see in later chapters.

Schroedinger noticed that starting from Newton's laws, it is possible to obtain an equation for the disturbance associated with a wave traveling along a string or in an acoustical medium. Similarly, from the laws of electricity and magnetism, one can find an equation for the electric and magnetic fields in an electromagnetic wave. Thus, for any classical wave there is a wave equation whose solutions are formulas for the possible wave disturbances or *wave functions*.

Schroedinger reasoned that an analogous wave equation must exist for the matter wave function, usually represented by ψ (psi). He proposed a procedure for constructing the wave equation for a particle or system of particles if the force or forces acting are known. For example, in a hydrogen atom consisting of a proton and an electron, this force is the electrical attraction given by Coulomb's law. The resulting equation relates derivatives of the wave function and involves mathematical complexities beyond the level of this book. In fact, the equations can be solved exactly only in a few special cases although very accurate approximation methods can often be devised. Fortunately, many of the results obtained by solving the wave equation may be understood using generalizations of the ideas of the preceding unit. These solutions also provide explanations of many interesting phenomena.

In Schroedinger's theory, the wave function ψ itself cannot be directly measured or observed. However, if one knows the wave function, measurable quantities such as the energy levels of an atom or a molecule can be computed. Also, from the appropriate wave functions, one can predict the outcome of experiments, such as the double-slit electron interference experiment discussed in the previous chapter. The results of all such analyses are consistent with the uncertainty principle and with de Broglie's wave hypothesis. Thus, these two concepts are embedded firmly in quantum mechanics.

A remarkable example of the parallel development of major scientific ideas is the fact that Werner Heisenberg developed an entirely different form of quantum mechanics at almost the same time that Schroedinger did his initial work. Heisenberg represented observable quantities, such as the position or momentum of an electron, by mathematical quantities called matrices and proposed certain equations relating these matrices. Although no wave functions appeared in his theory, it was soon realized that its predictions are always identical to those of Schroedinger's wave theory. Hence the matrix and wave forms of quantum mechanics are completely equiva-

531

lent. We restrict our discussions to Schroedinger's approach, although Heisenberg's methods are sometimes more convenient for detailed calculations.

30.1 | THE OUTLINE OF QUANTUM MECHANICS

In this chapter, we discuss those aspects of Schroedinger's theory which are important in understanding atomic structure. This will also set the foundation for our picture of how atoms combine to form molecules.

Some important ideas that we use can be described in terms of the results of the preceding chapter. In Bohr's model of one-electron atoms, we found that the electron has certain allowed orbital radii labeled by a *quantum number*. Each value of the quantum number designates a unique configuration or quantum state of the system. There are as many quantum states as there are values of the quantum number, and each state has a characteristic energy. The electron can be in any one of these states and can change configurations only by moving from one allowed state to another.

In Schroedinger's theory the full specification of an atomic state requires additional quantum numbers. Some of these are associated with the *orbital angular momentum* corresponding to the electronic motion about the nucleus. Others are associated with the *spin angular momentum* of the electrons or, more concisely, the electron *spin*. This spin can be pictured as due to a rotation somewhat like that of the earth about its axis. However, this classical analogy to the quantum mechanical concept of spin is of only limited applicability, since electrons have no discernable structure or size. Furthermore, all electrons have a spin angular momentum of exactly the same magnitude. Spin appears to be an intrinsic property of all the basic constituents of matter, the so-called *elementary particles*. These include not only the electron, proton, neutron, and photon but also some more recently discovered particles such as neutrinos, mesons, and hyperons. The fact that all particles of a given type have the same spin cannot be explained classically.

In addition to lacking a classical analog, the spin angular momentum does not appear in the original form of Schroedinger's theory. Although spin can be inserted into this theory, it appears naturally in the theory only when special relativity is included at the outset. The existence of the electron spin was first found experimentally in studies of the properties of atoms and plays a crucial role in our understanding of the structure of matter.

In the next two sections, we describe the quantum numbers and wave functions found from Schroedinger's theory for the hydrogen atom. We will then be able to make the step to atoms with more than one electron.

30.2 | HYDROGEN ATOM QUANTUM NUMBERS

Schroedinger's theory requires the solution of an equation for a wave function ψ. It is found that each allowed configuration of any system has a unique wave function labeled by certain values of the quantum numbers. We describe the quantum numbers for atomic hydrogen in this section and the wave functions in the following section. The quantum numbers for hydrogen are n, l, m_l, and m_s. These are associated with the energy, the total orbital angular momentum, a component of the orbital angular momentum, and a component of the spin angular momentum, respectively.

The Principal Quantum Number, n | In

Schroedinger's original theory of hydrogen, the electronic energy levels are completely determined by a *principal quantum number, n*, and satisfy the Bohr formula

$$E_n = -\frac{Z^2}{n^2} E_0 \qquad (30.1)$$

Using $Z = 1$ and $E_0 = 13.6$ eV,

$$E_n = -\frac{13.6}{n^2} \text{ eV}, \qquad n = 1, 2, 3, \dots$$

This formula is only approximately correct. The more complete relativistic theory of hydrogen developed shortly after Schroedinger's original work includes the small magnetic forces associated with the motion of the charges in the atom. This complete theory leads to energy levels that have a slight dependence on other quantum numbers. The energy levels in multielectron atoms depend even more strongly on other quantum numbers because of the additional forces among the electrons.

TABLE 30.1

Spectroscopic notation for the first seven angular momentum states in atoms. The sequence is alphabetic for $l \geqslant 3$.

$l =$	0	1	2	3	4	5	6
Notation	s	p	d	f	g	h	i

The Angular Momentum Quantum Number, l

In the Bohr model, one quantum number labeled both the angular momentum and the energy of each state. However, in solving Schroedinger's equation, it is found that the magnitude L of the total orbital angular momentum $\mathbf{L}$ is related to a separate quantum number l by,

$$L = \sqrt{l(l+1)}\,\hbar, \quad l = 0, 1, 2, \ldots, n-1 \quad (30.2)$$

Two important results should be noted. First, the angular momentum may be zero for any value of n. A classical picture of an electron with zero angular momentum would be one in which the electron moves back and forth directly through the nucleus! The second point is that for each value of n, l may be any integer up to and including $n - 1$.

A somewhat archaic nomenclature is still used to designate states with certain values of n and l. Table 30.1 shows the letters assigned to states with $l = 0$ through 6. With this notation, the state with $n = 1$, $l = 0$, is said to be the $1s$ state while $n = 2$, $l = 0$ is the $2s$ state and $n = 2$, $l = 1$ is the $2p$ state. Other states are denoted in a similar way.

The z Component of the Angular Momentum

The direction of the angular momentum of the one electron in hydrogen becomes important when a magnetic field is applied to the atom. An electron with an orbital angular momentum $\mathbf{L}$ is effectively a current loop or magnetic dipole. The energy of the loop depends on its orientation relative to the magnetic field $\mathbf{B}$ (Fig. 30.1).

In Section 21.5 on magnetic dipoles, we found that an electron of charge $-e$ and mass m moving in an orbit with angular momentum $\mathbf{L}$ has a magnetic dipole moment which is proportional to $\mathbf{L}$. In vector form, this relationship is

$$\boldsymbol{\mu} = -\frac{e}{2m}\mathbf{L} \quad (30.3)$$

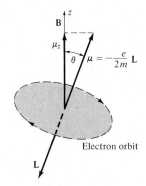

Figure 30.1. An electron with orbital angular momentum $\mathbf{L}$ has a magnetic dipole moment $\boldsymbol{\mu}$ opposite to $\mathbf{L}$ because the electron charge is negative. The energy of the dipole in the field $\mathbf{B}$ is $E = -\mu_z B$.

The minus sign is needed because the electronic charge is negative. The energy of the dipole in a magnetic field is $E = -\mu B \cos\theta$. If $\mathbf{B}$ is in the z direction, this can also be written in terms of the z component of $\boldsymbol{\mu}$ (Fig. 30.1),

$$E = -\mu_z B \quad (30.4)$$

Thus, we expect that the energy of an electron due to a magnetic field can have any value between $-\mu B$ and $+\mu B$.

The electron can change its energy by emitting or absorbing a photon. Since its energy can vary between $-\mu B$ and $+\mu B$, the photons observed might be expected to have a spread of energies; but in experiments only certain photon energies are seen. The analysis of the experiments and Schroedinger's theory both show that L_z, the z component of $\mathbf{L}$, can have only certain values:

$$L_z = m_l \hbar,$$
$$m_l = -l, -l+1, \ldots, 0, \ldots l-1, l \quad (30.5)$$

Thus m_l, the quantum number denoting the z component of $\mathbf{L}$, can have any of $2l + 1$ integer values for a given value of l (Fig. 30.2).

Since L_z can only have certain values, μ_z also has specific allowed values. The resulting energy levels are given (Fig. 30.3) by

$$E_{m_l} = -\mu_z B = -\left(-\frac{e}{2m}\right)m_l \hbar B = m_l \frac{e\hbar}{2m}B \quad (30.6)$$

The following example illustrates the enumeration of states possible for $n = 2$.

Figure 30.2. The allowed orientations of **L** for $l = 2$ correspond to five values of $L_z = m_l h$. Here m_l can have any of the values 0, ± 1, ± 2.

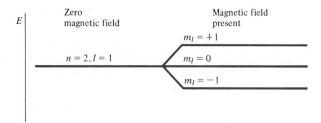

Figure 30.3. In the presence of a magnetic field the $2p$ state is broken up into three energy levels corresponding to the three allowed values of m_l. In the $m_l = +1$ state the magnetic dipole is parallel to the magnetic field; for $m_l = -1$, it is antiparallel, and for $m_l = 0$, it is perpendicular to the field.

Example 30.1

What are the allowed values of l and m_l for states of the hydrogen atom with $n = 2$?

For $n = 2$, the orbital angular momentum number l can be either 0 or 1. For $l = 0$, we can only have $m_l = 0$; for $l = 1$, m_l can be -1, 0, or $+1$. In tabular form,

State (n, l, m_l)	n	l	m_l
$2s$ state $(2, 0, 0)$	2	0	0
$2p$ states $(2, 1, -1)$	2	1	-1
$(2, 1, 0)$	2	1	0
$(2, 1, 1)$	2	1	1

We see that for $n = 2$, there are four possible combinations of the quantum numbers l and m_l. However, we see below that an additional quantum number is needed to fully specify all of the $n = 2$ states.

Spin Angular Momentum | When experiments were performed on atoms in a magnetic field the energy levels were found to be split, but more

Figure 30.4. For electrons, with spin $\frac{1}{2}$, the z component of the spin angular momentum is either $+\hbar/2$ or $-\hbar/2$. **S** is always at an angle with the z axis.

levels were observed than anticipated. Furthermore, even when no external magnetic field was present, it was found that many of the observed spectral lines were, in fact, closely spaced pairs of lines or *doublets*. These energy level splittings led to the conclusion that the electron has not only an orbital angular momentum but also an intrinsic spin angular momentum with an associated magnetic dipole moment. The observed effects were explained in terms of both the orbital and spin magnetic moments.

The magnitude of the spin angular momentum **S** of a particle is found to be (Fig. 30.4)

$$S = \sqrt{s(s + 1)}\,\hbar \qquad (30.7)$$

where s is the *spin quantum number*. For electrons, $s = \frac{1}{2}$, as it is for protons and neutrons; for photons, $s = 1$. These particles are said to have spin one-half and spin one, respectively.

In atomic structure, it is the z component of the spin angular momentum S_z that is important. For a spin one-half particle, S_z is

$$S_z = m_s \hbar, \qquad m_s = \frac{1}{2}, -\frac{1}{2} \qquad (30.8)$$

Despite the fact that **S** never points directly along the z axis, the two states $m_s = \pm\frac{1}{2}$ are often referred to as spin up ($m_s = \frac{1}{2}$) and spin down ($m_s = -\frac{1}{2}$). Thus, for every set of values for n, l, and m_l, there are two allowed values of m_s. This means that the results of Example 30.1 are incomplete. There are two $n = 1$ states and eight, rather than four, $n = 2$ states.

The electron spin accounts for the extra observed spectral lines. The electron has an orbital and a spin magnetic moment, and these two dipoles may have two possible relative orientations corresponding to the two values of S_z. Hence, all the energy levels are split into two, giving rise to the doublet spectral lines observed. This is called the *spin-orbit splitting* or *fine structure*.

Later Developments | In the two decades following Schroedinger's work, theory and experiment were found to be in excellent agreement for hydrogen and for more complex atoms. However, after World War Two, totally new tools were applied to the study of atomic structure. Most notable was the use of short-wavelength electromagnetic radiation (microwaves) employing technology developed in wartime radar research. It was found that the existing theory failed to predict small but significant details of the measurements.

Feynman, Schwinger, Tomonaga, and others showed that agreement with experiment could be restored if several new processes were taken into account. These processes included the emission and reabsorption of photons by electrons in atoms and the spontaneous production of short-lived pairs of electrons and *positrons*. The positron is a particle with the same mass as the electron but with a positive charge $+e$. This new theory is called *quantum electrodynamics* and is sufficient to explain all features of atomic structure that do not depend on the internal structure of the atomic nucleus. For most applications of quantum mechanics to problems in chemistry and biochemistry, quantum electrodynamic effects are negligibly small. Consequently, we do not consider them further.

In summary, the complete solution to the Schroedinger equation for hydrogen shows that neglecting spin, the energy levels are given correctly by the Bohr model if no external fields are present. The multiplicity of the states for a given energy level can be found from the splitting of a magnetic field. In the next section, we examine the wave functions corresponding to different sets of quantum numbers.

30.3 | HYDROGEN ATOM WAVE FUNCTIONS

As we mentioned earlier, the wave functions used to describe the states available to electrons are not measurable quantities. However, these wave functions are used to calculate measurable quantities such as the energy and angular momentum of a state. The wave function, denoted by ψ, is the analog of the displacement in the standing waves on strings and in air columns. In quantum mechanics the wave function is most directly related to the probability of finding an electron at a certain position when it is in a given state.

The square of the absolute value of the wave function, denoted by $|\psi|^2$, is a measure of the probability that an electron is at a given point in space.* For example, Fig. 30.5a shows a graph of ψ and $|\psi|^2$ for the $n = 1$, $l = 0$ state of hydrogen. We see that there is a high probability that the electron in this state will be found at small values of r or near the nucleus. All $l = 0$ wave functions are spherically symmetric and are largest near $r = 0$.

Notice that the wave function can be positive or negative but $|\psi|^2$ is always positive or zero. The shaded pictures can be thought of as a superposition of a large number of photographs of the electron position. Thus the electron is most likely to be found in the darker areas.

While there is a good probability that an $l = 0$ electron is near $r = 0$, the probability that it may be found in the nucleus is small. This is because the nuclear radius is about 10,000 times smaller than the Bohr radius. On the graphs of Fig. 30.5, the nuclear radius falls within the width of the line representing the vertical axis.

The wave functions for $l \neq 0$ are more complex (Fig. 30.6). They are not spherically symmetric and go to zero near $r = 0$. Thus, an electron in a 2p state is much less likely to be found near the nucleus than an electron in the 2s state.

The wave functions we have described are those that are most useful for atomic physics, where the magnetic interactions between the spin and orbital magnetic moments play a role. However, when atoms bind to one another, it is more convenient to use a different set of wave functions. In Fig. 30.7, we show representations of the three 2p states for hydrogen which are used in discussions of molecular binding. The wave functions labeled p_x, p_y, and p_z are superpositions of the three 2p states of Fig. 30.6. Both sets of wave functions are solutions of Schroedinger's

---

*Sometimes ψ contains $\sqrt{-1}$ and is therefore a complex number. However $|\psi|^2$ is a real number as we would expect for a probability. If ψ is real, then $|\psi|^2 = \psi^2$.

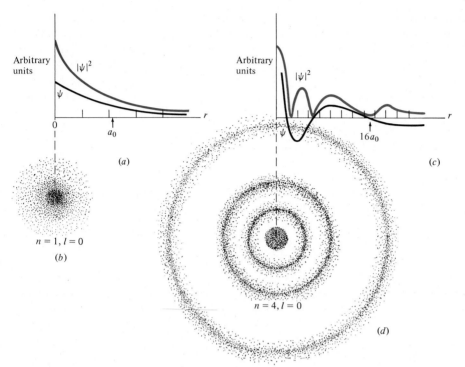

Figure 30.5. (a) A graph of ψ and $|\psi|^2$ (shown colored) versus r for the $n = 1$, $l = 0$ state of hydrogen. (b) A representation of $|\psi|^2$ for the same state. (c) Graphs of ψ and $|\psi|^2$ (shown colored) for the $n = 4$, $l = 0$ state of hydrogen. (d) The representation of $|\psi|^2$ for this state. a_0 is the radius of the first Bohr orbit.

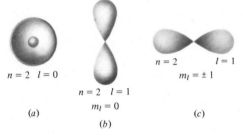

Figure 30.6. The representations of $|\psi|^2$ for the $n = 2$ states. (a) The $l = 0$ cloud is darkest at $r = 0$. (b) For $l = 1$ and $m_l = 0$ the representation of $|\psi|^2$ appears as two droplets above and below the origin. For this state $|\psi|^2 = 0$ at $r = 0$. (c) For $l = 1$, $m_l = \pm 1$, $|\psi|^2$ is doughnut shaped (here it is viewed in cross section from the side). Again $|\psi|^2 = 0$ at $r = 0$ for this state. While the pictures for $m_l = +1$ and $m_l = -1$ appear the same, the angular momentum is opposite. Thus one may think of the electron in the doughnut as circulating one way for $m_l = 1$ and oppositely for $m_l = -1$.

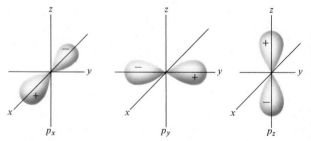

Figure 30.7. The three $2p$ wave functions most often used in chemistry. These wave functions are positive on one side of the origin and negative on the other side.

30.4 | THE PAULI EXCLUSION PRINCIPLE

We noted that all elementary particles have an intrinsic spin angular momentum. The spin plays a crucial role in atomic and molecular structure and in the behavior of collections of particles. For example, the two isotopes of helium, ^{3}He, with two protons and a neutron in the nucleus, and ^{4}He, with two protons

equation and the choice of which set to use depends on whether one is interested in magnetic effects or molecular structure. We use the states of Fig. 30.7 in discussing molecular binding in the next chapter.

and two neutrons, have very different properties at low temperatures because the net spins of the two nuclei are different. In ^{4}He the two protons have equal but oppositely directed spins, as do the two neutrons. Thus the net spin angular momentum is zero. This substance, which is a liquid at atmospheric pressure and at the lowest temperatures so far achieved (about 10^{-6} K), becomes a *superfluid* below about 2 K. Among other extraordinary characteristics, this superfluid will flow through openings that no other liquid or gas will. ^{3}He displays equally remarkable but very different behavior below about 5×10^{-3} K. Here the net nuclear spin is one-half due to the single unpaired neutron.

Another example of the effects of spin is in metals. At low temperatures the free electrons, each having a spin of one-half, match up to form pairs with a net spin of zero. When this happens the metal becomes a superconductor; its electrical resistance becomes zero!

All these characteristics depend on two observations:

1 *The Pauli exclusion principle.* Two or more indistinguishable or identical particles with spin one-half cannot be in the same quantum state. This means for example that in atoms with many electrons, no more than one electron can be in a level denoted by a given set of quantum numbers, n, l, m_l, and m_s.

2 *Indistinguishable or identical particles with a spin of zero or one* can be put in any quantum state whether it is occupied or not.

The word "indistinguishable" in these rules is important because in classical physics, individual particles are usually considered as being identifiable and distinguishable. That is, if we know the position and momentum of a particle at a certain moment, then using Newton's laws we can, in principle, follow the motion of that particle as long as we wish. In quantum systems, the uncertainty principle precludes an accurate determination of the position and momentum, so at various times we cannot tell if we are seeing the original particle or another that looks just like it. In the latter case the particles must be considered as indistinguishable. Atoms provide many illustrations of the effects of the Pauli exclusion principle.

Example 30.2

How many electrons can be in the $n = 2$ hydrogenlike

states of an atom?

In Example 30.1, we found that there were four possible values of l and m_l for $n = 2$. However, for each of these states, m_s can be $+\frac{1}{2}$ and $-\frac{1}{2}$. There are then a total of eight $n = 2$ states. The Pauli exclusion principle states that only one electron can be in each state, so the $n = 2$ states can accommodate eight electrons.

30.5 | ATOMIC STRUCTURE AND THE PERIODIC TABLE

We now have sufficient information to gain some insight into atoms with more than one electron. We will see that many of the physical and chemical properties of atoms can be explained by the *atomic shell model,* which is based on the idea that the electrons in atoms are in hydrogenlike states.

In the ground state of an atom the electrons occupy the lowest states allowed by the Pauli exclusion principle. When all the states corresponding to the same or nearly the same energy have been filled, an *electron shell* is said to be completed or *closed.* Atoms with all their electrons in closed shells are similar to each other in their physical and chemical properties and are very stable. For example, the minimum energy needed to remove one electron from a neutral ground state atom, the *ionization energy,* is very large in these closed shell atoms (Fig. 30.8). The regularities associated with successive shell closures are responsible for similarities among the elements in the same column of the periodic table (Appendix D.3).

We now consider individual atoms in more detail, starting with $Z = 2$.

Helium $(Z = 2)$ | Helium has two electrons and

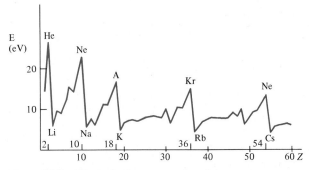

Figure 30.8. The ionization energy, the minimum energy required to remove one electron from a neutral atom in its ground state, versus Z for atoms up to $Z = 60$. Closed shell inert gases have $Z = 2$, 10, 18, 36, and 54.

537

a nucleus with a charge 2e. It is not presently possible to find exact solutions to the Schroedinger equation when there are three or more objects present, although very accurate approximate solutions to the helium atom problem can be obtained. These solutions show, as one might expect, that in the helium ground state the two electrons are both in $n = 1$, $l = 0$ states. In accordance with the Pauli principle, one electron has its spin directed up ($m_s = \frac{1}{2}$) and the other down ($m_s = -\frac{1}{2}$).

The fact that two electrons are present affects the energies of their states. Each electron has a probability distribution $|\psi|^2$ similar to that of Fig. 30.5a, so the electrons partially shield each other from the nucleus. Roughly half the time, each electron "sees" an attractive nuclear charge $Z = 2$, and half the time, it is shielded and sees a net charge $Z = 1$. Thus, we might suppose that in the Bohr formula we should use an effective nuclear charge $Z_{eff} = \frac{1}{2}(2 + 1) = 1.5$. However, the mutual repulsion of the electrons pushes them apart and reduces the shielding effect, so that Z_{eff} increases to approximately 1.7. Hence, when they are in $n = 1$ orbitals, using Eq. 30.1 with $Z = Z_{eff}$, the two electrons have a total energy

$$E_2 = -2\frac{Z_{eff}^2}{n^2}E_0 = -2\frac{(1.7)^2}{1^2}(13.6\text{ eV})$$

$$= -79\text{ eV}$$

If one electron is removed, then the remaining electron experiences the full nuclear charge $Z = 2$. The energy of this electron is then

$$E_1 = -\frac{2^2}{1^2}(13.6\text{ eV}) = -54\text{ eV}$$

The difference $E_1 - E_2 = -54 - (-79) = 25$ eV is the energy needed to remove one electron from the helium atom. This ionization energy is greater than the 13.6 eV needed to ionize hydrogen. We see then that the two electrons in helium are in the two 1s states, and despite their mutual repulsion, they are each more tightly bound than is the one electron in hydrogen. Helium is the lightest closed shell atom, since it has the $n = 1$ levels filled; chemically, it is the lightest inert or *noble* gas. In lithium, with $Z = 3$, the third electron will have to go into the higher energy $n = 2$ orbital.

Lithium—Neon($Z = 3$-10) | From our dis-

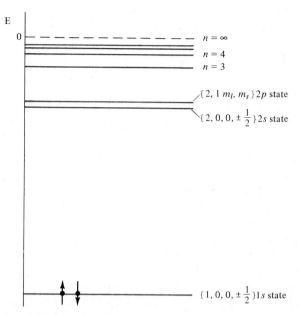

Figure 30.9. A schematic energy level diagram for the neutral helium atom. The two electrons have opposite spins and are both in 1s states. The states are labeled by $\{n, l, m_l, m_s\}$.

cussion of the hydrogen energy states, the $n = 2$ level contains a total of eight states with different values of l, m_l, and m_s. The $n = 2$ levels are significantly higher in energy than the $n = 1$ levels but are still well below the $n = 3$ levels (Fig. 30.9). In lithium ($Z = 3$) two electrons occupy the $n = 1$ states and one electron goes into a $n = 2$ level. Because this third electron is further from the nucleus on the average, it is shielded from the nuclear charge by the two inner electrons. This shielding is less complete in the $n = 2$, $l = 0$ state than in the $n = 2$, $l = 1$ state, since in the $l = 0$ state the electron has a higher probability of being close to the nucleus. Accordingly the energy is somewhat lower in the $l = 0$ state, and that level will be filled in the ground state of the atom. We find the effective nuclear charge for the $n = 2$, $l = 0$ electron in the next example.

Example 30.3

From the Bohr model, an $n = 2$ electron in the presence of an effective nuclear charge $Z_{eff} = 1$ will have an energy of $E_2 = -13.6/(2)^2$ eV $= -3.4$ eV. In lithium the energy of this electron is -5.39 eV. What is the effective nuclear charge for this state?

From the Bohr model with $Z = Z_{eff}$,

$$E_2 = -5.39\text{ eV} = -13.6\frac{(Z_{eff})^2}{(2)^2}$$

Solving, we find $Z_{eff} = 1.25$. Thus, because the electron can be close to the nucleus, it sees an effective nuclear charge which is 25 percent greater than that of a completely shielded electron.

For $Z = 4$, beryllium, the fourth electron can also be in an $n = 2$, $l = 0$ state with its spin opposite to that of the third electron. These two electrons shield each other somewhat. However, the nuclear charge is now 4 and the inner electrons only shield two of the protons. Thus the two outer electrons see an effective nuclear charge of between 1 and 2. The ionization energy is found to be 9.32 eV for beryllium.

With $Z = 5$, boron, the $n = 2$, $l = 1$ levels begin to fill. There are six $n = 2$, $l = 1$ states and electrons fill these states in going from boron to carbon ($Z = 6$), nitrogen ($Z = 7$), oxygen ($Z = 8$), fluorine ($Z = 9$), and finally neon ($Z = 10$). With neon, the $n = 2$ levels are filled, and the shell is completed; it is an inert gas. We note that for neon, the six outer electrons are competing with one another to share the portion of the nuclear charge that is not shielded by the four inner electrons. Because each of these six electrons can sample a nuclear charge as large as six part of the time, their effective nuclear charge is relatively large. For neon the ionization energy has the large value of 21.56 eV. This corresponds to a Z_{eff} of about 2.5.

The competition between the attractive effects of an effective nuclear charge and the repulsion of the electrons from each other is very sensitive to the number of electrons present. For example, fluorine, with five $n = 2$, $l = 1$ electrons, is one electron short of having a filled shell. The effect of adding an additional electron to fluorine to obtain the negative ion F^- is to cause six outer electrons to compete for five unshielded nuclear charges. The energy gained in this sharing is sufficient to offset the repulsive effects of the added electron, and F^- is quite stable. Its ionization energy is 4.2 eV. Fluorine is said to have a high *electron affinity*.

In a crude model of the molecule NaF, the fluorine atom takes one of the electrons from sodium. The molecule is then held together by the attractive electrical force between Na^+ and F^-. This is called *ionic binding*. Fluorine is the lightest of the *halogens*, which include chlorine, bromine, iodine, and astatine, all of which are one electron short of having a closed shell. All are highly reactive as they try to gain an electron.

We can also see how carbon ($Z = 6$), a constituent of so many organic compounds, combines with other elements. The $n = 2$ level of carbon has two electrons in the $l = 0$ states and two in the $l = 1$ states. Four more are needed to completely fill the $n = 2$ shell. Since atoms are most stable when they have closed shells, carbon will usually try to "share" four electrons from other elements. This is called *covalent binding*.

$Z = 11\text{-}18$ | As Z increases from 11, the $3s$ and then the $3p$ levels fill just as the $2s$ and $2p$ states were filled. The chemical characteristics of these elements are very similar to the corresponding elements with Z between 3 and 10. At first glance, we would expect that the next closed shell inert gas would appear at $Z = 28$, when all the $n = 3$ levels are filled (Table 30.2). However, the electrons in the $l = 2$ states spend much less time near the nucleus than those in $l = 1$ or 0 states and are therefore substantially higher in energy. In fact, the $4s$ levels ($n = 4$, $l = 0$) are also below the $3d$ levels ($n = 3$, $l = 2$). Accordingly, the $3s$ and $3p$ states form a complete shell by themselves. Argon ($Z = 18$) has filled $3s$ and $3d$ states and is therefore an inert gas. Similarly, chlorine ($Z = 17$) is a halogen like fluorine, and potassium ($Z = 19$) is an *alkali metal* similar to lithium and sodium, which also have one electron outside a closed shell.

$Z \geq 19$ | Above $Z = 18$ the order in which levels are filled becomes complicated. The zero angular momentum states in a shell are always filled first

TABLE 30.2

Atomic states for $n = 3$

State	l	m_l	m_s	
$3s$	0	0	$\pm\frac{1}{2}$	} 2 states
$3p$	1	-1	$\pm\frac{1}{2}$	
		0	$\pm\frac{1}{2}$	} 6 states
		$+1$	$\pm\frac{1}{2}$	
$3d$	2	-2	$\pm\frac{1}{2}$	
		-1	$\pm\frac{1}{2}$	
		0	$\pm\frac{1}{2}$	} 10 states
		$+1$	$\pm\frac{1}{2}$	
		$+2$	$\pm\frac{1}{2}$	

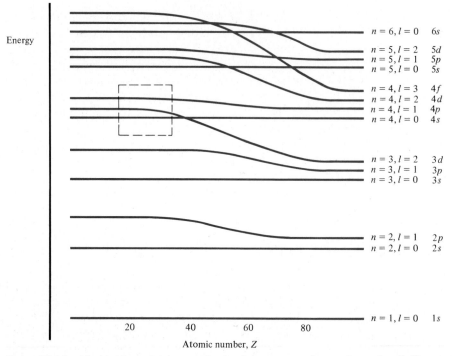

Energy

$n = 6, l = 0$ $6s$
$n = 5, l = 2$ $5d$
$n = 5, l = 1$ $5p$
$n = 5, l = 0$ $5s$
$n = 4, l = 3$ $4f$
$n = 4, l = 2$ $4d$
$n = 4, l = 1$ $4p$
$n = 4, l = 0$ $4s$
$n = 3, l = 2$ $3d$
$n = 3, l = 1$ $3p$
$n = 3, l = 0$ $3s$
$n = 2, l = 1$ $2p$
$n = 2, l = 0$ $2s$
$n = 1, l = 0$ $1s$

20   40   60   80

Atomic number, Z

Figure 30.10.   The positions of the energy levels of atoms versus the atomic number Z. The dotted square indicates the levels filled for $Z = 19$ through 36. The shells contain different levels in different parts of the periodic table. For example for $Z = 30$ the $3s$ and $3p$ levels comprise a shell but for $Z = 70$ the $3s$, $3p$, and $3d$ levels form a shell.

because electrons in these states penetrate closest to the nucleus. This effect is so strong that for $Z = 19$ the $n = 4$, $l = 0$ states begin to fill even though the $n = 3$, $l = 2$ states are empty. Thus in Fig. 30.10 we see that for intermediate values of Z, closed shells are not always associated with a single value of n. The one-to-one correspondence between shells and the quantum number n reappears for large values of Z.

Two general features of atomic structure are important. First, there is no obvious reason that elements with vary large values of Z could not exist. However naturally occurring or artificially produced elements have only been found for values of Z up to slightly over 100. Very heavy atoms do not exist because their nuclei tend to break up.

A second notable feature of atoms is the remarkable similarity in their radii. The alkali atoms, with a single, weakly bound s electron, have the largest radii. But the largest alkali atom, cesium, has a volume that is only about twice that of helium. This happens because, in heavy atoms, the innermost electrons see a very large Z_{eff} and consequently are much closer to the nucleus than in light atoms. This permits the

outer electrons to be closer to the nucleus as well, so that the average atomic radius does not increase much with Z.

30.6 | ATOMIC EMISSION AND ABSORPTION SPECTRA

We described earlier how the electrons in an atom can change their energies by moving from one state to another. This process is accompanied by the emission or absorption of photons.

The photon has a spin of 1. When an atom absorbs or emits a photon, the *total* angular momentum of the atom and photon must be constant. Thus, because the photon carries angular momentum, the electron that emits or absorbs that photon must change its angular momentum when making a transition.

Usually the electrons *change their orbital angular momentum quantum number l by* 1 when making transitions. For example, suppose a hydrogen atom in its ground state, which is a $l = 0$ state, absorbs a photon. It can only do so if its final orbital angular momentum quantum number is $l = 1$. During a

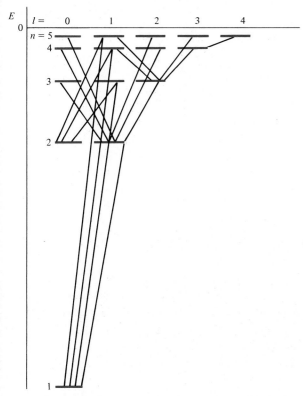

Figure 30.11. Allowed transitions for the electron in hydrogen. These are accompanied by the emission or absorption of one photon. The $n = 2$, $l = 0$ state is metastable because the electron cannot lower its energy by direct one photon emission.

downward transition the electron must also change its angular momentum by 1. Figure 30.11 shows these *allowed* transitions among the lowest five energy levels of hydrogen.

When atoms are excited, the electrons usually return to their ground states via allowed transitions. These excited states are very short-lived, lasting no more than about 10^{-8} seconds. However, when the electrons cannot return to their ground states via allowed transitions, they may do so by the emission of two photons. This is a relatively slow process, and such *metastable* states may last for times of the order of a second.

SUMMARY

The structure of atoms with more than one electron can be largely understood in terms of electronic energy levels similar to those of the hydrogen atom. The positions of the electronic energy levels are deter-

mined by the attractive electric force due to the nucleus and the repulsive electric forces among the electrons.

The Pauli exclusion principle requires that only one electron can be in a given state. These states are labeled by quantum numbers that indicate the energy, orbital angular momentum, and spin angular momentum of an electron in that state. In the ground states of atoms, electrons occupy the lowest energy states available consistent with the exclusion principle.

The square of the wave function is an indication of the position probability of an electron in a given state. This probability distribution can be used to understand the screening of the nucleus by other electrons and the relatively low energy of the zero angular momentum states.

The observed atomic spectra of complex atoms depend not only on the positions of the energy levels but also on angular momentum conservation.

Checklist

Define or explain:

wave function	closed electron shell
principle quantum number	ionization energy
quantum number l	effective nuclear charge
quantum number m_l	electron affinity
spin angular momentum	ionic binding
$|\psi|^2$	halogen
s states, p states	covalent binding
Pauli exclusion principle	alkali metal
	allowed transition
	metastable states

EXERCISES

Section 30.1 | The Outline of Quantum Mechanics, and Section 30.2 | Hydrogen Atom Quantum Numbers

30-1   An electron in hydrogen is in an $n = 4$ state. (a) What are the possible values of the orbital angular momentum quantum number l? (b) What is the spectroscopic notation for each state? (c) What is the maximum magnitude of the z component of the orbital angular momentum for each of these values of l?

30-2   Sketch a figure similar to Fig. 30.2 showing the possible orientations of the z component of the angular momentum for an $l = 3$ state. Label the values of m_l for each of the allowed orientations of **L**.

30-3 (a) What are the possible energy states of an electron in an $l = 2$ state of hydrogen when a magnetic field B is applied? (Neglect all effects of spin in this exercise.) (b) If the magnetic field has a magnitude of 10 T, what is the energy difference between the $m_l = 1$ and $m_l = 0$ states? (c) If an electron makes a transition from the $m_l = 0$ to the $m_l = 1$ state is a photon emitted or absorbed?

30-4 A spinning charge in a magnetic field is effectively a current loop. Quantum mechanics shows that an electron in a magnetic field B with a z component of spin S_z has an energy $\mathcal{U} = 2\mu_B B m_s$, where $\mu_B = e\hbar/2m$ is called the Bohr magneton; $\mu_B = 0.93 \times 10^{-23} \text{ A m}^2$. What is the energy difference between the spin up and spin down states of an electron in a magnetic field of magnitude 1 T?

30-5 How many $n = 4$ states are there in hydrogen? (Include spin effects.)

30-6 How many states in hydrogen have $n = 3$ and (a) $l = 0$; (b) $l = 1$; (c) $l = 2$? (Include spin effects.)

Section 30.5 | Atomic Structure and the Periodic Table

30-7 Why is potassium chemically similar to sodium and lithium? ($Z = 19$ for potassium.)

30-8 In $Z = 56$, barium, the $n = 6$, $l = 0$ state is filled and in succeeding atoms the electrons start filling the $n = 4$, $l = 3$ subshell. How many states are available in this subshell? (These are the rare earth elements.)

30-9 Using the ideas of screening described for the helium atom, estimate the energy required to remove the second electron from lithium, leaving Li^{++}. Compare your estimate of Z_{eff} with $Z = 3$, the nuclear charge of lithium.

30-10 One of the two electrons in a helium atom is excited to an $n = 2$ state while the other electron remains in the $n = 1$ state. Will the $n = 2$ electron have the same energy in the $l = 1$ and $l = 0$ states? If not, which state has the lower energy and why?

30-11 The alkali metals are strongly reactive; they readily interact with other atoms, particularly those that attract electrons. Why are these metals so chemically active?

30-12 In the water molecule, H_2O, the oxygen atom shares the electrons of the hydrogen atoms. Why is this such a stable compound?

30-13 Hydrogen combines with the halogens to form such acids as HF, HCl, HBr, and HI. Why do these compounds form so readily?

30-14 The Nobel prize work of Tomonaga, Feynman, and Schwinger on quantum electrodynamics was prompted, in part, by some unexplained features of atomic spectra. Techniques developed in the late 1940s enabled experimenters to observe that the splitting of energy levels in a magnetic field B due to the electron spin is not $\Delta\mathcal{U} = 2\mu_B B$ but is instead $\Delta\mathcal{U} = 2\mu_B(1 + \alpha/2\pi)B$. ($\alpha = e^2/\hbar c = 1/137$ is the fine structure constant. $\mu_B = e\hbar/2m = 0.93 \times 10^{-23} \text{ A m}^2$. See Exercise 30-4.) (a) In a magnetic field of 10 T, what is the magnitude of the discrepancy in the spectrum? (b) What is the percentage of the shift from the value $\Delta\mathcal{U} = 2\mu_B B$?

30-15 What are the quantum numbers for the outer electron in the ground state of sodium ($Z = 11$)?

30-16 Estimate the energy of a $1s$ electron in a lead atom ($Z = 82$).

Section 30.6 | Atomic Emission and Absorption Spectra

30-17 The effective nuclear charge for a $3s$ ($n = 3$, $l = 0$) electron is $Z_{eff} = 3.1$, and for the same electron in an excited state $n = 3$, $l = 1$, it is $Z_{eff} = 1.5$. What is the energy of the photon emitted when the electron returns to the $3s$ state?

30-18 When one of the innermost electrons of an atom is captured by the nucleus, an x-ray is emitted. Further x-rays are emitted as electrons from higher levels drop down to fill the vacant level or hole left by the captured electron. In copper, $Z = 29$. (a) Estimate Z_{eff} for the $2p$ and $1s$ states in copper. Explain your reasoning. (b) Estimate the wavelength of the x-ray emitted when an electron drops from the $2p$ state to the $1s$ state.

PROBLEMS

30-19 The energy necessary to remove a $1s$ electron in nitrogen ($Z = 7$) is 540 eV. (a) Using

$Z_{eff} = 6.5$, estimate the energy needed to remove a
1s electron. (b) Why does this calculation overestimate the energy required? (c) Might the presence
of the 2s electrons affect the energy of the 1s state?
Why?

30-20   Beryllium ($Z = 4$) has two 1s and two 2s
electrons in its ground state. The ionization energy
for the first 2s electron is 9.32 eV, and for the second it is 18.12 eV. (a) What is Z_{eff} for the single 2s
electron of Be⁺? (b) What is Z_{eff} for the two 2s
electrons in the neutral ground state of Be?

30-21   In singly ionized neon Ne⁺ ($Z = 10$), the
five outer 2p electrons see an effective nuclear
charge as high as 6 and as low as 2, depending on
the shielding due to the other four 2p electrons. By
a simple averaging process we find $Z_{eff} =$
$(6 + 5 + 4 + 3 + 2)/5 = 4$. (a) What is the total
energy of the five outer electrons of Ne⁺ using this
value of Z_{eff}? (b) The measured ionization energy
of neon is 21.56 eV. Assuming that the Z_{eff} of part
(a) is correct, compute Z_{eff} for the six 2p electrons
for neutral neon in its ground state.

30-22   Potassium ($Z = 19$) has one outer 4s electron and an ionization energy of 4.3 eV. (a) What is
Z_{eff} for this electron in potassium? (b) Calcium
($Z = 20$) has two outer 4s electrons. A rough estimate for the effective nuclear charge for the single
4s electron of Ca⁺ would be $Z_{eff} + 1$, where Z_{eff} is
the result from (a). The measured ionization energy
of calcium is 6.09 eV. Estimate Z_{eff} for the two 4s
electrons of neutral calcium.

SUPPLEMENTARY TOPICS

30.7 | MASERS AND LASERS

The ability to produce intense, coherent radio waves
at nearly a single frequency has made possible much
of modern communication. Masers and lasers now
produce very intense coherent monochromatic microwave and visible radiation in a narrow beam. The
word maser stands for *Microwave Amplification by
Stimulated Emission of Radiation*. The laser, which
relies on the same principles, produces visible light
rather than microwaves. In this section, we consider
some general features common to all masers and
lasers and then describe the helium-neon gas laser.

Two important conditions must occur in any maser

Figure 30.12.   Two energy levels of an atom. Under normal
conditions, a system of atoms will have more electrons in the
lower energy state, E_1. Heating will at best result in the nearly
equal population of the two states.

or laser. One is called *population inversion* and the
other is called *stimulated emission*. We describe these
ideas in terms of atoms that have two energy levels E_1
and E_2 (Fig. 30.12).

Under normal conditions most of the atoms will be
in the lower energy state, with a few in the upper
level. Atoms that are in the excited state spontaneously emit photons and drop into the lower state. This
process is random, and the emitted photons are not
coherent, that is, they are not in phase with one
another. In a laser, another type of emission process,
called stimulated emission, plays an important role. If
one atom emits a photon of energy $E_2 - E_1$, that
photon may collide with a second atom that is in the
state E_2. When this happens the second atom is stimulated to also emit a photon of the same energy.
These two photons are coherent with each other
(Fig. 30.13).

Stimulated emission may occur under normal conditions but is a rather minor effect because so few
atoms are in the higher energy state. However, in
lasers the upper energy level is caused to be overpopulated, with more atoms in the upper than the lower
state, and *population inversion* is said to occur. Under
these conditions the probability that stimulated emis-

Figure 30.13.   A photon with energy $hf = E_2 - E_1$ is incident
on an atom in the state E_2. The incident photon stimulates the
atom to drop into the lower energy state E_1, emitting a second
photon with the same energy and coherent with the first.

543

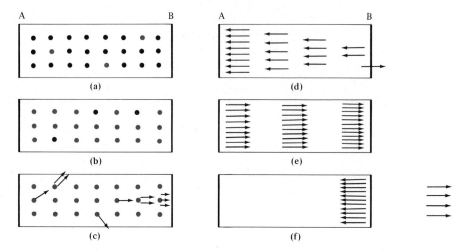

• atom in lower state
• atom in higher state

Figure 30.14. A schematic diagram of a laser. A material is placed between two parallel walls A and B. Wall A is a perfect reflector and B is 99% reflecting. (a) The material is in its normal state with most atoms, black circles, in a lower energy state and some, colored circles, are in a higher energy state. (b) Population inversion is achieved. This is done by feeding energy into the system and is often called *pumping*. (c) Photons are emitted by random atoms. Some photons are lost through the sides, but one is shown initiating stimulated emission perpendicular to the ends. (d) After partial reflection the buildup of coherent photons continues. (e) The buildup continues after total reflection from wall A and in (f) 1% of the very large number of photons present escape through wall B. The photons inside continue to sweep back and forth until the population becomes normal and all the photons escape.

sion occurs is large, and many coherent photons can be obtained.

We can now see how intense coherent radiation may be produced. If a system of atoms or molecules can be forced into having the higher of two energy states overpopulated, the emission of a single photon during a transition to the lower state by one atom will trigger the emission of photons by many more atoms. In practice, this is achieved by an arrangement like that in Fig. 30.14.

The laser diagrammed is said to be a two-state or pulsed laser. After the process of stimulated emission is completed and most of the atoms in the material have been deexcited, energy must again be fed into the system to invert the population. This energy is usually supplied by electromagnetic radiation.

Lasers that produce continuous radiation rather than a pulsed beam require the presence of at least three, rather than two, energy states. In a continuous laser, atoms are continuously pumped from a low energy state to a higher one. These excited atoms then do not return to the initial state, but instead fall into and overpopulate a third state of intermediate

energy. The laser action then takes place between this third level and some lower state.

One continuous laser, producing light at 632.8 nm, contains a mixture of helium and neon gases. These two atoms have filled $n = 1$ and $n = 2$ electronic shells, respectively. When these atoms are excited, the electrons must return to a $1s$ state in helium and to a $2p$ state in neon to reestablish the ground state (Fig. 30.15).

The $2s$ ($n = 2$, $l = 0$) level in helium is 20.61 eV above the ground state and is lower in energy than the $2p$ ($n = 2$, $l = 1$) state. When an electric discharge occurs in the gas, electrons that end up in the $2s$ state in helium cannot return to the $1s$ ground state because the orbital angular momentum of the two states is the same; the $2s$ state is metastable. However, the $4s$ state of neon happens to be 20.66 eV above its ground state. Thus one way that the electrons in helium can return to their ground states from the $2s$ state is by colliding with a neon atom and exciting a neon electron from the $2p$ state into the $4s$ state. The slight energy difference is made up by the thermal kinetic energy of the helium atom. In effect, the con-

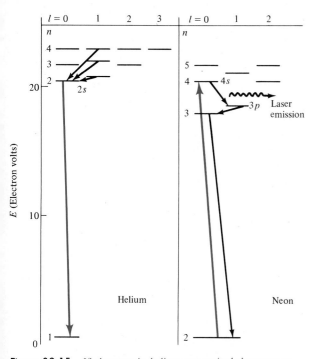

Figure 30.15. If electrons in helium are excited they may return to the ground state emitting photons in the process. However, many electrons end up in the metastable 2s state of helium. When such a helium atom collides with a neon atom, the electron in helium drops into the ground state (colored line) and an electron in neon is raised into its 4s state (also shown colored). Laser action occurs as the electron in neon drops into the 3p state.

tinuous fund of 2s electrons in helium is pumping electrons into the 4s state of neon. The laser action then occurs when the electrons of the 4s state of neon falls into the 3p state, emitting photons with wavelength 632.8 nm. Other coherent transitions can also occur, but none are in the visible spectrum.

A variety of lasers and masers are now available which produce radiation in large portions of the visible, infrared, and microwave regions. Although not all wavelengths in these regions are available, some devices are tunable; that is, their output wavelengths may be changed. Masers and lasers have already found a wide range of applications and more are being developed.

In surveying and construction work the intense narrow laser beam can be used as a reference for accurate alignment. Holography is almost totally reliant on the use of lasers. In medicine, the laser is used for repair of retinal damage and as a surgical tool.

Laser light scattering is an excellent tool for studying molecular structure and motion.

The same characteristics that make lasers a useful tool also pose hazards. While the total energy output of a laser is sometimes small, the rate at which it is released and the narrowness of the beam often result in extremely intense radiation. Available pulsed lasers can have power outputs of more than 10^{13} W with a beam area of less than 1 mm^2 = 10^{-6} m^2. However, the pulse may be as brief as about 10^{-12} seconds. Laser beams can vaporize metals and may cause severe damage to biological systems and tissue. For example the retina is a strong absorber of light. Thus, direct visual observation of even a relatively weak laser beam can cause severe retinal damage.

EXERCISES ON SUPPLEMENTARY TOPICS
Section 30.7 | Masers and Lasers

30-23 The angular spread of a particular laser beam is 10^{-5} rad. What is the diameter of the spot formed on the moon's surface if the laser is directed towards the moon from the earth? (The earth to moon distance is 3.8×10^5 km.)

30-24 (a) If a laser emits 10 J of energy in a pulse lasting 5×10^{-11} s, what power is emitted? (b) What is the intensity of the beam if it is 2×10^{-6} m^2 in area?

30-25 What is the length of a laser pulse in a vacuum if it is emitted in 10^{-11} s?

30-26 A ruby laser emits light at 693.4 nm. If the energy released in each 10^{-11} second pulse is 0.1 J, how many photons are in a pulse?

Additional Reading

R. Riuker, The Safe Use of Lasers, *Physics Teacher,* vol. 11, 1973, p. 455.

H. Weichel, W.A. Danne, and L.S. Pedroth, Laser Safety in the Laboratory, *American Journal of Physics,* vol. 42, 1974, p. 1006.

J.M. Coakley, Probability of Laser Injury to the Eye— Further Considerations, *Physics Teacher,* vol. 13, 1975, p. 388.

Scientific American articles:

Leslie Holliday, Early View on Forces Between Atoms, May 1970, p. 116.

H.R. Crane, The g Factor of the Electron, January 1968, p. 77.

Vernon W. Hughes, The Muonium Atom, April 1966, p. 93.

O.R. Frisch, Molecular Beams, May 1965, p. 58.

William C. Livingston, Magnetic Fields on the Quiet Sun, November 1966, p. 54.

J.P. Gordon, The Maser, December 1958, p. 42.

Arnold L. Bloom, Optical Pumping, October 1960, p. 72.

Arthur L. Schawlow, Optical Masers, June 1961, p. 52.

Arthur L. Schawlow, Advances in Optical Masers, July 1963, p. 34.

S.E. Miller, Communication by Laser, January 1966, p. 19.

George C. Pimentel, Chemical Lasers, April 1966, p. 32.

Alexander Lempick and Harold Samelson, Liquid Lasers, June 1967, p. 80.

Donald F. Nelson, The Modulation of Laser Light, June 1968, p. 17.

C.K.N. Patel, High Power Carbon Dioxide Lasers, August 1968, p. 22.

Arthur L. Schawlow, Laser Light, September 1968, p. 120.

Donald R. Herriott, Applications of Laser Light, September 1968, p. 141.

Peter Sorokin, Organic Lasers, February 1969, p. 30.

M.W. Berus and D.E. Rounds, Cell Surgery by Laser, February 1970, p. 98.

M.S. Field and V.S. Letokhov, Laser Spectroscopy, December 1970, p. 69.

M.B. Parish and I. Hayashi, A New Class of Diode Lasers, July 1971, p. 32.

Richard N. Zare, Laser Separation of Isotopes, February 1977, p. 86.

CHAPTER THIRTY-ONE

THE STRUCTURE OF MATTER

Molecules and bulk matter are aggregates of atoms held together by electrical forces. If the electrons and nuclei of a molecule have a lower energy than that of the separate neutral atoms, the molecule will be stable. Similarly in metals and semiconductors, the energy of the aggregate is less than that of its isolated constituents. The problem of understanding the structure of matter is one of finding the arrangement that has the lowest energy or greatest binding energy.

Fortunately, it is not necessary to solve Schroedinger's equation for all the electrons and nuclei of a system, because this can only be done approximately and is generally difficult. Instead, we use simplified models based on our knowledge of atoms which are very successful in accounting for the structure and properties of molecules and bulk materials. There are three models of primary interest to us here, which describe *ionic, covalent,* and *metallic* binding respectively. Weaker bonds such as *van der Waal's* and *hydrogen* bonds are also important where the stronger bonding mechanisms are not effective.

31. 1 | IONIC BINDING

When some molecules and crystals are formed, one or more electrons from one atom are completely transferred to another atom. The atom losing the electrons acquires a net positive charge, and the atom accepting these electrons becomes negatively charged. These two ions are then held together by the electric forces between them. This *ionic binding* usually involves one atom with one or more loosely bound electrons and a second with a nearly full outer shell. For example, the alkali metals, such as sodium and potassium with one valence electron, readily combine with the halides, such as chlorine and fluorine, which have one vacancy in their outer shells. This type of binding is responsible both for the formation of single molecules and of bulk crystals.

Ionic binding is illustrated by the ionic molecule potassium chloride, KCl. In order to understand why it is energetically favorable for potassium and chlorine atoms to form KCl molecules, we break up the formation of the molecule into three stages. These are the removal of an electron from the potassium atom, its binding to the chlorine atom, and the attraction of the resulting K^+ and Cl^- ions. These steps are not necessarily followed in the actual formation of the molecule.

1 The outer electron of potassium is removed leaving the K^+ ion and an electron. This requires an energy equal to the ionization energy of potassium which is 4.34 eV.
2 The extra electron binds to the neutral chlorine atom yielding Cl^-. This releases energy called the energy of formation or *electron affinity*. For chlorine this is 3.82 eV. The entire process of transferring one electron from potassium to chlorine then requires that $(4.34 - 3.82) = 0.52$ eV of energy be supplied.
3 The oppositely charged ions approach each other. Because of their mutual attraction, their potential energy initially decreases as they ap-

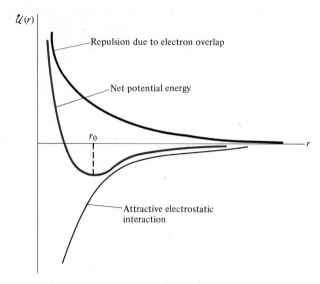

Figure 31.1. The potential energy of two ions versus the distance between their centers. The repulsive part due to the overlap of core electrons, which is positive, plus the attractive electrostatic part, which is negative, equals the net potential energy (shown colored). This energy is a minimum at a spacing r_0.

proach. However, when they are so close that their electron clouds begin to overlap, there is a repulsive force and the potential energy begins to increase (Fig. 31.1). The observed equilibrium separation of the ions in KCl is $r_0 = 2.79 \times 10^{-10}$ m. The electrical potential energy at this separation is

$$E = -\frac{ke^2}{r_0}$$

$$= -(9 \times 10^9 \text{ N m}^2 \text{ C}^{-2}) \frac{(1.6 \times 10^{-19} \text{ C})^2}{2.79 \times 10^{-10} \text{ m}}$$

$$= -8.26 \times 10^{-19} \text{ J} = -5.16 \text{ eV}$$

Ignoring other contributions to the potential energy momentarily, this means that the molecule has an energy of $(0.52 - 5.16)$eV $= -4.64$ eV compared to the separate neutral atoms. Conversely, 4.64 eV of energy would be required to break up the molecule into neutral potassium and chlorine. The measured value is 4.40 eV. The difference in the two results can be attributed to the repulsion of the inner electron clouds.

In ionic crystals the ions form a lattice. A number of crystal structures are possible, but the common feature of all of them is that ions of one charge have closest neighbors which are ions of the opposite charge. The electric attraction between neighbors is responsible for the binding of the crystals.

31.2 | COVALENT BINDING

The vast majority of molecules are formed by sharing the outermost or *valence* electrons of the constituent atoms. In such *covalent* binding, the distribution of electrons in the molecule may be quite different from that in the separated atoms. These electronic distributions and other molecular properties are studied by interpreting detailed experimental results with the aid of theoretical models. These models rely on two important concepts described in the preceding chapter.

1 The square of a wave function $|\psi|^2$ indicates the relative probability of an electron being at a particular location. Equivalently, $|\psi|^2$ determines how often an electron is at a given place. When two atoms share an electron, it effectively belongs to each atom part of the time. Thus we expect that in molecules the wavefunctions of the electronic states should overlap so that the shared electron is favorably positioned with respect to both atoms.

2 When the atoms combine, the Pauli principle requires that each spatial state contains at most two electrons, one with spin up and one with spin down.

We can illustrate these ideas with the diatomic hydrogen molecule, H_2. Neutral hydrogen atoms have one electron in the 1s shell, although this state could hold two electrons if their spins were opposite (Fig. 31.2a). When two such atoms move together, one of two things will occur. If the electron spins are parallel, the Pauli principle prevents the electrons from being very close together since they cannot be in the same spatial state. The electrons repel each other and their wavefunctions are distorted (Fig. 31.2b). Because the electrons are far from each other most of the time, the nuclear charges also repel each other, and the whole arrangement is energetically unfavorable.

On the other hand, if the spins are opposite each electron "fits" into the 1s state of the other atom. The electrons can now be close to each other, and the

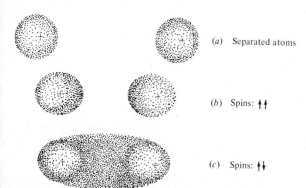

(a)   Separated atoms

(b)   Spins: ↑↑

(c)   Spins: ↑↓

Figure 31.2.   Representation of $|\psi|^2$ for the 1s states of two hydrogen atoms when they are far apart. (a) Contours representing the electron probability. (b) When the electrons have parallel spins, they repel each other because of the exclusion principle and the molecule does not form. (c) When the spins are opposite the two 1s electronic wave functions add. In a sense, each atom has a filled 1s shell. The energy of the H_2 molecule is lower than that of two separate neutral atoms and the molecule is stable.

molecular wave function looks like the sum of the two 1s wavefunctions of each atom (Fig. 31.2c). The electrons have a high probability of being between the two nuclei where they are attracted by both protons. This reduces the total energy of the molecule below that of the separated atoms and results in a stable molecule. The characteristic elongated wavefunction of Fig. 31.2c is called a σ (sigma) *orbital*. This bond is said to be a σ bond.

The spatial spreading of the wave functions, which occurs when the molecule is formed, plays an important role in reducing the electron energies. From the uncertainty principle, an increase in the position spread or uncertainty implies a decrease in the momentum uncertainty. This means the minimum possible momentum and the minimum possible kinetic energy are smaller, and the minimum total electron energy is reduced.

Hybridization
An approximate approach to the structure of complex covalent molecules which has proved very useful is called *hybridization*. This method employs wave functions that are combinations or *hybrids* of atomic states such as the 3s and 3p states of the atoms that have slightly different energies. These energy differences can usually be ignored in molecules because they are small compared to the molecular binding energy. To illustrate the method of hybridization we describe the scheme for magnesium fluoride, MgF_2.

Magnesium ($Z = 12$) has closed $n = 1$ and $n = 2$ shells, and two 3s valence electrons; its six 3p levels are empty. Fluorine ($Z = 9$) is one electron short of having a filled $n = 2$ shell and hence attracts electrons. Since the outer 3s wave functions of magnesium are spherically symmetric, their overlap with the 2p wave functions of fluorine is small (Fig. 31.3a). However, using combinations of magnesium wave functions, the overlap can be greatly increased, resulting in stronger binding (Fig. 31.3b,c).

We saw in the last chapter that the three p wave functions have pairs of lobes oriented along coordinate axes. If a p wave function and an s wave function are added and subtracted, the result is a pair of *sp hybridized* wave functions, or *orbitals*, each with a large lobe in one direction and a small lobe opposite (Fig. 31.3b). The two magnesium valence electrons can be thought of as being in these states, which overlap well with the single vacant p state of a fluorine atom. Consequently MgF_2 is a linear molecule with the magnesium atom between two fluorine atoms. Note that the sp orbital is a mixture of two atomic orbitals with slightly different energies, so a free magnesium atom would not ordinarily be found in such a state. However, the energy needed to occupy this sp orbital is much smaller than that gained in the formation of the molecule, so in the molecule it is energetically favorable for the electrons to be in this configuration.

In sp^3 *hybridization*, we consider combinations of one s state and all three p states. We illustrate this procedure for the water molecule H_2O. Oxygen ($Z = 8$) has two 2s electrons and four 2p electrons in its $n = 2$ shell. Thus it is two electrons short of having a closed $n = 2$ shell. Figure 31.4a shows the four sp^3 orbitals which can be formed from suitable combinations of the s and p states of oxygen. These states have a maximum angular separation, 109.5°. When occupied by electrons, this will minimize their repulsion.

Each sp^3 orbital can accommodate two electrons with opposite spins. In the water molecule, four of the $n = 2$ oxygen electrons fill two of the sp^3 orbitals. Each of the other two orbitals has one electron from the oxygen and one from a hydrogen atom (Fig. 31.4b,c).

A reminder that the hybridization model for water is not exact is provided by the experimental bond angle, 104.5°, which is slightly less than the predicted angle of 109.5°. The difference is due to the effects of

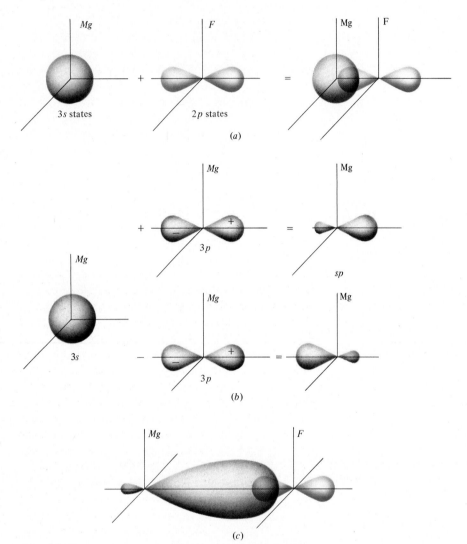

Figure 31.3.   (a) The 3s wave functions in magnesium and the 2p in fluorine have a small overlap area. (b) Adding and subtracting 3s and 2p magnesium wave functions produces sp wave functions with large lobes. (c) The sp wave function has a large overlap with the 2p fluorine wave function. In the MgF_2 molecule, a second fluorine atom binds on the left with the other sp wave function, forming a linear, symmetric molecule.

electrical forces among the bond electrons and the hydrogen nuclei in the asymmetric water molecule.

The structure of the water molecule makes it a *polar* molecule, one with a permanent electric dipole moment. The oxygen nucleus attracts the bond electrons somewhat more strongly than do the hydrogen nuclei. The result is an effective excess of positive charge near the hydrogen nuclei and an excess of negative charge (electrons) near the oxygen nucleus. This permanent electric dipole moment plays in important role in the properties of water. For example,

water as a solvent breaks ionic salts up into ions, as in $NaCl \rightarrow Na^+ + Cl^-$ (Fig. 31.5).

Carbon | Carbon ($Z = 6$), because it has only four out of a possible eight $n = 2$ electrons, can participate in covalent bonds in a variety of ways. When the s and p states are combined to form four sp^3 orbitals, carbon can, for example, bond with four hydrogen atoms forming methane CH_4. Each orbital has an electron from carbon and a second from a hydrogen atom. The carbon wave functions can also

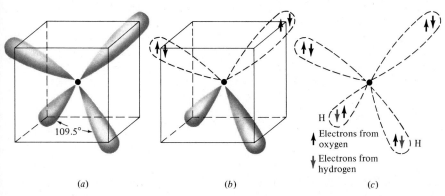

Figure 31.4. (a) The four sp^3 orbitals formed from combinations of s and p states of oxygen. (b) Two of the orbitals are full, containing a pair of electrons with opposite spins. The remaining two orbitals each have one of the oxygen electrons. (c) In the water molecule the hydrogen atoms provide the electrons to fill these two orbitals. The experimentally determined angle between the two hydrogen bonds is 104.5°.

Figure 31.5. (a) The water molecule is electrically neutral, but the electrons are closer to the oxygen nucleus. The result is a permanent electric dipole in the molecule. Water acts as a solvent because its dipole field can weaken and break the ionic bond in molecules such as NaCl. (b) The water molecules tend to surround the dissociated ions so that they are nearly electrically neutral and do not recombine.

be hybridized to form two sp orbitals or three sp^2 orbitals. The sp^2 orbitals are used for molecules such as ethylene $CH_2{=}CH_2$ (Fig. 31.6).

31.3 | THE METALLIC BOND

Covalent bonds occur in molecules containing two to several thousand atoms. By contrast, metallic binding holds billions of atoms together. Usually these substances are solid, with ionized atoms forming a rigid lattice and electrons contributed by the atoms that are free to move over the entire crystal. These delocalized electrons are the charge and energy carriers in metals and are responsible for their large electrical and thermal conductivities.

To see how the electrons become delocalized, consider lithium atoms, which have one $2s$ electron outside a filled $1s$ shell. This electron can have its spin up or down. When two lithium atoms are brought together, there are only two possible energy levels for the outer electrons. Their wave functions are similar to those for hydrogen (Fig. 31.2), the parallel spin state being higher in energy than the antiparallel state. Now, when a third atom is brought close to the first two, three separate energy levels become available for the outer electrons (Fig. 31.7). The energy levels are clustered near the original energy of the $2s$ electron in the free atom (Fig. 31.8).

Each time another atom is added, a new energy level results. However, the net effect of each new atom becomes smaller as the number of atoms already close together increases. Finally, with N atoms present, we have a *band* of N closely spaced energy levels. This band can accommodate $2N$ electrons. For example, for $N = 3$, there are three energy levels and six states (Fig. 31.8).

The electronic wave functions are spread over large distances and the electrons in this *conduction band* are free to move over the entire crystal. The reduction in kinetic energy due to the spreading out of the electronic wave function and the positive attraction of the electrons by the ions results in a stable configuration.

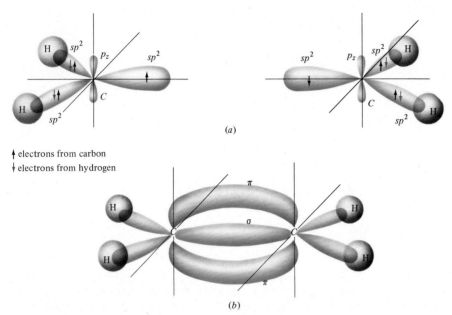

(a)

↑ electrons from carbon
↓ electrons from hydrogen

(b)

Figure 31.6. (*a*) The ethylene molecule $CH_2 = CH_2$ is formed by attaching two hydrogen atoms along two of the three sp^2 orbitals of each carbon atom and then (*b*) bringing the two complexes together. The σ bonds are formed along the C—C and C—H directions, and a bond is also formed by the overlap of the p_z wave functions of the two carbon atoms. The overlapping p_z states of the two complexes are called a π bond.

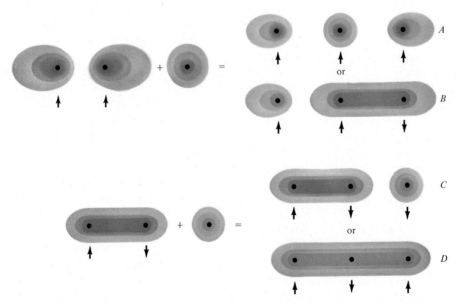

Figure 31.7. Two lithium atoms have their outer electron spins either parallel (above left) or antiparallel (below left). When a third atom is brought near, configurations like those at the right result. We expect that the configurations represented schematically as B and C would have the same energy. Reversing all the spins in A or D produces two more configurations A′ and D′ with the same energies as A and D. Thus the picture suggests three energy levels, each having two states. In fact this picture is oversimplified. Combining the wave functions of several particles is complicated and this diagram is only suggestive. Extending this type of diagram to four or more particles can lead to erroneous conclusions.

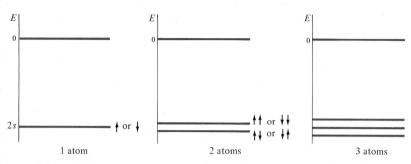

Figure 31.8. The lowest energy levels available to the outer electrons from one, two, and three lithium atoms placed close together.

Metals are good electrical and thermal conductors because of the presence of the conduction electrons. There are many unoccupied states in the conduction band into which electrons can move with only a small change in energy. Thus, if a metal is heated at one end, electrons in this region move into higher energy states and transport the heat energy away from this region. Similarly, a small voltage difference along a metal will lead to an increase in the energy of electrons which then move to regions of lower potential energy. It is the presence of the closely spaced unfilled energy levels of the band which facilitates these conduction processes. In a filled band in a semiconductor or insulator, electrons cannot easily change their states and conduction is much reduced.

31.4 | INSULATORS AND SEMICONDUCTORS

In metals the formation of the conduction band, resulting in a decrease in the kinetic energy of the electrons, plays a vital role in the binding of atoms. In materials such as diamond, germanium, and silicon, bands are also formed when a large number of atoms

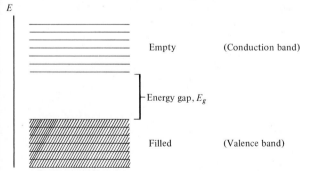

Figure 31.9. Energy bands of insulators and semiconductors.

are assembled. However, in these cases the valence electrons form a *band that is completely filled* at zero temperature.

Additional bands at higher energies are formed from higher atomic orbitals, but at zero temperature these are completely empty. There is an energy gap between the fully occupied *valence band* and the empty conduction band at a higher energy (Fig. 31.9).

In diamond the energy gap is about 6 eV. A temperature of 300 K, which is approximately room temperature, implies a mean thermal energy of $\frac{3}{2}k_BT = 0.04$ eV. Thus, virtually no electrons will have enough thermal energy to be excited from the filled band into the conduction band at any reasonable temperature. The motion of electrons due to electric forces or to temperature gradients requires that they increase their kinetic energy. Since the electrons in the filled band cannot change energy states, conduction is impossible and diamond is an excellent insulator.

In germanium ($E_g = 0.72$ eV), and in silicon ($E_g = 1.1$ eV), the energy gap is much smaller and a few electrons succeed in being thermally excited from the valence to the conduction band. Once in the conduction band they behave exactly as the conduction electrons in metals. The resistivity is higher in these *semiconductors* than in metals, because the number of electrons free to move is much smaller than in metals. For example, the room temperature resistivities of metals, semiconductors and insulators are on the order of 10^{-8}, 10^2, and 10^{13} ohm m, respectively.

A unique feature of semiconductors is the conduction by *holes* in the filled band. When an electron is excited to the conduction band, it leaves an empty state or hole in the valence band with an effective positive charge. If an electric field is applied, electrons from adjacent atoms jump into the hole and the

553

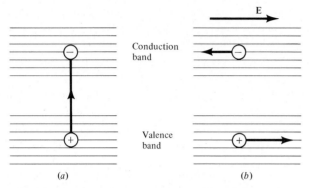

Figure 31.10. (*a*) An electron excited from the filled to the conduction band leaves a positively charged hole behind. (*b*) In an electric field both move, producing a net current.

hole moves. Thus, in an electric field, electrons in the conduction band move one way and holes in the valence band the other (Fig. 31.10).

The conductivity of a semiconductor can be increased by adding impurities to the sample. For example, if a few parts per million of arsenic is added to germanium, the conductivity increases a thousand-fold. Arsenic has one more electron in its outer shell than does germanium. When added as an impurity, it replaces a germanium atom, and its extra electron is freed for conduction. Similarly, gallium has one less outer electron than germanium, and when present as an impurity, it takes one electron from germanium, leaving a hole.

Semiconductors have many practical uses. A semiconductor can serve as a photocell for use in photography or in automatic door openers. The quanta of visible light have energies of about 2 or 3 eV, more than enough to excite electrons from the valence to the conduction band. In an electric field the semiconductor current increases greatly when the material is exposed to light and the number of charge carriers increases. Different types of semiconductors in contact are used in a wide variety of transistors and other devices. These devices have all but replaced the older vacuum tubes in electronics. They are desirable because they have low power requirements, no heating problems, are rugged, and last almost indefinitely.

31.5 | WEAKER BONDS

The bonds we have discussed typically have energies of a few electron volts. However, two bonds with

energies of at most one-tenth of an electron volt play an important role in nature: the *van der Waals bond* and the *hydrogen bond*. These bonds are especially important when the stronger types of bonds are not present. For example, the inert gases such as helium, neon, argon, and krypton condense to liquids and solids only because of attractive van der Waals forces among their atoms. Hydrogen bonding is often responsible for important structural features of molecules and solids such as the helical and pleated sheet structures of some organic molecules. In some instances, both types of bonding are present simultaneously.

van der Waals Attraction | The van der Waals attraction of molecules or atoms for each other is electric in nature just as are other types of bonds. However, this attraction is due to the presence of electric dipoles, either permanent or induced.

Molecules such as water have a permanent electric dipole moment corresponding to an effective separation of charge within the molecule (Fig. 31.5). As a result these polar molecules tend to cluster together (Fig. 31.11). This clustering is opposed by the normal thermal motion of the molecules. The motion diminishes as the temperature decreases, so the van der Waals attraction becomes more effective and may cause condensation or solidification.

A polar molecule can induce an electric dipole moment in a nearby nonpolar molecule. The positive end of the polar molecule repels the nucleus and attracts the electrons of the nonpolar molecule. Once this induced dipole moment is formed, the two molecules attract each other just as do two polar molecules.

A van der Waals attraction even occurs between nonpolar atoms and molecules. Although the centers

Figure 31.11. Polar molecules tend to cluster together due to the electric attraction of the positive and negative ends of neighboring molecules.

Figure 31.12. The instantaneous presence of a dipole moment in molecule *A* induces a dipole moment in molecule *B*, and the two nonpolar molecules attract one another.

of positive and negative charge are coincident on the average in a nonpolar atom, the electrons are constantly moving. At any instant the atom may have a temporary electric dipole moment in some random direction. This moment will induce a dipole moment in a neighboring atom, and the two dipoles are always aligned, so there is a net attraction. On the average then there is a net van der Waals attraction (Fig. 31.12).

Van der Waals forces are responsible for the attraction between inert gas atoms and for the behavior of liquids in capillary tubes. For example, water rises in a glass tube because the van der Waals attraction between the water and glass molecules is stronger than that among the water molecules themselves.

The Hydrogen Bond | The hydrogen atom

plays a unique role in compounds because of its small size and its single electron. A classic example of a hydrogen bond (H-bond) is found in ice. In liquid form, the primary attraction among water molecules is due to van der Waals forces among the polar mole-

cules, with H-bonding playing a secondary role. However, in the solid phase the H-bond is responsible for the specific structure of the ice crystal.

In water, two of the hybrid orbitals of oxygen hold paired spin electrons and do not participate in the binding of the molecule (Fig. 31.13). We refer to them as nonbonding orbitals. Since the bonding electrons are mainly between the oxygen and hydrogen nuclei, the region near the hydrogen nucleus is electrically positive. Each nonbonding orbital has a pair of electrons that is attracted to the positive hydrogen nucleus on an adjacent molecule. The hydrogen nucleus then is positioned with two electrons on either side (Fig. 31.13*b*). In ice, we find the hydrogen nuclei situated between the bonding and nonbonding orbitals of adjacent molecules. This is the hydrogen bond.

Hydrogen bonding plays an important role in the spatial arrangement of atoms in large organic molecules. For example, in DNA, there are two helical strands of alternate sugars and phosphate groups. Each sugar residue has a side chain, and the side chain strands are hydrogen bonded to each other. Thus the strands of the double helix have hydrogen bonds as their weakest link (Fig. 31.14).

SUMMARY

Molecules and bulk matter represent energetically favorable configurations of electrons and nuclei held

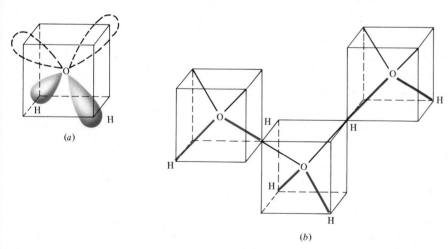

Figure 31.13   (*a*) The water molecule. Each orbital has a pair of electrons. The colored orbitals form the covalent bond with the hydrogen atoms. (*b*) In the crystal, water molecules join so that a hydrogen atom from one molecule connects to a nonbonding orbital (represented by a single black line) of an adjacent molecule. Bonding orbitals are shown as colored lines.

Figure 31.14. The DNA structure during replication. A, T, G, and C label side chain components attached to opposing side chains by hydrogen bonds as shown by the dashes. The double helix separates and rejoins at the hydrogen bonds.

Old   New   New   Old

together by electric forces. Ionic bonds form between atoms that have completely transferred an electron. The electric potential energy gained by having the ions close together more than compensates for the energy required to transfer the electron.

In covalent bonds, electrons are shared. In general, this sharing can be understood by examining the electronic structure of the constituent atoms. The wave functions of the atoms should overlap well, so that the electrons are favorably positioned with respect to the nuclei. Usually, such bonds contain two opposite spin electrons satisfying the Pauli principle. The pairs of bonding electrons tend to be as far apart as possible so the electric repulsion between electrons in different bonds is minimized. A model of the molecular wave function treats the constituent atoms as having hybrid electronic states in covalent bonding.

In metals the electrons become the common property of the metal as a whole. They are completely delocalized. This increased spatial uncertainty permits a major reduction in the kinetic energy of each electron. In semiconductors in their ground state, the electrons remain bound to the nuclei, although they may be shared by neighboring atoms. An electron

must increase its energy by at least a minimum amount to move into a conduction band. That is, excitation of an electron is equivalent to taking it away from an atom and allowing it to be free to move throughout the sample, just as in a metal. This excitation can be produced by heating or by the absorption of light.

Weak hydrogen bonds and van der Waals bonds often act between or within molecules. The van der Waals bond, which is caused by the mutual attraction of permanent or induced electric dipole moments, is responsible for the existence and properties of many liquids and solids. Hydrogen bonds are formed when a covalently bonded hydrogen atom attracts the electrons in a nearby orbital. This weak bond is responsible for much of the detailed structure in some solids and many organic molecules.

Checklist
Define or explain:

ionic bond	electric dipole moment
electron affinity	conduction band
covalent binding	valence band
Pauli principle	insulator
σ orbital	semiconductor
hybridization, sp, sp^2, sp^3	holes
	van der Waals forces
directional p state	hydrogen bond

EXERCISES

Section 31.1 | Ionic Binding

31-1   The ionization energy of neutral sodium (Na) is 5.12 eV, and the electron affinity of chlorine (Cl) is 3.82 electron volts. (Neglect the repulsions of the core electrons.) (a) How much energy is required to transfer one electron from Na to Cl? (b) If the ionic separation in the molecule NaCl is 2.36×10^{-10} m, what is the electric potential energy of the molecule? (c) What is the predicted energy required to separate NaCl into its separate neutral constituents, Na and Cl?

31-2   The ionization energy of lithium atoms is 5.39 eV, and the interatomic distance in the lithium fluoride molecule, LiF, is 1.51×10^{-10} m. The electron affinity of fluorine is 3.51 eV. (Neglect the repulsions of the core electrons.) (a) What is the potential energy of the ions Li^+ and F^- in the molecule? (b) What is the binding energy of LiF?

31-3   The binding energy of sodium bromide (NaBr) is 3.77 eV, the interatomic spacing is 2.50×10^{-10} m, and the ionization energy of Na is 5.12 eV. (Neglect the repulsions of the core electrons.) (a) What is the potential energy of the ions Na^+ and Br^- in the molecule? (b) What is the electron affinity of bromine?

Section 31.2 | Covalent Binding

31-4   Describe how magnesium and chlorine form the molecule $MgCl_2$.

31-5   Does the molecule H_2 have a permanent electric dipole moment? Explain.

31-6   The distance between the protons in a singly ionized H_2^+ molecule is 1.06×10^{-10} m. If the remaining electron is midway between the protons, what is its electric potential energy?

31-7   Describe the formation and structure of the methane molecule CH_4 using the sp^3 hybridization scheme for carbon.

31-8   What is the theoretical angle between the hydrogen-carbon bonds at one end of the ethylene molecule?

Section 31.3 | The Metallic Bond

31-9   When N atoms are brought together to form a certain material there are 2N electronic energy-levels formed in a band. If each atom contributes one valence electron to the band the band will be half filled and the material will behave like a metal. (a) Describe the occupation of the band for another material in which the atoms contribute two valence electrons. (b) How might the electrical and thermal conduction processes of this second material compare with those of the metal? Explain.

31-10   If the conduction band electrons are free to move about in the metal, why don't they leave the metal entirely?

Section 31.4 | Insulators and Semiconductors

31-11   What is the minimum frequency of light necessary to excite electrons from the valence to the conduction band in pure silicon? (The energy gap is 1.1 eV.)

31-12   Explain how a semiconductor might be used as a thermometer.

31-13   Is energy conserved when an electron drops from the conduction band to the valence band in a semiconductor? Explain.

Section 31.5 | Weaker Bonds

31-14   In Fig. 31.12, molecule A is polar and molecule B is nonpolar. Molecule A induces an electric dipole moment in B. Explain qualitatively why the net force between the molecules is attractive. Include the effects of the positive and negative charges of both molecules in your discussion.

31-15   Draw a schematic diagram of the boundary between a horizontal water surface and a vertical piece of glass. Describe qualitatively in terms of the van der Waals forces the shape of the water surface near the boundary.

Figure 31.15.   The structure of the methane molecule.

31-16   Figure 31.15 shows a methane molecule. Would you expect that hydrogen bonding is as important in solid methane as it is in ice? Explain.

PROBLEMS

31-17   The electric dipole moment of a molecule is the charge separation distance times the charge itself. What is the electric dipole moment of KCl? (The separation of the K^+ and Cl^- ions is 2.79×10^{-10} m.)

31-18   The neutral molecule H_3 is not stable although H_3^+ is. Using the exclusion principle, explain these results.

31-19   In a hydrogen atom the uncertainty in the position of an electron is about equal to the Bohr radius $a_0 = 5.29 \times 10^{-11}$ m. In the H_2 molecule the uncertainty is approximately equal to the distance between nuclei, 7.4×10^{-11} m. Using the uncertainty principle, estimate the reduction in the minimum kinetic energy of the electron in the molecule over that in the atom.

31-20   The ammonia molecule (NH_3) is usually described by an sp^3 hybridization of the nitrogen

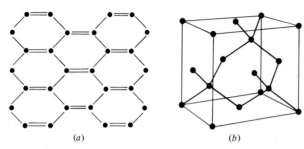

Figure 31.16. The positions of carbon atoms in (a) graphite and (b) diamond. The structure of graphite is nearly two dimensional, with each carbon bound to three neighbors in the same plane. In diamond, each carbon atom has four equidistant nearest neighbors.

Figure 31.17. (a) A crude model of a hydrogen bond in ice. The oxygen nuclei (shown colored) have effective charges $+e = 1.6 \times 10^{-19}$ C. The hydrogen nucleus (open circle) is midway between with a charge $+e$. The electrons with charges $-e$ are midway between the oxygen and hydrogen nuclei. (b) The same charges grouped as shown but separated by a large distance.

wavefunctions. Starting from the fact that the neutral nitrogen atom has two $2s$ and three $2p$ electrons in its $n = 2$ shell, describe the structure of NH_3. Do you expect this molecule to have a permanent electric dipole moment? Explain.

31-21   The hybridization model is a highly simplified one which sometimes leads to results which are not in agreement with experiment. For example, hydrogen sulfide (H_2S) might be expected to have the same shape as H_2O, since oxygen and sulfur are both two electrons short of having filled outer shells. However, the bond angle in H_2S is 92°, which is quite different from the bond angle in H_2O of 104.5°. Show that H_2S may be expected to have 90° bond angles if the hydrogen atoms overlap with the pure p states of sulfur. Sulfur has two $3s$ and four $3p$ electrons.

* **31-22**   Two common forms of pure solid carbon are graphite and diamond (Fig. 31.16). (a) Describe the hybridization scheme appropriate to the two structures. (b) Graphite can be used as a lubricant while diamond is extremely hard. Can this difference in properties be explained by their difference in crystal structure?

* **31-23**   In methane (CH_4) the carbon atom shares an electron from each of four hydrogen atoms. These four electrons would be sufficient to fill the $n = 2$ shell of carbon. Chlorine atoms need a p state electron to complete a shell. Carbon tetrachloride (CCl_4) is a covalently bound molecule. (a) What is the structure of CCl_4? (b) Does this molecule have a permanent electric dipole moment? (c) How might the positions of the bonding electrons differ from those in methane?

* **31-24**   Figure 31.17a shows a crude model of a hydrogen bond. Compute the difference between the total electric potential energy of the charges in Fig. 31.17a and Fig. 31.17b. (Do not include the repulsion between electrons which are in the same side of the hydrogen nucleus).

LINUS CARL PAULING

Born: 1901

(Wide World.)

Pauling belonged to the generation that developed the methods of quantum physics and applied them to atoms and molecules. After receiving a B.S. in chemical engineering from Oregon State College, Pauling obtained his Ph.D. in Chemistry at the California Institute of Technology in 1925. He has been a professor there and at Stanford University since 1931.

By the mid-1920s the quantum theory of atoms was relatively well developed, and enough was known so that one could try to understand molecules and molecular binding. The models then existing depicted molecules as composed of atoms containing fixed charges, somewhat like knobs on the surface of a spherical nucleus. Atoms combined when these knobs fit into place, held together by the electric attraction between the various electrons and nuclei. This model was unsatisfactory for several reasons and was completely incompatible with the wave-particle description of electrons in atoms.

From 1928 to 1932, Pauling published a series of papers in which he developed the quantum mechanical theory of molecular bonding. He was responsible for many of the ideas that are regarded today as fundamental. These included the hybridization scheme described in this chapter, the idea of resonance bonds such as those in benzene (see Chapter Thirty-two), and the correlation of interatomic distances with the electronic structure of molecules. The culmination of this work was the publication of *The Nature of the Chemical Bond* in 1939, a book that is still required reading for anyone doing research on the subject. Pauling received the Nobel prize in chemistry for this work.

In the mid-1930s Pauling turned his interest to biochemistry. He was the first to suggest and analyze the helical structure of some proteins. His work at the molecular level led to the idea that sickle-cell anemia is caused by a genetic defect in hemoglobin and also to a molecular theory of anesthesia.

Since World War Two, Pauling has been an outspoken critic of nuclear testing and of the large nuclear arsenals of the major powers. This ongoing activity resulted in his receiving the 1963 Nobel peace prize.

Pauling, who once picketed in front of the White House during the afternoon, and then dined inside with the president and other Nobel Laureates in the evening, has had a profound effect on physics, physical chemistry, and biochemistry. In his efforts for nuclear disarmament, Pauling has become a thoughtful and important advocate.

Additional Reading

Linus Pauling, *The Nature of the Chemical Bond,* Cornell University Press, Ithaca, New York, 1960.

Alan Holden, *Bonds Between Atoms,* Bell Telephone Laboratories, Murray Hill, N.J., 1966.

Gordon M. Barrow, *Physical Chemistry,* 3rd ed., McGraw-Hill Book Co., New York, 1973. Schematic representations of atomic and molecular orbitals.

Scientific American articles:

Boris V. Derjaguim, The Force Between Molecules, July 1960, p. 47.

Arnold C. Wahl, Chemistry by Computer, April 1970, p. 54.

Harry Selig et al., The Chemistry of the Noble Gases, May 1964, p. 66.

Ronald Breslow, The Nature of Aromatic Molecules, August 1972, p. 32.

The entire September 1967 issue is devoted to *Materials,* with a number of articles related to this chapter.

Gerald L. Pollack, Solid Nobel Gases, October 1966, p. 64.

L.K. Rumels, Ice, December 1966, p. 118.

J.D. Bernal, The Structure of Liquids, August 1960, p. 125.

A.R. Mackintosh, The Fermi Surface of Metals, July 1963, p. 110.

N.W. Ashcroft, Liquid Metals, July 1968, p. 72.

N.Ya. Azbel', M.I. Kaganov, and I.M. Lifshitz, Conduction Electrons in Metals, January 1973, p. 88.

Gordon A. Thomas, An Electron-Hole Liquid, June 1976, p. 28.

Bernard Bertman and Robert A. Guyer, Solid Helium, August 1967, p. 84.

James L. Fergason, Liquid Crystals, August 1964, p. 76.

Raymond Bowers, Plasmas in Solids, November 1963, p. 46.

R.D. Parks, Quantum Effects in Superconductors, October 1965, p. 57.

F. Reif, Superfluidity and "Quasi-Particles," November 1960, p. 138.

N. David Mermin and David M. Lee, Superfluid Helium 3, December 1976, p. 56.

Ednor M. Rowe and John H. Weaver, The Uses of Synchrotron Radiation, June 1977, p. 32.

David Adler, Amorphous-Semiconductor Devices, May 1977, p. 36.

James L. Dye, Anions of the Alkali Metals, July 1977, p. 92.

CHAPTER THIRTY-TWO

NUCLEAR MAGNETIC RESONANCE

A number of experimental techniques originally used by physicists are now in wide use by biochemists and others to study complex molecules. These studies yield information about the structure of molecules and their role in biological processes.

The experimental information necessary for any description of molecules includes the identification of the types of atoms present, their positions in the molecule, and the properties of the molecule as a whole. Usually the experimental evidence is acquired in bits and pieces and is very difficult to interpret, except in terms of models such as those of the preceding chapter.

In this chapter, we describe one of the most commonly applied and versatile techniques used to study molecular structure: *nuclear magnetic resonance or NMR.* NMR exploits the fact that many atomic nuclei behave in magnetic fields as small magnetic dipoles. Measuring the energy needed to reorient the nuclear magnetic moments determines the magnetic fields at the positions of the nuclei in the molecule. For example, if an external magnetic field $\mathbf{B}_e$ is applied to a molecule, the actual magnetic field at a dipole is the external field plus the fields due to the electrons and nuclei in the neighborhood of the dipole. We call this the total field $\mathbf{B}_T$. The experiment is performed to measure the difference $\mathbf{B}_T - \mathbf{B}_e$. As we will see, this yields detailed information about the structure of the molecule.

NMR can be used on any molecule that has nuclei with nonzero magnetic dipole moments. The single proton in a hydrogen nucleus has such a moment, and the presence of hydrogen in so many organic molecules makes them excellent subjects for study. Our discussion in this chapter focuses on the origin of this proton magnetic moment and the methods of NMR.

32.1 | THE BEHAVIOR OF A MAGNETIC DIPOLE IN A MAGNETIC FIELD

In Section 21.5, we found that a current loop can be considered as a magnetic dipole. The magnetic dipole moment was found to be the current times the area of the loop. In the case of a particle moving in a circular orbit this magnetic dipole moment turns out to be proportional to the particle's orbital angular momentum.

A spinning charge also has a magnetic dipole moment $\boldsymbol{\mu}$ proportional to its spin angular momentum, $\mathbf{S}$. Experimentally the relationship between these quantities for a proton is found to be

$$\boldsymbol{\mu} = 2.79 \frac{e}{m_p} \mathbf{S} \tag{32.1}$$

where m_p is the proton mass. In the remainder of this section, we describe how the energy of a proton depends on the magnetic field it experiences. We also find that the proton moment experiences a torque due to the magnetic field. This torque causes the proton moment and spin to precess. This motion is quite analogous to that of a precessing top (Section 8.5).

We saw in Section 21.5 that the energy of a dipole at an angle θ to a magnetic field $\mathbf{B}_T$ is $\mathcal{U} = -\mu B_T \cos\theta$. When the total magnetic field $\mathbf{B}_T$ is in the z direction, the potential energy of the dipole is $\mathcal{U} = -\mu_z B_T$. If S_z is the z component of the spin angular momentum, this energy is

$$\mathcal{U} = -2.79 \frac{e}{m_p} S_z B_T \qquad (32.2)$$

For the proton $S_z = +\hbar/2$ or $-\hbar/2$. It is conventional to define the *nuclear magneton* by

$$\mu_N = \frac{e\hbar}{2m_p} = 5.05 \times 10^{-27}\ \text{A m}^2$$

so the potential energy becomes

$$\mathcal{U} = \mp 2.79\,\mu_N B_T \qquad (32.3)$$

The minus sign holds when S_z and μ_z are parallel to the field (spin up) and the $+$ sign when they are opposite (spin down). The following example illustrates these energy relationships.

Example 32.1

A photon is absorbed when a proton moves from the spin up to the spin down state in a magnetic field of 1.4 T. (a) What is the energy of the absorbed photon in electron volts? (b) What is the photon frequency?

(a) Using $\mathcal{U} = -2.79\,\mu_N B_T$ for the spin up state and $\mathcal{U} = +2.79\,\mu_N B_T$ for the spin down state, the change in energy is,

$$\begin{aligned}\Delta\mathcal{U} &= 2(2.79)\mu_N B_T\\ &= 2(2.79)(5.05 \times 10^{-27}\ \text{Å m}^2)(1.4\ \text{T})\\ &= 3.94 \times 10^{-26}\ \text{J} = 2.47 \times 10^{-7}\ \text{eV}\end{aligned}$$

We see that the energy of the photon absorbed is small compared to the changes of 1 to 10 eV usually observed in atomic and molecular processes.

(b) The photon energy is hf, so

$$f = \frac{\Delta\mathcal{U}}{h} = \frac{2.47 \times 10^{-7}\ \text{eV}}{4.14 \times 10^{-15}\ \text{eV s}} = 5.96 \times 10^7\ \text{Hz}$$

It is important to notice that the spin angular momentum $\mathbf{S}$ (and also $\boldsymbol{\mu}$) is not parallel or antiparallel to $\mathbf{B}_T$. This follows from Chapter Thirty where we found that the magnitude of the spin angular momentum $\mathbf{S}$ for spin $\frac{1}{2}$ particles is given by $S = \sqrt{s(s+1)}\,\hbar = \sqrt{\frac{1}{2}(\frac{1}{2}+1)}\,\hbar = \sqrt{3}\,\hbar/2$. Since $S_z = \pm\hbar/2$, $\mathbf{S}$ must always be at an angle with the z axis (Fig. 32.1b).

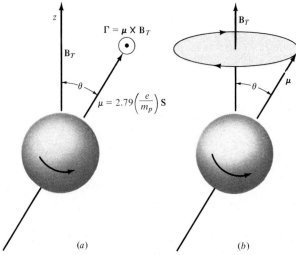

Figure 32.1. (a) A classical model of a spinning charge in a magnetic field $\mathbf{B}_T$. The spin angular momentum $\mathbf{S}$ and the magnetic moment $\boldsymbol{\mu}$ are parallel if the charge is positive. (b) The spin of a proton has a magnitude of $S = \sqrt{3}\,\hbar/2$, and the z component of $\mathbf{S}$ is $S_z = \pm\hbar/2$.

Figure 32.2. (a) The magnetic dipole moment $\boldsymbol{\mu}$ experiences a torque $\boldsymbol{\Gamma}$ due to the field $\mathbf{B}_T$. Here $\boldsymbol{\Gamma}$ is directed out of the page. (b) The magnetic moment $\boldsymbol{\mu}$ precesses around $\mathbf{B}_T$; the tip of the arrow representing $\boldsymbol{\mu}$ traces out a circle.

In a magnetic field, the proton moment experiences a torque $\boldsymbol{\Gamma}$ (Section 21.5) given by

$$\boldsymbol{\Gamma} = \boldsymbol{\mu} \times \mathbf{B}_T \qquad (32.4)$$

This torque is perpendicular to $\boldsymbol{\mu}$, $\mathbf{B}_T$ and also to $\mathbf{S}$ (Fig. 32.2a). The torque produces a change in the angular momentum $\mathbf{S}$ according to $\boldsymbol{\Gamma} = \Delta\mathbf{S}/\Delta t$ (Chapter Eight) and the result is a precession of $\mathbf{S}$ and $\boldsymbol{\mu}$ around the direction of the field $\mathbf{B}_T$ (Fig. 32.2b). We can see this more clearly if we write $\Delta\boldsymbol{\mu} = 2.79\,(e/m_p)\,\Delta\mathbf{S}$, so

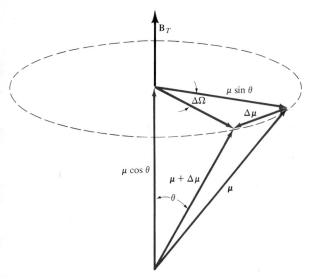

Figure 32.3.   The vector μ precesses around $\mathbf{B}_T$. The component of μ perpendicular to $\mathbf{B}_T$ changes by an amount $\Delta\mu = \mu \sin\theta \, \Delta\Omega$ in a time Δt.

$$\Gamma \, \Delta t = \Delta\mathbf{S} = \frac{m_p}{2.79 \, e} \, \Delta\mu \qquad (32.5)$$

Since Γ is always perpendicular to μ, the magnitude of μ is constant, but its direction changes. The vector μ is "pulled" around by the torque and μ precesses around $\mathbf{B}_T$ (Fig. 32.2b). This precession is analogous to the precession of a spinning top.

The rate of precession is an important parameter in a resonance experiment. From Fig. 32.3, we see that the z component of μ is constant, but the component of μ perpendicular to $\mathbf{B}_T$, $\mu \sin\theta$, turns through a small angle $\Delta\Omega$ in a time Δt. Since the arc AB is approximately equal to the chord $\Delta\mu$, $\Delta\mu = \mu \sin\theta \, \Delta\Omega$.

The torque is in the same direction as $\Delta\mu$, and from Fig. 32.2, we see that its magnitude is $\Gamma = \mu B_T \sin\theta$. Using these results for $\Delta\mu$ and Γ in Eq. 32.5, we find

$$\mu B_T \sin\theta \, \Delta t = \frac{m_p}{2.79 \, e} \, \mu \sin\theta \, \Delta\Omega$$

The precession frequency f_p is defined as $\Delta\Omega/2\pi \, \Delta t$, so we find

$$f_p = \frac{1}{2\pi} \frac{\Delta\Omega}{\Delta t} = 0.44 \, \frac{e}{m_p} B_T \qquad (32.6)$$

Note that the precession frequency is proportional to the magnetic field. This is illustrated in the next example.

Example 32.2

Calculate the precession frequency of a proton in a total magnetic field of 1.4 T.

Using Eq. 32.6, we have

$$f_p = 0.44 \, \frac{e}{m_p} B_T$$

$$= 0.44 \, \frac{(1.6 \times 10^{-19} \, \text{C})}{(1.67 \times 10^{-27} \, \text{kg})} (1.4 \, \text{T})$$

$$= 5.9 \times 10^7 \, \text{Hz}$$

This is a typical radio wave frequency.

In an NMR experiment the proton moments of the sample precess at a frequency $f_p = 0.44 \, eB_T/m_p$, where B_T is the total field at the position of the proton. A free proton precesses at a frequency $f_e = 0.44 \, eB_e/m_p$, where B_e is the magnitude of the external magnetic field. The difference between the two precession frequencies is

$$\Delta f = f_p - f_e = 0.44 \, \frac{e}{m_p} (B_T - B_e) \qquad (32.7)$$

The difference of $\mathbf{B}_T$ from the external field $\mathbf{B}_e$ indicates the effects of the surroundings. This information is obtained by measuring Δf.

32.2 | MEASURING THE PRECESSION FREQUENCY

The measurement of $\Delta f = f_p - f_e$ is the object of NMR experiments. To do this, a second magnetic field $\mathbf{B}_\perp$ is applied perpendicular to $\mathbf{B}_e$ and $\mathbf{B}_T$ (Fig. 32.4). This field changes with time, oscillating between the positive and negative x directions with a frequency $f_\perp$, and also produces a torque on the proton. When $f_\perp = f_p$, the resulting torque $\Gamma_\perp$ on the proton moment tends to increase the angle θ. At all other frequencies the effect of $\Gamma_\perp$ averages to zero.

As we noted earlier, the magnetic moment can only have two possible orientations with respect to the magnetic field $\mathbf{B}_T$. From Eq. 32.3, the energy difference between these states is $\Delta\mathfrak{U} = 2(2.79)\mu_N B_T = 5.58 \, \mu_N B_T$. When the magnetic field $\mathbf{B}_\perp$ is applied at the precession frequency, the torque $\Gamma_\perp$ will cause the dipole to "flip" (Fig. 32.5). Energy $\Delta\mathfrak{U}$ is absorbed by the dipole, and this absorption can be detected. The frequency at which energy absorption occurs determines the precession frequency. (*Note:* The magni-

563

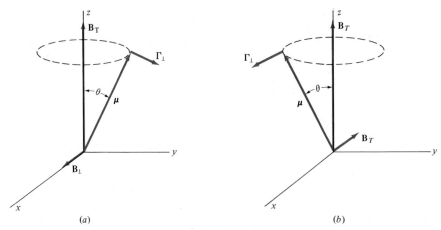

Figure 32.4. (*a*) An oscillating magnetic field $\mathbf{B}_\perp$, oriented as shown at a certain instant, produces a torque $\mathbf{\Gamma}_\perp = \boldsymbol{\mu} \times \mathbf{B}_\perp$ which tends to increase the angle θ. (*b*) If the field $\mathbf{B}_\perp$ varies at a frequency $f_\perp$ which is just equal to the precession frequency f_p, the torque $\mathbf{\Gamma}_\perp$ again tends to increase θ. If $f_\perp \neq f_p$, the effect of $\mathbf{B}_\perp$ averages to zero.

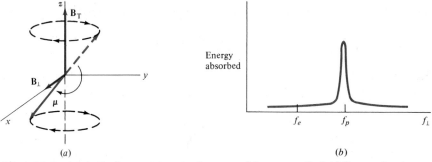

Figure 32.5. (*a*) A dipole precessing at a frequency f_p in a magnetic field $\mathbf{B}_T$ can absorb energy and "flip" over if a second field $\mathbf{B}_\perp$ is properly applied. (*b*) Absorption will occur when the frequency $f_\perp$ of $B_\perp$ is equal to f_p.

tude of $\mathbf{B}_\perp$ is always much smaller than that of $\mathbf{B}_T$ and is thus neglected in the calculation of $\Delta\mathcal{U}$.)

To summarize, when an external magnetic field is applied to a sample, the proton moments precess at the frequency f_p. The perpendicular time varying magnetic field acts to flip the dipoles over when $f_\perp = f_p$ (Fig. 32.5*b*).

32.3 | THE NMR APPARATUS

The NMR apparatus has four fundamental components:

1   A permanent magnet or electromagnet used to produce the large external field $\mathbf{B}_e$. Standard instruments produce a field of 1.4 T with the field uniform over the region of the sample to one part in 10^8.

2   A method for varying $\mathbf{B}_e$ over a small range. In Fig. 32.6, the sweep coils fill this function.

3   A radio frequency (R.F.) oscillator that produces electromagnetic radiation at a fixed frequency, normally $f_\perp = 6 \times 10^7$ Hz. The magnetic field of this radiation plays the role of $\mathbf{B}_\perp$.

4   A radio frequency receiver used as a detector for the absorption of energy from the oscillator. Sometimes a single coil serves as both the oscillator and detector. When energy is absorbed an induced EMF is produced in the coil. This induced EMF is slightly out of phase with the oscillator voltage and can be electronically detected.

Since the R.F. oscillator has a fixed frequency, variations in f_p are produced by varying B_e. This is done by altering the current in the sweep coils in Fig.

Figure 32.6. A schematic diagram of an NMR apparatus. The electronic instruments outside the box produce and monitor the energy absorbed. The sweep coils are used to make small variations in the field $\mathbf{B}_e$ produced by the permanent magnet. Other features are described in the text.

32.6. Resonance occurs when the external field is adjusted so that $f_p = f_\perp = 6 \times 10^7$ Hz. Each time a resonance appears energy is absorbed. As we will see, there are often many resonant frequencies for a given molecule.

The sample region is quite small, usually a tube of radius 0.2 cm filled about 3 cm deep. This is done to minimize the expense of the high precision magnet. Liquid solutions are usually used with the solvents chosen so that their resonant absorption frequencies do not overlap with those of the sample. An internal reference is commonly employed, normally tetramethylsilane (TMS), a molecule that does not interact strongly with other molecules. It has a single resonant frequency which shows up on the absorption graph and is used to calibrate the system.

Having established the physical basis for NMR measurements, we now consider what information can be obtained.

32.4 | THE CHEMICAL SHIFT

The *chemical shift* is the observed shift of the resonant frequency of protons in a given molecule from that of free protons in a field $\mathbf{B}_e$. For example, in benzene a single resonant frequency and chemical shift are observed. The measured frequency f_p is greater than f_e, so $\Delta f = f_p - f_e$ is positive. We can understand this result by examining the benzene molecule. Benzene has six carbon atoms in a ring with hydrogen atoms attached to them (Fig. 32.7a).

Figure 32.7. (a) The chemical diagram of benzene. (b) Electrons in benzene move freely in two rings above and below the plane of the molecule. (c) When $\mathbf{B}_e$ is applied, an induced electron current produces a magnetic field which opposes $\mathbf{B}_e$ inside the rings and adds to it outside the rings. (d) The field at the protons is larger than $\mathbf{B}_e$, so the resonant frequency is larger than f_e.

565

Figure 32.8.   When $\mathbf{B}_e$ is applied to a hydrogen atom, the electron circulates as shown, producing the induced field $\Delta\mathbf{B}$. The field at the proton is then $B_T = B_e - \Delta B$.

Above and below the carbon ring, electrons are free to move in doughnut shaped regions (Fig. 32.7*b*). As an external magnetic field $\mathbf{B}_e$ is applied, the electrons circulate in accordance with Lenz's law (Chapter Twenty-two). They form a current whose magnetic field opposes the initial increase in $\mathbf{B}_e$, reducing the flux inside the ring (Fig. 32.7). However, *outside the ring,* this induced current produces an increase in the magnetic field, and the total magnetic field there is greater than B_e by ΔB. Since the hydrogen atoms in benzene are outside the ring, they have a precession frequency which is greater than f_e by Δf.

Notice that at least two qualitative pieces of information about benzene have been confirmed. First, there exists a closed path around which the electrons can circulate and second, the hydrogen nuclei are outside this path. If the hydrogen nuclei were inside the ring, the resonant frequency would be less than

f_e; the chemical shift would be in the opposite direction.

Subtle differences in the positions of electrons in different bonds can be seen using NMR. When an external field $\mathbf{B}_e$ is applied to a neutral hydrogen atom in its ground state, the *s* electron has its energy changed slightly, and it circulates according to Lenz's law (Fig. 32.8). The local field at the proton is less than B_T and the precession frequency is smaller than f_e. This is called the *diamagnetic* effect.

When hydrogen occurs in a covalent bond, the electron wave function is no longer symmetric about the nucleus. The more the electron is pulled away from the hydrogen nucleus, the smaller the diamagnetic effect, since its ability to produce an effective induced field at the nucleus is reduced. For example, in an OH group the oxygen atom strongly attracts the bond electron from the hydrogen atom and the precession frequency of the proton in hydrogen is only slightly smaller than f_e. The bond electrons in CH_2 are not as far from the hydrogen as in OH. Hence the diamagnetic effect is larger and the precession frequency of these two protons is less than in OH. Finally, electrons in CH_3 are only weakly attracted by the carbon so the diamagnetic effect and chemical shifts are even larger. The precession frequency of these three protons is well below f_e. Thus in ethanol, CH_3—CH_2—OH, we see three chemical shifts (Fig. 32.9).

The identification of the three resonance peaks in ethanol is also facilitated by the fact that for every proton in an OH group which absorbs energy, there are two protons in CH_2 and three in CH_3. Energy is

Figure 32.9.   (*a*) Schematic diagram of the center of charge of bond electrons in the OH, CH_2, and CH_3 groups of ethanol. (*b*) The changes in the electron density and current at the hydrogen nucleus are evidenced by the different chemical shifts or precession frequencies of each group.

Figure 32.10. The NMR absorption spectrum of ethanol (CH_3—CH_2—OH) under high resolution at room temperature. (The effect of temperature will be discussed below.)

absorbed in the ratio of 1 to 2 to 3 in the three groups as seen in Fig. 32.9b.

When an NMR experiment of higher resolution is performed on ethanol, the broad spectrum of Fig. 32.9 is resolved into a number of sharp peaks (Fig. 32.10). The explanation of these peaks is the subject of the next section.

32.5 | SPIN-SPIN SPLITTING

The same proton magnetic moments that precess and absorb energy in molecules give rise to magnetic fields within the molecule. These dipole magnetic fields produce modifications in the electronic configuration of the molecule which, in turn, affect other dipoles. The net result is that the orientation of each dipole may influence the energy of the other dipoles.

When the effects of interactions among spins within a group such as CH_3 are taken into account, the energy required to flip any one spin does not depend on the orientation of the others. On the other hand, the effect of proton spins on adjacent groups is significant and measureable. Consider first the effect of the CH_3 spins on the protons of an adjacent group.

The three proton spins in CH_3 can have four possible types of spin configurations:

1. ↑ ↑ ↑
2. ↑ ↑ ↓ or ↑ ↓ ↑ or ↓ ↑ ↑
3. ↓ ↓ ↑ or ↓ ↑ ↓ or ↑ ↓ ↓
4. ↓ ↓ ↓.

The configurations with one spin opposite to the other two are three times as probable as those where all the spins are parallel.

Each of the configurations shown produces a different effect on dipoles of a neighboring group. Thus

the protons of the group adjacent to CH_3 will experience four slightly different magnetic fields and in a sample four slightly different precession frequencies will be observed. The fourfold splitting of the CH_2 line in Fig. 32.10 shows this. It is also seen that the two central peaks are about three times as high as the outermost peaks, corresponding to the relative frequency of occurrence of the spin configurations in CH_3.

The two proton spins in CH_2 can have three different configurations:

1. ↑ ↑
2. ↑ ↓ or ↓ ↑
3. ↓ ↓

These three configurations produce three different fields in adjacent groups, and the CH_3 resonance is split into three peaks. The central peak is very close to twice as intense as the two side peaks (Fig. 32.10). Using the same reasoning we expect to see a threefold splitting of the OH resonance due to the adjacent CH_2 group. This splitting is not seen in the room temperature spectrum of Fig. 32.10. However, if the temperature of the sample is lowered, the threefold splitting is finally observed (Fig. 32.11). Furthermore, the CH_2 resonance undergoes an additional splitting so that eight peaks are seen.

The disappearance of this additional structure at room temperature is explained by *chemical exchange*. The hydroxyl group OH is rather weakly bound in ethanol and can be pictured as moving from molecule to molecule, binding briefly at each site and then moving again. This process is so rapid that during the time it takes for the resonant absorption of energy to occur, about the time it takes for the dipole to precess once, several OH groups have been attached to the

Figure 32.11. At low temperatures, further structure is observed in the spectrum of ethanol. There are eight peaks near the CH_2 resonance, and the OH line has three peaks.

567

same molecule. Thus the resonance experiment detects an average effect from the presence of the OH groups and the detailed structure is not seen. As the temperature is lowered, the thermal energy that facilitates this random exchange motion is decreased, and the OH group remains attached to each molecule for a longer period of time. When the exchange frequency is lower than the precession frequency, the effects of the OH group are seen. The OH resonance splits into three peaks due to the spin-spin coupling with the adjacent CH_2 group, and there are two possible spin alignments of OH:

1. ↑
2. ↓

These lead to a twofold splitting of the CH_2 resonance. Since the CH_2 resonance already had four peaks, it now has eight.

When experiments are performed on complex molecules, it may be difficult to distinguish the resonance of a given group from the spin-spin splitting of another group. In such cases the experiment can be performed at several different values of the external field. The size of the chemical shift varies with the applied field but the spin-spin coupling does not. The chemical shift depends on the magnitudes of induced fields which, in turn, depend on the applied field. On the other hand, the spin-spin coupling is produced by the effects of spins on each other. Because these spins have the same magnitude, regardless of the applied field, the separation of the spin-spin splittings is always the same.

SUMMARY

In NMR experiments, an external field is applied that causes the precession of the nuclear magnetic dipole moments in the molecule. When a second time varying field is applied perpendicular to the first, energy is absorbed by the precessing dipoles if the frequency of this field is equal to the precession frequency. Since the precession frequency is very sensitive to the local magnetic field produced by the molecule itself, measurement of this frequency yields detailed information about the molecule.

Two of the most important effects seen in NMR experiments are the chemical shift and spin-spin coupling. The chemical shift is the result of magnetic

fields produced by induced currents in molecules. The existence and magnitude of these currents are closely tied to the constituents, positions, and types of bonds in the molecule. The spin-spin coupling is independent of the external field applied and measures the effect of spin magnetic moments on one another. The orientation of one spin produces changes in the electron distribution which then effects other spins. Much of the detailed structure in NMR experiments is due to the spin-spin coupling.

Checklist

Define or explain:

magnetic dipole moment	radio frequency field
	chemical shift
total magnetic field	resonant frequency
dipole energy	diamagnetic effect
precession frequency	spin-spin coupling
spin flip	

EXERCISES

Section 32.1 | The Behavior of a Magnetic Dipole in a Magnetic Field

32-1 What is the energy of a proton in a magnetic field of 1.2 T if the z component of the spin angular momentum is (a) along the field; (b) opposite to the field?

32-2 A photon of frequency 5×10^7 Hz is emitted when a proton spin flips. (a) What is the orientation of the proton spin angular momentum in the final state after the spin flips? (b) How large is the magnetic field?

32-3 Indicate the direction of the torque and the path of precession of the proton magnetic moment μ in the field $\mathbf{B}_T$ of Figs. 32.12 *a* and *b*.

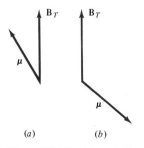

(a) (b)

Figure 32.12. At the instant shown, the proton magnetic moment μ is in the plane of the page as is $\mathbf{B}_T$.

32-4   What is the precession frequency f_p of a proton in a magnetic field $B_T = 1.2$ T?

32-5   When a magnetic field is applied to protons in a molecule, the difference between the precession frequencies $\Delta f = f_p - f_e$ is found to be 300 Hz. What is the difference between the actual and applied magnetic fields at the position of the dipole?

32-6   Show that the energy absorbed when a proton spin flips in a magnetic field $\mathbf{B}_T$ can be written as $\Delta \mathfrak{U} = h f_p$, where f_p is the precession frequency.

32-7   (a) Using the results of Exercise 32-6, compute the frequency of the photons absorbed by protons in a magnetic field of 5 T. (b) What is the energy change when absorption occurs?

Section 32.2 | Measuring the Precession Frequency

32-8   In a typical NMR experiment the difference between the external field and the field at the position of the proton is $B_e - B_T = 1.5 \times 10^{-6}$ T. What is the measured frequency shift Δf?

32-9   A free proton precesses with a frequency f_e in an external field B_e. At resonance in a molecule the total field is $B_T > B_e$. Is the frequency at which the perpendicular magnetic field is applied greater or less than f_e?

32-10   In Fig. 32.13 the proton magnetic moment precesses with a frequency f_p. The period, the time necessary for the dipole to precess once, is $T = 1/f_p$. At resonance the perpendicular magnetic field $\mathbf{B}_\perp$ has a maximum magnitude and is oriented as shown at $t = 0$. Sketch the field $\mathbf{B}_\perp$ and show the direction of the torque on $\boldsymbol{\mu}$ due to $\mathbf{B}_\perp$ at $t = 0$, $T/4$, $T/2$, $3T/4$, and T.

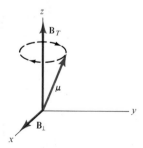

Figure 32.13.   At $t = 0$, $\boldsymbol{\mu}$ and $\mathbf{B}_T$ are in the yz plane, and $\mathbf{B}_\perp$ is along the x axis. Exercise 32-10.

Section 32.3 | The NMR Apparatus

32-11   Standard instruments produce a field $\mathbf{B}_\perp$ at a frequency of $f_\perp = 6 \times 10^7$ Hz. At what value of B_e will resonance occur for a free proton?

32-12   Describe briefly what might be observed if the applied field $\mathbf{B}_e$ in an NMR apparatus did not have exactly the same value everywhere in the sample.

Section 32.4 | The Chemical Shift

32-13   Sketch a graph of energy absorbed versus $f_\perp$ for a molecule shaped like benzene but with the hydrogen atom protons located inside the carbon ring.

32-14   In the compound CH_3—CH_2—CH_2—I, the two CH_2 groups experience slightly different local fields because their environments in the molecule are not identical. (a) Assuming the chemical shift in the CH_2 group beside the iodine (I) atom is smaller than in the adjacent CH_2 group, sketch a graph of energy absorbed versus $f_\perp$ for this molecule analogous to that in Fig. 32.9. Do not include the effects of spin-spin coupling. (b) What is the predicted ratio of the peak heights of the resonances which appear for this molecule?

32-15   When molecules such as ethanol, CH_3—CH_2—OH, are in certain solutions, the hydrogen in the OH group may form the center for a hydrogen bond with a solvent molecule. In such a bond, the hydrogen proton has four electrons near it rather than two as when the hydrogen bond is not present. The OH resonance when hydrogen bonds are present is observed to shift so that it appears between the CH_3 and CH_2 resonances in Fig. 32.9. Describe why this shift occurs in terms of the diamagnetic effect.

Section 32.5 | Spin-Spin Splitting

32-16   Sketch and explain the NMR spectrum of 1,1,2-trichloroethane,

$$Cl—CH_2—CH\begin{array}{l} Cl \\ \diagdown \\ Cl \end{array} .$$

(The chemical shift of CH is smaller than that of CH_2.)

Figure 32-14. Exercise 32-17.

Figure 32.15. An electron current is induced in the C—O group of the aldehyde molecule when an external field is applied. *R* stands for the remainder of the molecule. Problem 32-18.

32-17 Figure 32.14 shows the NMR absorption spectrum of a molecule containing one carbon atom, one oxygen atom, and four hydrogen atoms. What is the formula for the molecule?

PROBLEMS

32-18 Figure 32.15 shows the circulating electron current of the C—O bond in aldehyde. What will the effect of this induced current be on the precession frequency of the proton dipole moment in the hydrogen atom?

32-19 Sketch the NMR absorption spectrum of acetaldehyde:

$$CH_3-\overset{\displaystyle O}{\underset{\displaystyle H}{C}}$$

(Use the fact that the chemical shift of the CH group is smaller than that of CH_3.)

32-20 When an external magnetic field is applied to acetylene the electrons of the C—C bond can circulate around the surface of a cylinder (Fig.

Figure 32.16. Problem 32-20.

Figure 32.17. Problem 32-21.

32.16). (a) What is the direction of the induced electron current if viewed looking along the $\mathbf{B}_e$ direction? (b) What is the effect of this current on the precession frequency of the protons in the hydrogen atoms?

32-21 The NMR absorption spectrum of $CH_3-CH_2-CH_2-I$ is shown in Fig. 32.17. The different chemical shifts of the two CH_2 groups are due to the different environments of the two groups in the molecule. (a) Explain the threefold splitting of the CH_3 peak. (b) Why is the central CH_2 line split into 12 peaks?

Additional Reading

R.B. Setlow and E.C. Pollard, *Molecular Biophysics*, Addison-Wesley, Reading, Mass., 1962, chapter 7.

John R. Dyer, *Applications of Absorption Spectroscopy of Organic Compounds*, Prentice-Hall, Inc., Englewood Cliffs, N.J., 1965. Paperback.

Dudley H. Williams and Ian Fleming, *Spectroscopic Methods in Organic Chemistry*, McGraw-Hill Publishing Co., London, 1966. Paperback.

D. Chapman and P.D. Magnus, *Introduction to Practical High Resolution Nuclear Magnetic Resonance Spectroscopy*, Academic Press, London and New York, 1966. Paperback.

Stanley J. Opella, Biological Nuclear Magnetic Resonance, *Science,* vol. 198, 1977, p. 158.

Scientific American article:

George B. Benedek, Magnetic Resonance at High Pressure, January 1965, p. 102.

UNIT NINE

THE ATOMIC NUCLEUS

We have seen that the development of quantum physics in the early decades of this century made possible an understanding of atomic and molecular phenomena. By the early 1930s the frontier of physics and the search for the ultimate structure of matter had moved on to the much smaller scale of the atomic nucleus.

Until 1939, nuclear physics was a scholarly activity of little practical importance. In that year O. Hahn (1879–1968) and F. Strassman, working in Germany, discovered that a uranium nucleus can fission into two lighter nuclei. This process releases millions of times more energy than any chemical reaction among atoms or molecules and represents an energy source of unprecedented magnitude.

The great contemporary impact of nuclear physics is demonstrated by a few examples. Nuclear weapons brought a gruesome but rapid end to World War Two, and the threat of their possible use has become a central factor in international politics. The world's reserves of fossil fuels are rapidly diminishing, and of all the possible alternative energy sources, only nuclear fission reactors have as yet become available for widespread use. Many kinds of radioactive materials have been made available; their use in medicine has saved many lives, and important applications have been developed in research, agriculture, and industry. Major advances in geology and archeology have resulted from radioactive dating and other nuclear physics tools.

This unit is divided into two chapters. The first covers the fundamental physics of the nucleus, and the second deals with the effects and uses of ionizing radiation.

CHAPTER THIRTY-THREE

NUCLEAR PHYSICS

The atomic nucleus is a very small, dense object made up of two kinds of nucleons: protons and neutrons. A proton has a positive electrical charge equal in magnitude to the electronic charge, and a mass about 1840 times that of the electron. Neutrons are about 0.1 percent more massive than protons. As their name suggests, they bear no electrical charge.

A nucleus is specified by its *atomic number Z* and its *mass number A*. Z is the number of protons, and A is the total number of nucleons, so the *neutron number N* equals $A - Z$. The standard notation for nuclei is illustrated by $^{238}_{92}U$. This nucleus has 238 nucleons, of which 92 are protons and $238 - 92 = 146$ are neutrons. U is the chemical symbol for the 92nd element, uranium. Sometimes the atomic number is omitted, since it is implicitly given by the name of the element.

Nuclear species, or *nuclides,* which have the same atomic number but different neutron numbers are called *isotopes*. Because the electronic structure of atoms depends mainly on the total positive charge of the nucleus, different isotopes of an element are nearly identical chemically. Slightly over 100 naturally occurring or artificially produced elements and about 300 stable nuclides are known.

Three distinct types of forces play important roles in nuclei. Nuclei are held together by very strong, short-ranged *nuclear forces* among the nucleons. *Electrical forces* are smaller in magnitude but they become progressively more important as the number of protons in the nucleus increases. The *weak interactions* are much weaker than either the strong nuclear forces or the electromagnetic interactions, but they are responsible for the *beta decay* processes in which, for example, neutrons in nuclei are converted into protons as they emit electrons and neutrinos. Gravitational forces are weaker still, and are unimportant in nuclear physics.

The detailed motions of the individual nucleons inside a nucleus are very complex. However, their average orbits and the resulting nuclear structure can be fairly well described by a nuclear shell model, which closely resembles the atomic shell model.

In this chapter, we first explore some basic properties of nuclei and then discuss nuclear forces and the nuclear shell model. We conclude with an examination of nuclear decay processes. Nuclear energy is discussed in the supplementary topics.

33.1 | RADIOACTIVITY

Antoine Henri Becquerel (1852–1908) accidentally made the first observation of a purely nuclear phenomenon in 1896, 15 years before Rutherford inferred the existence of the nucleus. Becquerel noted that uranium compounds produce invisible rays or *radiation* which can penetrate an opaque container and expose a photographic emulsion. Soon thereafter, Pierre and Marie Curie showed that uranium ores also contain traces of polonium ($Z = 84$) and radium ($Z = 88$), both much more intensely *radioactive* than uranium. Many other radioactive nuclear species or *radionuclides* were subsequently found.

Before long, some important properties of the radi-

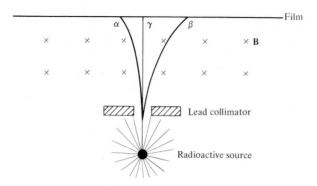

Figure 33.1. Radiation from a radioactive source splits into three components in a magnetic field. The α particles are $^4_2$He nuclei, the β particles are electrons and the uncharged γ rays are photons.

ation were discovered. A lead plate an inch or so thick stops most of the radiation from a uranium source, so a plate with a small hole can be used to form a narrow collimated beam of radiation. In the presence of a magnetic field, this beam splits into three components, labeled *alpha* (α), *beta* (β), and *gamma* (γ) Fig (33.1). *Alpha particles* are positively charged and have a very short range in matter; they are now known to be helium nuclei ($^4_2$He). The negatively charged *beta particles* have a longer range in matter and are electrons. The neutral *gamma rays* penetrate furthest in matter. They are photons whose energies are usually greater than those of x-rays.

The energies of α, β, and γ radiation are as much as several million electron volts (MeV) per particle. Since atomic and molecular processes typically involve energies of a few electron volts, radioactivity represented a totally new kind of phenomenon and suggested the existence of forces much stronger than electrical forces.

In 1903, Rutherford and Soddy showed that when a uranium nucleus emits an alpha particle, it is changed into a thorium nucleus. Symbolically,

$$^{238}_{92}\text{U} \rightarrow {}^{234}_{90}\text{Th} + {}^4_2\text{He}$$

Thus a transmutation of the elements occurs in α *decay processes*. This is a change that cannot be produced by chemical means, with or without medieval incantations. Similar transmutations occur in β *decays,* but not in γ *decays.*

Along with the early discoveries of the basic physical properties of radiation, there was progress in understanding the biological effects of radiation. By the

start of this century, it was realized that both x-rays and nuclear radiation can cause skin burns. However, it was not then known that they also can induce cancer. Research scientists, physicians, and industrial workers often received massive radiation doses, and many developed malignancies, sometimes decades after the exposure had ceased. Not until the 1920s were the first governmental limitations on radiation exposures formulated.

33.2 | HALF-LIFE

Consider a large group of students at a dull lecture. If they all toss coins, saying "heads I stay, tails I leave," about half will remain after one toss. Roughly half of these will remain after the next toss, and so on. We cannot predict in any way when a specific student will depart, and the process is said to be *random*. Because about half the students leave after each toss, the time between tosses is called the *half-life* of the class.

A nuclear decay is a similar random process and is characterized by a half-life T, the time required for half the nuclei present to decay. If at time $t = 0$, there are N_0 nuclei, then one half-life later at $t = T$, an average of $N_0/2$ will remain. At $t = 2T$, when two half-lives have elapsed, half of these, or $N_0/4$, nuclei will be left; at $t = 3T$, $N_0/8$ will be left, and so on. Depending on the nuclide, the half-life may vary from a small fraction of a second to billions of years.

When the elapsed time is not an exact integer multiple of the half-life, we can find the number of nuclei remaining as follows. The change ΔN in the number of nuclei N present occurring in a short time Δt is proportional to N and to Δt, so

$$\Delta N = -\lambda N \Delta t \qquad (33.1)$$

The minus sign is needed because N is decreasing and ΔN is negative; the proportionality constant λ is called the *decay constant*. As we will show in Section 33.13, this equation implies that if at time $t = 0$ there are N_0 nuclei, then at a later time t the number of nuclei remaining is determined by

$$\boxed{\frac{N}{N_0} = e^{-\lambda t}} \qquad (33.2)$$

This equation is called the *exponential decay formula*. We will also show that the decay constant is related to the half-life by

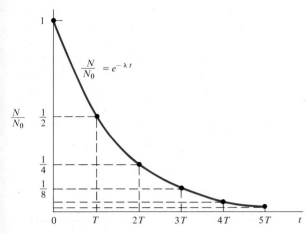

Figure 33.2. The fraction of radioactive nuclei N/N_0 remaining after time t. N_0 nuclei are present at $t = 0$, and T is the half-life.

TABLE 33.1

Half-lives of some radionuclides used in medicine and biology

Nuclide	Organ Where Concentrated	Half-Life (days) Physical	Half-Life (days) Biological
$^{3}_{1}H$	Total body	4.6×10^3	19
$^{14}_{6}C$	Fat	2.09×10^6	35
	Bone	2.09×10^6	180
$^{24}_{11}Na$	Total body	0.62	29
$^{32}_{15}P$	Bone	14.3	1200
$^{35}_{16}S$	Skin	87.1	22
$^{36}_{17}Cl$	Total body	1.6×10^8	29
$^{42}_{19}K$	Muscle	0.52	43
$^{45}_{20}Ca$	Bone	152	18,000
$^{59}_{26}Fe$	Blood	46.3	65
$^{64}_{29}Cu$	Liver	0.54	39
$^{131}_{51}I$	Thyroid	8.1	180

$$\lambda = \frac{\ln 2}{T} = \frac{0.693}{T} \qquad (33.3)$$

Here $\ln 2 = 0.693\ldots$ is the natural logarithm of 2, that is, the logarithm of 2 to the base e. Note that a short half-life implies a large decay constant and a rapid decay, as we might expect.

The exponential decay formula is plotted versus time in Fig. 33.2. When $t = T$, $N/N_0 = \frac{1}{2}$, in accordance with our discussion of the half-life; when $t = 2T$, $N/N_0 = \frac{1}{4}$, and so forth. Values of N/N_0 for any time can be read from the graph or calculated using tabulated values of e^{-x}, as in the next example.

Example 33.1

Iodine 131 is used in the treatment of thyroid disorders. Its half-life is 8.1 days. If a patient ingests a small quantity of ^{131}I and none is excreted from the body, what fraction N/N_0 remains after 8.1 days, 16.2 days, 60 days?

Since 8.1 days is the half-life, the fraction remaining at this time is $\frac{1}{2}$. Similarly, 16.2 days is $2T$, so $(\frac{1}{2})(\frac{1}{2}) = \frac{1}{4}$ remains. Sixty days is not an exact integer multiple of the half-life, so to find the fraction remaining we use the exponential decay formula:

$$\frac{N}{N_0} = e^{-\lambda t} = e^{-0.693\,t/T}$$

$$= e^{-0.693(60\ \text{d})/8.1\ \text{d}}$$

$$= e^{-5.13} = 0.0059$$

Thus only 0.59 percent of the radioactive iodine remains after 60 days.

The assumption made in the above example that no ^{131}I is lost from the body by biological processes is not quite correct; ^{131}I is excreted steadily but slowly with a biological half-life of 180 days. Thus if nonradioactive iodine were ingested, only $\frac{1}{2}$ would remain in the body after 180 days, $\frac{1}{4}$ after 360 days, and so on. The effective half-life T_{eff} is obtained by combining the biological half-life T_b and the radioactive or physical half-life T_p according to the formula

$$\frac{1}{T_{\text{eff}}} = \frac{1}{T_b} + \frac{1}{T_p} \qquad (33.4)$$

Some representative half-lives are given in Table 33.1.

Example 33.2

^{59}Fe is administered to a patient to diagnose blood anomalies. Find its effective half-life.

Using Table 33.1, $T_b = 65$ days and $T_p = 46.3$ days. Thus

$$\frac{1}{T_{\text{eff}}} = \frac{1}{T_b} + \frac{1}{T_p} = \frac{1}{65\ \text{d}} + \frac{1}{46.3\ \text{d}}$$

$$= 0.037\ \text{d}^{-1}$$

$$T_{\text{eff}} = 27\ \text{days}$$

Note that the effective half-life is shorter than either the biological or physical half-life. This happens because both processes are depleting the supply of the radionuclide.

577

For many applications, it is convenient to plot ln N versus t because the resulting graph is a straight line. This result follows from taking the natural logarithm of

$$\frac{N}{N_0} = e^{-\lambda t}$$

to obtain

$$\ln N = \ln N_0 - \lambda t$$

This equation is of the form $\ln N = a + bt$, so $\ln N$ and t are linearly related and have a straight line graph. If values of N versus t are plotted on semilog paper, a graph of $\ln N$ versus t is obtained directly.

This kind of graph is useful when we wish to determine the half-life of a radioactive sample. We can measure its activity with one of several detectors described in the next chapter. From Eq. 33.1, _the number of decays observed per second, the count rate_ $\Delta N/\Delta t$, _is proportional to the number N of radioactive nuclei present and will diminish at the same rate_. Consequently, we can deduce the half-life from observations of the count rate by noting when the count rate drops to about half its initial value. However, a better value is obtained by plotting the data on semilog paper and drawing a straight line through the points. This procedure is more accurate since it uses all the observations, not just two data points. It is illustrated by the following example.

Example 33.3

The measured count rate for a radioactive sample is given in Table 33.2. The initial count rate is 400 per second, the rate after 2 minutes is 336 per second, and so on. What is the half-life of the sample?

Inspecting the table, we can immediately estimate the half-life by noting that the count rate dropped to 200 per second or half the initial rate of 400 per second sometime between $t = 6$ minutes and $t = 8$ minutes. Hence, T is between 6 and 8 minutes. To find a more exact value, we plot the data in the table on semilog paper as

TABLE 33.2

t (min)	Count Rate (s^{-1})	t (min)	Count Rate (s^{-1})
0	400	8	194
2	336	10	162
4	280	12	131
6	230	14	110

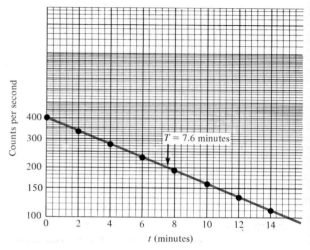

Figure 33.3. Data of Table 33.2 plotted on semilogarithmic paper.

in Fig. 33.3. The straight line drawn through the data points crosses 200 counts per second at 7.6 minutes, so this is the half-life.

It is worth noting that the half-life is determined without using any information about the absolute number of nuclei present. Also, one does not need to know what fraction of the radiation is observed by the detector as long as it is a constant fraction.

33.3 | DATING IN ARCHAEOLOGY AND GEOLOGY

The ratio of the original and present-day quantities of a radionuclide in an object indicates the time elapsed since that object was created or formed. While the present-day concentration of a radionuclide may be measured directly, the original amount of the radionuclide must be determined indirectly.

Carbon 14 Dating | Radiocarbon dating of seeds, wooden artifacts, human and animal remains, and other objects containing plant or animal materials provides a reliable way of dating events occurring within the past 60,000 to 70,000 years. Developed by Willard Libby in the 1940s, this technique has had a tremendous impact on archaeology and related fields.

Carbon 14, which has a half-life of 5692 years, is always present in the environment because of the effects of cosmic ray particles arriving in the atmosphere from outer space. These energetic particles interact with the atomic nuclei of the upper atmosphere to produce neutrons (n) which subsequently collide with nitrogen nuclei to produce ^{14}C and protons (p) in the reaction

$$n + {}^{14}_{7}N \rightarrow {}^{14}_{6}C + p$$

The radiocarbon mixes thoroughly with the ordinary carbon in the environment and is ingested by all living organisms. Once an organism dies, the input of radiocarbon stops, and the ratio of radiocarbon to ordinary carbon decreases steadily as the ^{14}C decays. Thus the quantity of ^{14}C remaining indicates the date of death.

Although this basic concept is simple, the actual application of radiocarbon dating is complex. One reason for this is that the ^{14}C is a minute fraction of the total carbon, only about one part in 10^{12}. Consequently the radioactivity produced by the ^{14}C as it beta decays to ^{14}N is very small compared to the total background radiation from sources normally present in the environment. To reduce this background, it is necessary to perform a careful chemical separation of the carbon and to use elaborate shielding arrangements during the counting. A second complication arises from uncertainties about the levels of radiocarbon in antiquity. The plausible assumption that radiocarbon levels have remained constant leads to predicted radioactivities that are close to the measured values for objects of known age (Fig. 33.4). However, studies made on tree rings have shown that minor variations have occurred in the ^{14}C level. These studies make use of the fact that only the outer portion of a tree is living material, so the ring formed in any one year records the radiocarbon level at that time. Using older trees which have long since fallen, but whose lifetimes overlapped appropriately, tree ring chronologies have been constructed extending back some 8000 years. These data permit small corrections to be made in doing accurate work.

The fraction of ^{14}C in the atmosphere decreased by about 3 percent over the past century because of the additional carbon dioxide produced by the large scale burning of fossil fuels which no longer contain ^{14}C. Hydrogen bomb testing, which began in 1954, reversed the trend, doubling the radiocarbon level by 1963. For example, whiskies made after 1954 can be dated accurately by their radiocarbon levels.

Figure 33.4. Predicted and observed radioactivities of samples of known age, assuming radiocarbon levels were constant in past centuries. "Bible" refers to Dead Sea scrolls, "Tayinat" to wood from the floor of a Syrian palace, and the other names refer to wooden objects from Egyptian tombs. (Adapted from W.F. Libby, *Radiocarbon Dating*, 2nd ed., University of Chicago Press, Chicago, 1955.)

Geochronology

The radionuclides used for dating rocks have half-lives comparable to geological times (Table 33.3). The oldest rocks on the surface of the earth are 3.3 billion years old, while the earth itself is estimated to be about 4.5 billion years old.

The techniques used in geochronology depend on the particular kind of rock or mineral under study. For example, ordinary lead of nonradioactive origin is a mixture of ^{204}Pb, ^{206}Pb, ^{207}Pb, and ^{208}Pb. From Table 33.3, we see that the radioactive decays of the uranium isotopes and thorium produce all of these isotopes except ^{204}Pb. If the lead in a sample does not

TABLE 33.3

Radionuclides used in geochronology		
Natural Radionuclide	Stable Nuclide Produced	Half-Life (Billion Years)
^{238}U	^{206}Pb	4.49
^{235}U	^{207}Pb	0.71
^{232}Th	^{208}Pb	14.1
^{87}Rb	^{87}Sr	50
^{40}K	^{40}A	1.3

contain any ^{204}Pb, this indicates that the lead present was produced by radioactive decay, and the sample can be used for dating. According to the table, ^{238}U decays into ^{206}Pb with a half-life of 4.49 billion years. Suppose a sample with no ^{204}Pb contains equal numbers of ^{238}U and ^{206}Pb nuclei. Then exactly one half-life or 4.49 billion years must have elapsed since the specimen was formed. The ^{232}Th to ^{208}Pb ratio can be employed in the same way. Alternatively the ^{206}Pb/^{207}Pb ratio can be used, since ^{235}U and ^{238}U decay at different rates. The ^{87}Rb/^{87}Sr and ^{40}K/^{40}Ar ratios are also sometimes employed.

33.4 | NUCLEAR SIZES

Beginning in 1907, Rutherford conducted a series of experiments in which he bombarded various atoms with alpha particles. As we saw in Chapter Twenty-nine, he found that an atom contains a small positive nucleus with a radius of less than 10^{-14} m, which is about 10^{-4} times the radius of an atom. From further experiments with alpha particles, nucleons, and other projectiles, considerable information has emerged about the spatial distribution of matter in the nucleus.

Roughly speaking, a nucleus containing A nucleons is a uniformly dense sphere of radius (Fig. 33.5)

$$R = 1.4\,A^{1/3} \times 10^{-15} \text{ m}$$
$$= 1.4\,A^{1/3} \text{ femtometers} \qquad (33.5)$$

(1 femtometer = 1 fm = 10^{-15} m). The radius is proportional to the cube root of the mass number, and therefore increases very slowly as is seen in this next example.

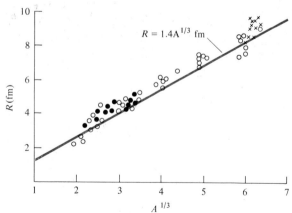

Figure 33.5. Experimental values of the radius of the nucleus versus $A^{1/3}$.

Example 33.4

Find the nuclear radii of ^{27}Al and ^{64}Zn.
The radius of ^{27}Al is

$$R = 1.4\,A^{1/3} \text{ fm} = 1.4(27)^{1/3} \text{ fm}$$
$$= (1.4)(3) \text{ fm}$$
$$= 4.2 \text{ fm}$$

Similarly, the radius of ^{64}Zn is

$$R = 1.4(64)^{1/3} \text{ fm} = (1.4)(4) \text{ fm}$$
$$= 5.6 \text{ fm}$$

The radius has increased only by a third although A has more than doubled.

^{238}U is the largest naturally occurring nuclide, with a radius of about 9×10^{-15} m. By comparison, atomic radii are about 10^{-10} m, more than 10,000 times larger. The fraction of the atomic volume occupied by the nucleus is much smaller than the corresponding fraction occupied by the sun in the solar system.

Because the nuclear radius varies as $A^{1/3}$, it follows that the nuclear volume $\frac{4}{3}\pi R^3$ varies as $(A^{1/3})^3$ or A. Thus the volume is proportional to the number of nucleons. This is similar to the situation in ordinary bulk matter; if we double the number of water molecules, we double the volume of water present.

33.5 | PROTONS AND NEUTRONS

In 1921, Rutherford produced hydrogen nuclei or protons ($^1_1$H or p) by bombarding nitrogen with alpha particles ($^4_2$He or α). This reaction,

$$^4_2\text{He} + {}^{14}_7\text{N} \rightarrow {}^{17}_8\text{O} + {}^1_1\text{H}$$

was the first artificially induced transmutation of elements, and suggested that protons are a constituent of nuclei. Since it was known that electrons are emitted by nuclei in β decay, it seems plausible that a nucleus has A protons and $A - Z$ electrons.

Two problems arose with this idea. First, it was found that $^{14}_7$N has spin 1. However, 14 protons and 7 electrons, each with spin $\frac{1}{2}$, can only combine to give a *half integer* spin ($\frac{1}{2}, \frac{3}{2}, \ldots$). Second, if an electron is confined to a region of nuclear dimensions the uncertainty principle can be used to show that its energy levels must differ by roughly 100 MeV. However, the energy needed to excite a nucleus is usually 5 MeV or less. Thus the observed spacing of energy levels is

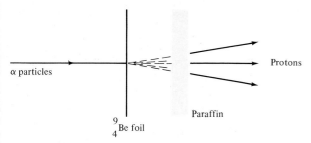

Figure 33.6. Chadwick's discovery of the neutron. A beam of α particles struck a thin beryllium foil, producing uncharged particles (shown dashed). When a block of paraffin (a hydrocarbon containing many hydrogen nuclei or protons) was placed in front of the foil, protons were ejected. From measurements of the energy and momentum of the protons, he found the velocity and mass of the neutral particles striking them. The mass of these neutral particles turned out to be slightly greater than that of the proton.

much smaller than would be predicted by having electrons in the nucleus.

James Chadwick's (1891–1974) discovery of the neutron in 1932 began the modern era of nuclear physics (Fig. 33.6). Bombarding beryllium with alphas, he produced neutrons in the reaction

$$ {}^{4}_{2}\text{He} + {}^{9}_{4}\text{Be} \rightarrow {}^{12}_{6}\text{C} + \text{n} $$

Neutrons have spin $\frac{1}{2}$, no electric charge, and a mass about 0.1 percent greater than the proton mass. This discovery suggested that nuclei are made up of protons and neutrons, which has now been well verified.

We have mentioned that nuclei that differ only in their neutron number are called isotopes and are chemically very similar. For example, hydrogen exists in three forms: ordinary hydrogen, ${}^{1}_{1}\text{H}$; deuterium or heavy hydrogen, ${}^{2}_{1}\text{H}$; and tritium, ${}^{3}_{1}\text{H}$. The first two nuclei are stable, but tritium β decays into ${}^{3}_{2}\text{He}$ with a half-life of 12.3 years. For a given atomic number, it is found that from zero to nine stable isotopes may exist, and several naturally occurring or artificially produced radioactive isotopes may also exist. Each chemical element as it is found on the earth normally contains a mixture of its stable isotopes with nearly constant relative abundances.

33.6 | NUCLEAR MASSES AND BINDING ENERGIES

The masses of many nuclei have been accurately measured (Table 33.4) using mass spectrometers (Chapter Twenty-one). A little arithmetic shows that the mass of a nucleus is less than the sum of the masses of its constituents. For example, $6m_p + 6m_n + 6m_e = 12.0989$ amu, while a ${}^{12}_{6}\text{C}$ atom has a mass of only 12.000 amu. This *mass defect* tends to increase with the mass number A.

The significance of the mass defect is made clear by Einstein's principle of the equivalence of mass and energy (Chapter Twenty-seven).

$$ E = mc^2 \tag{33.6} $$

In words, a mass m of matter can be converted into an amount of energy, E; the quantity c is the speed of

TABLE 33.4

Atomic masses in amu. Except for e, p, and n, the masses given are for the *neutral atom and include the electronic masses.* The atomic mass unit is defined so that the mass of a ${}^{12}\text{C}$ atom is exactly 12 amu

Nuclide	Mass, m (amu)	$Zm_p + Nm_n + Zm_e$ (amu)	Difference (amu)	Binding Energy per Nucleon (MeV)
e	5.48×10^{-4}			
p	1.00728			
n	1.00866			
${}^{1}_{1}\text{H}$	1.00783			
${}^{2}_{1}\text{H}$	2.0141	2.0165	0.0024	1.1
${}^{4}_{2}\text{He}$	4.0026	4.0330	0.0304	7.1
${}^{12}_{6}\text{C}$	12.0000	12.0989	0.0989	7.7
${}^{13}_{6}\text{C}$	13.0034	13.1078	0.1044	7.5
${}^{56}_{26}\text{Fe}$	55.9349	56.4633	0.5284	8.8
${}^{238}_{92}\text{U}$	238.0508	239.9845	1.9337	7.6

light in vacuum. The mass of a ^{12}C nucleus is less than that of its constituent nucleons because it is a bound system. One must supply energy equal to the *nuclear binding energy* to pull apart the protons and neutrons. According to Einstein's principle, this energy is equal to the mass defect times c^2.

To relate the mass defect and binding energy, we first calculate the energy associated with a mass of 1 amu = 1.66×10^{-27} kg:

$$E = (1 \text{ amu})(c^2)$$
$$= (1.66 \times 10^{-27} \text{ kg})(3 \times 10^8 \text{ m s}^{-1})^2$$
$$= 1.49 \times 10^{-10} \text{ J}$$

Since 1 eV = 1.602×10^{-19} J, this energy can be expressed in electron volts as 931×10^6 eV = 931 MeV. Effectively,

$$1 \text{ amu} = 931 \text{ MeV} \qquad (33.7)$$

The ^{12}C mass defect of 0.0989 amu corresponds to a total binding energy of 0.0989×931 MeV = 92.1 MeV. Dividing by the mass number $A = 12$, the *binding energy per nucleon* in ^{12}C is 7.7 MeV.

The binding energy per nucleon for the stable nuclei is plotted versus A in Fig. 33.7. It is about 8 MeV per nucleon except for the lightest nuclei. There is a broad maximum in the region of medium size nuclei, with a peak of 8.8 MeV per nucleon at

^{56}Fe. Above $A \simeq 100$, the curve gradually declines, reaching 7.6 MeV per nucleon for uranium.

The initial increase and later decrease in the binding energy per nucleon can readily be explained. The strong *nuclear forces* among the nucleons which hold the nucleus together have a very short range; these forces are zero at distances greater than a few femtometers. Accordingly a nucleon is attracted only to its closest neighboring nucleons. A nucleon near the surface has fewer neighbors than one in the interior of the nucleus and is less tightly bound. This *surface energy* effect implies that the binding energy per nucleon will rise as the nucleus increases in size and proportionately fewer nucleons are near the surface. This explains the initial rise in the binding energy per nucleon for the light elements.

To explain the decrease in the binding energy per nucleon for large A, we must take into account the fact that the electrical repulsions among the protons are proportional to the number of proton pairs, or to Z^2. The potential energy of two protons at a separation r is ke^2/r, so the total electrical potential energy due to the proton charges varies as Z^2e^2/R, where R is the nuclear radius. This energy grows rapidly as the number of protons increases. In the region near ^{56}Fe, the changes in the surface and electrical energies are roughly equal in magnitude but opposite in sign. Above $A \simeq 100$, the electrical repulsion gradually outstrips the surface effects, leading to the observed gradual decline in the binding energy per nucleon.

The fact that intermediate size nuclei have the greatest binding energy per nucleon has some important consequences. If a heavy nucleus splits or *fissions* into two intermediate-size nuclei, the binding energy increases by close to 1 MeV per nucleon. The extra energy is released as kinetic energy of the fission products or as γ rays. Similarly, if two very light nuclei such as ^{2}H or ^{3}H combine, this *fusion* releases several MeV. Fission and fusion will be discussed in the supplementary topics at the end of this chapter, but we can already see that both processes release large amounts of energy.

33.7 | NUCLEAR FORCES

We now discuss the strong nuclear forces that hold the nucleus together. We have seen that the nuclear volume is proportional to A; as more nucleons are

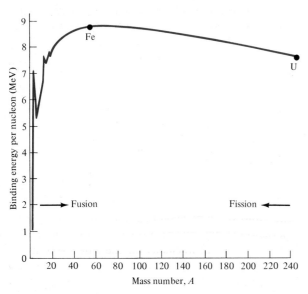

Figure 33.7. Binding energy per nucleon versus A for stable nuclei.

added, the nucleus grows in size but its density remains constant. We also saw that the binding energy of each additional nucleon is roughly constant at 8 MeV. These properties both occur because the *nuclear forces are short ranged* and are zero at distances greater than a few femtometers. Note that if the forces were long ranged as is true for electrical and gravitational forces, then every pair of nucleons would interact, leading to an increase in the density and mean binding energy with A. These do not happen because each nucleon interacts with only a few neighbors.

Complex nuclei are tightly bound despite the large electrical repulsions among the pairs of closely spaced protons. This indicates that nuclear forces are much stronger than electric forces. They must also be attractive, at least at the average internucleon separation in nuclei. However, it is found by studying collisions between nucleons that the force actually becomes repulsive at very short internucleon distances (Fig. 33.8). It has also been established that, except for the electrical repulsion between two protons, the proton-proton, proton-neutron, and neutron-neutron forces are the same; *nuclear forces do not depend on the electric charge.*

Since the nuclear radius is $1.4\,A^{1/3}$ fm, the average internucleon separation turns out to be over a femtometer. At this distance the nucleon-nucleon potential energy is of moderate size and slowly varying. Consequently, the nucleus should not be viewed as a rigid arrangement of nucleons. To a surprisingly good approximation, each nucleon in the interior of the nucleus can be viewed as moving fairly freely with a

constant potential energy arising from the effects of its neighbors. This idea is the basis for the nuclear shell model, which is discussed in the next section.

33.8 | NUCLEAR ENERGY LEVELS AND NUCLEAR STABILITY

A nucleus, like an atom, has a ground state and excited states. These states can be studied by bombarding nuclei with energetic protons, alpha particles, and other projectiles from particle accelerators. Many of the observed energy levels and other properties of nuclei can be predicted using the *nuclear shell model.* Initially developed in 1949 by Maria Mayer (1906–1972) and J.H. Jensen, this model resembles the atomic shell model.

Successive nuclear energy levels are usually separated by a few million electron volts or less, and the average binding energy of a nucleon in a nucleus is 8 MeV. Accordingly, most nuclear physics research has been done with projectile kinetic energies from a few million electron volts up to about 1000 MeV. At the higher energies the de Broglie wavelength is shorter, so the nucleus can be probed on a finer scale.

Nuclear processes studied with accelerators are of several types. A projectile may scatter *elastically,* so that the nucleus remains in its original state. It can scatter *inelastically,* exciting the nucleus and giving up part of its kinetic energy. In collisions, there may also be a *transfer* of one or more nucleons to or from the target increasing or decreasing its mass number. Measurements of the kinetic energies and relative directions of the incident and outgoing particles can be used to study the processes that occur. Experiments of these and other types have yielded a great deal of information about the energies and wave functions of nuclear states.

Nuclear Shell Model | In the preceding section, we noted that when a nucleon is deep inside the nucleus, its potential energy is approximately constant. However, a nucleon near the edge of a nucleus experiences a net inward attraction, since there are fewer neighboring nucleons on the side away from the nuclear center. Thus near the edge of a nucleus the potential energy increases as r increases (Fig. 33.9). Such a potential energy curve is referred to as a *potential well.*

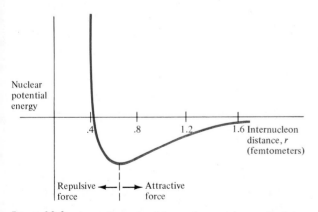

Figure 33.8. A rough graph of the nucleon-nucleon potential energy. The force becomes large and repulsive for $r \lesssim 0.4$ fm.

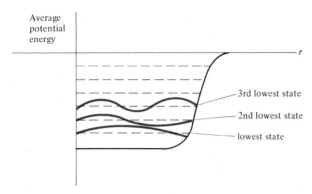

Figure 33.9. Average potential energy of a nucleon in the nucleus versus the distance r from the nuclear center. The possible nucleon energy levels are shown, along with the wave functions for the lowest three states.

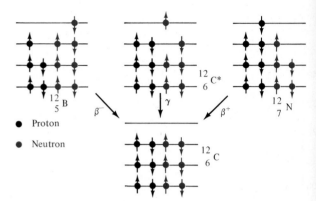

Figure 33.10. Lowest states of three $A = 12$ nuclei, and the first excited state of $^{12}_6C$. The $^{12}_5B$, $^{12}_6C^*$, and $^{12}_7N$ nuclei have the same energy. All three decay to the ground state of $^{12}_6C$ as described in the text. (*Indicates excited state.)

The possible energy levels for a nucleon are obtained by solving the basic equation of quantum mechanics, the Schroedinger equation. Its solutions for stable states correspond to fitting waves into the potential well. The longest waves that fit into the potential well have nodes at $r = 0$ and at the edge. Now using the de Broglie relation $p = h/\lambda$, the kinetic energy of a nucleon $\frac{1}{2}mv^2 = p^2/2m$ can be rewritten as $h^2/2m\lambda^2$. Thus the longest wavelength wave functions represent the lowest energy states. The longest waves fitting into the well are shown in Fig. 33.9. When angular momentum is taken into account, the level scheme is more complex, but it is found that, as in atoms, the levels fall into closely spaced groups or shells.

Nucleons are spin $\frac{1}{2}$ particles, so the Pauli principle applies, and two identical nucleons cannot occupy a single quantum state. Any energy level may contain at most two protons, one with spin up and one with spin down, the two possible orientations for a spin $\frac{1}{2}$ particle. Since there can also be spin up and spin down neutrons in an energy level, it can have at most four nucleons.

The lowest energy configuration for a given number of protons and neutrons is obtained by filling the lowest level with two protons and two neutrons, then the next level, and so on, until all the nucleons are used. A schematic diagram for three $A = 12$ nuclei is given in Fig. 33.10. Because of the Pauli principle, the lowest energy ground state occurs for the $Z = N$ case, $^{12}_6C$. Neglecting the small differences due to the electrical repulsions among the protons and to the neu-

tron-proton mass difference, the *excited state* of $^{12}_6C$ and the *ground states* of $^{12}_5B$ and $^{12}_7N$ have the same energy. The latter two nuclei will tend to β decay into the ground state of $^{12}_6C$, since that is a state of lower total energy. (In β^+ decay, a proton is converted into a neutron and in β^- decay, a neutron is converted into a proton. See Section 33.9.) Also, the excited $^{12}_6C$ state will tend to emit a γ ray and undergo a transition to the ground state.

In general, the ground state of a nucleus is found by filling the lowest states in accord with the Pauli principle. Excited states correspond to one or more nucleons in higher states.

From our $A = 12$ example, we expect the stable nuclei to have $N \simeq Z$, which is in fact observed for nuclei with values of Z up to about 20 (Fig. 33.11). This equality maximizes the number of nucleons in the lowest nuclear energy states. If N differs substantially from Z, the nucleus tends to β decay. As Z increases beyond 20, the known stable and radioactive nuclei tend to have increasing neutron excesses. A neutron excess occurs because the energy that would be released by converting a neutron to a proton (β^- decay) and placing it in a lower nuclear energy level is less than the increase that would occur in the electrical potential energy as a result of the increase in Z. Since the electrical energy is proportional to Z^2e^2/R, it becomes increasingly important as Z becomes large. Consequently, in the heaviest nuclei there are about 50 percent more neutrons than protons.

Figure 33.11. The known stable and radioactive nuclei. Nuclei beyond those shown in most cases rapidly break up into fragments.

We saw in Chapter Thirty that atoms with a completely filled outermost shell are inert gases and are very stable. Similarly, nuclei with closed outer shells are more tightly bound than neighboring nuclei, and are called *magic nuclei*. Nuclei with both proton and neutron shells closed are especially stable.

33.9 | RADIOACTIVE DECAYS

At the beginning of this chapter, we noted that many nuclei undergo alpha, beta, or gamma decay. In these decays, as in all nuclear processes, the following quantities must be conserved:

1   energy (including mass energy)

2   momentum (both linear and angular)
3   electric charge
4   number of nucleons

Notice in particular that the total charge and number of nucleons does not change. We now discuss these decays individually.

γ Decay | Gamma rays are electromagnetic quanta or photons emitted when a nucleus undergoes a transition from a higher to a lower energy level. They are completely equivalent to the light quanta and x-rays emitted by excited atoms, but their energies are usually much greater. The half-lives for γ decay are usually very short, typically 10^{-13} seconds; however, in a few special cases the half-lives may be as long as several years.

Closely related to γ decay is *internal conversion*. Here an excited nucleus gives up its excess energy to an electron in one of the inner atomic shells, ejecting it from the atom. No gamma ray is emitted. Some radionuclides used in nuclear medicine decay in this fashion.

β Decay | In β decay, an electron (e^-) or in a few cases a positron (e^+) is emitted by a nucleus. In 1928, P.A.M. Dirac proposed a theory of the electron that included the effects of special relativity and accurately predicted some fine details of the hydrogen atom spectrum not given by Schroedinger's equation. Dirac's equations also had solutions corresponding to positrons, *positively* charged particles with the same mass and spin as the electron. This *antiparticle* was first discovered in the products of cosmic ray reactions in 1932. Today, antiprotons, antineutrons, and many other antiparticles have been seen, and all particles are believed to have antiparticles. Particle-

TABLE 33.5

Typical beta decays

Decay	Half-Life	Maximum Kinetic Energy (MeV)
$^3_1H \rightarrow {}^3_2He + e^-$	12.3 years	0.186
$^{14}_6C \rightarrow {}^{14}_7N + e^-$	5692 years	0.156
$n \rightarrow p + e^-$	11.0 minutes	0.782
$^{34}_{15}P \rightarrow {}^{34}_{16}S + e^-$	12.4 seconds	5.1
$^{22}_{11}Na \rightarrow {}^{22}_{10}Ne + e^+$	2.60 years	0.546
$^{13}_7N \rightarrow {}^{13}_6C + e^+$	9.99 minutes	1.19

585

Figure 33.12. (*a*) The ground state of $^{14}_{6}$C decays into $^{14}_{7}$N as a neutron is converted into a proton. An electron and a neutrino are emitted, and they share the energy given up as the nucleus goes into a lower energy state. (*b*) The number of electrons emitted versus kinetic energy for a large number of decays. Point (1) represents a decay in which the electron has less energy than that represented by point (2). The total energy available is constant so the neutrino in decay (1) has correspondingly more energy than in decay (2).

antiparticle pairs can annihilate in a burst of gamma rays, converting all their mass to energy.

Some typical β decays are listed in Table 33.5. The half-lives are very long compared to γ decay half-lives, varying from seconds to many years. This indicates that the force responsible for the β decay is weak compared to the electromagnetic forces that are responsible for γ decay.

A striking feature of β decay is that the emitted betas are variable in kinetic energy (Fig. 33.12). For example, in $^{3}_{1}$H $\rightarrow$ $^{3}_{2}$He $+ e^{-}$, the mass energy of the $^{3}_{1}$H nucleus exceeds that of a $^{3}_{2}$He nucleus plus an electron by 0.0186 MeV, the maximum electron kinetic energy. However, sometimes the electron energy is less than 0.0186 MeV. What has happened to the missing energy?

Enrico Fermi (1901–1954) supplied the answer in 1933. He proposed that when a nucleus β decays, it creates not only an electron but also a *neutrino* (ν): a massless, uncharged, spin $\frac{1}{2}$ particle. Then the $^{3}_{1}$H decay is more completely written as

$$^{3}_{1}\text{H} \rightarrow {}^{3}_{2}\text{He} + e^{-} + \nu$$

Since the decaying nucleus emits two particles they can share the decay energy in various combinations.*

The $\beta^{\pm}$ decay processes *in nuclei* result in the following conversions

$$\text{n} \rightarrow \text{p} + e^{-} + \nu \qquad (\beta^{-} \text{ decay}) \qquad (33.8)$$
$$\text{p} \rightarrow \text{n} + e^{+} + \nu \qquad (\beta^{+} \text{ decay}) \qquad (33.9)$$

Both processes tend to occur whenever they produce a nucleus with more binding energy. The β^{+} decay can never happen for a free proton, since the proton mass is less than that of the neutron. However, the free neutron does β^{-} decay with a half-life of 11 minutes.

Another closely related way in which protons are changed into neutrons in some nuclei is *electron capture*. Here an atomic electron and a proton interact via

$$\text{p} + e^{-} \rightarrow \text{n} + \nu \qquad (\text{electron capture}) \quad (33.10)$$

Two additional aspects of beta decay are worth noting. First, the forces responsible for the beta decay

---

*According to the modern form of Fermi's theory, an *antineutrino* ($\bar{\nu}$) is emitted in a β^{-} decay, and a *neutrino* (ν) in a β^{+} decay. We will ignore the differences between the neutrino and antineutrino.

processes, including electron capture, are very weak compared both to the strong nuclear forces and to the electromagnetic forces among the nucleons. Accordingly, these β decay forces are referred to as *weak interactions;* they are the weakest of the fundamental forces in nature, except the gravitational forces. Second, as we mentioned earlier, beta decay is responsible for the fact that in stable nuclei the proton and neutron levels are filled to about the same maximum energy. When electrical forces are taken into account, this implies $N \simeq Z$ for light nuclei and $N \simeq 1.5\, Z$ for the heaviest nuclei.

α Decay

Unlike β decay, α decay always produces particles with the same kinetic energy. The emitted particle is a $^4_2$He nucleus, with two protons and two neutrons. Alpha decay is usually observed in the heavier unstable nuclei. All known nuclei above $Z = 83$ are unstable; those that do not β decay have been observed to α decay with half-lives ranging from about 10^{-3} seconds to 10^{10} years.

In β decay, the weak interaction determines the decay time. We might expect α decay to be much faster, since it is due to the strong nuclear interaction. However, typical decay processes are much slower than β decays. Why then does α decay take so long if it is energetically favorable for the strong nuclear forces to eject four nucleons in the form of an α particle?

The explanation of why this process is so slow involves a quantum mechanical phenomenon called *tunnelling*. In classical physics, a ball bouncing elastically inside an open container can bounce over the walls and escape from the container only if its energy is great enough. Specifically, its energy must exceed the gravitational potential energy at the walls (Fig. 33.13). Quantum mechanically, if a particle reaches a

Figure 33.14. The quantum mechanical wave associated with a particle approaching a barrier, a region where the potential energy exceeds the energy of the particle. The wave function ψ is small but nonzero to the right of the barrier, so the probability $|\psi|^2$ that the particle may be found beyond the barrier is greater than zero. This probability diminishes rapidly as the barrier becomes taller or wider.

region of high potential energy without sufficient energy to pass over this *barrier,* it has a small but nonzero probability of passing through the barrier (Fig. 33.14). This means that each time a particle strikes the barrier, it has a small probability of getting past the barrier. After enough impacts, the odds are that it will have passed through the barrier and escaped. It is then said to have *tunnelled* through the region where, according to classical physics, it lacks sufficient energy to ever penetrate.

To see how this argument applies to α decay, we must consider the origin of the energy barrier for α particles. We do this by examining the potential energy curve for an alpha particle and a nucleus (Fig. 33.15). If we bring an α particle up to a nucleus, we must do work to overcome the electrical repulsion. Thus the potential energy increases as the particles approach and r decreases. At the nuclear surface, the α particle begins to feel the attractive nuclear force, which corresponds to a negative potential energy.

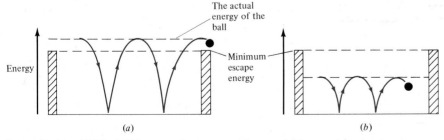

Figure 33.13. (*a*) The ball has enough energy and bounces high enough so that it can escape from the container. (*b*) The ball is trapped according to classical mechanics.

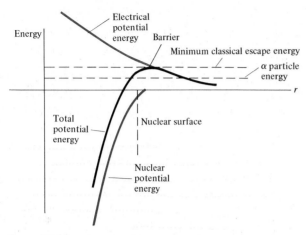

Figure 33.15. The potential energy of an α particle and a nucleus.

The sum of the two potential energies has a positive maximum value somewhere near the nuclear surface. This peak corresponds to the barrier in the preceding discussion.

   When the Schroedinger equation for the α particle inside this potential well is solved, it is found that the lowest α particle state has an energy greater than zero but below the barrier peak. Thus the alpha can reach a lower energy state by leaving the nucleus, and if it strikes the barrier often enough, it will eventually escape. This will take a long time if the barrier is very high or wide. Detailed calculations based on this theory are in excellent agreement with the measured α decay half-lives and energies.

SUMMARY

The half-life of a radioactive or unstable nucleus is the time required for half the nuclei to spontaneously decay. Nuclei are approximately spheres of radius $1.4\,A^{1/3}$ fm. The mass of a nucleus is less than the mass of its constituent protons and neutrons by an amount equal to its mass defect. According to the Einstein mass-energy relationship, this represents the nuclear binding energy.

   Using beams of particles produced by accelerators, the energy levels and structure of nuclei can be studied in detail. The nuclear shell model, which is similar to the atomic shell model, accounts for much of this data. Light nuclei either have $Z \simeq N$ or are unstable against β decay; because of electrical forces, very heavy stable nuclei have $N \simeq 1.5\,Z$. Excited states of nuclei tend to γ decay to their ground states. All nuclei above $Z = 83$ are unstable; those which do not β decay are all α particle emitters, and some will also spontaneously fission.

Checklist
Define or explain:

mass number	equivalence of mass
atomic number	and energy
neutron number	fission
isotope	fusion
alpha, beta, gamma	nuclear force
radiation	nuclear shell model
half-life	internal conversion
nuclear radius	positron
mass defect	electron capture
binding energy	tunnelling

EXERCISES

Section 33.2 | Half-Life

33-1   After 24 hours the radioactivity of a nuclide is 1/8 times its original level. What is its half-life?

33-2   How many half-lives are required for the activity of a radionuclide to decrease by a factor of 64?

33-3   Using the data in Table 33.1, find the effective half-life for ^{35}S.

33-4   A radionuclide with a physical half-life of 10 days is observed to have an effective half-life of 6 days when administered to a patient. What is the biological half-life of the nuclide?

33-5   Estimate the half-life of the radioactive substance whose count rate is given in Table 33.6 by inspection of the table.

TABLE 33.6

Time (days)	Count Rate (counts per minute)
0	455
1	402
2	356
3	315
4	278
5	246
6	218
7	193
8	171
9	151
10	133

33-6   Find the half-life of the radioactive substance whose count rate is given in Table 33.6 by plotting the data on semilog paper.

Section 33.3 | Dating in Archaeology and Geology

33-7   A wooden bowl has one-fourth the ^{14}C activity observed in contemporary wooden objects. Estimate its age. (Assume ^{14}C levels in the atmosphere have remained the same.)

33-8   A rock contains three ^{207}Pb nuclei for each ^{235}U nucleus. How old is the rock, assuming all the ^{207}Pb is from the uranium decay?

33-9   Why is there no radiocarbon in fossil fuels?

Section 33.4 | Nuclear Sizes

33-10   (a) Find the nuclear radii for ^{4}He, ^{27}Al, ^{64}Cu, ^{125}I and ^{216}Po. (b) Draw a rough graph of the nuclear radius versus A.

33-11   What fraction of the volume of a helium atom is occupied by its nucleus, assuming an atomic radius of 10^{-10} m?

33-12   (a) Calculate the density of an oxygen nucleus in kilograms per cubic meter. (b) Find the ratio of this density to the density of water, 10^3 kg m^{-3}.

Section 33.5 | Protons and Neutrons

33-13   How many neutrons are there in $^{14}_{6}$C, $^{36}_{17}$Cl, $^{64}_{29}$Cu, and $^{208}_{82}$Pb?

33-14   (a) Of the nuclides $^1_1$H, $^2_1$H, $^3_1$H, $^3_2$He, and $^4_2$He, which have the same neutron numbers? (b) Which nuclides have similar chemical properties?

Section 33.6 | Nuclear Masses and Binding Energies

33-15   The atomic mass of $^{208}_{82}$Pb is 207.9766 amu. What is its average binding energy per nucleon?

33-16   The atomic mass of $^{207}_{82}$Pb is 206.9759 amu, and the atomic mass of $^{208}_{82}$Pb is 207.9766 amu. (a) Find the difference of their mass defects. (b) What is the minimum energy needed to remove a neutron from $^{208}_{82}$Pb?

33-17   A nucleus has a mass defect of 1.5 amu. (a) What is its binding energy in MeV? (b) If the mass number is 200, find the binding energy per nucleon.

Section 33.7 | Nuclear Forces

33-18   (a) Find the magnitude of the electric force between two protons separated by a typical nuclear distance, 10^{-15} m. (b) A typical atomic dimension is about 10^{-10} m. How large is the electrical force between a proton and an electron with this separation? (c) Calculate the ratio of the two forces in parts (a) and (b).

33-19   When a proton moves toward a nucleus, its initial kinetic energy must be great enough to overcome the electrical repulsion if it is to experience nuclear forces. For the largest naturally occurring nucleus, $^{238}_{92}$U, this minimum energy is about 15 MeV. Find the corresponding energies for incident (a) alpha particles ($^4_2$He nuclei); (b) neutrons.

Section 33.8 | Nuclear Energy Levels and Nuclear Stability

33-20   What is X in the following reactions? ($d = {}^2_1$H) (a) $p + {}^{12}_{6}$C $\to d + X$. (b) ${}^3_2$He $+ {}^3_2$He $\to {}^4_2$He $+ p + X$. (C) $p + {}^{14}_{7}$N $\to {}^{12}_{7}$N $+ X$.

33-21   What is X in the following reactions? ($d = {}^2_1$H, $\alpha = {}^4_2$He) (a) $d + d \to X + p$. (b) ${}^6_3$Li $+ X \to \alpha + \alpha$. (c) $X + {}^{16}_{8}$O $\to {}^{19}_{9}$F $+ p$.

33-22   In Fig. 33.10, the ground states of $^{12}_{5}$B and $^{12}_{7}$N and an excited state of $^{12}_{6}$C have the same states filled. (a) Which of these nuclides has the greatest binding energy if electrical forces are taken into account? (b) What effect does the neutron-proton mass difference have on the relative energies? Explain.

Section 33.9 | Radioactive Decays

33-23   An excited nucleus decays, losing 2 MeV. Find (a) the frequency of the gamma ray photon emitted; (b) the wavelength of the photon.

33-24   Complete the following decay processes by adding the missing decay particles (α, γ, or $\beta^{\pm} + \nu$): (a) ${}^{11}_{6}$C $\to {}^{11}_{5}$B $+$? (b) ${}^{32}_{15}$P $\to {}^{32}_{16}$S $+$? (c) ${}^{12}_{6}$C$^* \to {}^{12}_{6}$C $+$? (d) ${}^{240}_{94}$Pu $\to {}^{236}_{92}$U $+$?

PROBLEMS

33-25   How many half-lives have elapsed if the activity of a radionuclide has diminished to one percent of its initial value?

33-26   An excavated wooden beam has 20 percent of the ^{14}C found in atmospheric carbon. How old is it? (Assume the atmospheric ^{14}C levels have remained the same.)

33-27 ^{90}Sr has a half-life of 28 years. How many years must this material be stored before its activity drops to $1/e$ times its original value?

33-28   A radioactive material contains two radionuclides, one with a half-life of 1 day and the other with a half-life of 8 days. Initially the radioactivity of the shortlived nuclide is $2^7 = 128$ times greater than that of the longlived nuclide. When will their activities be equal?

* **33-29**   In the preceding problem, suppose one cannot distinguish the radiation from the two radionuclides and plots the *total* count rate on semilog paper. (a) What is the shape of the graph for the first few days? (b) Why does the shape change when the two activities are nearly equal? (c) What is the shape of the graph when the shortlived nuclide has largely disappeared? Is there any difference from the initial curve?

33-30   The μ meson (muon) is a short-lived particle which, like the electron, has electrical interactions with nuclei but does not experience the strong nuclear force. The muon mass is 207 times that of an electron. Both $\mu^{\pm}$ have been observed and the magnitude of their charge is equal to that of the electron. (a) What is the radius of the first Bohr orbit of a negative muon in orbit around a $^{238}_{92}$U nucleus? (b) Find the ratio of that radius to the nuclear radius. (Actually the radii and energy levels of muonic atoms are changed somewhat from the Bohr values because a significant fraction of the wave function is inside the nucleus for lowlying atomic states. These changes provide information about the distribution of protons in the nucleus.)

33-31   The energy of the sun is derived from reactions which convert four protons into an alpha particle ($^4_2$He) with the emission of two positrons and two massless neutrinos. (a) How much energy in MeV is released in this process? (See Table 33.4 for the required atomic masses.) (b) When H_2 and O_2 gases combine to produce H_2O, 6.2 eV is released for each molecule formed. What mass of hydrogen would have to be burned to equal the energy released by the nuclear fusion of 1 kg of hydrogen into alphas?

33-32   How can reactions in which a projectile

transfers nucleons to a target nucleus be used to find nuclear energy levels?

33-33 $^9_4$Be is stable but $^9_5$B is unstable as is $^9_3$Li. (a) Draw nuclear energy level diagrams similar to those of Fig. 33.10 for these three $A = 9$ nuclides. (b) Suggest why two of these three nuclides are unstable. (c) What decay process would be expected for $^9_3$Li?

33-34   After a heavy nucleus α decays, a β^- decay often follows but never a β^+ decay. What is the reason for this?

SUPPLEMENTARY TOPICS
33.10 | NUCLEAR FISSION

We saw earlier that the binding energy per nucleon decreases gradually as A increases beyond 100. Consequently, about 1 MeV per nucleon is released if a large nucleus such as uranium splits into two smaller fragments. Such fission processes are the energy sources in nuclear power reactors and in the fission weapons first developed during World War Two.

A simple model of the $^{235}_{92}$U nucleus shows how fission occurs. The model pictures a $^{235}_{92}$U nucleus as composed of two parts, each containing a large number of protons and neutrons. The potential energy curve for the two parts is much like that of an α particle and a nucleus. If the two positively charged parts are initially far apart and gradually approach, their electrical potential energy rises. When they become close enough so the strong attractive nuclear forces play a role, the total potential energy begins to diminish (Fig. 33.16).

The ground state of the ^{235}U nucleus turns out to have an energy considerably greater than zero, but the system is trapped in its bound state below the peak in the potential energy curve. The region that is inaccessible according to classical physics is called the *fission barrier*. As in α decay, tunneling through the barrier makes it possible for this nucleus to *spontaneously fission*. As the two parts of the nucleus separate, a large amount of electrical potential energy is converted into kinetic energy. The spontaneous fission half-life is about 700 million years for ^{235}U and over 4 billion years for ^{238}U.

Some nuclides can also undergo *induced fission* when they are bombarded by *thermal neutrons,* very slow neutrons with kinetic energies comparable to kT

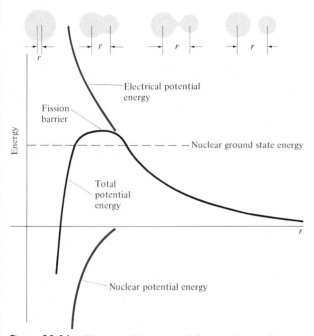

Figure 33.16. The potential energy of the uranium nucleus versus the separation r of the two parts of the nucleus. The corresponding shape of the nucleus is shown above the graph.

(less than 1 eV). Thermal neutrons have a large de Broglie wavelength and therefore a large effective size for an interaction. Many nuclei have high probabilities (large *cross sections*) for capturing thermal neutrons that pass close by and incorporating them into their structure. When this happens, a nucleus with A nucleons is transformed into one with $A + 1$ nucleons. The average binding energy of the last nucleon in a heavy nucleus is roughly 7 MeV. Hence when the neutron is captured the $A + 1$ particle nucleus will have an energy excess of about 7 MeV, and it will be in an excited state. If this excitation energy is sufficient to put the nucleus above the fission barrier, it will immediately fission, releasing a large amount of energy.

Precisely this happens when ^{235}U is bombarded by thermal neutrons. An excited state of ^{236}U is formed, with an energy 6.8 MeV above the ground state. Since the ^{236}U fission barrier height is also 6.8 MeV, the nucleus immediately fissions. On the other hand, ^{238}U does not fission when it captures a slow neutron, because the resulting excitation energy in the ^{239}U nucleus is 5.3 MeV, while the fission barrier is 7.1 MeV. Consequently, induced fission can occur with ^{238}U only if the incoming neutron has at least 1.8 MeV of kinetic energy. Such fast neutrons have shorter de Broglie wavelengths and smaller effective sizes and are hard to capture. The fast neutron fission cross section for ^{238}U is 1/2000 times the thermal neutron fission cross section for ^{235}U. For this reason, ^{235}U is much more useful in a fission energy source.

Natural uranium is 99.3 percent ^{238}U and only 0.7 percent ^{235}U. However, it is necessary to use nuclear fuels with a larger percentage of ^{235}U in most types of power reactors and in nuclear weapons. Since separation of these isotopes is costly and difficult, this both retards nuclear power development and tends to stabilize some aspects of international politics.

The energy released in nuclear fission can be estimated by noting that the binding energy per nucleon is 7.6 MeV for ^{235}U and about 8.5 MeV for the two $A \simeq 100$ fission fragments. (See Fig. 33.7). Consequently the energy released is 235(8.5 MeV—7.6 MeV) $\simeq$ 200 MeV per nucleus. To put that in perspective, we may note that chemical reactions typically liberate about 1 eV per atom, so nuclear fission is 200 million times more powerful! If all the nuclei in one kilogram of ^{235}U fission, the energy released is equivalent to that of 20,000 tons of TNT. The World War Two fission bombs released approximately this much energy.

Usually the fission fragments have proportionately too many neutrons for medium A nuclei, so they eject

Figure 33.17. Induced fission of ^{235}U. A thermal neutron approaching a ^{235}U nucleus is captured, producing an excited ^{236}U nucleus which fissions. In this example the two fission fragments that are produced emit three prompt neutrons.

591

one or more *prompt neutrons* almost immediately in 10^{-13} seconds or less (Fig. 33.17). They then undergo a series of three or four β^- decays, each of which converts one neutron into a proton and brings the nucleus closer to stability. Accordingly, the first β^- decay proceeds in a few seconds, while the next has a half-life of minutes or hours, and the last may take days or years. It is these relatively long-lived radioactive nuclides that comprise the *radioactive wastes* produced in reactors and that must be safely stored until the radioactivity is negligible. Sometimes a *delayed neutron* is ejected after the first β^- decay. On the average, there are 2.6 prompt neutrons for each fission; about 1 percent of the fission events lead to the emission of a delayed neutron after a delay averaging some 10 seconds.

The neutrons released in an induced ^{235}U fission can be captured by other ^{235}U nuclei, causing them to fission in turn in a *chain reaction.* Suppose, on the average, the 2.6 neutrons from one fission induce more than one additional fission. The *fission ratio* is then greater than 1, and the reaction will grow exponentially. Before long an explosion will occur if the growth proceeds unchecked. On the other hand, if enough neutrons escape from the ^{235}U or are absorbed by other materials, the fission ratio is less than 1. The process will then not be self-sustaining and will die out after being initiated by a stray neutron. When each fission produces exactly one more fission, the fission ratio is one and the reaction continues at a constant rate. The ^{235}U is then said to have a *critical mass.* In a nuclear reactor the fission ratio is controlled so that fission reactions occur at a desired rate.

33.11 | FISSION REACTORS AND EXPLOSIVES

To achieve a critical mass, explosives and most fission reactors require enriched uranium containing more than the 0.7 percent of ^{235}U found in natural uranium. In the most common reactors, enrichment to 3 percent is required, while explosives use about 90 percent enriched material. The two uranium isotopes cannot be separated chemically, since their chemical properties are almost identical. However, the average molecular kinetic energy is determined by the temperature, so molecules containing a heavier isotope move more slowly on the average. Hence physical processes, such as diffusion, that depend on the aver-

age velocity can be used to separate isotopes. The masses of the uranium isotopes are so similar that although the separation may be accomplished by diffusion, it is slow and costly. Electromagnetic separators, which are similar to mass spectrometers, are sometimes used to further concentrate partially enriched uranium.

Reactors | All the commercial electric power reactors in the United States employ thermal neutrons to induce fission and have several common features. They have many *fuel rods,* tubes containing enriched uranium, along with a *moderator* and *control rods.* The moderator serves to slow down the neutrons. This is necessary because the neutrons are produced with an initial energy of several MeV, and the ^{235}U fission cross section is largest below 1 eV. The control rods are made of a substance such as boron, which has a large neutron capture probability, and are used to control the fission rate. As these rods are withdrawn from the reactor assembly, the chain reaction grows until the desired rate is reached. The approximately 1 percent of the neutrons that are delayed play a key role in this process. They permit the operators to gradually adjust the control rods and to make readjustments if the reaction grows too rapidly.

The majority of the commercial reactors in opera-

Figure 33.18. The boiling water reactor. The water serves as the moderator and also carries off the heat energy produced. Steam heated in the reactor drives a turbine and is then condensed before returning to the reactor.

Figure 33.19. The reactor at Vernon, Vermont, is a typical boiling water reactor. Located on the Connecticut River, its cooling system normally requires 800 ft³ of water per second, which exceeds the flow of the river during dry periods. Hence, it is equipped with cooling towers that require evaporation of much smaller quantities of water. The reactor is designed to produce 515 MW of electrical power, which is about 0.1% of the U.S. electrical generating capacity as of 1978. It has 68 tons of uranium enriched to 2.7% ^{235}U contained in 23,000 fuel rods. (ERDA photo.)

tion or under construction in the U.S. are *boiling water reactors* (Figs. 33.18 and 33.19). The water in the reactor serves as the moderator; neutrons slow down when they collide with the hydrogen nuclei (protons) of the water molecules. The energy released in the fission process is converted into heat when the fast moving fission fragments collide with other atoms in the fuel rods. The heated water carries off this heat energy and then in the form of steam is used to drive turbines and produce electricity.

An important feature of boiling water reactors is their intrinsic stability against runaway chain reaction accidents. Any unintended rise in temperature increases the rate of boiling and bubble formation, thereby reducing the amount of water available to moderate the neutrons and slowing the chain reaction. Thus, such a reactor cannot explode like a nuclear bomb, and only an energy equivalent to a minor chemical explosion could be released in an accident. The controversy surrounding the use of nuclear power reactors centers not so much on the risk of a large-scale explosion as on the possible accidental dispersal of radioactive wastes into the environment.

Breeder Reactors | If the number of nuclear power reactors expands rapidly, uranium supplies are likely to be depleted in a few decades. A possible solution to this problem is the *breeder reactor,* which produces more fissionable material than it consumes. One type of breeder reactor produces ^{239}Pu, which can be fissioned like ^{235}U by thermal neutrons. In this reactor a fast neutron causes the reaction

$$^{238}_{92}\text{U} + \text{n} \rightarrow {}^{239}_{92}\text{U} + \gamma$$

This is followed by two β^- decays, which produce first neptunium and then plutonium:

$$^{239}_{92}\text{U} \rightarrow {}^{239}_{93}\text{Np} + \beta^-, \qquad {}^{239}_{93}\text{Np} \rightarrow {}^{239}_{94}\text{Pu} + \beta^-$$

The plutonium α decays with a half-life of 24,000 years, so it is relatively stable. It is a suitable material for thermal neutron reactors as well as for bomb construction but has the unfortunate property of being highly toxic.

Since fast neutrons are needed for breeding, a breeder reactor contains no moderator. The design currently receiving the greatest study in the United States uses as a coolant liquid sodium metal, which is notoriously reactive chemically. The reactor fuel is highly enriched, and the reactor core is very compact and hot. This places very great demands on the components and materials.

Fission Explosives | In a fission or atomic bomb, a subcritical sphere of ^{235}U or ^{239}Pu is surrounded by a chemical explosive. When this is detonated, shock waves implode the sphere, substantially reducing its volume and increasing its density. This makes the fission ratio greater than 1, and the chain reaction grows rapidly. Because of the inertia of the imploding material, the fission ratio stays above 1 long enough for a large total energy release to occur before the bomb material disperses. Such a bomb is limited in size by the requirements that it must not be critical before it is detonated, and it must be brought to critical size quickly and held together long enough for a substantial amount of fission to occur. The largest fission explosives are equivalent to about 250 kilotons of TNT, and are about 12 times the size of the World War Two atomic bombs.

33.12 | NUCLEAR FUSION

Nuclear fusion is potentially a much greater source of energy than fission because the supplies of suitable materials are nearly inexhaustible. Fusion is also attractive because its final products are stable and do not present radioactive waste disposal problems. Nevertheless, the achievement of controlled fusion and the extraction of useful power is a very difficult challenge.

The source of the difficulty is seen by considering two fusion reactions among the hydrogen isotopes, deuterium ($d = {}^2_1\text{H}$) and tritium ($t = {}^3_1\text{H}$):

$$d + d \rightarrow t + p + (4 \text{ MeV kinetic energy})$$
$$t + d \rightarrow {}^4_2\text{He} + \text{n} + (17.6 \text{ MeV kinetic energy})$$

In both reactions, there is a strong electrical repulsion keeping the two positively charged nuclei apart. Unless their total initial kinetic energy is at least 0.1 MeV, they cannot come close enough together for the nuclear force to cause them to fuse. It is easy to supply 0.1 MeV or more to a nucleus in an accelerator, but such a machine consumes much more energy than is released, since fusion occurs in only a relatively few nuclei. Thus the only practical way to attain large-scale fusion is in a *thermonuclear reaction,* that is, by heating the materials until at least a small fraction of the nuclei have enough kinetic energy to fuse. This requires temperatures of about one million degrees Celsius.

The necessary temperatures are attained in a fission-fusion bomb by using a fission bomb to heat the hydrogen isotopes. Tritium is desirable in a thermonuclear device, since the *d-t* fusion process starts at a lower temperature than does the *d-d* reaction. However, tritium is radioactive, with a half-life of 12 years, so it is costly to produce and difficult to work with. These problems are avoided in hydrogen bombs by effectively storing the needed hydrogen isotopes in a solid, nonradioactive form in the chemical compound ${}^6\text{Li}$ d, lithium deuteride. When struck by a fast neutron from the fission explosion, the ${}^6\text{Li}$ nuclide produces tritium in the reaction

$$ {}^6_3\text{Li} + \text{n} \rightarrow {}^4_2\text{He} + t + (4.8 \text{ MeV kinetic energy})$$

The tritium produced in this reaction then undergoes fusion with the deuterium. The required ${}^6\text{Li}$ is separated from natural lithium, which also contains the more abundant ${}^7\text{Li}$, at a much lower cost than that of producing tritium.

The basic problem in achieving controlled fusion is containing the materials, since no solid can exist at the required temperatures. Since the hot deuterium or tritium gases are completely ionized, they form an electrically conducting mixture of positive and negative charges, which is called a *plasma* and can be contained by magnetic fields. After the plasma is heated by bombardment from an electron gun, the magnetic field is suddenly increased. This compresses the plasma, raises its temperature still further and causes some nuclei to fuse.

Although thermonuclear reactions have been produced in this way, it is not yet possible to produce them for a long enough time and in a large enough volume to be useful. It may turn out that an approach quite different from the magnetic containment schemes will work best. For example, considerable progress has been made recently using intense laser beams or electron beams focused on tiny spheres containing hydrogen isotopes. In general, progress has been slower than was hoped when the fusion program began in the 1950s, and practical power seems to be decades away at best. However, deuterium, unlike fossil fuels and uranium, is available in almost unlimited amounts. This is because one hydrogen nucleus out of 20,000 is a deuteron, making the oceans giant reservoirs of this fuel. Thus the potential benefits from the successful generation of power from deuterium fusion are very great.

33.13 | DERIVATION OF THE EXPONENTIAL DECAY FORMULA

We saw in Section 33.1 that the change ΔN in the number of nuclei N in a radioactive sample in a short time interval Δt is

$$\Delta N = -\lambda N \Delta t \qquad (33.1)$$

Dividing by Δt and letting Δt become arbitrarily small,

$$\frac{dN}{dt} = -\lambda N$$

If at $t = 0$ there are N_0 nuclei, then as may be verified by differentiation, the solution of this equation is

$$\frac{N}{N_0} = e^{-\lambda t} \qquad (33.2)$$

This is the exponential decay formula stated earlier.

To relate λ to T, we note that when one half-life has elapsed, N/N_0 is $\frac{1}{2}$. Substituting $t = T$ in Eq. 33.2 gives

$$e^{-\lambda T} = \frac{1}{2}$$

or

$$e^{\lambda T} = 2$$

Taking the natural logarithm of both sides, we find $\lambda T = \ln 2 = 0.693$. Thus we obtain the result given earlier,

$$\lambda = \frac{0.693}{T}$$

EXERCISES ON SUPPLEMENTARY TOPICS

Section 33.10 | Nuclear Fission, and
Section 33.11 | Fission Reactors and
Explosives

33-35   Why are β^-'s and γ's emitted in the decay of fission fragments but not β^+'s?

33-36   If control rods are inserted so that the chain reaction in a reactor stops, the reactor still produces a substantial amount of heat for a long time. Why?

33-37   Geologists spend a great deal of time inspecting potential nuclear reactor sites. What do you think they are looking for?

33-38   It is easier to separate fissionable ^{239}Pu bred in a reactor from the uranium and other materials present than to separate ^{235}U from the more abundant ^{238}U. Why? (This plutonium isotope is suitable for bombmaking. One reason for international inspection of reactors sold to the non-nuclear powers is that any thermal reactor breeds a small amount of ^{239}Pu. Over a period of years a reactor could make enough plutonium to assemble a bomb.)

33-39   Suppose an object of mass m_1 strikes a stationary object of mass m_2 squarely so that object 2 moves straight ahead. If the collision is elastic the final and initial kinetic energies of object 1 satisfy

$$K_f = \left(\frac{m_1 - m_2}{m_1 + m_2} \right)^2 K_i$$

Use this result to explain why the hydrogen nuclei are more effective than the oxygen nuclei in water in slowing down neutrons.

Section 33.12 | Nuclear Fusion

33-40   The $^3_1$H-$^2_1$H fusion process releases 17.4 MeV, while $^{235}_{92}$U fission releases about 200 MeV. Find the energy released per nucleon for each process.

33-41   The reaction ^{6}Li $+ d \rightarrow \alpha + \alpha + 22$ MeV occurs in a hydrogen bomb. Why does this reaction need higher temperatures to occur than does the $d + d$ or $d + t$ fusion reactions? ($d = ^2_1$H; $t = ^3_1$H.)

33-42   2.2 MeV γ rays are produced when a proton captures a slow neutron in the reaction p + n $\rightarrow$ $d + \gamma$. ($d = ^2_1$H.) Why can this fusion process occur at very low temperatures? (This process is not usable as a fusion power source since suitable neutron sources do not exist.)

PROBLEMS ON SUPPLEMENTARY TOPICS

33-43   (a) Calculate the energy released in joules when 1 kg of ^{235}U fissions, assuming 200 MeV per nucleus is released. (b) The electric power capacity of the U.S. is about 5×10^{11} W. Assuming 30 percent efficiency in producing electricity from nuclear energy, at what rate must ^{235}U be consumed to provide this power?

33-44   In a pressurized water reactor, as in a boiling water reactor, water cools the reactor core and serves as a moderator. However, the water is kept at high pressure so that it is well below its boiling point. Why is this reactor less stable than a boiling water reactor? (*Hint:* What is the effect, if any, of a local increase in the water temperature?)

33-45   (a) At what temperature is the average thermal kinetic energy of a nucleus, $\frac{3}{2}kT$, equal to 0.1 MeV? (b) Why is it not necessary to achieve this temperature to initiate a fusion reaction which requires a kinetic energy of 0.1 MeV?

(Los Alamos Laboratory.)

ENRICO FERMI

(1901 – 1954)

Physics has become sufficiently specialized so that most physicists concentrate their efforts either on experimental work or theoretical calculations. Enrico Fermi was a notable exception to this rule. He made many important contributions of both types to physics.

Born in Rome, Fermi received his doctorate in 1922 from the University of Pisa for his research on x-rays. As a student he taught himself the new quantum theories that were being developed elsewhere but were not yet known in Italy. His efforts to explain these concepts to his fellow students and his professors helped to establish modern physics in Italy and also developed his abilities as a teacher.

After a brief period of further study in Germany and Holland, Fermi returned to Italy in 1924 and became a professor at the University of Rome in 1926. In that year, he developed the theory of an ideal gas whose atoms obey the exclusion principle introduced in 1925 by Pauli. He found striking departures from the behavior predicted using classical physics, particularly at low temperatures and high densities. An example of such a *Fermi gas* is provided by the conduction electrons in a metal, which are nearly free and obey the exclusion principle. Fermi showed that many previously unexplained features of the electrical and thermal properties of metals were correctly predicted by this theory.

Turning his attention to nuclear physics, Fermi proposed a theory of beta decay in 1933, which in a slightly modified form remains our present-day basis for understanding this process. In 1934, he began a series of experiments in which he systematically bombarded a variety of targets with neutrons. Soon he discovered that placing water or a hydrocarbon between the source and the target increased the rate of production of artificial radioactivity. Fermi realized that the light atoms had absorbed some of the kinetic energy of the neutrons in a series of collisions and that the resulting slow neutrons were more readily captured by the target nuclei.

When a neutron is captured by a nucleus, its mass number A is increased by one; if a subsequent β^- decay occurs, the atomic number Z also is increased by one. Fermi and his collaborators therefore tried, in 1934, to go beyond the last known element by bombarding uranium ($Z = 92$) with neutrons. They thought mistakenly that they succeeded in producing the first transuranic element

$(Z = 93)$ when in fact they had caused uranium nuclei to fission. This was not realized until further work was done by Hahn and Strassman in Germany in 1939 which conclusively identified the fission products. Shortly before Fermi and his family had fled the Fascist regime in Italy by traveling to Stockholm where he accepted the Nobel prize and then proceeding to New York. It was there that he learned of the work done in Germany.

Fermi immediately realized the importance of the discovery of fission and the possibility of a nuclear chain reaction. Working initially at Columbia University and later at the University of Chicago, he supervised the construction of the first nuclear reactor. When it was first operated on December 2, 1942, a now famous but then secret telegram announced that "the Italian navigator has entered the new world."

Fermi worked on the development of the atomic bomb during World War Two and then returned to academic life at the University of Chicago. In 1949, he joined several other leading scientists in opposing the development of the hydrogen bomb on ethical grounds. His postwar research centered on neutron studies, the properties of newly discovered particles called pi mesons, and the origin of cosmic rays. Shortly after his death in 1954, the artificially produced element with atomic number 100 was named fermium in his honor.

Additional Reading

W.A. Blanpied, *Physics: Its Structure and Evolution,* Ginn Blaisdell, Waltham, 1969, chapters 20–23.

W.F. Libby, *Radiocarbon Dating,* 2nd ed., University of Chicago Press, Chicago, 1955.

I. Perlman, F. Asara, and H.V. Michel, Nuclear Applications in Art and Archaeology, *Annual Reviews of Nuclear Science,* vol. 22, 1972, p. 383.

D.E. Nelson, R.G. Korteling, and W.R. Stott, Carbon 14: Direct Detection at Natural Concentrations, *Science,* vol. 198, 1977, p. 807.

C.L. Bennett et al., Radiocarbon Dating Using Electrostatic Accelerators: Negative Ions Provide the Key, *Science,* vol. 198, 1977, p. 808. A new method is described which will permit dating of much smaller and older samples.

D.R. Inglis, *Nuclear Energy: Its Physics and Its Social Challenge,* Addison-Wesley, Reading, Mass., 1973.

B.N. DaCosta Andrade, *Rutherford and the Nature of the Atom,* Science Study Series, Doubleday and Company, Garden City, N.Y., 1964.

Enrico Fermi, The Nucleus, *Physics Today,* vol. 5, March 1952, p. 6.

C.F. Tsang, Superheavy Elements, *Physics Teacher,* vol. 13, 1975, p. 279.

R.F. Post and F.L. Ribe, Fusion Reactors as Future Energy Sources, *Science,* vol. 186, 1974, p. 397.

Otto R. Frisch and John A. Wheeler, the Discovery of Fission, *Physics Today,* vol. 20, Nov. 1967, p. 43.

F.L. Culler, Jr. and W.O. Harms, Energy from Breeder Reactors, *Physics Today,* vol. 25, May 1972, p. 28.

K.G. McNeill, Photonuclear Reactions in Medicine, *Physics Today,* vol. 27, April 1974, p. 75. See also the printing corrections in vol. 27, May 1974, p. 71.

Scientific American articles:

O. Hahn, The Discovery of Fission, February 1958, p. 76.

Lawrence Bodash, How the Newer Alchemy Was Received, August 1966, p. 88.

Victor F. Weisskopf, The Three Spectroscopies, May 1968, p. 15.

G.T. Seaborg and A.R. Fritsch, The Synthetic Elements: III, April 1963, p. 68.

Glenn T. Seaborg and Justin L. Bloom, The Synthetic Elements: IV, April 1969, p. 56.

Clyde E. Wiegand, Exotic Atoms, November 1972, p. 102.

Vernon D. Barger and David B. Cline, High-Energy Scattering, December 1967, p. 76.

G. Feinberg and M. Goldhaber, Conservation Laws, October 1963, p. 36.

M.D. Kameu, Tracers, January 1949, p. 31.

C. Emiliani, Ancient Temperatures, February 1958, p. 54.

C. Renfrew, Carbon 14 and the Prehistory of Europe, October 1971, p. 63.

Stephen Moorbath, The Oldest Rocks and the Growth of Continents, March 1977, p. 92.

J.D. Macdougall, Fission-Track Dating, December 1976, p. 114.

Maria G. Mayer, The Structure of the Nucleus, March 1951, p. 228.

Hans Bethe, What Holds the Nucleus Together, September 1953, p. 201.

Victor F. Weisskopf and E.P. Rosenbaum, A Model of The Nucleus, December 1955, p. 261.

R.H. Hofstadter, The Atomic Nucleus, July 1956, p. 55.

R.B. Peierls, Models of the Nucleus, January 1959, p. 235.

Robert B. Marshak, The Nuclear Force, March 1960, p. 98.

Lawrence Cranberg, Fast-Neutron Spectroscopy, March 1964, p. 79.

Michael Baranger and Raymond A. Sorenson, The Size and Shape of Atomic Nuclei, August 1969, p. 58.

H.W. Kendall and W. Panofsky, The Structure of the Proton and the Neutron, June 1971, p. 66.

Chris D. Zafiratus, The Texture of the Nuclear Surface, October 1972, p. 100.

Sheldon Penman, The Muon, July 1961, p. 46.

Leon M. Lederman, The Two-Neutrino Experiment, March 1963, p. 60.

W.B. Fowler and N.P. Samios, The Omega-Minus Experiment, October 1964, p. 36.

John N. Bahcall, Neutrinos From the Sun, July 1969, p. 29.

Steven Weinberg, Unified Theories of Elementary Particle Interactions, July 1974, p. 50.

Sidney D. Drell, Electron-Positron Annihilation and the New Particles, June 1975, p. 50.

Sheldon Lee Glashow, Quarks with Charm and Color, October 1975, p. 38.

David B. Cline, Alfred K. Mann, and Carlo Rubbia, The Search for New Families of Elementary Particles, January 1976, p. 44.

S.B. Treimann, The Weak Interactions, March 1959, p. 72.

D.J. Hughes, The Reactor as a Research Instrument, August 1953, p. 23.

A. M. Weinberg, Power Reactors, December 1954, p. 33.

Alvin M. Weinberg, Breeder Reactors, January 1960, p. 82.

R.B. Leachman, Nuclear Fission, August 1965, p. 49.

Glenn T. Seaborg and Justin L. Bloom, Fast Breeder Reactors, November 1970, p. 13.

Hugh C. McIntyre, Natural-Uranium Heavy-Water Reactors, October 1975, p. 17.

George A. Cowan, A Natural Fission Reactor, July 1976, p. 36.

H.A. Bethe, The Necessity of Fission Power, January 1976, p. 21. See also April 1976, pp. 8–21 for letters to the editor.

Bernard L. Cohen, The Disposal of Radioactive Wastes From Fission Reactors, June 1977, p. 21.

William P. Bebbington, The Reprocessing of Nuclear Fuels, December 1976, p. 30.

George A. Vendryes, Superphénix: A Full Scale Breeder Reactor, March 1977, p. 26.

R.F. Post, Fusion Power, December 1957, p. 73.

T.K. Fowler and R.F. Post, Progress Toward Fusion Power, December 1966, p. 21.

Francis F. Chen, The Leakage Problem in Fusion Reactors, July 1967, p. 76.

William C. Gough and Bernard J. Eastlund, The Prospects of Fusion Power, February 1971, p. 50.

Moshe J. Lubin and Arthur P. Fraas, Fusion by Laser, June 1971, p. 21.

Bruno Coppi and Jan Rems, The Tokamek Approach in Fusion Research, July 1972, p. 65.

John L. Emmett, John Nuckolls, and Lowell Wood, Fusion Power by Laser Implosion, July 1975, p. 24.

CHAPTER THIRTY-FOUR

IONIZING RADIATION

The radioactive decay of nuclei produces several kinds of *ionizing radiation* with energies that are typically several million electron volts per particle or quantum. When this radiation passes through matter, it leaves a trail of ionized atoms along its path. Even a small amount of ionization can seriously disrupt a sensitive system such as a living cell or a transistor.

The term "ionizing radiation" includes both nuclear radiation and atomic x-rays. The less energetic quanta of lower frequency electromagnetic waves such as visible light and microwaves do not ordinarily cause appreciable ionization. In general, radiation refers only to ionizing radiation in this chapter.

Radiation is an excellent example of an area of science that has been studied intensively by physicists because of its intrinsic interest and that has also become invaluable in applications to many other fields including biology and medicine. Radiation also illustrates with unusual clarity how a scientific advance may, despite its great benefits, have a very large potential for harm. For example, x-ray pictures are often essential in diagnosing a serious illness, but even one x-ray exposure slightly increases the chance of developing cancer. Consequently, all those working with radiation, especially those in the health sciences, have an obligation to understand the physics and biology of radiation and to use it wisely and carefully.

In this chapter we first consider the interaction of radiation with matter. We then discuss common radiation sources and the biological effects of radiation. We conclude with a discussion of some applications where radiation is used.

34.1 | THE INTERACTION OF RADIATION WITH MATTER

There are four major categories of radiation of interest to us. In order of increasing range in matter, these are:

1 positive ions, such as alpha particles
2 electrons and positrons
3 gamma rays
4 neutrons

Positive Ions | Alpha particles, protons, and other positive ions have very short ranges in matter. Roughly speaking, the average range or stopping distance varies inversely with the density of the medium, so a 5-MeV alpha that can travel about 4 cm in air cannot penetrate a sheet of paper or a layer of skin (Fig. 34.1).

A fast alpha particle undergoes frequent collisions with atomic electrons as it passes through matter, leaving a wake of excited and ionized atoms. About 100 eV is transferred in a single collision, so many collisions occur as the alpha slows. When its kinetic energy decreases to about 1 MeV, the alpha acquires two electrons and becomes a neutral helium atom. The neutral atom comes to rest in a short distance after a few more collisions.

Because an alpha is so much more massive than an electron, it is scarcely deflected in the collisions, and its path is nearly a straight line. Although individual alphas do not all travel exactly the same distance before stopping, the spread in stopping distances or

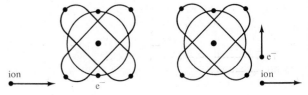

Figure 34.3. (a) An ion approaches an atomic electron which is moving slowly compared to the ion. (b) The ion collides with the electron, transferring momentum and energy to it. In this case the electron is removed, so the atom it came from is now ionized. Collisions can also excite atoms without ionizing them.

Figure 34.1. Range of alpha particles versus energy for air, water, aluminum, and lead. Water is a good approximation to soft animal tissue. (Adapted from G. S. Hurst and J. E. Turner, *Elementary Radiation Physics*, Copyright © 1970, John Wiley, New York.

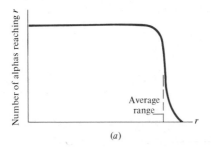

Figure 34.2. Alpha particles from a monoenergetic point source enter a medium. (a) The number of alphas passing a point a distance r into the medium. (b) The number of alphas stopping versus distance r traveled. Note most alphas travel a distance within a few percent of the average range.

straggling is only a few percent of the average range (Fig. 34.2).

Energy Loss Rate | The fact that alpha particles have a short range in matter compared to elec-

trons of the same energy can be understood from basic physical principles. An ion traveling through a medium collides repeatedly with atomic electrons and gradually loses energy (Fig. 34.3). We show in Section 34.8 that in a given medium, the energy ΔK lost per unit distance by an ion with charge q and velocity v is proportional to q^2/v^2, or

$$\Delta K \propto -\frac{q^2}{v^2} \qquad (34.1)$$

The minus sign indicates that the ion is losing kinetic energy. ΔK is also approximately proportional to the density of atomic electrons in the medium. It is convenient to rewrite this result for the rate of energy loss with distance in terms of the kinetic energy instead of v^2. If the ion is moving slowly compared to the speed of light, its kinetic energy is $K = \frac{1}{2}mv^2$, where m is the mass of the ion. Thus $v^2 = 2K/m$ and Eq. 34.1 can be rewritten as

$$\Delta K \propto \frac{-mq^2}{2K} \qquad (v \ll c) \qquad (34.2)$$

This shows that for a given kinetic energy, the energy loss rate is proportional to ion mass, so that massive particles such as alphas lose energy rapidly and come to rest in a short distance.

These equations are in good agreement with observed energy loss rates except at very low ion velocities. They show that the rate of energy loss increases as the ion slows and is a maximum near the end of the path. To the extent that they are correct, these equations allow one to compare the energy loss rates of various ions in a given medium. This is illustrated in the following example.

Example 34.1

Compare the rates of energy loss in a given material

for protons and alphas having the same initial kinetic energies, assuming $v \ll c$.

From Eq. 34.2, the energy loss rate for a given particle is proportional to mq^2. A proton ($^1_1$H) has a charge e, while an alpha ($^4_2$He) has a charge $2e$. Thus q^2 for the two particles is related by $q_p{}^2 = q_\alpha{}^2/4$. The mass of a proton is close to 1 amu and that of an alpha is close to 4 amu, so $m_p = m_\alpha/4$. The energy loss rate for the proton is then proportional to

$$m_p q_p{}^2 = \frac{m_\alpha}{4} \frac{q_\alpha{}^2}{4} = \frac{m_\alpha q_\alpha{}^2}{16}$$

Thus the rate of energy loss with distance for a proton is 1/16 times that of an alpha. This suggests that a proton has a range 16 times that of an alpha with the same initial energy. Experimentally, this is only roughly correct; the proton range is typically about 10 times the alpha range. This happens because, as noted, Eq. 34.2 does not hold for very low velocities.

Electrons and Positrons | These products of nuclear beta decays have ranges that are typically a hundred times greater than those of alpha particles. For example, a 1-MeV electron has a range in water or soft tissues of 0.4 cm (Fig. 34.4). Like positive ions, electrons lose energy mainly by ionizing or exciting atoms. However, for a given kinetic energy the velocity will be much larger than for a proton or an alpha, because the electron mass is so small. Thus the rate of

Figure 34.4. Average range of electrons in water, aluminum, and lead. The range of alpha particles in water is shown again for comparison. (Adapted from Hurst and Turner.)

energy loss given by Eq. 34.1 will be much smaller. This accounts for the large range of the electrons. Also, because of the small electron mass, a large deflection occurs in each collision with an atomic electron. Hence the electrons do not travel in a straight line, but instead wander randomly, and the straggling is large.

The range of positrons is approximately the same as that of electrons. Eventually a positron slows and comes close enough to an electron, so that they annihilate producing gamma rays.

Gamma Rays | Gamma rays do not produce appreciable ionization directly. Instead they lose energy to electrons which, in turn, cause ionization. Consequently, gamma rays have a long range in matter. For example, a 1-MeV gamma ray in water has a mean range of roughly 10 cm.

Gamma rays transfer energy to electrons by three processes (Fig. 34.5). For gamma rays with energies below 1 MeV, as well as x-rays, the *photoelectric effect* ① is most important; here a gamma ray photon is absorbed by an atom, and an atomic electron is ejected. This process is most likely for inner shell electrons and for atoms with large atomic numbers. At about ② 1 MeV, *Compton scattering* dominates; this is photon-electron scattering, with the γ transferring some but not all of its energy to an atomic electron. At ③ higher energies, it becomes possible for a gamma ray to produce an *electron-positron pair*. This can occur near a nucleus when the gamma-ray energy is greater than the total mass energy of the two particles, $2m_e c^2 = 1.02$ MeV. Above a few million electron volts this is the most likely fate of a gamma ray.

Gamma ray absorption probabilities usually diminish as the energy rises. Consequently, as the gamma ray energy increases, the absorption becomes more gradual, and the radiation becomes more penetrating or *harder*.

Neutrons | Neutrons are uncharged and produce ionization only indirectly. Since they interact primarily with the small atomic nuclei rather than the atomic electrons, they have a very long range in matter. Neutrons with energies of a few million electron volts may travel a meter or so in water or in animal tissues.

Neutrons are slowed by elastic scattering from nu-

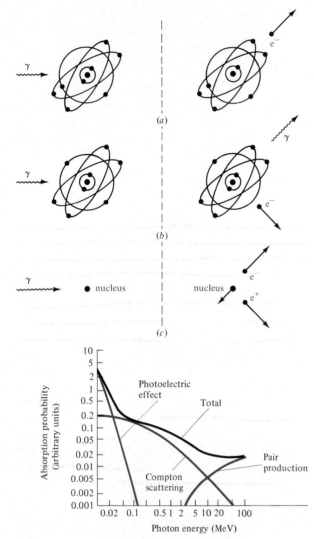

Figure 34.5.   Energy loss mechanisms for gamma rays. (*a*) Photoelectric effect. The gamma ray is absorbed by an atom, and an inner shell electron is ejected. (*b*) Compton scattering. Some energy and momentum is transferred to an atomic electron. (*c*) Pair production. The gamma ray disappears and an electron-positron pair is created. Some momentum is transferred to a nucleus. (*d*) Relative importance in water of the three mechanisms versus energy. [(*d*) is adapted from Gloyna and Ledbetter.]

clei and by nuclear reactions. Some of these reactions lead to proton or gamma-ray emission; the protons in particular are significant in causing biological effects. Once a neutron has slowed to thermal energies—less than 1 eV—it has a high probability of being captured by a nucleus. Such a capture is often followed by gamma-ray emission.

34.2 | RADIATION UNITS

Four types of radiation measurements are used in various applications: *source activity, exposure, absorbed dose,* and *biologically equivalent dose.* We discuss each of these briefly and define their most common units.

The *source activity* α is the disintegration rate of a radioactive material, or the rate of decrease in the number of radioactive nuclei present. It is measured in *curies,* where the curie is defined by the *exact* relationship

$$1 \text{ curie} = 1 \text{ Ci}$$
$$= 3.7 \times 10^{10} \text{ disintegrations}$$
$$\text{per second} \qquad (34.3)$$

One gram of radium has approximately 3.7×10^{10} disintegrations per second, so that 1 Ci of radium has a mass of about 1 g.

The activity of a sample is related to its half-life, T. From Chapter Thirty-three, $\Delta N = -\lambda N \Delta t$ and $\lambda = 0.693/T$, so

$$\alpha = \frac{-\Delta N}{\Delta t} = \lambda N = \frac{0.693}{T} N$$

The minus sign is needed because ΔN is the change in the number of nuclei present and is negative, while the disintegration rate or activity is positive. If there are n moles in the sample, then the number of atoms is $N = n N_A$, where Avogadro's number $N_A = 6.02 \times 10^{23}$ is the number of atoms in a mole. The activity of n moles of a sample is then

$$\alpha = \frac{0.693}{T} n N_A \qquad (34.4)$$

Note that the activity varies inversely with the lifetime. This result is used in the following example.

Example 34.2

^{60}Co beta decays with a half-life of 5.27 years = 1.66×10^8 seconds into ^{60}Ni, which then promptly emits two gamma rays. These gamma rays are widely used in treating cancer. What is the mass of a 1000-Ci cobalt source?

Solving Eq. 34.4 for the number of moles and using $\alpha = 1000$ Ci,

$$n = \frac{\alpha T}{0.693 \, N_A}$$
$$= \frac{1000(3.7 \times 10^{10} \text{ s}^{-1})(1.66 \times 10^8 \text{ s})}{0.693(6.02 \times 10^{23} \text{ mole}^{-1})}$$

$$= 0.0147 \text{ moles}$$

Since a mole of ^{60}Co has a mass of 60 g, the mass of the sample is

$$m = (0.0147 \text{ moles})(60 \text{ g mole}^{-1}) = 0.883 \text{ g}$$

Exposure and Absorbed Dose | *Exposure*

indicates the amount of radiation reaching a material, while *absorbed dose* indicates the energy absorbed in the material from the beam. Thus the absorbed dose depends on the properties of the material and the beam, while the exposure depends on the characteristics of the beam alone.

Exposure is defined only for x-rays and gamma rays with energies up to 3 MeV and not for other forms of radiation. It is defined as the amount of ionization produced in a unit mass of dry air at standard conditions. The conventional unit is

$$1 \text{ roentgen} = 1 \text{ R}$$
$$= 2.58 \times 10^{-4} \text{ coulombs per} $$
$$\text{kilogram} \quad (34.5)$$

Thus 1 roentgen of x-rays will produce 2.58×10^{-4} coulombs of positive ions in a kilogram of air.

The *absorbed dose* is the energy imparted by ionizing radiation to a unit mass of absorbing tissue. It is measured in *rads*, where

$$1 \text{ rad} = 0.01 \text{ joules per kilogram} \quad (34.6)$$

A 1 roentgen exposure to x-rays or gamma rays produces a soft tissue absorbed dose of approximately 1 rad. The *gray* (Gy), which is 1 joule per kilogram or 100 rads, is the official S.I. unit and is gradually coming into use.

Unlike exposure, the absorbed dose is used with all kinds of ionizing radiation, not just x-rays and gamma rays with energies below 3 MeV. Its use is illustrated in the following example, which also shows the great effectiveness of energy in the form of ionizing radiation in altering biological structures.

Example 34.3

Living tissues exposed to 10,000 rads are completely destroyed. By how much will this absorbed dose raise the temperature of the tissues if none of the heat is lost? (Assume that the specific heat of the tissue is the same as that of water, $c = 1 \text{ kcal kg}^{-1} \text{ K}^{-1} = 4180 \text{ J kg}^{-1} \text{ K}^{-1}$.)

According to Chapter Twelve, the heat ΔQ needed to produce a temperature change ΔT in a mass m is $\Delta Q = mc \, \Delta T$. Now 10,000 rads corresponds to an ab-

sorbed energy per unit mass $\Delta Q/m = 10,000(0.01 \text{ J kg}^{-1}) = 100 \text{ J kg}^{-1}$. Thus

$$\Delta T = \frac{\Delta Q}{m} \frac{1}{c} = (100 \text{ J kg}^{-1}) \left(\frac{1}{4180 \text{ J kg}^{-1} \text{ K}^{-1}} \right)$$
$$= 0.0239 \text{ K}$$

Such a small temperature rise would have a negligible effect if it were achieved by simply heating the tissue. Radiation is so lethal to living tissue because it does not deposit the energy uniformly at all points. Instead it imparts energy in relatively large amounts to single atoms at random locations, and this may disrupt critical biological molecules.

BIOLOGICAL QUANTITIES | The absorbed

dose refers to a physical effect: the transfer of energy to a material. However, the effects of radiation on biological systems also depend on the type of radiation and its energy. The *relative biological effectiveness* (RBE) of a particular radiation is defined by comparing its effects to those of a standard kind of radiation, which is usually taken to be 200 keV x-rays. For example, fast neutrons (with energies above 0.1 MeV) have an RBE of about 10 for causing cataracts. Hence the absorbed dose (in rads) of 200 keV x-rays needed to produce cataracts is 10 times the dose required for neutrons. The RBE varies with the radiation type and energy, with the animal species, and the biological effect under consideration (Table 34.1). Positive ions, which deposit more energy per unit length than beta or gamma rays, generally do more biological damage than the same absorbed dose of betas or gammas. However, their effects are often limited to the surface tissue because they have short ranges.

The *rem* and the millirem $= 10^{-3}$ rem are the units used in discussions of biological effects. For example,

TABLE 34.1

Typical RBE values. By definition, the RBE is exactly 1 for 200 keV x-rays

Radiation	Typical RBE
^{60}Co γ rays (1.17 and 1.33 MeV)	0.7
4 MeV γ rays	0.6
β particles	1.0
Protons (1 to 10 MeV)	2
Neutrons	2–10
α particles	10–20

TABLE 34.2

Radiation units

Quantity	Unit	Definition
Source activity	curie (Ci)	3.70×10^{10} disintegrations per second
Exposure (x and γ rays)	roentgen (R)	2.58×10^{-4} C kg^{-1} in dry air
Absorbed dose	rad	10^{-2} J kg^{-1}
Biologically equivalent dose	rem	RBE $\times$ (dose in rads)

in the case of cataract formation, 1 rad of 200 keV x-rays and 0.1 rads of fast neutrons each produce 1 rem of damage. In any situation the *biologically equivalent dose* (in rems) is the physical absorbed dose (in rads) times the RBE. *One rem of any kind of radiation* produces the same biological effect, namely the effect of *one rad of 200 keV x-rays*.

A summary of various radiation units is given in Table 34.2. An illustration of the use of these units is given in the next example.

Example 34.5

A cancer is irradiated with 1000 rads of ^{60}Co gamma rays, which have an RBE of 0.7. Find the exposure in roentgens and the biologically equivalent dose in rems.

For gamma rays, the soft tissue dose in rads is approximately equal to the exposure in roentgens, so the exposure is approximately 1000 R. The biologically equivalent dose is the product of the RBE and the dose in rads, or (0.7)(1000) = 700 rems.

34.3 | HARMFUL EFFECTS OF RADIATION

When radiation passes through living cells, it can alter or damage the structure of important molecules. This may lead to the malfunctioning or death of the cells and ultimately to the death of the organism. Usually the cells that are growing or dividing most rapidly are the most radiosensitive. Often cancer cells are rapidly growing; in such cases, they are highly vulnerable to radiation. Similarly, fetuses and infants are much more readily harmed by radiation than are adults. For example, in one study it was found that children whose mothers received pelvic x-rays while pregnant had a 30 to 40 percent increase in the incidence of cancer.

The limited knowledge of the immediate effects on humans of large radiation doses comes from studies of victims of atomic bomb explosions and of occasional accidents. Whole body doses under 25 rems have no observable effect. As the dose increases above 100 rems, damage to the blood forming tissues becomes evident, and above 800 rems severe gastrointestinal disorders occur. Death usually follows in a period of days or weeks if the dose is much more than about 500 rems.

Sublethal, short-term doses and doses acquired gradually over a long period may lead to cancer after a latent period of many years during which no ill effects are discernible. The chance of dying of cancer is doubled by a dose somewhere between 100 and 500 rems. Over a wide range of exposures, both in animal experiments and in human data, the increase in the cancer rate is directly proportional to the total accumulated dose.

Studies of low level doses are difficult and inconclusive, because of the much larger incidence of cancer from other causes. Consequently, it is possible that the damage resulting from doses below some threshold is repaired, so that no increase in the cancer rate occurs. However, most experts favor the less

Figure 34.6. Leukemia rates in humans. Both the linear and threshold hypotheses are consistent with the high dose data. Measurements of the very small rates at low doses such as 1 to 10 rads would require huge populations to be statistically meaningful. (Adapted from Hurst and Turner.)

optimistic *linear response hypothesis,* which assumes that the effects of radiation in causing cancer are proportional to the dose at all levels (Fig. 34.6).

Genetic Effects | Although genetic mutations have helped humans to evolve to their present state, most mutations are harmful. Any increase in the mutation rate means more prenatal deaths and more people born with serious defects. In addition to the personal tragedy this implies, there is also the problem that the increasing ability of medical science to keep such people alive until they reproduce tends to keep defective genes in the human genetic pool.

Mutations can be increased above the normal rate by elevated temperatures, some chemicals, and by ionizing radiation. The mutations caused by radiation are similar to those occurring naturally. It is generally believed that their rate is proportional to the absorbed dose, no matter how small, and that there is no threshold or repair mechanism. The dose that will double the mutation rate is probably between 25 and 150 rems.

34.4 | CHRONIC RADIATION EXPOSURE

Living things have always been exposed to chronic low-level radiation arising from cosmic rays and from radionuclides naturally present in the environment. To this natural background, we have added a nearly equal amount of artificially produced radiation, mostly from medical diagnostic x-rays (Table 34.3). Some individuals receive much more than the average chronic doses from natural sources. For example, some villages in Brazil and India happened to be built on soil with a high thorium content; doses in these villages are up to 100 times the norm. Also, the water from some wells and mineral springs in Italy, Austria, and elsewhere contains up to a million times the usual radioactivity. In these localities the cancer rates are abnormally high.

The National Council on Radiation Protection establishes a maximum permissible dose (MPD) for radiation workers and the general public from all artificial radiation sources *except* medical diagnosis and treatment. (The council does not regulate medical exposures.) The MPD has been lowered repeatedly as the hazards of radiation have become better understood. For example, in the late 1930s the MPD was 0.1 rems per day. The current MPD for radiation workers is a seventh of this, or 5 rems per year (Table 34.4). The average permitted for the entire population averaged over all individuals is a thirtieth of the occupational MPD, or 170 millirems per year.

Since any radiation exposure implies some hazard, these standards represent a compromise involving judgments of risks versus benefits. No single radiation-producing device or activity, such as watches, TV sets, or nuclear reactors, is permitted to expose the public to more than a small fraction of the MPD. Total present-day general population exposures are much less than the limit.

Nevertheless, the 1972 National Academy of Sciences report quoted in Table 34.4 urged that the allowed doses be lowered substantially. It is estimated that *if* everyone received the present MPD from non-medical sources, between 3000 and 15,000 *additional* U.S. cancer deaths would occur annually. To see how this estimate comes about, let us assume that 250 rems will double the cancer rate. At 170 millirems per year, the average American who is about 30 years old will have accumulated 30(170 mrem) $\simeq$ 5000 mrem = 5 rems. This dose is 5/250 or 2 percent of the amount needed to double the cancer rate. The total death rate is roughly the U.S. population of 200 million divided by 70 years, or about 3 million, of which one-tenth or 300,000 deaths are due to cancer. A 2 percent increase would mean 6000 more deaths per year. (The range of additional deaths quoted in the Academy report corresponds to the uncertainty in the doubling dose which is somewhere between 100 and 500 rems.) At present, nonmedical artificial radiation sources other than bomb testing presently contribute less than 1 percent of the MPD. Thus the actual number of deaths due to these sources is small.

Medical diagnostic x-rays are currently the major source of artificial radiation exposure, and they are increasing annually. These average 40 percent of the nonmedical MPD and are estimated to cause 1500 to 3000 cancer deaths annually, plus a possibly greater number of genetic deaths in future generations via lethal mutations. It is estimated that *medical exposures could be greatly reduced* without any loss of the benefits of diagnostic x-rays. Clearly the medical and dental professions have a great responsibility to minimize exposures. This means avoiding unneeded di-

TABLE 34.3

Chronic sources of radiation in the United States. All values are approximate or average (1 millirem = 10^{-3} rems)

Source	Dose (millirems per year except as noted)	
Cosmic rays:		
Sea level	41	
Denver (5000 ft)	70	
Leadville, Colorado area (10,500 ft)	160	
20,000 ft	400	
Commercial jet (35,000 ft)	0.7 millirems per hour	
U.S. average from cosmic rays		44
γ Rays from rocks, soil (Ra, U, Th, K, etc.):		
Atlantic Coastal Plains	22.8	
Colorado Front Range	89.7	
U.S. average from external radionuclides	40	
Calculation of gonadal dose:		
Correction factor due to housing shielding	0.8	
Correction factor due to biological shielding	0.8	
U.S. average gonadal dose from external radionuclides	(40)(0.8)(0.8)	26
Internal radionuclides:		
^{40}K	16	
^{14}C, Ra and decay products	2	
Total		18
Total U.S. environmental average gonadal dose: (Oakley, 1972)		88 ± 11
Fallout (1970)	4	
Nuclear power	0.003	
Medical diagnostic	72	
Medical radiopharmaceuticals	1	
Occupational	0.8	
Miscellaneous	2	

(Sources: Natural Radiation Exposure in the U.S., by D. T. Oakley, U.S. Environmental Protection Agency, June 1972; National Academy of Sciences Committee on Biological Effects of Ionizing Radiation Report, November 1972.)

agnostic x-rays and assuring the use of modern, well-shielded machines and high-sensitivity films. It also means educating the public, so that people do not equate diagnostic x-rays with good medical practice in evaluating the care they receive.

The importance of using the proper x-ray devices and techniques is illustrated by the range of doses received from chest x-rays. The best procedures give only 6 millirems, although the average is 200 millirems. By contrast, the x-ray machines used in mobile units to screen large populations for tuberculosis, which are relatively inexpensive to operate, gave about 1000 millirems to the chest area. Public health authorities no longer recommend their general use, since they believe the risks of radiation exposure outweigh those of lung diseases for the average person.

To summarize, any radiation exposure carries with it a small but very real risk. The benefits of such exposures must be weighed against the hazards, both for the individual and for society overall.

TABLE 34.4

Maximum permissible doses (MPD) for whole body exposures as set in 1971 by the National Council on Radiation Protection. Larger doses are permitted for some parts of the body

	MPD
General Population	
MPD for any individual[a]	500 millirems per year
Average for whole U.S. population[b]	170 millirems per year
Radiation Workers	
Annual	5000 millirems per year
3-month period	3000 millirems per 3 months
Pregnant workers	500 millirems per 9 months

[a] Facilities using radiation may occasionally expose some individuals to this dose.
[b] This is the average for everyone, including radiation workers.

Figure 34.7   An example of the misuse of x-rays was the shoe-fitting fluoroscopy machine, a popular selling feature of many children's shoe stores from 1946 until the late 1950s. The first machines were largely unshielded. Doses to the feet sometimes were as high as 40 rems per minute and averaged 8 rems per minute. The machines were eventually eliminated by legislation that restricted x-ray use to licensed medical practitioners. Here, an expert is measuring x-ray levels while a little girl watches the bones in her toes wiggle. (Courtesy of Ira A. Paul, Bureau of Radiation Control, New York City Department of Health.)

34.5 | RADIATION IN MEDICINE

Despite its hazards, the use of ionizing radiation in medical research, diagnosis, and therapy has been invaluable and has saved many lives. Within a few weeks of their discovery by Roentgen in 1895, the value of x-rays for medical diagnosis was realized. By early in this century, x-rays and naturally occurring radionuclides (radium, radon) were being used in cancer therapy. Today, many different radionuclides are produced in reactors and accelerators and are available for use in nuclear medicine.

Medical Research | Amino acids, sugars, DNA, and penicillin are a few of the hundreds of biological compounds commercially available today containing ^{14}C, ^{3}H, ^{35}S, ^{32}P, or other radionuclides. The radioactivity of these *tracers* makes it possible to follow their pathways and metabolism very accurately and conveniently. The active transport of sodium in nerve fibers, the metabolism of starches and sugars, the building up of proteins from amino acids, and the action of hormones and drugs are examples of fundamental biological processes that have been studied with tracers.

Diagnosis | Most diagnostic studies done with radionuclides share some common features. A radioactively labeled compound is administered which is absorbed in the organ of interest, and a detector outside the body measures the radioactivity at various times or locations. The radiation doses are normally comparable to those from diagnostic x-rays. To minimize the dose, short-lived nuclides are selected which emit gamma rays with just enough energy to be detected. For many purposes, a long-lived excited state of technicium ($Z = 45$) referred to as ^{99m}Tc is almost ideal. Formed when ^{99}Mo beta decays, this nuclide decays to its stable ground state ^{99}Tc with a half-life of 6.0 hours by emitting a 140 keV gamma ray. In suitable chemical compounds ⁹⁹ᵐTc can be directed at many different organs.

A typical *volume study* is the determination of the total amount of blood plasma in the body. A radioactively labeled blood protein is injected into the bloodstream. Within about 10 minutes, this material is thoroughly mixed with the plasma. Measurement of the radioactivity of a plasma sample then indicates the amount of dilution and hence the total plasma volume. Similarly, the total number of red blood cells can be determined if some of the patient's own red cells are labeled with ^{51}Cr or ^{59}Fe and returned to the bloodstream.

A *dynamic function study* measures the rate at which a thyroid, kidney, or other organ absorbs or eliminates a material. For example, the amount of a ⁹⁹ᵐTc labeled compound taken up by the thyroid as measured by a counter held against the throat is an indicator of its condition. Similarly, when diuretics containing radioactive compounds are concentrated in the kidneys, alterations from the usual patterns of uptake and excretion signal possible abnormalities.

In a *gamma ray-imaging study* or *scan,* a picture is produced by detecting the radiation from an absorbed radionuclide. The radiation is detected with the aid of a *scintillating crystal,* which emits flashes of light when beta or gamma rays pass through it. (Scintillation detectors are discussed in more detail in the supplementary topics at the end of this chapter.) The scintillation counter is equipped with a lead collimator containing several tapered holes which transmit the gamma rays from a small region of the organ

which has absorbed the radionuclide. The counter is slowly moved or *scanned* over the region of interest, and the pulses generated are amplified and recorded on paper, photographic film, or video tape (Fig. 34.8). For example, a thyroid scan is done by administering a suitable compound containing ⁹⁹ᵐTc. Since nonfunctioning tissue in the thyroid does not absorb these materials, it appears as a less radioactive region in the thyroid. Other common scans include those of the liver, kidneys, and brain.

The *Anger* or *gamma camera* is an ingenious device developed by H.O. Anger in 1957. It measures the radioactivity everywhere in a large area simultaneously and displays the results on an oscilloscope screen. This makes it possible to perform dynamic studies in which one observes a radionuclide enter or leave an organ. It also permits static examinations to be completed in a minute or so. The older scanners, by contrast, required up to an hour.

The gamma camera has a single, large, thin scintillating crystal shielded by a lead collimator containing hundreds of holes (Fig. 34.9). This collimator is placed over the organ under study. Nineteen light-sensitive *photomultiplier tubes* behind the crystal observe flashes of light produced when a gamma ray passes through the crystal and send signals to a small computer. The amount of light entering a particular tube depends on its distance from the point where the gamma ray passes through the crystal. The computer locates this point from the relative light intensities and sends appropriate signals to the oscilloscope plates. The oscilloscope then displays a two-dimen-

Photomultiplier

Crystal

Pb

Focusing collimator

Thyroid

L

R

(a)

(b)

Figure 34.8. Diagnostic scans. (*a*) A scintillation counter shielded with a lead collimator. (*b*) A kidney scan. Note the difference in appearance between the healthy right kidney and the diseased left kidney.

Figure 34.9. The Anger or gamma camera.

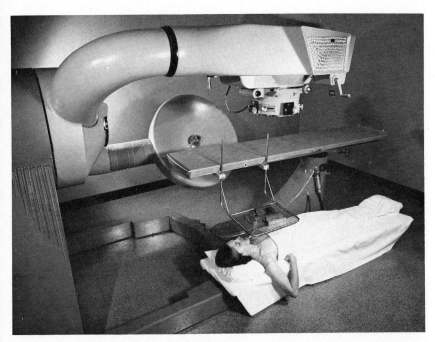

Figure 34.10. A facility for ^{60}Co radiation therapy. The cobalt is in a heavily shielded container that can be moved near various parts of the patient's body. Since the cobalt produces gamma rays that have a greater energy than is available with standard x-ray machines, they penetrate further into the body and are more effective in treating many forms of cancer. (Courtesy of the University of Texas, Houston, M.D. Anderson Hospital & Tumor Institute.)

sional picture whose intensity at each point is proportional to the level of radioactivity at the corresponding point in the organ. Gamma cameras are used to diagnose diseases of the thyroid, liver, brain, kidneys, lungs, spleen, heart and circulatory system.

Therapy | As we noted earlier, rapidly dividing cancer cells are highly vulnerable to radiation. Accordingly, at least half of all cancer patients receive *radiation therapy,* often in combination with surgery. The radiation may be administered externally, with x-ray machines, ^{60}Co, or occasionally electron accelerators serving as the source (Fig. 34.10). Alternatively, radium-filled needles, small seeds containing radon gas, or wires or ribbons containing artificial radionuclides may be implanted in the tumor. Occasionally, radioactive compounds are administered which are absorbed by the affected part of the body. Still in the experimental stage is the use of accelerator-produced beams of alpha particles and other particles with fairly high energies. Because these particles deposit much of their energy in a small region near the end of their range, it is hoped that they will do

less damage to the tissues surrounding the cancer being irradiated and therefore permit larger doses.

34.6 | OTHER USES OF RADIATION

Radiation has hundreds of uses in agriculture and industry. Sometimes radiation offers the cheapest or most convenient way to perform some function, while in other situations, no other technique is available. For example, there is often no substitute for tracer studies of metabolic processes or complex reactions.

In agriculture, tracers have been used to study how fertilizers, hormones, weed killers, and pesticides perform their functions. With labeled growth hormones or pesticides, it is even possible to measure residues in the final food products which are too minute to be detected chemically.

Radiation has been used to induce plant mutations, which have led to improved varieties of many crops, such as wheat, peas, and beans, with higher yields and more resistance to diseases. It has also been used to eliminate a major agricultural pest, the screwworm fly, without employing pesticides. This was done by

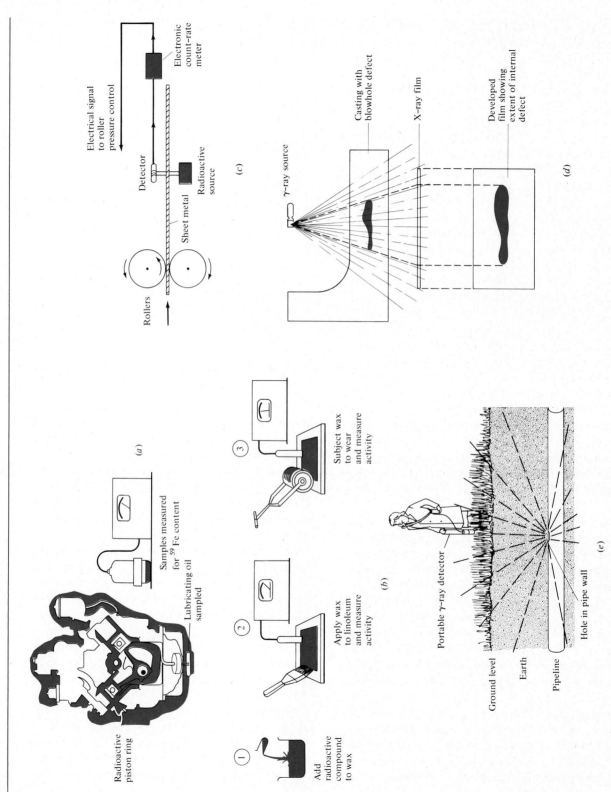

Figure 34.11. Typical industrial uses of radiation. Measurement of wear rates of (*a*) engine parts and (*b*) floor wax. (*c*) A thickness gauge. (*d*) Inspection of castings. (*e*) Locating pipeline leaks with tracers.

using radiation to sterilize hordes of male flies and releasing them to compete with normal males for mates. The females mate only once, and unfertilized eggs do not hatch. Hence, repeatedly saturating an area with millions of sterile males soon wiped out the population completely.

The range of uses that radiation has found in industry can be seen with some typical examples (Fig. 34.11). Gamma rays or x-rays are used much like medical diagnostic x-rays to reveal potentially dangerous but invisible defects in metal castings and welds. The thickness of rolled sheets of plastic, paper, or metal, or of thin coatings, can be controlled by measuring the transmission of radiation. Densities and fluid levels are easily measured. Tracers are used to locate leaks in underground pipelines, to measure the rate of wear of automobile tires and engine parts, and to determine the effectiveness of detergents.

SUMMARY

Charged particles passing through matter ionize atoms along their paths by colliding with atomic electrons. For a given energy, the rate of energy loss with distance is greatest for the most massive particles. Hence alpha particles penetrate about 0.01 cm in water, while electrons travel about 1 cm. Gamma rays lose energy to electrons via the photoelectric, Compton, and pair production processes; their ranges are typically 10 cm. Neutrons lose energy only in nuclear processes, and their ranges are of the order of a meter.

The source activity refers to the number of nuclear disintegrations occurring per second in a source and is measured in curies. The exposure indicates the strength of an x-ray or gamma-ray beam and is measured in roentgens. The absorbed dose for any kind of radiation is the energy absorbed per unit mass of material; its unit is the rad. One roentgen of x-rays or gamma rays produces a dose of about one rad. The biologically equivalent dose in rems is the absorbed dose in rads times the relative biological effectiveness, which is defined to be one for 200 keV x-rays,

Radiation doses from cosmic rays and naturally occurring radionuclides average about 88 millirems per year in the United States. Medical diagnostic x-rays contribute almost as large a dose on the average. The effects of low-level radiation in causing cancer and inducing mutations are apparently proportional to the total dose accumulated. Rapidly growing cells are highly radiosensitive, so many kinds of cancer cells can be killed with radiation.

Checklist
Define or explain:

ionizing radiation	rad
range	RBE
energy loss rate	rem
photoelectric effect	doubling dose
Compton scattering	maximum permissible
pair production	dose
source activity	linear hypothesis
curie	latent period
exposure	tracers
roentgen	Anger camera
absorbed dose	

EXERCISES

Section 34-1 | The Interaction of Radiation with Matter

34-1 If a microwave beam is sufficiently intense, it can heat a material sufficiently to ionize some of its atoms. How does this differ qualitatively from the way gamma-ray photons cause ionization?

34-2 The rate of energy loss for a charged particle in a medium with atomic number Z is almost exactly proportional to nZ, where n is the number of atoms per unit volume. Give an argument for this.

34-3 In a laboratory experiment to study the properties of radioactive sources, a student finds that the count rate drops sharply to zero when a Geiger-Mueller counter in the air is moved from 9 to 10 cm from the source. (a) What is the most likely nature of the radiation? Explain. (b) Estimate the range of this radiation in aluminum, assuming the range varies inversely with the density. (The density of aluminum is 2700 kg m^{-3}, and that of air is 1.29 kg m^{-3}.)

34-4 Using Fig. 34.4, find the ranges of 10-MeV alpha particles and 10-MeV electrons in water.

34-5 In the photoelectric process, a gamma ray is absorbed and an electron is ejected from an atom. Usually x-rays are then produced. Explain why.

34-6 When a gamma ray is absorbed in a photo-electric process, an electron is ejected from an atom, and the gamma ray disappears. Is the kinetic energy of the electron equal to the gamma ray energy? Explain your answer.

34-7 In electron-positron pair production by gamma rays, some momentum is transferred to a nearby nucleus. Why is the energy transferred to the nucleus very small?

34-8 According to Fig. 34.5, when gamma rays are absorbed in water, at what energy are pair production and Compton scattering equally important?

34-9 Protons and deuterons ($^1_1$H and $^2_1$H) lose the same amount of energy in a thin sheet of material. How are their energies related?

34-10 If protons and deuterons ($^1_1$H and $^2_1$H) have the same kinetic energy when they enter a thin sheet of material, how are the energies they lose related?

34-11 How could thin foils and a proton detector be used to determine the energy of a beam of protons produced by an accelerator?

Section 34.2 | Radiation Units

34-12 What is the mass of a 1 microcurie (10^{-6} Ci) ^{131}I source? (The half-life of ^{131}I is 8.1 days.)

34-13 How many positive ions are produced in 1 kg of air at standard conditions by a 1-R x-ray exposure? (Assume each ion has one electronic charge, $+e$.)

34-14 Treating fish or meat with 200,000 rads kills many of the bacteria present and increases the refrigerated shelf life five to seven times. (a) How much energy is absorbed per kilogram of food? (b) Neglecting heat losses, estimate the temperature rise. (Assume that the specific heat is equal to that of water.)

34-15 A ^{60}Co source produces an absorbed dose of 4000 rads per hour in tissue. The RBE is 0.7 for cobalt gamma rays. (a) How much time is required for an absorbed dose of 300 rads? (b) How much time is required for a biologically equivalent dose of 300 rems?

34-16 A mouse receives an absorbed dose of 200 rads of 10 MeV protons, which have an RBE of 2. What is the biologically equivalent dose?

Section 34.3 | Harmful Effects of Radiation, and Section 34.4 | Chronic Radiation Exposure

34-17 A nurse receives a whole body dose of 0.1 rems every time she loads a radium source. How many times is she permitted to load the source in (a) a month; (b) a year? (The nurse is considered a radiation worker.)

34-18 The lethal dose of radiation for mammals and birds is under 1000 rads, while it takes about 100,000 rads to kill amoebas and over a million rads to kill viruses. What is the likely reason for this?

34-19 Suppose 100,000 radiation workers each receive 5 rems. Is this more or less important genetically than the entire world population of about 3 billion people receiving 5 millirems? Why?

34-20 A stewardess flies at an average height of 35,000 feet for 20 hours per week. If she receives 0.7 millirems per hour, what is her annual biologically equivalent dose? Compare this with the MPD for the general population and for radiation workers.

34-21 Estimate the death rate due to diagnostic x-rays, assuming the U.S. cancer death rate is 300,000 per year, that the doubling dose for cancer is 250 rems, and that the average person of age 30 has received 72 millirems per year.

34-22 A radiation worker accidentally receives 50 rems. Using the assumptions in the preceding exercise, estimate his chances of developing cancer because of this accident.

34-23 The earliest safety rules for radiological personnel limited their dose to the equivalent of 100 rems per year, a level low enough to avoid skin burns. Estimate the cancer risk and genetic effects of such doses.

Section 34.5 | Radiation in Medicine

34-24 In *rotation therapy* the patient or the radiation source is rotated about an axis through the tumor being irradiated. What is the advantage of this procedure?

34-25 How do you explain the apparent paradox that x-rays induce cancer and that x-rays are used to treat cancer?

Section 34.6 | Other Uses of Radiation

34-26 How can you use radioactively labeled dirt to test the effectiveness of a detergent?

PROBLEMS

34-27 X-ray machines produce a continuous distribution of x-ray photon energies up to a maximum determined by the voltage of the machine. A thin sheet of aluminum will absorb the lowest energy x-rays without appreciably absorbing the higher energy photons. (a) What is the reason for this? (b) What is the advantage of such an aluminum filter in medical uses of x-rays?

34-28 A 1000-Ci alpha emitter is in a lead container. (a) How many disintegrations occur per second? (b) The alpha particles have an energy of 2.5 MeV, and all the alphas are stopped in the lead. At what rate is heat generated in the lead?

34-29 Cobalt 60 produces two gamma rays per disintegration. (a) How many gammas are emitted per second by a 10-Ci source? (b) What is the flux of gammas per square meter at a distance of 1 m from the source? (Assume the radiation is uniform in all directions.)

34-30 A 10-Ci β^- emitter is 2 m from your hand. How many electrons strike each square centimeter of your hand per second? (Neglect absorption in air.)

34-31 A radiation worker is accidentally exposed to 100 R of 200 keV x-rays. Estimate the absorbed and biologically equivalent doses.

34-32 A milligram of radium produces a dose of 8.2 rads per hour at a distance of 0.01 m. The RBE for radium gamma rays is 0.965 for the tissue irradiated. (a) Explain why the exposure varies inversely with the square of the distance from the source. (b) How long is required for a 60-mg radium source to give a dose of 100 rems at a distance of 0.01 m? (c) How long would this take at 0.05 m?

34.33 A small quantity of radioactive serum albumin is administered to a patient, and an equal quantity is placed in 2000 cm³ of water to serve as a standard. After 10 minutes, a blood sample is taken from the patient and the red cells removed by centrifugation, leaving the serum. The activity of 10 cm³ of the serum is measured with a scintillation counter and found to be 2600 counts per minute. A 10-cm³ sample from the standard gives 1892 counts per minute. In both cases, 155 counts per minute are due to background from cosmic rays and other environmental sources. What is the total volume of the patient's blood serum?

SUPPLEMENTARY TOPICS

34.7 | RADIATION DETECTION AND MEASUREMENT

Radiation is detected by observing the ionization it produces in matter. In this section, we describe the radiation detectors most commonly used in biomedical applications of radiation.

Gas Ionization Counters | These detectors, which are easy to use, are very sensitive to beta particles. They can also be used for gamma rays, but are not very effective in stopping this more penetrating radiation. A typical arrangement has argon gas in a cylinder with a fine wire along its axis (Fig. 34.12). When ionization is produced by radiation, a brief current pulse results which is metered or recorded.

Depending on the applied voltage V_0, the counter operates in basically different ways (Fig. 34.13). A small voltage drives all the ions and electrons produced to the cathode or anode, causing a small cur-

Figure 34.12. Gas ionization counter. When radiation ionizes one of the argon atoms, the electrons are attracted to the positively charged wire (anode) and the positive ions to the negatively charged outer cylinder (cathode). This produces a current through R and a voltage drop across it which is measured by the voltmeter.

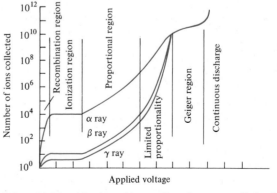

Figure 34.13. Number of ions collected versus applied voltage in a gas ionization counter.

Figure 34.14. Gamma ray spectra obtained with a multichannel analyzer. The gamma ray energy is the horizontal coordinate and the vertical coordinate gives the intensity at that energy. (*a*) ^{58}Co. (*b*) ^{46}Sc. (*c*) A mixture of ^{58}Co and ^{46}Sc. Mixtures of radionuclides can be analyzed if the energy resolution of the apparatus is good enough to distinguish the characteristic peaks. Sometimes samples are made radioactive artificially by exposing them to beams of neutrons, gamma rays, or charged particles that induce various nuclear reactions. Such *activation* techniques require only small samples and can be used to detect and measure extremely small traces of many elements.

rent pulse. Increasing the voltage further, we reach the *proportional region,* where each electron released in the original ionization process acquires sufficient kinetic energy to ionize one or more additional atoms before it reaches the anode. The total charge in the current pulse now is proportional to the initial amount of ionization, and it indicates the energy of the beta particle or gamma ray which passed through the detector. Often these current pulses are fed to a *multichannel analyzer,* which records the number of pulses produced versus the total charge in the pulse. This information can be used to identify the nuclide producing the radiation (Fig 34.14).

If the applied voltage is increased still more, the counter operates in the *Geiger region.* Here each electron produces several secondary electrons, which in turn produce others, and so on. A current pulse develops which is large enough so that it can be observed or recorded with very simple circuits. Such a detector is called a *Geiger-Mueller counter,* and it can be quite compact and portable. However, the total amount of charge in a pulse is the same for any incident ionizing particle, since the tube becomes fully ionized. Consequently a Geiger-Mueller counter cannot be used to identify the nature or energy of the radiation being detected.

Scintillation Counters | These counters have relatively high efficiencies for detecting gamma rays and can be used to measure their energies. They are the detectors most widely used in biomedical applications.

A scintillation counter consists of a crystal such as sodium iodide and a *photomultiplier.* When it is irradiated, the crystal *scintillates* or emits flashes of visible light as atoms excited by the radiation emit photons. The photons enter the photomultiplier and strike a light sensitive cathode, causing the emission of electrons (Fig. 34.15). These electrons are accelerated by a potential difference and strike an electrode called a *dynode,* which emits approximately four secondary electrons for each incident electron. A series of 10 dynodes gives an amplification factor of 4^{10} or approximately a million. As in the proportional gas counter, the total charge in the current pulse is proportional to the amount of ionization, and a multichannel analyzer can be used to identify the source.

Figure 34.15. Scintillation counter.

Semiconducting Detectors | When radiation deposits energy in a semiconducting material, electrons can be excited from filled valence bands to conduction bands. This produces pairs of conduction electrons and electron vacancies or *positive holes*. If a potential difference is maintained across the semiconductor, these charge carriers will move, and a current pulse will be observed. The energy needed to produce an electron-hole pair is typically about 1 eV, considerably less than the energy needed to ionize gas atoms or to excite atoms in a scintillating crystal. Accordingly, a relatively large number of charge carriers is produced for a given energy loss. The statistical fluctuations in the number of atoms excited or ionized limit the energy resolution of any detector. The large number of charge carriers produced in the semiconductor reduces these statistical effects and enables such a detector to determine energies very accurately when used with a multichannel analyzer.

At the present time, these detectors are expensive and difficult to make unless they are quite small. Small semiconducting detectors have the disadvantage of low gamma-ray efficiencies.

Photographic Emulsions | Film badges are used to monitor the beta and gamma ray exposure of radiation personnel. Thin metal strips shield parts of the film from beta rays so that they indicate the gamma flux; the unshielded area gives the total dose from beta and gamma rays. The radiation exposure is determined by examining the processed film. Neutron doses can be measured if the film is impregnated or covered with a material having a high probability for neutron reactions.

Thermoluminescent dosimeters or TLDs are relatively new, but they are widely used in calibrating radiation sources in hospitals and are rapidly replacing film badges in personnel monitoring. A TLD is a crystal such as lithium fluoride (LiF) or calcium fluoride (CaF_2) containing trace impurities. After a TLD is exposed to radiation, it will emit visible light upon being heated. The total amount of light emitted is directly proportional to the dose received by the crystal, even if the crystal is stored for many months after exposure. TLDs are more accurate than film badges and are usable over a much larger range of radiation levels.

The principle of the TLD is illustrated in Fig.

Figure 34.16. Energy levels of a LiF crystal. A perfect crystal has a filled band and an empty conduction band. Isolated levels below the conduction band are caused by impurities. If radiation excites an electron to the conduction band, leaving a valence band hole, the electron can fall into an isolated level and become trapped. Heating will provide the energy needed for the electron to return to the conduction band. It will emit visible light when it drops to the vacant valence band level.

34.16, which shows the energy levels of a LiF crystal. A perfect crystal will have a completely filled valence band and an empty conduction band. Electrons cannot be in energy levels between these bands. Electrons can be excited from the valence band to the conduction band, leaving empty valence levels or holes if they absorb enough energy from the radiation passing through the crystal. The impurity atoms present in the crystal add isolated energy levels to this band structure which lie just below the conduction band. If an electron is excited to the conduction band and falls into one of these isolated levels, it is "trapped." It cannot migrate and thus cannot return to an empty valence level. When the crystal is heated at a later time, the trapped electron can receive enough thermal energy to return to the conduction band and migrate until it reaches a hole. When the electron loses energy by filling the hole, it emits a photon of visible light. The total light emitted as the crystal is heated is a measure of the number of trapped electrons and therefore of the total absorbed radiation.

34.8 | DERIVATION OF THE ENERGY LOSS RATE FORMULA

We now derive Eq. 34.1 for the rate of energy loss with distance for an ion in matter.

The velocity $\mathbf{v}_e$ of an atomic electron in the medium is negligible compared to its final velocity $\mathbf{v}'_e$ after it is struck by the ion. The momentum transferred by the ion to the electron is therefore $\mathbf{F}\,\Delta t = m_e(\mathbf{v}'_e - \mathbf{v}_e) \simeq m_e\mathbf{v}'_e$, where $\mathbf{F}$ is the force exerted by the ion and Δt is the time duration of the collision. The energy transferred to the electron is approximately $\frac{1}{2}m_e v_e'^2$, which is proportional to $(\mathbf{F}\,\Delta t)^2$. Now the electrical force between the ion and the electron is proportional to the ion charge q, and the collision time varies inversely with the ion velocity v. Thus $\mathbf{F}\,\Delta t$ must vary as q/v, and the energy transfer as q^2/v^2, or as we stated earlier,

$$\Delta K \propto -\frac{q^2}{v^2} \qquad (34.1)$$

EXERCISES ON SUPPLEMENTARY TOPICS

Section 34.7 | Radiation Detection and Measurement

34-34 In gamma ray scans, the scintillation counter pulses are sometimes simply counted, and the count rate is then used to construct a picture. Alternatively, the pulses may be processed by a multichannel analyzer before constructing a picture. How can this latter procedure reduce the effects of background radiation from other sources?

34-35 Given a Geiger-Mueller counter and several thin slabs of a material containing a large amount of hydrogen, such as a hydrocarbon, how can you tell whether a given radioactive source is emitting neutrons?

Additional Reading

Steward C. Barshong, Radiation Exposure in Our Daily Lives, *The Physics Teacher*, vol. 15, 1977, p. 135. An up-to-date survey of environmental, occupational, and medical exposures.

G.S. Hurst and J.E. Turner, *Elementary Radiation Physics*, John Wiley and Sons, New York, 1970. An excellent brief introduction at the level of this book.

E.F. Gloyna and J.E. Ledbetter, *Principles of Radiological Health*, Marcel Dekker, New York, 1969. Emphasis on simplified calculational methods and engineering estimates.

Louis E. Etter (ed.), *The Science of Ionizing Radiation*, Charles C Thomas, Springfield, Ill., 1965. Interesting chapters on history and on medical and other applications.

J.N. Gregory, *The World of Radioisotopes*, Angus and Robertson, Sydney, 1966. Directed at nontechnical audiences; many interesting applications.

H.E. Johns and J.R. Cunningham, *The Physics of Radiology*, 3rd ed., Charles C Thomas, Springfield, Ill., 1969.

Victor Arena, *Ionizing Radiation and Life*, C.V. Mosby Company, St. Louis, 1971.

Bacq and Alexander, *Fundamentals of Radiobiology*, Pergamon, New York, 1961.

P.N. Goodwin, E.H. Quimby, and R.H. Morgan, *Physical Foundations of Radiology*, 4th ed., Harper and Row, New York, 1970.

G.E.M. Jauncey, The Early Days of Radioactivity, *American Journal of Physics*, vol. 14, 1946, p. 226.

Gordon L. Brownell and Robert J. Shalek, Nuclear Physics of Medicine, *Physics Today*, August 1970, p. 33.

Advisory Committee on the Biological Effects of Ionizing Radiations, *The Effects on Populations of Exposure to Low Levels of Ionizing Radiation*, National Academy of Sciences, National Research Council, Washington D.C., 1972.

Arthur Robinson, Image Reconstruction (I) Computerized X-ray Scanners, *Science*, vol. 190, 1975, p. 542; (II) Computerized Scanner Explosion, *Science*, vol. 190, 1975, p. 647.

Scientific American articles:

S.A. Korff, Counters, July 1950, p. 40.

G.B. Collins, Scintillation Counters, November 1953, p. 36.

D.A. Glaser, The Bubble Chamber, February 1955, p. 46.

H. Yagoda, The Tracks of Nuclear Particles, May 1956, p. 40.

Gerard K. O'Neill, The Spark Chamber, August 1962, p. 36.

Olexa-Myron Bilamick, Semiconductor Particle Detectors, October 1962, p. 78.

R.L. Fleischer, P.B. Price, and R.M. Walker, Nuclear Tracks in Solids, June 1969, p. 30.

W.H. Wahl and H.H. Kramer, Neutron Activation Analysis, April 1967, p. 68.

The September 1959 issue was devoted to ionizing radiation.

H.J. Muller, Radiation and Human Mutations, November 1955, p. 58.

E.F. Knipling, The Eradication of the Screw-Worm Fly, October 1960, p. 54.

Theodore T. Puck, Radiation and the Human Cell, April 1960, p. 142.

Renato Baserga and W.B. Kisielski, Autobiographies of Cells, August 1963, p. 103.

Richard Gordon, Gabor T. Herman, and Steven A. Johnson, Image Reconstruction from Projection, October 1975, p. 56.

EPILOGUE

PHYSICS AND THE FUTURE

Many textbooks simply stop at a suitable point, without an epilogue or a postscript or sometimes even a concluding paragraph. However, because physics is still a rapidly developing science with many exciting discoveries being made, we would like to conclude with some brief remarks about the directions in which physics seems to be going.

We have at least touched on some of the currently active subfields of physics, such as atomic, nuclear, and solid-state physics. Physicists working in these areas are continuing to explore new frontiers. For example, nuclear physicists are trying to produce *superheavy nuclei,* with mass and atomic numbers much greater than those presently known. Solid state physicists are studying the curious behavior of many materials at temperatures as low as a millidegree (0.001 K). It has been known since 1956 that the weak interactions responsible for processes such as nuclear beta decay violate the principle called *parity conservation.* This means that a physical process would not be the same in a mirror-image world as in our world if weak interactions are involved. Atomic and nuclear physicists are performing a variety of experiments designed to probe this and other symmetry breakdowns that have been observed.

A subfield that presently involves a large number of research physicists, high energy or particle physics, deals with phenomena at the scale of the nucleon diameter (about 1 fm = 10^{-15} m) or smaller. When two particles collide at high speeds, some of their kinetic energy may be converted into mass energy with the creation of other particles. Using particle accelerators of ever-increasing energy, a vast number of short-lived mesons, hyperons, and other particles have been created, and their masses, lifetimes, decays, and interactions have been studied. This research has led to a subnuclear spectroscopy, characterized by new quantum numbers with intriguing names such as *strangeness* and *charm.* Evidence has accumulated suggesting that all these particles, in turn, may be made up of objects called *quarks,* which bear electric charges that are a fraction of the electronic charge.

Because much of particle physics is tentative and remote from applications, we have not discussed particle physics in this book. For similar reasons, we have excluded any discussion of astrophysics, even though many striking discoveries have occurred in recent years. Many of these have been made using *radio telescopes,* which detect the electromagnetic waves emitted at radio frequencies by stars just as ordinary telescopes detect the light they emit. Among the remarkable objects found with radio telescopes are *pulsars,* which emit rapid pulses in the radio and optical regions. Pulsars are believed to be *neutron stars,* objects first predicted theoretically, with about the same mass as our sun but a radius of only 10 kilometers! Also, there are several double star systems in which one member may be a *black hole:* an object so large and dense that the gravitational force bends all light rays back into it, and no light ever escapes! Since new kinds of astronomical tools—millimeter radio wave, infrared, and x-ray telescopes—are just now being developed, we can expect more exciting advances in astronomy and astrophysics in the decade ahead.

Returning to the earth, physicists are currently active in many applied and interdisciplinary fields. One of these is plasma physics, the physics of highly ionized matter. Thermonuclear reactions occur in a plasma, so advances in this field are crucial for achieving controlled fusion. Increasing numbers of physicists are working in biophysics and medical physics, using physical tools to probe basic biological processes and to develop new kinds of instrumentation. For example, light scattering using laser sources is now being used to study aspects of nerve conduction. Light scattering is also employed in a system being developed which will automatically sort biological cells.

To summarize, physicists are continuing the process of learning the basic laws of the universe, as they have for centuries. Although we cannot predict the future course with any certainty, we can be sure that many surprises lie ahead. We can also be confident that physics will continue to make important contributions to many fields, including the life sciences.

APPENDIX A

TRIGONOMETRIC TABLES

Angle		Sine	Cosine	Tangent	Angle		Sine	Cosine	Tangent
Degree	Radian				Degree	Radian			
0°	0.000	0.000	1.000	0.000					
1°	.017	.017	1.000	.017	46°	0.803	0.719	0.695	1.036
2°	.035	.035	0.999	.035	47°	.820	.731	.682	1.072
3°	.052	.052	.999	.052	48°	.838	.743	.669	1.111
4°	.070	.070	.998	.070	49°	.855	.755	.656	1.150
5°	.087	.087	.996	.087	50°	.873	.766	.643	1.192
6°	.105	.104	.994	.105	51°	.890	.777	.629	1.235
7°	.122	.122	.992	.123	52°	.908	.788	.616	1.280
8°	.140	.139	.990	.140	53°	.925	.799	.602	1.327
9°	.157	.156	.988	.158	54°	.942	.809	.588	1.376
10°	.174	.174	.985	.176	55°	.960	.819	.574	1.428
11°	.192	.191	.982	.194	56°	.977	.829	.559	1.483
12°	.209	.208	.978	.212	57°	.995	.839	.545	1.540
13°	.227	.225	.974	.231	58°	1.012	.848	.530	1.600
14°	.244	.242	.970	.249	59°	1.030	.857	.515	1.664
15°	.262	.259	.966	.268	60°	1.047	.866	.500	1.732
16°	.279	.276	.961	.287	61°	1.065	.875	.485	1.804
17°	.297	.292	.956	.306	62°	1.082	.883	.470	1.881
18°	.314	.309	.951	.325	63°	1.100	.891	.454	1.963
19°	.332	.326	.946	.344	64°	1.117	.899	.438	2.050
20°	.349	.342	.940	.364	65°	1.134	.906	.423	2.145
21°	.366	.358	.934	.384	66°	1.152	.914	.407	2.246
22°	.384	.375	.927	.404	67°	1.169	.920	.391	2.356
23°	.401	.391	.920	.424	68°	1.187	.927	.375	2.475
24°	.419	.407	.914	.445	69°	1.204	.934	.358	2.605
25°	.436	.423	.906	.466	70°	1.222	.940	.342	2.747
26°	.454	.438	.899	.488	71°	1.239	.946	.326	2.904
27°	.471	.454	.891	.510	72°	1.257	.951	.309	3.078
28°	.489	.470	.883	.532	73°	1.274	.956	.292	3.271
29°	.506	.485	.875	.554	74°	1.292	.961	.276	3.487
30°	.524	.500	.866	.577	75°	1.309	.966	.259	3.732
31°	.541	.515	.857	.601	76°	1.326	.970	.242	4.011
32°	.558	.530	.848	.625	77°	1.344	.974	.225	4.331
33°	.576	.545	.839	.649	78°	1.361	.978	.208	4.705
34°	.593	.559	.829	.674	79°	1.379	.982	.191	5.145
35°	.611	.574	.819	.700	80°	1.396	.985	.174	5.671
36°	.628	.588	.809	.726	81°	1.414	.988	.156	6.314
37°	.646	.602	.799	.754	82°	1.431	.990	.139	7.115
38°	.663	.616	.788	.781	83°	1.449	.992	.122	8.144
39°	.681	.629	.777	.810	84°	1.466	.994	.104	9.514
40°	.698	.643	.766	.839	85°	1.484	.996	.087	11.43
41°	.716	.656	.755	.869	86°	1.501	.998	.070	14.30
42°	.733	.669	.743	.900	87°	1.518	.999	.052	19.08
43°	.750	.682	.731	.933	88°	1.536	.999	.035	28.64
44°	.768	.695	.719	.966	89°	1.553	1.000	.017	57.29
45°	.785	.707	.707	1.000	90°	1.571	1.000	.000	∞

APPENDIX B

MATHEMATICAL REVIEW

This appendix reviews topics covered in high school or introductory college mathematics courses which are needed in various parts of this book. Students who are somewhat rusty in basic algebra, geometry, and trigonometry may find it helpful to study Sections B.1 to B.6 in detail; the remaining sections are provided mainly for reference at specific points in the book. A list of paperback mathematics review books is given at the end of the appendix for students who need additional preparation in this area. Answers to all the review problems are also given at the end of this appendix.

B.1 | POWERS AND ROOTS

A quantity x multiplied by itself n times is written as x^n. For example, $(2)(2)(2) = 2^3$; in words, this is 2 raised to the *exponent* or *power* 3. The basic rule in manipulating powers of a given number is that exponents add. For example, $(2^2)(2^3) = (2)(2) \cdot (2)(2)(2) = 2^5$. In symbols, the rule is

$$(x^n)(x^m) = x^{n+m} \qquad (B.1)$$

From this rule, we see that $x^n x^0 = x^n$, so $x^0 = 1$ for any value of x. Also, $x^n x^{-n} = x^0 = 1$, so x^{-n} is the inverse of x^n:

$$x^{-n} = \frac{1}{x^n} \qquad (B.2)$$

For example, $10^{-2} = 1/10^2 = 1/100 = 0.01$. Another useful rule is

$$(x^n)^m = x^{nm} \qquad (B.3)$$

For example, $(10^2)^3 = (10)(10) \cdot (10)(10) \cdot (10)(10) = 10^6$; similarly, $(10^2)^{-3} = 10^{-6}$.

When two numbers are raised to the same power, their products and quotients obey simple rules:

$$(x^n)(y^n) = (xy)^n \qquad (B.4)$$

$$\frac{x^n}{y^n} = \left(\frac{x}{y}\right)^n \qquad (B.5)$$

For example, $(4)^3(2)^3 = (4 \cdot 2)^3 = (8)^3$, and $(4)^3(2)^{-3} = (4/2)^3 = (2)^3$.

A fractional exponent means that a root of a number is involved. For example, $x^{1/2}x^{1/2} = x^1 = x$, so $x^{1/2}$ is the square root of x. The quantity $x^{1/n}$ is the nth root of x:

$$x^{1/n} = \sqrt[n]{x} \qquad (B.6)$$

For example, $(64)^{1/3} = \sqrt[3]{64} = 4$. More complicated fractional powers can be evaluated with the aid of Eq. B.3. For example, $(27)^{2/3} = (27^{1/3})^2 = (3)^2 = 9$. All the other rules listed above also apply to fractional powers.

REVIEW PROBLEMS

Evaluate or simplify the following quantities:

1. 2^4
2. 3^2
3. $(2^2)(2^3)$
4. $(x^5)(x^3)(x)$
5. 5^{-2}
6. $(5^{-3})(5^4)$
7. $(x^4)(x)(x^{-3})$
8. x^4/x^2
9. x^4/y^4
10. $(a^2 x^4)^{1/2}$
11. $(a^3 x^6)^{1/2}$
12. $(x^2 y^6)^{1/2}$
13. $(x^4 y^4)^{-1/2}$
14. $(1000)^{1/3}$

15. $(10,000)^{-1/4}$

16. $x^2(x^6)^{-1/3}$

17. $(125)^{-1/3}$

18. $\left(\dfrac{x^2}{64}\right)^{1/2}$

19. $(x^4 y^{-8})^{1/2}$

20. $(10^4)^{3/4}$

B.2 | SCIENTIFIC NOTATION

A number is said to be in scientific notation when it is written as a number between 1 and 10 times a power of 10. For example, 376 can be written as $3.76 \times 100 = 3.76 \times 10^2$, since $10^2 = 10 \times 10 = 100$. One advantage of this notation is compactness; 376,000,000 can be written as 3.76×10^8. Note that the power of 10 is the number of places the decimal point has been shifted to the left. Similarly, $0.000,037,6 = 3.76 \times 0.000,01 = 3.76 \times 10^{-5}$. Here the number in this negative exponent indicates how many places the decimal point has been shifted to the right.

Scientific notation facilitates many kinds of numerical calculations. It is especially useful in manipulations involving very large or small numbers. As an illustration, consider 2×10^{20} times 3×10^{-15} divided by 8×10^8:

$$\frac{(2 \times 10^{20})(3 \times 10^{-15})}{8 \times 10^8} = \frac{(2)(3)}{8} \times 10^{20-15-8}$$
$$= 0.75 \times 10^{-3} = 7.5 \times 10^{-4}$$

The use of scientific notation also aids in the evaluation of roots, as in the following illustrations:

$$(2.32 \times 10^8)^{1/2} = (2.32)^{1/2}(10^8)^{1/2} = \sqrt{2.32} \times 10^4$$
$$= 1.52 \times 10^4$$
$$(2.32 \times 10^8)^{1/3} = (232 \times 10^6)^{1/3}$$
$$= (232)^{1/3}(10^6)^{1/3}$$
$$= \sqrt[3]{232} \times 10^2 = 6.14 \times 10^2$$
$$(9.37 \times 10^{-4})^{1/3} = (937 \times 10^{-6})^{1/3}$$
$$= (937)^{1/3}(10^{-6})^{1/3}$$
$$= \sqrt[3]{937} \times 10^{-2}$$
$$= 9.79 \times 10^{-2}$$

In the last two examples, we have rewritten the power of 10 so that it leads to an integer power of 10 when the root is calculated. Notice also from the first two examples that the cube root of a number greater than 1 is less than the square root; the fourth root is smaller still. The converse statement holds for numbers less than 1.

REVIEW PROBLEMS

Write the following numbers in scientific notation:

21. 27,631

22. 2,763,100

23. 15,000

24. 0.000,000,034

25. 1,600

26. 4,329.76

27. 0.003,902

28. 0.080,02

Express the following numbers in ordinary notation:

29. 2.34×10^{-3}

30. 1.76×10^6

31. 5.799×10^{-5}

32. 4.5×10^7

33. 0.067×10^4

34. 27.2×10^5

35. 0.0272×10^8

Evaluate the following expressions:

36. $(3 \times 10^6)(5 \times 10^4)$

37. $\dfrac{4 \times 10^8}{8 \times 10^6}$

38. $(5 \times 10^{10})(3 \times 10^{-8})(4 \times 10^6)$

39. $\dfrac{(4.4 \times 10^6)(3 \times 10^3)^2}{6 \times 10^{-4}}$

40. $\dfrac{(8.25 \times 10^4)(3.14)(5.2 \times 10^3)^2}{(6.25 \times 10^{-3})}$

41. $(4 \times 10^4)^{1/2}$

42. $(90,000)^{1/2}$

43. $(2.7 \times 10^7)^{1/3}$

44. $(8,000)^{1/3}$

45. $(4 \times 10^{-6})^{1/2}$

46. $(160,000)^{1/4}$

47. $(10^{10})^{1/2}$

48. $(10^{10})^{-1/2}$

49. $(10^{10})^{1/3}$

50. $(3.2 \times 10^8)^{1/3}$

B.3 | SIGNIFICANT FIGURES

The accuracy of any measurement is limited by errors of various types (see Section 1.1). It is important to keep track of these errors at least approximately in using or manipulating experimentally determined numbers. This is accomplished most readily with the rules for significant figures.

The principle involved is illustrated by the problem of determining the area A of a rectangular sheet of paper using a ruler whose smallest spacing is 0.1 cm. If we place one end of the ruler at the edge of the paper, the other edge might lie between the markings indicating 8.4 and 8.5 cm. We can, at best, then judge its position to one-tenth of a spacing, so we might report our reading as 8.43 cm. However, a more elaborate measuring arrangement might well give a length closer to 8.44 or 8.42 cm; the last digit we report is somewhat uncertain. The number 8.43 is said to have three *significant figures*. In the same way we might find 6.77 cm for the other dimension of the rectangle. The area is then the product

$$A = (8.43 \text{ cm})(6.77 \text{ cm}) = 57.0711 \text{ cm}^2 = 57.1 \text{ cm}^2$$

Each of the factors in the product is uncertain in the third place, so only three places on the right have any meaning. Hence, A is given to three significant figures. To clarify the reason for this, suppose the first factor is found to be closer to 8.42 cm when more careful measurements are made. Then the area becomes $A = (8.42 \text{ cm})(6.77 \text{ cm}) = 57.0034 \text{ cm}^2$, and the digits beyond 57.0 are changed. Clearly, these digits in the product are meaningless, and the area A is somewhat uncertain in the third digit. Note that our answer for A has been rounded up from $57.07\cdots$ to 57.1; a number below 57.05 would be rounded down to 57.0.

In all computations involving multiplication and division, the factor with the fewest significant figures determines the number of significant figures in the answer. For example, in

$$\frac{(8.2239)(2.7)(98.35)\pi^2}{2764} = 7.797899\cdots$$

the first three factors in the numerator have five, two, and four significant figures, respectively; $\pi^2 = (3.1415926\cdots)^2$ is known to an arbitrarily great accuracy; and the denominator is known to four significant figures. Accordingly, the answer obtained for this expression should be rounded to two figures, that is, to 7.8. However, it is a good idea to retain one or more extra places in *intermediate steps* of the calculation in order to avoid introducing additional errors in the process of rounding off the numbers. This is important in complex multistep calculations and is easy to do with an electronic calculator.

The significant figures procedure used in addition and subtraction differs from that for multiplication and division. It is illustrated by the sum

$$\begin{array}{r} 45.76 \\ +\ 0.123 \\ \hline 45.883 \end{array}$$

Here, the 6 in the first number is somewhat uncertain, and the next place is completely unknown. Accordingly, the 3 in the sum is meaningless, and the answer is rounded to 45.88. *The answer contains as many places relative to the decimal point as the "least accurate number" in the sum.* Notice that in this example the least accurate number that limits the accuracy is 45.76, which has four significant figures; 0.123 has only three significant figures but is more accurate in the sense meant here.

Since the same ideas apply to subtraction, the difference of two nearly equal numbers may have very few significant figures. For example, consider

$$\begin{array}{r} 35.179 \\ -35.17813 \\ \hline 0.001 \end{array}$$

This result has essentially no accuracy, since it is uncertain by approximately 1 in the last place. If a new set of measurements changed the numbers slightly, their difference could well be 0.002 or -0.001.

Adding or subtracting numbers expressed in scientific notation requires that they be written with the same power of 10. For example,

$$\begin{aligned} 2.25 \times 10^6 + 6.4 \times 10^7 &= 2.25 \times 10^6 + 64 \times 10^6 \\ &= 66.25 \times 10^6 \\ &= 6.6 \times 10^7 \end{aligned}$$

Note that we have rounded 66.25 to 66 in accordance with the rules.

Significant Zeros

The number 1200 may have two, three, or four significant figures, depending upon whether the zeros represent measurements or are merely used to locate the decimal point. Scientific notation avoids this ambiguity; 1.2×10^3, 1.20×10^3, and 1.200×10^3 have two, three, and four significant figures, respectively.

REVIEW PROBLEMS

Round off the following quantities to three significant figures and write them in scientific notation:

51. 27632.0
52. 0.3729
53. 4.6667
54. 3.33333
55. 2.45558×10^4
56. 0.000,034,567

How many significant figures are there in the final result for each of the following expressions?

57. $(3.2)(8.67)/(3.008)$
58. $(0.0002)(45.6)$
59. $(2.0 \times 10^5)(3.777 \times 10^{-4})$
60. $17.2 + 2.35 + 4.3333$
61. $88.45 + 9.24 - 6.05043$
62. $186.45 - 186.12$

Evaluate the following expressions in accordance with the rules for significant figures:

63. $3.28 \times 10^5 + 4.25 \times 10^7$
64. $3.7 \times 10^6 + 2.91 \times 10^7$
65. $1.91 \times 10^{-3} - 1.7 \times 10^{-5}$

B.4 | SOLUTION OF ALGEBRAIC EQUATIONS

The application of physical laws often leads to one or more algebraic equations that must be solved for the desired quantities. To do this, we must have as many equations as there are unknowns; for example, if we want to find two forces acting on an object, we must have two different equations relating them.

Equations in One Unknown | The basic rule in manipulating any algebraic equation is that both sides of the equation must be treated in the same way. If we add a number to one side, or multiply it by some factor, or square it, we must do the same thing to the other side.

To illustrate how a simple equation is solved, we first consider a *linear* equation, one in which the unknown quantity x appears only to the first power:

$$5x - 10 = 30$$

To solve for x, we first add 10 to both sides, giving

$$5x = 40$$

Dividing by 5 leads to the desired solution,

$$x = 8$$

A *quadratic* equation is one in which the highest power of the unknown is 2. Quadratic equations may or may not have a term linear in the unknown. An example of the latter situation is

$$16t^2 = 64$$

Dividing by 16 gives $t^2 = 4$. Taking the square root then gives two possible answers, $t = +2$ and $t = -2$, since either of these numbers squared is equal to 4. (Quadratic equations generally have two solutions.) Since t represents some number such as a time that is measured, it can only have one correct value. The circumstances of the specific problem will indicate which is the appropriate solution. For example, in Chapter One, in determining when a ball thrown straight up is at a specific height, we encounter equations in the time variable t. The two solutions of these equations correspond to the time when the ball is at that height on the way up and on the way down.

The quadratic equation

$$t^2 - 6t + 8 = 0$$

is somewhat more complicated than the preceding one, since it contains t to the first power as well as t^2. It can be solved by *factoring*, rewriting the equation as a product of two factors that is equal to zero:

$$(t - 2)(t - 4) = 0$$

This factoring can be verified by noting that when we multiply out the two factors, we obtain four terms, $t^2 - 4t - 2t + 8 = t^2 - 6t + 8$. Clearly the product can be zero only if either the first *or* the second factor is 0. Thus

$$t - 2 = 0, \qquad t = 2$$

or

$$t - 4 = 0, \qquad t = 4$$

The two solutions of this equation are $t = 2$ and $t = 4$.

The disadvantage of factoring is that one must guess the factors somehow. The *quadratic formula* is a general solution that can always be used without any guesswork. The equation

$$at^2 + bt + c = 0$$

has two solutions,

$$t = \frac{-b + \sqrt{b^2 - 4ac}}{2a}$$

and

$$t = \frac{-b - \sqrt{b^2 - 4ac}}{2a} \tag{B.7}$$

This can be applied to the example $t^2 - 6t + 8 = 0$ by setting $a = 1$, $b = -6$, and $c = 8$. We find then

$$t = \frac{-(-6) + \sqrt{(-6)^2 - 4(1)(8)}}{2(1)}$$

$$= \frac{6 + \sqrt{4}}{2} = 4$$

and

$$t = \frac{-(-6) - \sqrt{(-6)^2 - 4(1)(8)}}{2(1)}$$

$$= \frac{6 - \sqrt{4}}{2} = 2$$

These are the same solutions as found before.

Simultaneous Equations | Two different
equations containing the same two unknowns are called *simultaneous equations*. The unknowns are found by combining the equations in such a way that a single equation is obtained which contains just one unknown. For example, consider these two equations for a force F and an acceleration a, $F - 6a = 20$ and $-F + 8a = 0$. If we add these equations, the F and $-F$ cancel:

$$\begin{array}{r} F - 6a = 20 \\ -F + 8a = 0 \\ \hline F - F - 6a + 8a = 20 + 0 \end{array}$$

This reduces to

$$2a = 20$$

or

$$a = 10$$

F is then found by substituting the value found for a into the first equation:

$$F - 6(10) = 20, \qquad F = 20 + 60 = 80$$

Sometimes one equation must be multiplied by a factor before the equations are added or subtracted in order to eliminate one unknown. For example, consider $x + 3y = 6$ and $2x - y = 5$. Neither the x's nor y's will completely cancel if these equations are added. However, if we multiply the first equation by 2 and the second by -1, we find upon adding them

$$\begin{array}{r} 2x + 6y = 12 \\ -2x + y = -5 \\ \hline 7y = 7 \end{array}$$

and $y = 1$. Substituting this value of y into either equation then gives $x = 3$.

The procedure used above is readily extended to three simultaneous equations containing three unknowns, x, y, and z. Two of the equations are combined to yield an equation containing only two unknowns, say x and y, and another combination of two of the original equations is formed which again contains only x and y. The problem has then been reduced to two equations in two unknowns, and one then proceeds as in the examples above to find x and y.

REVIEW PROBLEMS

Solve the following equations for the unknown quantities.

66. $x - 7 = 3$
67. $3x + 7 = 4 + 6x$
68. $1 + 0.2x = 7$
69. $x^2 + 4 = 13$
70. $x^{1/2} + 4 = 13$
71. $-4x + 7 = 2x + 15$
72. $(x/3)^{1/2} = 2$
73. $0 = 64 - 16t^2$
74. $x^3 - 1 = 63$
75. $(x + 2)(x + 4) = 0$
76. $x^2 + 3x + 2 = 0$
77. $3x^2 + 2x - 5 = 0$
78. $x^2 + 4 = -4x$
79. $2x^2 = -3x$
80. $-3x + 2x^2 - 5 = 0$
81. $x + y = 5, \quad x - y = 1$
82. $2 - T = 3a, \quad T = 4a$
83. $x + 3y = 9, \quad x - 2y = 10$
84. $2x - y = 10, \quad x + y = 6$
85. $3x - 7y = 2, \quad 3x - 2y = 4$

B.5 | GRAPHS

Just as pictures are often more informative than words, graphs are often more useful than algebraic formulas in understanding what is happening in a physical system. To illustrate how a graph is constructed, consider the equation for the position coordinate x of an object moving at a constant velocity at a time t:

$$x = 5 + 2t$$

We start by making a table of the values of x ob-

Figure B.1. The graph of $x = 5 + 2t$.

tained from this equation for several values of t between -4 and $+4$:

time, t	-4	-2	0	2	4
position, x	-3	1	5	9	13

Using ordinary (Cartesian) graph paper, we draw a horizontal axis for the independent variable, t, and a vertical axis for the dependent variable, x. The tabulated values are then used to locate the points marked on the graph (Fig. B.1).

In this example, all the points fall on a single straight line because t appears only to the first power. If other powers are present the graph of the equation will be a curve. Some examples of such curves are shown in Fig. B.2.

The choice of the dependent and independent variables is not fixed, and it depends on what one is trying to do. For example, the equation

$$x = 16t^2$$

gives the position of an object dropped from $x = 0$ at time $t = 0$. If we want to know when it will be at a position x, we must solve for t in terms of x:

$$t = \tfrac{1}{4}\sqrt{x}$$

Now x has become the independent variable. The graphs of the two forms of the equation look quite different (Fig. B.3), although both are curves rather than straight lines. We can also get a straight line graph if we consider t^2 rather than t as a variable. This is particularly useful in analyzing experimental measurements, since it is easy to see if the data points all fall on a straight line as expected.

REVIEW PROBLEMS

Draw the graphs of the following equations:
86. $y = 3x - 7$
87. $x = 2t^2$
88. $y = 2x^4 - 3$

B.6 | PLANE GEOMETRY AND TRIGONOMETRIC FUNCTIONS

Plane Geometry | The following results from plane geometry are often useful.

1 The sum of the internal angles of any triangle is $180°$. In a right triangle, where one angle is $90°$, the other two angles must add up to $90°$.

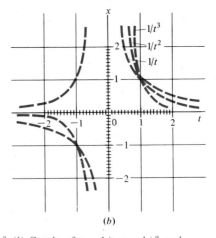

(a)   (b)

Figure B.2. (a) Graphs of $x = t$, $x = t^2$, and $x = t^3$. (b) Graphs of $x = 1/t$, $x = 1/t^2$, and $x = 1/t^3$. Notice that $1/t^3$ becomes very small most rapidly as t increases and also grows most rapidly as t approaches zero.

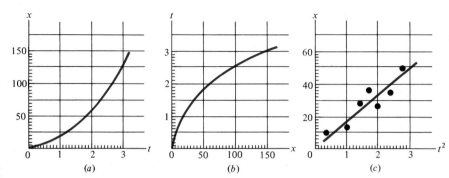

Figure B.3. (*a*) A graph of $x = 16t^2$. (*b*) A graph of $t = \frac{1}{4}\sqrt{x}$. (*c*) Measured pairs of x and t^2 values are plotted. They fall close to a straight line drawn through them with a ruler. On a plot of x versus t, they would fall near a curved line which would be harder to draw accurately and to analyze numerically.

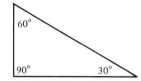

2 Two triangles are *similar* if two of their angles are equal. The corresponding sides of similar triangles are proportional. For example, $a/A = b/B$ in this figure:

3 Two angles are equal if their sides are parallel.

4 Two angles are equal if their sides are mutually perpendicular. For example, the angle between the weight vector **w** and the line perpendicular to the inclined plane equals the angle between the plane and the horizontal direction.

5 Two angles are equal if they are *vertical* angles.

6 Two angles are said to be *complementary* if they add up to 90° and *supplementary* if they add up to 180°. In the figure above θ and ϕ are supplementary.

Trigonometric Functions | The sine, cosine, and tangent of an angle are abbreviated as sin, cos, and tan, respectively, and are defined in terms of the right triangle below.

$$\sin \theta = \frac{\text{opposite side}}{\text{hypotenuse}} = \frac{a}{c} \qquad \text{(B.8)}$$

$$\cos \theta = \frac{\text{adjacent side}}{\text{hypotenuse}} = \frac{b}{c} \qquad \text{(B.9)}$$

$$\tan \theta = \frac{\text{opposite side}}{\text{adjacent side}} = \frac{a}{b} \qquad \text{(B.10)}$$

The Pythagorean theorem states that

$$a^2 + b^2 = c^2 \qquad \text{(B.11)}$$

Dividing by c^2,

$$\frac{a^2}{c^2} + \frac{b^2}{c^2} = 1$$

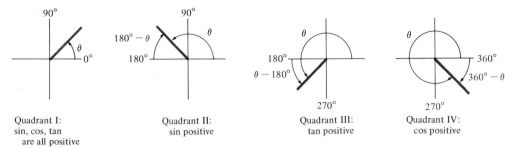

Quadrant I:
sin, cos, tan
are all positive

Quadrant II:
sin positive

Quadrant III:
tan positive

Quadrant IV:
cos positive

Figure B.4. A summary of the rules for finding the trigonometric functions. For example, if θ is in the third quadrant, one looks up $\theta - 180°$ in the tables; a plus sign is used for $\tan \theta$, and a minus sign for $\sin \theta$ and $\cos \theta$.

or

$$\sin^2 \theta + \cos^2 \theta = 1 \qquad (B.12)$$

The trigonometric functions of angles larger than 90° may be either positive or negative, depending on the angle. In the notation of the diagram below, they are defined by

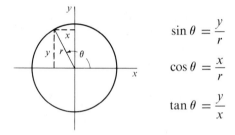

$$\sin \theta = \frac{y}{r}$$

$$\cos \theta = \frac{x}{r}$$

$$\tan \theta = \frac{y}{x}$$

By convention, r is always positive. However, for the angle shown, which is said to be in the second *quadrant*, x is negative and y is positive. Accordingly the sine is positive, and the tangent and cosine are negative. In the third quadrant only the tangent is posi-

tive, and in the fourth only the cosine is positive.

Appendix A contains a table of the trigonometric functions for angles up to 90°. For angles between 90° and 180°, one looks up the supplement, $180° - \theta$. Between 180° and 270°, one looks up $\theta - 180°$; and from 270° to 360°, one looks up $360° - \theta$. Minus signs are inserted wherever required in accordance with the discussion above. For example, $\sin 150° = \sin 30°$, $\cos 150° = -\cos 30°$, and $\tan 150° = -\tan 30°$. These rules are summarized in Fig. B.4.

Graphs of the sine, cosine, and tangent (Fig. B.5) are useful in understanding the general behavior of these quantities. They all repeat after one full circle or cycle; their *period* is 360° or 2π radians. (Measuring angles in radians is discussed in Chapter Eight.)

The average of the sine or cosine over a full cycle is zero, since for every positive value, there is a corresponding negative value. Notice that the $\sin \theta$ and $\cos \theta$ curves are identical if the $\cos \theta$ curve is shifted 90° to the right. The same shift also makes the $\sin^2 \theta$

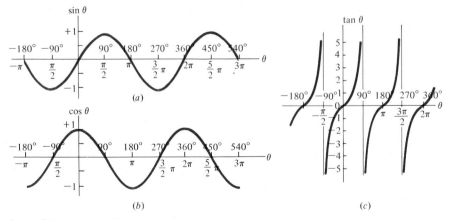

Figure B.5. Graphs of (*a*) $\sin \theta$; (*b*) $\cos \theta$; (*c*) $\tan \theta$.

628

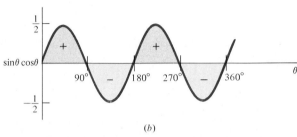

Figure B.6. (a) $\sin^2 \theta$ and $\cos^2 \theta$ have the same average value, $\frac{1}{2}$. (b) The average value of $\sin \theta \cos \theta$ is zero.

and $\cos^2 \theta$ curves (Fig. B.6) identical, so the average values of these quantities over a full cycle must be equal. Denoting the average by a bar, this means

$$\overline{\sin^2 \theta} = \overline{\cos^2 \theta}$$

Since we know from Eq. B.12 that $\sin^2 \theta + \cos^2 \theta = 1$, it follows that

$$\overline{\sin^2 \theta} + \overline{\cos^2 \theta} = 1$$

Therefore

$$\overline{\sin^2 \theta} = \overline{\cos^2 \theta} = \tfrac{1}{2} \qquad \text{(B.13)}$$

Also, we can see from Fig. B.6 that the product, $\sin \theta \cos \theta$, has a corresponding negative value for each positive value, so averaging the product over a full cycle must give zero:

$$\overline{\sin \theta \cos \theta} = 0 \qquad \text{(B.14)}$$

REVIEW PROBLEMS

89. If two angles of a triangle are 29° and 111°, what is the third angle?

90. Find sin 120°; cos 120°; tan 120°.

91. Find sin 270°; cos 270°; tan 270°.

92. At what angle or angles between 0° and 360° is $\sin \theta$ equal to 0, $+1$, -1? Give the corresponding angles for the cosine and tangent.

B.7 | SERIES EXPANSIONS

We are often interested in the value of a trigonometric function or algebraic expression when the variable is much less than 1. In these cases, it is often convenient to approximate the exact formula by a series expansion involving successively higher powers of the variable, since these terms quickly become very small.

As an example, we may consider the series expansion for $(1 - x)^{-1}$ derived in algebra texts:

$$\frac{1}{1 - x} = 1 + x + x^2 + x^3 + \cdots$$

When $x = 0.1$, $(1 - x)^{-1} = 1/0.9 = 1.11$ to two decimal places. Since $x = 0.1$, $x^2 = 0.01$, and $x^3 = 0.001$. Thus

$$1 + x + x^2 = 1.11$$

is sufficient to approximate the formula to two decimal places. If $x = 0.01$, then $1 + x$ is sufficient to give two place accuracy.

Many series expansions involve the quantity $n!$ (read "n factorial") defined by

$$n! = n(n - 1)(n - 2) \cdots (2)(1)$$

For example, $4! = 4 \times 3 \times 2 \times 1 = 24$. By definition,

$$1! = 1, \qquad 0! = 1$$

The following series expansions are often useful:

$$(1 \pm x)^{-1} = 1 \mp x + x^2 \mp x^3 + \cdots$$
$$(-1 < x < 1) \quad \text{(B.15)}$$

$$(1 \pm x)^n = 1 \pm nx + \frac{n(n - 1)x^2}{2!}$$
$$\pm \frac{n(n - 1)(n - 2)x^3}{3!} + \cdots$$
$$(-1 < x < 1) \quad \text{(B.16)}$$

$$(1 \pm x)^{-n} = 1 \mp nx + \frac{n(n + 1)x^2}{2!}$$
$$\mp \frac{n(n + 1)(n + 2)x^3}{3!} + \cdots$$
$$(-1 < x < 1) \quad \text{(B.17)}$$

$$e^x = 1 + x + \frac{x^2}{2!} + \frac{x^3}{3!} + \cdots \qquad \text{(B.18)}$$

In the following series for the trigonometric functions,

the angles must be measured in radians, where $1 \text{ rad} = 180°/\pi = 57.3°$.

$$\sin x = x - \frac{x^3}{3!} + \frac{x^5}{5!} - \cdots \quad \text{(B.19)}$$

$$\cos x = 1 - \frac{x^2}{2!} + \frac{x^4}{4!} - \cdots \quad \text{(B.20)}$$

$$\tan x = x + \frac{x^3}{3} + \frac{2x^5}{15} + \frac{17x^7}{314} + \cdots \quad \text{(B.21)}$$

REVIEW PROBLEMS

93. What is the percentage error in using $\sin x = x$, when $x = 10° = 0.1745$ rad? What is the error at $30°$?

94. For $x = 0.1$, $e^x = e^{0.1} = 1.105$. How many terms in the series expansion for e^x are needed to obtain this accuracy?

95. (a) Write out the first three terms in the series for $(1 + x)^{1/2} = \sqrt{1 + x}$. (b) What does this approximation give for $x = 0.1$? Compare this with the exact answer to five decimal places, 1.04881.

B.8 | DERIVATIVES

At the end of several chapters in this book, we present derivations of various equations which are based on arguments using differentiation. Here we list the derivatives used. Note that the derivative of a constant is zero. Also, when an expression is multiplied by a number, its derivative is multiplied by the same number. For example,

$$\frac{d}{dt}(3t^2) = 3\frac{d}{dt}(t^2)$$

In the following expressions a and n are constants.

$$\frac{d}{dt}(t^n) = nt^{n-1} \quad \text{(B.22)}$$

$$\frac{d}{dt}\left(\frac{1}{t}\right) = -\frac{1}{t^2} \quad \text{(B.23)}$$

$$\frac{d}{dt}\left(\frac{1}{t^n}\right) = \frac{-n}{t^{n+1}} \quad \text{(B.24)}$$

$$\frac{d}{dt}(e^{at}) = ae^{at} \quad \text{(B.25)}$$

$$\frac{d}{dt}\sin at = a\cos at \quad \text{(B.26)}$$

$$\frac{d}{dt}\cos at = -a\sin at \quad \text{(B.27)}$$

REVIEW PROBLEMS

Find the derivatives of the following expressions.

96. $3t + 7$
97. $4t^3$
98. $1 - 1/t$
99. $4e^{-3t}$
100. $10 \sin 2\pi t$

Additional Reading

Students needing additional mathematical preparation may find the following books helpful.

Jerry B. Marion and Ronald C. Davidson, *Mathematical Preparation for General Physics,* W.B. Saunders, Philadelphia, 1972. A paperback; same level as this appendix.

Clifford E. Swartz, *Used Math,* Prentice-Hall, Englewood Cliffs, N.J., 1973. A paperback; includes more advanced material.

Answers to Review Problems in Appendix B

1. 16	**21.** 2.7631×10^4
2. 9	**22.** 2.7631×10^6
3. $2^5 = 32$	**23.** 1.5×10^4
4. x^9	**24.** 3.4×10^{-8}
5. $\frac{1}{5^2} = 0.04$	**25.** 1.6×10^3
	26. 4.32976×10^3
6. 5	**27.** 3.902×10^{-3}
7. x^2	**28.** 8.002×10^{-2}
8. x^2	**29.** 0.00234
9. $\left(\frac{x}{y}\right)^4$	**30.** 1,760,000
10. ax^2	**31.** 0.000,057,99
11. $a^{3/2}x^3$	**32.** 45,000,000
12. xy^3	**33.** 670
13. $\frac{1}{x^2y^2}$	**34.** 2,720,000
	35. 2,720,000
14. 10	**36.** 1.5×10^{11}
15. 10	**37.** 50
16. 1	**38.** 6×10^9
17. $\frac{1}{5} = 0.2$	**39.** 2.2×10^{16}
18. $\frac{x}{8}$	**40.** 1.12×10^{15}
19. $x^2y^{-4} = \frac{x^2}{y^4}$	**41.** 200
20. 10^3	**42.** 300

43. 300

44. 20

45. 2×10^{-3}

46. 20

47. 10^5

48. 10^{-5}

49. 2.15×10^3

50. 684

51. 2.76×10^4

52. 3.73×10^{-1}

53. 4.67

54. 3.33

55. 2.46×10^4

56. 3.46×10^{-5}

57. two

58. one

59. two

60. three

61. four

62. two

63. 4.28×10^7

64. 3.28×10^7

65. 1.89×10^{-3}

66. $x = 10$

67. $x = 1$

68. $x = 30$

69. $x = \pm 3$

70. $x = 81$

71. $x = -\frac{4}{3}$

72. $x = 12$

73. $t = \pm 2$

74. $x = 4$

75. $x = -2, x = -4$

76. $x = -1, x = -2$

77. $x = 1, x = -\frac{5}{3}$

78. $x = -2$

79. $x = -\frac{3}{2}, x = 0$

80. $x = -1, x = \frac{5}{2}$

81. $x = 3, y = 2$

82. $a = \frac{2}{7}, T = \frac{8}{7}$

83. $y = -\frac{1}{5}, x = 9\frac{3}{5}$

84. $x = 5\frac{1}{3}, y = \frac{2}{3}$

85. $y = \frac{2}{5}, x = 1\frac{3}{5}$

86.

87.

88.

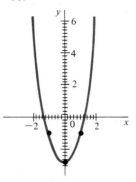

89. $40°$

90. $\sin 120° = 0.866$, $\cos 120° = -0.5$,
$\tan 120° = 1.732$

91. $\sin 270° = -1$, $\cos 270° = 0$, $\tan 270° = \infty$

92. $\sin \theta = 0$ at $0°$, $180°$, $360°$
$\sin \theta = 1$ at $90°$
$\sin \theta = -1$ at $270°$
$\cos \theta = 0$ at $90°$, $270°$
$\cos \theta = 1$ at $0°$, $360°$
$\cos \theta = -1$ at $180°$
$\tan \theta = 0$ at $0°$, $180°$, $360°$
$\tan \theta = 1$ at $45°$, $225°$
$\tan \theta = -1$ at $135°$, $315°$

93. 0.52 percent at $10°$, 4.7 percent at $30°$

94. Three terms

95. (a) $1 + \frac{1}{2}x - \frac{1}{8}x^2$
(b) 1.04875, which is 0.006 percent less than the exact answer.

96. 3

97. $12t^2$

98. $\dfrac{1}{t^2}$

99. $-12e^{-3t}$

100. $20\pi \cos 2\pi t$

APPENDIX C

CONVERSION OF UNITS

Equalities denoted by $\equiv$ are exact.
Metric prefixes are given in Table 1.3 on p. 6.

Length
1 meter $\equiv$ 39.37 inches = 3.281 feet
1 inch $\equiv$ 2.54 centimeters
1 foot $\equiv$ 30.48 centimeters
1 kilometer $\equiv$ 10^3 meters = 0.6214 miles
1 mile $\equiv$ 5280 feet = 1.609 kilometers
1 angstrom $\equiv$ 10^{-10} meters
1 nanometer $\equiv$ 10^{-9} meters

Area
1 in.^{2} $\equiv$ 6.4516 cm^2
1 ft^2 = 9.29 $\times$ 10^{-2} m^2
1 cm^2 $\equiv$ 10^{-4} m^2 = 0.155 in.^{2} = 1.076 $\times$ 10^{-3} ft^2
1 m^2 $\equiv$ 10^4 cm^2 = 10.76 ft^2

Volume
1 in.^{3} = 16.39 cm^3
1 ft^3 = 2.832 $\times$ 10^{-2} m^3
1 cm^3 $\equiv$ 10^{-6} m^3 = 6.102 $\times$ 10^{-2} in.^{3}
$\quad$ = 3.531 $\times$ 10^{-5} ft^3
1 m^3 $\equiv$ 10^6 cm^3 = 35.31 ft^3
1 liter $\equiv$ 10^{-3} m^3 = 0.264 gallons
1 gallon = 3.786 liters = 231 in.^{3}

Time
1 hour $\equiv$ 60 minutes $\equiv$ 3600 seconds
1 day $\equiv$ 24 hours $\equiv$ 1440 minutes
$\quad$ $\equiv$ 8.64 $\times$ 10^4 seconds
1 year = 365.24 days = 3.156 $\times$ 10^7 seconds

Mass
1 gram $\equiv$ 10^{-3} kilograms = 6.852 $\times$ 10^{-5} slugs
$\quad$ = 6.024 $\times$ 10^{23} amu

1 kilogram $\equiv$ 10^3 grams = 6.852 $\times$ 10^{-2} slugs
$\quad$ = 6.024 $\times$ 10^{26} amu
1 slug = 14.59 kilograms
1 amu = 1.66 $\times$ 10^{-27} kilograms

Mass density
1 gm cm^{-3} $\equiv$ 10^3 kg m^{-3} = 1.94 slug ft^{-3}
1 slug ft^{-3} = 0.5153 gm cm^{-3}
$\quad$ = 5.153 $\times$ 10^2 kg m^{-3}

Speed
1 cm s^{-1} $\equiv$ 10^{-2} m s^{-1} $\equiv$ 3.6 $\times$ 10^{-2} km h^{-1}
1 m s^{-1} $\equiv$ 3.6 km h^{-1} = 2.24 mi h^{-1}
1 ft s^{-1} $\equiv$ 30.48 cm s^{-1} $\equiv$ 0.3048 m s^{-1}
$\quad$ = 1.097 km h^{-1}
1 mi h^{-1} = 0.447 m s^{-1} = 1.609 km h^{-1}
$\quad$ = 1.467 ft s^{-1}
1 km h^{-1} = 0.2778 m s^{-1} = 0.6214 mi h^{-1}

Angle and Angular Speed
180 degrees $\equiv$ π radians
1 radian = 57.3 degrees
1 degree = 1.745 $\times$ 10^{-2} radians
1 rad s^{-1} = 0.159 rev s^{-1} = 9.549 rev min^{-1}
1 rev min^{-1} = 0.0167 rev s^{-1} = 0.1047 rad s^{-1}

Force
1 pound = 4.448 newtons = 4.448 $\times$ 10^5 dynes
1 newton $\equiv$ 10^5 dynes = 0.2248 pounds
1 dyne $\equiv$ 10^{-5} newtons = 2.248 $\times$ 10^{-6} pounds

Pressure
1 atmosphere = 1.013 $\times$ 10^5 pascals
$\quad$ = 14.7 lb in.$^{-2}$
1 pascal $\equiv$ 10 dyn cm^{-2} = 1.450 $\times$ 10^{-4} lb in.$^{-2}$
$\quad$ = 7.501 $\times$ 10^{-4} cm Hg

1 cm Hg $= 1.333 \times 10^4$ dyn cm^{-2}
   $= 1.316 \times 10^{-2}$ atmosphere $= 1.333 \times 10^3$
   pascals
1 in. H$_2$O $= 1.868$ mm Hg $= 249.1$ pascals
1 lb in.$^{-2} = 6.895 \times 10^3$ pascals
   $= 6.805 \times 10^{-2}$ atmosphere
1 lb ft$^{-2} = 47.88$ pascals
1 torr $\equiv 1$ mm Hg $= 133.3$ pascals

Viscosity and Flow Resistance
1 Pa s $\equiv 10$ poise
1 Pa s m$^{-3} \equiv 0.750 \times 10^{-8}$ torr s cm^{-3}

Energy
1 joule $\equiv 10^7$ ergs $= 0.2390$ calorie
   $= 0.7376$ ft lb
1 calorie $= 4.184$ joules

1 kcal $\equiv 10^3$ calories
1 joule $= 6.24 \times 10^{18}$ electron volts
1 electron volt $= 1.602 \times 10^{-19}$ joule
1 kWh $\equiv 3.6 \times 10^6$ joules
1 Btu $= 1.054 \times 10^3$ joules
1 ft lb $= 1.356$ joules

Mass Energy Conversion
1 amu $= 931 \times 10^6$ electron volts $\equiv 931$ MeV

Power
1 watt $\equiv 10^{-3}$ kilowatts $= 0.7376$ ft lb s^{-1}
   $= 1.341 \times 10^{-3}$ horsepower
1 horsepower $\equiv 550$ ft lb s$^{-1} = 7.457 \times 10^2$ watts
1 kilowatt $\equiv 10^3$ watts $= 1.341$ horsepower

Magnetic Field
1 gauss $= 10^{-4}$ tesla

APPENDIX D

PHYSICAL CONSTANTS

D.1 | FUNDAMENTAL CONSTANTS

The numerical values of most constants have been rounded off to three significant figures for convenience.

Quantity	Symbol	Numerical Value
Speed of light (in vacuum)	c	3.00×10^8 m s^{-1}
Gravitational constant	G	6.67×10^{-11} N m^2 kg^{-2}
Avogadro's number	N_A	6.02×10^{23} molecules mole^{-1}
Universal gas constant	R	8.31 J K^{-1} mole^{-1}
Boltzmann constant	k_B	1.38×10^{-23} J K^{-1}
		8.62×10^{-5} eV K^{-1}
Stefan's constant	σ	5.67×10^{-8} W m^{-2} K^{-4}
Atomic mass unit	amu	1.66×10^{-27} kilograms
Coulomb constant	k	9.00×10^9 N m^2 C^{-2}
	$\varepsilon_0 = 1/4\pi k$	8.85×10^{-12} C^2 N^{-1} m^2
Biot-Savart constant	k'	10^{-7} T m A^{-1}
Electron charge	$-e$	-1.60×10^{-19} coulombs
Electron mass	m_e	9.11×10^{-31} kilograms
Proton charge	e	1.60×10^{-19} coulombs
Proton mass	m_p	1.673×10^{-27} kilograms
Neutron mass	m_n	1.675×10^{-27} kilograms
Planck's constant	h	6.63×10^{-34} J s
		4.14×10^{-15} eV s
	$\hbar = h/2\pi$	1.055×10^{-34} J s
		6.58×10^{-16} eV s
Rydberg constant	R_H	1.10×10^7 meters^{-1}
Bohr radius	a_0	5.29×10^{-11} meters
Bohr magneton	μ_B	9.27×10^{-24} J T^{-1}

D.2 | SOLAR AND TERRESTIAL DATA

The numerical values given have been rounded off to three significant figures for convenience.

Standard atmospheric pressure	1 atm
	1.01×10^5 Pa
	14.7 lb in.$^{-2}$
	760 mm Hg
	760 torr
Acceleration of gravity, g	9.81 m s^{-2}
	32.2 ft s^{-2}
Magnetic field (Washington, D.C.)	5.7×10^{-5} teslas
Speed of sound (dry air)	331 m s^{-1}
Mass of earth	5.98×10^{24} kilograms
Volume of earth	1.09×10^{21} m^3

Mean radius of earth	6.38×10^6 meters
Mean density of earth	5.52×10^3 kg m^{-3}
Mean angular rotational speed of earth	7.29×10^{-5} rad s^{-1}
Earth to sun, mean distance	1.50×10^{11} meters
Earth to moon, mean distance	3.84×10^8 meters
Mean orbital speed of earth about sun	2.98×10^4 m s^{-1}
Sun, mean radius	6.95×10^8 meters
mass	1.99×10^{30} kilograms
Moon, mean radius	1.74×10^6 meters
volume	2.20×10^{19} m^3
mass	7.35×10^{22} kilograms
mean density	3.34×10^3 kg m^{-3}
acceleration of gravity	1.62 m s^{-2}

D.3 | PERIODIC TABLE OF ELEMENTS

Each entry lists the atomic number before the symbol for the element. Below the element symbol is the atomic weight in atomic mass units of the element as it occurs naturally on earth. For example, the atomic mass of carbon 12 is defined as exactly 12.0 amu. In the table, the atomic mass of carbon is listed as 12.01115, because naturally occurring carbon is 98.89 percent ^{12}C and 1.11 percent ^{13}C. More accurate atomic masses of some pure isotopes are given in Table 33.4 on p. 581.

For artifically produced elements the approximate atomic mass of the most stable isotope is given in brackets.

Period	Series	I	II	III	IV	V	VI	VII	VIII			O
1	1	1 H 1.00797										2 He 4.0026
2	2	3 Li 6.939	4 Be 9.0122	5 B 10.811	6 C 12.01115	7 N 14.0067	8 O 15.9994	9 F 18.9984				10 Ne 20.183
3	3	11 Na 22.9898	12 Mg 24.312	13 Al 26.9815	14 Si 28.086	15 P 30.9738	16 S 32.064	17 Cl 35.453				18 A 39.948
4	4	19 K 39.102	20 Ca 40.08	21 Sc 44.956	22 Ti 47.90	23 V 50.942	24 Cr 51.996	25 Mn 54.9380	26 Fe 55.847	27 Co 58.9332	28 Ni 58.71	
	5	29 Cu 63.54	30 Zn 65.37	31 Ga 69.72	32 Ge 72.59	33 As 74.9216	34 Se 78.96	35 Br 79.909				36 Kr 83.80
5	6	37 Rb 85.47	38 Sr 87.62	39 Y 88.905	40 Zr 91.22	41 Nb 92.906	42 Mo 95.94	43 Tc [99]	44 Ru 101.07	45 Rh 102.905	46 Pd 106.4	
	7	47 Ag 107.870	48 Cd 112.40	49 In 114.82	50 Sn 118.69	51 Sb 121.75	52 Te 127.60	53 I 126.9044				54 Xe 131.30
6	8	55 Cs 132.905	56 Ba 137.34	57–71 Lanthanide series*	72 Hf 178.49	73 Ta 180.948	74 W 183.85	75 Re 186.2	76 Os 190.2	77 Ir 192.2	78 Pt 195.09	
	9	79 Au 196.967	80 Hg 200.59	81 Tl 204.37	82 Pb 207.19	83 Bi 208.980	84 Po [210]	85 At [210]				86 Rn [222]
7	10	87 Fr [223]	88 Ra [226.05]	89–Actinide series**								

* Lanthanide series: 57 La 138.91 | 58 Ce 140.12 | 59 Pr 140.907 | 60 Nd 144.24 | 61 Pm [147] | 62 Sm 150.35 | 63 Eu 151.96 | 64 Gd 157.25 | 65 Tb 158.924 | 66 Dy 162.50 | 67 Ho 164.930 | 68 Er 167.26 | 69 Tm 168.934 | 70 Yb 173.04 | 71 Lu 174.97

** Actinide series: 89 Ac [227] | 90 Th 232.038 | 91 Pa [231] | 92 U 238.03 | 93 Np [237] | 94 Pu [242] | 95 Am [243] | 96 Cm [245] | 97 Bk [249] | 98 Cf [249] | 99 Es [253] | 100 Fm [255] | 101 Md [256] | 102 No [254] | 103 Lw [257]

APPENDIX E

THE GREEK ALPHABET

A	α	alpha
B	β	beta
Γ	γ	gamma
Δ	δ	delta
E	ϵ	epsilon
Z	ζ	zeta
H	η	eta
Θ	θ	theta
I	ι	iota
K	κ	kappa
Λ	λ	lambda
M	μ	mu
N	ν	nu
Ξ	ξ	xi
O	o	omicron
Π	π	pi
P	ρ	rho
Σ	σ	sigma
T	τ	tau
Υ	υ	upsilon
Φ	ϕ	phi
X	χ	chi
Ψ	ψ	psi
Ω	ω	omega

ANSWERS TO ODD NUMBERED EXERCISES AND PROBLEMS

Chapter 1

1-1 4.05×10^3 m^2

1-3 3.79 liters

1-5 2.76×10^{-7} in.

1-7 40 mi h^{-1}

1-11 80 mi h^{-1}

1-13 10.2 m s^{-1}

1-15 (a) -32 ft s^{-1} (b) -96 ft s^{-1}

1-17 (a) 10 ft s^{-1} (b) 5 ft s^{-1} (c) 0

1-21 positive at 0, τ; negative $\tau/2$

1-25 (a) $T_0 < t < 3T_0$ (b) $0 < t < T_0, 3T_0 < t < 5T_0$ (c) $t > 5T_0$

1-27 5 m s^{-2}

1-29 (a) 40 m s^{-1} (b) 0

1-31 225 m

1-33 (a) 48.8 ft s^{-2} (b) 7.35 s

1-35 (a) 1.28×10^4 m (b) 51.1 s (c) 100 s, 2.04 s

1-37 (a) 14.7 m s^{-1} (b) 1.50 s (c) 3.50 s

1-39 (a) 39.2 m s^{-1} (b) 78.4 m

1-41 (a) 49.0 m s^{-1} (b) 123 m

1-43 (a) 1.10×10^3 m (b) 1.40×10^3 m

1-45 7 m s^{-1}

1-51 6 s

1-53 (a) 3.93 yd s^{-2} (b) 3.51 s (c) 13.8 yd s^{-1} (d) 220, yes; mile, no

1-55 (a) 1.84 m (b) 1.22 s

1-57 $v = 2.42$ m s^{-1}, $a = 98.0$ m s^{-2}

1-59 62.5 m

Chapter 2

2-1 (a) **D** (b) **G** (c) **C** (d) 0 (e) **A** (f) **G** (g) **C** (h) **D**

2-3 (b) 10.4, 17° above $-x$ axis

2-5 (a) 7.21, 56° above **A** (b) 7.21, 56° above $-$**A** (c) 7.21, 56° below **A**

2-7 24.9, 36° above $-x$ axis

2-9 224 mi, 27° N of W

2-11 (a) 8.54, 69° above **C** (b) 17.0, 28° below **A**

2-13 (a) 39.3 s (b) 12.7 m s^{-1}

2-15 $v_x = 28.2$ m s^{-1}, $v_y = 10.3$ m s^{-1}

2-17 (a) 78.5 s (b) 0.510 m s^{-2} toward south

2-21 (a) 1.75 s (b) 35 m

2-23 (a) x = 8.67 m, y = 2.10 m (b) $v_x = 8.67$ m s^{-1}, $v_y = -4.8$ m s^{-2}

2-25 (a) 20.4 m (b) 2.04 s (c) 70.7 m

2-27 (a) $v_x = 8.32$ m s^{-1}, $v_y = 5.55$ m s^{-1} (b) 0.180 s (c) 0.840 m

2-29 121 ft s^{-1}, 8° above horizontal

2-31 (a) 75.4 ft s^{-1} (b) 13.3 ft past net

2-33 (a) 35.3 m (b) 35.3 m (c) 40.8 m

2-35 4 m

2-39 44.8 m s^{-1}

2-41 (a) Center of body is lowered (b) Less.

2-43 (a) 0.676 m s^{-1} (b) 286 m s^{-2}

Chapter 3

3-1 37° below the 20 N force, 25 N

3-3 15° from either force, 19.3 N

3-5 490 N

3-7 (a) 4.90×10^5 dyn (b) 4.90 N (c) 1.10 lb

3-9 Holds for (a), (c)

3-11 (a) Yes (b) Gravity, normal, friction, air resistance

3-13 (a) No (b) Gravity, normal, friction
(c) Forward
3-19 (a) 1490 N (b) 3.04
3-21 3000 N
3-23 3.52×10^5 km
3-25 73.9 lb
3-27 $m_E/8$
3-29 5.98×10^{24} kg
3-31 (a) 0 (b) 1.63 m s^{-2}
3-33 (a) 888 N (b) 288 N
3-35 2.53 m s^{-2}
3-37 (a) 100 m s^{-2} (b) 10^5 N
3-39 (a) 200 N (b) $\mu_s \geq 0.2$
3-41 6.40 ft s^{-2}
3-43 (a) 7200 lb (b) 4800 lb
3-45 313 lb
3-47 (a) 5 slug (b) 20 ft s^{-2} (c) 100 lb
3-49 (a) 4.43 m s^{-1} (b) 3600 N (c) 1080 N
3-51 (a) 1.23 m s^{-2} (b) 32,000 N (c) 16,000 N
3-53 (a) 0.647 m s^{-2} (b) 30.6 kg
3-55 (a) 1000 N (b) 2010 N
3-57 5.12 m s^{-2}
3-59 (a) $1.55mg$, $1.22mg$ (b) $0.225mg$
3-61 (a) 23.5 N (b) 1.96 m s^{-2} (c) 0.98 m s^{-1},
0.245 m
3-63 190 kg
3-65 (a) 6.24 m s^{-2} (b) 22.3 m s^{-1}
3-67 3.20 m s^{-2}

Chapter 4

4-1 $\mathbf{w}_1$, 0; $\mathbf{w}_2$, -6 N m; $\mathbf{w}_3$, -40 N m; $\mathbf{w}_4$, -75 N m
4-3 (a) $\mathbf{A} \times \mathbf{A}$, $\mathbf{A} \times \mathbf{C}$, (b) $\mathbf{A} \times \mathbf{D}$, $\mathbf{A} \times \mathbf{E}$
(c) $\mathbf{A} \times \mathbf{B}$ (d) $\mathbf{A} \times \mathbf{D}$, $\mathbf{A} \times \mathbf{E}$
4-5 (a) 30 N m, into page; 24 N m, into page;
21.2 N m, out of page (b) position (a)
4-7 $T_1 = 25.7$ N, $T_2 = 40.3$ N
4-9 $F_1 = 0.75$ N, $F_2 = 0.25$ N
4-11 (a) $w + w_1 + w_2$ (b) w_1/w_2
4-13 40 N
4-15 1.14 m
4-17 Lower C. G.
4-19 500 N
4-21 0.667
4-25 (a) I (b) III (c) Splenius
4-27 $T = 2000$ N, $H = 1730$ N
4-29 500 N
4-31 (a) 400 lb (b) $X = 0.36$ m, $Y = 0.15$ m

4-33 (a) $X = M/(100 + M)$,
$Y = (110 + 1.5\,M)/(100 + M)$ (b) 17.6 kg
4-35 32°
4-37 0.126 m
4-41 (a) 0.075 m (b) 3 N
4-43 3 m from scale 1
4-45 $X = 0.229h$, $Y = 0.443h$
4-47 75°
4-49 42.4 N

Chapter 5

5-1 0.64 m s^{-2}
5-3 367 m
5-5 (a) 19,700 N (b) 2010 kg
5-7 5100 m
5-9 0.546
5-11 31.3 m s^{-1}
5-13 1°
5-15 48.0 m s^{-2}
5-17 (a) 21.6 m s^{-2} (b) 66°
5-19 16,300 m
5-21 $1.83w$
5-23 6.25×10^{12} must be removed
5-25 3.33×10^{-6} C
5-27 (a) -9.63×10^7 C (b) $+9.63 \times 10^7$ C
(c) 8.35×10^{25} N
5-29 9.22×10^{-10} N
5-31 (a) 18.9 m s^{-2} (b) 1.06
5-33 (a) 22.8 m s^{-1} (b) 57.8 m s^{-1}, 0
5-35 (a) 0.0338 m s^{-2} (b) 698 N (c) Further reduces
weight
5-37 16.0 s
5-39 4 AU
5-41 (a) 2.56×10^{-3} m s^{-2} (b) 2.78×10^{-3} m s^{-2}
(c) They differ by 8% (due to inaccuracies in
constants)

Chapter 6

6-1 56.4 J
6-3 76°
6-5 80 J
6-7 12.5 J
6-9 (a) 24,200 ft lb (b) 8070 lb
6-11 (a) 2500 J (b) 2500 J
6-13 13.1 J
6-15 Less energy per second required
6-17 $mgl \sin \theta$

6-19 225 ft

6-21 (*a*) 6.50 J (*b*) 2.25 J (*c*) 8.75 J

6-23 $(2gh)^{\frac{1}{2}}$

6-25 0.918

6-27 0.592

6-29 (*a*) 49,000 J (*b*) 6.45×10^{-3} kg

6-31 $\frac{3}{4}gmR_E$

6-33 2.37×10^3 m s^{-1}

6-35 (*a*) Right (*b*) Left

6-37 $13.6\,(1 - 1/n^2)$ eV

6-39 $v_y = 0.816$ m s^{-1}

6-41 0.204 m s^{-1}

6-43 5.10 kg

6-45 12 cents

6-47 Most, kangaroo; least, flea

6-49 l^{-1}

6-51 5 (including pool at top)

6-55 $v_0 = CR$

6-57 441 lb

6-59 193 ft s^{-1} = 132 mi h^{-1}

6-65 0.549 g s^{-1}

6-67 (*a*) 732 N (*b*) 447 N (*c*) Force is limited by rate of metabolic energy conversion

Chapter 7

7-1 69.1 m s^{-1}

7-3 (*a*) 36,000 kg m s^{-1} (*b*) 36.0 m s^{-1}

7-9 $mv_0/(m + M)$

7-11 (*b*) and (*c*)

7-13 19,500 kg m s^{-1}

7-15 (*a*) 10.4 kg m s^{-1} (*b*) 52 N (*c*) 11.6 m s^{-1}

7-17 4.62 m s^{-1}

7-21 (*a*) 2.48 m s^{-1} (*b*) 0.314 m

7-25 104 N

7-27 2.24 kg

7-29 1200 N

7-35 0.892 m s^{-1}

Chapter 8

8-1 (*a*) 2.5 (*b*) 143°

8-3 (*a*) 60° (*b*) 135° (*c*) 405°

8-5 (*a*) Up (*b*) Down

8-7 (*a*) 12.5 rad s^{-1}; along axle, away from us (*b*) 2.5 rad s^{-2}

8-9 7670 rad s^{-1}; 73,200 rev min^{-1}

8-11 (*a*) 7.27×10^{-5} rad s^{-1} (*b*) 0.0338 m s^{-2}

8-13 9.83×10^{37} kg m^2

8-15 $g/11$

8-17 (*a*) 66.7 rad s^{-1} (*b*) 4.44 rad s^{-1} (*c*) 500 rad

8-19 (*a*) 16.7 rad s^{-1} (*b*) 36.1 J (*c*) 4.33 kg m^2 s^{-1}

8-21 (*a*) 0.141 m (*b*) 0.251 kg m^2 s^{-1} (*c*) 0.790 J

8-23 To increase I, hence, stability

8-25 (*a*) 2.5×10^8 J (*b*) 12,500 s

8-29 0.0251 rad

8-33 $n^2h^2/8\pi^2 I$

8-35 (*a*) 1.71 rad s^{-1} (*b*) 1.71 rad s^{-1}

8-39 (*a*) 2.95×10^{32} kg m^2 s^{-1} (*b*) 6.42×10^{21} N

Chapter 9

9-1 1000 N m^{-2}

9-3 0.0125

9-5 (*a*) 9800 N m^{-2}, 4.90×10^{-8} (*b*) 9.80×10^{-8} m (*c*) 5.10×10^6 kg

9-7 77,400 N

9-9 (*a*) 3.18×10^6 N m^{-2}, 1.59×10^{-5} (*b*) 4.77×10^{-5} m

9-11 2.25×10^7 N m^{-1}

9-13 78.5 N m

9-15 8.73 m

9-17 37.6 m, compared to 8.73 m

9-19 (*a*) 10^6 N m^{-2}, 1.19×10^{-5} (*b*) 1.19×10^{-7} m

9-21 (*a*) 8.80×10^{-5} m^4 (*b*) 7.41×10^{-7} m^4

9-23 $R(\text{solid})/R(\text{hollow}) = 0.0914$

9-25 1.73 cm

9-27 $m^{1/4}$

9-29 Largest

9-31 (*a*) Infants (*b*) $m^{5/4}$

9-35 0.01 m

Chapter 10

10-1 (*a*) 0.0796 Hz (*b*) 12.6 s

10-3 (*a*) 0.1 m, 0 (*b*) 0, -0.314 m s^{-1} (*c*) -0.1 m, 0

10-5 -148 m s^{-2}

10-7 13.2 N m^{-1}

10-9 39.2 m s^{-2}

10-11 (*a*) $\simeq 0.2$ kg m^2 (*b*) $\simeq 1$ Hz

10-13 (*a*) 219 N m^{-1} (*b*) 1.10 J (*c*) 2.10 m s^{-1}

10-15 (*a*) 58.8 Hz (*b*) Yes

10-17 (*a*) 27.6 Hz (*b*) 0.0725 s

10-19 (*a*) x_0, 0, $-4\pi^2 f^2 x_0$ (*b*) $-x_0$, 0, $4\pi^2 f^2 x_0$

10-21 (*a*) 19.7 N m^{-1} (*b*) 0.995 m

10-23 (*a*) $\simeq 0.7$ Hz (*b*) $\simeq 4$ km

10-25 5.95 m s^{-1}

10-27 11.7 Hz

10-29 11,100 N m^{-1}

Chapter 11

11-1 $37.8\,°C$

11-3 $-40°$

11-5   36.461 amu

11-7   4.032 grams

11-9 1.806×10^{24}

11-11   1.58 moles

11-13   127 grams

11-15 5.07×10^8 N

11-17   2030 N; no

11-19   (a) $\simeq 10^5$ Pa (b) $\simeq 1000$ Pa

11-21   5

11-23   1.37

11-25   10,970 K

11-27   Pressure is doubled

11-29   0.9937

11-31   (a) 4.25×10^{-21} J (b) 205 K

11-33   746 s

11-35 8.79×10^{23}

11-37 1.31×10^{24}

11-39   Kr

11-41   (a) 6.52×10^{-7} moles m^{-3}, 1.79×10^{-7} moles m^{-3} (b) 1.68×10^{-3} Pa, 4.61×10^{-4} Pa (c) 1.71×10^{-7} m, 4.71×10^{-8} m

11-43   447 moles m^{-3}

11-45   Expand

Chapter 12

12-1 5.08×10^{-3} m

12-3 4.6×10^{-4} m

12-5   Lid expands more than glass

12-7   11,200 cal

12-9 $-9.91\,°C$

12-11   0.157 kcal kg^{-1} K^{-1}

12-13   847 kcal

12-15   (a) No (b) $0\,°C$

12-17   0.096 kcal s^{-1}

12-19   0.15 kcal s^{-1}

12-21 $21.2\,°C$

12-23 9.35×10^{-6} m

12-25 1.27×10^{-6} m

12-27   2%

12-29   Clouds reflect radiation

12-31   (a) 3.1×10^{-4} m (b) Slow (c) 1.27×10^{-4} (c) 11.0 s

12-33 3.41×10^{-3} K^{-1}

12-35   1.2 K h^{-1}

12-37   (a) 14.9 kcal (b) 24.9 kcal

12-39   171 cal

12-41   (a) $102\,°C$ (b) 0.76 kcal s^{-1}

12-43 $100.8\,°C$

12-45   22%

12-47   (a) 418 W m^{-2}

12-49   (a) 0.0792 kcal s^{-1} (b) No

12-51   0.259 kg h^{-1}

Chapter 13

13-1   (a) $P_1(V_3 - V_1)$ (b) $-P_1(V_3 - V_1)$

13-3   (a) $P_3(V_2 - V_1) + P_2(V_3 - V_2)$ (b) $P_1(V_3 - V_1)$ (c) $-P_3(V_2 - V_1) - P_2(V_3 - V_2)$

13-5   (a) 1500 J (b) 1500 J

13-9   (a) No (b) Increased

13-11   (a) 58.1% (b) 3.3×10^6 J (c) 41.8%

13-13   500 K, 429 K

13-15   (a) 70,000 J (b) 80,000 J (c) 20,000 J

13-17   50%

13-19   (a) 0 or 6 white, 1 way; 1 or 5, 6 ways; 2 or 4, 15 ways; 3, 20 ways (b) 3 yellow, 3 white

13-21   (a) 148 s

13-23   (a) 43.9% (b) 1.46×10^8 W

13-25   (a) 7.43×10^{-3} liters s^{-1} (b) 214 liters

13-27   (a) 8.83 liters (b) 34.6 liters

13-29   (a) 25% (b) 231 W

13-31 1.24×10^6 J

13-33   2.69 grams

13-35   No

13-37   0.375 W kg^{-1}

13-39   18.6 d

Chapter 14

14-1   680 kg m^{-3}

14-3   0.775 m^3

14-5   850 lb

14-7   (a) 3.04×10^5 Pa (b) 9.55×10^7 N

14-9 4.95×10^{-8} m^3

14-11   4.00

14-13   1.09 N

14-15   12 m s^{-1}

14-17 $9v/4$

14-19   (a) Yes (b) No, flow is not steady state

14-21   (a) 8.05×10^7 Pa (b) 1.21×10^6 N

14-23   6300 Pa $\simeq$ 0.06 atm

14-25   9.92 atm = 1.005×10^6 Pa

14-27   8540 Pa = 64 torr

14-29 -9.09×10^{-3} m

14-31 750 kg m^{-3}

14-35 99.986 atm

14-37 0.193 m

14-39 (a) 3.13 m s^{-1} (b) 1.25×10^{-3} m^3 s^{-1}

14-41 9.98×10^4 Pa = 0.986 atm

14-43 160 mi h^{-1}

14-45 12 m s^{-1}

14-47 (b) no

14-49 (a) 0.0594 W (b) 79.2 W kg^{-1}

Chapter 15

15-1 Halfway

15-3 1.8 W

15-5 (a) 1.13×10^{-7} m^3 s^{-1} (b) 0.0720 m s^{-1}

15-7 (a) 0.0314 N (b) 4.71×10^{-4} W

15-9 (a) 0.354 m s^{-1} (b) 632 Pa

15-11 (a) 7.96 m s^{-1} (b) Turbulent (c) No

15-13 (a) 0.983 m s^{-1} (b) 1.24×10^{-5} m^3 s^{-1}

15-15 2.35×10^{-5} m s^{-1}

15-17 (a) 3.47×10^{-5} m (b) 6.14×10^{-5} m

15-19 1.45 m s^{-1}

15-21 (a) 0.099 N (b) 0.0297 W

15-23 (a) 1.22×10^5 m s^{-1} (b) 24.7 m s^{-1} (c) Stokes' law does not hold

15-25 42,300 rev min^{-1}

15-27 (a) 10,000 s (b) 0.1 s

15-29 (a) 3.66×10^{-11} N s m^{-1} (b) 17,400 amu

15-31 10^{-5} m s^{-1}

15-37 (a) 2.29×10^{-24} m^3 (b) 8.18×10^{-9} m
(c) 1.55×10^{-10} N s m^{-1}
(d) 4.35×10^{-10} N s m^{-1} (e) no

Chapter 16

16-1 0.02 N m^{-1}

16-3 Minimize film area outside

16-5 -0.0535 m

16-7 0.0703 m

16-9 (a) 1000 Pa (b) 1.26×10^{-4} N

16-13 1262 Pa

16-15 9800 Pa

16-19 (a) 3.96×10^{-4} N (b) 2.38×10^{-3} N

Chapter 17

17-1 Skin

17-3 5.33 cm^3, 12.96 cm^3

17-5 Lower pressures in veins

17-7 Increased blood pressure and volume in head

17-9 -25 torr

17-11 (a) 38 torr, 303 torr (b) 100 torr, 100 torr

17-13 No

17-15 (a) Tertiary branches (b) Mesenteric artery

17-17 (a) 1.12×10^5 torr s cm^{-3}
(b) 8.93×10^{-5} cm^3 s^{-1}

17-19 21%

17-21 0.16%

17-23 (a) Kidneys (b) Bronchial

17-25 (a) 7.24 cm^3 s^{-1} (b) 7.72 cm^3 s^{-1}
(c) 2.54 cm^3 s^{-1}

17-29 (a) 10^5 Pa s m^{-3} (b) 10 W (c) No

Chapter 18

18-1 $(2kQ^2/9b^2)\widehat{\mathbf{y}}$

18-3 $-(kQ^2/2b^2)\widehat{\mathbf{y}}$

18-5 4.36×10^{-9} N

18-7 (a) 1.32×10^{13} N C^{-1}, outward
(b) 2.11×10^{-6} N, toward nucleus

18-9 5.69×10^{-4} N C^{-1}, opposite to **a**

18-11 $-(kQ/2b^2)\widehat{\mathbf{y}}$

18-13 -1.77×10^{-8} C

18-15 (a) 86.4 V (b) -1.38×10^{-17} J

18-17 (a) 8.84×10^{-8} C m^{-2} (b) 400 V

18-19 3.13×10^{-11} m

18-21 $2qE$

18-23 8.85×10^{-11} F

18-25 5

18-27 (a) 894 V (b) 0.0447 C

18-29 No

18-31 14.2 N C^{-1}

18-33 (a) $-(9000$ N C$^{-1})\widehat{\mathbf{x}}$ (b) $(2250$ N C$^{-1})\widehat{\mathbf{y}}$
(c) 1130 N C^{-1}, 45° below $-x$ direction

18-35 $-(0.144$ N C$^{-1})\widehat{\mathbf{r}}$ (toward center of plate)

18-37 2.71 kq/a

18-39 (a) $\alpha ke/r^2$ (b) $-\alpha k^2 e^2/r^4$

18-41 (a) 3.83×10^{-11} m, $0.724r(\mathrm{H})$
(b) 7.69×10^{-5} eV

18-43 (a) 3.27×10^{-3} C (b) 2.94×10^{-4} J

18-45 6 μF

Chapter 19

19-1 6.25×10^{12}

19-3 3.43×10^{-4} m s^{-1}

19-5 4.75×10^{-4} kg

19-7 2.5 ohms

19-9 34.2 ohms

19-11 1.05×10^{-3} m

19-13 20 ohms

19-15 (*a*) 6 A (*b*) 60 C (*c*) −720 J (*d*) 720 J (*e*) 0 (*f*) 720 J (*g*) Chemical energy in the battery

19-17 (*a*) 0.15 A (*b*) −1.5 V, 1.5 V, 1.5 V, −0.6 V, −0.9 V

19-19 (*a*) 3000 A, 3.6×10^4 W (*b*) 0.146 ohm (*c*) 934 W (*d*) 25.6 W

19-21 (*a*) 144 ohms (*b*) 0.833 A

19-23 (*a*) 0.4 A (*b*) 4.8 W, −3.2 W (*c*) 0.64 W, 0.96 W

19-25 64.8 cents

19-27 (*a*) 4.17 A (*b*) 1.15 ohms

19-29 460

19-31 (*a*) 240 ohms (*b*) 40 ohms

19-33 1 ohm

19-35 90-ohm series resistor

19-37 (*a*) 0.1 ohm (*b*) 10^{-7} s

19-39 0.02 s

19-41 (*a*) 0.012 A (*b*) 0.005 s

19-43 (*a*) 10 V (*b*) 100 V

19-47 (*a*) 7200 C (*b*) 4.50×10^{22} (*c*) 6.15×10^{23}

19-49 (*a*) 2.09×10^{-3} ohm (*b*) 0.00162 m (*c*) $w(\text{Al})/w(\text{Cu}) = 0.462$

19-51 (*a*) 4600 W (*b*) 37.7 min

19-53 (*a*) 1.51×10^4 J (*b*) 5.16 K

19-55 (*a*) Positive (*b*) 241 W (*c*) 12.06 V

19-57 (*a*) 0.1 C (*b*) 10 ohms (*c*) 37 A

19-61 $\mathcal{E} = \mathcal{E}_1 R/(R + R_1)$

19-63 1.295 ohm m

19-65 (*a*) 13.6 ohm m (*b*) 7.37×10^{-2} ohm^{-1} m^{-1}

Chapter 20

20-1 10^6

20-3 1.59×10^9 ohms

20-5 (*a*) 1.26×10^{-9} F (*b*) 1.59×10^6 ohms

20-7 (*a*) 1.2×10^7 N C^{-1}, into axon (*b*) 8.48

20-9 (*a*) 4.5×10^{-6} C m^{-2} (*b*) 9.0×10^{-4} C m^{-2}

20-11 10 microns

20-13 0.8 microns

20-15 8.17×10^{12}

20-17 74.1 mV

20-19 (*a*) 0.0289 A (*b*) 0.00260 W

20-21 1.5 mm

20-23 (*a*) −86.3 mV (*b*) −88.6 mV

20-25 (*a*) 0.2 s (*b*) 0.01 s

20-27 (*a*) 2.67×10^7 ohm m (*b*) 1.46×10^{10} m^{-2} (*c*) 8.28 microns

20-29 $\rho_a = 0.812$ ohm m, $\rho_{if} = 0.599$ ohm m

20-31 (*a*) 3.9×10^{-10} C (*b*) 2.44×10^9

20-37 (*a*) 75 (*b*) 75,000

Chapter 21

21-1 (*a*) P_4 (*b*) P_1

21-3 No; can have **v** ∥ **B**

21-5 5×10^{-4} N, 0.05 m s^{-2}; into page

21-7 (*a*) 30 N

21-9 (*a*) 2 N, 2 N, 0 (*b*) 0

21-11 (*a*) 0.1 A m^2 (*b*) 0.01 N m (*c*) **B** in plane of loop

21-13 Yes; −x direction

21-15 3.77×10^{-7} T

21-17 6.28×10^{-3} T

21-19 (*a*) 8×10^{-6} T (*b*) 8×10^{-7} T

21-21 (*a*) 2×10^{-4} m (*b*) Yes

21-23 (*a*) $ev_d B$, upward (*b*) $v_d B$, upward (*c*) $av_d B$ (*d*) $av_d B$ (*e*) 2×10^{-5} V (*f*) Out of page (*g*) Opposite in sign for a given current direction

21-25 (*a*) 0.24 A m^2 (*b*) 0.12 N m (*c*) $\boldsymbol{\mu} \parallel \pm$ **B**

21-27 2×10^{-6} N, toward left

21-29 (*a*) 10^{-5} T, into page (*b*) 1.67×10^5 T, into page (*c*) 2.33×10^5 T, out of page (*d*) $x = 0.0333$ m

21-31 15.9 A

21-33 (*a*) 1.09×10^{-3} A (*b*) 13.4 T

21-35 $\pi k'I/a$

21-37 (*a*) 2.85×10^{-5} T (*b*) 5×10^{13} m s^{-2}

21-39 (*a*) 0.7 T (*b*) 8.35×10^{-3} m

21-41 (*a*) 1.31 T (*b*) 1.26×10^8 m s^{-1}

21-43 0.209 m, 2.29×10^7 Hz

21-45 (*a*) 8.38×10^7 m s^{-1} (*b*) 8.05×10^{-16} N (*c*) 1.42×10^{-3} m

Chapter 22

22-1 (*a*) Yes (*b*) No (*c*) Yes

22-3 (*a*) Same (*b*) Large in copper, almost zero in rubber.

22-5 3.14 A

22-7 (*a*) 48 V (*b*) 0

22-9 Clockwise

22-11 $2\omega BA/\pi$

22-13 417

22-15 Use a movable ferromagnetic core in a coil

22-17 −20 V

22-19 1.97 H

22-21 0.32 H

22-23 (*a*) qvB (*b*) $qvBl$ (*c*) $-Blv$; same

22-25 Viewed from above: (*a*) counterclockwise;
(*b*) zero; (*c*) clockwise

22-27 (*a*) Porportional to ω (*b*) Reduces it
(*c*) Increases I (*d*) I^2R becomes large

22-31 (*a*) $4\pi K_m k'I\,N/l$ (*b*) $2\pi K_m k'N^2I^2/l^2$
(*c*) $1/(8\pi k')$

22-37 (*a*) 60 ohms (*b*) 0.01 s (*c*) 0.6 H

23-29 (*a*) 0.5 A (*b*) 0.707 A (*c*) 120

22-41 (*a*) 1800 W (*b*) 3600 W

22-43 (*a*) 39.8 μF (*b*) 66.7 ohms (*c*) 240 V

22-45 (*a*) 754 ohms (*b*) 120 V (*c*) 170 V

22-49 1.27×10^{-9} F

22-51 (*a*) 10.05 ohms (*b*) 1430 W

22-53 (*a*) 1000 rad s^{-1} (*b*) 159 Hz (*c*) 20 W

22-55 Less energy lost in heating battery

22-57 (*a*) Decrease (*b*) Wastes energy due to heating

Chapter 23

23-1 0.344 m

23-3 (*a*) 10^{10} Hz (*b*) 5×10^9 waves (*c*) 1.5×10^8 m

23-5 173 m s^{-1}

23-13 (*a*) 435 m s^{-1} (*b*) 2.91×10^{-4} kg m^{-1}

23-15 (*a*) 3.5 m (*b*) 1.75 m

23-17 6.51 m

23-19 (*a*) 0.25 m (*b*) 0.132 m

23-23 5 Hz

23-25 0.0866 m

23-27 0.25

23-31 3.42 m s^{-1}

23-33 40.0 s

23-35 (*a*) 1.52×10^{-3} s

23-37 (*a*) 80% (*b*) 20%

23-39 (*a*) 2263 Hz (*b*) 1792 Hz

23-41 1.91 Hz

23-43 3.41 m s^{-1}

23-45 5.43×10^{14} Hz

23-47 (*a*) 15.7

23-51 (*a*) 99,997 Hz (*b*) 1.16×10^{-6} m^3 s^{-1}

23-53 4.60×10^{-7} m

Chapter 24

24-1 365 Hz

24-3 41.3 m

24-5 1480 m

24-7 0.0516 m

24-9 In water, c is larger, λ larger

24-11 132 Hz, 265 Hz, 397 Hz, 529 Hz

24-13 2.65 m, 0.0823 m

24-15 8.50 Pa

24-17 0.0156

24-19 4

24-21 14.7

24-23 1.72×10^{-3} Pa

24-25 17,200 Hz

24-31 10

24-33 60 db

24-35 8×10^{-13} W

24-41 (*a*) 0.103 m (*b*) 4.4×10^{-3} s

24-43 (*a*) 86.0 Hz (*b*) 1.828 m

24-45 (*a*) 1.26×10^4 W (*b*) 1.26×10^3 J

24-47 49.5 db

24-49 (*a*) 10^{-8} W m^{-2} (*b*) 50.5 db (*c*) 6.28×10^{-4} W

24-51 (*a*) 99.9% (*b*) 99.9%

24-53 $\simeq 2$ mm

Chapter 25

25-1 2.25×10^8 m s^{-1}

25-3 (*a*) 450 nm (*b*) Yellow

25-5 0.0204

25-7 0.997

25-9 (*a*) 15° (*b*) 11°

25-11 24°

25-13 1.91×10^{-5} m

25-15 5.89×10^{-5} m

25-17 (*a*) 17° (*b*) 7 lines ($m = 0$, ±1, ±2, ±3)

25-19 (*a*) 32° (*b*) 3280

25-21 13°

25-25 1.75°

25-27 1.37 Å

25-29 77°

25-37 (*a*) 2 (*b*) 4

25-39 Wavelets from each point on edge are in phase

25-41 0.050 m

25-43 $r = 3.66 \times 10^{-3}$ m

25-47 (*a*) All of it (*b*) Resolution is reduced in both
cases

25-49 150 nm

25-53 Spot is 0.117 m from plate

25-55 All wavelengths interfere destructively

Chapter 26

26-1 4 m

26-3 -2.92 m

26-5 1.50 m

26-7 Blue, 0.775 m; red, 0.795 m

26-9 −0.333 m

26-11 (a) −0.4 m (b) 5; erect

26-13 (a) 0.0508 m (b) −0.0169 (c) 1.42 m

26-15 4 times larger

26-17 10^{-3} m

26-19 (a) 0.133 m from objective (b) 0.00412 m
(c) −277

26-21 −0.125 m, −0.167 m

26-23 0.5 diopters

26-25 6 m

26-27 Edges of retina have most rods

26-29 0.75 m

26-31 (a) 0, −f, ∞, 3f, 3f/2

26-35 −1.20 m

26-37 0.01 s

26-39 To reduce diameter and cost

26-41 (a) 1.38 (b) 145 nm

26-43 (a) −0.5 m (b) 0.167 m (c) 0.25 m

26-45 (a) 0.125 m (b) 0.5 m

26-47 −20

Chapter 27

27-1 2000 km h^{-1}

27-3 100 s

27-5 (a) 3.75×10^{-8} s (b) 6.75 m

27-7 0.872 m

27-9 $0.866c$

27-11 $0.866c$

27-13 (a) 5.35×10^{-9} amu (b) 1.78×10^{-10}

27-15 8.37×10^{-4} amu

27-17 (a) 2150 MeV (b) 1210 MeV

27-19 (a) 1.005 y (b) 10.05 y (c) Both

27-21 (a) 5 h (b) 1.25 h (c) 6.25 h (d) 4 h

27-23 3.50×10^{-7} kg

27-25 (a) 1.12 (b) 1.35×10^{8} m s^{-1}
(c) 1.45×10^{8} m s^{-1}

27-29 (a) 167 m (b) A, by 6.68 s

27-31 (a) $0.882c$ (b) 0

27-33 (a) $0.806c$ (b) $-0.263c$

27-35 (a) 4 m (b) 10^{-8} s (c) 5.8 m

27-37 $v_x = 0.8c$, $v_y = 0.54c$, $v = 0.965c$

Chapter 28

28-1 4.53×10^{14} Hz

28-3 Removal of electrons leaves positive charge

28-5 48.6 m

28-7 (a) 3.10 eV (b) 1.77 eV

28-9 5.32×10^{20} Hz

28-11 39,300 V

28-13 2.5×10^{-9} m^2

28-15 (a) 2.77×10^{18} s^{-1} (b) 892 m
(c) 2.77×10^{11} s^{-1} m^{-2}

28-17 (a) 8.35×10^{-28} kg (b) 14.6 N m^{-1}

28-19 (a) 1.57×10^{-27} kg (b) 1.79×10^{14} Hz

28-21 15.3 s

28-23 (a) 1.07×10^{-23} kg m s^{-1} (b) 6.20×10^{-11} m

Chapter 29

29-1 4.73×10^{-14} m

29-3 486 nm

29-5 6.63×10^{-24} kg m s^{-1} (same for both)

29-7 3.70×10^{-63} m

29-9 (a) 6.63×10^{-34} m (b) No; λ is too small

29-11 (a) 0.821 Å (b) 224 eV

29-13 Yes

29-15 (a) 13.6 eV (b) 54.4 eV (c) 122 eV

29-17 0.348 eV

29-19 (a) 6.8 eV (b) $2a_0 = 1.06 \times 10^{-10}$ m

29-21 (a) -7.21×10^{5} eV (b) 1.60×10^{-14} m (c) 4

29-23 12.75 eV, 12.1 eV, 10.2 eV, 2.55 eV, 1.89 eV,
0.661 eV

29-25 (a) $n^2\hbar^2/2I$, $n = 0, 1, 2, \ldots$

29-27 12.1 eV, 10.2 eV, 1.89 eV

Chapter 30

30-1 (a) 0, 1, 2, 3 (b) 4s, 4p, 4d, 4f (c) 0, $\hbar$, $2\hbar$, $3\hbar$

30-3 (a) 0, $\pm e\hbar B/2m$, $\pm 2e\hbar B/2m$
(b) 5.79×10^{-4} eV (c) absorbed

30-5 32

30-7 All have one s electron outside closed shells

30-9 48 eV

30-11 Small ionization energies

30-13 H electron closes shell in halogen atoms

30-15 $n = 3$, $l = 0$, $m_l = 0$, $s = 1/2$, $m_s = \pm 1/2$

30-17 11.1 eV

30-19 (a) 483 eV (b) Mutual repulsion of 1s
electrons (c) Yes; additional shielding

30-21 (a) −272 eV (b) 3.79

30-23 3.8 km

30-25 0.003 m

Chapter 31

31-1 (a) 1.30 eV (b) −6.10 eV (c) 4.80 eV

31-3 (a) −5.76 eV (b) 3.13 eV

31-5   No; it is symmetrical
31-9   (*b*) 4   (*c*) 2
31-11 2.66×10^{14} Hz
31-13   Yes; photons are emitted or lattice excited
31-17 4.46×10^{-29} C m
31-19   6.74 eV
31-23   (*b*) No; it is symmetrical   (*c*) Closer to C atom

Chapter 32

32-1   (*a*) -1.06×10^{-7} eV   (*b*) 1.06×10^{-7} eV
32-5 7.12×10^{-6} T
32-7   (*a*) 2.12×10^{8} Hz   (*b*) 8.81×10^{-7} eV
32-9   Greater
32-11   1.42 T
32-17 CH_3—OH
32-21   (*a*) 3 spin orientations in CH_2 (*b*) CH_3 group causes 4-fold splitting, CH_2 causes 3-fold; $4 \times 3 = 12$

Chapter 33

33-1   8 h
33-3   17.6 d
33-5   5.6 to 5.7 d
33-7   11,400 y
33-9   too old
33-11 1.1×10^{-14}
33-13   8, 19, 35, 126
33-15   7.86 MeV
33-17   (*a*) 1397 MeV   (*b*) 6.98 MeV
33-19   (*a*) 30 MeV   (*b*) 0
33-21   (*a*) $^3_1$H   (*b*) $^2_1$H   (*c*) α
33-23   (*a*) 4.84×10^{20} Hz   (*b*) 6.20×10^{-13} m
33-25   6.64

33-27   40.4 y
33-29   (*a*) Straight line   (*b*) Decay rate is not a simple exponential   (*c*) Straight line, less rapid decrease
33-31   (*a*) 24.7 MeV   (*b*) 1.99×10^6 kg
33-33   (*b*) $^9_5$B and $^9_3$Li have unfilled lower levels   (*c*) β
33-35   Fragments have too many neutrons
33-37   Geological stability (against earthquakes)
33-41   Electrical repulsion is 3 times larger
33-43   (*a*) 8.21×10^3 J kg^{-1} (*b*) 0.0203 kg s^{-1}
33-45   (*a*) 7.72×10^8 K   (*b*) Some atoms have above average energies

Chapter 34

34-1   A single γ can ionize atoms; one microwave quantum cannot
34-3   (*a*) α particles   (*b*) 0.0045 cm
34-7   Nuclear mass is much larger than electron mass
34-9 $K(d) = 2K(p)$
34-11   Determine range
34-13 1.61×10^{15}
34-15   (*a*) 4.50 min   (*b*) 6.43 min
34-17   (*a*) 30   (*b*) 50
34-19   Less important, because of smaller total dose to genetic pool
34-21   2600 y^{-1}
34-23   1 to 5 years doubles cancer rate, 0.7 to 4 years doubles mutation rate
34-27   (*a*) Low energy x-rays have shorter ranges   (*b*) Reduce skin dose
34-29   (*a*) 7.40×10^{11} s^{-1} (*b*) 5.89×10^{10} m^{-2} s^{-1}
34-31   100 rad, 100 rem
34-33   1420 cm^3

INDEX

Index

Index

Index

Index

Index

Index